ENVIRONMENTAL TRACKING FOR PUBLIC HEALTH SURVEILLANCE

T0382890

International Society for Photogrammetry and Remote Sensing (ISPRS) Book Series

Book Series Editor

Paul Aplin
School of Geography
The University of Nottingham
Nottingham, UK

information from imagery

Environmental Tracking for Public Health Surveillance

Editors

Stanley A. Morain & Amelia M. Budge

Earth Data Analysis Center, University of New Mexico, Albuquerque, New Mexico, USA

CRC Press
Taylor & Francis Group
Boca Raton London New York

CRC Press is an imprint of the
Taylor & Francis Group, an **informa** business

A BALKEMA BOOK

Cover image:
Environmental tracking involves a host of international space agency satellites and their on-board sensors. The upper portion of the image illustrates one of several satellite constellations (the *A-Train*) carrying land, ocean, and atmospheric sensors. The bottom three elements of the image show a pollen burst resulting in a respiratory reaction (left), typical disease carriers like mosquitoes, hookworms and ticks (center), and public health interventions like vaccinations (right). The challenge is to improve current environmental tracking capabilities with next generation sensor systems to predict disease threats and mitigate their outcomes.

Image credits:
A-Train: Courtesy US National Aeronautics and Space Administration (NASA)
Girl sneezing: Courtesy frogblog
Mosquito: Photo by James Gathany courtesy Centers for Disease Control and Prevention (CDC) Tick: Photo by James Gathany courtesy CDC
Hook worm: Courtesy CDC
Boy being vaccinated: Photo by James Gathany courtesy CDC

First published in paperback 2024

First published 2013
by CRC Press/Balkema
4 Park Square, Milton Park, Abingdon, Oxon, OX14 4RN

and by CRC Press/Balkema
2385 NW Executive Center Drive, Suite 320, Boca Raton FL 33431

CRC Press/Balkema is an imprint of the Taylor & Francis Group, an informa business

Library of Congress Cataloging-in-Publication Data

Environmental tracking for public health surveillance / editors, Stanley A. Morain & Amelia M. Budge.
    p. cm.
  Includes bibliographical references and index.
  ISBN 978-0-415-58471-5 (hbk.) – ISBN 978-0-203-09327-6 (eBook)
  I. Morain, Stanley A.   II. Budge, Amelia M.
  [DNLM: 1. Disease Outbreaks–prevention & control.   2. Environment.   3. Information
Systems.   4. Population Surveillance.   5. Satellite Communications. WA 105]
  362.1—dc23
                                                                          2012025102

ISBN 13: 978-0-41-558471-5 (hbk)
ISBN 13: 978-1-03-291940-9 (pbk)
ISBN 13: 978-0-42-910668-2 (ebk)

DOI: 10.1201/b12680

Typeset by MPS Limited

Visit the Taylor & Francis Web site at
http://www.taylorandfrancis.com

and the CRC Press Web site at
http://www.crcpress.com

Table of contents

Preface

The volume editors and Chapter author/editors have compiled a survey of health science research and application developments linking satellite Earth observations with types of diseases, their potential transmission pathways, and prospects for early warning of outbreaks or epidemics. Applying Earth science as a way to advance health science is both relevant and timely, given that in 2011 Earth's human population passed seven billion; and especially since health science includes broader social and economic factors that drive wellbeing and quality-of-life issues. The predominant focus is on global satellite observations that show how modelling space-based environmental measurements actually improve assessments of future health outcomes in scientifically valid ways. Its authors/editors and chapter contributors demonstrate that Earth's environments not only affect human exposures to disease but also serve as triggers for massive outbreaks and deadly epidemics.

The book was compiled by Chapter author/editors who addressed major categories of health and environment issues. Most lead author/editors invited additional international contributors to develop the Chapter content. The result is a collection representing the current state-of-the-art for environmental tracking for health surveillance, mitigation strategies, and policy decisions.

Some themes and topics recur in more than one Chapter. Information systems, for example are described in Chapters 6 and 10, but they describe different systems. In Chapter 6, *information systems* refer to data extraction from an extensive list of satellite sensors and their orbiting platforms, and the datasets and other products developed from them. In Chapter 10, *information systems* refer to tracking environmental parameters and suites of parameters for assessing transmission pathways and for health monitoring and forecasting. The Chapter also addresses semantic approaches to *information systems* as tools for identifying health threats rapidly based on global networks of media sources, the potential for crowd sourcing, and other forms of social networking for identifying evolving health threats that could slide from order to chaos in a matter of hours.

The reference section at the end of each Chapter is a compendium of citations for readers to extract the history and development of thought surrounding air-, water-, vector-, and soil-borne diseases and zoonotic diseases. By and large the author/editors and their contributors have strengthened convergence of the two scientific philosophies. We are confident that this volume will become a stimulus for greater cooperation among the sciences, and become a source reference for several years to come.

The following two conventions have been adopted to provide consistency between Chapters. First, while recognizing that Universal Record Locators (URLs) can be ephemeral, they nevertheless point to important resources for online data and information. The speed of communication is such that on-line publishing results in URLs as the best means for retrieving information. Evolving search engines will no doubt lead inquirers to a wealth of additional resources. Chapters that cite many URLs include tables in their text to relevant Sections or sub-Sections that point to them. Additional URLs are cited in the reference sections.

The second convention recognizes that readers will be interested in specific sensors, sensor systems, services, diseases, or medical and health terms that recur frequently in Chapters. All scientific terms are spelled out at the place of their first use in the book; and thereafter, by acronym only. Acronyms for proper titles (satellite platforms, satellite missions and experiments, international organizations, programmes, commissions, and centres) are capitalized at their first use in the text and thereafter by acronym only; those referring to orbiting sensors, government or university systems, online systems, services, models, or processes are *not* capitalized at their first use in the text, but are given as acronyms in subsequent references; and those acronyms used only locally by authors pertinent to their material, but not otherwise recognized widely as acronyms, are not included in the list of acronyms. Readers interested mainly in material in the later Chapters will find acronyms spelled-out in earlier Chapters, and will need to refer to the comprehensive list of acronyms provided at the end of the book. Lastly, the United Kingdom and the United States are given as UK and US. All other countries are spelled out.

This work is a product of the International Society for Photogrammetry and Remote Sensing (ISPRS): Commission VIII – *Remote Sensing Applications and Policy*/ Working Group-2 – *Health*. The working group had an international membership of over fifty active and passive remote sensing and health science experts during its four-year commission from July 2008 to September 2012.

Albuquerque April 30, 2012
Amelia Budge, Chair
Richard Kiang, Co-Chair
Stan Morain, Technical Secretary

Foreword

Justinian's Flea (Rosen 2007) explores a systems approach to understanding the decline of the Roman Empire, a centuries-long process involving political, religious, economic, social, military, and personality threads interacting over a gradually changing physical environment. The arrival of the plague bacterium (*Yersinia pestis*) to the lower Nile valley around 542 AD, and its subsequent spread throughout the Mediterranean region, reshaped the political and social orders of Europe, and in no small way represented the final straw that destroyed the Empire. In the context of Environmental Tracking for Public Health Surveillance, *Y. pestis*, the flea that hosted it, and the black rat that transported it, became possible through a minor lowering of temperatures that *brought the coast within the [bacterium's] active temperature range of 59–68 degrees Fahrenheit* (Brown 1999). According to tree ring analyses and historical evidence, such a cooling took place around 530 AD. *Y. pestis* migrated from its East African focus where it had been active for hundreds, if not thousands, of years, and along the way apparently jumped to human populations. Justinian nor anyone else in the mid-sixth century could have foretold the horrifying health consequences of a slightly cooler temperature along the southern Mediterranean coast.

Estimating the moment when climate, flea behaviour, food availability [wheat for the rats], and a dozen other variables [would] combine to cause a rat population explosion is not impossible, but very nearly so (Rosen 2007, 291). These words are a reminder that complexity is, at best, able to balance *order* and *chaos* only temporarily. While physics and math provide precise solutions to two-body problems whose parameters are measurable, they provide only vague approximations to problems governed by multiple interacting influences whose parameters are not known as well, and whose impacts on system function are nonlinear.

Twenty-first century environmental health tracking, made possible by vastly superior science and technology capabilities (compared to the early centuries AD), suggest that we *can* foretell health scenarios, and do so at almost any geographic or human scale we choose. Why, then, do we not? According to Sterman (2006) it is because we do not convince policy and decision makers. He argues in favor of scientific methods and formal modelling to guide policy decisions for otherwise intractable system dynamics, all the while knowing that the greater the number of interacting components, the more complex the system will become. Complexity hinders the generation of evidence, learning from that evidence, and implementing policies based on that evidence. *System thinking requires us to examine issues from multiple perspectives, to expand the boundaries of our mental models, [and] to consider the longer-term consequences of our actions, including their environmental, cultural, and moral implications* (Sterman 2006, 511).

It is safe to argue that early civilizations made little, if any, connection between diseases and their possible underlying causes either at an individual or societal level. The empires of Greece, Rome, Macedonia, China, Peru, Mexico and elsewhere could not connect cause and effect (aetiological) relationships. Yet, at the everyday level, early populations had fundamental beliefs that the common cold was somehow rooted in cold weather; that some plants were poisonous; that fevers were brought on by something in the victim's daily experiences; or that there were supernatural forces at work.

Not being a history of health and medicine, this book fast-forwards to the last quarter of the 20th Century and the advent of satellite observations of Earth. These digital data and the products derived from them not only provide the means for monitoring ever-changing environmental parameters, but also contain clues to identify conditions that trigger health consequences. The Chapters focus on a variety of observations that are known to affect health. Some of these reveal slowly changing or emerging environments; others represent dramatic events having quick and often catastrophic health consequences.

On the cusp of the 21st Century it became apparent to world bodies that environment and health were tightly linked. The first principle to emerge from the 1992 United Nations Conference on Environment and Development held in Rio de Janeiro was that *Human beings are at the center of concerns for sustainable development. They are entitled to a healthy and productive life in harmony with nature.* Here, perhaps, is an early convergence of environmental health with public health. Ten years later at the 2002 World Summit on Sustainable Development (WSSD), the Johannesburg Plan of Implementation (POI) was adopted (UN 2004). In this Plan, paragraphs 53–57 refer specifically to human health issues. It is stated that *there is an urgent need to address the causes of ill health, including environmental causes, and their impact on development, and to reduce environmental health threats* (UN 2004, 31).

Many challenges in Earth system science require not only integrating complex physical processes into system models, but also coupling environmental biogeochemical and chemical phenomena that trigger human health responses. The next generation of modelers will be required to form teams that partner members from the biogeophysical realm with those from the medical realm to assess quickly changing and highly vulnerable situations.

People and Pixels (Liverman *et al.*, 1998) was among the early publications to draw humankind into the arena of satellite remote sensing. Early scientific literature focused on physical and natural applications in agriculture, forestry, rangeland, hydrology, and mineral exploration. After *People and Pixels*, interest migrated to people-oriented issues like food security, environmental health, public health, disasters and hazards, and most recently on security and antiterrorism. Because of their immense humanitarian and policy implications, remote sensing and geospatial programmes are moving quickly to address the consequences of weather and climate cycles on human health, air and water quality degradation, and diseases following natural disasters. For the photogrammetry, remote sensing, and geospatial information sciences, the key language in Paragraphs 54 and 56 in the POI includes the following. For the most part they are general aims and goals, but a few are quite specific.

§54: Integrate health concerns into strategies, policies, and programs for sustainable development; provide technical and financial assistance for health information systems and integrated databases; target research efforts and apply research results to priority public health issues and reduce exposures to public health risks; start international initiatives that assess health and environment linkages; and, develop preventive, promotive, and curative programs for non-communicable diseases.

§56: Reduce respiratory diseases and other health impacts resulting from air pollution.

Once articulated, it was inevitable that genetic and molecular systems would eventually be linked with the broader biogeochemical forces of nature.

Stan Morain, May 9, 2012

REFERENCES

Brown, N.G. 1999. *Challenge of Climate Change.* New York: Routledge.
Liverman, E.F. Moran, R.R. Rindfuss & P.C. Stern (eds.), *People & Pixels: Linking Remote Sensing & Social Science.* Washington DC: National Academy Press.
Rosen, W. 2007. *Justinian's flea.* New York: Penguin Books.
Sterman, J.D. 2006. Learning from evidence in a complex world. *Amer. J. Pub. Health* 96(3):505–514.
UN 2004 Available from: http://www.un.org/esa/sustdev/documents/WSSD_POI_PD/English/POIChapter6.htm [Accessed 6th April 2012].

List of contributors

Abdel-Dayem, Mahmoud, S.
Cairo University & US Naval Medical Research Unit, Cairo, Egypt, FPO AE 098-0007, E-mail: Mahmoud.Abdel-Dayem.eg@med.navy.mil

Achee, Nicole L.
Uniformed Services University, Bethesda, MD, US 20814, E-mail: nachee@usuhs.mil

Anyamba, Assaf
Universities Space Research Association, Columbia, MD, US 21044 & NASA Goddard Space Flight Center, Greenbelt, MD, US 20771, E-mail: assaf.anyamba@nasa.gov

Barker, Christopher M.
School of Veterinary Medicine, University of California, Davis, CA, US 95616, E-mail: cmbarker @ucdavis.edu

Benedict, Karl
Earth Data Analysis Center, University of New Mexico, Albuquerque, NM, US 87131-0001, E-mail: kbene@edac.unm.edu

Brown, Christopher W.
Satellite & Information Service, National Oceanic and Atmospheric Administration, College Park, MD, US 20740, E-Mail: christopher.w.brown@noaa.gov

Brown, Heidi E.
School of Geography & Development, University of Arizona, AZ, US 85721-0076, E-mail: HeidiBrown@email.arizona.edu

Brownstein, John S.
Massachusetts Institute of Technology & Children's Hospital, Boston, MA, US 02215, E-mail: john.brownstein@harvard.edu

Budge, Amelia M.
Earth Data Analysis Center, University of New Mexico, Albuquerque, NM, US 87131-0001, E-mail: abudge@edac.unm.edu

Caian, Mihaela
National Meteorological Administration, Bucharest, Romania 013686, E-mail: mihaela.caian@ gmail.com

Castronovo, Denise
Mapping Sustainability LLC, Palm City, FL, US 34990, E-mail: denise@mappingsustain ability.com

Ceccato, Pietro
Environmental Monitoring Div., International Research Institute for Climate & Society, Columbia University, Palisades, NY, US 10964, E-mail: pceccato@iri.columbia.edu

Charland, Katia M.
Children's Hospital Informatics Program, Boston, MA, US 02115, E-mail: not available

Chen, Robert S.
Socioeconomic Data & Applications Center, CIESIN, Columbia University, Palisades, NY, US 10964, E-mail: bchen@ciesin.columbia.edu

Colacicco-Mayhugh, Michelle G.
Division of Entomology, Walter Reed Army Institute of Research, Silver Spring, MD, US 20910-7500, E-mail: Michelle.Colacicco@us.army.mil

Comrie, Andrew C.
School of Geography and Development, University of Arizona, AZ, US 85721-0076, E-mail: comrie@arizona.edu

Connor, Stephen J.
School of Environmental Sciences, University of Liverpool, Liverpool L69 3BX, UK, E-mail: sjconnor@liv.ac.uk

Crǎciunescu, Vasile
Remote Sensing and GIS Laboratory, National Meteorological Administration, Bucharest, Sos. Bucuresti-Ploiesti 97, Bucharest, Romania, E-mail: vasile.craciunescu@meteoromania.ro

Daszak, Peter
EcoHealth Alliance, New York, NY, US 10001, E-mail: daszak@ecohealthalliance.org

Deng, Zhi
Civil and Environmental Engineering, Louisiana State University, Baton Rouge, LA, US 70803, E-mail: zdeng@lsu.edu

Durant, John L.
Department of Civil & Environmental Engineering, Tufts University, Medford, MA, US 02155, E-mail: john.durant@tufts.edu

Epstein, Jonathan H.
EcoHealth Alliance, New York, NY, US 10001, E-mail: epstein@ecohealthalliance.org

Estes, Sue
Marshall Space Flight Center, Universities Space Research Association, Huntsville, AL, US 35812, E-mail: sue.m.cstcs@nasa.gov

Faruque, Fazlay S.
The University of Mississippi Medical Center, Jackson, MS, US 39216-4505, E-mail: FFaruque@umc.edu

Fearnley, Emily
South Australian Department of Health, Adelaide, South Australia 5000, Australia, E-mail: Emily.Fearnley@health.sa.gov.au

Ford, Timothy
University of New England, Portland, ME, US 04103, E-mail: tford@une.edu

Gibbons, Robert V.
Armed Forces Research Institute of Medical Sciences, Bangkok, Thailand 10400, E-mail: robert.gibbons@afrims.org

Golden, Meredith L.
Socioeconomic Data & Applications Center, CIESIN, Columbia University, Palisades, NY, US 10964, E-mail: mgolden@ciesin.columbia.edu

Green, David
NOAA Weather Service, Silver Spring, MD, US 20910, E-mail: david.green@noaa.gov

Grieco, John P.
Uniformed Services University, Bethesda, MD, US 20814, E-mail: jgrieco@usuhs.mil

Griffin, Dale W.
Geology Division, US Geological Survey, 2639 North Monroe St., Tallahassee, FL, US 32303, E-mail: dgriffin@usgs.gov

Gurley, Emily S.
International Centre for Diarrhoeal Disease Research, Dhaka, Bangladesh 1212, GPO 128, E-mail: egurley@icddrb.org

Hamner, Steven
Department of Microbiology, Montana State University, Bozeman, MT, US 59717, E-mail: shiva_dancing@yahoo.com

Harrington, Laura C.
Cornell University, Ithaca, NY, US 14853, E-mail: lch27@cornell.edu

Hickey, Barbara M.
School of Oceanography, University of Washington, Seattle, WA, US 98195, E-mail: bhickey@u.washington.edu

Hossain, M. Jahangir
International Centre for Diarrhoeal Disease Research, Dhaka, Bangladesh 1212, GPO 128, Email: jhossain@icddbr.org

Huang, Q.
George Mason University, Fairfax, VA US 22030, Email: not available

Hudspeth, William
Earth Data Analysis Center, MSC01-1110, University of New Mexico, Albuquerque, NM, US 87131-0001, E-mail: bhudspeth@edac.unm.edu

Huff, Amy
Battelle Memorial Institute, Columbus, OH, US 43201, E-mail: not available

Irwin, Daniel
NASA Marshall Space Flight Center, Huntsville, AL, US 35812, E-mail: dan.irwin@nasa.gov

Jacobs, John M.
NOAA Ocean Service, Oxford, MD, US 21654, E-mail: john.jacobs@noaa.gov

Jacquez, Geoffrey M.
BioMedware, Ann Arbor, MI US 48104-1382, E-mail: Jacquez@biomedware.com

Jagai, Jyotsna S.
Office of Research & Development, US Environmental Protection Agency, Chapel Hill, NC, US 27599, E-mail: jagai.jyotsna@epamail.epa.gov

Kass-Hout, Taha A.
Public Health Surveillance Program Office Centers for Disease Control & Prevention, Atlanta, GA, US 30333, E-mail: kasshout@gmail.com

Kempler, Steven
NASA Goddard Space Flight Center, Greenbelt, MD, US 20771, E-mail: Steven.J.Kempler@nasa.gov

Kiang, Richard K.
NASA/Goddard Space Flight Center, Greenbelt, MD, US 20771, E-mail: richard.kiang@nasa.gov

Koch, Magaly
Center for Remote Sensing, Boston University, Boston, MA, US 02215, E-mail: mkoch@bu.edu

Kramer, Vicki
Vector-borne Disease Section, California Department of Public Health, Sacramento, CA, US 95899-7377, Vicki.Kramer@cdph.ca.gov

Kumar, Sunil
Department of Ecosystem Science and Sustainability, Natural Resource Ecology Laboratory, Colorado State University, Fort Collins, CO, US 80523-1499, E-mail: Sunil.Kumar@colostate.edu

Lanerolle, Lyon W.J.
NOAA Ocean Service, Silver Spring, MD, US 20910, E-mail: lyon.lanerolle@noaa.gov

Lary, David J.
William B. Hanson Center for Space Science, University of Texas, Richardson, TX, US 75080-3021, E-mail: djl101000@utdallas.edu

Leptoukh, G.G.
NASA Goddard Space Flight Center, Greenbelt, MD, US (deceased).

Linthicum, Kenneth J.
Agricultural Research Service, US Department of Agriculture, Gainesville, FL, US 32608, E-mail: kenneth.linthicum@ars.usda.gov

Liu, Yang
School of Public Health, Emory University, Atlanta, GA, US 30322, E-mail: yang.liu@emory.edu

Lo, Martin
Navigation and Mission Design Section, Jet Propulsion Laboratory, California Institute of Technology, Pasadena, CA, US 91109-8099, E-mail: martin.lo@jpl.nasa.gov

Luby, Stephen P.
International Centre for Diarrhoeal Disease Research, Bangladesh Centre for Health and Population Research, Mohakali, Dhaka, Bangladesh 1212, GPO 128, E-mail: sluby@icddrb.org

Luvall, Jeff
NASA Marshall Space Flight Center, Huntsville, AL, US 35812, E-mail: jluvall@nasa.gov

Lyles, Mark B.
Medical Sciences & Biotechnology, Center for Naval Warfare Studies, US Naval War College, Newport, RI, US 02841-1207, E-mail: mark.lyles@usnwc.edu

Manibusan, Pedro A.
Tripler Army Medical Center, Honolulu, HI, US, 96859, E-mail: pedro.manibusanjr@us.army.mil

Maxwell, Susan
BioMedware, Ann Arbor, MI, US 48104, E-mail: susan.maxwell@biomedware.com

McClure, Leslie
University of Alabama at Birmingham, Birmingham, AL, US 35294, E-mail: LMcClure@ms.soph.uab.edu

McEntee, Jesse C.
Department of Urban & Environmental Policy & Planning, Tufts University, Medford, MA, US 02155, E-mail: jesse.mcentee@gmail.com

Moore, Stephanie
NOAA National Marine Fisheries Service, Seattle, WA, US 98112, E-mail: stephanie.moore@noaa.gov

Morain, Stanley A.
Earth Data Analysis Center, MSC01-1110, University of New Mexico, Albuquerque, NM, US 87131-0001, E-mail: smorain@edac.unm.edu

Myers, Todd E.
Naval Medical Research Center, Silver Spring, MD, US 20910-7500, E-mail: todd.myers@med.navy.mil

Naumova, Elena N.
Department of Civil & Environmental Engineering, Tufts University, Medford, MA, US 02155, E-mail: elena.naumova@tufts.edu

Pavlin, Julie A.
Armed Forces Research Institute of Medical Sciences, Bangkok, Thailand, 10400, E-mail: julie.pavlin@us.army.mil

Pinzon, Jorge E.
Science Systems & Applications, Inc., Lanham, MD, US 20706 & Goddard Space Flight Center, Greenbelt, MD, US 20771, E-mail: jorge.e.pinzon@nasa.gov

Pulliam, Juliet R.C.
College of Veterinary Medicine, University of Florida, Gainesville, FL, US 32610, E-mail: pulliam@ufl.edu

Ragain, R. Michael
Saint Louis University School of Public Health, St Louis, MO, US 63103,E-mail: Ragain@slu.edu

Reisen, William K.
School of Veterinary Medicine, University of California, Davis, CA, US 95616, E-mail: wkreisen@ucdavis.edu

Richards, Allen L.
Naval Medical Research Center, Silver Spring, MD, US 10001, E-mail: allen.richards@mad.navy.mil

Rosenberg, Mark
Department of Geography, Queen's University, Kingston, Ontario, Canada K7L 3N6code, E-mail: mark.rosenberg@queensu.ca

Rommel, Robert G.
BioMedware, Ann Arbor, MI US 48104-1382, Email: Robert.rommel@biomedware.com

Scharl, Arno
Department of New Media Technology, MODUL Am Kahlenberg 1, 1190, University of Vienna, WIEN, Austria, E-mail: arno.scharl@modul.ac.at

Schwab, David J.
NOAA, Office of Oceanic & Atmospheric Research, Ann Arbor, MI, US 48108, E-mail: david.schwab@noaa.gov

Selinus, Olle
Geological Survey of Sweden (retired), Linnaeus University, Kalmar, Sweden 39233, E-mail: olle.selinus@gmail.com

Simpson, Gary
New Mexico Department of Health (retired), E-mail: garyl.simpson@comcast.net

Skelly, Chris
University of Queensland, St. Lucia, Brisbane, Queensland 4072, Australia, E-mail: wchris.skelly@googlemail.com

Soebiyanto, Radina P.
Universities Space Research Association, Columbia, MD, US 21044 & NASA/Goddard Space Flight Center, MD 20771, E-mail: radina.p.soebiyanto@nasa.gov

Sonricker, Amy L.
Massachusetts Institute of Technology & Children's Hospital, Boston, MA, US 02215, E-mail: amy.sonricker@childrens.harvard.edu

Stanhope, William
Institute for Biosecurity, Saint Louis University, School of Public health, St. Louis, MO, US 63104, E-mail: Stanhope@slu.edu

Steinnes, Eiliv
Department of Chemistry, Norwegian University of Science and Technology, Trondheim, Norway, NO-7491, E-mail: eiliv.steinnes@chem.ntnu.no

Stohlgren, Thomas J.
US Geological Survey, Fort Collins Science Center, Fort Collins, CO, US 80526-8118, E-mail: stohlgrent@usgs.gov

Stumpf, R.P.
National Oceanic and Atmospheric Administration, National Ocean Service, Silver Spring, MD, US 20910, E-mail richard.stumpf@noaa.gov

Tilburg, Charles E.
University of New England, Biddeford, ME, US 04093, E-mail: ctilburg@une.edu

Tong, Daniel Q.
Center for Spatial Information Science & Systems, George Mason University, Fairfax, VA, US 22030, E-mail: quansong.tong@nasa.gov

Trainer, Vera L.
NOAA, National Marine Fisheries Service, Seattle, WA, US 98112, E-mail: vera.l.trainer@noaa.gov

Trtanj, Julie
NOAA Ocean Service, Silver Spring, MD, US 20910, E-mail: juli.trtanj@noaa.gov

Turner, Elizabeth J.
NOAA, National Ocean Service, Durham, NH, US 03824, E-mail: elizabeth.turner@noaa.gov

Ward, T.G.
City of Lubbock Health Department (retired), Lubbock, TX, US 79411, E-mail: not available

Weinstein, Philip
Graduate Research Centre, University of South Australia, Adelaide, South Australia 5001, Australia, E-mail: philip.weinstein@unisa.edu.au

Wimberly, Michael
Geographic Information Science Center of Excellence, South Dakota State University, Brookings, SD, US 57007, E-mail: Michael.Wimberly@sdstate.edu

Witt, Clara J.
USPHS, Center for Disaster & Humanitarian Assistance Medicine, MD, US 20814, E-mail: cwitt@cdham.org

Wood, Robert J.
NOAA, National Ocean Service, Oxford, MD, US 21654, E-mail: bob.wood@noaa.gov

Wynne, Timothy T.
NOAA, National Ocean Service, Silver Spring, MD, US 20910, E-mail: timothy.wynne@noaa.gov

Xiao, Xiangming
Center for Spatial Analysis, University of Oklahoma, Norman, OK, US 73019, E-mail: xiangming.xiao@ou.edu

Yang, Phil
George Mason University, Fairfax, VA, US 22030, E-mail: not available

Zaitchik, Benjamin
Johns Hopkins University, Baltimore, MD, US 21218, E-mail: zaitchik@jhu.edu

Zayed, Alia
Cairo University & US Naval Medical Research Unit, Cairo, Egypt, FPO AE 098-0007, E-mail: Alia.Zayed.eg@med.navy.mil

Zeeman, Stephan I.
Department of Marine Sciences, University of New England, Biddeford, ME, US 04005, E-mail: szeeman@une.edu

Zelicoff, Alan
Saint Louis University School of Public Health, St. Louis, MO, US 63103, E-mail: Zelicoff@slu.edu

Zollner, Gabriela
Walter Reed Army Institute of Research, Silver Spring, MD, US 20910, E-mail: gabriela.zollner@us.army.mil

Environmental Tracking for Public Health Surveillance – Morain & Budge (eds)
© 2013 Taylor & Francis Group, London, ISBN 978-0-415-58471-5

Acronyms

ACE	Atmospheric Chemistry Experiment
ACES	Applied Climate for Environment & Society
ADAM	Asian Dust Aerosol Models
ADDS	Africa Data Dissemination Service
ADHS	Arizona Department of Health Services
AFO	Animal Feeding Operations
AFRIMS	Armed Forces Research Institute of Medical Sciences
AI	Avian Influenza
AIC	Akaike Information Criterion
AIDS	Acquired Immune Deficiency Syndrome
AIRS	Atmospheric Infrared Sounder
Aka	Also Known As
ALADIN	Aire Limitee Adaptation Dynamique INitialisation
ALI	Advanced Land Imager
AMSR-E	Advanced Microwave Scanning Radiometer-EOS
AOD	Aerosol Optical Depth *aka* Aerosol Optical Thickness
API	Annual Parasite Indices
APRHB	Air Pollution and Respiratory Health Branch
AQS	Air Quality System
ARC	Ames Research Center
ARIMA	Autoregressive Integrated Moving Average
AROME	Applications of Research to Operations at Mesoscale
ASDC	Atmospheric Science Data Center
ASP	Amnesic Shellfish Poisoning
ASTER	Advanced Spaceborne Thermal Emission & Reflection Radiometer
ATSDR	Agency for Toxic Substances & Disease Registry
AVHRR	Advanced Very High Resolution Radiometer
AVIRIS	Airborne Visible/Infrared Imaging Spectrometer
BEBOV	Bundibugyo Ebola Vrus
BFU	Bacterial-colony Forming Unit
BI	Business Intelligence
BIDSS	Border Infectious Disease Surveillance System
BRT	Boosted Regression Trees
BSA	Black-Sky Albedo
BTD	Brightness Temperature Difference
CABLE	CSIRO Atmosphere Biosphere Land Exchange
CAFO	Concentrated Animal Feeding Operations
CALIOP	Cloud & Aerosol LiDAR Orthogonal Polarization
CART	Classification & Regression Tree
CAS	Complex Adaptive System
CASTNET	Clean Air Status & Trends Network
CATHALAC	Water Center for the Humid Tropics of Latin America & the Caribbean
CBEPS	Chesapeake Bay Ecological Prediction System
CCAD	Central American Commission for Environment & Development

CCC	Climate Change Collaboratory
CDC	Centers for Disease Control & Prevention
CDPH	California Department of Public Health
CEGLHH	Center of Excellence for Great Lakes & Human Health
CERES	Cloud & Earth Radiation Energy Sensor
CEV	California Encephalitis Virus
CFU	Colony-Forming Unit
CI	Cryptosporidium Infections *also* Cyanobacterial Index
CIEBOV	Cote d'Ivoire Ebola Virus
CIESIN	Center for International Earth Science Information Network
CL	Cutaneous Leishmaniasis
CLM	Common Land Model
CMAQ	Community Multi-scale Air Quality
CMAVE	Center for Medical, Agricultural & Veterinary Entomology
CMORPH	Climate Prediction Center Morphing Technique
CMS	Content Management System
CMVSRP	California State Mosquito-borne Virus Surveillance & Response Plan
CNES	Centre National d'Etudes Spatiales
CNSA	Chinese National Space Agencies
COADS	Comprehensive Ocean-Atmosphere Data Set
CONAE	Comisión Nacional de Actividades Espaciales
COP	Community-of-Practice
COPD	Chronic Obstructive Pulmonary Disorder
CORINAIR	CORe INventory AIR
CPI	Cryptosporidium Infections
CSA	Canadian Space Agency
CTM	Chemical Transport Model
CSR	Canine Seroprevalence Rate
CST	Cross-Species Transmission
CSV	Comma Separated Values
CT	Computerized Tomography
CVEC	Center for Vectorborne Diseases
CyanoHABs	Cyanobacterial Harmful Algal Blooms
DA	Domoic Acid
DAAC	Distributed Active Archive Center
DALY	Disability-Adjusted Life Years
DBPs	Disinfectant By-Products
DCL	Diffuse Cutaneous Leishmaniasis
DDT	Dichlorodiphenyltrichloroethane
DEM	Digital Elevation Model
DENV	Dengue Virus
DF	Dengue Fever
DHF	Dengue Haemorrhagic Fever
DHS	Demographic & Health Surveys
DINEOF	Data Interpolation with Empirical Orthogonal Functions
DLR	Deutsches Zentrum für Luft und Raumfahrt
DPSEEA	Driving force, Pressure, State, Exposure, Effects & Action
DQSS	Data Quality Screening Service
DREAM	Dust Regional Atmospheric Model
DSS	Decision Support System
DUST-DISC	Deep Blue Utilization of SeaWiFS/Data & Information Services Center
EBOV	Ebola Virus
E. coli	*Escherichia coli*
ECMWF	European Centre for Medium Range Weather Forecasts

ECOHAB PNW	Ecology & Oceanography in the Pacific Northwest
ECOWAS	Economic Community of West African States
EDAC	Earth Data Analysis Center
EEEV	Eastern Equine Encephalomyelitis
EHEC	Enterohaemorrhagic *Escherichia coli*
EIA	Enzyme Immunoassay
EIP	Extrinsic Incubation Period
ELISA	Enzyme-Linked Immunosorbent Assays
EMD	Empirical Mode Decomposition
EMPRES	Emergency Prevention System
EMS	Emergency Management Systems e.g. Ambulance services
ENFA	Environmental Niche Factor Analysis
ENM	Ecological Niche Modelling
ENSO	El Niño Southern Oscillation
ENVISAT	Environmental Satellite
EO	Earth Observation
EOS	Earth Observing System
EOSDIS	Earth Observing System Data & Information System
EPA	Environmental Protection Agency
EPHTN	Environmental Public Health Tracking Network
EPHTS	Environmental Public Health Tracking System
EPI	Environmental Performance Index
ER	Emergency Room
ERDAS	Earth Resources Data Analysis System
EROS	Earth Resources Observation & Science Centre
ESA	European Space Agency
ESMF	Earth System Modeling Framework
ESRI	Environmental Systems Research Institute
EVI	Enhanced Vegetation Index
FAO	Food & Agriculture Organization
FDA	Food and Drug Administration
FEWS	Famine Early Warning System
FOSS	Free & Open Source GIS
FTP	File Transfer Protocol
GAM	Generalized Additive Model
GAMS	General Algebraic Modeling System
GARP	Genetic Algorithm for Rule-set Production
GASP	GOES Aerosol & Smoke Product
GBD	Global Burden of Disease
GCMs	Global Climate Models
GDAL	Geospatial Data Abstraction Library
GDD-WB	Growing Degree Day-Water Budget
GDP	Gross Domestic Product
GDS	GrADS Data Server
GEIS	Global Emerging Infectious Surveillance and Response System
GEO	GEostationary-Orbit (GEO); *also* Group on Earth-Observations
GEOSS	Global Earth-Observing System of Systems
GES-DISC	Goddard Earth Sciences - Data & Information Service Center
GFATM	Global Fund to Fight AIDS, Tuberculosis & Malaria
GFCS	Global Framework for Climate Services
GFDL	Geophysical Fluid Dynamics Laboratory
GFS	Global Forecast System
GHCN	Global Historical Climatology Network
GHMF	Global Hazard Model-Flood

GIMMS	Global Inventory Monitoring & Mapping Studies
GIOVANNI	GES-DISC Interactive Online Visualization and Analysis Infrastructure
GIS	Geographic Information System
GISN	Global Influenza Surveillance Network
GLAS	Geoscience Laser Altimeter System
GLCF	Global Land Cover Facility
GLCFS	Great Lakes Coastal Forecasting System
GLDAS	Global Land Data Assimilation System
GLM	Generalized Linear Model
GMAO	Global Modeling & Assimilation Office
GMS	Greater Mekong Sub-region
GMU	George Mason University
GNOME	General NOAA Operational Modeling Environment
GOARN	Global Outbreak Alert & Response Network
GOCART	Goddard Chemistry Aerosol Radiation & Transport
GOES	Geostationary Operational Environmental Satellite
GOME	Global Ozone Monitoring Experiment
GPCP	Global Precipitation Climatology Project
GPM	Global Precipitation Measurement
GPS	Global Positioning System
GPW	Gridded Population of the World
GPWfe	Gridded Population of the World future estimates
GRACE	Gravity Recovery & Climate Experiment
GRUMP	Global Rural-Urban Mapping Project
GSFC	Goddard Space Flight Center
GSSTF	Goddard Satellite-based Surface Turbulent Fluxes
HAB	Harmful Algal Bloom
HAB-OFS	Harmful Algal Bloom Operational Forecast System
HACCP	Hazard Analysis Critical Control Point
HAPEM	Hazardous Air Pollution Exposure Model
HBM	Hierarchical Bayesian Model
HAN	Health Alert Network
HPS	Hantavirus Pulmonary Syndrome
HEM	Human Exposure Model
HF	Haemorrhagic Fever
HIV	Human Immunodeficiency Virus
HPC	High-Performance Computing
HTTP	Hypertext Transfer Protocol
HUS	Haemolytic Uraemic Syndrome
HWRF	Hurricane Weather Research Model
HYSPLIT	HYbrid Single-Particle Lagrangian Integrated Trajectory
IAEA	International Atomic Energy Agency
IASI	Infrared Atmospheric Sounding Interferometer
IC	International Charter
ICA	Independent Component Analysis
ICDDRB	International Centre for Diarrhoeal Disease Research
ICSU	International Council for Science
IDEL	Infectious Disease Eco-climatic Link
IDL	Interface Description Language
IDNDR	International Decade for Natural Disaster Reduction
IDODSS	Infectious Disease Outbreak Decision Support System
IDV	Integrated Data Viewer
IEDCR	Epidemiology Disease Control and Research
IFA	Indirect Immunofluorescence Assay

IGAD	Intergovernmental Authority for Development
IGBP	International Geosphere/Biosphere Programme
IgG	immunoglobulin-G; *hence also* IgA, IgM
IHD	Ischaemic Heart Disease
IHR	International Health Regulations
I-HEAT	Internet-based Heat Evaluation & Assessment Tool
IMF	Intrinsic Mode Function
IOC	Intergovernmental Oceanographic Commission
IOOS	Integrated Ocean Observing System
IPCC	Intergovernmental Panel on Climate Change
IRI	International Research Institute
IRS	Indoor Residual Spraying; *also* Indian Remote Sensing Satellite
ISID	International Society for Infectious Diseases
ISRO	Indian Space Research Organization
IT	Information Technology
IUBS	International Union of Biological Sciences
IUSS	International Union of Soil Sciences
JAXA	Japan Aerospace Exploration Agency
JEV	Japanese Encephalitis Virus
JPL	Jet Propulsion Laboratory
KML	Keyhole Mark-up Language
LAADS	Level-1 Atmosphere Archive & Distribution System
LACV	La Crosse Virus
LaRC	Langley Research Center
LCLU	Land Cover/Land Use
LCLUC	Land cover/Land Use Change
LDAS	Land Data Assimilation System
LDCM	Landsat Data Continuity Mission
LECZ-URE	Low Elevation Coastal Zone Rural Urban Estimates
LEO	Low-Earth-Orbit
LiDAR	Light Detection & Ranging
LIS	Lighting Image Scanner
LISS	Linear Imaging & Self Scanning
LMM	Liverpool Malaria Model
LPRM	Land Parameter Retrieval Model
LR	Logistic Regression
LSMS	Living Standards Measurement Surveys
LST	Land Surface Temperature
MARV	Marburg virus
MASINGAR	Model of Aerosol Species in the Global Atmosphere
Maxent	Maximum Entropy
McIDAS	Man computer Interactive Data Access System
MCL	Muco-Cutaneous Leishmaniasis
MCP	Minimum Convex Polygon
MDDNR	Maryland Department of Natural Resources
MDG	Millennium Development Goals
MEI	Multivariate ENSO Index
MENTOR	Monitoring Environment for Total Risk
MERIS	Medium-spectral Resolution Imaging Spectrometer
MERIT	Meningitis Environmental Risk Information Technologies
MERRA	Modern Era Retrospective-Analysis for Research & Applications
MEWS	Malaria Early Warning System
MFS	Modelling & Forecasting System
MGET	Marine Geospatial Ecology Toolset

MIPAS	Michelson Interferometer for Passive Atmospheric Sounding
MISR	Multi-angle Imaging Spectroradiometer
MLE	Maximum Likelihood Estimate
MNC	Multiple-Nested Coupled
MOCAGE	Multi-scale Chemistry & Transport Model
MODIS	Moderate Resolution Imaging Spectroradiometer
MRI	Magnetic Resonance Imaging
MSFC	Marshall Space Flight Center
MVEV	Murray Valley Encephalitis Virus
MWCC	Media Watch on Climate Change
NAAPS	Navy Aerosol Analysis & Prediction System
NAAR	North American Regional Reanalysis
NAHQP	NASA Health & Air Quality Program
NAO	North Atlantic Oscillation
NASA	National Aeronautics & Space Administration
NASS	National Agricultural Statistics Service
NCAR	National Center for Atmospheric Research
NCDC	National Climate Data Center
NCEP	National Centers for Environmental Prediction
NCEP/eta	National Centers for Environmental Prediction, eta version
NCEP/nmm	NCEP Non-hydrostatic Mesoscale Model
NDVI	Normalized Difference Vegetation Index
NEESPI	Northern Eurasia Earth Science Partnership Initiative
NEPA	National Environmental Policy Act
NFSC	Northwest Fisheries Science Center
NGO	Non-Government Organization
NICED	National Institute for Cholera & Enteric Diseases
NIDIS	National Integrated Drought Information System
NIEHS	National Institute of Environmental Health Sciences
NIH	National Institutes of Health
NIR	Near-infrared
NiV	Nipah Virus
NLDAS	North American Land Data Assimilation System
NMVOC	Non-Methane Volatile Organic Compound
NN	Neural Network
NOAA	National Oceanic & Atmospheric Administration
NOGAPS	Navy Operational Global Prediction Center System
NOS	National Ocean Service
NPL	National Priorities List
NPP	NPOESS Preparatory Project
NSF	National Science Foundation
NWS	National Weather Service
ODE	Ordinary Differential Equations
OGC	Open Geospatial Consortium
OGR	Simple Feature Library
OHHI	Oceans & Human Health Initiative
OIE	World Organization for Animal Health
OMI	Ozone Monitoring Instrument
OML	Operationelle Meteorologiske Luftkvalitsmodeller
OPeNDAP	Open Source Project for a Network Data Access Protocol
ORHAB	Olympic Region Harmful Algal Bloom
OSPM	Operational Street Pollution Model
OTC	Open Top Chamber
PAF-LDAS	Peruvian Amazon Frontier Land Data Assimilation System

PALSAR	Phased Array L-band Synthetic Aperture Radar
PAM	Primary Amoebic Meningoencephalitis
PAR	Population-at-Risk
PCB	Polychlorinated biphenyls
PCR	Polymerase Chain Reaction
PD	Participatory Design
PDA	Personal Digital Assistant
PDE	Partial Differential Equations
PDO	Pacific Decadal Oscillation
PDVI	Paediatric Dengue Vaccine Initiative
PEAM	Potential Epizootic Area Mask
PES	Population Estimation Service
PET	Positron-Emission Tomography
PHAiRS	Public Health Applications in Remote Sensing
PHO	Public Health Official
PI	Pandemic Influenza
PKDL	Post-Kala-azar Dermal Leishmaniasis
PLACE	Population, Landscape & Climate Estimates
PM_{10}	Particulate Matter ($10\,\mu m$)
$PM_{2.5}$	Particulate Matter ($2.5\,\mu m$)
PNWTOX	Pacific Northwest Toxins
POPs	Persistent Organic Pollutants.
POW	Powassan Encephalitis
PR	Precipitation Radar
ProMED	Program for Monitoring Emerging Diseases
PVDI	Paediatric Dengue Vaccine Initiative
RACMO	Regional Atmospheric Climate Model
R&D	Research & Development
RAMS	Regional Atmospheric Modelling System
RBFNN	Radial Basis Function Neural Network
RCMRD	Regional Center for Mapping of Resources for Development
REGARDS	REasons for Geographic And Racial Differences in Stroke
REST	Representational State Transfer
RF	Random Forest
RFLP	Restriction Fragment Length Polymorphism
RH	Relative Humidity
RI	Rotavirus Infections
RMSE	Root Mean Square Error
ROMS	Regional Ocean Modeling System
RVF	Rift Valley Fever
RVFV	Rift Valley Fever Virus
RWQE	Reduced Water Quality Events
SADC	Southern African Development Community
SAMC	Southern Africa Inter-Country Programme for Malaria Control
SARS	Severe Acute Respiratory Syndrome
SAVI	Soil Adjusted Vegetation Index
SBDSS	Syndrome-based Decision Support System
SCIAMACHY	SCanning Imaging Absorption spectroMetre for Atmospheric CHartographY
SDG	Sulphur Dioxide Group
SDSWAS	Sand & Dust Storm Warning Advisory & Assessment System
SE	Susceptible & Exposed (disease condition model *see* SEIR)
SeaWiFS	Sea-viewing Wide Field-of-view Sensor
SEBOV	Sudan Ebola virus
SEDAC	Socioeconomic Data & Applications Center

SEIR	Susceptible, Exposed, Infectious, Recovered (disease condition models)
SERVIR	Regional Visualization & Monitoring System
SHWB	Science for Health & Wellbeing
SI	Solar Insolation *also* Susceptible & Infectious (disease condition model *see* SEIR)
SIM	Simple Arithmetic Averaging
SIR	Susceptible, Infectious, Recovered (disease condition model *see* SEIR)
SLEV	St Louis Encephalitis Virus
SMAP	Soil Moisture Active-Passive
SMS	Sentinel Monitoring Sites
SOAP	Simple Object Access Protocol
SOHC	Stepwise Optimal Hierarchical Clustering
SOI	Southern Oscillation Index
SOM	Self-organizing Map
SORCE	Solar Radiation & Climate Experiment
SPECT	Single Photon Emission Computed Tomography
SPOT	Satellite Pour l'Observation de la Terre
SRTM	Shuttle Radar Topography Mission
SSC	Stennis Space Center
SSEC	Space Science & Engineering Center
SST	Sea Surface Temperature
SSW	Simple Subset Wizard
STEC	Shiga-toxin producing *Escherichia coli*
SVD	Singular Value Decomposition
SVG	Scalable Vector Graphics
SVM	Support Vector Machine
SVR	Support Vector Regression
SWHC	Status Warning Health Codes
SWIR	Short-Wave Infrared
SYRIS	Syndrome Reporting Information System
TBE	Tick-borne Encephalitis
TC	Tasseled Cap
TCI	Tasseled Cap Index
TCDD	Tetrachlorodibenzodioxin
TES	Troposheric Emissions Spectrometer
TESSEL	Tiled ECMWF Scheme for Surface Exchanges over Land
THORPEX	The Observing system Research & Predictability Experiment
THREDDS	Thematic Real-time Environmental Distributed Data Services
TIROS	Television Infrared Observation Satellite
TM	Thematic Mapper
TMI	TRMM Microwave Imager
TNF	Tumour-Necrosis Factor
TOMS	Total Ozone Mapping Spectrometer
TOPS	Terrestrial Observation & Prediction System
TOVS	TIROS Operational Vertical Sounder
TRMM	Tropical Rainfall Measuring Mission
TSR	Task & Stay Resident
TTHM	Trihalomethane
UAV	Unoccupied (or unManned) Aerial Vehicle
UCS	Union of Concerned Scientists
US/AID	US Agency for International Development
USGCRP	United States Global Change Research Programme
USGS	United States Geological Survey
USLE	Universal Soil Loss Equation
USRA	Universities Space Research Association

UV	UltraViolet
VAAA	Volcano Ash Advisory Archive
V&V	Verification & Validation
VEEV	Venezuelan Equine Encephalomyelitis
VFCE	Valley Fever Center for Excellence
VI	Vegetation Index
VIC	Variable Infiltration Capacity
VIIRS	Visible & Infrared Scanner
VL	Visceral Leishmaniasis
VNIR	Visible & Near Infrared
WCS	Web Coverage Services
WebCGM	Web Computer Graphics Metafile
WEEV	Western Equine Encephalomyelitis
WELD	Web-Enabled Landsat Data
WHO	World Health Organization
WIO	Western Indian Ocean
WIST	Warehouse Inventory Search Tool
WMO	World Meteorological Organization
WMS	Web Mapping Service
WNV	West Nile Virus
WONDER	Wide-ranging Online Data for Epidemiological Research
WPC	Weighting by Pixel Counts
WPS	Web Processing Service
WSA	White Sky Albedo
YSAWS	Yellow Sand Activity Warning System
YCELP	Yale Center for Environmental Law & Policy
ZCL	Zoonotic Cutaneous Leishmaniasis
ZEBOV	Zaire Ebola virus

Introduction

Environmental Tracking for Public Health Surveillance – Morain & Budge (eds)
© 2013 Taylor & Francis Group, London, ISBN 978-0-415-58471-5

Chapter 1

Earth observing data for health applications

S.A. Morain & A.M. Budge (Auth./eds.)
Earth Data Analysis Center, University of New Mexico, NM, US

ABSTRACT: This chapter summarizes more than forty years of scientific, technological and societal trends leading to current practices for Earth observations from space. All sectors of society are participating in a global common cause to use near-Earth space environments to collect and warehouse whatever data are needed to better understand disease transmission mechanisms, human exposures to air, water, soil, and vector-borne diseases, and to promote health and wellbeing. Subsequent chapters show how this common cause has stimulated environmental tracking for public health surveillance.

1 TECHNICAL & SOCIAL TRENDS

This book is based on technical and scientific literature beginning in the 1960s and that, by the turn of the Century, had transformed advanced applications of Earth-observing satellites into quantifiable economic and social benefits. Space-based Earth science applications began with Geostationary Operational Environmental Satellites (GEOS) and Television Infrared Observation Satellites (TIROS) to record changing cloud patterns and to infer surface wind speeds and directions. Monitoring cloud patterns to forecast hurricanes and typhoons were among the first civilian objectives because of their devastating impacts on life and property. Since the 1960s, hundreds of satellites have been launched carrying sensors with increasingly better combinations of spatial, spectral, and temporal resolutions. These sensor systems have produced long-term records of physical environmental cycles that are now being used in concert with knowledge from medical science communities to examine physical earth environments that have human health consequences.

In the 1970s low-earth-orbiting (LEO) satellites (namely, Landsat) added the ability to record fluxes in surface phenomena (Morain 1998). Since then, phenomenal progress has been made on a host of cross-cutting applications, among which human health has emerged as a defining thread (Epstein 1986; Norboo *et al.*, 1991; Colwell 1996; King *et al.*, 1999; Goudie *et al.*, 2001; Yates *et al.*, 2002; Wiggs *et al.*, 2003; Kaya *et al.*, 2004; Deckers & Steinnes 2004; Oxford *et al.*, 2005; Selinus 2005; Taubenberger *et al.*, 2005; Sterman 2006; Morain & Budge 2010; Budge *et al.*, 2010). Drivers of these changes included advances in spectral sensor design and digital image processing technology; the advent of data discovery, access, and retrieval systems; advanced forms of information extraction using data fusion and assimilation; and, migration from one-dimensional to multi-dimensional modelling systems.

1.1 *Sensor technology trends and space-based health applications*

Table 1 highlights important technology trends contributing to international interest in space-based health applications. Since Landsat was launched in 1972, each line in the list had its own pace for progress; and each stimulated growth in a variety of communities-of-practice (COPs). Vertical convergence of technologies has also accelerated, leading to complex systems for health applications

Table 1. Progress in Earth-observing technology from pre-1960s (top) to present (bottom).

(1860s–1960s)	(1970s-present)
Balloons	Satellites
Visible spectrum (aerial & land cameras)	Total EM spectrum (active & passive sensors)
Analogue film images	Digital image data ($2^6 - 2^{8+}$ bit) sensitivity
Static photographs & multi-spectral images	Dynamic, time series images & short revisit cycles
Manual analyses	High-speed machine-assisted analyses
Photogrammetric reconnaissance	Complex situational problem solving
Commercial firms & government archives	Multiple national government & commercial archives
Site-specific project areas	Continuous global surveillance & surveys
Experimental aerial and satellite sensors	Multi-purpose sensor webs
Long image delivery times	Nearly real-time, dynamic forecasting
Uni-dimensional models	Interactive modelling frameworks
Govt. & private project-specific applications	International cooperating research teams
National/regional/local/site-oriented research	Multi-national problem solving

that were only beginning to emerge in the first decade of the 21st Century. An excellent, but now out-dated bibliography relating these trends was produced by Beck (2004). It lists hallmark efforts published between 1970 and 2004. Among the seminal citations is one by Beck and her associates outlining roles for remote sensing and health in the context of new surveillance opportunities (Beck *et al.*, 2004).

Temporal and spatial resolution of satellite and air-borne sensors, and sensor detectivity are keys to promoting and evolving health applications. An ability to observe details in proximal environments enables one to predict health outcomes as they develop locally; and an ability to monitor distal environments enables one to forecast health risks migrating over larger areas and longer distances. An example of the former might be the Haitian earthquake of 2009; and, of the latter, downwind migration of nuclear particles from Chernobyl in 1986 or Fukushima in 2011. Improvements in sensor design parameters are actually driving new human health applications. However, one of the most contentious issues of sub-orbiting and orbiting sensors is how fast these improvements have taken place against a backdrop of national security concerns that began with Landsat-1.

Figure 1 shows how improved sensor resolution and detectivity have progressed since 1972. A vignette from the Cold War between the US and former USSR illustrates a technology issue that, at the time, controlled the pace of civilian space-based applications. Through logic derived from sensor design parameters available publicly in the 1980s, information was published in 1986 that compromised US security and caused the book *Deep Black* to be temporarily banned (Burrows 1986). Burrows described how precise measurements of missile lengths from spy satellites were measured during the time they were being moved to launch sites and were thus *visible* to international compliance monitoring (Tsipis 1987). At the heart of this monitoring issue was the difference between *reconnaissance* and *surveillance*. Reconnaissance is intelligence gathered by fine resolution sensors or cameras that collect imagery on-demand over specific locations and for specific purposes, or by essentially staring at an area of interest. Surveillance is routine data gathering set by orbital design parameters that determine revisit cycles, and sensor parameters that record specific spectral regions. Their purpose is to monitor natural and human-caused phenomena and longer term changes in forestry, agriculture, directions of urban sprawl, shorelines and similar spectral and spatial changes. In Figure 1, improvements in sensor resolution are shown from right (1000 m) to left (0.5 m). The grey areas between secret and civilian uses represent the zone of this contention. In the US this zone is defined by official space policies revised periodically at Presidential and Congressional levels. Through time, technologies advanced enough that decades-old photographs from formerly secret satellite programs in the US have subsequently been made available for civilian uses.

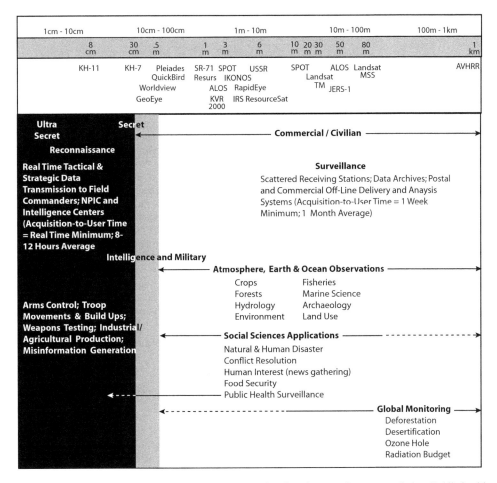

1cm - 10cm	10cm - 100cm	1m - 10m	10m - 100m	100m - 1km

| 8 cm | 30 cm | .5 m | 1 m | 3 m | 6 m | 10 m | 20 m | 30 m | 50 m | 80 m | 1 km |

KH-11   KH-7   Pleiades   SR-71 SPOT   USSR   SPOT   ALOS Landsat   AVHRR
            QuickBird   Resurs   IKONOS   Landsat   MSS
         Worldview   ALOS   RapidEye   TM   JERS-1
         GeoEye   KVR   IRS ResourceSat
                  2000

Ultra Secret   **Secret**   ← ——————————— **Commercial / Civilian** ——————————— →

Reconnaissance

Real Time Tactical & Strategic Data Transmission to Field Commanders; NPIC and Intelligence Centers (Acquisition-to-User Time = Real Time Minimum; 8-12 Hours Average

                              Surveillance
                   Scattered Receiving Stations; Data Archives; Postal
                   and Commercial Off-Line Delivery and Anaysis
                   Systems (Acquisition-to-User Time = 1 Week
                   Minimum; 1 Month Average)

Intelligence and Military

              ← ——— **Atmosphere, Earth & Ocean Observations** ——— →
                   Crops   Fisheries
                   Forests   Marine Science

Arms Control; Troop Movements & Build Ups; Weapons Testing; Industrial/ Agricultural Production; Misinformation Generation

                   Hydrology   Archaeology
                   Environment   Land Use

              ← ——— **Social Sciences Applications** ——— - - - - →
                   Natural & Human Disaster
                   Conflict Resolution
                   Human Interest (news gathering)
                   Food Security
  ← - - -   Public Health Surveillance

                                       Global Monitoring →
  ← - →
                                       Deforestation
                                       Desertification
                                       Ozone Hole
                                       Radiation Budget

Figure 1. The scope of social science applications resulting from improved sensor resolution. Public health surveillance is identified in this book as an application that would benefit significantly from higher resolution environmental data.

Relaxation of policies related to sensor resolution has expanded the range of both air-borne and satellite surveillance into realms that are relevant to the social sciences in general, and the health sciences in particular. Today, sub-metre spatial resolution digital image data are supplied routinely by commercial vendors. There are many health applications that use these satellite data both in the commercial/civilian area and the intelligence/military operations area that were once classified or difficult for research communities to obtain. This book reports on many direct applications, and touches on ancillary applications associated with disaster relief management.

Since Landsat-1 was launched, over a thousand satellites have been launched worldwide (UCS 2011). The overwhelming majority of these have been communication satellites sponsored by a growing number of international government and commercial entities. Many of these (e.g. communication satellites and global positioning satellites) are playing valuable roles for health and wellbeing through telemedicine and rescue operations. Of special interest here, however, is the nearly 240 satellites launched specifically for Earth observations that support and enhance earth and social sciences. The Union of Concerned Scientists (UCS) categories for satellites are not altogether exclusive. Among them are: earth observations, earth sciences, meteorology, global positioning and navigation, remote sensing, remote sensing research, remote sensing technology

Table 2. Selected list of societal awareness and collaborations linking environment and health.

Year	Event
1984	CEOS adopts principle on international cooperation to develop Earth-observing systems
1988	UNEP/WMO creates Intergovernmental Panel on Climate Change (IPCC)
1992	1st World Summit on Sustainable Development (WSSD) places human health at centre stage
1999	Recognition that core data sets are needed for international environmental policies
1999	ESA/CNES creates the International Charter for disaster relief assistance using EO data
2002	NASA includes human health as a cross-cutting issue for space science applications
2003	Group on Earth-Observations (GEO) created to develop an international *system-of-systems*
2005	WMO/WWRP inaugurates *The Observing System Research and Predictability Experiment*
2007	NSF supports creation of an *International Society for Disease Surveillance*
2007	ICSU initiates interdisciplinary group to explore health and wellbeing issues in urban areas
2008	ISPRS adopts *Beijing Declaration* to support disaster and health monitoring from space
2010	UNOOSA holds 1st international workshop on economic and societal benefits from space sensors
2011	ISPRS holds international symposium on *Advances in Geospatial Technologies for Health*
2011	ICSU approves SHWB initiative for systems approach to health in urban environments

and development, and remote sensing reconnaissance. At least 115 countries are now space-faring nations that design, launch, operate, or collect data from satellites, many of which are multinational and commercial efforts. Most of the last four decades have been devoted to sensors gathering data about weather and climate, natural resources, and disaster detection (that is hurricanes, tsunamis, floods, droughts, and fires, among others).

1.2   *Societal trends and global awareness of health monitoring*

Table 2 lists major social trends that recognize society's need at local, national, regional, and global scales to monitor individual and population health. Given the speed at which viruses can mutate and spread, and the growing size of populations being affected, the global economy requires multinational and intergovernmental collaborations to insure public safety. Human health and disaster mitigation represent the right and left hands for establishing a global safety net for both the humanitarian and economic wellbeing of nations. Significant among these hand-shakes are the expanding collaborations between civilian and military research and development (R&D) communities-of-health practice; but this is traditionally truer on the health side than on the technology side. There are hundreds of regional, national, provincial, and local efforts that are engaging satellite technology in their research and health care programmes. Many of these in the developing world are looking at the next generation of nano-satellites as their entry into independent local and regional health care infrastructures that include telemedicine, environmental monitoring and health surveillance.

Subsequent Chapters describe and demonstrate how data returned from these systems can be analysed mathematically, fused or assimilated into both simple and complex modelling systems, and incorporated into sophisticated information systems for health tracking and intervention decisions. The technology has matured rapidly and is being adopted and incorporated into routine health monitoring programmes. As capabilities increase over coming decades, and their respective technologies are integrated, one can imagine that health systems will evolve to link molecular medicine with macro physical and chemical processes at the global level.

1.2.1   *Rise of international health programmes*

Earth observations from space matured independently from concerns for human health; but not in an unconscious way. Discussions that follow show clearly that public health, individual health, and human wellbeing have intrigued and motivated a variety of COPs almost from the beginning of EO image and data collection. In the Modern era one can cite the series of World Summits on Sustainable Development (WSSD), the Committee on Earth Observation Satellites (CEOS), and

Table 3. Responses to environmental events having health/wellbeing consequences (IC-2000–2011).

World Region*	NA	C/SA	EU	RU	AF	ME	IN	AS	AU	OC	Total by event
Floods	12	35	20	3	26	11	6	18	1		132
Volcanoes	1	10	2					2		1	16
Earthquakes		6	2	1	5	10	2	8		3	37
Landslides		8	4	1	2		1	9			25
Ocean storms	1	3			1		1	2			8
Wildfires	3	6	7			1			1		18
Snow								2			2
Tropical cyclones					1		2	6	1	2	12
Tsunami		1						2	1	1	5
Hurricanes	4	9	2				1			2	18
Tornadoes	1										1
Human caused**	1	2	4	1	1	3		2		1	15
Total by region	23	80	41	6	36	25	13	51	4	1	289

*NA = North America; C/SA = Central/South America; EU = Europe; RU = Russia; AF = Africa; ME = Middle East; IN = India; AS = Asia; AU = Australia; OC = Oceania (includes New Zealand)
**Includes oil spills

the creation of the Intergovernmental Panel on Climate Change (IPCC) as transformational stimuli to convert thoughts into actions. Examples from international programmes are provided below to frame the general course of progress.

1.2.1.1 International Charter

The International Charter (IC) was created after the UNISPACE III conference. The European and French space agencies (ESA and CNES) were the founders. In subsequent years, it was joined by space agencies representing the US, Canada, Germany, India, Argentina, Japan, and China (NASA, CSA, DLR, ISRO, CONAE, JAXA and CNSA respectively). The US is represented by both the National Oceanic and Atmospheric Administration (NOAA) and the US Geological Survey (USGS).

The IC provides imagery and data through its growing list of affiliates to any country impacted by a disaster, and which officially requests technical assistance. Between November 2000 and October 2011, the IC delivered assistance for over 282 disasters worldwide, most of which had associated public health and wellbeing effects (IC 2012). By the end of 2011, almost 300 disastrous weather events had sparked national and regional decision makers to request synoptic satellite data and imagery from the IC to inform local authorities about the extent of damage, and to direct their mitigation efforts. Table 3 is a tabulation of these events by type and by world region obtained from its website. Each event carried explicit human wellbeing and health issues needing immediate attention; and, implicit consequences for cascading economic and social impacts.

Table 3 demonstrates that floods, earthquakes, and landslides account for two-thirds of the world's catastrophes. Almost thirty per cent are caused by volcanic eruptions, wildfires, tropical cyclones, hurricanes, and human-caused accidents (mainly oil spills). All but the last are accompanied typically by injuries, diseases and mortality. It is also true that two-thirds of the events represent extreme weather events. Floods, ocean storms, snow, tropical cyclones, hurricanes and tornadoes are associated directly with atmospheric conditions. Some wildfires, landslides and oil spills are also attributable indirectly to weather conditions, discounting those not directly attributable to human causes. Extreme weather events affect human wellbeing and have adverse impacts on the young, the elderly and diseased individuals. Major disease outbreaks usually occur days or weeks after an event. The pervasive roles of weather on health are a repeating theme in subsequent Chapters, from both the environmental health and human health perspectives.

Health communities rely on the bio-geosciences to predict daily and annual weather cycles with increasing accuracy for reliable health interventions and preparedness. For epidemiology, they also are interested in decadal and longer term climate trends if they have acceptable levels of uncertainty against which to correlate health outcomes in populations. Both of these needs can be assisted by advanced sensors and data processing strategies, and by more robust interactive numerical environment/health modelling systems.

Aside from the IC, a growing number of other national and international programmes are underway to combine Earth-observations and geospatial analyses with human health R&D. A few of those listed in Table 2 are highlighted below.

1.2.1.2   Human health and global urbanization

By 2008, most people lived in urban or rapidly urbanizing environments, many of which are informal slum settlements lacking basic health and wellbeing infrastructures. They exist in virtually all countries. Recognizing this trend, the International Council for Science (ICSU) formulated a ten year initiative focused on science for health and wellbeing (SHWB) in urban areas, beginning with those developing in sub-Saharan Africa (Budge *et al.*, 2009; ICSU 2011). This programme is based on a systems analysis approach directed at health policy and decision-making. Consequently, it requires EO data from existing and evolving international sensor systems; mathematical modelling of contextual factors that address *a priori* needs in defined urban areas and their inhabitants; and open access to evolving data systems for biodiversity and human health. SHWB will involve mapping processes for human-environment interactions, and learning how these dynamics lead to measurable outcomes. Appropriate systems analysis models are emerging that integrate data types and data sets that can characterize these interactions to identify knowledge gaps, while at the same time providing scientifically valid decision-making and policy-making information.

Health and wellbeing COPs will require clinician-based and data mining systems for diagnoses; routine environmental forecasts for alerting authorities to weather events or zoonotic outbreaks that trigger health outcomes; reliable epidemiology from cohort and longitudinal analyses that identify cause and effect relationships; and a system of systems to inform decision makers. All of these represent infrastructural elements of an end-to-end health forecasting and preparedness system for a global sustainable and responsible social and economic future.

1.2.1.3   GEOSS

The global Earth-observing system of systems (GEOSS) is the end goal of the GEO. It was created at a ministerial level meeting in 2003, stimulated by the 2002 World Summit on Sustainable Development. It grew from a relatively small number of member nations to a 2011 membership of eighty-eight countries, the European Commission, and several national and international scientific bodies. Human health is recognised as a cross-cutting element in the 2012–2015 Work Plan adopted at the GEO-VIII plenary in Istanbul (GEO 2011a). The work plan's scope focuses on target driven approaches to facilitate the GEOSS ten year implementation plan. Of interest here is the topical approach for members and users to obtain information for societal benefits.

Sixteen tasks were approved by the 8th Plenary. In the broadest sense, they all have ties to human health as a cross-cutting societal benefit; but three have direct ties. One of these (*DI-01*) was designed to reduce loss of life and property from natural and human induced disasters. In collaboration with the IC, it is being implemented by the geo-hazards and coastal zone COPs, among others. In addition, there are two health tasks designated as HE-01 and HE-02. HE-01 focuses on developing tools and information for health decision-making. Its implementation has four sectors supported by implementation teams, some of which are listed in Table 2. These sectors are air-borne, water-borne, vector-borne diseases, and holistic approaches. Task HE-02 is focused on tracking pollutants and has two components: developing a global Mercury observation system; and, tracking global persistent organic pollutants (POPS) as global change indicators. These activities are integrated technologically with GEOSS member organizations to ensure comprehensive development and application (GEO 2011b).

1.2.1.4 THORPEX/MERIT

The World Meteorology Organization (WMO) organized a ten year experimental programme called THORPEX to extend weather forecasting capabilities in Africa. The core objectives of this experiment are to: 1) build a knowledge base for understanding the variability of high-impact weather systems over Africa to explore the limits of predictability; 2) identify the optimal design of observing and communicating systems needed to disseminate data and information; 3) facilitate exchange of information for THORPEX-related forecasts and research outputs; and 4) develop socio-economic applications using weather forecast and climate change indicators for decision-making communities. An element of THORPEX is the meningitis environmental risk information technologies, or MERIT project (MERIT 2011). Its aim is to initiate projects that relate real-time processing of historical EO data and models to *in-situ* socioeconomic and epidemiological data for detecting disease trends, and to provide early warning systems for meningitis epidemics.

1.2.2 *Weather, climate, climate change and health*

There is a growing body of evidence that extreme weather events are increasing in frequency; and, according to some sources, severity. In the US alone there were nearly 3000 monthly weather records broken in communities around the nation, reportedly because of heat waves, floods, or fires (Spatialnews 2011). Many invoke these numbers to confirm that climate is changing. In a geologic context, climate is indeed changing, and the speed of global warming seems to be exacerbated by human impacts. Nevertheless the phrase *climate change*, as most often used in the following Chapters, is better categorized as *climate variability* or *climate cycles* associated with Earth and Sun cycles. Severe climate episodes may also be occurring because of land use changes related to rapid population growth, and to human concentration into urban and urbanizing areas. When struck by extreme weather events, health and wellbeing impacts are reported instantly to the world (for example, the March 2011 earthquake and tsunami in Japan). One is compelled to ask whether the events are becoming more extreme, or more pronounced. Or whether they seem to be more frequent and widespread because of better reporting and communications. Gradual climate warming is well documented in the scientific literature by many earth science disciplines, and is perhaps most easily demonstrated by physical evidence of glacial and interglacial oscillations over the past two million years (Morain 1984). These changes and their impacts on human populations cannot be disputed and can even be traced in situations like desertification in the American southwest and the associated retreat of forests in the Sky Islands of Arizona and New Mexico (Zimmer 1995; Morain 2000). However, use of the phrase *climate change* in the context of health is problematic because the accumulated evidence for environment and health interactions only spans the last few centuries in historical documents (e.g. Justinian's plague); the last 100–150 years of weather records; and only the last few decades of actual synoptic measurements from satellite observations.

Table 4 lists examples of how weather events lead to observed health outcomes (Rose *et al.*, 1999). Table 5 lists examples of vector-borne, water-borne, food-borne, and air-borne diseases that were known in 1999 to be sensitive to weather parameters. Rose and his associates use the phrase *climate change*, but here it is hoped the reader will interpret the phrase as *weather events*

Table 4.   Key effects of global climate change (Courtesy American Academy of Microbiology).

Potential effect of climate change	Direct health effect	Indirect health effect
Extreme heat	Heat-related mortality	
Extreme storms	Flood-related mortality	Increased prevalence of enteric pathogens in surface waters; lack of adequate potable water
Air pollution	Respiratory effects	Increase in allergies and asthma
Ecological shifts		Changes in pathogen and/or vector geographical range; changes in virulence; changes in incidence

Table 5. Climate sensitive diseases (Courtesy American Academy of Microbiology).

Vector-borne	Water-and food-borne	Air-borne
Malaria	Cholera	Meningococcal meningitis
Dengue fever	Other non-cholera vibrio *spp.**	Coccidioidomycosis
Lyme disease	Leptospirosis	Respiratory syncytial virus (colds)
Rocky Mountain spotted fever	Schistosomiasis	Legionnaires disease
Encephalitis; Murray Valley	Sea-bather's eruption	Influenza
Western Equine	Giardiasis	
Rift Valley fever	Cryptosporidiosis	
Ross River fever	Human-enteric viruses	
Ehrlichiosis	Campylobacteriosis	
Hantavirus Pulmonary	Cyclospora cayetanensis	
Syndrome	Salmonella enteritidis	
Leishmaniasis		
African Trypanosomiasis		
Tularaemia		
Plague		
Onchoceriasis (river blindness)		

*e.g. *V. vulnificus*; *V. parahaemolyticus*

and climate cycles. In the Chapters to follow, more diseases are described, along with their weather and climate associations.

2 SCOPE OF EARTH OBSERVATION AND HEALTH

Science and society are driven to address health issues by any means possible. However, it is not immediately clear to those outside the Earth observing community just how sensors can play functional roles in the health sciences. Medical doctors have identified many types of diseases that have identifiable environmental determinants. There are also areas of health concern for which Earth observations do not at first appear to be applicable. Table 6 lists categories of diseases generally recognized by medical communities. Use of EO data for some categories is self-evident, but not necessarily mature in the applied sciences; other applications lag far behind or are obscure.

Within the past few years, Earth scientists have been incorporating environmental measurements into investigations of degenerative, neoplastic, and metabolic diseases. Many of these diseases and debilitating conditions actually have direct and indirect links to one's proximal environment. Consequently, malnutrition, post-traumatic stress disorder, micronutrient deficiencies and other conditions are of interest to geoscientists, more so if environmental links are known to exist. These conditions and diseases arise through an absence of appropriate enzymes needed to digest substances in food or drink, or their inability to reassemble into nutritional building-blocks. Under these circumstances, intermediate toxic substances may accumulate in a body's system, leading to a loss of wellbeing.

Even human immunodeficiency virus (HIV) acquired immune deficiency syndrome (AIDS), which is classified as a lifestyle disease with no apparent geophysical basis, has related environmental determinants that promote or retard its spread. Immuno-compromised individuals whose systems are already weakened can be overcome directly by the soil fungus *Aspergillus fumigatus*, which then causes a range of additional health complications for AIDS and leukaemia victims. An indirect environmental factor that spreads HIV/AIDS is the transportation net. New transportation corridors introduce the disease into districts formerly isolated by topography, relief, or other geophysical factors.

Table 6. Categories of disease.

Category	Examples	Transmission pathways
Environmental	Asthma, cholera, malaria	Natural & built environments
Infectious & zoonotic	HIV/AIDS, TB, plague, influenza	Transmitted by contact
Degenerative	Arthritis, atherosclerosis	Non-contagious
Neoplastic	Cancers, tumours, moles	Abnormal cell proliferation
Metabolic	Diabetes, muscle disorders	Abnormal body reactions

Table 7. Elements of happiness & dimensions of wellbeing (adapted from Wahlqvist 2004).

Happiness	Dimensions
Personal security	Social
Expectations	Occupational
Connectedness	Spiritual
Recognition	Physical (diet & physical activity)
Closure (conflict resolution)	Intellectual
Problem solving	Emotional

Wahlqvist (2004) described a new science for measuring happiness in context of wellness dimensions (Table 7). The two elements are intertwined in complex ways and may be exacerbated by environment and nutrition.

Finally, the focus of this book lies predominantly on natural and anthropogenic environments that infect or affect humans. However, in pursuing disease aetiology and transmission pathways one cannot discount broader issues of environmental health and ecosystem health; i.e. nutrition derived from domestic plants and animals, or those wild forms in close contact with human populations.

3 ENVIRONMENTAL DISEASES

Synopses of diseases transmitted by air-, water-, soil- and vector-borne pathogens are presented below as a point of departure for those who are not conversant with the major categories of disease, or how their carriers might be linked to environmental parameters. Every human disease agent has one or more transmission pathways for entering the human system and exposure to these agents, whether biological or chemical, cause the system to react.

3.1 *Air-borne diseases*

Diseases carried through the air are both infectious and contagious. They include acute respiratory infections like anthrax, chickenpox, smallpox, influenza, tuberculosis, Hantavirus, pertussis, pneumonia, diphtheria, common cold, and bronchitis. They also include allergens and chemical pollutants that exacerbate asthma and chronic obstructive pulmonary disorders (COPD); and air-borne vector diseases like trypanosomiasis (tsetse fly) and leishmaniasis (sand fly). Air-borne viral and bacterial diseases are found worldwide. They are often associated with poverty, overcrowding, or unsanitary conditions. Exacerbating circumstances include indoor and outdoor particulates, aerosols, chemical pollutants, radiation and biomass fuels. Allergens and anthropogenic emissions are worldwide but are predominant in the Northern Hemisphere. Diseases caused by air-borne vectors are found primarily in Africa, Latin America, and the Middle East.

Infectious air-borne diseases are caused by bacterial and viral infections. Some (chickenpox, smallpox) are very contagious during the pre-crusted stage of blisters. Influenza is extremely contagious and has such a short incubation period that it can infect whole communities at once. For others like tuberculosis, prolonged or repeated exposures are needed, and even then may not be evident for years. Allergic and chemical reactions (also known as [aka] hay fever) are caused by inhaling pollen, soil mould, and/or atmospheric pollutants. Allergic reactions caused by food that lead to forms of eczema are not considered here; neither are reactions to poisonous plants. Air-borne vector diseases are caused by flies and flying insects (see also Section 3.4).

Air-borne diseases are transmitted on atmospheric dust particles; or, as nearly invisible aerosolized droplets when infected people sneeze, cough, laugh, or exhale. Some can be transmitted also by touch. Allergens and chemical pollutants are released as ordinary natural processes in plant and soil ecosystems. Air-borne anthropogenic emissions of ozone, NO_X, SO_X, and other chemical gases irritate the nose and throat. Other air-borne chemicals such as pesticides, organophosphates, PCBs, dioxin, arsenic, cadmium, lead, and mercury contaminate food and water supplies and enter the digestive system.

Chronic diseases arising from life-long or prolonged exposure to dusty environments (particles in the size range between PM_{10} and $PM_{2.5}$) are managed best by wearing masks or staying indoors during dust events. Diseases and conditions arising from acute respiratory infections require treatments targeting specific pathogens; those exacerbated by air-borne allergens (pollen) can be managed by over-the-counter antihistamines, through prescriptions, or by patient interventions. Zoonotic infections that enter the cardiovascular system from soils require treatment and usually are prevented through access to cleaner water, improved personal hygiene, or interventions to avoid contaminated areas. Diseases transmitted by mosquitoes and flies can be managed by removing or chemically treating their breeding habitats. Personal habits such as smoking exacerbate respiratory diseases and conditions.

3.2 Soil-borne diseases

There are many types of soil-borne diseases from parasitic, to fungal and bacterial, and those related to basic soil contamination and nutrient deficiencies. A few of the biotic diseases are: salmonellosis, cholera, dysentery, tetanus, anthrax, hepatitis-A, hepatitis-E, and *Escherichiacoli (E. coli)*. There also are intestinal diseases caused by bacteria, viruses and moulds; and parasitic diseases caused by nematodes such as ascariasis, hookworm, and pinworm, strongyloidiasis, and trichinosis. Abiotic diseases, or medical conditions derived from soils, include lead, mercury, and other heavy metal poisoning from mining operations; cancers from natural asbestos; and conditions arising from soil nutrient deficiencies like iodine, selenium, and molybdenum.

Soil-borne diseases make contact with humans directly through the soil, or indirectly via food, water, and air. Soil is a primary place of origin for organisms that cause disease in humans or their food supplies. Diseases associated with soils are classified as soil-associated, soil-related, soil-based, or soil-borne. Soil-associated diseases are caused by opportunistic or emerging pathogens that are part of the soil micro-biota. *Aspergillus fumigatus* is a common soil fungus that attacks the human immune system, especially those who have leukaemia or HIV/AIDS. Soil-related diseases enter humans via food contaminated with entero- or neurotoxins. Soil-based diseases are caused by pathogens indigenous to soil (e.g. tetanus, anthrax); and soil-borne diseases are enteric pathogens arising from human or animal excreta. These include bacteria, virus, protozoa, and parasitic roundworms. When they reach the food chain, some can withstand freezing and desiccating environments. *Ascarislumbricoides*, a giant roundworm infection in humans, is transmitted most often through human faeces in close proximity to dense populations. Severe cases can be fatal. Hookworm infections also are associated with human faeces. Under cool conditions, the worms migrate to soil surfaces waiting to attach to humans walking through contaminated areas.

These diseases are major aetiological agents in countries with poor sanitary conditions, but also in developed and affluent societies. Hookworms are common worldwide in warm, moist climates, and have complex life cycles focused on the small intestine. Pinworm is a parasite in temperate regions, even in areas of good sanitation. It can be a persistent or re-infectious disease within families because eggs survive outside the host for several weeks. Strongyloidiasis is widespread from temperate to tropical areas.

In areas of known occurrence in the soil, people should not walk barefooted or touch the soil with their hands. Environmental conditions that support the transmission of soil-borne disease include absence of clean water, inadequate sanitation, and crowded living conditions. Lifestyle factors include poor personal hygiene, inadequately cooked food, and squalid living conditions. Children in particular are vulnerable to parasitic worms that deprive them of nutrition.

3.3 *Water-borne diseases*

Water-borne diseases are divided into four subcategories: water-borne; water-washed; water-based; and water-related. Common water-borne diseases include cholera, typhoid, amoebic and bacillary dysentery, diarrhoea diseases, cryptosporidiosis, and infectious hepatitis. Water-washed diseases include scabies, trachoma, typhus, leprosy, tuberculosis, tetanus, diphtheria, and flea, lice, and tick-borne diseases. Water-based diseases include dracunculiasis, schistosomiasis, Guinea worm, and other helminths like flukes, tapeworms, roundworms, and tissue nematodes. Water-related diseases include those caused by vectors that breed or bite in water.

Water-borne diseases are caused by drinking water contaminated by human, animal, or chemical wastes. These include a wide variety of viruses, bacteria, and protozoan parasites that differ in size, structure, composition, and excretion. Their incidence and behaviour in water environments also differ, making it difficult to test the efficiency of water treatment processes and water testing. Water-washed diseases (aka water-scarce diseases) are those caused by poor personal hygiene and skin or eye contact with contaminated water. Water-based are those caused by parasites found in host organisms living in contaminated water. Water-related are those caused by vectors breeding in water.

Water-borne diseases are found everywhere because of improper sanitary waste disposal and consequent degradation of clean water for drinking, cooking, and washing. Cryptosporidiosis is an emerging pathogen in developed countries, and is particularly life-threatening for persons with HIV/AIDS. Water-washed diseases thrive where fresh water is scarce and sanitation is poor. They occur mainly in hot and arid environments; but are global wherever dirty water is used for multiple purposes, including drinking; but they can be controlled through better hygiene. Water-based diseases are found in tropical and subtropical areas, often in stagnant water behind impoundments and in irrigation systems. Water-related diseases are found primarily in the tropics and subtropics where precipitation and humidity are high.

Best disease management practices for water-borne diseases include: improving water quality; preventing casual use of unimproved sources; constructing sanitary drinking water systems; improving water quantity and water accessibility; improving personal hygiene; decreasing the need for water contact; controlling snail populations; improving water quality; improving surface water management; destroying breeding sites of insects; and decreasing the need to visit breeding sites. Many of these management practices can be addressed with current state-of-the-art satellite sensors by accessing fine resolution data and imagery made available routinely and cost-free from government and government-sponsored online databases. Others might require developing step-wise procedures, models and algorithms before they can be implemented.

3.4 *Diseases transmitted through vectors*

Most vector-borne diseases are found in the tropics and subtropics. During the 20th Century many of them were eradicated through pesticides and other means, but reductions in financial support and loss of public health infrastructure have allowed some of them to re-emerge. In the past, they have been described as rare in temperate regions; but continued global warming promotes vectors and their pathogens to expand into larger territories. Temperature and humidity are considered to be the key environmental variables along with attendant effects on hydrology, agriculture, and related factors.

Vector-borne diseases are numerous. They include malaria, dengue fever, yellow fever, plague, West Nile and Japanese encephalitis, Lyme disease, Rift Valley fever, haemorrhagic fever, louse-borne typhus, sleeping sickness, trypanosomiasis, filariasis, Rocky Mountain spotted fever, and tularaemia. Several are analysed and described in subsequent Chapters in context of their environmental determinants and transmission pathways. Most vector-borne diseases are zoonoses that survive by using mammals like foxes, bats, raccoons, and skunks to spread rabies; or by humans to spread malaria and dengue fever. Arthropods like fleas, mosquitos and ticks, however, are the most prevalent reservoirs for pathogens. Bites from these vectors transmit pathogens when they take a blood meal from humans. Transmission depends on three factors: the pathologic agent; the arthropod vector; and the number of human hosts. There are hundreds of viruses, bacteria, protozoa, and helminths that require vertebrates and arthropods for transmission into human populations. Discovering transmission pathways by employing EO data is a major area for earth and medical scientists, as is detailed in Chapters 2 through 5.

Most environmental interventions focus on controlling arthropod vectors rather than the pathogens by modifying or eliminating breeding areas, stalling their entry into new areas by monitoring ports and airports, or reducing the population sizes of host reservoirs in areas of human habitation. There are some efforts to control pathogens by vaccinating a sufficient number of people to interrupt further spread of the disease. Management practices are numerous and intermixed. They are comprised of natural, cultural, and economic factors that are determined largely at national and local levels. They include: insecticide and drug resistance; changes in public health policy; emphasis on emergency response; de-emphasis of prevention programs; demographic and societal changes; and, genetic changes in pathogens. Mosquito-borne, rodent-borne, and water-borne diseases are often associated with urbanization in developing countries.

4 PROLEGOMENON

This Chapter has summarized the last four decades of satellite Earth observations R&D linking environmental and health sciences. Arguably, the next forty years will show even greater strides linking quantitative Earth science *macro*-system models that monitor plant, animal, and geophysical processes known to transmit diseases with progress in medical *micro*-systems research. Results from stem cell biology, genomics, proteomics, radio biomarkers, and pharmaceutical chemistry that trace the aetiology and evolution of diseases in humans will revolutionize healthcare information systems. It is conceivable that basic and applied research on environmental pollutants will be driven more by human health impacts than by concerns over climate change, as such; or, at least confirm direct cause-and-effect climate change relationships that are still anecdotal. It is also conceivable that future generations of sensing systems will provide reliable and scientifically sound measurements using nano-detectors of internal biological processes driving degenerative, neoplastic, and metabolic diseases. As the capabilities evolve, it may be possible to link internal disease biology at the organic level with external environments at the geographic level.

4.1 *Physical basis for environment and health interactions*

There is a rich literature in Order and Chaos that demonstrates the concept of self-adaptive responses to environment (Kauffman 1993). Self-adaptation refers to an organism's response to stimuli; a phenomenon well recognized in microorganisms that build immunity to pesticides and poisons. In higher organisms, the process is slower but well recorded in the geological record; and it includes humans adapting to long-term glacial and interglacial episodes, shorter term drought and economic cycles, and related environmental impacts not yet well understood or incorporated into empirical or deterministic models (e.g. solar output). Self-adaptive models emulate the individual and collective responses of genetic systems whose physiological, ecological and geographic tolerances can be characterized. Organisms are tied to their biology and controlled by the Earth-Sun system.

As a specific example, it is common knowledge that ultraviolet (UV) and shorter wavelength solar radiation can cause genetic mutations in life forms that evolve into newer life forms. Drilling through the elements of the Earth system both horizontally and vertically shows the theoretical pathway for how environmental tolerances, mutations, and new forms might evolve. Variations in solar irradiance account for about 0.1 per cent over the sun-spot cycle, but most of this is found in the UV spectrum. These variations are measureable in the stratosphere, and create heat gradients in the troposphere. In turn, changes in UV dosage can be traced through genetic networks to monitor how incoming UV radiation affects organisms (Loomis & Sternberg 1995). In 1995, these pathways were based on laboratory data only. Since then considerable progress has been made in understanding these networks.

There are several high-level roles for spectral environmental data applied to health monitoring. Some of these include 1) change detection and time series analyses for assessing human factors; 2) inputs to genetic networks (i.e. time and space components of environmental attributes that control genetic triggering mechanisms); 3) drivers for rule-based decision-making; 4) using pixel windows for assessing fractal dimensions of environmental change; and 5) identifying environmental common denominators (i.e. vegetation indices). Perhaps the most often employed index is the normalized difference vegetation index (NDVI). It works because it measures proven biophysical phenomena controlling plant vigour (Sellers *et al.*, 1997). According to the authors, *land surface parameterizations have evolved from simple, unrealistic schemes into credible representations of the global soil-vegetation-atmosphere transfer system as advances in plant physiological and hydrological research, advances in satellite data interpretation, and the results of large-scale field experiments have been exploited.* NDVI and several other indices are described in subsequent Chapters. As years have lapsed, science has recognized that multidimensional analysis methods can be applied whether one is studying thousands of gene expressions, or sensor data from thousands of points around the globe (George Michaels, pers. comm.).

4.2   *From images to data*

Readers will recognize several recurring themes across the following Chapters. Three are most important. The first is that environmental tracking data can be used to forecast health issues that will occur statistically months, or even years, into the future. The second is that there are emerging statistical approaches in numerical models that incorporate sensor data to forecast health impacts. While these approaches seem to be identical, they are not. The first emphasizes anecdotal links between environment and health that are still being tested and examined. The linkages are probably true, but need further verification. The second seeks to establish the proverbial *smoking gun* for environmental cause-and-effect relationships by employing models that have verifiable statistical limits of accuracy for health outcomes. At publication the approach has acquired a considerable following and the second is beginning to focus on verification and validation of health outcomes using EO data. The aim, of course, is to move beyond the fascination of satellite images as seductive, thought-provoking pictures and to learn how the digital data *behind* those images can be used directly to model complex real-world health outcome scenarios. The third theme runs full-circle back to incorporating digital image data into numerical environmental modelling systems that are interoperable with, and capable of, solving complex environmental/biophysical interactions. To accomplish this, microsystems must emerge for understanding cell and tissue biology and their interactions with environmental stimuli.

4.3   *Medical and biological microsystems*

Microsystems are defined as integrated, intelligent, miniaturized devices and systems that are fabricated using processes compatible with semiconductor technology, and combining these technologies with non-invasive sensing technologies, high-speed computation and actuation. Microsystems normally combine two or more of the following elements: electrical, mechanical, optical, chemical, biological, or magnetic properties within a single or multi-chip integrated

15

system. Such systems are emerging rapidly in health care fields and are beginning to emerge in environmental management fields. Enabling technologies include miniaturization through nanotechnology, and integration of devices to produce systems with intelligent processing in real time using low power communication via wireless sensor networks. Among the many types of microsystems are those being developed for metabolomics, proteomics, proton therapy, stem cell therapy, genomics, and bio-markers. Uses of microsystems might include:1) environmental tracking and impacts of persistent organic pollutants on health; 2) mapping environmental exposures that signal disease pathways perturbed by single and multiple toxicants in model cells; 3) determining the role these pathways play in cells, organs and animals during development, and in the adult; and 4) assessing perturbations in the regulation of gene expression and how these changes feed-back to alter cellular signalling.

Metabolomics focuses on small molecules arising from complex protein, gene, and environmental interactions. *Environmental* in this context refers to metabolic environments. Measurements are facilitated by high resolution mass spectrometers, nuclear magnetic resonance spectrometers, and other techniques. Information is kept in databases containing information about constituent chemicals. Metabolic changes can be measured in seconds or minutes compared to days or weeks for analyses made through gene chips, protein analyses or tissue biopsies.

Proteomics focuses on protein interaction networks involving disease-causing proteins that aid development of bio-markers as diagnostic tools and potential new therapeutic drugs, intelligent drugs, nano-medicine, and personal health management systems. Through proteomics one might build a picture of the molecular networks that form the cell's operating system. Microorganisms and the environment play key roles in human health, and both influence risks for developing cancers, asthma, obesity and other complex diseases. Scientists are studying the molecular basis of host-pathogen interactions. Since many human pathogens have free living life stages in water, soil and alternative hosts, they are using meta-genomic approaches to describe microbial communities, and to identify novel viruses and bacteria in the environment.

Proton therapy and radiation oncology reduce the radiation dose required by high-energy x-rays and deliver more concentrated doses at more precise locations in the body. Bio-markers have become popular ways to unravel the pathways for signal transduction and protein interactions through medical imaging. Technologies are developing for making short half-life tracers that are sufficiently *hot* radiologically that, when made in appropriate quantities, can be transported over distances of two to four half-lives and still have enough material for medical use.

Structural imaging methods such as computerized tomography (CT) and conventional magnetic resonance imaging (MRI) provide anatomical information, while positron emission tomography (PET) and single photon emission computed tomography (SPECT) use radiotracers to image biochemical processes that allow tissues to be distinguished according to function rather than structure. Combining these emerging medical imaging technologies (CT, MRI, PET and SPECT) might allow chemical and biological processes to be both localized and viewed spatially *in vivo* to provide previously unattainable information. Finally, genomics make it possible to compare molecular differences in several cancer sub-types found in different geographic areas.

REFERENCES

Beck, L.R. 2004. Remote sensing/GIS and human health: A partial bibliography. Available from: http://geo.arc.nasa.gov/sge/health/rsgisbib.html [Accessed 13th February 2010].
Beck, L.R., Lobitz, B.M. & Wood, B.L. 2004. Remote sensing and human health: New sensors and new opportunities. Available from: http://spacejournal.ohio.edu/issue14/outlook_remote.html [Accessed 13th February 2010].
Budge, A.M., Grobicki, A.M., Rosenberg, M.R., Selinus, O., Steinnes, E. & Enow, A. 2009. Mapping GeoUnions to the ICSU framework for sustainable health & wellbeing: Focus on sub-Saharan African cities. Joint Science Project Team for Health (JSPT-H). Contractor report for ICSU Committee on Scientific Planning & Review.
Burrows, P. 1986. *Deep black: Space espionage and national security*. New York: Random House.

Colwell, R.R. 1996. Global climate and infectious diseases: The cholera paradigm. *Science* 274: 2025–2031.

Cowen, D.J. & Jensen, J.R. 1988. Extracting & modeling of urban attributes using remote sensing attributes. In D. Liverman, E.F. Moran, R.R. Rindfuss & P.C. Stern (eds.), *People & pixels: Linking remote sensing & social science*: 164–188. Washington DC: National Academy Press.

Deckers, J. & Steinnes, E. 2004. State of the art on soil-related geo-medical issues in the world. *Advanc. Agron.* 24: 1–35.

Epstein, P.R. 1986. Health applications of remote sensing and climate modeling. In D. Liverman, E.F. Moran, R.R. Rindfuss & P.C. Stern (eds.), *People & Pixels: Linking remote sensing & social science*: 197–207. Washington DC: National Academy Press.

Goudie, A.S. & Middleton, N.J. 2001. Saharan dust storms: Nature and consequences. *Earth Sci. Rev.* 56: 179–204.

GEO 2011a. GEO 2012–2015 Work Plan.Document 20. Geneva: Geo Secretariat.

GEO 2011b. Available from: http://www.earthobservations.org/geoss_imp.php [Accessed 17th January 2011].

IC 2011. Available at http://www.disasterscharter.org/web/ charter/activations [Accessed 27th December 2011].

ICSU 2011. Report of the ICSU planning group on health and wellbeing in the changing urban environment: A systems analysis approach. Paris: International Council for Science.

Kauffman, S.A. 1993 *The origins of order: Self organization and selection in evolution*. New York: Oxford UP.

Kaya, S., Sokol, J. & Pultz, T.J. 2004. Monitoring environmental indicators of vector-borne disease from space: A new opportunity for RADARSAT-2. *Can. J. Rem. Sens.* 30(3): 560–565.

King, M.D., Kaufman, Y.J., Tanre, D. & Nakajima, T. 1999. Remote sensing of tropospheric aerosols from space: Past, present, future. *Bull. Amer. Met. Soc.* 80(11): 2229–2259.

Lauer, D.T., Morain, S.A. & Salomonson, V.V. 1997. The Landsat program: Its origins, evolution & impacts. *Photog. Eng. & Rem. Sens.* 63(7): 831–838.

Loomis, W.F. & Sternberg, P.W. 1995. Genetic networks. *Science* 269(5224): 649. doi:10.1126/science. 7624792.

MERIT 2011. Available at http://merit.hc-foundation.org/ [Accessed 18th December 2011].

Morain, S.A. 1984. *Systematic and regional biogeography*. New York: Van Nostrand-Reinhold.

Morain, S.A. 1992. From Columbus to Columbia. *ISPRS J. Photogram. & Rem. Sens.* 47: 285–305.

Morain, S.A. 1993. Emerging technology for biological data collection and analysis. *Ann. Missouri Bot. Gard.* 80(2): 309–316.

Morain, S.A. 1998. A brief history of remote sensing applications with emphasis on Landsat. In D. Liverman, E.F. Moran, R.R. Rindfuss & P.C. Stern (eds.), *People & pixels: Linking remote sensing & social science*: 28–50. Washington DC: National Academy Press.

Morain, S.A. 2000. Receding forests of the Sky Islands: Impacts of potential climate change and a vision for future prediction. Proceedings 8th biennial forest service remote sensing applications conference. Bethesda: ASPRS (CD ROM).

Morain, S.A. & Budge A.M. 2010. Suggested practices for forecasting dust storms and intervening their health effects. In O. Altan, R. Backhaus, P. Boccardo, & S. Zlatanova (eds.),*Geoinformation for disaster and risk management*: 45–50. Copenhagen: Joint Board of Geospatial Information Societies & United Nations Office for Outer Space Affairs.

Norboo, T., Angchuk, P.T., Yahya, M., Kamat, S.R., Pooley, F.D., Corrin, B., Kerr, I.H., Bruce, N. & Ball, K.P. 1991. Silicosis in a Himalayan village population: Role of environmental dust. *Thorax* 46: 341–343.

Oxford, J.S., Lambkin, R., Sefton, A., Daniels, R., Elliot, A., Brown, R. & Gill, D. 2005. A hypothesis: the conjunction of soldiers, gas, pigs, ducks, geese, and horses in Northern France during the Great War provided the conditions for the emergence of the Spanish influenza pandemic of 1918–1919. *Vaccine* 23(7): 940–945.

Rose, J.B., Huq, A. & Lipp, E.K. 1999. Health, climate and infectious disease: A global perspective. Colloquim report. Washington DC: American Academyof Microbiology

Sclinus, O. (ed.) 2005. *Essentials of medical geology: Impacts of the natural environment on public health*. London: Elsevier.

Sellers, P.J., Dickinson, R.E, Betts, D.A., Hall, F.G., Berry, J.A., Collatz, G.J., Denning, A.S., Mooney, H.A., Nobre, C.A., Sato, N., Field, C.B. & Hendsserson-Sellers, A. 1997. Modeling the exchanges of energy, water, and carbon between continents and the atmosphere. *Science* 275: 502–509.

Spatialnews 2011. Available from: http://spatialnews.geocomm.com/daily news/2011 /Dec/09/ news2.html [Accessed 24th December 2011].

Sterman, J.D. 2006. Learning from evidence in a complex world. *Amer. J. Pub. Health* 96(3):505–514.

Taubenberger, J.K., Reid, A.H. & Fanning, T.G. 2005. Capturing a killer flu virus. *Sci. Amer.* (January): 62–71.

Tsipis, K. 1987. Arms control treaties can be verified. *Discover* 8(4): 78–91.

Wahlqvist, M.L. 2004.Sciences for health & well-being. *Int. Union Biol. Sci.* Cairo: 18–22 Jan.

Wiggs, G.F.S., O'Hara, S.I., Wegerdt, J., Meers, J. van der, Small, I. & Hubbard, R. 2003. The dynamics and characteristics of aeolian dust in dryland central Asia: Possible impacts on human exposure and respiratory health in the Aral Sea Basin. *Geog. J.* 169: 142–157.

Yates, T.L., Mills, J.N., Parmenter, C.A., Ksiazek, T.G., Parmenter, R.R., Vande Castle, J.R., Calisher, C.H., Nichol, S.T., Abbott, K.D., Young, J.C., Morrison, M.L., Beaty, B.J. & Dunnam, J.L. 2002. The ecology and evolutionary history of an emergent disease: Hantavirus pulmonary syndrome. *BioSci.* 52(11): 989–997.

Zimmer, C. 1995. How to make a desert. *Discover* (February). Available from: http://discovermagazine.com/1995/feb/howtomakeadesert467 [Accessed 10th April 2012].

Infectious and contagious diseases in the environment

Chapter 2

Vector-borne infectious diseases and influenza

R.K. Kiang (Auth./ed.)[1]; with R.P. Soebiyanto[1,2], J.P. Grieco[3], N.L. Achee[3],
L.C. Harrington[4], W.K. Reisen[5], A. Anyamba[1,2], K.J. Linthicum[6], J.E. Pinzon[1,7],
G. Zollner[8], & M.G. Collacicco-Mayhugh[8]
[1] *NASA Goddard Space Flight Center, Greenbelt, MD, US*
[2] *Universities Space Research Association, Columbia, MD, US*
[3] *Uniformed Services University, Bethesda, MD, US*
[4] *Cornell University, Ithaca, NY, US*
[5] *School of Veterinary Medicine, University of California, Davis, CA, US*
[6] *US Department of Agriculture, Agricultural Research Service, Gainesville, Florida, US*
[7] *Science Systems and Applications, Inc., Greenbelt, MD, US*
[8] *Walter Reed Army Institute of Research, Silver Spring, MD, US*

ABSTRACT: Breeding, propagation and survivorship of disease-transmitting arthropod vectors often depend on meteorological and environmental parameters. Statistical and biological modelling that incorporate satellite-acquired data provide useful tools for assessing and forecasting these diseases, assessing possible outbreak areas and providing vital information for policies and decision making.

1   INTRODUCTION

The role of environment and climate in propagating infectious disease has been recognized since early human history. The effect is particularly evident in vector-borne diseases such as malaria where temperature, precipitation and humidity influence the abundance and lifecycle of mosquitoes and pathogens. When environmental conditions are favourable for vectors and pathogens, disease outbreaks typically become more intense. Consequently, changes in climate and ecosystem, whether human caused or due to natural variability, will alter the burden and spread of the disease. For example, the rise of Lyme disease in the US has been linked to human encroachment of wildlife habitats and reforestation of farmlands. Meanwhile, global warming can potentially expand the geographic ranges of vectors and pathogens, hence bringing emerging diseases to new areas. In spite of advances in modern medicine, vector-borne and zoonotic infectious diseases remain a serious threat to human and animal health. The appearance of new viral strains and drug resistant microorganisms present a difficult challenge for prevention and treatment. Personal protection, vector control, and disease surveillance and response remain the most effective measures. However, advances in modelling techniques combined with rapid progress in remote sensing technology, have provided researchers with tools to link environment and climate quantitatively with infectious disease. This has enhanced spatiotemporal disease risk assessment that can be used to aid public health and agricultural agencies in prevention and control efforts.

In a broad sense, remote sensing imagery data and derived products can either be used to characterize and identify suitable habitat for vector, pathogens, reservoir hosts, and in some cases, the disease itself; or to estimate the timing and magnitude of the outbreak. Hence remote sensing data and products are important components in decision support tools for public health agencies in mitigating infectious disease risks. Furthermore, the successful use of remote sensing to aid public health efforts in prevention and control of infectious diseases not only hinges upon accurate measures of climatic and environment indicators, but also on a multitude of other factors. This includes the availability and completeness of historical disease surveillance data that are appropriate for use in modelling with remote sensing data. More importantly, success will depend on minimizing the current gap in spatiotemporal resolution between remote sensing and disease data.

In this chapter, we hope to give the readers different perspectives on the broad spectrum of remote sensing use in vector-borne infectious diseases and seasonal influenza. Specifically, seven diseases are discussed, including malaria, dengue, several mosquito-borne encephalitides, Rift Valley fever, plague, leishmaniasis, and seasonal influenza. State-of-the-art analytic, mathematical, statistical and modelling methods for applications concerning these diseases are also discussed. For example, this chapter presents various methods to be used in conjunction with EO data, from regression, clustering, ecological niche modelling, neural network, to Hilbert-Huang-transform. More importantly, most of the examples presented here not only use real epidemiological data but also real life applications where the resulting disease maps or risk assessments are implemented as public health agency decision tools.

This chapter is organized as follows. In Section 2, Doctors Grieco and Achee discuss malaria in the Americas with case studies drawn from Belize, Mexico and South America. The major *Anopheles* species transmitting malaria in these regions include *An. darlingi, An. albimanus* and *An. vestitipennis*. Although their larval habitats may be too small to be detected from satellite altitudes with medium or lower spatial resolutions, the vegetation or the land cover associated with their breeding sites can often be detected and used as risk indicators. Remote sensing data used in these studies include those from Landsat Thematic Mapper (TM), Satellite Pour l'Observation de la Terre (SPOT) and Radarsat.

In Section 3, Doctors Kiang and Soebiyanto discuss malaria in Asia focusing on appropriate use of remote sensing data in detecting, predicting and reducing malaria risks. Specifically, textual-contextual techniques can be used to identify the vegetation types associated with larval habitats for certain mosquito species; statistical models can estimate malaria prevalence and predict future transmission; and agent-based biological models can identify the strategy to reduce malaria transmission efficiently and cost-effectively. Examples are given for three countries: Thailand, Indonesia and Afghanistan.

In Section 4, Doctor Harrington describes the ecology of the *Aedes aegypti* mosquito, which is the most important vector of dengue and yellow fever viruses worldwide, and discusses the major environmental determinants for dengue transmission such as temperature, precipitation and humidity. The satellite data products that provide such parameters or their proxies are discussed, and examples are given. Doctor Harrington also discusses the difficulties of inferring environmental parameters from satellite to the mosquito microhabitats that are often in-door or peridomestic.

In Section 5, Doctor Reisen discusses the epidemiology, invasion and health significance of representative encephalitis viruses from the *Flaviviridae, Togaviridae* and *Bunyaviridae* families. These zoonoses are maintained and amplified among wildlife or domestic animals by *Culex* or *Aedes* mosquitoes. One of the emerging *Flaviviridae* is West Nile virus that has caused public, veterinary and wildlife health problems on five continents. Doctor Reisen further demonstrates how EO data can help to identify regions and areas of elevated risks based on the landscape, environmental, meteorological, climatic and socio economical characteristics.

In Section 6, Doctors Anyamba and Linthicum discuss Rift Valley fever, a mosquito-borne viral disease with pronounced health and economic impact to domestic animals and humans in sub-Saharan Africa and the Middle East. Infected animal species, including sheep, cattle and goats, suffer significant mortality and abortion. Most outbreaks occurred during the warming phase of the El Niño Southern Oscillation (ENSO). The authors have developed a predictive model based on anomalies observed in NDVI, sea surface temperature (NINO 3.4 and western Indian Ocean) and outgoing long-wave radiation that is able to provide early warning several months in advance.

In Section 7, Doctor Pinzon discusses plague, which is a highly virulent flea-borne zoonotic disease maintained in nature as an infection of rodents. The typical reservoir species include ground squirrels, prairie dogs, chipmunks and wood rats. Plague's aetiological agent, *Yersinia pestis*, is one of the most pathogenic bacteria for humans. NDVI, vegetation type, land cover, surface temperature and rainfall are some of the parameters that can be used for predictive modelling. Doctor Pinzon employed several mathematical techniques including regression, empirical mode decomposition, and stepwise optimal hierarchical clustering.

In Section 8, Doctors Zollner and Colacicco-Mayhugh discuss leishmaniasis and how EO data can be used to identify regions and areas with leishmaniasis risks. The disease is caused by *Leishmania* parasites and transmitted by one or more sand-fly species, and can be either zoonotic or anthroponotic. Several rodent species, dogs, certain wild animals, and humans can serve as reservoirs. Vegetation type, land use and land cover, precipitation, house construction, and settlements are some of the environmental and socioeconomic determinants. Ecological niche modelling has been one of the main techniques to estimate the presence of vectors, reservoir and disease risks.

In Section 9, Doctors Soebiyanto and Kiang discuss the seasonality of influenza for various parts of the world. The fact that seasonality is latitude dependent suggests that climate may be a principal driver for influenza transmission. They then discuss the empirical and biological factors that are known to affect seasonality. The subsets of meteorological, environmental and climatic parameters that can be remotely sensed are further used to model the transmission intensity of seasonal influenza. Regression, time series analysis and neural network are some of the techniques for modelling influenza.

2   ENVIRONMENTAL DETERMINANTS OF MALARIA IN THE AMERICAS

2.1   *Introduction*

Malaria transmission occurs when an infected anopheline mosquito bites a human host. Prevention of malaria, therefore, can be focused on breaking human – vector contact by protecting humans from mosquito bites in their homes as well as reducing vector populations through larval habitat management and/or adult control. The key component to successful disease prevention strategies is to understand clearly the risk of man-vector contact at a given space and time. This includes knowledge of species-specific ecologies such as flight distance, larval habitat preference, seasonal adult density fluctuations, time of biting, house entry behaviour and an understanding of the human population at risk to include work location, housing design, travel history and current vector control practices. The former data sets can be evaluated in the field using standard entomological approaches and correlated to environmental factors using geographical information system (GIS) and remote sensing technologies to identify predictors of disease to guide implementation of intervention strategies for maximum success.

2.2   *Burden of disease in the Americas*

Malaria cases within the Americas have historically been maintained through organized vector control strategies such as indoor residual spraying with chemicals like dichlorodiphenyltrichloroethane (DDT), as well as performing larval habitat management through engineering techniques like drainage and irrigation (Roberts *et al.*, 2002a, 2002b). Despite decades of vested interest in malaria prevention, however, the burden of disease within Central and South America still remains because malaria is endemic from Brazil, Paraguay and Bolivia north to Mexico. Current estimates range from forty-nine cases in El Salvador to over 457,000 in Brazil (PAHO 2007). The countries of French Guiana, Guyana and Colombia share the highest burden with annual parasite indices (API) of 132.1, 43.7 and 33.2, respectively. The predominate parasite species in Central America and portions of South America is *Plasmodium vivax* while *Plasmodium falciparum* is more widely found in countries within the Guiana shield. Although mortality within the Americas resulting from malaria is lower than in other regions of the world, specifically Africa, morbidity rates continue to affect daily life activities within endemic populations and contribute to a general lack of country-wide economic development due to the inability of infected persons to perform normal workloads.

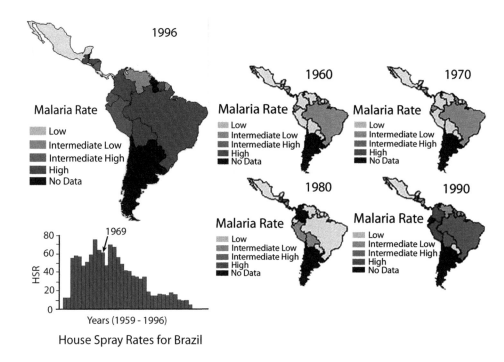

Figure 1.   Changes in malaria prevalence distribution within South America from the 1960s–1990 (Re-created from Roberts *et al.*, 2002b). (see colour plate 1)

The reasons why malaria transmission continues to be a burden are complex but are exacerbated by a de-emphasis of preventive measures (e.g. the virtual elimination of organized vector control approaches to disease control). In addition, environmental changes in each locality influence the presence and abundance of vector species or human populations, thereby affecting the probability of malaria. Many developing countries are in fact experiencing a trend toward increasing numbers of cases and geographical distribution of malaria (Figure 1) (Mouchet & Manguin 1998; Curtis & Lines 2000; Roberts *et al.*, 1997), as well as other vector-borne diseases (Pinheiro & Corber 1997; Davies *et al.*, 1994). Indeed, these diseases now seem to be expanding their range, as defined by climate, ecology and inter-relationships between environmental conditions and human activities. Human activities can work for or against disease occurrence. In the case of preventive measures, human activities can exert powerful control over incidence of disease. In the absence of preventive measures, human modifications of the environment may actually improve conditions for disease vectors and exacerbate the problem of disease burden. These anthropogenic changes and their potential influence on malaria vectors and disease occurrence are the subjects of most remote sensing research programs.

2.3   *Malaria disease transmission*

Malaria is a protozoan disease transmitted to humans by the bite of female mosquitoes within the genus *Anopheles*. There are four species of human malaria: *Plasmodium falciparum*, *P. vivax*, *P. malariae* and *P. ovale*. Variations exist in manifestations and intensity of clinical symptoms, duration and intensity of disease and the probability of relapse among the different species with *P. falciparum* causing the highest mortality. Most cases and deaths are in sub-Saharan Africa. However, Asia, Latin America, the Middle East and parts of Europe are also affected. In 2008, malaria was present in 108 countries and territories (WHO 2010a). Although each species of parasite is found in large regions of the world, and several parasite species can occur within the same region, broad generalities can be made. *P. falciparum* predominates in sub-Saharan Africa and south-east Asia; *P. malariae* in other regions of Africa; *P. vivax* in Europe and the Americas; and *P. ovale* in Asia and Indonesia (Warrell & Gilles 2002).

All four human malaria species follow the same life cycle and require both the human and mosquito host to complete development (Warrell & Gillies 2002). The sexual stage occurs with infected humans, thus humans are known as the definitive host, while asexual reproduction occurs in the mosquito vector, or intermediate host. Malaria transmission begins when the anopheline mosquito takes a blood meal and transfers sporozoites from its salivary glands into the victim. The sporozoites immediately enter the human liver and remain there for seven to ten days, depending on the species of parasite. The parasite then moves actively out of the liver and enters the bloodstream to invade red blood cells. Within the red blood cell, the parasite undergoes asexual reproduction when at final development will burst the red blood cell. The newly formed parasites subsequently reinvade other red blood cells to continue the cycle. It is during this erythrocytic cycle that clinical manifestations become apparent in the infected human, including high fever and rigors, followed by a period of general malaise. Depending on the immunity of the host and species of parasite, the disease can range from general influenza-like symptoms to severe infections, and potential death. During the blood phase, some of the parasites develop into male or female gametocytes and circulate within the blood stream. These gametocytes are ingested by the anopheline vector during feeding and undergo sexual reproduction within the mid-gut of the mosquito where eventually sporozoites will be formed and enter the anopheline salivary glands. The life cycle is repeated when the mosquito takes another blood meal.

Once the female anopheline is infected with the parasite, she remains infected for life, although transmission of disease will not occur unless the female bites another human host. Because the average number of days required for sporozoites to develop following ingestion of gametocytes is more or less ten days, there is a high probability of mosquito mortality due to environmental factors experienced during normal behaviour patterns. These include resting, oviposition and/or mating prior to the mosquito taking a subsequent blood meal. Knowledge of which of these environmental factors is most influential in affecting species-specific behaviours is critical for assessing risk and defining interventions for disease transmission control.

2.4 *Environmental and contextual determinants*

Associations between the environment and mosquito vectors are core to the production, survival and development of both the anopheline mosquito and malaria parasite. Temperature, rainfall, elevation, distance to human habitations, land cover and habitat modification (whether natural or through anthropogenic changes) are some of the determinants that influence the location, abundance, area, and duration of viable vector habitats and therefore vector population densities available for disease transmission. The same determinants can also be used to predict probability of parasite development in the vector and which human populations are most at risk.

It cannot be overstated that a full understanding of basic vector behaviour is important to identify which environmental determinants are most important in a given area. This usually includes evaluating mosquito breeding habitats, flight ranges, seasonal density fluctuations, and biting patterns over time. Of the more than 430 anopheline species found throughout the world, about seventy have been established as vectors capable of transmitting the malaria parasite. Even fewer are considered to be of major importance (Warrell & Gilles 2002). In the Americas, *Anopheles darlingi* is the primary vector, with *An. albimanus*, *An. pseudopunctipennis*, *An. aquasalis* and *An. nuneztovari* serving secondary roles (Zimmerman 1992; PAHO 1994). The larval ecologies of these vectors are far-ranging such that some breeding habitats consist of large areas of wetland marshes, cropped fields, or primary river bodies, while others use a flooded area no larger than a human footprint for egg deposition. Regardless of habitat-type, each of these water sources is dependent on precipitation to maintain a water level sufficient to develop larvae and maintain nutritional thresholds for larvae to survive to pupae, the last immature stage of development prior to adult eclosion (i.e. emergence of an adult insect from a pupal case or an insect larva from an egg).

Of course, temperature variations influence rainfall and both are typically characterized by seasonal patterns. By following seasonal fluctuations over time, routine patterns can be characterized and used to identify predictors of larval habitat formation as well as provide risk assessment under aberrant conditions. In the same manner, elevation is directly related to lowering ambient temperature, which may disrupt the normal life cycle of the mosquito population. This results in delays or even prevention of larval development to adults or reduces the life span of the adult vector and subsequent parasite development within an infected mosquito. Both scenarios modify disease transmission dynamics.

Location and densities of human populations are also linked to environmental determinants. Land cover within endemic regions can be used to predict risk of disease transmission. In many malaria endemic countries, rivers provide water for drinking, cooking and washing, as well as for irrigating crops. This necessitates people living in close proximity to rivers and in turn places them at risk for contacting vectors able to breed and fly within the inhabited range of a village. In addition, cash crops such as citrus and logging industries, promote the congregation of migratory human populations that may introduce malaria transmission by exposing indigenous anophelines to imported parasites.

2.5   Remote sensing for malaria surveillance and control

The connection between environment and the abundance, presence and distribution of malaria vectors, and consequently disease transmission risk, offers a unique opportunity to apply knowledge of vector ecology to monitor and control this disease using aerial and satellite imagery, data and products (Roberts & Rodriguez 1994). Remote sensor data for studying and controlling malaria rests on an important relationship; namely, malaria cases appear when humans, vectors and the malaria parasites occur together. The disease is associated with natural environments to the extent that *Anopheles* mosquitoes are also associated with those natural environments. The existence of this fundamental relationship is the basis for using GIS and sensor data to study malaria epidemiology and control. EO data can be used in a GIS to perform rapid integration and analysis with maps and field data. GIS technology can be used to create maps quickly to determine spatial patterns in the field and relate them to environmental features such as streams or marshes. Geographic analysis and modelling within a GIS can be used to produce risk maps based on criteria such as proximity of villages to mosquito habitat, the type of housing and other factors such as weather data. For example, a ten year average for weather data can be combined with a 1992–1993 derived land cover map to predict malaria prevalence for all of South America (Figure 2).

For appropriate use of satellite observations, one must first know the specific relationships between different vector species and their various environmental conditions. These relationships can be quantified through systematic field studies and refined with properly designed experiments. Once completed, satellite data might be used to develop predictions about where vectors might be present or abundant. In essence, mapping key environmental determinants aids in identifying target areas at risk for malaria and subsequently guides resource management for implementing control interventions. The range of remote sensing products available is wide and each has specific benefits depending on the disease. A more detailed description of these products is provided in Chapter 6.

2.6   Case studies in the Americas

Several malaria projects have demonstrated the potential of satellite imagery for identifying breeding sites of anopheline mosquitoes. The utility of predictive capabilities depends on validating the accuracy of predictions through follow-up field surveys.

2.6.1   Belize

Previous investigations in Belize have characterized the roles and efficiency of some anopheline vectors found throughout the country, including: *Anopheles albimanus* Weidemann;

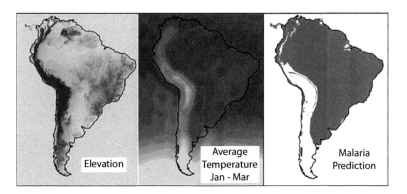

| Elevation | Average Temperature Jan - Mar | Malaria Prediction |

Figure 2.   Sensor data for elevation (left) and temperature (middle) can be used to predict malaria prevalence (right), assuming that transmission only occurs at elevations lower than 2600 metres in areas where average January to March temperature is lower than 15°C (Courtesy Roberts *et al.*, 2002b).

An. pseudopunctipennis Theobald; *An. vestitipennis* Dyar & Knab; and, *An. darlingi* Root. These species are known to be competent vectors for transmitting malaria in the Americas (Lourenco-de-Oliveira 1989; Klein *et al.*, 1991; Loyola *et al.*, 1991; Padilla *et al.*, 1992; Ramsey *et al.*, 1994), and specifically in Belize (Roberts *et al.*, 1993; Achee *et al.*, 2000; Grieco *et al.*, 2005). Developing and evaluating EO data and products to predict high-risk areas for individual vector breeding sites have been an important component of malaria control in Belize.

2.6.1.1   *Anopheles darling*

The environmental determinants of *An. darlingi* have been identified as floating mats of detritus in shaded areas within freshwater river systems in association with overhanging spiny bamboo (Manguin *et al.*, 1996). In 2006, a study was performed to quantify the contribution of riverbank clearance to bamboo growth in an effort to guide policies for land use that would prevent or reduce *An. darlingi* habitat formation (Achee *et al.*, 2006). Two river systems were studied: the Belize River and the Sibun River, both located in the central region of the country and productive for the target vector. For each river system, the total amount of bamboo present along the riverbanks was measured by overlaying hand-held global positioning system (GPS) points collected while on canoe trips onto SPOT twenty meter resolution and IKONOS one meter resolution products, respectively. The study used general land cover categories (i.e. forest, orchard, and pasture) because the vegetation diversity and interspersion of land cover is high in the humid tropics, and because spectral reflectance characteristics of mixed vegetation often are not distinct, causing problems in digital classification (Roy *et al.*, 1991; Sader *et al.*, 1991). This approach was acceptable given that the goal of the research was to determine the association between *cleared* and *undisturbed* land cover categories with bamboo growth. Results from both study areas indicated that bamboo growth along riverbanks is not associated with cleared land cover. Even though there existed a significantly greater percentage of cleared land cover classes in the disturbed transect of the Belize River, the total length of bamboo was similar, if not slightly less, than the length of mapped bamboo in the undisturbed transect. In addition, no difference in the percentage of cleared land cover classes existed between transects mapped with and without bamboo along the Sibun River. These results were the same using four, ten and twenty meter buffer zones. The only indicator of *An. darlingi* habitats in the Belize study was the distance of detritus patches to surrounding homes. Those debris mats not containing larvae of the target species were an average 162 m further away from mapped house locations compared to positive habitats. In addition, *An. darlingi* positive patches had twenty-three per cent more homes within a 1000 m radius than *An. darlingi* negative patches.

These conclusions are similar to those of other studies in Belize using satellite measurements and GIS to predict malaria risk based on distance of homes to rivers. Examination of malaria

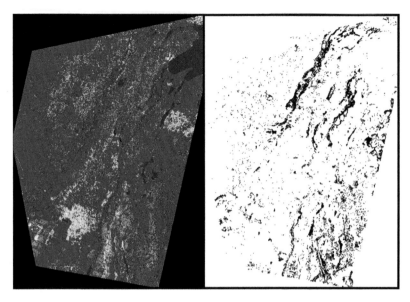

Figure 3.   Classified SPOT and Radarsat imagery of marshes in northern Belize (Permission *ESA.*, 2005). (see colour plate 2)

prevalence in Belize during 1989–1999 found that proximity to a stream less than one kilometre from a household to be predictive for malaria (Hakre *et al.*, 2004). In addition, Roberts (*et al.*, 2002a) found the distance from homes to rivers to be a good predictor of adult *An. dar*lingi during the wet season. In a study that investigated the ability of multispectral satellite data to predict the location of adults of another vector, *An. pseudopunctipennis*, *An. darlingi* adults were also collected at all high probability sites and none were captured at houses predicted as low probability (Roberts *et al.*, 1996). The criteria for site selection included distance of houses from waterways.

2.6.1.2   *Anopheles albimanus* and *An. vestitipennis*

Malaria research employing satellite imagery in areas of Central America has benefited from the fact that the main anopheline species utilize large vector breeding habitats that are detectable directly from space. It has been demonstrated that using dry season SPOT multispectral satellite imagery could produce accurate predictions of dry season adult *An. albimanus* populations in areas of northern Belize that were verified through human landing collections (Rejmankova *et al.*, 1995). In these studies, predictions were based on distance to marshes with low sparse macrophyte vegetation (mostly *Eleocharis spp.*) that included algal (cyanobacteria) mats which were detectable in SPOT images (Figure 3). Habitats containing these algal mats cover large areas in northern Belize, where they have been documented to be the primary larval habitat for *An. albimanus* (Rejmankova *et al.*, 1993; 1996). SPOT imagery was also found to be useful in the study of breeding habitats of other important malaria vectors in Belize, such as *An. vestitipennis*, (Rejmankova *et al.*, 1998). *Anopheles vestitipennis* larval habitats are found in marshes with tall dense macrophyte vegetation (typically *Typha domingensis* and *Cladium jamaicense*) and large expanses of flooded forest. The SPOT image analyses identified large areas with *An. vestitipennis* larval habitat, but could not separate tall dense macrophyte habitat accurately from some cropland and pasture. In addition, flooded forests could not be identified accurately in the SPOT imagery due to the limited availability of images from the dry season when most forests were not flooded. Nevertheless, the SPOT image analyses were useful for guiding malaria control efforts by targeting of interventions in villages at risk for malaria that could be identified based on their proximity to *An. vestitipennis* larval habitat. More recent research in Belize with multi-temporal Radarsat C-band data have indicated that this product may provide better discrimination of anopheline mosquito breeding habitats than do analyses of dry season SPOT imagery (Figure 3). The image on the left shows marsh as red, purple, and pink; cattail marsh as black; forest as green; and agricultural areas as brown, yellow and orange. The marsh image on the right can be used to calculate the area and cost to treat with larvicide or other control methods. Radarsat data analyses can provide information relevant to seasonal changes within the marshes, and findings indicate that both the expansion of algal mats in low sparse macrophyte marshes in conjunction with changes in the extent of flooding in tall dense macrophyte marshes, can be monitored.

2.6.2 Mexico

Research in Mexico with thematic mapper (TM) imagery focused on identifying land cover types associated with *An. albimanus* breeding sites (Beck *et al.*, 1994; Pope *et al.*, 1994). Actual breeding habitats were small and many were near or below the resolution of the TM sensor, which is approximately thirty meters. Nevertheless, *An. albimanus* larval habitats could be defined by specific plant associations (Savage *et al.*, 1990; Rejmankova *et al.*, 1992; Rodriguez *et al.*, 1993), which were then correlated with larger vegetation associations detectable with the TM sensor. This approach produced successful predictions of high and low adult *An. albimanus* population densities in villages, based on the proximity of larval habitats inferred from the TM image analyses (Beck *et al.*, 1994). A similar indirect approach was used in central Belize, where larval habitats of *An. pseudopuntipennis* were inferred on the basis of the presence of sunlit streams detected as open water in the dry season by SPOT multispectral twenty meter resolution satellite imagery (Roberts *et al.*, 1996). This approach was developed from field studies that confirmed the presence of *An. pseudopuntipennis* larval habitats in dry season streams where sunlight supported the growth of filamentous algae (Rejmankova *et al.*, 1993). SPOT image-based predictions of the presence or absence of adult *An. pseudopuntipennis* populations in settlements located near or far from the inferred larval habitat were highly accurate.

2.6.3 South America
2.6.3.1 *Anopheles darlingi*

There are vast areas of South America that maintain a high capacity for malaria transmission based on their environmental variables. These areas are the equatorial, para-equatorial, and subtropical regions of the Americas (Casman & Dowlatabadi 2002). Within these zones, there is the potential for increasing the environmental capacity for malaria transmission due to anthropogenic changes such as through agriculture or deforestation. These associations are most commonly achieved by looking for associations between vector breeding habitat and landscape types or ecological features. Vittor (*et al.*, 2009) utilized a classified TM image in conjunction with human population census data to select appropriate adult *An. darlingi* sampling areas. The sites were stratified on vegetation type and population densities. The researchers ran a one kilometre transect in a north south orientation at the point of the adult collection along which larval collections were conducted. The field data were then evaluated using eight land cover types with an unsupervised classification of a 2001 TM image. A maximum flight range for *An. darlingi* of seven kilometres (Charlwood & Alecrim 1989) was used as the selection criteria for the grid size to be evaluated. The percentage of each landscape class was calculated in each grid. The study indicated that there are positive associations between ecological features that correlate with deforestation and the presence of positive *An. darlingi* breeding sites. This was particularly true for secondary growth that results after forest clearing.

In addition to evaluating the impact of anthropogenic changes on vector populations, GIS and remote sensing tools also have been used in South America to focus control efforts. Bautista (*et al.*, 2006) conducted a spatial point pattern analysis to examine the distribution of malaria incidence in the Northern Peruvian Amazon city of Iquitos utilizing retrospective surveillance data. Spatial scan statistics were employed to identify the relative risk of spatial clusters (i.e. areas with excess cases) and spatial-temporal clusters (i.e. areas with excess cases in a period of time) (Hjalmars *et al.*, 1996; Kulldorff 1997). It was found that there were consistent, micro-high risk areas and spatial malaria clusters (hot spots) where transmission was occurring. These same areas were in proximity to land-cover types classified as secondary growth forest where *An. darlingi* breeding sites are found often.

2.7 Conclusion

It is reasonable to conclude that EO data and imagery in conjunction with GIS technologies offer considerable promise for cost-effective applications in malaria control and prevention. Rapid growth in these technologies will bring more refined multispectral satellite data and computing facilities at lower cost in the future. Regardless of which product is utilized, full knowledge of the target vector ecology is critical to maximize success in predicting malaria risk and managing disease. As with any tool, the utility of GIS output products is completely dependent on the data chosen for analysis and the end user.

3 MALARIA IN ASIA

3.1 Introduction

Malaria is a parasitic disease that infects both humans and primates. For humans, it is caused by a number of plasmodium species, such as *Plasmodium vivax, P. falciparum, P. malariae, P. ovali*, and the more recently discovered *P. knowlesi* (Singh *et al.*, 2004). The disease is transmitted by infected female anophelines after taking blood meals from infectious humans. Malaria is endemic in most parts of the tropics, especially in developing countries. Half of the world's population is at risk for malaria infection (RBM 2011). Worldwide, there are approximately 250 million cases annually with 0.9 million deaths. The gravity of the disease is reflected often in the statistics that on average one death occurs every thirty seconds. Africa has nearly ninety per cent of the cases and deaths, but it is also a significant problem in South and Southeast Asia.

Malaria may still become a significant health issue for countries outside of tropics where public health support is inadequate because of economic constraints or military conflicts. For example, since 1993 vivax malaria re-emerged in North Korea (Feighner *et al.*, 1998). Similarly, situated around 34°N with an arid climate, Afghanistan has approximately 0.41 to 0.6 million cases annually (Youssef 2008; WHO-EMRO 2007), and is the country most endemic with malaria in the WHO's Eastern Mediterranean Region.

In this section, malaria modelling and surveillance are discussed in context of remote sensing data. Examples are drawn from malaria in South Korea, Thailand, Indonesia and Afghanistan.

3.2 Determinants for malaria transmission

Many factors contribute to malaria transmission, including meteorological and environmental conditions, socioeconomic status, military conflicts and natural disasters. Among these, meteorological and environmental factors are perhaps the most measureable. For example, malaria transmission may increase with the arrival or the end of a rainy season, and living near forest or water bodies may pose greater risk of mosquito bites and getting malaria. The ENSO is a quasi-periodic climatic cycle that occurs every three to seven years across the tropical Pacific Ocean, and which causes excessive precipitation or droughts. It has been shown to promote malaria transmission (Bouma & van der Kaay 1996; Kovats 2003). Aside from precipitation, temperature and humidity are also important factors. Warmer temperatures hasten larval and vector development, and prolong mosquito life span and its consequent ability to transmit malaria. Warmer air retains more moisture and improves mosquito survivorship, so has higher humidity.

Malaria transmission is also linked to vegetation type at their breeding sites. *Anopheles dirus* is a forest breeder while *An. maculates* and *An. sawadwongpori* are rice field breeders. NDVI is one of the indices for vegetation condition (Tucker 1979). It is defined as the difference between the responses from the infrared and the red bands normalized by their mean. Vegetation type and plant species, however, generally cannot be derived from NDVI, but need to be identified using classification methods based on multispectral data. Modelling infectious diseases using NDVI most often infer precipitation amounts received in an area before satellite measurements were taken. The spatial distribution of NDVI can also be used to differentiate among urban, peri-urban, suburban and rural areas. Such information on the nature of the area is useful for malaria prevention and control.

3.3 Remote sensing measurements for environmental determinants

Satellite measurements of precipitation for estimating disease risks are most often derived from the Tropical Rainfall Monitoring Mission (TRMM) (Kummerow 1998). TRMM is a collaboration between the US and Japan. Japan built the precipitation radar (PR) and launched the spacecraft in 1999. There are five instruments on board: PR, TRMM microwave imager (TMI), Visible and Infrared Scanner (VIIRS), Cloud and Earth Radiation Energy Sensor (CERES), and Lighting Image Scanner (LIS). The main sensor that measures precipitation is TMI. The follow-on mission for TRMM is the Global Precipitation Measuring (GPM) mission. Instead of a single satellite, GPM consists of a constellation of satellites; a core satellite, and another eight to achieve more complete and more frequent coverage.

Because the spatial resolution of the TRMM measurement is relatively coarse, NDVI measured from a medium resolution satellite are sometimes used to infer recent rainfall at a higher spatial resolution. Such inferences are more effective in arid regions where little rainfall is received and vegetation growth is sensitive to rainfall, but are less effective for regions with plentiful rainfall.

Land surface temperature and NDVI are both provided by the moderate resolution imaging spectro-radiometer (MODIS). This sensor has thirty-six spectral bands spanning the visible and near-infrared wavelengths. Both the Terra (EOS-AM) and Aqua (EOS-PM) observatories are equipped with MODIS. NDVI can be computed from any satellite instruments with red and infrared channels. However, because of the differences in band definitions, instrument characteristics, satellite orbits and measuring conditions, NDVI from different sensors must be cross-calibrated before data from different sensors can be compared.

In addition to the data sets described above, several other satellites also provide data for ground cover classification, identification of potential larval habitats, and modelling malaria risks. For example, the well-known Landsat and SPOT series of satellites, the advanced space-borne thermal emission and reflection radiometer (ASTER), and the advanced land imager (ALI) are among the multispectral sensors used for health monitoring purposes (JPL 2009; USGS 2009). Microwave sensors like Radarsat and the phased array L-band synthetic aperture radar (PALSAR) are used over areas that are obscured frequently by clouds (CSA 2011; JAXA 2011). The geoscience laser altimeter system (GLAS) is a light detection and ranging (LiDAR) sensor that is useful for differentiating vegetation types (NASA 2011). For high spatial resolution imagery, commercial data from IKONOS, QuickBird, or WorldView can be used (GeoEye 2011; DigitalGlobe 2011).

3.4 Modelling and analytical techniques that use remote sensing data

Remote sensing data can be used in a number of ways to model disease risks. For malaria in general, the data are useful for identifying potential larval habitats and for estimating the present and future malaria risks. In the following, several examples are given to illustrate the methodology. Specific examples are drawn from South Korea, Thailand, Indonesia and Afghanistan.

3.4.1 Techniques for ground-cover classification

The commonly used techniques include: the parallelepiped method, the maximum likelihood estimator method (Duda *et al.*, 2000), the neural network methods (Principe *et al.*, 2000; Debeir *et al.*, 2001; Bocco *et al.*, 2007), and decision or regression trees (Breiman *et al.*, 1984). Classification can be performed on a per-pixel basis, at sub-pixel level or spatially on a group of pixels. Per-pixel classification is the most commonly found. Sub-pixel classification can be used to estimate the classes within a pixel. When a class can only be identified relative to the pixel's surroundings, spatial classification approach is taken. In this approach, several windows around each pixel in study are classified at a time on the basis of the pixels' textural and contextual information. A similar approach to textural classification, though it is less used, is object-oriented classification. Here, the image is first segmented into objects and the homogenous objects are then classified.

3.4.1.1 Potential larval habitats of *Anopheles sinensis* in South Korea

Carefully planned applications of larvicide and insecticide are an effective method for vector control. They minimize the damage to the environment and the possibility of inducing insecticide resistance. Because mosquito larval habitats are often small, special techniques beyond the usual per-pixel spectral-only classification techniques are often needed to achieve reasonable classification accuracy. Vivax malaria is the indigenous form in Korea. It was eradicated in 1979, but has reappeared since 1993 throughout the Korean peninsula. By 2000, there were approximately 4000 cases in South Korea. One of the primary vector species transmitting vivax malaria is *Anopheles sinensis*. It was discovered previously that malaria transmission through *An. sinensis* is due to its population density instead of its transmission potential (Lee *et al.*, 2001). Hence larval and vector control, as well as reducing the possibility of biting, should be the emphasis of malaria control.

Irrigation and drainage ditches along rice fields are common larval habitats for *An. sinensis*. Because the width of the irrigation and drainage ditches is much narrower than the footprint of common civilian satellites like Landsat, commercial satellite data of higher spatial resolution must be used. It has been shown that the native four metre spatial resolution IKONOS data is not adequate for differentiating ditches; hence, pan-sharpened IKONOS data must be used to identify these surface features.

Pan-sharpening enhances spatial resolution and makes narrow linear features visible. At the same time, it has been shown that pan-sharpening makes clusters in spectral space more defused as the spatial scale becomes finer (Kiang *et al.*, 2003). Therefore, one would expect lower classification accuracy in pan-sharpened data if the same classification method is used. To maintain reasonably good classification accuracy after the imagery data are sharpened, it is necessary to use spatial features, such as textural and contextual information, in the classification. For a 3.2 km^2 by 3.2 km^2 test site in South Korea, the average accuracy for ten classes of forest/trees for the following configurations were obtained using maximum likelihood method: (a) 4-band native resolution (4 m) 88.4 per cent; (b) the preceding configuration plus panchromatic band 88.6 per cent; (c) 3 pan-sharpened bands from visible and near infrared (VNIR), red and blue 71.8 per cent; (d) the preceding configuration plus 4 bands at native resolution 92.1 per cent; and (e) 3 pan-sharpened bands plus 3 × 3 neighbourhood mean and variance 82.9 per cent. Hence, the 7-band configuration (4 native bands plus 3 pan-sharpened bands) gives the highest classification accuracy.

3.4.2 *Statistical methods for modelling malaria risks*

In general, statistical and mechanistic, or processed-based approaches are used to model malaria risks based on satellite observations of meteorological and environmental parameters. In the statistical approach, epidemiological data are correlated with satellite data. The unknown parameters in the models are determined using statistical goodness of fit criteria, such as mean squared errors or Akaike information criteria (AIC). Once the model is trained, it can then be applied to other situations than those used for deriving the model parameters. How the pathogens actually transmit the disease under different meteorological and environmental conditions is not explicitly modelled in this approach. The common methods in this category include regression, time series analysis, and neural network. Examples are given in the following sub-sections for using this approach to model malaria cases in Thailand, Indonesia and Afghanistan.

3.4.2.1 Thailand

Thailand is one of the countries in the greater Mekong sub-region (GMS). Other countries and regions in GMS include Myanmar, Laos, Cambodia, Vietnam and China's Yunan Province. GMS is the world's epicentre of multidrug resistant falciparum malaria. The Thai provinces that are endemic

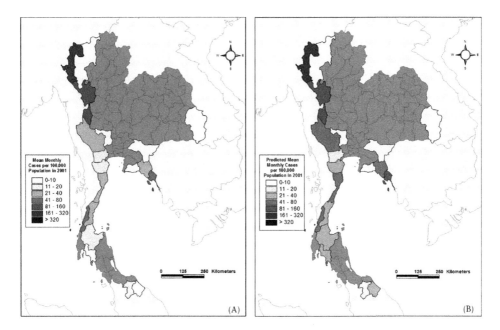

Figure 4. Comparison of actual and modelled malaria case rates (per 100,000 population) in 2001. Training of neural networks was performed using 1994–2000 data: (a) actual; (b) modelled case rates. (see colour plate 3)

with malaria are mainly the provinces at the borders with other countries. The provinces bordering with Myanmar especially have significant transmission due to non-permanent populations going back and forth across the border to seek better economic opportunities or escape from military conflicts.

The main malaria vector species in Thailand include *Anopheles minimus, An. dirus,* and *An. maculatus.* The typical breeding habitat of *An. minimus* is shaded small streams. Typical habitats for *An. dirus* and *An. maculates* are shaded ground pools and rice fields, respectively. Figure 4 illustrates the result of using a multi-level perceptron neural network to model provincial malaria case rates in 2001. The meteorological and environmental parameters used for modelling include TRMM precipitation, MODIS NDVI, and MODIS land surface temperature (LST). The provincial malaria case rates are shown in Figure 4(a), and the modelled case rates are shown in Figure 4(b). Overall, the modelled case rates resemble the actual distribution closely (Kiang *et al.*, 2006).

3.4.2.2   Indonesia
In Southeast Asia, Indonesia has the third highest malaria incidence after Myanmar and India (WHO-SEARO 2011). Approximately forty per cent of Indonesia's population lives in malarial areas. Malaria is hypoendemic on Java and Bali, but on the outer islands, malaria ranges from hypoendemic to hyperendemic. Malaria endemicity in Indonesia had been in decline in the early 1990s, but most malaria control efforts were stopped for economic reasons in 1997. Currently vivax malaria is the main species on Java and Bali. On the outer islands, there are approximately equal numbers of vivax and falciparum cases. Since the decentralization policy was implemented in 2001, malaria control efforts have been carried out at district or regency levels. It is challenging to reduce malaria transmission efficiently for a country with vast and diverse territories and more than 17,500 islands.

The major malaria vector species are *An. balabacensis, An. minimus* and *An. maculatus.* All three are primarily anthropophilic and exophilic. Menoreh Hills in Central Java is one region in Indonesia

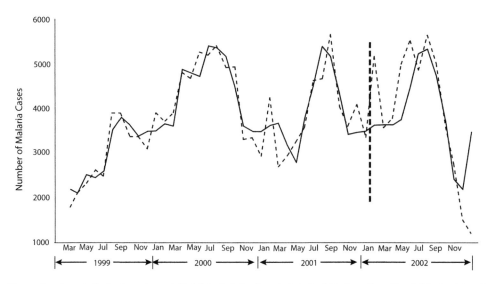

Figure 5. Actual (dashed grey), modelled (black, left of vertical dashed line), and predicted (black, right of vertical dashed line) malaria cases in Jawa Tengah Province, Indonesia.

that has persistent malaria transmission. The hilly topography and numerous streams running through the steep hills presents a particular challenge for effective larval control (Barcus *et al.*, 2002). Menoreh Hills is also the main part of the Jawa Tengah Province that is endemic with malaria. Figure 5 shows the monthly malaria case distribution for Jawa Tengah as well as the modelled and predicted distributions. Both the modelled and predicted distributions are reasonably close to the actual time series. Neural network was used for modelling. The same set of environmental parameters as described in the Thailand study are also used here.

3.4.2.3 Afghanistan
As the result of the three-decade long military conflicts, malaria control and public health service nearly disappeared from Afghanistan. It was only since 2004 that an effort to rebuild the public health infrastructure began. Currently, nearly sixty per cent of the population lives in areas endemic with malaria. Vivax malaria is the dominant malaria species. Nowadays there are only a small number of falciparum malaria cases. Malaria distribution is heterogeneous in Afghanistan because of the diverse landscapes. In general, endemic areas are below 610 m. Rice growing areas in river valleys are common mosquito breeding grounds. For this analysis, a linear regression on precipitation, NDVI, and surface temperature with corrections of auto-regression, seasonal variation and linear trend were used. Provincial malaria data for twenty-three provinces from 2004 to June 2007 were used for training, and the trained models were used to predict malaria cases for July to December 2007. As shown in Figure 6, the predicted cases follow the real distribution of malaria cases closely (Adimi *et al.*, 2010).

3.4.3 *Process-based modelling*
The studies described for Thailand, Indonesia and Afghanistan use a statistical modelling approach. A biological, process-based, mechanistic model describes how plasmodium parasites are transmitted among residents and households through infected mosquitoes. The particular approach used here is discrete-event, or agent-based simulation. In essence, it simulates the detailed interactions among vectors, parasites and human hosts at individual vector, dwelling, animal, and human host. The human host can be at a number of distinct stages, parasite-free, pre-patient, and incubation, among others. Likewise, the mosquito has a number of stages in its lifecycle, and a number of

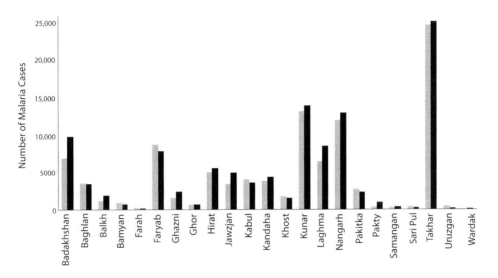

Figure 6. Actual (grey) and predicted (black) malaria cases between July and December 2007 for twenty-three Afghanistan provinces.

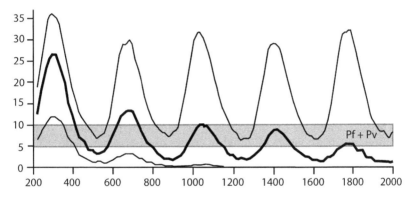

Figure 7. Modelled prevalence over 2000 days for three levels of larval loadings (Kiang, unpublished data). The shaded band represents observed malaria prevalence for both falciparum and vivax malaria (Zollner, unpublished data).

stages in its sporogonic cycle, the stages in a mosquitoes' stomach when parasites multiply. The transition from one stage to the next is partially influenced by environmental, meteorological, and socioeconomic factors. Figure 7 shows a simulation of malaria transmission among twenty-three houses in a setting that resembles a test site that used active case detection in Thailand. The top curve is for heavier than normal larval loading, the middle curve is for normal loading, and the lower curve is for lighter than normal loading. Because active case detection was used and infected individuals were always treated, the percentage of infected residents decreased over time. Overall, there is reasonably good agreement between the modelled result and the field observations.

3.5 *Discussion*

This section demonstrates that remote sensing data can be used to help detect potential larval habitats, estimate current endemicity, and predict future transmission. Malaria transmission is known

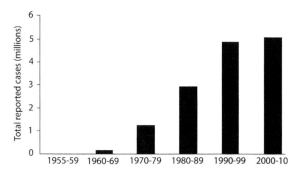

Figure 8.   Total reported cases of dengue since 1955 (Source data: WHO Dengue Net 2010).

to be spatially heterogeneous. The results in this section indicate that using low resolution, provincial epidemiological data paired with medium resolution remotely sensed data can still achieve sufficient accuracy for malaria surveillance and control.

4   DENGUE FEVER

Dengue infections cause more human morbidity and mortality worldwide than any other arthropodal viral disease (Gubler 2002; Kuno 1997). It is estimated that 2.5–3.0 billion people are at risk of infection each year, and millions have been infected during recent epidemics. Dengue is caused by four different viral serotypes in the family Flaviviridae. It occurs throughout the tropics but the number of serotypes circulating can vary by geographic region. Incidence rates have increased steadily since the 1950s. Between 1990 and 2010 dengue has emerged as a major public health threat throughout the world, including Latin America, with 50–100 million fever cases and 250,000–500,000 dengue haemorrhagic fever/dengue shock syndrome cases worldwide. (Guzman & Kouri 2003; Isturiz *et al.*, 2000) (Figure 8). In urban centres of Southeast Asia, dengue haemorrhagic fever is among the leading causes of paediatric hospitalization (WHO 2008). The burden of illness across the spectrum of clinical outcomes for dengue can be expressed in terms of disability adjusted life years (DALYs), which are cumulative years lost prematurely (see also Chapter 3). The WHO estimated the global burden of dengue to be 670,000 DALYs, or healthy life years lost in 2004 (WHO 2008). The burden of disease may vary by dengue serotype. For example dengue virus-2 is estimated to cause more DALYs in Thailand compared with dengue-1, -3, and -4 .(Anderson *et al.*, 2007). Dengue is the most important arboviral infection of humans worldwide with 2.5 to 3.0 billion people at risk of infection each year. A critical tool for dengue surveillance and control is the ability to predict dengue seasonality, incidence, and risk. Although difficult to measure directly in field studies, temperature, precipitation and humidity may be important environmental drivers influencing transmission. Using EO sensor data to understand environmentally driven and spatial patterns of dengue incidence is on the rise and represents a valuable new approach for understanding patterns and risk for this major vector-borne disease. Ideally, studies should utilize surface climate conditions; land cover and land use (LCLU) classification data with high spatial resolution and frequent coarse resolution environmental satellite data to map disease distribution and risk. Some of the challenges to employing remotely sensed data in dengue studies are discussed here.

4.1   *Dengue transmission and vector biology*

The yellow fever mosquito, *Aedes aegypti,* is the most important vector of dengue and yellow fever viruses worldwide (Gubler 1997; Gubler 2002) (Figure 9). This species is uniquely adapted to a close association with humans and is a highly efficient vector of human pathogens. Immature *Ae. aegypti*

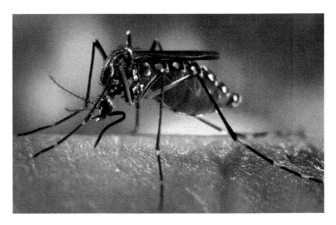

Figure 9. *Aedes aegypti*, the primary vector of dengue viruses (Source, CDC Public Health Image Library; photo credit, James Gathany). (see colour plate 4)

Figure 10. Larval breeding sites of *Aedes aegypti* in Thailand: (a) indoor bathroom basins; (b) ant trap under table leg; (c) outdoor animal water vessel; (d) indoor and outdoor water storage jars. (see colour plate 5)

develop primarily in human-made water storage containers, other water holding containers and rain-filled debris (Figure 10), laying their eggs in containers during the late afternoon. (Harrington *et al.*, 2008a; Kuno 1997). Females are highly anthropophilic, resting inside houses and feeding frequently and preferentially on human blood (Scott *et al.*, 1993a; Scott *et al.*, 1993b). Ingesting blood from humans confers a fitness advantage (Harrington *et al.*, 2001). Few *Ae. aegypti* disperse beyond 100 m of houses in which they reside, because food, mates, and oviposition sites are readily available within human habitations (Edman 1998; Harrington 2005). Today, with no clinical cure or commercially available vaccine, dengue control is dependent on the reduction or elimination of its vector, *Ae. aegypti* (Gubler 2002b).

The vast majority of dengue virus transmission is horizontal between mosquito and human hosts, although there is some evidence of sylvatic transmission cycles that include non-human primates. In addition, there is some evidence for vertical transmission of dengue viruses by mosquitoes (Rodhain & Rosen 1997; Gunther *et al.*, 2007). Horizontal transmission begins when a female mosquito takes a viremic human blood meal. Extrinsic incubation requires at least twelve days at ambient temperatures in the tropics (Scott *et al.*, 2000; Watts *et al.*, 1987). Once infective, *Ae. aegypti* females can transmit virus each time she probes a host or ingests a blood meal (Putnam & Scott 1995). Dengue virus incubation in humans is typically four to seven days. Symptoms, if present, are first observed at the end of the viremic period. Viremia often precedes fever and typically lasts about five days, subsiding in concert with detection of virus in the blood (Vaughn *et al.*, 2000).

4.2 *Environmental and contextual determinants of dengue transmission*

The ecology of dengue is complex, with three key components in the transmission cycle (host, mosquito vector, and virus). Several factors may influence these key components and, ultimately, the number and severity of dengue cases. Environmental factors affect the vector, virus and host in the transmission cycle directly and indirectly. Although difficult to measure directly in field studies, it is logical that temperature, precipitation, and humidity play roles in dengue transmission. Conflicting results on the influence of these variables appear in the literature with little rigorous support from field studies. However, a few recent studies have made significant progress towards identifying the association between weather and dengue cases. Johansson (*et al.*, 2009a) found a significant relationship between precipitation and temperature with dengue transmission in Puerto Rico. In their study Johansson (*et al.*, 2009b) emphasized the role of variation in local climate on patterns of dengue transmission, a challenge when using remotely sensed proxies for environmental variables. In another study Johansson (*et al.*, 2009a) used wavelet analysis to assess multiannual patterns related to the ENSO with varying results. Only local precipitation was correlated with dengue in Puerto Rico. No associations for temperature or precipitation with dengue were detected for Mexico (Johansson *et al.*, 2009a).

4.2.1 *Temperature*

Temperature is an important physiological determinant for development, physiological ageing and survival of *Ae. Aegypti*, all of which may influence dengue transmission by the vector (Christophers 1960; Keirans & Fay 1968; Gerade *et al.*, 2004; Harrington *et al.*, 2008b;). Temperature also drives the extrinsic incubation period (EIP) of the dengue virus, which is the interval of time for a female mosquito to ingest a viremic blood meal until she is capable of transmitting virus to a new host. The higher the ambient temperature, the faster a vector undergoes the EIP up to a certain limit. The impact of temperature on EIP is likely to fluctuate due to varying temperatures in the microclimates found in nature. Most estimates of EIP used in vector-borne disease models are made with laboratory reared and infected mosquitoes held under constant conditions (Watts *et al.*, 1987). Few studies have investigated the impact of fluctuating temperatures on dengue virus EIP in *Ae. aegypti* under more natural conditions that they may experience in the field. Differences in vector competence among geographic mosquito strains as well as differences in virus genotype also may affect the length of the EIP (Salazar *et al.*, 2007). As a consequence it can be difficult to estimate EIPs accurately in the field from surface temperature data.

Temperature also increases blood feeding frequency and egg production of *Ae. aegypti* (Pant & Yasuno 1973). Once a female is infectious, she may transmit virus for her life. Higher ambient temperatures can increase the frequency of blood meals and potential transmission events, but also will decrease longevity or the number of survival days. Higher temperature can influence mosquito body size through faster larval development which can result in smaller body size. Body size can influence important aspects of mosquito biology epidemiologically; for example, Scott (*et al.*, 2000b) demonstrated greater multiple feeding rates among smaller *Ae. aegypti* females in Thailand. The effect of temperature on the physiology and development of *Ae. aegypti* has been reviewed (Focks *et al.*, 1993).

4.2.2 *Precipitation*

As a water-filled container breeding species, precipitation has the potential to play a significant role in supporting *Ae. aegypti* populations (Christophers 1960). However, this effect can be complex. For example, in many tropical cultures with distinct rainy seasons, stored rainwater provides essential drinking and washing water throughout the dry season. During times with little to no rainfall, this practice may inadvertently provide a large aquatic resource to sustain *Ae. aegypti* larvae. In contrast, rainfall can create additional breeding sites beyond managed storage containers in the environment. In a study in Peru, large outdoor and unmanaged, rain-filled containers accounted for seventy-two per cent of total *Ae. aegypti* positive containers and seventy-eight per cent of all pupae (Morrison *et al.*, 2004). The impact of rainfall can be detrimental if it is excessive and leads to flushing of immature mosquitoes from breeding sites. However, evidence for the flushing effect of rain has only been reported anecdotally and is based on the association between heavy rain and lower mosquito catch numbers without direct observations (Geery & Holub 1989; DeGaetano 2005;). Excessive rainfall may not be an issue for *Ae. aegypti* populations. This species appears to be adapted to periods of torrential rainfall with the ability to dive to the bottom of containers and remain for long periods of time during major rainfall events (Koenraadt & Harrington 2008). Some claim that dengue vectors are not impacted by rainfall in the same way as other species like *Anopheles* because of differences in their larval ecology (Martens *et al.*, 1997).

4.2.3 *Humidity*

Humidity is difficult to disentangle from precipitation effects. It is known that low humidity can reduce the survival rate of *Ae. aegypti* (Machado-Allison & Craig 1972), and Hales (*et al.*, 2002) used long term average vapour pressure successfully as a proxy for humidity and were able to model the geographical limits of dengue transmission with high accuracy. It is clear that temperature, precipitation and humidity have strong potential for influencing vector and virus biology in nature. These determinants can be used for estimating risk and for surveillance and control strategies. Other non-environmental factors probably also influence dengue transmission in such ways as to reduce model accuracies that are based solely on satellite sensor data of atmospheric conditions. For example, personal practices such as cleaning water storage containers, or religious practices such as the use of spiritual vases in homes may decrease or increase vector populations (Bohra & Andrianasolo 2001; Spiegel *et al.*, 2007). In addition, host immunity, malnutrition, and human genetic determinants may influence the outcome of dengue infection (Nguyen *et al.*, 2005; Coffey *et al.*, 2009).

4.3 *Dengue surveillance and control using Earth observations data*

A review of remote sensing applications for vector-borne disease surveillance has been provided by Kalluri (*et al.*, 2007). Only a few studies have utilized sensor data for understanding dengue transmission and risk compared with studies of malaria, Rift Valley fever, tick-borne and water-borne infections (Hales *et al.*, 2002; Tran & Raffy 2006; Rotela *et al.*, 2007; Vanwambeke *et al.*, 2007; Johansson *et al.*, 2009a); however, the use of these data for dengue is increasing and represents a valuable new approach for estimating dengue risk. Satellite observations of terrestrial environments for vector-borne diseases have ranged from studies that correlate land use and land cover types with vector populations to spectral signatures and more complex approaches that link seasonal environmental variables to disease vectors. Satellite sensors by themselves cannot estimate vector populations, but they can provide data about environmental drivers that influence these populations. Several scientific studies have used remote sensing technology to investigate the association between multivariate environmental data and mosquito abundance or vector-borne disease incidences to create risk maps. One theoretical study used a non-static diffusion model with environmental data obtained from sensor imagery to model the spatial and temporal dynamics of dengue fever (Tran & Raffy 2006).

Proxy variables to estimate land surface temperatures have been employed in vector-borne disease models to estimate EIP and entomological inoculation rates. NDVI and other vegetation indices are commonly used. NDVI is correlated with rainfall and has been used to model Rift Valley fever incidence in Africa successfully (Linthicum *et al.*, 1991). Rogers (*et al.*, 2006) used thermal Fourier-processed measurements of NDVI to map global distribution of dengue. Cold cloud duration also has been used as an indirect measure of precipitation for vector-borne disease studies (Hay *et al.*, 1996; Hay 2000).

Land use changes also can impact vector-borne diseases (Patz & Olson 1996) including dengue. Vanwambeke (*et al.*, 2007) used satellite data to investigate the impact of land use changes on dengue in northern Thailand. In another study conducted in Argentina, a predictive map for dengue was built on Landsat-5 TM data to develop a synthetic multiband image including variables such as distances to main roads, water and other potentially important landscape factors (Rotela *et al.*, 2007).

Ideally, Kalluri (*et al.*, 2007) suggest that vector-borne disease studies should use high spatial resolution surface climate conditions and land cover data combined with frequent coarser resolution environmental satellite data to map vector-borne disease distribution and risk. Although remotely sensed data for dengue epidemiological studies are increasing, challenges remain. Traditionally there has been a major trade-off between resolution (temporal and spatial) and spatial coverage (Johansson & Glass 2008). One major issue is scaling regional temperature data to the microhabitats within a house where dengue vectors may rest, feed, and oviposit. Another major challenge for satellite data as a means for predicting accurate distribution and risk of dengue is learning more about the relationships between disease and environmental drivers. On the positive side, difficulties that have traditionally hindered access to appropriate data in useful formats at low or no cost (Achee *et al.*, 2006; Herbreteau *et al.*, 2007) have, for the most part, disappeared. Another major challenge when using EO data for predicting accurate distribution and risk of dengue is our lack of understanding about the relationship between this disease and environmental drivers. More detailed field studies are required to understand these relationships.

5 MOSQUITO-BORNE ENCEPHALITIS

Neuro-invasive viruses transmitted by mosquitoes may cause clinical diseases ranging in severity from acute encephalitis (infection of the brain) and death to mild febrile illnesses. These viruses are zoonoses maintained and amplified among wildlife or domestic animals by culicine mosquitoes (those of the genus *Culex)*, with human and domestic animal infection resulting from tangential or spill-over transmission from the basic cycle. These viruses arose independently in the tropics within three divergent arbovirus families where they are maintained typically at low enzootic levels. Periodically virulent strains have invaded temperate latitudes leading to wide-spread epizootics and epidemics among susceptible vertebrate host and human populations. The epidemiology, invasion, and health significance of representative encephalitis viruses are discussed, showing the value of remote sensing in understanding distributions in time and space.

Mosquito-borne encephalitis is a disease of the brain caused by swelling and pathology associated with or as sequellae to infection by a virus transmitted by a mosquito bite (see Chapter 10 Section 2 for a discussion on the West Nile virus [WNV] detection system in California). Frequently neuro-invasive disease is preceded by or associated with cases of febrile illness with influenza-like symptoms and infection of the meninges leading to cervical rigor. These viral encephalitides are caused by a genetically diverse assemblage of zoonotic viruses within three taxonomically diverse and antigenically distinct viral families, the *Flaviviridae, Togaviridae* and *Bunyaviridae*. These zoonoses are distributed globally and are maintained and amplified among wildlife or domestic animals by a variety of culicine mosquitoes, mostly within the genera *Culex* and *Aedes* (Figure 11). Historically in North America, encephalitides included: St. Louis encephalitis (SLE), eastern equine encephalitis (EEE), western equine encephalitis (WEE), LaCross encephalitis (LAC), and Powassan

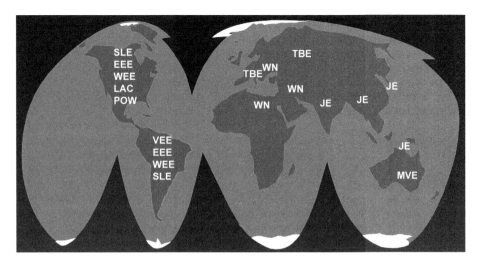

Figure 11.   Global distribution of the arboviral encephalitides (Courtesy CDC).

encephalitis (POW). Venezuelan equine encephalitis (VEE) is limited to the neotropics of Central and South America. In the Eastern Hemisphere, West Nile encephalitis (WN) and Japanese encephalitis (JE) are widespread, but tick-borne encephalitis (TBE) and Murray Valley encephalitis (MVE) are relatively more local in their distribution.

These arboviruses are diverse genetically and are maintained typically at low levels within enzootic cycles in the tropics, but occasionally produce highly infectious genotypes that invade or emerge at northern latitudes where they cause major epidemics among immunologically naive populations (Weaver & Reisen 2009). Typically, transmission involves one or more susceptible mosquito and vertebrate hosts that amplify the virus enzootically to levels where infection of other hosts or vectors occurs tangentially. Tangential hosts such as humans or domestic animals usually are a dead end for the virus, but may develop serious disease and death.

5.1   *Flaviviridae*

Viruses that cause encephalitis are grouped within the Japanese encephalitis virus (JEV) serocomplex within the genus *Flavivirus* and are transmitted mostly among birds by mosquitoes in the subgenus *Culex* of the genus *Culex*. JEV, the type virus, remains the greatest cause of neurological human illness and is widely distributed throughout eastern and central Asia. Currently, more than three billion people are at risk from JEV disease, with an estimated 30,000–50,000 clinical cases occurring annually (Erlanger *et al.*, 2009), although laboratory confirmation and case reporting are limited, especially in rural settings. JEV circulates in nature by horizontal transmission among wading birds in the family Ardeidae, including Black-crowned Night Herons (*Nycticorax nycticorax*) and Asian Cattle Egrets (*Bubulcus ibis coromandus*). Infection in domestic and perhaps feral swine seems necessary for amplification to levels where tangential transmission frequently spreads to equids and humans (Burke & Leake 1988). *Culex tritaeniorhynchus* is the primary vector throughout Asia, but other members of the *vishnui* complex, *bitaeniorhynchus* complex, *Cx. gelidus* and *Cx. fuscocephalus* may be important locally. Transmission occurs at low levels throughout the year at southern latitudes in Sri Lanka and Indonesia, but subsides during winter at northern latitudes in Korea, China, India and Nepal when *Culex* vectors enter diapause or become regionally extinct (Wang *et al.*, 1989; Min & Xue 1996). Risk factors include rice production that provides extensive habitat for larval mosquito development and pig farming, especially at family farms where the pigs are kept near to the home. Intervention includes centralizing swine farming to remove the amplifying host from the peridomestic environment, vaccinating swine to interrupt amplification

41

and preclude losses due to abortion, and vaccinating school age children. In combination, these programs have limited human infection in Japan to fewer than ten cases per year; whereas in areas like Uttar Pradesh in India and the adjacent Terrai of Nepal epidemics recur annually during mid to late summer. Areas at greatest risk may be determined remotely by delineating the ecotones or intersects among rice culture, swine production and human habitation.

West Nile virus within the JEV serocomplex is an emerging virus that has caused public, veterinary, and wildlife health problems on five continents (Kramer *et al.*, 2008). Disease in naive human populations ranges from in-apparent infection, to influenza-like symptoms to neuro-invasive disease and death, especially among the elderly. Sequellae can include permanent neurological complications. Within developing countries, repeated in apparent infection within the wide range of childhood febrile disease imparts acquired immunity levels that seem to protect the elderly from neuro-invasive disease (Hayes *et al.*, 1982). In contrast, periodic invasion of northern latitudes has resulted in self-limiting outbreaks in northern Africa and Europe, but has been the cause of a massive on-going epidemic in North America since 1999 that is the largest mosquito-borne encephalitis virus epidemic ever recorded in the New World. There were more than 25,000 cases and more than 1000 deaths documented globally (Kramer *et al.*, 2008). Equids also develop a similar broad spectrum of disease, including fatal neurological infection, but now are largely protected in the US by annual vaccination and naturally acquired immunity. Although a wide range of mosquitoes and birds has been found infected naturally (Komar 2003), mosquitoes within the *Culex pipiens* complex and *Cx. tarsalis* and birds within the order Passeriformes seem to be the primary vectors and host species, respectively. Different bird taxa produce an extremely varied response to infection, ranging from high viremia and death in the Corvidae such as American crows, to very low viremia and in apparent infection in galliforms such as domestic chickens and quail (Kilpatrick *et al.*, 2007; Wheeler *et al.*, 2009). Most enzootic transmission among roosting/nesting birds and tangential transmission to humans occurs in urban or rural residential settings where *Culex* vectors blood feed at night and serve as both amplification and bridge vectors. At temperate and northern latitudes in North America, outbreaks seem closely tied to climate variation, where hot dry summers extend the transmission season, shorten the extrinsic incubation period of the virus in the vector, and lead to more frequent vector-host contact (Reisen *et al.*, 2006). Although equids now are protected by vaccination, there is no vaccine currently approved for humans. Neither are there effective therapeutics. Therefore, intervention relies on personal protection, integrated vector management programs that attempt to keep vector populations below thresholds where humans become infected; and emergency adulticide applications that attempt to break the chain of amplification and tangential transmission by reducing or eliminating the number of infected and infectious vectors. Remotely acquired data for weather events and climate cycles can be used to track outbreak risk skilfully (Reisen 2010), especially in northern latitude dry land agricultural ecosystems where elevated temperatures are required for efficient transmission (Reisen *et al.*, 2006). Remotely sensed data of landscapes, temperatures and socioeconomic conditions have been linked to WNV risk in urban and peri-urban settings (Liu *et al.*, 2008; Liu & Weng 2009).

Other members of the JEV serocomplex of public health concern include St. Louis encephalitis virus (SLEV) in the New World and Murray Valley encephalitis virus (MVEV) in Australia (Mackenzie *et al.*, 2002). The spectrum of human disease attributed to these viruses generally is similar to JEV and WNV. SLEV is endemic in North America where, historically, it was a major cause of neurological disease, especially during the 1970s in the Ohio River drainage (Monath 1980). Transmission cycles, vectors and avian hosts generally are similar to WNV, except that the passeriformes rarely succumb to infection, produce lower titred viremias, and the main vectors (*Cx. pipiens* complex, *Cx. tarsalis* and *Cx. nigripalpis*) become infected at comparatively lower avian host viremias (Reisen 2003). Since the 1970s, the number of SLEV cases has decreased markedly and now SLEV largely seems to have been competitively displaced by WNV (Reisen 2003; Reisen & Brault 2007). Similar to WNV, outbreaks of SLEV have been linked strongly to La Niña conditions of high temperature and low rainfall (Lumsden 1958; Day 2001).

MVEV causes intermittent outbreaks of encephalitis in rural Australia, frequently among aboriginal populations in arid biomes (Broom *et al.*, 2002). Transmission cycles seem very similar to JEV, principally involving the mosquito *Culex annulirostris* and ardeid birds, although the inter-epidemic persistence mechanisms in arid Australia are not well understood (Boyle *et al.*, 1983; Kay *et al.*, 1984; Broom *et al.*, 1995). Outbreaks in areas such as the Kimberly seem closely tied to the ENSO warm wet periods, and therefore would seem to have a strong correlation with EO-data from vegetative change similar to that of Rift Valley fever virus discussed below (Nicholls 1986).

5.2 *Togaviridae*

Three mosquito-borne viruses within the genus *Alphavirus* produce neuro-invasive diseases in the New World: western equine (WEEV), eastern equine (EEEV) and Venezuelan equine (VEEV) encephalomyelitis viruses. In North America these viruses have spatially distinct distributions, different enzootic amplification cycles, and a different suite of vectors and hosts. Historically, WEEV produced large epizootics among equines in western North America during the early 1900s where it crippled horse/mule-drawn agriculture at that time and was a frequent cause of encephalitis in children residing in rural environments (Reeves & Hammon 1962; Reisen & Monath 1989). A large proportion of the adult population at that time was protected by acquired immunity that has been lost in recent years (Reeves 1990; Reisen & Chiles 1997). Although an effective vaccine largely has eliminated equine disease, reasons for the decline in human cases remain obscure, but do not appear to be related to changes in viral fitness (Reisen *et al.*, 2008b). WEEV is amplified primarily among passeriform birds by *Cx. tarsalis*, with increased enzootic transmission linked strongly with antecedent increases in *Cx. tarsalis* abundance associated with mild winter temperatures and wet spring conditions (Reeves & Hammon 1962; Reisen *et al.*, 2008a; Barker *et al.*, 2009). Flood-water *Aedes* in the *campestris* complex and *Ae. trivittatus* also transmit WEEV among lagomorphs and rodents in a secondary cycle, and may have been responsible for some of the equine and human infection (Hardy 1987). *Cx. tarsalis* abundance patterns have been related spatially to landscape and agriculture using remotely sensed data (Barker 2008).

EEEV causes severe encephalitis in humans and equines in the eastern US, with frequent mortality among apparent infections (Morris 1988). EEEV is amplified within cypress bogs in an enooztic cycle involving summer resident song birds and the primary vector, *Culiseta melanura*, with transmission to equines and humans associated with *Aedes* and *Coquillettidia* bridge vectors that carry the virus from these wetlands to adjacent pastures and human residential areas (Morris 1988; Crans *et al.*, 1994). Outbreaks are associated typically with cool wet summers that increase *Cs. melanura* populations and consequently increase mosquito-bird contact. Intervention is through equine vaccination, mosquito control, repellent use, and public education. Unique wetland settings required for the enzootic amplification of EEEV have enabled development of risk maps using remote sensing (Moncayo *et al.*, 2000; Jacob *et al.*, 2010).

VEEV is endemic in the New World and is the type species of the VEE complex that includes seven different viral species and multiple subtypes and varieties (Weaver & Reisen 2009). Enzoonotic cycles (i.e. transmission cycles of a pathogen) of VEEV include several rodent groups and mosquitoes in the subgenus *Melanoconion* of *Culex*. They occur typically in humid tropical forest or swamp habitats in Central and South America. VEEV infection overlaps considerably in signs and symptoms with many acute, tropical febrile infectious diseases such as dengue fever, so most cases probably are under-reported. Major epidemics occur sporadically when VEEV subtypes IAB and IC arise and amplify in equids utilizing *Aedes* and *Psorophora* bridge vectors, which are abundant in agricultural settings and lead to spill-over transmission to humans (Shope & Woodall 1973). Epidemic VEEV causes a highly incapacitating febrile illness that occasionally leads to fatal encephalitis in children. The emergence of epidemic VEEV relies on a combination of ecological and viral genetic factors that must intersect in time and space to produce rolling epidemics that spread rapidly (Weaver & Reisen 2009). The creation of pasture within or adjacent to fragmented tropical forests create ideal ecological settings for the emergence of epidemic VEEV that are definable through remote sensing and NDVI (Barrera *et al.*, 2001).

5.3 Bunyaviridae

Several viruses within the California encephalitis virus (CEV) serocomplex cause neuro-invasive disease in humans in the US (Calisher 1983). CEV is maintained in western North America vertically within *Aedes* mosquitoes in the *campestris* complex and is amplified horizontally by transmission among rabbits, but rarely causes disease in humans, even though infection appears to be frequent based on antibody surveys (Reeves 1990; Eldridge *et al.*, 2001). In contrast, La Crosse virus (LACV) is a major cause of encephalitis in children residing within, or near, forested areas in the eastern US, with epicentres in Wisconsin, Ohio and West Virginia (CDC 2011). The virus is maintained vertically through transovarial transmission from stabilized infections in the tree-hole breeding mosquito, *Aedes triseriatus*, but amplified horizontally by transmission among small mammals, especially chipmunks. The recent invasion of LACV endemic areas by *Aedes albopictus* may facilitate horizontal transmission (Kitron *et al.*, 1998). Humans become infected within and near woodlots during the day by this diurnal biting species. Remote sensing coupled with GIS have been useful in relating landscape to disease occurrence (Barker *et al.*, 2003).

Rift Valley fever virus (RVFV) in the genus *Phlebovirus* causes intermittent epizootics and epidemics in Africa (Bird *et al.*, 2008). Infection causes abortion and high mortality in sheep and abortion in cattle, with occasional spill-over to include other domestic animals and humans. Camels frequently become infected, but seem to suffer less illness than other livestock. In one to two per cent of affected humans, RVF progresses to severe diseases that include hepatitis, encephalitis, retinitis, and/or haemorrhagic fever. The case fatality rate among these severe cases is approximately ten to twenty per cent (Madani *et al.*, 2003). Historically, RVFV was restricted to sub-Saharan eastern Africa, especially the Rift Valley of Kenya and Tanzania, where outbreaks were caused by multiple lineages of the virus (Meegan & Bailey 1989; Bird *et al.*, 2008). Subsequent outbreaks with human involvement have been documented in South Africa, the Nile Valley from Sudan to the Egyptian delta, and the Saudi Arabian peninsula. The 1977–79 outbreak in Egypt was especially devastating, with more than 200,000 human infections, 600 deaths, and costly losses in livestock exceeding $100M (Meegan *et al.*, 1978). Transmission is enzootic during most years, but epizootic during wet years, especially those following droughts when there has been an accumulation of infected *Aedes* eggs in the soil and animal populations have low herd immunity. RVFV apparently is maintained between rainy seasons and between high rainfall years by vertically infected, desiccation resistant eggs of several flood-water *Aedes* species, especially *Aedes macintoshi* (Linthicum *et al.*, 1985a). High water years inundate eggs, giving rise to large numbers of mosquitoes. Infection rates among *Aedes* adults reared from field-collected larvae typically are low and therefore extensive horizontal amplification is required for maintenance (Linthicum *et al.*, 1985a). The large variety of wildlife naturally infected with RVFV would seem to support this notion. Warming in the Indian Ocean is linked closely to the ENSO in the Pacific and is followed by high rainfall events in the Rift Valley that are predictive of RVFV outbreaks (Linthicum *et al.*, 1999). Widespread flooding is apparent from LEO satellite altitudes by examining the NDVI data that trigger *Aedes* egg hatching and produce a cohort of blood-feeding adults rapidly, some of which are infectious with RVFV and able to transmit this infection to ruminants. Both wild and domestic ruminants are drawn or driven by herdsmen to these new sources of water and grass, bringing susceptible hosts into close proximity with infectious mammal-feeding *Aedes* mosquitoes (Linthicum *et al.*, 1985b). Work in Senegal using NDVI and GIS to estimate livestock density are in agreement with this scenario (Pin-Diop *et al.*, 2007). The concordance between the rainy season and calving of wild and domestic ruminants assures that highly susceptible host populations are available to be fed upon by infectious mosquito vectors during warm humid weather conditions conducive for transmission (Vignolles *et al.*, 2009).

5.4 Remote sensing

In summary, remote spectral and temporal measurements of landscape and climate environments, coupled with geographic information systems and statistical analyses, have enhanced epidemiological research greatly by enabling production of risk maps and information for decision support systems (Rogers & Randolph 2003; Clements & Pfeiffer 2009; Reisen 2010). Landscape dynamics are an especially useful surrogate for mosquito, wildlife host, and other surveillance data and often can be used to interpolate among sampling points or to extrapolate real or potential distributional patterns of viruses or hosts. Frequently, landscape and associated vegetative dynamics can be linked closely with transmission dynamics and used to predict skilfully from variation in climate cycles such as ENSO. WNV incidence in the North American prairie biomes, for example, is linked closely with warm temperatures that extend the transmission season, increase the frequency of vector-host contact, and shorten the extrinsic incubation period (Reisen *et al.*, 2006a). RVFV outbreaks in Kenya have been predicted skilfully from ENSO cycles that are linked to the warming of the Indian Ocean and rainfall over eastern Africa and then verified in real time by NDVI (Linthicum *et al.*, 1987; Anyamba *et al.*, 2002).

6 RIFT VALLEY FEVER

Section 6 addresses Rift Valley fever (RVF) as a disease transmitted by mosquitoes that are a serious viral threat to animals and humans in Africa and the Middle East. The disease was isolated first in Kenya during an outbreak in 1930 (Daubney *et al.*, 1931). Subsequent outbreaks have had significant impacts on animal and human health and associated national economies. Outbreaks are a major concern for international agricultural and public health communities. RVF outbreaks have been linked closely to unusually heavy rainfall resulting from local and regional climate variability in the Horn of Africa. Concurrent elevation of SSTs associated with the ENSO in the equatorial east-central Pacific and western equatorial Indian Ocean produce above normal and widespread rainfall that is the primary ecological cause of RVF outbreaks in the region. Retrospective analyses of satellite derived time series from NDVI, in combination with other climate variables like cloudiness, rainfall, and SST are being used to map potential RVF outbreak areas. A Rift Valley fever mapping and prediction system focused on sub-Saharan Africa, the Nile Basin in Egypt, and the western Arabian Peninsula, has been developed and has been shown to be an important tool for local, national, and international organizations involved in disease control and prevention. The risk monitoring and mapping system permits focused and timely implementation of disease control strategies several months before an outbreak. If used appropriately, this would allow for timely, targeted implementation of mosquito control, animal quarantine, vaccine strategies, and public education to reduce or prevent animal and human disease.

RVF is a mosquito-borne viral disease with pronounced health and economic impacts to domestic animals and humans in much of sub-Saharan Africa, particularly the savannah regions (Meegan & Bailey 1989). The virus belongs to the family Bunyaviridae and is a member of the genus *Phlebovirus*, a genus associated primarily with sand flies. The documented expansion of RVF's range has extended beyond sub-Saharan Africa into Egypt in 1977 and more recently to its emergence in Saudi Arabia and Yemen in 2000. This expansion has led some to speculate further globalization (Peters & Linthicum 1994; CDC 2000). Rift Valley fever virus (RVFV) is thought to be maintained in an endemic cycle that depends on intermittent heavy rainfall events and periodic short term flooding of low lying habitats known as dambos or pans and on the vertical transovarial transmission of the virus by flood-water *Aedes* mosquitoes (Figure 12). RVF epizootics/epidemics occur during exceptional years of above normal and prolonged rainfall that flood dambos that allow virus-infected mosquito eggs to hatch and become large populations. Many mosquito species worldwide are capable of biologically transmitting RVFV.

In humans the disease includes fever, chills, and myalgias and normally resolves without further consequences after several days. In less than three per cent of cases, more severe symptoms occur, including: retinal lesions, haemorrhagic fever, or encephalitis. Human infections are caused by a bite from an infected mosquito, from close contact with infected animals, or from exposure to infectious aerosols. Humans may be protected by diligent use of insect repellents, bed nets, and insect control measures. Vaccination of domestic animals would likely prevent epizootic epidemics

Figure 12. Aerial photograph of dambo habitats in Eastern Free State province, South Africa (Photo courtesy R. Swanepoel). (see colour plate 6)

since these animals serve as the primary amplifying hosts of the virus. Vector control is a viable option, especially with the recent development and operational use of accurate predictions of epidemic disease and innovative approaches focusing on the specific habitats of immature flood-water *Aedes* mosquito reservoirs/vectors (Anyamba *et al.*, 2010). Human vaccination could prevent human disease in endemic areas; however, effective human vaccines remain investigational and are available only in limited quantities.

Infection of domestic animals also is caused by an infected mosquito bite. The most important animals in RVF epidemics are sheep, cattle, and goats; and these species suffer significant nearly 100 per cent mortality and abortion after infection. Their high viremias are sufficient to infect many arthropod vector species (Peters & Linthicum 1994). Vaccination of domestic animals could reduce epidemic transmission of RVFV; and, an effective vaccine would also protect vaccinated animals from disease, reduce abortion, and produce antibodies that would be distinguishable from natural infection of the virus. Currently, both inactivated and live attenuated vaccines are available in endemic areas of sub-Saharan Africa; but they are not ideal.

RVF epidemics over the last few decades in Egypt, West Africa, Sudan, Madagascar, and the Arabian Peninsula have emphasized the importance of land-use, irrigation, and deforestation, in the continuing emergence of this virus. The occurrence of either prolonged or intense transmission in East Africa in 1997–1998 and again in 2006–2007, Sudan in 2007, South Africa in 2008–2010, and Madagascar in 2007–2008 have increased the potential for exploitation of RVFV to suitable but immunologically naive areas such as may exist in the Middle East, the Mediterranean, and the Americas. The effect of an RVF outbreak on the economy in the US or other non-endemic country would be substantial. Livestock feed suppliers, health care insurers, and the loss of confidence in food-service industries and food suppliers would be enormous. In 2003, the US had beef-related exports of $5.7 billion, and the World Organisation for Animal Health (OIE) imposed a four year trade ban on any country with confirmed RVF transmission (Linthicum *et al.*, 2007).

6.1 History, ecology and mosquito vectors

Examination of historical annual reports by the Kenya Department of Veterinary Services suggest that the disease had been occurring for some time previous to this outbreak in 1931, and Shimshony (1979) speculated that RVF virus was a cause of biblical plagues. A large RVF outbreak occurred in sheep in South Africa in 1950–1951 (Weiss 1957). In the 1950s, '60s and '70s periodic outbreaks with significant impact on human and domestic animal health occurred in eastern and southern Africa during years with very high rainfall. RVF virus was first detected in West Africa in 1967, associated with deaths of cattle imported into Nigeria. Human epidemic disease was first observed in West Africa in 1987 when completion of the Diama Dam on the Senegal River resulted in a large outbreak in northern Senegal and southern Mauritania (Digoutte & Peters 1989; Linthicum *et al.*, 1994). There have been subsequent outbreaks in this region since then. In 1990, RVF virus caused an epidemic in Madagascar, the first outside of continental Africa, and outbreaks have continued through 2008 (Morvan *et al.*, 1992; Anyamba *et al.*, 2010).

A landmark epizootic/epidemic occurred in 1977–1979 in Egypt when RVF occurred for the first time outside sub-Saharan Africa after it was most likely introduced from Sudan. During this outbreak there were very high human and animal mortalities highlighting the need to consider the ramifications should the disease be introduced into immunologically naive populations. In 2000 RVF expanded for the second time outside of Africa when an animal and human outbreak occurred on the West coast of the Arabian Peninsula in parts of Saudi Arabia and Yemen along the Red Sea into another immunologically naive area. It produced severe human and animal disease outbreaks (CDC 2000).

Inter-epizootic circulation of RVF virus has been documented through serosurveys in humans, and ecological studies from South Africa, Kenya, and Zimbabwe have led further to discovering the participation of flood-water *Aedes* mosquitoes in maintaining RVF virus in nature through transovarial transmission (Peters & Meegan 1981; Linthicum *et al.*, 1985a). RVF epizootics and epidemics are closely linked to the occurrence of the warm phase of the ENSO phenomenon (Linthicum *et al.*, 1999). These conditions can lead to unusually heavy rainfall and eventual flooding of low lying habitats that are suitable for producing immature *Aedes* and *Culex* mosquitoes that serve as the primary and secondary RVF vectors (Anyamba *et al.*, 2010). A monitoring and risk mapping system has been developed using NDVI time series data derived from the advanced very high resolution radiometer (AVHRR) instrument to map potential areas for RVF outbreaks (Anyamba *et al.*, 2002a; Linthicum *et al.*, 2007). This forecasting system operates within a three to five month lead time to predict RVF risk (Linthicum *et al.*, 1999, 2007; Anyamba *et al.*, 2006b, 2010). Satellite systems permit one to identify eco-climatic conditions associated with disease outbreaks over a large area. This monitoring and risk mapping system has proved to be an important tool for local, national, and international organizations involved in preventing and controlling animal and human disease to focus on timely disease control strategies before an outbreak occurs (Anyamba *et al.*, 2010).

The RVF outbreak on the Arabian Peninsula in 2000 demonstrated that regions other than those in sub-Saharan Africa might be at risk (CDC 2000). To address this prospect, the forecasting system developed for Africa was adapted for the Arabian Peninsula and to assess the risk of RVF and other arthropod-borne disease outbreaks in new ecological settings (Anyamba 2006a). This system has been further modified to include a GIS-based early warning system for RVF vectors in the US (Linthicum *et al.*, 2007). This system uses a variety of mosquito surveillance data collected by public health agencies and mosquito control authorities, and combined them with climate data derived from satellite measurements and terrestrial weather stations into a GIS framework. Information will be disseminated throughout the US that should provide several months' warning before conditions are suitable for elevated RVF mosquito populations to emerge. The system also should permit timely, targeted implementation of mosquito control, animal quarantine and vaccine strategies to reduce or prevent animal and human disease, should RVF virus enter the US. The infrastructure and systems developed for RVF will be laterally transferred to inform strategies against any other introduced mosquito-borne disease threat.

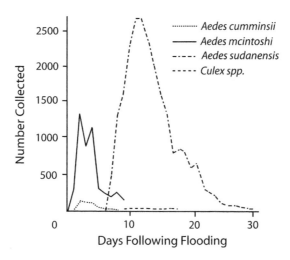

Figure 13.   Evolution of mosquito vector species at a flooded dambo site near Ruiru, Kenya (Linthicum *et al.*, 1984; 1985a,b).

6.2   *Climatic influences*

The first findings that illustrated explicit links between epizootics, RVF epizootics, and rainfall were by Davies (*et al.*, 1985). Through an analysis of time series rainfall records from numerous stations in Kenya between 1950 and 1982, a composite rainfall index was developed to compare the number of rainy days and rainfall totals for each month as a cumulated function over twelve months. It was found that periods with positive surplus rainfall corresponded to periods when RVF epizootics occurred. Widespread, frequent, and persistent rainfall was shown to be a prominent feature of epizootic periods. Heavy rainfall raises the level of the water table in certain areas, flooding the dambo habitats that support the immature stages of certain flood-water *Aedes* mosquitoes (Figure 13). These findings have been collaborated by findings in Southern Africa (Swanepoel 1976; 1981) and West Africa (Bicout & Sabatier 2004). RVF virus is probably transmitted transovarially in these species, very large numbers of which emerge under these damp and flooded conditions. This is when clinical signs of the disease are first seen in animals and subsequently in humans.

   This pattern of rainfall was confirmed by field studies at Ruiru, Kenya (Linthicum *et al.*, 1985a,b). In an examination of mosquito vector dynamics following the artificial flooding of a dambo, it was found that species populations emerged in a cascading sequence over time, exhibiting a batch-hatching pattern. This continued for up to thirty to fifty days as the dambo was sequentially flooded (Figure 14). RVF virus was isolated from several species of mosquito species. These data confirm that prolonged rainfall is required to create appropriate flood conditions for vectors to emerge in sufficient populations to trigger an outbreak. When this process is replicated on a large regional scale it leads to an epizootic/epidemic.

   Several studies examining the history of outbreaks in Kenya since 1951 have established that episodic outbreaks are coupled closely with above normal rainfall associated with an inter-annual time scale coincident with warm phase of ENSO and warm events in the equatorial western Indian Ocean (Cane 1983; Nicholson 1986; Ropelewski & Halpert 1987; Birkett *et al.*, 1999; Linthicum *et al.*, 1999; Saji *et al.*, 1999; Anyamba *et al.*, 2002b). Such warm events are associated with above normal and extended rainfall over East Africa and are further enhanced when both the sea surface anomalies in the western Indian Ocean and equatorial central-eastern Pacific are synchronized. More that ninety per cent of RVF outbreak events since 1950 have occurred during warm ENSO events (Linthicum *et al.*, 1999) (Figure 14). The inter-epizootic period is dominated by La Niña events (the cold phase of ENSO), which results in drought in East Africa and wetter than normal conditions in southern Africa (Nicholson & Entakhabi 1986; Anyamba *et al.*, 2002b). Recent

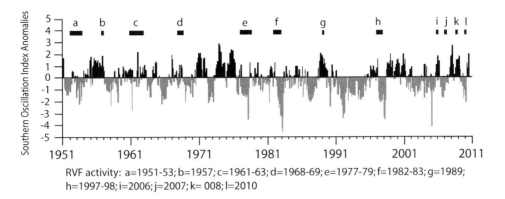

RVF activity: a=1951-53; b=1957; c=1961-63; d=1968-69; e=1977-79; f=1982-83; g=1989; h=1997-98; i=2006; j=2007; k= 008; l=2010

Figure 14. RVF outbreak events plotted against time and the southern oscillation index (SOI), a measure of the phase of El Niño southern oscillation events. Most RVF outbreak events have occurred during the warm phase of ENSO shown in black; negative SOIs are shown in grey (Linthicum *et al.*, 1999).

evidence shows that RVF outbreaks in southern Africa are coupled to La Niña patterns (Anyamba *et al.*, 2010). Inter-annual variability, in part driven by ENSO events with differential impacts on rainfall anomaly patterns on eastern and southern Africa, largely influence the temporal outbreak patterns of RVF.

6.3   *Remote sensing*

Remote sensing data for studying ecological conditions associated with RVF rely primarily on NDVI (Tucker 1979). Rainfall and green vegetation dynamics are a major determinant of life cycles of arthropods in many semi-arid lands, globally. Research indicates a close relationship between the seasonal trace of green vegetation development with breeding and upsurge patterns of particular insect vectors, including locusts and mosquitoes (Tucker *et al.*, 1985; Hielkema *et al.*, 1986; Linthicum *et al.*, 1987; Linthicum *et al.*, 1990). Therefore, early attempts to utilize satellite data to detect emergent habitats of mosquito vectors were made in the late 1980s using a combination of aircraft mounted radar and TM data (Pope *et al.*, 1992). However, while dambo habitats could be identified easily by using cloud penetrating radar, this technique was not cost effective on an operational basis over large areas. High resolution data are more meaningful if used to identify vector habitats for pre-treatment and vector control during periods of high risk. For operational monitoring, attempts were made to use NDVI data from NOAA satellites to detect and monitor flooded, man-made vector mosquito habitats in West Africa (Linthicum *et al.*, 1989; 1994) and natural habitats in Kenya (Linthicum *et al.*, 1987, 1990). It was found that time series NDVI data could be used to infer and identify areas of abundant vegetation growth associated with flooded mosquito vector habitats. The use of climate data sets including sea surface temperatures and vegetation index as a proxy for ecological dynamics in tying together the coupling of ocean-atmosphere and land system as a driver of RVF outbreaks is also documented (Linthicum *et al.*, 1999). Their findings show that all known RVF virus outbreaks in East Africa from 1950 to May 1998, and probably earlier, followed periods of abnormally high rainfall. Furthermore, these outbreaks were associated with the time variability of sea surface temperature anomalies in the equatorial Pacific and equatorial Indian Oceans. Using data during the 1997–1998 RVF epizootic/epidemic, it was found that such outbreaks could be predicted with a lead time of up to five months, and concurrent near real-time monitoring with satellite NDVI data could identify actual affected areas.

6.4   *Early warning*

Following the first effort to design an RVF environmental monitoring risk mapping and prediction system, a second version was attempted (Linthicum *et al.*, 1999; Anyamba *et al.*, 2002a). This

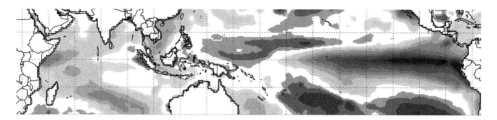

Figure 15a. Sea surface temperature anomalies during December 1997–February 1998. (See colour plate 7a)

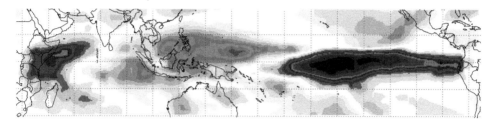

Figure 15b. Outgoing long wavelength radiation anomalies during December 1997–February 1998. (See colour plate 7b)

second version is informed by several factors including: 1) monitoring global SST with reference to the phase and amplitude of anomalies and equatorial western Indian Ocean SST anomalies bounded by the geographic area from 5S-5N, 170W-120W (aka the NINO 3.4) (Figure 15a); 2) monitoring patterns of outgoing long wavelength radiation anomalies to infer and detect large scale changes and shift in the major atmospheric centres of tropical convection as resulting from ENSO (Figure 15b); and 3) monitoring anomalous NDVI patterns over Africa as a proxy for ecological dynamics. In Figure 15a, temperature anomalies range from minus 5°C (deep blue) to plus 5°C (deep red); and in Figure 15b, outgoing long wavelength radiation is measured in watts/m² ranging from minus eighty watts/m² (deep purple) to plus eighty watts/m² (red).

The risk mapping component is based on two factors: 1) mapping epizootic/epidemic regions as informed by areas of reported historical outbreaks combined with thresholding long-term NDVI and rainfall to identify areas with a high inter-annual variability signal to create a potential epizootic area mask/map; and 2) temporal dynamics of mosquito vector populations in a dambo habitat. The model detects areas having persistent positive NDVI anomalies (greater than + 0.1 NDVI units) using a three-month moving window to flag regions at greatest epizootic risk potential area mask. The system is designed to account for periods of extended above normal NDVI (by inference rainfall) and to consider the complex life cycle of mosquitoes that maintain and transmit RVF virus to domestic animals and people. An historical reconstruction of areas at risk illustrated that regions of potential outbreaks have occurred predominantly during warm ENSO events in east Africa (Figure 14) and during cold ENSO events in southern Africa; and that there is a close agreement between confirmed outbreaks between 1981 and 2000, particularly in east Africa. This has been adapted and is now used for near real-time monitoring and risk mapping on a monthly basis as a useful tool in RVF disease surveillance. Figure 16 shows elevated RVF risk depicted in dark grey for much of east Africa. Medium grey represents the savannah mask or the area for potential epizootic activity. The largest RVF outbreak in the last twenty years occurred during the time period between December 1997 and February 1998. It covered an area about 1.28 Mkm².

The first successful prediction using this system was made in 2006 (Anyamba *et al.*, 2006b; 2009). The prediction followed the steps outlined above. The developments of the ENSO warm event in August 2006, and associated anomalous warming in the equatorial Indian Ocean resulted in above normal and widespread rainfall over the endemic areas in east Africa during the period

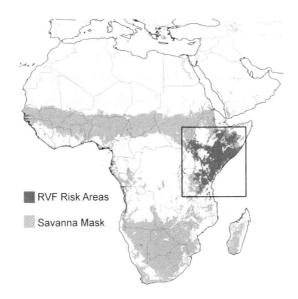

Figure 16.   RVF risk map for December 1997 to February 1998.

September to December 2006. This resulted in dambo habitats being flooded and anomalous high growth in vegetation, creating ideal conditions for mosquito vector production. Early warnings were issued every month from September 2006 through May 2007 (Figure 17). This system provided a lead time of three to four months to respond, but response and mitigation activities only started one month prior to the first reported outbreak. The predictions were subsequently confirmed by entomological and epidemiological field investigations of virus activity in the areas mapped to be at risk in Kenya, Somalia and Tanzania. Following the outbreak in east Africa, this system provided further predictions of the outbreak in Sudan in late 2007 and January 2008, 2009 and 2010 in southern Africa. These predictions and outbreak assessments are provided in detail in Anyamba (*et al.*, 2010). In Figure 17 areas shown in green represent RVF potential epizootic areas; and those in red represent pixels that were mapped by the prediction system to be at risk for RVF activity during the respective time periods. Blue dots indicate human cases identified to be within the RVF risk areas; and yellow dots represent human cases in areas not mapped to be at risk.

Risk maps are generated and are interpreted in relation to inter-annual variability patterns in SST anomalies in the western Indian Ocean (WIO) and NINO 3.4 regions. RVF risk results are presented as binary images. Areas flagged as red within the savannah mask represent elevated risk for RVF for that period of time (Figures 16, 17). Areas shown in green are within the RVF endemic mask and show reduced risk. Areas shown in yellow represent desert and dense tropical forest but are not included in the analysis.

6.5   *RVF risk assessment system for Africa*

In east Africa the RVF risk model retrospectively detected the last three RVF outbreaks in 1982–1983, 1989, and 1997–1998, and each of these events was correlated to positive SST anomalies. The 1982 to 1983 outbreak corresponded to a warm ENSO event with peak positive anomalies in the NINO 3.4 and WIO regions of 3.0°C and 0.8°C, respectively (Figure 14). Positive NDVI anomalies persisted for several months and covered most of the semi-arid lands of east Africa (Figure 16).

In 1988–1989 warm SST conditions prevailed in the WIO region but the NINO 3.4 changed to cold conditions (Figure 15a). Positive NDVI anomalies persisted in southern African and east Africa. We identified areas of potential RVF risk in northwest Kenya extending to the central

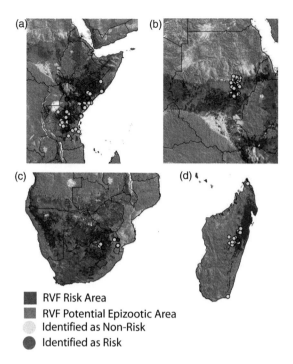

RVF Risk Area
RVF Potential Epizootic Area
Identified as Non-Risk
Identified as Risk

Figure 17.   Summary RVF risk areas for: (a) eastern Africa, Sep. 2006 to May 2007; (b) Sudan, May 2007 to Dec. 2007; (c) southern Africa, Sep. 2007 to May 2008; and (d) Madagascar, Sep. 2007 to May 2008 (Courtesy Anyamba). (see colour plate 8)

Rift Valley region around Lakes Nakuru and Naivasha in Kenya, and other areas of South Africa (Figure 14). Localized RVF activity in mosquitoes, domestic animals and humans was detected in the area around Lake Naivasha, Kenya (Logan *et al.*, 1992).

Warming SST conditions started in May 1997 in both the WIO and NINO 3.4, reaching a peak of 1.2°C in the WIO and 4.0°C in the NINO 3.4 (Figures 14, 15a). Large areas of East Africa received widespread and heavy rainfall during the short rainy season of September through November 1997 and this rainfall extended into the dry season of December 1997 through February 1998, and is shown by the negative anomalies in the outgoing long wavelength radiation data (depicted in blue) in Figure 15a. Significantly elevated NDVI values occurred in much of east Africa. This led to an increased risk for RVF transmission over an area of 1.2 million km^2, including Kenya, Somalia, Tanzania, Sudan, Uganda and parts of Ethiopia (Figure 16). The largest RVF outbreak in the last 20 years occurred during this period over the same area determined to be at risk (CDC 1998).

In the Sahel, the largest RVF outbreak in Senegal and Mauritania was detected in 1987; however, limited RVF activity in Senegal in 1993, Burkina Faso in 1983, and the Central African Republic in 1985 was not detected. In southern Africa, elevated RVF activity was detected in Zambia in 1985–1986 but RVF activity between 1982 and 1985 was variable. The initial stages of the RVF outbreak in Madagascar in 1992 were detected.

In November 2006, positive NDVI anomalies, forty to sixty per cent above normal in east Africa, indicated heavy rainfall. Elevated risk of RVF activity is noted in Somalia, and Namibia and Botswana in southern Africa. It had been ten years since the 1997–1998 RVF outbreak in east Africa. Elevated rainfall, NDVI, and RVF risk suggest that if those conditions persisted, there would be a high likelihood of another RVF outbreak. The forecast for this outbreak was provided in October and November 2006 (Anyamba 2006b).

The RVF risk model uses population dynamics of mosquito vectors and eco-climatic indicators to reconstruct likely historical patterns of RVF outbreaks and monitor eco-climatic conditions associated with disease outbreaks over large areas with satellite measurements (Anyamba 2006a). Disease mapping might provide public health authorities with information with which to target disease surveillance and control teams. Targeted response increases efficacy and minimizes the cost of surveillance over large areas.

6.6   Temporal and spatial products

The temporal and spatial products of the RVF early warning system identify regions at risk to disease transmission in sub-Saharan Africa, the Middle East, including Egypt and the Arabian Peninsula, and Madagascar. This information has been used to permit early disease detection and implementation of multiple control strategies. Spatially-specific recommendations allow limited resources to be distributed more effectively, for instance freeing some areas from unnecessary blanket precautions by spatially targeting vector control, distribution of vaccines and diagnostics, and diffusion of information to stakeholders based on the status of mosquito populations. Since climate in North America is linked to phenomena such as the ENSO, climate activity elsewhere in the world could be used to estimate risk indices months into the future and provide ample time for preparation and prevention across the US.

6.7   Conclusions

In 2012, the monitoring and risk mapping system based on rainfall, NDVI, and global SST data is effective for assessing the potential spatial and temporal distribution of RVF transmission in Africa, as demonstrated by having predicted the RVF outbreaks in 2006–2010. Many of the systems were developed in preparation for RVF and can be transferred to inform strategies against other mosquito-borne disease threats.

The RVF monitoring and prediction system (Anyamba et al., 2009) produced forecasting information that was used operationally during the most recent RVF outbreaks in east Africa, Sudan, southern Africa, and Madagascar. This information provided significantly improved spatial and temporal warnings of imminent RVF transmission, and permitted early disease detection and implementation of multiple control strategies. Forecasting RVF activity in these regions of Africa could have been used also to enhance various preparedness activities such as targeting both adult and immature stages of the most important mosquito vector species. Costs of mosquito control are significant but have been shown to be effective in suppressing arbovirus transmission (Carney et al., 2008), potentially reducing human and animal morbidity and mortality and reducing economic impacts. Theoretically, the use of sustained release methoprene or other immature mosquito larval control products would be more costly than post outbreak control measures but massive applications at the earliest indications of elevated rainfall and before flooding would decrease the quantity of RVF virus introduced into the environment by killing the majority of the mosquito reservoir before it is able to transmit the virus to domestic animals.

While for the most part the predictions were correct on a regional scale, there are a number of elements in the model that need to be improved going into the future. The model described here uses NDVI as the primary data input as a proxy for both ecological dynamics and rainfall. The addition of real-time rainfall data that are now readily available into the risk mapping model would provide a back-up check on the NDVI anomalies and would improve risk mapping based on accumulated rainfall.

The RVF epizootic area mask is based on an RVF literature survey to identify countries where there have been episodes of RVF activity (Anyamba et al., 2002a). These maps are then improved through climate variables (rainfall and NDVI) and thresholding to derive a potential epizootic area mask (PEAM) (Anyamba et al., 2002a). Some of the RVF outbreaks along coastal Kenya in 2006–2007, in South Africa (Jan.–Feb. 2008), in Sudan within the Gezira irrigation scheme and some areas in Madagascar were outside of the PEAM area. Improvement the PEAM will involve either a change in the rainfall and NDVI thresholding values, or incorporating detailed land cover maps into the model using inputs from the Food and Agriculture Organization (FAO), WHO, and in-country experts.

Inclusion of livestock and human-population data would enable improved risk ranking of potential areas of RVF activity and improve early targeting of locations of vector-virus surveillance by entomological and veterinary teams. For real-time monitoring of rainfall and ecological conditions, a sentinel monitoring site (SMS) database could be created that contains locations of foci or epicentres of recent and previous RVF outbreaks, and could be used also to identify and monitor rainfall and NDVI to serve as area specific indicators of early warning information. The plotted series from the SMS locations could be published along with risk maps on a monthly basis to provide additional value-added information to the response planning component teams at country and international levels. In addition, application and refinement of early warning models outside of east Africa based on the improvements above would make it possible to take specific regional landscape characteristics into consideration. This would require building a regional model for each target region.

6.8 *Acknowledgments*

Work presented in this Section was supported by the Department of Defense, Armed Forces Health Surveillance Center, Division of Global Emerging Infections Surveillance and Response System (GEIS) Operations and the US Department of Agriculture's Agricultural Research Service, Center for Medical, Agricultural & Veterinary Entomology (CMAVE).

7 RISK AND INCIDENCE OF HUMAN PLAGUE

Plague is a highly virulent flea-borne, zoonotic disease maintained in nature as an infection of rodents, typically ground squirrels, prairie dogs, chipmunks and wood rats. Its aetiological agent, *Yersinia pestis*, is one of the most pathogenic bacteria for humans. Plague is transmitted usually to humans by the bite of an infected rodent flea, less often by handling infected animals, only rarely by direct air-borne spread (Dennis & Meier 1997; Gage 1998; Parmenter *et al.*, 1999; Gage & Kosoy 2005). It remains endemic in many parts of the Americas, Asia and Africa and it is characterized by quiescent and epizootic periods that usually precede the majority of human cases with increased depopulation of affected rodents, probably forcing the dispersal of infected fleas in search of new hosts (Perry & Fetherston 1997; Levy & Gage 1999; Stapp *et al.*, 2004). Phenotypically, there are three *Y. pestis* biotypes that caused three infamous pandemics: 1) Antiqua that caused the Justinian plague, the first pandemic in the 6th Century AD; 2) Mediavalis, the Black Death of the 14th Century; and 3), Orientalis in the 19th Century (Clem & Galwankar 2005; Torrea *et al.*, 2006). The Orientalis pathogen was introduced to the US around 1900 by rats and fleas arriving on ships from plague-endemic regions of Asia that then spread across North and South America. Today, Orientalis is the dominant strain with almost worldwide distribution and has likely been the only source of all endemic cases on the North American continent. The two other biotypes have a geographically restricted distribution: Mediavalis in Asia, and Antiqua in some parts of Africa and Central Asia. In any case, since it has been reappearing in areas formerly plague-free, WHO listed it as an international quarantine re-emerging disease with the status of biological weapon, category-1 (Daszak *et al.*, 2000; WHO 2004; Clem & Galwankar 2005; Wolf *et al.*, 2007).

In the US there are still occasional human cases of plague diagnosed in the western states with high fatality rates (Craven *et al.*, 1993; CDC 2003). Most of the human cases occur in a peridomestic environment and are reported from the Four-Corners States (Arizona, Colorado, New Mexico and Utah), with eighty-three per cent of 416 total cases reported since 1950 (CDC 2006; Eisen *et al.*, 2007a.). Mortality rates for untreated bubonic plague infections range from forty to seventy per cent, compared to almost 100 per cent for untreated pneumonic or septicemic infections (Levy & Gage 1999). Little is known about the dynamics of plague in its natural reservoirs, particularly about the survival strategies of *Y. pestis* (Perry & Fetherston 1997; Gage & Kosoy 2005). Yet, effective rodent control and prompt diagnosis followed by appropriate antibiotic treatment have reduced morbidity and mortality from the disease greatly (Levy & Gage 1999; CDC 2003; Gage & Kosoy 2005; CDC 2006). Thus, informed preemptive decisions about plague management and prevention preceding outbreaks would be more sustainable and cost-beneficial than "fire-fighting" approaches that wait for an epidemic before implementing emergency response plans that are usually too late to mitigate transmission (Stenseth *et al.*, 2008). Natural cycles of plague are conditioned by ecology, environment, host, and agent; that is, factors that vary temporally and spatially. Recent studies have identified local climatic factors and landscape features associated with increased plague activity. In Arizona and New Mexico, epizootic activity intensifies when cool summer temperatures follow wetter winter/spring seasons (Parmenter *et al.*, 1999; Enscore *et al.*, 2002). When favourable conditions occur, assuming a trophic cascade of climatological, ecological and demographic events, they can lead to increases in numbers of rodent hosts and flea vectors that in turn lead to higher risks for human exposure (Yates 2002). Several limitations hinder use of these temporal models as predictive models. One of the most limiting factors is their sensitivity to spatial heterogeneity that makes the ability of a good generalization unlikely.

7.1 *Earth observations of plague*

Satellite sensing over nearly thirty years has enabled systematic analysis and mapping of the relationships between disease vectors and indicators of climate variability on a global scale at high-temporal and moderate spatial resolutions. These variables include ENSO, rainfall, temperature and vegetation (Linthicum *et al.*, 1999; Beck *et al.*, 2000; Pinzon *et al.*, 2005a; Anyamba *et al.*, 2009). These tools have been used to expand knowledge of the coupling between plague and climate. Logistic regression models have been used in the Four Corners region to identify local landscape features associated with human plague cases, and to create a predictive geographical model of high-risk habitats for human exposure to *Y. pestis* (Eisen *et al.*, 2007a, b). Although these models were not intended to be predictive in the temporal domain, they provide fine-scale modelling of adjacent spatial interactions between landscape elements and the ability to predict where infections are likely to occur, albeit in limited areas. These models relate the risk of exposure to *Y. pestis* to suitable environmental indicators using occurrence-only data from the contact rates between humans and the pathogen. Even though the epidemiological data are of high quality, one cannot equate absence to unsuitability. Nonetheless, these studies provide useful insights into those aspects of incorporating explicit ecological features to improve understanding and prediction of disease risk that could not be treated in previous models. Prior to classification, the landscape and climatic features associated with human risk to exposure to *Y. pestis* included elevation and four quarterly means of NDVI seasonal profiles for each eight kilometre grid cell. The stepwise optimal hierarchical clustering (SOHC) approach can be applied to identify training samples that contribute the most in the cluster classification and optimize criterion functions (Duda *et al.*, 2000).

The climatic, landscape, and ecological properties of these samples can be used to identify and generalize temporal characteristics of plague outbreaks in the four corners region (Figure 18). These features were selected for being the most valuable and robust from a larger set of candidate features such as monthly variance of NDVI seasonal profiles, land-cover habitat type, and profiles of monthly mean temperature and rainfall (Pinzon *et al.*, 2005a). Moreover, on a seasonal scale, NDVI is well correlated with rainfall, evapotranspiration and surface temperature variables across a wide range of environmental conditions. When these variables are viewed from different timescales using the method of empirical mode decomposition (EMD), it is possible to identify linkages between climate, ENSO and plague epizootics in the region (Huang *et al.*, 1998; Pinzon *et al.*, 2005b; Huang *et al.*, 2009).

The most commonly used ENSO index is the SOI computed from the Darwin and Tahiti pressure differences, and indices based on sea surface temperature in several regions of the equatorial Pacific Ocean: R12 (0-10S,80-90W); R3 (5S-5N,150W-90W); R4 (5S-5N,160E-150W); and R3.4 (5S-5N,170W-120W). One can monitor ENSO through the multivariate ENSO index (MEI). This is a multivariate measure of the ENSO signal expressed in the first principal component of six observed variables from the tropical Pacific Comprehensive Ocean-Atmosphere data set (COADS): sea level

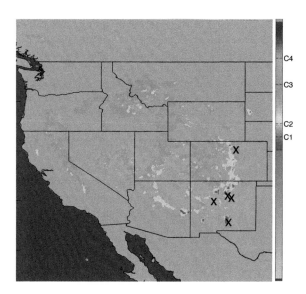

Figure 18. A stepwise optimal hierarchical clustering approach was applied to identify training samples that contribute the most in the cluster classification and optimize criterion functions. The climatic, landscape and ecological properties of these samples were used to identify and generalize temporal characteristics of plague outbreaks in the region (Pinzon *et al.*, unpublished). (See colour plate 9)

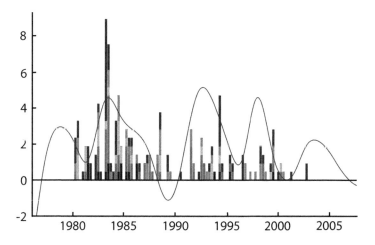

Figure 19. Three intrinsic mode functions (IMF_{4-6}) of the EMD of the multivariate ENSO index (MEI) are combined into an almost ten year-cycle wave (blue line), and lagged twelve months to overlap the number of human cases of plague in each quarter of the year. The number of cases is clustered according to the five SOHC classes. Notice that incidence of plague occurred when IMF_{4-6} is positive (Pinzon *et al.*, unpublished). (see colour plate 10)

pressure, surface zonal and meridional wind components, sea surface temperature, surface air temperature and cloudiness (Wolter & Timlin 1998). MEI is re-computed every month to monitor the strength of ENSO conditions since 1950. Correlations between the MEI and other most common indices range from 0.8 to 0.9 (Wolter &Timlin 1998).

To start, the last three intrinsic mode functions (IMF) components (i.e. IMF_{4-6}) are combined into a ten year-cycle wave and lagged twelve months to overlap the time series for the number of human plague cases in each quarter of the year (Figure 19). In this Figure, the incidence of plague

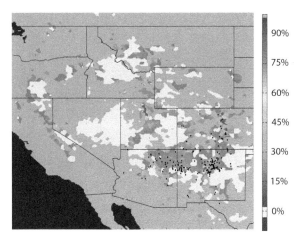

Figure 20.   Plague risk maps masked with a density population map of the Four Corners region. The plague endemic area is thus concentrated on peridomestic regions that constitute about eighty-five per cent of the cases (Pinzon *et al.*, unpublished). (see colour plate 11)

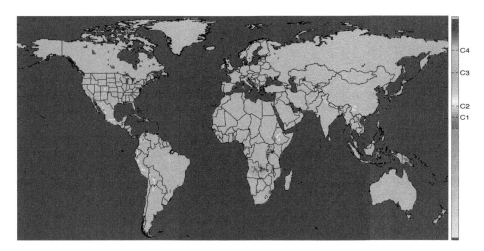

Figure 21.   Global extension of the SOHC model based on five clusters. (see colour plate 12)

is explained by positive ENSO years. The first two principal components of the correspondence analysis explain ninety-eight per cent of the variance of the category matrix (Pinzon *et al.*, unpublished results). Two salient features are apparent: 1) a distinct contribution to the total number of cases; and 2) the time when this contribution peaks in each cluster. Each SOHC cluster contributes different numbers of cases to the total number, making available a new tool to rank plague risk: C1 (eight per cent), C2 (twenty-six per cent), C3 (thirteen per cent), C4 (twenty per cent), and C5 (thirty-five per cent). Targeting limited prevention resources would further improve if the risk maps were masked with a density population map of the region (Figure 20). The plague endemic area is thus concentrated on peridomestic regions that constitute about eighty-five per cent of the cases. Moreover, using the climatic and ecological features to extend the SOHC model globally, one can identify and validate common characteristics of endemic plague regions (Figure 21). Seventy-nine per cent of the cases (C5+C2+C4) occur in regions where NDVI is lower than 0.4 with an average elevation less than 2000m. Thus, the early recognition and improved management of plague provided by these models has underscored the importance of an even better understanding of the spatial

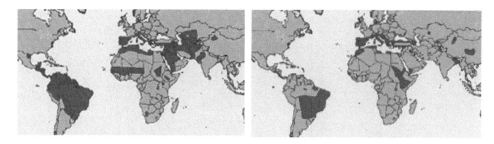

Figure 22.   Global distribution of CL (left) and VL (right) (Courtesy WHO). (see colour plate 13)

and temporal conditions that promote emergence of plague in both human and animal populations. Further, by developing critical tools to make plague surveillance more comprehensive and timely, it is possible to minimize the importance, or to prevent the potential for human outbreaks. This kind of system could operate in near real-time to monitor plague risk on a monthly basis and could offer an opportunity to identify eco-climatic conditions associated with potential vector-borne disease outbreaks over large areas. Still, many aspects of this subject are ripe for further investigation and improvement.

8   LEISHMANIASIS

Leishmaniasis is caused by obligate intra-macrophage protozoan parasites in the genus *Leishmania* (Kinetoplastida: Trypanosomatidae). The leishmaniases of human importance include twenty-one species that can be categorized into four clinical syndromes depending on where the parasite replicates in the body: (1) cutaneous leishmaniasis (CL) or (2) diffuse cutaneous leishmaniasis (DCL), both of which are dermal infections that cause skin ulcers at the site of the sand fly bite or chronic skin lesions resembling leprosy, respectively; (3) muco-cutaneous leishmaniasis (MCL), an infection of the naso-oro-pharyngeal mucosa that causes facial disfigurement; and (4) visceral leishmaniasis (VL) or kala-azar, an infection of macrophages in the internal organs (e.g. liver and spleen), which can be fatal.

8.1   *Burden of leishmaniasis*

Leishmaniasis is a serious public health problem globally and hampers socioeconomic progress in many areas. The estimated global prevalence is fourteen million cases, with a yearly incidence of approximately 1.5 million CL cases and 500,000 VL cases, respectively (WHO 2007) (Figure 22). Leishmaniasis is endemic in eighty-eight countries, including seventy-two developing countries, located in central and South America, the Mediterranean region, sub-Saharan Africa and Asia. The geographical distribution of VL has increased to include areas that were previously non-endemic (e.g. where an aggressive form of explosive outbreaks of VL is associated with HIV co-infections), or from which the disease has been eradicated previously (e.g. Sri Lanka) and Italy (Fuzibet *et al.*, 1988; Surendran *et al.*, 2007; ECDC 2009). The burden of leishmaniasis falls on the poorest segments of the population (Alvar *et al.*, 2006). Approximately ninety per cent of all cases are reported from just eleven developing countries: VL in Bangladesh, Brazil, India, Nepal and Sudan; MCL in Bolivia, Brazil and Peru; and CL in Afghanistan, Brazil, Iran, Peru, Saudi Arabia and Syria (WHO 2009a).

Poor nutrition, infection, and other stressors predispose patients to increased morbidity and mortality rates for leishmaniasis. CL, DCL and MCL have a considerable impact on morbidity, which can lead to social isolation, and mortality can occur due to secondary infection. Mortality rates in untreated VL cases are approximately seventy-five to ninety-five per cent, or as high as 100

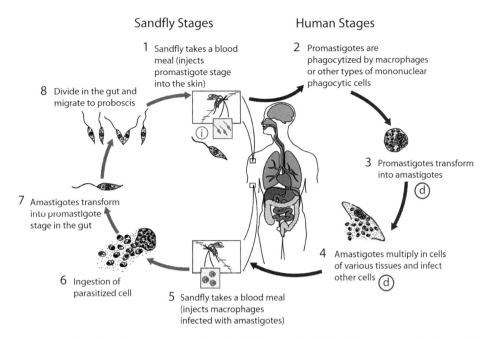

Sandfly Stages

1 Sandfly takes a blood meal (injects promastigote stage into the skin)

8 Divide in the gut and migrate to proboscis

7 Amastigotes transform into promastigote stage in the gut

6 Ingestion of parasitized cell

5 Sandfly takes a blood meal (injects macrophages infected with amastigotes)

Human Stages

2 Promastigotes are phagocytized by macrophages or other types of mononuclear phagocytic cells

3 Promastigotes transform into amastigotes

4 Amastigotes multiply in cells of various tissues and infect other cells

Figure 23. Life cycle of *Leishmania* parasites (Modified from CDC). The infectious stage (i) in red circle occurs in stage 1; diagnostic stages (d) in blue circles occur in stages 3 and 4 (Courtesy CDC). (see colour plate 14)

per cent within two years in developing countries, particularly in children aged one to four years (DCPP 2009). In southern Asia, post kala-azar dermal leishmaniasis (PKDL) may occur after VL treatment, which enables patients to act as a durable infection reservoir (Rahman *et al.*, 2010). VL also has emerged as an AIDS-associated opportunistic infection, for example in intravenous drug users in southern Europe (Herwaldt 1999). In 2001, the estimated mortality from leishmaniasis was 51,000 deaths, including 40,000 in South Asia and 8,000 in sub-Saharan Africa, with estimated 1757 DALYs lost. In the developing world, the high cost of anti-leishmanial drugs prevents many people from seeking treatment (DCPP 2009), and increasingly widespread resistance of *Leishmania spp.* to anti-leishmanial drugs compromises their efficacy (Maltezou 2010). Currently there are no vaccines or prophylactic drugs to prevent leishmaniasis. Prevention is based on reducing sand fly and host-vector contact by insecticide indoor residual spraying (IRS) in houses and animal shelters, and by using insecticide-treated bed nets and topical or area repellents (Alexander & Maroli 2003).

8.2 *Leishmaniasis transmission*

Leishmania parasites are transmitted from one mammalian host to another by the bite of an infectious female phlebotomine sand fly (Diptera: Psychodidae) that was previously infected by an animal or human host (Figure 23). Sand flies ingest the amastigote stages of the parasite when they feed on infected animal or human blood, and it takes approximately seven days for the parasites to develop into the promastigote stage; that is the infectious form that may be transmitted to humans during a subsequent blood meal. The parasite development rate in the sand fly is dependent on ambient temperature.

The leishmaniases cause either zoonotic or anthroponotic diseases as characterized by their eco-epidemiology in urban, peri-urban or rural environments. In most cases, CL (for example, *Le. major*) is a zoonosis where several rodent species serve as the infectious reservoir. Susceptible humans acquire CL when they move into enzootic forest areas or co-exist with wild reservoir hosts.

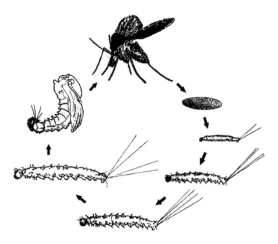

Figure 24.   Life cycle of the sand fly (*Phlebotomus* and *Lutzomyia* spp) (Permission: R. Dillon, Lancaster University).

One species (*Le. tropica*) can also be anthroponotic (human to human transmission) in urban areas like Kabul in Afghanistan, and in the Middle East. VL caused by *Le. infantum* or *Le. chagasi* is a zoonosis in China, the Mediterranean basin, and the Americas, respectively. Dogs are the main animal reservoir, and wild canines and other animals in a peri-domestic environment also may serve as reservoirs. Canine leishmaniasis is an important veterinary issue in most areas, though it is outside the scope of this book. VL caused by *Le. donovani* is anthroponotic and is a serious public health problem in India, Nepal and Bangladesh.

Different *Leishmania spp.* may be vectored by one or more known or suspected sand fly species; that is, *Lutzomyia spp.* in the Americas and *Phlebotomus spp.* in Africa, Asia and Europe. Only females feed on blood, but both sexes derive energy in the form of sugar meals from plant sources. Whereas most hematophagous flies have an aquatic life cycle, sand flies have a terrestrial life cycle. Larvae breed in moist, dark soil containing a high organic content in forests, caves or rodent burrows; and they are notoriously difficult to locate. In arid desert environments, *P. papatasi* breeds in rodent burrows near clumps of vegetation. The life cycle is approximately thirty-five to forty days from the time a gravid sand fly takes a blood meal and oviposits her eggs until the first flies emerge (Figure 24). The developmental time ranges from twenty-five to more than fifty days depending on the temperature of the surrounding environment (Killick-Kendrick *et al.*, 1977; Rangel *et al.*, 1986). Some sand fly species (e.g. *Lu. longipalpis*) may live for two to four weeks, though few female flies survive to oviposit more than once (Killick-Kendrick *et al.*, 1977). Sand flies are most active during crepuscular (i.e. low light dusk and dawn hours) and nocturnal hours.

8.3   Environmental determinants

The distribution and abundance of sand fly vectors and human and/or reservoir hosts are affected by various physical factors (temperature, precipitation, humidity, surface water and wind) and biotic factors (vegetation, host species, predators, competitors, parasites and human interventions) (WHO 2005). Thus, the determinants of leishmaniasis, both stable levels of infection or disease outbreaks, can be categorized into environmental determinants that include natural factors like vector ecology and rainfall; human activities like urban development, irrigation, soil moisture and insecticide use; and contextual determinants like socio economic factors (poverty) and human physiological factors (human immune response to leishmaniasis). According to WHO, the most important determinants include environmental changes such as deforestation, creating water bodies by building dams, installing new irrigation schemes, urbanization and migration of non-immune people to endemic areas (WHO 2009a). Neto (*et al.*, 2009) describe several environmental and socioeconomic factors associated with visceral leishmaniasis in Brazil. Climate cycles are considered a separate risk factor, as described below.

8.3.1 *Natural environmental determinants*

The epidemiology of leishmaniasis in endemic areas is strongly correlated with the ecology, behaviour, temporal and geographical distribution of the sand fly vector, and the reservoir host for zoonotic leishmaniases. Sand fly activity patterns are seasonal depending on temperature, latitude and altitude. In addition, sand fly vector distribution and abundance are strongly correlated with rainfall (Nieto *et al.*, 2006), which may shift leishmaniasis distribution patterns both temporally and spatially (Elnaiem *et al.*, 2003; WHO 2005; Ben-Ahmed *et al.*, 2009). The estimated flight range of sand flies (*Lutzomyia spp.*) is less than 100m/night depending on the surrounding environment, prevailing meteorological conditions and proximity to a blood meal or sugar sources (Chaniotis *et al.*, 1974; Alexander 1987). Killick-Kendrick (*et al.*, 1984; 1986) estimated a flight speed of 2.3–2.5 km/h and a dispersal range greater than one kilometre for female *P. ariasi* sand flies. In most temperate climates, sand flies are active during the summer season and generally have a wider range than the respective *Leishmania* species they transmit (e.g. *Le. infantum* (ECDC 2009).

In some areas, sand fly species have preferred types of vegetation that provide food sources. Some feed on honeydew produced by aphids on oak trees (Killick-Kendrick & Killick-Kendrick 1987); others feed directly on plants (e.g. Syrian mesquite) (Schlein & Jacobson 1999). Vegetation also may be associated with the food preferences of rodent reservoir hosts. For example, *Atriplex spp.* is a staple for the fat sand rat (*Psammomys obesus*), a rodent reservoir of *Le. major*. In hot, arid climates, vegetation may also provide a microhabitat with cooler temperatures and higher humidity for sand flies. For instance, sand fly population densities were significantly higher in traditional (shaded) than intensified (non-shaded) coffee plantations in Colombia (Alexander *et al.*, 2001). In addition, populations of *P. orientalis* sand flies were found preferentially resting in termite mounds constructed in *A. seyal* wood, and were uncommon in villages with poor vegetation at a VL site in the eastern Sudan (Elnaiem *et al.*, 1999). Vegetation may also provide suitable sites for mating or breeding sand flies (Elnaiem *et al.*, 1999). Several studies have used sensor imagery to identify the relevant vegetation in areas where CL or VL is present (e.g. in Brazil) (Miranda *et al.*, 1996).

8.3.2 *Human environmental determinants*

Human activities have tremendous impacts on their environments that result in creating favourable conditions for leishmaniasis transmission (Wilson 2001). The density and distribution of sand fly vector species may be indirectly associated with environmental modification caused by urbanization, deforestation and agrarian practices, which in turn changes land-use patterns and the types of vegetation present in endemic areas (WHO 2005). One study used satellite sensor data and imagery to correlate the occurrence of epidemic outbreaks of CL cases with areas of logging in Paraná, Brazil (Arraes *et al.*, 2008). A second study identified land use characteristics and analysed environmental factors for VL in the urban areas of Teresina, Piauí, Brazil (Correia *et al.*, 2007). A third study correlated an increased infection prevalence of *Le. major* in the reservoir host (*Psammomys obesus*) with the development of human dwellings in previously undisturbed areas in southern Israel (Wasserberg *et al.*, 2003).

In zoonotic CL foci, environmental modification resulting from new irrigation practices, the construction of dams or extensive land reclamation can change the temperature and humidity of the soil and vegetation, which can result in changes in the composition of sand fly species and increased densities of sand fly and rodent reservoir populations (Aytekin *et al.*, 2006; Ben-Rachid & Ben-Ismail 1987). In southern Israel, the density of *P. papatasi* sand flies was correlated positively with soil moisture, especially in habitats subject to anthropogenic disturbances in which the addition of new water sources not only increased the abundance of lush vegetation, an important food source for the rodent reservoir host, but also improved breeding conditions for both the sand fly vector and rodent species (Wasserberg *et al.*, 2002; 2003). Furthermore, the superior water retaining capability of alluvial soil samples collected from larval habitats in VL-endemic areas in Bihar State, India, was compared with samples containing non-porous granular particles collected from non-endemic areas (Sudhakar *et al.*, 2006).

Historically, sand fly control to reduce leishmaniasis transmission has been a by-product of mosquito control in public health campaigns to reduce malaria transmission. However, the indiscriminate use or inconsistent application of insecticides has led to insecticide resistance in sand flies in some areas (Ehrenberg & Ault 2005). For instance, the increasing incidence of CL cases in Sri Lanka has been linked to the emerging resistance of *P. argentipes* (main sand fly vector) to insecticides, particularly malathion used in anti-malarial campaigns (Surendran *et al.*, 2007).

In zoonotic CL or VL foci, disease epidemiology may be correlated strongly with the proximity of the human population to the reservoir host population, primarily in zoonotic VL foci where a large canine population constitutes the primary reservoir of infection (Sherlock 1996). In a VL focus in north-east Brazil, the presence of sand flies (*Lutzomyia spp.*) in dogs and chickens, either in the house or in the neighbourhood, were possible risk factors for leishmaniasis infection (Caldas *et al.*, 2002). However, in other analyses they appeared to protect from infection, perhaps because the animals provided zooprophylaxis by deflecting sand flies from biting humans. In a CL focus in Turkey, hot temperatures in the summer caused people to sleep next to rodents on the roofs of their houses (Aytekin *et al.*, 2006).

In endemic areas, risk of infection may also be increased where housing construction attracts sand flies. Houses with thatched, mud-plastered roofs or walls encourage sand flies to rest in the cracks and crevices during the day (Sudhakar *et al.*, 2006; Hogsette *et al.*, 2008). In addition, rich organic debris on the ground assists the development of immature stages (Sudhakar *et al.*, 2006).

8.4   *Contextual determinants*

8.4.1   *Socio economic factors*
Whereas environmental determinants may be directly correlated with leishmaniasis transmission, socio economic factors and other contextual determinants are more difficult to quantify with respect to disease transmission. Poverty that results from uncontrolled urban development impacts human health and increases vulnerability to diseases, especially in areas with limited or no access to high quality health care, education, water resources, nutrition, adequate housing and personal protection methods against sand flies (Sherlock 1996; Larrea *et al.*, 2002; Ehrenberg & Ault 2005; WHO 2009). Efforts to conduct infectious disease surveillance and to prevent transmission are compromised or halted where community resources are limited or non-existent; and people who cannot afford proper treatment may improperly treat themselves with alternative medications (Ehrenberg & Ault 2005). On the other hand, widespread misuse of drugs available cheaply can lead to emerging parasite resistance, as has been noted for acquired resistance of *Le. donovani* (causative agent of VL) to pentavalent antimonials in Bihar State, India during the past sixty years (Croft *et al.*, 2006). The increasing number and geographic distribution of VL patients with *Le infantum*/HIV co-infections in Europe and elsewhere is of particular concern, especially in cases where intravenous drug users can acquire the parasite (and virus) by needle (Cruz *et al.*, 2002). Use amphotericin B (antibiotic) to treat leishmaniasis has increased, and there is some evidence for the emergence of resistance of *Le. infantum* to the drug in VL/HIV co-infection cases in France (Durand *et al.*, 1998; Croft *et al.*, 2006).

8.4.2   *Human physiological factors*
Human physiological factors play an important role in leishmaniasis epidemiology, particularly the status of the human immune response and the genetic makeup of the human population in endemic areas (Ehrenberg & Ault 2005). The formation of new settlements with non-immune populations facilitates the outbreak of leishmaniasis in endemic areas; for example, the outbreak of zoonotic cutaneous leishmaniasis (ZCL) in the central and southern governorates of Tunisia in 1982–83 occurred following construction of the Sidi Saad Dam (Ben Rachid & Ben-Ismail 1987). Refugees and displaced people are particularly vulnerable in war-torn areas or following natural disasters. Since 1983, the civil war in Sudan has displaced millions of people who have been affected by VL

outbreaks in the conflict-torn regions in the east and south (Ritmeijer & Davidson 2003). In the Western Upper Nile Province, one ferocious VL epidemic in a previously non-endemic region was linked to regenerating forest vegetation in which the sand fly vector (*P. orientalis*) is associated, and also to introducing the parasite (*Le. donovani*) from endemic areas further east (Ashford & Thomson 1991).

8.5 Remote surveillance of leishmaniasis

8.5.1 Targeting surveillance and control efforts

Knowledge of the temporal and spatial distribution of leishmaniasis cases, and the density and abundance of vector/reservoir host species are important for predicting where disease transmission may or may not occur (Gebre-Michael *et al.*, 2004). Public health planners and vector surveillance programs can take advantage of sensor data and imagery to target vector and disease surveillance and control efforts in endemic areas where the sand flies are most likely to occur. Imagery is used to characterize local environmental conditions that can be incorporated into a model or GIS along with parasite (leishmaniasis) case data, and vector (sand fly) distribution and abundance data, to predict risk of transmission. The resulting data can be used to derive vegetation cover, landscape structure and water bodies to address questions about relationships between disease transmission and environmental variables in a risk model (Beck *et al.*, 2000).

Sufficient knowledge about vector ecology and leishmaniasis epidemiology is crucial, particularly where vector species are strongly associated with environmental variables (Beck *et al.*, 2000; Seto *et al.*, 2007). For instance, sand flies are not associated ecologically with water bodies due to their terrestrial life cycle. However, soil moisture may be used to identify suitable habitats for rodent reservoir hosts, and hence the location of developing immature stages of sand flies. The availability of high resolution EO soil moisture data is still in the early developmental stages (Schneider *et al.*, 2008). For zoonotic leishmaniases associated with rural or peri-urban areas, efforts are focused on identifying habitats with particular types of vegetation that are preferred by the vector and/or reservoir host animal where applicable. Landscape structure may also be used to assess environmental changes or in geographical areas where economic resources or infrastructure do not exist to support disease surveillance systems (Beck *et al.*, 2000).

Since the 1990s, much of the research using satellite-acquired data has focused on identifying environmental conditions associated with habitats of sand fly vector species that are implicated in leishmaniasis transmission. One of the first published studies to incorporate satellite imagery to model disease risk used NDVI to identify the probable distribution of the sand fly vector (*P. papatasi*) of *Le. major* in Southwest Asia, though the model was never validated (Cross *et al.*, 1996). It has been suggested that NDVI may be a useful tool to identify areas that are suitable for sand fly populations and where there is higher incidence of visceral leishmaniasis in Bahia, Brazil (Bavia *et al.*, 2005). Ecological niche modelling has also been used to develop distribution models for *P. papatasi* and *P. alexandri* across the Middle East (Colacicco-Mayhugh *et al.*, 2010). Distributions were associated strongly with urban and woody savannah land cover, and weakly with temperature and precipitation. Adding NDVI to model development had no effect. At present, NDVI appears to be a more useful tool in modelling distributions of New World species of *Lutzomyia* than Old World species of *Phlebotomus spp*. However, because of ecological differences between the genera, further research needs to be conducted into the relationship between NDVI products and leishmaniasis.

Incorporating remote sensing data into public health planning may result in more effective disease prevention programs and targeted resources for optimum disease intervention. The usefulness of risk maps for VL distribution, together with rainfall and altitude data in eastern Sudan, has been attempted to plan the location of VL treatment centres and limits of sand fly control programs (Elnaiem *et al.*, 2003). In addition, efforts have sought to identify high-risk areas for VL transmission in Brazil to facilitate efficient targeting of control measures (Werneck *et al.*, 2002). In some endemic areas, sand fly control for CL often has been a by-product of mosquito control for malaria. Where multiple parasitic diseases are prevalent, spatially-explicit databases may be created to target multiple species effectively in a cost-effective manner (Malone *et al.*, 2006).

Functional surveillance systems should be shared among countries with common borders that are affected by outbreaks of leishmaniasis. GIS and EO data are being developed continuously through many global partnerships to improve the accuracy of risk mapping and prediction of disease transmission. Following sustained collaborative control efforts to modify environmental risk factors in endemic areas, EO data can also be used to assess the efficacy of such control efforts on the incidence of leishmaniasis in retrospective studies, particularly in areas with high population densities exposed to the anthroponotic cycle of CL.

8.5.2 *Conditions signalling leishmaniasis transmission risk*

Environmental disturbances like flooding, medium-term fluxes in precipitation and longer-term climate variability create conditions that allow leishmaniasis transmission to increase. This results from periodic increases in vector species populations or socio economic changes that promote greater contact between people and the vector in native settlements that also are zoonotic foci (Ready 2008; WHO 2009a). Ready provides a detailed review of climate impacts on leishmaniasis emergence, including case studies of spatial-temporal modelling of CL and VL emergence, and an excellent discussion of control strategies to prevent leishmaniasis emergence. Climate variability affects leishmaniasis transmission directly through the effect of temperature on parasite life-cycle dynamics, vector competence (incubation changes), and the range, abundance and behaviour of sand fly vector species (WHO 2005; Ready 2008). Various integrated modelling studies have forecasted that climate fluxes, caused by an increase in ambient temperature, may contribute to expanding the geographical ranges of sand fly vector species upslope to higher altitudes or poleward to higher latitudes. The opposite might occur in a period of decreased temperature, if conditions become too cool or dry for sand fly survival (Hunter 2003; McMichael *et al.*, 2003; ECDC 2009;). Leishmaniasis caused by *Le. infantum* in Europe has a wider geographical distribution than that of *Le. infantum*, though imported cases of canine leishmaniasis as a reservoir host in central and northern Europe are common (Maier 2003; Naucke & Schmitt 2004; ECDC 2009). If ambient temperatures continue to rise in northern Europe, making conditions for transmission more suitable, these imported cases could serve as a reservoir of infection and lead to establishing new endemic foci (Ready 2008; ECDC 2009).

Developing an early warning system is important for effective public health planning. An early warning system might consider a GIS that includes: 1) leishmaniasis incidence in humans and animal reservoirs; 2) infection levels of vectors; 3) number of humans and animal reservoirs in the region; and 4) spatial analyses that determine the spatial coincidence between sand fly habitats and human or animal reservoir populations (Seto *et al.*, 2007). A climate-based early warning system would be useful to plan disease interventions for controlling leishmaniasis outbreaks (Chaves & Pascual 2006). A past study has confirmed the role flooding plays in outbreaks of CL in Argentina, and that sand fly vector abundance is correlated with rainfall in the previous year (Salomon *et al.*, 2006). Their discussion includes sensor data to trigger an early warning system for outbreaks. Another climate-based study used climatic data and ten years of CL cases in Costa Rica to build advanced statistical models to investigate how sea surface temperature influences resulting from ENSO affect leishmaniasis transmission (Chaves & Pascual 2006). In that study, CL cases tended to peak in May with cases rising and falling in three-year cycles corresponding to climate cycles.

For prediction scenarios that incorporate both remote sensing data and disease case data, limitations of environmental data to predict leishmaniasis outbreaks exist where leishmaniasis case data may not be reported in the place of transmission; that is, they are geographically incorrect; or where case data may be under-reported if the poorest segments of the population do not seek treatment due to the prohibitive cost of anti-leishmanial drugs. In addition, no studies to date have used sensor data to predict an increased incidence of leishmaniasis based on the resistance of sand flies to insecticides for vector control, or the resistance of parasites to pentavalent antimonials or other drugs for treating *Leishmania* parasitic infections. While these control efforts cannot be captured directly from orbiting platforms, they should not be ignored in context of an early warning system for leishmaniasis control. Predictive maps generated from temporal and spatial distribution of leishmaniasis cases, as well as the density and abundance of vector/reservoir host species, need to be updated periodically to reflect environmental changes and possible shifts in the distribution of the disease, vector, and reservoir host species (Gebre-Michael *et al.*, 2004).

8.6 *Remote sensing case studies*

Three case studies are presented that incorporate environmental sensor data to predict the incidence of VL cases or risk of infection in endemic parts of Brazil, India and Italy, based on VL case data or sand fly collection data. The first study used enhanced imagery to achieve greater accuracy in a retrospective eco-epidemiological analysis to determine the relative risk of VL infection in children in a region in Bahia State, Brazil (Thompson *et al.*, 2002). The second study employed sensor data to develop two climate-based risk models using different approaches to predict the distribution of the sand fly vector species and VL risk across Bahia State (Nieto *et al.*, 2006). The third study used sensor data in a GIS context to develop landscape predictors of sand fly densities in VL-endemic areas of Bihar State, India (Sukhakar *et al.*, 2006). They sought to identify high-risk areas for VL to better define disease control and prevention efforts at regional and national levels. The fourth study used a variety of sensor data from aerial photography and satellite images to examine ecological factors affecting population levels of sand flies on opposite sides of Mt. Vesuvius in Italy (Rossi *et al.*, 2007).

8.6.1 *Bahia, Brazil*

A retrospective seventeen year cohort study in Bahia used remote sensing applications to examine the effects of precipitation and region on the relative risk of infection of VL (*Le. chagasi*) to children in Canindé, Ceará, a hot, semi-arid area in northern Brazil. The authors used commercial software to overlay a grid of 873 four square kilometre cells containing juvenile VL case data (1981–1997) and house locations on a Landsat thirty metre resolution TM image obtained during the 1986 dry season. They classified the image pixels into 100 spectral classes and removed cloud, cloud shadows and other shadow areas using a combination of near infrared, mid-infrared and visible green spectral bands. The resulting images were processed by the Earth Resources Data Analysis System (ERDAS) NDVI and tasseled cap (TC) algorithms to determine the mean index values for active vegetative growth in each four square kilometre area. A supervised classification scheme was used to reclassify areas of 200 pixels into four classes: foothills, plains, city and other (including water bodies and mining areas). Six models were developed to estimate population-at-risk (PAR) values for each of these four regions, based on different estimates of the yearly rate of change in the population during the seventeen-year study. The results showed that VL incidence was higher in the foothills than in other regions, which might be due to the presence of the sand fly vector (*Lu. longipalpis*) in brush land and areas of high humidity and cooler temperatures that are typical of the Canindé foothills. In addition, VL incidence was also negatively correlated with a rolling three-year rainfall average, which might be due to a concentration of people and reservoir hosts around existing water supplies in the foothills during times of drought, bringing malnourished children into close proximity with potentially infected sand flies.

Two predictive models were developed to assess the distribution of *Lu. longipalpis* sand flies and the potential risk of infection by VL (*Le. chagasi*) in Bahia State (Nieto *et al.*, 2006). First, an ecological niche model (see Chapter 8) was developed using genetic algorithm for rule-set production (GARP) software within a GIS that included annual VL prevalence data, monthly average climate data and one kilometre resolution Shuttle Radar Topography Mission (SRTM) data. Second, a growing degree day-water budget (GDD-WB) model was developed to estimate the suitability limits for *Le. chagasi-Lu. longipalpis* transmission in relation to rainfall and optimal temperature for propagating *Lu. longipalpis* sand flies. Both models predicted similar distribution and abundance patterns for vector-parasite transmission, with the highest transmission risk in the Caatinga region described below (Figure 25a,b,c). The authors discuss the usefulness of GARP and GDD-WB prediction models of VL risk in disease control programs and the requirement for more detailed sand fly distribution and parasite prevalence data to further develop disease risk models. It should be noted that the two modelling approaches described work well for leishmaniasis

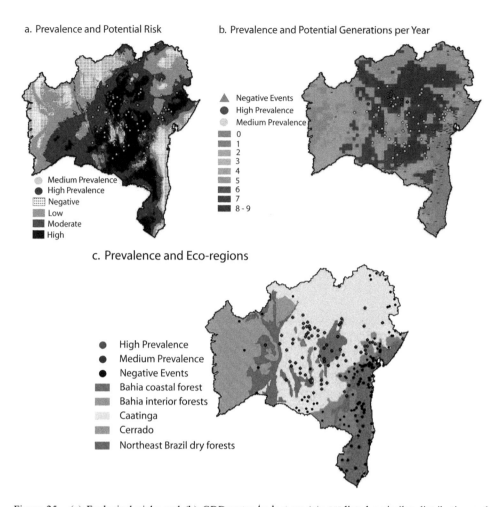

a. Prevalence and Potential Risk

Medium Prevalence
High Prevalence
Negative
Low
Moderate
High

b. Prevalence and Potential Generations per Year

Negative Events
High Prevalence
Medium Prevalence
0
1
2
3
4
5
6
7
8 - 9

c. Prevalence and Eco-regions

High Prevalence
Medium Prevalence
Negative Events
Bahia coastal forest
Bahia interior forests
Caatinga
Cerrado
Northeast Brazil dry forests

Figure 25. (a) Ecological niche and (b) GDD-water budget models predicted a similar distribution and abundance pattern for vector-parasite transmission. Highest transmission risk was predicted in the Caatinga region, and no risk was predicted in the coastal forest for (c) ecological regions of Bahia State (Modified from Nieto *et al.*, 2006). (see colour plate 15)

epidemiology systems in which the sand fly vector is closely associated with a human host, or both human and reservoir hosts. These approaches may not work where leishmaniasis is primarily a zoonosis (i.e. for *Le. major* in the eastern Mediterranean region), or where infected sand flies or rodent reservoir hosts may be present in areas uninhabited by people.

8.6.2 *Bihar State, India*
In Bihar State, remotely sensed land cover data were used to detect and map differences in land cover between endemic and non-endemic areas where population levels of *P. argentipes* sand flies and VL (*Le. donovani*) incidence rates were high or low (Sudhakar *et al.*, 2006). Indian Remote Sensing (IRS)-1D Linear Imaging and Self Scanning (LISS) III satellite data were corrected and segregated into seven primary land use categories, including water bodies, marshy areas, agricultural areas, fruit orchards and other vegetation types. For each study area, the satellite images were fused with sand fly vector collection data and locations of ten villages, each surrounded by a five square kilometre buffer zone. Soil type, soil chemistry, crop type and house flooring material were also

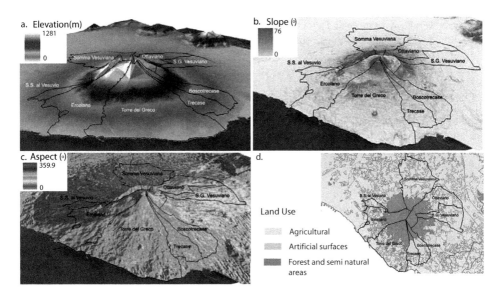

Figure 26.   Elevation, slope, aspect and land use classes formed the data layers used to construct a GIS database (Modified from Rossi *et al.*, 2007). (see colour plate 16)

included as environmental factors in the analysis. Sand fly densities were higher in the endemic study area and positively correlated with water- or irrigation-based land cover features, including water bodies, marshy areas and certain agricultural crops (e.g. banana, sugarcane and rice paddies). The authors discuss the usefulness of image spectral data for generating information about land cover to develop landscape predictors of sand fly vector abundance, and hence high-risk areas of VL transmission. While this study describes the importance of including geomorphology and soil type in an analysis of sand fly vector density, the absence of VL incidence data or vector infection rates in the study raise the question of accuracy in targeting effective disease prevention or control efforts.

8.6.3   *Vesuvius, Italy*

For Mt. Vesuvius, sensor imagery was employed to examine ecological factors thought to affect populations of *P. perniciosus* sand flies on two opposite slopes of Mt. Vesuvius (Rossi *et al.*, 2007). Zoonotic VL (*Le. infantum*) is endemic around the volcano, and dogs are the primary reservoir host. The authors constructed a GIS using digital aerial photographic data having one metre resolution; elevation, slope and aspect data from a DEM having forty metre resolution; municipal boundaries; and land use based on a 1:50,000 scale agricultural map having five categories and thirty-eight classes. The five categories were urban, agriculture, forest, wetlands and water (Figure 26). GIS and municipal boundaries were used to plan sand fly collection sites at similar elevations and surfaces on opposite slopes of the volcano. The density of sand flies was higher on the coastal side compared with the inland (Apennine) side. This might be explained by a steeper aspect and a higher proportion of forested areas on the coastal side. The authors suggest that descriptive maps of vector density and environmental features arc an important tool to generate predictive maps of leishmaniasis risk by statistical analysis and ecological niche modelling methods.

9   CLIMATE AND INFLUENZA

Influenza not only continues to be a significant public health burden, it is also a threat for pandemic. The spatiotemporal variation of influenza across latitudes often suggests, though remains arguable, that climate and environmental factors determine transmission and pathogenesis. This Section reviews several environmental factors that are implicated frequently in influenza transmission. Modelling and forecasting influenza using climate is also discussed.

9.1 *Characteristics of influenza*

Influenza is an acute viral respiratory disease that infects five to fifteen per cent of the world population and causes 250,000–300,000 deaths each year (WHO 2009). Despite vaccination and the largely mild cases, the burden of influenza remains significant. This is in part due to health care cost and the loss of productivity resulting from infections. For example in the in the US, where the annual influenza epidemic can cause up to 200,000 hospitalizations and more than 30,000 deaths (CDC 2010), the estimated economic burden based on the 2003 population is around $87.1 billion USD (Molinari *et al.*, 2007). In addition to the significant burden of seasonal influenza, the rapid and continuous mutational changes of the virus can cause unexpected pandemic.

There are three types of influenza virus circulating in the world: A, B and C. Type A and B are the most commonly found in humans, and it is type A, which is further classified into subtypes (i.e. H1N1), that is the most virulent. As an intrinsic evolutionary mechanism that enables effective host invasion, influenza virus undergoes two types of mutational change, namely, antigenic drift and shift that generally impact the ability of existing antibody to recognize the virus. The more danger-ous of the two, antigenic shift, may result in a novel virus to which humans have no immunity. Such mutations are responsible for the influenza pandemics in 1918, 1957 and 1968 (Carrat & Flahault, 2007). Meanwhile, antigenic drift occurs more often. In order to accommodate the ever-changing circulating influenza strain, vaccine composition recommendations are altered twice a year. The rec-ommendations are made by the WHO through its Global Influenza Surveillance Network (GISN). By monitoring the currently circulating strains, the WHO not only estimates the strain that will circulate but also serves as the global alert for emerging novel viruses that may cause pandemic.

It is well understood that influenza transmission in temperate climates is seasonal and peaks in the winter months (Figure 27). In the US, for example, influenza outbreaks often start as early as October, peak in February, and diminish by April or May; thus forming a distinct inter-annual oscillation pattern. In the tropics, there are significant influenza cases throughout the year, with one or two less distinct peak(s) whose timing varies geographically (Figure 28). As seen in Figure 28 and shown by others (Viboud *et al.*, 2006), influenza seasonal patterns vary with latitude, forming a traveling wave across the globe. Several studies that have explored the global migration pattern of influenza show varying travel patterns of influenza virus A (Russell *et al.*, 2008): migration out of the tropics and China; and migration between northern and southern hemispheres (Russell *et al.*, 2008). Another study in Brazil showed that influenza starts in a low-population state near the equator during March-April, and travels southward towards temperate and more populous states (Alonso *et al.*, 2007).

The spatiotemporal variation of influenza across latitude suggests that environmental factors determine its seasonality. However, the role of climate and environment in influenza transmission are arguable. This is because factors associated with influenza outbreaks in one region may not be valid in others. Moreover, there are few biological experiments that explore and validate climate's role in influenza transmission.

9.2 *Climate and seasonality in influenza transmission*

Influenza can be transmitted via contact, droplets or aerosols (Tellier 2009). Contact transmission can be a direct physical contact between humans, or through contact with a contaminated object. Pathogenic droplets are formed when an infected person sneezes or coughs and the virus is propelled through the air over a short distance. The size of a droplet is usually more than 10μm, though this varies in the literature. Droplet particles are not as effective as aerosols for inducing infection because there is an inverse relationship between the size of a particle and its penetration into the respiratory tract. Droplets smaller than 10μm have a better chance of inducing infection (Bridges

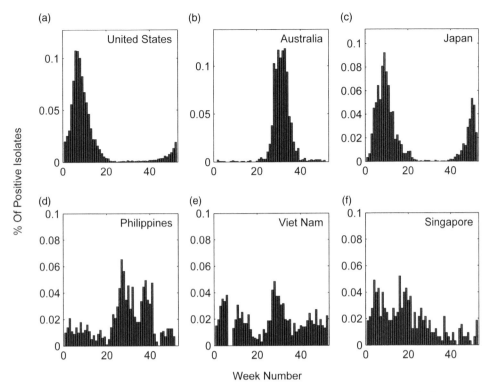

Figure 27.   Proportion of influenza positive isolates (Y-axis) by calendar week in 2007 (X-axis) for countries representing the Northern Hemisphere (a), Southern Hemisphere (b), and East Asia from temperate (c) to tropical (f) countries (data from FluNet, WHO 2010b). (see colour plate 17)

et al., 2003; Tellier 2009). On the other hand, aerosols smaller than 5μm are more contagious, remain suspended in the air longer, and are dispersed farther.

Any of the aforementioned transmission routes can be affected directly and indirectly by climatic factors. In addition to transmission modes, climate has also been attributed to virus survivorship and host susceptibility to infection. Temperature and humidity are the two most-studied determinants of influenza transmission. Virus survivorship outside of a host is prolonged at colder temperatures. At about 22°C, the outer membrane of influenza virus is in gel phase, providing protection during air-borne transmission; whereas at higher temperatures, the outer membrane melts and the virus dries out or is weakened (NIH 2008; Polozov *et al.*, 2008). It has been shown that influenza virus can survive at 28°C and 35–49 per cent humidity on a range of surfaces, thus enabling indirect transmission through fomites – contaminated inanimate objects (Bean *et al.*, 1982; Bridges *et al.*, 2003). In aerosolized transmission, temperature and humidity also influence influenza transmission efficiency. Lowen (*et al.*, 2007) provided supporting evidence for this using guinea pigs. They found that dry and cold conditions (5–20°C and 20–35 per cent relative humidity) were ideal for transmission. This is consistent with the observed winter influenza peak in temperate regions. Further studies reveal that temperatures approaching 30°C can block aerosol transmission. This suggests that contact transmission routes prevail in the tropics where influenza rates are high year round (Lowen *et al.*, 2008). Although results from Shaman & Kohn (2009) also show that humidity is a determinant for influenza transmission, they argue that absolute humidity impacts influenza virus survival and transmission more significantly than relative humidity. The role of absolute humidity is further integrated into a population-based mathematical model as a factor

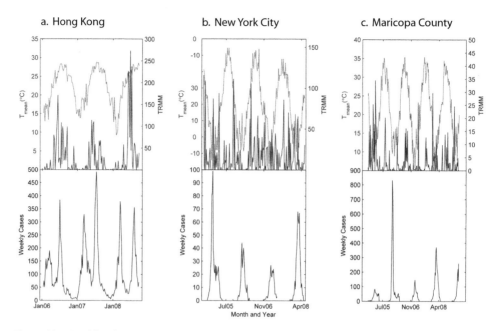

a. Hong Kong   b. New York City   c. Maricopa County

Figure 28. Weekly climatic factors and lab-confirmed influenza positive isolates for (a) Hong Kong, (b) New York City and (c) Maricopa County. In the top panel for each location, the green line mean temperature (°C) compared to the blue line showing TRMM precipitation rate (mm/hr.); bottom panel for each location is the weekly number of influenza cases for each location. (see colour plate 18)

in the transmission rate (Shaman *et al.*, 2009). The model was applied to the continental US and the results showed that anomalously low absolute humidity levels precede the onset of wintertime influenza-related mortality.

In addition to biological experiments, a number of empirical studies show significant correlations between influenza incidence and both temperature and humidity. In Japan, influenza evidently increases when there are fewer days with temperatures exceeding 10°C and more days of relative humidity less than sixty per cent per week (Urashima *et al.*, 2003). A study in Maricopa County Arizona shows that influenza is correlated with temperature, and that influenza in Hong Kong is correlated with LST and relative humidity data (Soebiyanto *et al.*, 2010). Another study on influenza in Hong Kong also shows its correlation with temperature and relative humidity that characterize cold and dry conditions (Chan *et al.*, 2009).

Unlike temperature and humidity, there is no evidence suggesting that rainfall effects influenza transmission directly. Instead, rainfall is commonly seen as an indirect effector by driving people to congregate indoors on rainy days, thus increasing the rate of contact transmission. Regions that have reported rainfall as influenza determinant include Brazil (Moura *et al.*, 2009), Singapore (Chew *et al.*, 1998) and Thailand (Chumkiew *et al.*, 2007).

The susceptibility of the host to influenza virus depends on the capability of the immune system to safeguard against it. The immune system is partly dependent on the level of melatonin and Vitamin D, which in turn is determined by photoperiod (Lofgren *et al.*, 2007). Light-dark cycles regulate the hormone melatonin, whereas the amount of sunlight determines Vitamin D level directly. As discussed in Dowell (2001) and Lofgren (*et al.*, 2007), there are several experiments that expose influenza virus to humans and mice at different times of the year, and the results show higher influenza prevalence during winter. Besides sunlight's contribution to the strength of the host's immune system, empirical evidence shows that sunlight's ultraviolet radiation could possibly increase the inactivation of the viruses in the environment (Sagripanti & Lytle (2007)).

70

9.3 Modelling influenza using climate variables

Empirical models provide an attractive complement to biological studies on the climate-related causes of influenza transmission since such models can provide predictions even where biological mechanisms are not fully understood. Empirical models may also identify patterns for further study that do not agree with known causal mechanisms. Regression analysis has become the most popular tool in understanding climate's role in influenza incidences (Chew *et al.*, 1998; Urashima *et al.*, 2003; Youthao *et al.*, 2007; Liao *et al.*, 2009; Moura *et al.*, 2009;). In addition, time series regression such as autoregressive integrated moving average (ARIMA) has been employed (Soebiyanto *et al.*, 2010; Tang *et al.*, 2010). A more biological approach was developed by Shaman (*et al.*, 2009), where absolute humidity was factored into the transmission rate in a classical compartmental model.

In the following, a neural network (NN) model was developed to predict influenza outbreaks in Hong Kong, New York City, and Maricopa County, Arizona, with latitudes of 22°N, 40°N, 33°N, respectively (Figure 28). Each area has different weather and climate attributes. Hot and humid conditions prevail during summer in Hong Kong with temperatures that exceed 31°C during the day. Winters are mild, with temperatures as low as 6°C. New York City, on the other hand, has colder winters with average low temperatures dipping to 2°C, and average highs in the summer of about 29°C. Dry conditions prevail in Maricopa County, Arizona, where the mean winter low is 5°C and the mean summer high is 41°C.

9.3.1 Materials and methods

The weekly count of lab-confirmed positive isolates was obtained from influenza surveillance issued by each region's public health department (Hong Kong Department of Health 2009; Maricopa Department of Public Health 2009; New York State Department of Health 2009). Since the number of specimens tested for New York City was available, the data were further assessed for the proportion of positive isolates (over total specimens tested in that week) in the analysis; otherwise, the weekly total positive isolates were used.

Climatic and meteorological parameters were obtained from ground stations and satellite-derived measurements. Daily data were collected for the three study areas from: Hong Kong Observatory (2009), National Climatic Data Center (2009) and the flood control district of Maricopa County (2009). The ground data include temperature (maximum, mean, minimum), dew point temperature, relative humidity, global solar radiation, total evaporation, and air pressure, among others. Station data from ground observation stations were aggregated as averages for each study area by week. For the remotely-sensed climate parameters in the three regions, LSTs were retrieved from MODIS at 0.05° (approximately 7.5 km spatial resolution). In addition, daily precipitation as measured from the instruments mounted on TRMM were downloaded via NASA's Goddard Earth Sciences GES-DISC interactive online visualization and analysis infrastructure (GIOVANNI) (Acker & Leptoukh 2007) (see also Chapter 6). Similarly, the remote sensing data were averaged across region and week.

In this study, influenza incidences were modelled and forecasted using NN, a computational method that emulates human brain function. It consists of interconnected weighted nodes, where each node carries a threshold function such as a sigmoid function. The nodes are typically arranged in layers: input, hidden, and output. The nodes in the input layer take the predictors through the weighted connection, and subsequently calculate the threshold function. The outputs from the first layer nodes are consequently fed into the next layer for further evaluation on the threshold function, and so forth. The weights and the parameters in each threshold function are calculated during a training phase, typically using least squares estimators. A more rigorous treatment on neural networks can be found in Pao (1989).

For the three areas in question, a radial basis function neural network (RBFNN) was used in which the threshold function was a linear combination of basis function such as Gaussian. RBFNN has been widely used for various applications, especially in function approximation. In developing RBFNN, the inputs and the hidden nodes were limited to three each. This choice was made to reduce the complexity of the model and the number of estimated parameters. Lagged environmental

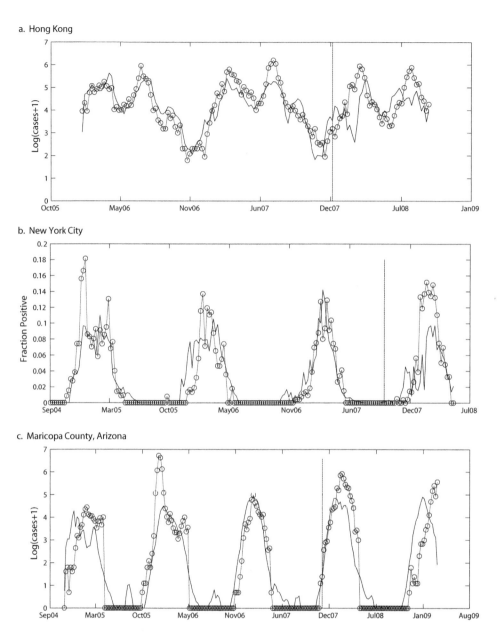

a. Hong Kong

b. New York City

c. Maricopa County, Arizona

Figure 29.   RBFNN model outputs for three study sites. The line with circles represents observations and the adjoining solid line represents model output. Fitted and predicted values are separated by the vertical line located near December 2007 at each study site.

variables up to lag 4 were included also as inputs. All inputs were smoothed before applying to the network. The resulting models were evaluated based on their root mean square error (Figure 29).

9.3.2   *Results*
All possible sets of three input combinations were explored. Results showed that the input combinations with best root mean square error (RMSE) calculated for training data were: mean dew point (lag 1), mean temperature (lag 3) and land surface temperature (lag 4) for Hong Kong;

maximum and minimum temperature (lag 1 & 2, respectively) and precipitation (lag 3) for New York City; minimum relative humidity (lag 1), mean temperature (lag 4) and maximum solar radiation (lag 4) for Maricopa County. The resulting models were further used to predict the dynamics of the forthcoming influenza season, as shown in Figure 29.

The NN models can generally predict the timing of the influenza peak reasonably well, particularly the influenza cases in New York City and Maricopa County. For Maricopa County, the model can well-predict the height of influenza peak. Inputs that result in the best training data RMSE for the three regions include temperature, humidity, dew point, precipitation and solar radiation. Most of these environmental factors have been implicated in influenza transmission process as previously described: temperature and humidity impacts aerosol borne transmission and virus survivorship; rainfall indirectly promotes contact transmission; and solar radiation regulates the human (host) immune system that in turn guards against influenza virus. Dew point, on the other hand, is a proxy to relative humidity. It should be noted that although there are biological evidences supporting the roles of the aforementioned climatic factors in influenza transmission and pathogenesis, the causative relationship cannot be justified through the NN models developed here. Instead, a biologically-based, mechanistic model such as developed by Shaman (*et al.* 2009) is needed. However, the NN model is sufficient for forecasting purposes. Integration of an influenza forecasting model with a surveillance system can further advance public health preparedness that in turn may reduce the associated economic burden.

REFERENCES

Achee, N.L., Korves, C.T., Bangs, M.L., Rejmankova, E., Lege, M., Curtin, D., Lenares, H., Alonzo, Y., Andre, R.G. & Roberts, D.R. 2000. *Plasmodium vivax* polymorphs and *Plasmodium falciparum* circumsporozoite proteins in *Anopheles* (Diptera: Culicidae) from Belize, Central America. *J. Vector Ecol.* 25: 203–211.

Achee, N., Grieco, J., Masuoka, P., Andre R. & Roberts. D. 2006. Use of remote sensing and geographic information systems to predict locations of *Anopheles darlingi*-positive breeding sites within the Sibun River in Belize Central America. *J. Med. Entomol.* 43: 382–392.

Acker, J.G. & Leptoukh, G. 2007. Online analysis enhances use of NASA Earth Science Data. *Trans. Amer. Geophys. Un.* 88(2): 14–17.

Adimi, F., Soebiyanto, R.P., Safi, N. & Kiang, R. 2010. Towards malaria risk prediction in Afghanistan using remote sensing. *Malaria J.* 9: 125.

Alexander, B. 1987. Dispersal of phlebotomine sand flies (Diptera: Psychodidae) in a Columbian coffee plantation. *J. Med. Entomol.* 24: 552–558.

Alexander, B., Agudelo, L.A., Navarro, F., Ruiz, F., Molina, J., Aguilera, G. & Quiñones, M.L. 2001. Phlebotomine sandflies and leishmaniasis risks in Colombian coffee plantations under two systems of cultivation. *Med. & Vet. Entomol.* 15(4): 364–373.

Alonso, W.J., Viboud, C., Simonsen, L., Hirano, E.W., Daufenbach, L.Z., Miller, M.A. 2007. Seasonality of influenza in Brazil: a traveling wave from the Amazon to the subtropics. *Amer. J. Epidemiol.* 165(12): 1434–1442.

Alvar, J., Yactayo, S. & Bern, C. 2006. Leishmaniasis and poverty. *Trends in Parasit.* 22(12): 552–557.

Anderson, K., Chunsuttiwat, S., Nisalak, A., Mammen, M., Libraty, D., Rothman, A., Green, S., Vaughn, D., Ennis, F. & Endy, T. 2007. Burden of symptomatic dengue infection in children at primary school in Thailand: A prospective study. *Lancet* 369: 1452–1459.

Anyamba, A., Linthicum, K.J., Mahoney, R., Tucker, C.J. & Kelley P.W. 2002a. Mapping potential risk of Rift Valley fever outbreaks in African savannas using vegetation index time series data. *Photog. Engin. & Rem Sens* 68: 137–145

Anyamba, A., Tucker, C.J. & Mahoney, R. 2002b. From El Niño to La Niña: Vegetation response patterns over east and southern Africa during the 1997–2000 period. *J. Climate* 15: 3096–3103.

Anyamba, A., Chretien, J-P., Formenty, P.B.H., Small, J., Tucker, C.J., Malone, J.L., El Bushra, H., Martin, V. & Linthicum, K.J. 2006a. Rift Valley fever potential, Arabian Peninsula [letter]. *Emerg. Infect. Dis.* 12(3): 518–520.

Anyamba, A., Chretien, J.-P., Small, J., Tucker, C.J. & Linthicum, K.J. 2006b. Developing global climate anomalies suggest potential disease risks for 2006–2007. *Int. J. Health Geogr.* 5: 60. doi:10.1186/1476-072X-5-60.

Anyamba, A., Chretien, J.-P., Small, J., Tucker, C.J., Formenty, P., Richardson, J.H., Britch, S.C., Schnabel, D.C., Erickson R.L. & Linthicum. K.J. 2009. Prediction of the Rift Valley fever outbreak in the Horn of Africa 2006–2007. *Proc. Nat. Acad. Sci.* 106: 955–959.

Anyamba, A., Linthicum, K.J., Small, J., Britch, S.C., Pak, E., de La Rocque, S., Formenty, P., Hightower, A.W., Breiman, R.F., Chretien,J.-P., Tucker, C.J., Schnabel, D., Sang, R., Haagsma, K., Latham, M., Lewandowski, H.B., Magdi, S.O., Mohamed, M.A., Nguku, P.M., Reynes, J.-M. & Swanepoel, R. 2010. Prediction, assessment of the Rift Valley fever activity in East and Southern Africa 2006–2008 and possible vector control strategies. *Amer. J. Trop. Med. & Hyg.* 8: 43–51.

Arraes, S.M., Veit, R.T., Bernal, M.V., Becker, T.C & Nanni, M.R. 2008. American cutaneous leishmaniasis in municipalities in the northwestern region of Paraná State: Use of remote sensing for analysis of vegetation types and places with disease occurrence. [in Portuguese]*Rev. Soc. Bras. Med. Trop.* 41(6): 642–647.

Aytekin, S., Ertem, M., Yaðdiran, O. & Aytekin, N. 2006. Clinico-epidemiologic study of cutaneous leishmaniasis in Diyarbakir Turkey. *Derm. Online J.* 12(3): 14.

Barcus, M.J., Laihad, F., Sururi, M., Sismadi, P., Marwoto, H., Bangs, M.J. & Baird, J.K. 2002. Epidemic malaria in the Menoreh Hills of central Java. *Amer. J. Trop. Med. Hyg.* 66(3): 287–292.

Barker, C.M. 2008. Spatial and temporal patterns in mosquito abundance and virus transmission in California. Ph.D. Dissertation, University of California, Davis.

Barker, C.M., Brewster, C.C. & Paulson, S.L. 2003. Spatiotemporal oviposition and habitat preferences of *Ochlerotatus triseriatus* and *Aedes albopictus* in an emerging focus of La Crosse virus. *J. Amer. Mosq. Cont. Assoc.* 19: 382–391.

Barrera, R., Torres, N., Freier, J.E., Navarro, J.C., Garcia, C.Z., Salas, R., Vasquez, C. & Weaver, S.C. 2001. Characterization of enzootic foci of Venezuelan equine encephalitis virus in western Venezuela. *Vect.-borne Zoon. Dis.* 1: 219–230.

Bautista C.T., Chan, A.S.T., Ryan, J., Calampa, C., Roper, M.H., Hightower, A.W. & Magill, A.J. 2006. Epidemiology and spatial analysis of malaria in the Northern Peruvian Amazon. *Am. J. Trop. Med. Hyg.* 75: 1216–1222.

Bavia, M.E., Carneiro, D.D., Gurgel Hda, C., Madureira Filho, C. & Barbosa, M.G. 2005. Remote sensing and geographic information systems and risk of American visceral leishmaniasis in Bahia, Brazil. *Parassitologia* 47(1): 165–169.

Bean, B., Moore, B.M., Sterner, B., Peterson, L.R., Gerding, D.N. & Balfour, H.H. Jr. 1982. Survival of influenza viruses on environmental surfaces. *J. Infect. Dis.* 146: 47–51.

Beck, L., Rodriguez, M., Dister, S., Hacker, C., Paris, J., Rejmankova, E., Roberts, D., Rodriguez, A., Spanner, M., Washino, R., Wood, B. & Legters, L. 1994. Remote sensing as a landscape epidemiological tool to identify villages at high risk of malaria transmission. *Amer. J. Trop. Med. & Hyg.* 51: 271–280.

Ben Rachid, M.S. & Ben-Ismail, R. 1987. Current situation in regard to leishmaniasis in Tunisia. In *Research on control strategies for the leishmaniases:* Manuscript 184e. *Proc. Int. Workshop.* Ottawa. IDRC-CIID.

Ben-Ahmed, K., Aoun, K., Jeddi, F., Ghrab, J., El-Aroui, M.A. & Bouratbine, A. 2009. Visceral leishmaniasis in Tunisia: Spatial distribution and association with climatic factors. *Amer. J. Trop. Med. & Hyg.* 81(1): 40–45.

Bicout, D.J. & Sabatier, P. 2004. Mapping Rift Valley fever vectors and prevalence using rainfall variations. *Vect.-borne & Zoon. Dis.* 4(1): 33–42.

Bird, B.H., Githinji, J.W., Macharia, J.M., Kasiiti, J.L., Muriithi, R.M., Gacheru, S.G., Musaa, J.O., Towner, J.S., Reeder, S.A., Oliver, J.B., Stevens, T.L., Erickson, B.R., Morgan, L.T., Khristova, M.L., Hartman, A.L., Comer, J.A., Rollin, P.E., Ksiazek, T.G. & Nichol, S.T. 2008. Multiple virus lineages sharing recent common ancestry were associated with a large Rift Valley fever outbreak among livestock in Kenya during 2006–2007. *J. Virol.* 82: 11152–11166.

Birkett, C., Murtugudde, R. & Allan, T. 1999. Indian Ocean climate event brings floods to East Africa's lakes and the Sudd Marsh. *Geophys. Res. Letts.* 26: 1031–1034.

Bocco, M., Gustavo Ovando, S.S. & Willington, E. 2007. Neural network models for land cover classification from satellite images. *Agricultura Tecnica* 67: 414–421.

Bohra, A. & Andrianasolo, H. 2001. Application of GIS in modeling of dengue risk based on sociocultural data: Case of Jalore, Rajasthan, India. *Dengue Bull.* 25: 92–102.

Bouma, M. & van der Kaay, H. 1996. The El Niño Southern Oscillation and the historic malaria epidemics on the Indian subcontinent and Sri Lanka: An early warning system. *Trop. Med. & Int. Health* 1(1): 86–96.

Boyle, D.B., Dickerman, R.W. & Marshall, I.D. 1983. Primary viraemia responses of herons to experimental infection with Murray Valley encephalitis, Kunjin and Japanese encephalitis viruses. *Aust. J. Exp. Biol. Med. Sci.* 61(Pt6): 655–664.

Breiman, L., Friedman, J.H., Olshen, R.A. & Stone, C.J. 1984. *Classification and regression trees*. Boca Raton: CRC Press.

Bridges, C.B., Kuehnert, M.J. & Hall, C.B. 2003. Transmission of influenza: Implications for control in health care settings. *Clinic. Infect. Dis.* 37: 1094–1101.

Broom, A.K., Lindsay, M.D., Johansen, C.A., Wright, A.E. & Mackenzie, J.S. 1995. Two possible mechanisms for survival and initiation of Murray Valley encephalitis virus activity in the Kimberley region of Western Australia. *Amer. J. Trop. Med. & Hyg.* 53: 95–99.

Broom, A.K., Lindsay, M.D., Plant, A.J., Wright, A.E., Condon, R.J. & Mackenzie, J.S. 2002. Epizootic activity of Murray Valley encephalitis virus in an aboriginal community in the southeast Kimberley region of Western Australia: Results of cross-sectional and longitudinal serologic studies. *Amer. J. Trop. Med. & Hyg.* 67: 319–323.

Burke, D.S. & Leake, C.J. 1988. Japanese encephalitis. In T.P. Monath (ed.), *The arboviruses: epidemiology and ecology*: 63–92. Boca Raton: CRC Press.

Caldas, A.J.M., Costa, J.M.L., Silva, A.A.M., Vinhas, V. & Barral, A. 2002. Risk factors associated with asymptomatic infection by *Leishmania chagasi* in north-east Brazil. *Trans. Roy. Soc. Trop. Med. & Hyg.* 96(1): 21–28.

Calisher, C.H. 1983. Taxonomy, classification, and geographic distribution of California serogroup bunyaviruses. *Prog. Clin. Biol. Res.* 123: 1–16.

Cane, M.A., 1983. Oceanographic events during El Niño. *Science* 222: 77–90.

Carney, R.M., Husted, S., Jean, S., Glaser, C. & Kramer, V. 2008. Efficacy of aerial spraying of mosquito adulticide in reducing incidence of West Nile virus, California, 2005. *Emer. Infect. Dis.* 14(5): 747–754.

Carrat, F. & Flahault, A., 2007. Influenza vaccine: The challenge of antigenic drift. *Vaccine* 25: 6852–6862.

Casman, E.A. & Dowlatabadi, H. 2002. *The contextual determinants of malaria*. Washington DC: Resources for the Future.

CDC 2003. Imported plague: New York City, 2002. *MMWR* 52: 725–728.

CDC 2006. Human plague- Four states*MMWR* 55: 1–3.

CDC 2010. Key facts about seasonal influenza. Available from: *http://www.cdc.gov/influenza/keyfacts.htm.* [Accessed 18th January 2012].

CDC 2011. Available from: http://www.cdc.gov/lac/tech/epi.html [Accessed 22nd January 2012].

CDC 1998. Rift Valley fever: East Africa, 1997–1998. *MMWR* 47: 261–264.

CDC 2000. Outbreak of Rift Valley fever-Saudi Arabia. *MMWR* 49: 905–908.

Chan, P.K.S., Mok, H.Y., Lee, T.C., Chiu, I.M.T., Lam, W.Y. & Sung, J.L.Y. 2009. Seasonal influenza activity in Hong Kong and its association with meteorological variations. *J. Med. Virol.* 81(10): 1797–1806.

Chaniotis, B.N., Correa, M.A., Tesh, R.B. & Johnson, K.M. 1974. Horizontal and vertical movements of phlebotomine sandflies in a Panamanian rain forest. *J. Med. Ent.* 11: 369–375.

Charlwood, J.D. & Alecrim, W.A.1989. Capture-recapture studies with the South American malaria vector *Anopheles darlingi*, Root. *Ann. Trop. Med. Parasitol.* 83: 569–576.

Chaves, L.F. & Pascual, P. 2006. Climate cycles and forecasts of cutaneous leishmaniasis, a nonstationary vector-borne disease. PLoS Med 3: e295. DOI:10.1371/journal.pmed.0030295.

Chew, F.T., Doraisingham, S., Ling, A.E., Kumarasinghe, G. & Lee, B.W. 1998. Seasonal trends of viral respiratory tract infections in the tropics. *Epidemiol. Infect.* 121(1): 121–128.

Christophers, S. 1960. *Aedes aegypti (L.): The yellow fever mosquito*. Cambridge: Cambridge UP.

Chumkiew, S., Srisang, W., Jaroensutasinee, M. & Jaroensutasinee, K. 2007. Climatic factors affecting on influenza cases in Nakhon Si Thammarat. *Proc. World Acad. Sci. Engin. & Tech.* 21: 364–367.

Clem, A. & Galwankar, S. 2005. Plague: A decade since the 1994 outbreaks in India. *J. Assoc. Physicians, India* 53: 457–464.

Clements, A.C. & Pfeiffer, D.U. 2009. Emerging viral zoonoses: Frameworks for spatial and spatiotemporal risk assessment and resource planning. *Vet. J.* 182: 21–30.

Colacicco-Mayhugh, M.G., Masuoka, P.M. & Grieco, J.P. 2010. Ecological niche model of *Phlebotomus alexandri* and *P. papatasi* (Diptera: Psychodidae) in the Middle East. *Int. J. Health Geogr.* 9: 2.

Correia, V.R., Monteiro, A.M., Carvalho, M.S. & Werneck, G.L. 2007. A remote sensing application to investigate urban endemics. [in Portuguese] *Cad Saude Publica.* 23(5): 1015–1028.

Crans, W.J., Cassamise, D.F., McNelly, J.R. 1994. Eastern equine encephalomyelitis in relation to the avian community of a coastal cedar swamp. *J. Med. Entomol.* 31: 711–728.

Craven, R.B., Maupin, G.O., Beard, M.L., Quan, T.J. & Barnes, A.M. 1993. Reported cases of human plague infections in the United States, 1970-1991. *J. Med. Entomol.* 30: 758–761.

Cross, E.R., Newcomb, W.W. & Tucker, C.J. 1996. Use of weather data and remote sensing to predict the geographic and seasonal distribution of *Phlebotomus papatasi* in Southwest Asia. *Am. J. Trop. Med. Hyg.* 54(5): 530–536.

CSA. 2009. RADARSAT-1. Available from: *http://www.asc-csa.gc.ca/eng/satellites/radarsat1 /default.asp.* [Accessed 12th March 2012].

Curtis, C.F. & Lines. J.D. 2000. Should DDT be banned by international treaty? *Parasit. Today* 16(3): 119–121.

Daszak, P., Cunningham, A.A. & Hyatt, A.D. 2000. Emerging infectious diseases of wildlife: Threats to biodiversity and human health. *Science* 287: 443–449.

Daubney, R., Hudson, J.R. & Garnham, P.C. 1931. Enzootic hepatitis or Rift Valley Fever: An undescribed virus disease of sheep, cattle and man from East Africa. *J. Path. & Bacteriol.* 34: 545–579.

Davies, F.G., Linthicum, K.J. & James, A.D. 1985. Rainfall and epizootic Rift Valley Fever. *Bull. WHO* 63: 941–943.

Davies, C.R., Llanos-Cuentas, A., Canales, J., Leon, E., Alvarez, E., Monge, J., Tolentino, E., Gomero, Q., Pyke, S. & Dye, C. 1994. The fall and rise of Andean cutaneous leishmaniasis: Transient impact of the DDT campaign in Peru. *Trans Roy. Soc. Trop. Med. & Hyg.* 88: 389–393.

Day, J.F. 2001. Predicting St. Louis Encephalitis virus epidemics: Lessons from recent, and not so recent, outbreaks. *Annu. Rev. Entomol.* 46: 111–138.

DCPP 2009. Avalable from: http://www.dcp2.org/pubs/DCP/23/ Section/3154 [Accessed 6th February 2012].

Debeir, O., Latinne, P. & Van Den Steen, I. 2001. Remote sensing classification of spectral, spatial, and contextual data using multiple classifier systems. *Image Anal. Stereol.* 20(Suppl1): 584–89.

DeGaetano, A.T. 2005. Meteorological effects on adult mosquito (*Culex*) populations in metropolitan New Jersey. *J. Biometeor.* 49: 345–353.

Dennis, D. & Meier, F. 1997. Plague. In C.R. Horsburgh & A.M. Nelson (eds.), *Pathology of emerging infections:* 21–47. Washington DC: ASM Press.

DigitalGlobe Inc. 2011. QuickBird and WorldView. Available from: *http://www.digitalglobe.com* [Accessed 12th March 2012].

Digoutte, J.P. & Peters, L.J. 1989. General aspects of the 1987 Rift Valley fever epidemic in Mauritania. *Res. Virol.* 140: 27–30.

Dowell, S.F. 2001. Seasonal variation in host susceptibility and cycles of certain infectious diseases. *Emerg. Infect. Dis.* 7(3): 369–374.

Duda, R.O., Hart, P.E. & Stork, D.G. 2000. *Pattern classification.* New York: Wiley Interscience.

ECDC. 2009. Climate change vector-borne diseases. Report of the European Centre for Disease Prevention and Control . Available from: http://ecdc.europa.eu/en/healthtopics/Pages/Climate Change_Vector Borne_Diseases.aspx [Accessed 6th February 2012].

Edman, J.D., Scott, T.W., Costero, A., Morrison, A.C., Harrington, L.C. & Clark, G.G. 1998. *Aedes aegypti* (Diptera: Culicidae) movement influenced by availability of oviposition sites. *J. Med. Entomol.* 35: 578–583.

Ehrenberg, J.P. & Ault, S.K. 2005. Neglected diseases of neglected populations: Thinking to reshape the determinants of health in Latin America and the Caribbean. *BMC Public Health* 5: 119.

Eisen, R.J., Enscore, R.E., Biggerstaff, B.J., Reynolds, P.J., Ettestad, P., Brown, T., Pape, J., Tanda, D., Levy, C.E., Engelthaler, D.M., Cheek, J., Bueno, R., Targhetta, J., Montenieri, J.A. & Gage, K.L. 2007a. Human plague in the southwestern United States, 1957–2004: Spatial models of elevated risk of human exposure to Yersinia pestis. *J. Med. Entomol.* 44(3): 530–537.

Eisen, R.J., Reynolds, .P.J, Ettestad, P., Brown, T., Enscore, R.E., Biggerstaff, B.J., Cheek, J., Bueno, R., Targhetta, J, Montenieri, J.A. & Gage, K.L. 2007b. Residence-linked human plague in New Mexico: A habitat-suitability model. *Am. J. Trop. Med. & Hyg.* 77: 121–125.

Eldridge, B.F., Glaser, C., Pedrin, R.E. & Chiles, R.E. 2001. The first reported case of California encephalitis in more than 50 years. *Emerg. Infect. Dis.* 7: 451–452.

Elnaiem, D.E., Schorscher, J., Bendall, A., Obsomer, V., Osman, M.E., Mekkawi, A.M., Connor, S.J., Ashford, R.W. & Thomson, M.C. 2003. Risk mapping of visceral leishmaniasis: the role of local variation in rainfall and altitude on the presence and incidence of kala-azar in eastern Sudan. *Amer. J. Trop. Med. & Hyg.* 68(1): 10–17.

Enscore, R.E., Biggerstaff, B.J., Brown, T.L., Fulgham, R.F., Reynolds, P.J., Engelthaler, D.M., Levy, C.E., Parmenter, R.R., Montenieri, J.A., Cheek, J.E., Grinnell, R.K., Ettestad, P.J. & Gage, K.L. 2002. Modeling relationships between climate and the frequency of human plague cases in the southwestern United States, 1960–1997. *Am. J. Trop. Med. & Hyg.* 66: 186–196.

Erlanger, T.E., Weiss, S., Keiser, J., Utzinger, J. & Wiedenmayer, K. 2009. Past, present, and future of Japanese encephalitis. *Emerg. Infect. Dis.* 15: 1–7.

Feighner, B.H., Pak, S.I., Novakoski, W.L. & Kelsey, L.L. 1998. Re-emergence of plasmodium vivax malaria in the Republic of Korea. *Emer. Infect. Dis.* 4(2): 295–297.

Flood Control District of Maricopa 2009. Weather information. Available from: http://www.fcd.maricopa.gov/ Rainfall/ *Weather/weather.aspx* [Accessed 12th March 2012].

Focks, D., Haile, D. Daniels, E. & Mount, G. 1993. Dynamic life table model for *Aedes aegypti* (Diptera: Culicidae): Analysis of the literature and model development. *J. Med. Entomol.* 30: 1003–1017.

Fuzibet, J.G., Marty, P., Taillan, B., Bertrand, F., Pras, P., Pesce, A., LeFichoux, Y. & Dujarin, P. 1988. Is *Leishmania infantum* an opportunistic parasite in patients with anti-human immunodeficiency virus antibodies? *Arch. Internal Med.* 148(5): 1228.

Gage, K.L. & Kosoy, M.Y. 2005. Natural history of plague: Perspectives from more than a century of research. *Annu. Rev. Entomol.* 50: 505–528.

Gage, K.L. 1998. Plague. In L. Colier, A. Balows & M. Sussman (eds.), *Topley and Wilson's microbiology and microbial infections:* 885–904. Oxford: Oxford UP.

Gebre-Michael, T., Balkew, M., Ali, A., Ludovisi, A. & Gramiccia, M. 2004. The isolation of *Leishmania tropica* and *aethiopica* from *Phlebotomus* (*Paraphlebotomus)* species (Diptera: Psychodidae) in the Awash Valley, northeastern Ethiopia. *Trans. Roy. Soc. Trop. Med. Hyg.* 98: 64–70.

Geery, P.R. & R.E. Holub. 1989. Seasonal abundance and control of *Culex spp.* in catch basins in Illinois. *J. Amer. Mosq. Cont. Assoc.* 5: 537–540.

GeoEye 2011. Ikonos products and specifications. Available from: *http://www.geoeye.com* [Accessed 12th March 2012].

Gerade, B.B., Lee, S.H., Scott, T.W., Edman, J.D., Harrington, L.C., Kitthawee, S., Jones, J.W. & Clark, J.M. 2004. Field validation of *Aedes aegypti* (Diptera: Culicidae) age estimation by analysis of cuticular hydrocarbons. *J. Med. Entomol.* 41: 231–238.

Grieco, J.P., Achee, N.L., Roberts, D.R. & Andre, R.G. 2005. Comparative susceptibility of three species of *Anopheles* from Belize, Central America, to *Plasmodium falciparum* (NF-54). *J. Am. Mosq. Cont. Assoc.* 21: 279–90.

Gubler, D. 1997. Dengue and dengue haemorrhagic fever: A global public health problem in the 21st Century. In W.M. Scheld, D. Armstrong & J.M. Hughes (eds.), *Emerging infections:* 1–14. Washington DC: ASM Press.

Gubler, D.J. 2002. Epidemic dengue/dengue haemorrhagic fever as a public health, social and economic problem in the 21st century. *Trends in Microbiol.* 10: 100-103.

Gunther, J., Martinez-Munoz, J.P., Perez-Ishiwara, D.G. & Salas-Benito, J. 2007. Evidence of vertical trans-mission of dengue virus in two endemic localities in the state of Oaxaca, Mexico. *Intervirology* 50: 347–52.

Guzman, M. & Kouri, G. 2003. Dengue and dengue haemorrhagic fever in the Americas: Lessons and challenges. *J. Clinic. Virol.* 27: 1–13.

Hakre, S., Masuoka, P., Vanzie, E. & Roberts, D.R. 2004. Spatial correlations of mapped malaria rates with environmental factors in Belize, Central America. *Int. J. Health Geograp.* 3: 6 doi:10.1186/1476-072X-3-6.

Hales, S., Wet, N.D., Maindonald, J. & Woodward, A. 2002. Potential effect of population and climate changes on global distribution of dengue fever: An empirical model. *Lancet* 360: 830-834.

Hardy, J.L. 1987. The ecology of western equine encephalomyelitis virus in the Central Valley of California, 1945–1985. *Amer. J. Trop. Med. & Hyg.* 37: 18–32.

Harrington, L.C., Edman, J.D. & Scott, T.W. 2001. Why do female *Aedes aegypti* (Diptera: Culicidae) feed preferentially and frequently on human blood? *J. Med. Entomol.* 38: 411–422.

Harrington, L.C., Scott, T.W., Lerdthusnee, K., Coleman, R.C., Costero, A., Clark, G.G., Jones, J.J., Kitthawee, S., Kittayapong, P., Sithiprasasna, R. & Edman, J.D. 2005. Dispersal of the dengue vector *Aedes aegypti* within and between rural communities. *Amer. J. Trop. Med. & Hyg.* 72: 290-220.

Harrington, L., Ponlawat, A., Edman, J. & Scott, T. 2008a. Physical container traits influence oviposition behavior of the *Aedes aegypti* mosquito in Thailand. *Vect-borne. & Zoon. Dis.* 8: 415–423.

Harrington, L., Vermeylen, F., Jones, J., Kitthawee, S., Sithiprasasna, R., Edman, J. & Scott, T. 2008b. Age-dependent survival of the dengue vector,*Ae. aegypti,* demonstrated by simultaneous release and recapture of different age cohorts. *J. Med. Entomol* 7: 307–313.

Hay, S. 2000. An overview of remote sensing and geodesy for epidemiology and public health application. *Advan Parasit.* 47: 1–35.

Hay, S., Tucker, C., Rogers, D. & Packer. M. 1996. Remotely sensed surrogates of meteorological data for the study of the distribution and abundance of arthropod vectors of disease. *Ann. Trop. Med. & Parasit.* 90: 1–19.

Hayes, C.G., Baqar, S., Ahmed, A., Chowdhry, M.A. & Reisen, W.K. 1982. West Nile virus in Pakistan. 1. Sero-epidemiological studies in Punjab Province. *Trans. Roy. Soc. Trop. Med. Hyg.* 76: 431–436.

Herbreteau, V., Salem, G., Souris, M., Hugot, J. & Gonzalez, J. 2007. Thirty years of use and improvement of remote sensing applied to epidemiology: From early promises to lasting frustration. *Health Place* 13.

Herwaldt. 1999. Available from: http://www.ncbi.nlm.nih.gov/pubmed/10513726 [Accessed 13th March 2012].

Hielkema, J.U., Roffey, J. & Tucker, C.J. 1986. Assessment of ecological conditions associated with the 1980/81 desert locust plague upsurge in West Africa using environmental satellite data. *Int. J. Rem. Sens.* 7: 1609–1622.

Hjalmars, U., Kulldorff, M., Gustafsson, G. & Nagarwalla, N. 1996. Childhood leukaemia in Sweden: using GIS and a spatial scan statistic for cluster detection. *Stat. Med.* 15: 707–715.

Hogsette, J.A., Hanafi, H.A., Bernier, U.R., Kline, D.L., Fawaz, E.Y., Furman, B.D. & Hoel, D.F. 2008. Discovery of diurnal resting sites of phlebotomine sand flies in a village in southern Egypt. *J. Amer. Mosq. Cont. Assoc.* 24(4): 601–603.

Hong Kong Department of Health. 2009. Flu express. Available from: http://www.chp.gov.hk/en/guideline1_year/29/134/ 304.html [Accessed 12th March 2012].

Hong Kong Observatory. 2009. Extract of meteorological observations for Hong Kong. Available from: *http://www.hko.gov.hk/wxinfo/pastwx/extract.htm* [Accessed 12th March 2012.

Huang, N.E., Shen, Z., Long, S.R., Wu, M.C., Shih, E.H., Zheng, Q., Tung, C.C. & Liu, H.H. 1998. The empirical mode decomposition method and the Hilbert spectrum for non-stationary time series analysis, *Proc. Roy. Soc. Lond.* 454A: 903–995.

Hunter, P.R. 2003. Climate change and waterborne and vector-borne diseases. *J. Appl. Microbiol.* 94: 37S-46S.

Isturiz, R., Gubler, D. & del Castillo, J. 2000. Emerging and re-emerging diseases in Latin America: Dengue and dengue hemorrhagic in Latin America and the Caribbean *Infect. Dis. Clin. North Amer.* 14: 121–140.

Jacob, B.G., Burkett-Cadena, N.D., Luvall, J.C., Parcak, S.H., McClure, C.J., Estep, L.K., Hill, G.E., Cupp, E.W., Novak, R.J. & Unnasch, T.R. 2010. Developing GIS-based eastern equine encephalitis vector-host models in Tuskegee, Alabama. *Int. J. Health Geogr.* 9: 12

JAXA. 2011. PALSAR. Available from: *http://www.eorc.jaxa.jp/ALOS/en/about/palsar.htm* [Accessed 12th March 2012].

Johansson, M. & Glass, G. 2008. High-resolution spatiotemporal weather models for climate studies. *Int. J. Health Geogr.* 7: 52.

Johansson, M., Cummings, D. & Glass, G. 2009a. Multiyear climate variability and Dengue El Nino Southern Oscillation, weather, and Dengue incidence in Puerto Rico, Mexico, and Thailand: A longitudinal data analysis. *PloS Med.* 6: e1000168. doi:100110.1001371/journal.pmed.1000168.

Johansson, M., Dominici, F. & Glass, G. 2009b. Local and global effects of climate on Dengue transmission in Puerto Rico. *PLoS Negl. Trop. Dis.* 3: doi:10.1371/journal.pntd.0000382.

Kalluri, S., Gilruth, P., Rogers, D. & Szczur. M. 2007. Surveillance of arthropod vector-borne infectious diseases using remote sensing techniques: A review. *PLoS Path.* 3: 1361–1371.

Kay, B.H., Fanning, I.D. & Carley, J.G. 1984. The vector competence of Australian *Culex annulirostris* with Murray Valley encephalitis and Kunjin viruses. *Aust. J. Exp. Biol. Med. Sci.* 62(5): 641–650.

Keirans, J. & Fay, R. 1968. Effect of food and temperature on *Aedes aegypti* (L.) and *Aedes triseriatus* (Say) larval development. *Mosq. News* 28: 338–341.

Kiang, R.K., Hulina, S.M., Masuoka, P.M. & Claborn, D.M. 2003. Identification of mosquito larval habitats in high resolution satellite data. *Proc. SPIE 5093*, 353; doi:10.1117/12.487016.

Kiang, R., Adimi, F., Soika, V., Nigro, J., Singhasivanon, P., Sirichaisinthop, J., Leemingsawat, S., Apiwathnasorn, C. & Looareesuwan, S. 2006. Meteorological, environmental remote sensing and neural network analysis of the epidemiology of malaria transmission in Thailand. *Geospat. Health* 1: 71–84.

Killick-Kendrick, R., Leaney, A.J. & Ready, P.D. 1977. The establishment, maintenance and productivity of a laboratory colony of *Lutzomyia longipalpis* (Diptera: Psychodidae). *J. Med. Entomol.* 13(4–5): 429–440.

Killick-Kendrick, R., Rioux, J.A., Bailly, M., Guy, M.W., Wilkes, T.J., Guy, F.M., Davidson, I., Knechtli, R., Ward, R.D. & Guilvard, E. 1984. Ecology of leishmaniasis in the south of France. 20. Dispersal of *Phlebotomus ariasi* Tonnoir, 1921 as a factor in the spread of visceral leishmaniasis in the Cévennes. *Ann. Parasitol. Hum. Comp.* 59(6): 555–572.

Killick-Kendrick, R., Wilkes, T.J., Bailly, M. & Righton, L.A. 1986. Preliminary field observations on the flight speed of a phlebotomine sandfly. *Trans. Roy. Soc. Trop. Med. & Hyg.* 80(1): 138–142.

Killick-Kendrick, R. & Killick-Kendrick, M. 1987. Honeydew of aphids as a source of sugar for *Phlebotomus ariasi*. *Med. & Vet. Entomol.* 1(3): 297–302.

Kilpatrick, A.M., LaDeau, S.L & Marra, P.P. 2007. Ecology of West Nile virus transmission and its impact on birds in the western hemisphere. *Auk* 124: 1121–1136.

Kitron, U., Swanson, J., Crandell, M., Sullivan, P.J., Anderson, J., Garro, R., Haramis, L.D. & Grimstad, P.R. 1998. Introduction of *Aedes albopictus* into a LaCrosse Virus-enzootic site in Illinois. *Emerg. Infect. Dis.* 4: 627–630.

Klein, T.A., Tada, M.S. & Lima, L.B.P. 1991. Comparative susceptibility of anopheline mosquitoes to *Plasmodium falciparum* in Rodonia, Brazil. *Amer. J. Trop. Med. & Hyg.* 44: 598–603.

Komar, N. 2003. West Nile virus: Epidemiology and ecology in North America. *Adv. Virus Res*. 61: 185–234.

Kramer, L.D., Styer, L.M. & Ebel, G.D. 2008. A gobal perspective on the epidemiology of West Nile virus. *Ann. Rev. Entomol.* 53: 61–81.

Kulldorff, M. 1997. A spatial scan statistic. *Communica. Stat. Theor. Methods* 27: 1481–1496.

Kummerow, C., Barnes, W., Kozu, T., Shiue, J. & Simpson, J. 1998. The tropical rainfall measuring mission (TRMM) sensor package. *J. Atmos. & Oceanic Tech*. 15: 809–817.

Kuno, G. 1997. *Dengue and dengue hemorrhagic fever*. New York: CAB International.

Lcc, II.I, Lee, J.S., Shin, E.H., Lee, W.J., Kim, Y.Y. & Lee, K.R. 2001. Malaria transmission potential by *Anopheles sinensis* in the Republic of Korea. *Korean J. Parasit*. 39(2): 1–10.

Levy, C.E. & Gage, K.L. 1999. Plague in the United States, 1995–1997. *Infect. Med.* 16: 54–64.

Liao, C.-M., Chan, S.-Y., Chen, S.-C. & Chio, C.-P. 2009. Influenza-associated morbidity in subtropical Taiwan. *Int. J. Infect. Dis.* 13: 589–599.

Linthicum, K.J., Davies, F.G., Bailey, C.L. & Kairo, A. 1984. Mosquito species encountered in a flooded grassland dambo in Kenya. *Mosq. News* 44: 228–232.

Linthicum, K.J., Davies, F.G., Kairo, A. & Bailey, C.L. 1985a. Rift Valley fever virus (family Bunyaviridae, genus *Phlebovirus*). Isolations from diptera collected during an interepizootic period in Kenya. *J. Hyg* 95:197–209.

Linthicum, K.J., Kaburia, H.F., Davies, F.G. & Lindqvist, K.J. 1985b. A blood meal analysis of engorged mosquitoes found in Rift Valley fever epizootics area in Kenya. *J. Am. Mosq. Cont. Assoc*. 1: 93–95.

Linthicum, K.J., Bailey, C.L., Davies, F.G. & Tucker, C.J. 1987. Detection of Rift Valley fever viral activity in Kenya by satellite remote sensing imagery. *Science* 235: 1656–1659.

Linthicum, K.J., Bailey, C.L., Tucker, C.J., Mitchell, K., Gordon, S.W., Logan, T.M., Peters, C.J. & Digoutte, J.P. 1989. Polar orbiting satellite monitoring of man-made alterations in the ecology of the Senegal River basin, as they relate to a Rift Valley fever epidemic, in arbovirus research in Australia. Proc. 5th Symp. *CSIRO, Div. Anim. Health & Queensland Inst. Med. Res*: 116–118.

Linthicum, K.J., Bailey, C.L., Tucker, C.J., Mitchell, K.D., Logan, T.M., Davies, F.G., Kamau, C.W., Thande, P.C. & Wagateh, J.N. 1990. Application of polar-orbiting, meteorological satellite data to detect flooding of Rift Valley fever virus vector mosquito habitats in Kenya. *Med.& Vet. Entomol.* 4: 433–438.

Linthicum, K., Bailey, C., Tucker, C., Angleberger, D. & Cannon, T. 1991. Towards real-time prediction of Rift-Valley fever epidemics in Africa. *Prevent. Vet. Med.* 11: 325–334.

Linthicum, K.J., Bailey, C.L., Tucker, C.J., Gordon, S.W., Logan, T.M., Peters, C.J. & Digoutte, J.P. 1994. Observations with NOAA and SPOT satellites on the effect of man-made alterations in the ecology of the Senegal River basin in Mauritania on Rift Valley fever virus transmission. *Sistema Terra* 3: 44–47.

Linthicum, K.J., Anyamba, A., Tucker, C.J., Kelley, P.W., Myers, M.F. & Peters, C.J. 1999. Climate and satellite indicators to forecast Rift Valley fever epidemics in Kenya. *Science* 285: 397–400.

Linthicum, K.J., Anyamba, A., Britch, S., Chretien, J.-P., Erickson, R.L., Small, J., Tucker, C.J., Bennett, K.E., Mayer, R.T., Schmidtmann, E.T., Andreadis, T.G., Anderson, J.F., Wilson, W.C., Freier, J., James, A., Miller, R., Drolet, B.S., Miller, S., Tedrow, C., Bailey, C., Strickman, D.A., Barnard, D.R., Clark, G.G. & Zou, L. 2007. A Rift Valley fever risk surveillance system for Africa using remotely sensed data: Potential for use on other continents. *Vet. Italiana* 43: 663–674.

Liu, H., Weng, Q. & Gaines, D. 2008. Spatio-temporal analysis of the relationship between WNV dissemination and environmental variables in Indianapolis, USA. *Int. J. Health Geogr*. 7: 66.

Liu, H. & Weng, Q. 2009. An examination of the effect of landscape pattern, land surface temperature, and socioeconomic conditions on WNV dissemination in Chicago. *Environ. Monit. Assess.* 159: 143–161.

Lofgren, E., Fefferman, N.H., Naumova, Y.N., Gorski, J. & Naumova, E.N. 2007. Influenza seasonality: Underlying causes and modeling theories. *J. Virol*. 81(11): 5429–5436.

Logan, T.M., Linthicum, K.J. & Ksiazek, T.G. 1992. Isolation of Rift Valley fever virus from mosquitoes collected during an outbreak in domestic animals in Kenya. *J. Med. Entomol.* 28: 293–295.

Lourenco-de-Oliveira, R., Guimaraes, A.E., Arle, M., da Silva, T.F., Castro, M.G., Motta, M.A. & Deane, L.M. 1989. Anopheline species, some of their habitats and relation to malaria in endemic areas of Rondonia State, Amazon region of Brazil. *Rev. Bras. Biol.* 49: 393–397.

Lowen, A.C., Mubareka, S., Steel, J. & Palese, P. 2007. Influenza virus transmission is dependent on relative humidity. *PLoS Path*. 3(10): e151.

Lowen, A.C., Steel, J., Mubareka, S. & Palese, P. 2008. High Temperature (30°C) blocks aerosol but not contact transmission of influenza virus. *J. Virol*. 82(11): 5650-5652.

Loyola, E.G., Arredondo, J.I., Rodriguez, M.H., Brown, D.N. & Vaca-Marin, M.A. 1991. *Anopheles vesitipennis*, the probable vector of *Plasmodium vivax* in the Lacandon forest of Chiapas, Mexico. *Trans. Roy. Soc. Trop. Med. Hyg*. 85: 171–174.

Lumsden, L.L. 1958. St. Louis encephalitis in 1933. Observations on epidemiological features. *Publ. Health Rep*. 73: 340-353.

Machado-Allison, C. & Craig. G. 1972. Geographic variation in resistance to desiccation in *Aedes aegypti* and *A. atropalpus* (Diptera: Culicidae). *Ann. Entomol. Soc. Amer*. 65: 542–547.

Mackenzie, J.S., Barrett, A.D. & Deubel, V. 2002. The Japanese encephalitis serological group of flaviviruses: A brief introduction to the group. *Curr. Top. Microbiol. Immunol*. 267: 1–10.

Madani, T.A., Al-Mazrou, Y.Y., Al-Jeffri, M.H., Mishkhas, A.A., Al-Rabeah, A.M., Turkistani, A.M., Al-Sayed, M.O., Abodahish, A.A., Khan, A.S., Ksiazek, T.G. & Shobokshi, O. 2003. Rift Valley fever epidemic in Saudi Arabia: Epidemiological, clinical, and laboratory characteristics. *Clinic. Infect. Dis*. 37: 1084–1092.

Maier, W.A. 2003. Possible effect of climate change on the distribution of arthropode vector-borne infectious diseases and human parasites in Germany. *Umweltbundesamt* 1–386.

Malone, J.B., Nieto, P. & Tadesse, A. 2006. Biology-based mapping of vector-borne parasites by geographic information systems and remote sensing. *Parassitologia*. 48(1–2): 77–79.

Maltezou, H.C. 2010. Drug resistance in visceral leishmaniasis. *J. Biomed. & Biotech*. 617521. Epub Nov 1 2009.

Manguin, S., Roberts, D.R., Andre, R.G., Rejmankova, E. & Hakre, S. 1996. Characterization of *Anopheles darlingi* (Diptera: Culicidae) larval habitats in Belize, Central America. *J. Med. Ent*. 33: 205–211.

Maricopa Department of Public Health, Maricopa County Weekly Influenza Summary. 2009. Available from: http://www.maricopa.gov/Public_Health/epi/influenza.aspx [Accessed 5th February 2012].

Martens, W., Jetten, T. & Focks, D. 1997. Sensitivity of malaria, schistosomiasis and dengue to global warming *Climate Change* 35: 145–156.

McMichael, A.J., Campbell-Lendrum, D.H., Corvalán, C.F., Ebi, K.L., Githeko, A., Scheraga, J.D. & Woodward, A. (eds.). 2003. *Climate change and human health: Risks and responses*. Geneva: WHO.

Meegan, J.M. & Bailey, C.L. 1989. Rift Valley fever. In T.P Monath (ed.), *The Arboviruses: Epidemiology and Ecology:* 51–76. Boca Raton: CRC Press.

Meegan, J.M., Moussa, M.I., el-Mour, A.F., Toppouzzada, R.H. & Wyess, R.N. 1978. Ecological and epidemiological studies of Rift Valley fever in Egypt. *J. Egypt Publ. Health Assoc*. 53: 173–175.

Min, J.G. & Xue, M. 1996. Progress in studies on the overwintering of the mosquito *Culex tritaeniorhynchus*. *SE Asian J. Trop. Med. & Publ. Health* 27: 810-817.

Miranda, C., Massa, J.L. & Marques, C.C. 1996. Occurrence of American cutaneous leishmaniasis by remote sensing satellite imagery in an urban area of Southeastern Brazil. [in Portuguese] *Rev. Saude Publica*. 30(5): 433–437.

Molinari, N.A., Ortega-Sanchez, I.R., Messonnier, M.L., Thompson, W.W., Wortley, P.M., Weintraub, E. & Bridges, C.B.2007. The annual impact of seasonal influenza in the US: Measuring disease burden and costs. *Vaccine* 25(27): 5086–5096.

Monath, T.P. 1980. Epidemiology. In T.P. Monath (ed.), *St. Louis:* 239–312. Washington DC: Amer. Pub. Health Assoc.

Moncayo, A.C., Edman, J.D. & Finn, J.T. 2000. Application of geographic information technology in determining risk of eastern equine encephalomyelitis virus transmission. *J. Amer. Mosq. Cont. Assoc*. 16: 28–35.

Morris, C.D. 1988. Eastern equine encephalomyelitis. In T.P. Monath (ed.), *The arboviruses: Epidemiology and ecology:* 1–20. Boca Raton: CRC Press.

Morrison, A., Gray, K., Getis, A., Astete, H., Sihuincha, M., Focks, D., Watts, D., Stancil, J., Olson, J., Blair, P. & Scott, T. 2004. Temporal and geographic patterns of *Aedes aegypti* (Diptera: Culicidae) production in Iquitos, Peru. *J. Med. Entomol*. 41: 1123–1142.

Morvan, J., Rollin, P.E., Laventure, S., Rakotoarivony, I. & Roux, J. 1992. Rift Valley fever epizootic in the central highlands of Madagascar. *Res. Virol*. 143: 407–415.

Mouchet, J. & Manguin, S. 1998. Evolution of malaria in Africa for the past 40 years: Impact of climatic and human factors. *J. Amer. Mosq. Cont. Assoc*. 14(2): 121–130.

Moura, F.E., Perdigao, A.C. & Siqueira, M.M. 2009. Seasonality of influenza in the tropics: A distinct pattern in northeastern Brazil. *Amer. J. Trop. Med. & Hyg.* 81(1): 180–3.

NASA 2011. ICESat. Available from: *http://icesat.gsfc.nasa.gov* [Accessed 12th March 2012].

Naucke, T.J. & Schmitt, C. 2004. Is leishmaniasis becoming endemic in Germany? *Int. J. Med. Microbiol.* 293(Supp37): 179–181.

NCDC. 2009. Climate data online: Global summary of the day. Available from: *http://www7.ncdc.noaa.gov/ CDO/cdoselect.cmd?datasetabbv=GSOD&countryabbv=& georegionabbv=*L Accessed 5th February 2012].

Neto, J.C., Werneck, G.L. & Costa, C.H.N. 2009. Factors associated with the incidence of urban visceral leishmaniasis: An ecological study in Teresina, Piauí State, Brazil. *Cadernos de Saúde Públ.* 25(7): 1543–1551.

New York State Department of Health. 2009. Influenza activity, surveillance and reports. Available from: http://www.health.state.ny.us/diseases/communicable/influenza/surveillance/ [Accessed 5th February 2012.

Nicholls, N. 1986. A method for predicting Murray Valley encephalitis in southeast Austrlaia using the Southern Oscillation. *Aust. J. Exp. Biol. Med. Sci.* 64: 587–594.

Nicholson, S.E. & Entekhabi, D. 1986. The quasi-periodic behavior of rainfall variability in Africa and its relationship to the Southern Oscillation, *Arch. Meteorol. Geophysik & Bioklimat.* 34: 311–348.

Nieto, P., Malone, J.B. & Bavia, M.E. 2006. Ecological niche modeling for visceral leishmaniasis in the state of Bahia, Brazil, using genetic algorithm for rule-set prediction and growing degree day-water budget analysis. *Geospat. Health* 1(1): 115–126.

NIH. 2008. Why the influenza virus is more infectious in cold winter temperatures. Available from: http://www.sciencedaily.com/releases/2008/03/080330203401.htm [Accessed 5th February 2012].

Padilla N., Molina P., Juarez J., Brown D. & Cordon-Rosales C. 1992. Potential malaria vectors in northern Guatemala. *J. Am. Mosq. Cont. Assoc.* 7: 456–461.

PAHO 1994. Status of malaria programs in the Americas. XLII Report. Washington DC: PAHO.

PAHO 2007. Status of malaria in the Americas, 1994–2007: A series of data tables. Available from: http://www.paho.org/english/ad/dpc/cd/mal-americas-2007.pdf [Accessed 1st February 2012].

Pant, C. & Yasuno, M. 1973. Field studies on the gonotrophic cycle of *Aedes aegypti* in Bangkok, Thailand. *J. Med. Entomol.* 10: 219–223.

Pao, Y.-H. 1989. *Adaptive pattern recognition and neural networks.* Sydney: Addison-Wesley.

Parmenter, R.R., Yadav, E.P., Parmenter, C.A., Ettestad, P.J. & Gage, K.L. 1999. Incidence of plague associated with increased winter-spring precipitation in New Mexico. *Am. J. Trop. Med. Hyg.* 61: 814–821.

Patz, J. & Olson, S. 1996. Climate change and health: global to local influences on disease risk. *Ann. Trop. Med. & Parasit.* 100: 535–549.

Perry, R.D. & Fetherston, J.D. 1997. *Yersina pestis:* etiologic agent of plague. *Clin. Microbiol Rev.* 10(1): 35–66.

Peters, C.J. & Meegan, J.M. 1981. Rift Valley fever. In G.B. Beran (ed.), *Handbook Series in Zoonoses. Section B: Viral Zoonoses:* 403–420. Boca Raton: CRC Press.

Peters, C.J. & Linthicum, K.J. 1994. Rift Valley fever, In G.B. Beran (ed.), *Handbook of Zoonoses:* 125–138. Boca Raton: CRC Press.

Pin-Diop, R, Toure, I, Lancelot, R, Ndiaye, M, Chavernac, D. 2007. Remote sensing and geographic information systems to predict the density of ruminants, hosts of Rift Valley fever virus in the Sahel. *Vet. Ital.* 43: 675–686.

Pinheiro, F.P. & Corber, S.J. 1997. Global situation of dengue and dengue haemorrhagic fever, and its emergence in the Americas. *Wld. Hlth. Statist. Quart.* 50: 161–169.

Pinzón, J.E., Brown, M.E. & Tucker, C.J. 2005b. Empirical mode decomposition correction of orbital drift artifacts in satellite data stream., In N.E. Huang & S.S. Shen (eds.), *Hilbert-Huang Transform and Applications:* 167–186. Singapore: World Scientific.

Pinzón, J.E., Wilson, J.M. &Tucker, C.J. 2005a. Climate-based health monitoring for eco-climatic conditions associated with infectious diseases.*Bull. Société de Pathol. Exotiq.* 98(3): 239–243.

Polozov, I.V., Bezrukov, L., Gawrisch, K. & Zimmerberg, J. 2008. Progressive ordering with decreasing temperature of the phospholipids of influenza virus. *Nature Chem. Biol.* 4(4): 248–55.

Ponlawat, A. & Harrington, L.C. 2005. Blood feeding patterns of *Aedes aegypti* and *Aedes albopictus* in Thailand. *J. Med. Entomol.* 42(5): 844–849

Pope, K.O., Sheffner, E.J., Linthicum, K.J., Bailey, C.L., Logan, T.M., Kasischke, E.S., Birney, K., Njogu, A.R. & Roberts, C.R. 1992. Identification of central Kenyan Rift Valley fever virus vector habitats with Landsat TM and evaluation of their flooding status with airborne imaging radar. *Rem. Sens. Environ.* 40: 185–196.

Pope, K., Rejmankova, E., Savage, H., Arredondo-Jimenez, J., Rodriguez, M., Roberts, D. 1994. Remote sensing of tropical wetlands for malaria control in Chiapas, Mexico. *Ecol. Appl.* 4: 81–90.

Pope, K., Masuoka, P., Rejmankova, E., Grieco, J., Johnson, S. & Roberts, D. 2005. Mosquito habitats, land use, and malaria risk in Belize from satellite imagery. *Ecolog. Appl.* 15 (4): 1223–1232.

Principe, J., Euliano, N. & Lefebvre, W.C. (eds.) 2000. *Neural and adaptive systems: Fundamentals through simulations.* New York: John Wiley & Sons.

Putnam, J.L. & Scott, T.W. 1995. Blood-feeding behavior of dengue-2 virus-infected *Aedes aegypti. Amer. J. Trop. Med. & Hyg.* 52: 225–227.

Rahman, K.M., Islam, S., Rahman, M.W., Kenah, E., Galive, C.M., Zahid, M.M., Maguire, J., Rahman, M., Haque, R., Luby, S.P. & Bern, C. 2010. Increasing incidence of post-kala-azar dermal leishmaniasis in a population-based study in Bangladesh. *Clinic. Infect. Dis.* 50(1): 73–76.

Ramsey, J.M., Salinas, E., Brown, D.M. & Rodriguez, M.H. 1994. *Plasmodium vivax* sporozoite rates from *Anopheles albimanus* in southern Chiapas, Mexico *J. Parasitol.* 80: 489–493.

Rangel, E.F., Souza, N.A. & Wermelinger, E.D. 1986. Biology of *Lutzomyia intermedia* Lutz & Neiva, 1912 and *Lutzomyia longipalpis* Lutz & Neiva, 1912 (Diptera: Psychodidae) in the laboratory: I. Some aspects of feeding in larvae and adults. *Memór. Instit. Oswaldo Cruz* 81(4): 431–438.

RBM 2011. Available from: Http://www.rbm.who.int [Accessed 12th March 2012].

Ready, P.D. 2008. Leishmaniasis emergence and climate change. *Rev. Scient. et Tech. (International Office of Epizootics).* 27(2): 399–412.

Reeves, W.C. 1990. Clinical and subclinical disease in man. In: W.C. Reeves (ed.), *Epidemiology and control of mosquito-borne arboviruses in California, 1943–1987:1–25.* Sacramento: California Mosq. Vect. Cont. Assoc.

Reeves, W.C. & Hammon, W.M. 1962. Epidemiology of the arthropod-borne viral encephalitides in Kern County, California, 1943–1952. Berkeley: UC Berkeley Publ. Hlth.

Reisen, W.K. 2003. Epidemiology of St. Louis encephalitis virus. In T.J. Chambers & T.P. Monath (eds), *The Flaviviruses: Detection, diagnosis and vaccine development: 139–183.* San Diego: Elsevier Academic Press.

Reisen, W.K. 2010. Landscape epidemiology of vector-borne diseases. *Ann. Rev. Entomol.* 55: 461–483.

Reisen, W.K. & Monath, T.P. 1989. Western equine encephalomyelitis. In T.P. Monath (ed.), *The arboviruses: epidemiology and ecology:* 89–138. Boca Raton: CRC Press.

Reisen, W.K. & Chiles, R.E. 1997. Prevalence of antibodies to western equine encephalomyelitis and St. Louis encephalitis viruses in residents of California exposed to sporadic and consistent enzootic transmission. *Amer. J. Trop. Med. & Hyg.* 57: 526–529.

Reisen, W.K., Fang, Y, & Martinez, V.M. 2006. Effects of temperature on the transmission of West Nile virus by *Culex tarsalis* (Diptera: Culicidae). *J Med. Entomol.* 43: 309–317.

Reisen, W.K. & Brault, A.C. 2007. West Nile virus in North America: Perspectives on epidemiology and intervention. *Pest Manag. Sci.* 63: 641–646.

Reisen, W.K., Cayan, D., Tyree, M., Barker, C.M., Eldridge, B.F. & Dettinger, M. 2008a. Impact of climate variation on mosquito abundance in California. *J. Soc. Vector Ecol.* 33: 89–98.

Reisen, W.K., Fang, Y. & Brault, A.C. 2008b. Limited interdecadal variation in mosquito (Diptera: Culicidae) and avian host competence for Western equine encephalomyelitis virus (Togaviridae: Alphavirus). *Amer. J. Trop. Med. & Hyg.* 78: 681–686.

Rejmankova, E., Savage, H., Rodriguez, M. & Roberts, D. 1992. Aquatic vegetation as a basis for classification of *Anopheles albimanus* Wiedemann (Diptera: Culicidae) larval habitats. *Environ. Entomol.* 21: 598–603.

Rejmankova, E., Roberts, D., Harbach, R., Pecor, J., Peyton, E., Manguin, S., Krieg, R., Polanco, J. & Legters, L. 1993. Environmental and regional determinants of *Anopheles* larval distribution in northern Belize. *Environ. Entomol.* 22: 978–992.

Rejmankova, E., Roberts, D., Pawley, A., Manguin, S. & Polanco, J. 1995 Predictions of adult *Anopheles albimanus* in villages based on distance from remotely sensed larval habitats. *Amer. J. Trop. Med. & Hyg.* 53: 482–488.

Rejmankova, E., Pope, K., Post, R. & Maltby, E. 1996. Herbaceous wetlands of the Yucatan Peninsula: Communities at extreme ends of environmental gradients. *Inter. Rev. Ges. Hydrobiol* 81: 223–252.

Rejmankova, E., Pope, K., Roberts, D., Lege, M., Andre, R., Greico, J. & Alonzo, Y. 1998 Characterization and detection of *Anopheles vestitipennis* and *Anopheles punctimacula* (Diptera: Culicidae) larval habitats in Belize with field survey and SPOT satellite imagery. *J. Vect. Ecol.* 23: 74–88.

Roberts, D.R., Chan, O., Pecor, J., Rejmankova, E., Manguin, S., Polanco, J. & Legters, L. 1993. Preliminary observations on the changing roles of malaria vectors in southern Belize. *J Am Mosq Cont. Assoc* 9: 456–459.

Roberts, D.R. & Rodriguez, M.H. 1994. The environment, remote sensing, and malaria control. In *Disease in Evolution, 740:* 396–402. New York: Annals of New York Academy of Sciences.

Roberts, D.R., Paris, J., Manguin, S., Harbach, R., Woodruff, R., Rejmankova, E., Polanco, J., Wullschleger, W. & Legters, L. 1996. Predictions of malaria vector distributions in Belize using multispectral satellite data. *Amer. J. Trop. Med. & Hyg.* 54: 304–308.

Roberts, D.R., Laughlin, L., Hsheih, P. & Legters, L. 1997. DDT, global strategies and malaria control crisis in South America. *Emerg. Infect. Dis.* 3: 1–9.

Roberts, D.R., Manguin, S., Rejmankova, E., Andre, R., Harbach, R.E., Vanzie, E., Hakre, S. & Polanco, J. 2002a. Spatial distribution of adult *Anopheles darlingi* and *Anopheles albimanus* in relation to riparian habitats in Belize, Central America. *J. Vector Ecol.* 27(1): 21–30.

Roberts, D.R., Masuoka, P. & Au, A.Y. 2002b. Determinants of Malaria in the Americas. In E.A. Casman & H. Dowlatabadi (eds.), *The contextual determinants of malaria*: 35–58. Washington DC: Resources for the Future.

Rodhain, F. & Rosen, L. 1997. Mosquito vectors and dengue virus-vector relationships. In D. Gubler & G. Kuno (eds.), *Dengue and dengue haemorrhagic fever.* Wallingford: CAB International.

Rodriguez, A.D., Rodriguez, M.H., Hernandez, J.E. Meza, R.A., Rejmankova, E., Savage, H.M., Roberts, D.R., Pope, K.O. & Legters, L. 1993. Dynamics of population densities and vegetation associations of *Anopheles albimanus* larvae in a coastal area of southern Chiapas, Mexico. *J. Amer.Mosq. Cont. Assoc.* 9: 46–58.

Rodriguez, A.D., Rodriguez, M.H., Hernandez, J.E., Meza, R.A., Rejmankova, E., Savage, H.M., Rogers, D.R., Wilson, D.A., Hay, S. & Graham, A. 2006. The global distribution of yellow fever and dengue. *Adv. Parasit.* 62: 181–120.

Rogers, D.J. & Randolph, S.E. 2003. Studying the global distribution of infectious diseases using GIS and RS. *Nature Revs. Microbiol.* 1: 231–236.

Rogers, D.R., Wilson, D.A., Hay, S. & Graham, A. 2006. The global distribution of yellow fever and dengue. *Adv. Parasit.* 62: 181–120.

Rogers, D.J. & Randolph, S.E. 2006. Climate change and vector-borne diseases. *Adv. Parasit.* 62: 345–381.

Ropelewski, C.F. & Halpert, M.S. 1987. Global and regional scale precipitation patterns associated with El Niño/Southern Oscillation. *Month. Weath. Rev.* 115: 1606–1626.

Rossi, E., Rinaldi, L., Musella, V., Veneziano, V., Carbone, S., Gradoni, L., Cringoli, G. & Maroli, M. 2007. Mapping the main Leishmania phlebotomine vector in the endemic focus of the Mt. Vesuvius in southern Italy. *Geospat. Health* 1(2): 191–198.

Rotela, C., Fouque, F., Lamfri, M., Sabatier, P., Introini, V., Zaidenberg, M. & Scavuzzo, C. 2007. Space–time analysis of the dengue spreading dynamics in the 2004 Tartagal outbreak, Northern Argentina. *Acta. Tropica.* 103: 1–13.

Roy, R.S., Ranganath, B.K., Diwakar, P.G., Vohra, T.P.S., Bhan, S.K., Singh, J.J. & Pandian, V.C. 1991. Tropical forest type mapping and monitoring using remote sensing. *Int. J. Rem. Sens.* 12: 2205–2225.

Sader, S.A., Powell, G.V.N. & Rappole, J.H. 1991. Migratory bird habitat monitoring through remote sensing. *Int. J. Rem. Sens.* 12: 363–372.

Sagripanti, J.L. & Lytle, C.D. 2007. Inactivation of influenza virus by solar radiation. *Photochem. Photobiol.* 83(5): 1278–1282.

Saji, N.H., Goswami, B.N., Vinayachandran, P.N. & Yamagata, T. 1999. A dipole mode in the tropical Indian Ocean. *Nature* 401: 360–363.

Salazar, M., Richardson, J., Sánchez-Vargas, I., Olson, K. & Beaty, B. 2007. Dengue virus type 2: replication and tropisms in orally infected *Aedes aegypti* mosquitoes. *BMC Microbiol.* 30(7): 9.

Salomón, O.D., Orellano, P.W., Lamfri, M., Scavuzzo, M., Dri, L., Farace, M.I. & Quintana, D.O. 2006. Phlebotominae spatial distribution associated with a focus of tegumentary leishmaniasis in Las Lomitas, Formosa, Argentina, 2002. *Mem Inst Oswaldo Cruz.* 101(3): 295–299.

Savage, H. M., Rejmankova, E., Arredondo-Jimenez, J.I., Roberts, D.R. & Rodriguez, M.II. 1990. Limnological and botanical characterization of larval habitats for two primary malarial vectors, *Anopheles albimanus* and Anophe*les pseudopunctipennis*, in coastal Chiapas, Mexico. *J. Am. Mosq. Cont. Assoc* 6: 612–620.

Schlein, Y. & Jacobson, R.L. 1999. Sugar meals and longevity of the sandfly *Phlebotomus papatasi* in an arid focus of *Leishmania major* in the Jordan Valley. *Med.& Vet. Entomol.* 13(1): 65–71.

Scott, T.W., Chow, E., Strickman, D., Kittayapong, P., Wirtz, R.A., Lorenz, L.H. & Edman, J.D. 1993a. Blood-feeding patterns of *Aedes aegypti* (Diptera: Culicidae) collected in a rural Thai village. *J. Med. Entomol.* 30: 922–927.

Scott, T.W., Clark, G.G., Lorenz, L.H., Amerasinghe, P.H., Reiter, P. & Edman, J.D. 1993b. Detection of multiple blood feeding in *Aedes aegypti* (Diptera: Culicidae) during a single gonotrophic cycle using a histologic technique. *J. Med. Entomol.* 30: 94–99.

Scott, T.W., Morrison, A.C., Lorenz, L.H., Clark, G.G., Strickman D., Kittayapong, P., Zhou, H. & Edman, J.D. 2000. Longitudinal studies of *Aedes aegypti* (Diptera: Culicidae) in Thailand and Puerto Rico: population dynamics. *J. Med. Entomol.* 37: 77–88.

Scott, T.W., Priyanie, H., Amerasinghe, P.H., Morrison, A.C., Lorenz, L.H., Clark, G.G. Strickman, D., Kittayapong, P. & Edman, J.D. 2000. Longitudinal Studies of *Aedes aegypti (*Diptera: Culicidae) in Thailand and Puerto Rico: Blood feeding frequency. *J. Med. Entomol.* 37: 89–101.

Seto, E.Y.W., Moore, C.G. & Hoskins, R. 2007. Geographic information systems and remote sensing for infectious disease surveillance. In N. M'ikanatha & R. Lynfield (eds.), *Infectious Disease Surveillance*: 408–418. London: Blackwell.

Shaman, J. & Kohn, M. 2009. Absolute humidity modulates influenza survival, transmission, and seasonality. *Proc Natl Acad Sci* 106(9): 3243–3248.

Shaman, J., Pitzer, J., Viboud, C., Lipsitch, M. & Grenfell, B., 2009. Absolute humidity and the seasonal onset of influenza in the Continental US. *PLoS Curr. Influenza* Dec 18: RRN1138.

Sherlock, I.A. 1996. Ecological interactions of Visceral Leishmaniasis in the State of Bahia, Brazil. *Memorias do Inst. Oswaldo Cruz* 91(6): 671–683.

Shimshony, A. 1979. RVF outbreak echoes biblical plague. *Vet. Rec.* 104: 511.

Shope, R.E. & Woodall, J.P. 1973. Ecological interaction of wildlife, man and virus of the Venezuelan equine encephalomyelitis complex in a tropical forest. *J. Wildlife Dis.* 9: 198–203.

Singh, B., Sung, L.K., Matusop, A., Radhakrishnan, A., Shamsul, S.S., Cox-Singh, J., Thomas, A. & Conway, D.J. 2004. A large focus of naturally acquired Plasmodium knowlesi infections in human beings. *Lancet* 363(9414): 1017–1024.

Soebiyanto, R.P., Adimi, F. & Kiang, R.K. 2010. Modeling and predicting seasonal influenza transmission in warm regions using climatological parameters. *PLoS ONE* 5(3): e9450.

Spiegel, J., Bonet, M., Ibarra, A., Pagliccia, N., Ouellette V. & Yassi, A. 2007. Social and environmental determinants of Aedes aegypti infestation in Central Havana: results of a case–control study nested in an integrated dengue surveillance programme in Cuba. *Trop. Med. Int. Health* 12: 503–510.

Stapp, P., Antolin, M.F. & Ball, M. 2004. Patterns of extinction in prairie dog metapopulations: Plague outbreaks follow El Niño events. *Front. Ecol.* 2: 235–240.

Stenseth, N.C., Atshabar, B.B., Begon, M., Belmain, S.R., Bertherat, E., Carniel, E., Gage, K.L., Leirs, H. & Rahalison, L. 2008. Plague: Past, present, and future. *PLoS* 5: 9–13.

Sudhakar, S., Srinivas ,T., Palit, A., Kar, S.K. & Battacharya, S.K. 2006. Mapping of risk prone areas of kala-azar (Visceral leishmaniasis) in parts of Bihar State, India: An RS and GIS approach. *J. Vect. Borne Dis.* 43(3): 115–122.

Surendran, S.N., Kajatheepan, A. & Ramasamy, R. 2007. Socio-environmental factors and sandfly prevalence in Delft Island, Sri Lanka: implications for leishmaniasis vector control. *J. Vect.-borne Dis.* 44(1): 65–68.

Swanepoel, R. 1976. Studies on the epidemiology of Rift Valley fever. *J. So. Afri. Vet. Assoc.* 47: 93–94.

Swanepoel, R. 1981. Observations on Rift Valley fever virus in Zimbabwe. *Contribs Epidemiol. & Biostats.* 3: 83–91.

Tang, J.W., Lai, F.Y.L. & Hon, K.L.E. 2010. Incidence of common respiratory viral infections related to climate factors in hospitalized children in Hong Kong. *Epidemiol. Infect.* 138: 226–235.

Tellier, R. 2009. Aerosol transmission of influenza A virus: A review of new studies. *J. Roy. Soc. Interface.* 6: S783–S790.

Thompson, R.A., Wellington de Oliveira Lima, J., Maguire, J.H., Braud, D.H. & Scholl, D.T. 2002. Climatic and demographic determinants of American visceral leishmaniasis in northeastern Brazil using remote sensing technology for environmental categorization of rain and region influences on leishmaniasis. *Amer. J. Trop. Med. & Hyg.* 67(6): 648–655.

Torrea, G., Chenal-Francisque, V., Leclercq, A. & Carniel, E. 2006. Efficient tracing of global isolates of *Yersinia pestis* by restriction fragment length polymorphism analysis using three insertion sequences as probes. *J. Clinic. Microbiol.* 44(6): 2084–2092.

Tran, A. & Raffy, M. 2006. On the dynamics of dengue epidemics from large-scale information. *Theoret. Popul. Biol.* 69: 3–12.

Tucker, C.J. 1979. Red and photographic infrared linear combinations for monitoring vegetation. *Rem Sens. Environ.* 8: 127–150.

Tucker, C.J., Hielkema, J.U. & Roffey, J. 1985. The potential of satellite remote sensing of ecological conditions for survey and forecasting desert-locust activity. *Int. J. Rem. Sens.* 6: 127–138.

84

Urashima, M., Shindo, N. & Okabe, N. 2003. A seasonal model to simulate influenza oscillation in Tokyo. *Japan J. Infect. Dis.* 56(2): 43–47.

USGS. 2009. Earth Observing 1. Available from: http://eo1.usgs.gov [Accessed 12th March 2012].

Vanwambeke, S., Lambin, E., Eichhorn, M., Flasse, S., Harbach, R., Oskam, L., Somboon, P., vanBeers, S., vanBenthem, B., Walton, C. & Butlin, R. 2007. Impact of land-use change on dengue and malaria in Northern Thailand. *EcoHealth* 4: 37–51.

Vaughn, D.W., Green, S., Kalayanarooj, S., Innis, B.L., Nimmannitya, S., Suntayakorn, S., Endy, T.P,. Raengsakulrach, B., Rothman, A.L., Ennis, F.A. & Nisalak,. A. 2000. Dengue viremia titer, antibody response pattern, and virus serotype correlate with disease severity. *J. Infect. Dis.* 181: 2–9.

Viboud, C., Alonso, W.J. & Simonsen, L. 2006. Influenza in tropical regions. *PLoS Med* 3(4): e89.

Vignolles, C., Lacaux, J.P., Tourre, Y.M., Bigeard, G., Ndione, J.A. & Lafaye, M. 2009. Rift Valley fever in a zone potentially occupied by Aedes vexans in Senegal: dynamics and risk mapping. *Geospat. Health* 3: 211–220.

Vittor, A.Y., Pan, W., Gilman, R.H., Tielsch, J., Glass, G., Shields, T., Sanchez-Lozano, W., Pinedo, V.V., Salas-Cobos, E., Flores, S. & Patz, J.A. 2009. Linking deforestation to malaria in the Amazon: Characterization of the breeding habitat of the principal malaria vector, *Anopheles darlingi*. *Amer. J. Trop. Med. & Hyg.* 81: 5–12.

Wang, R.L, Zhang, E.Y. & Min, J.G. 1989. Comparative studies on diapause responses between *Culex tritaeniorhynchus* and *Culex pipiens* pallens. *Zhongguo Ji. Sheng Chong. Xue. Yu Ji. Sheng Chong. Bing. Za Zhi.* 7:35–39.

Warrell, D.A. & Gilles, H.M. (eds.) 2002. *Essential malariology*. London: Arnold.

Wasserberg, G., Abramsky, Z., Anders, G., El-Fari, M., Schoenian, G., Schnur, L., Kotler, B.P., Kabalo, I. & Warburg, A. 2002. The ecology of cutaneous leishmaniasis in Nizzana, Israel: infection patterns in the reservoir host, and epidemiological implications. *Int. J. Parasitol.* 32: 133–143.

Wasserberg, G., Abramsky, Z., Kotler, B.P., Ostfeld, R.S., Yarom, I. &Warburg, A. 2003. Anthropogenic disturbances enhance occurrence of cutaneous leishmaniasis in Israel deserts: Patterns and mechanisms. *Ecolog. Appl.* 13(3): 868–881.

Watts, D.M., Burke, D.S., Harrison, B.A., Whitmire R.W. & Nisalak, A. 1987. Effect of temperature on the vector efficiency of *Ae. aegypti* for dengue-2 virus. *Amer.J. Trop. Med. & Hyg.* 36: 143–152.

Weaver, S.C. & Reisen, W.K. 2009. Present and future arboviral threats. *Antiviral Res*: 85(2): 328–345.

Weiss, K.E. 1957. Rift Valley fever: A review. *Bull. Epizootic Dis. Africa* 5: 431–458.

Werneck, G.L., Rodrigues, L. Jr., Santos, M.V., Araújo, I.B., Moura, L.S., Lima, S.S., Gomes, R.B.B., Maguire, J.H. & Costa, C.H.N. 2002. The burden of *Leishmania chagasi* infection during an urban outbreak of visceral leishmaniasis in Brazil. *Acta Trop.* 83(1): 13–18

Wheeler, S.S., Barker, C.M., Armijos, M.V., Carroll, B.D., Husted, S.R. & Reisen, W.K. 2009. Differential impact of West Nile virus on California birds. *The Condor* 111: 1–20.

WHO 2002. *WHO Wkly Epidemiol. Rec.* 77: 365–372.

WHO 2004. Human plague in 2002 and 2003. *Wkly. Epidemiol. Rec.* 79: 301–306.

WHO 2005. WHO Tech. Paper *Vector-borne diseases: Addressing a re-emerging public health problem (EM/RC52/3)*.

WHO 2007. Control of leishmaniasis. Available from: http://apps.who.int/gb/ebwha/pdf_files/WHA60/A60_10-en.pdf [Accessed 2nd February 2012].

WHO 2007. Control of leishmaniasis. Available from: http://apps.who.int/gb/ebwha/pdf_files/WHA60/A60_10-en.pdf [Accessed 26th January 2012].

WHO 2008. *The Global Burden of Disease*. Geneva: WHO Press.

WHO 2009. Influenza (Seasonal) -Fact Sheet. Available from: http://www.who.int/mediacentre/fact sheets/fs211/en/ [Accessed 26th January 2012.

WHO 2009. Available from: http://www.who.int/entity/leishmaniasis/resources/Interventions_old_world_cutaneous_lcish.pdf [Accessed 2nd February 2012].

WHO 2010a. Malaria fact sheet Number 94. Available from: http:// www.who.int/mediacentre/factsheets/fs094/en/index.html [Accessed 1st February 2012].

WHO 2010b. FluNet. Available from: http://gamapserver.who.int/GlobalAtlas/home.asp [Accessed 5th February 2012].

WHO DengueNet. 2010. Available from: http://www.who.int/csr/disease/dengue/denguenet/en/index.html [Accessed 13th March 2012].

WHO-EMRO 2007. Strategic plan for malaria control and elimination in the WHO Eastern Mediterranean region 2006–2010. Cairo: EMRO.

WHO-SEARO 2011. Malaria Situation in SEAR Countries. Available from: http://www.searo.who.int/en/Section10/Section21/Section340_4022.htm [Accessed 30th June 2011].

Wilson, M.L. 2001. Ecology and infectious disease. In J.L. Aron & J.A. Patz (eds.), *Ecosystem change and public health: A global perspective*: 283–324. Baltimore: Johns Hopkins UP.

Wolter, K. & Timlin, M.S. 1998. Measuring the strength of ENSO events: How does 1997/98 rank? *Weather* 53: 315–324.

Yates, T. L., Mills, J., Parmenter, C., Ksiazek, T., Parmenter, R., Calisher, C., Nichol, S., Abbot, K., Young, J., Morrison, M., Beaty, B., Dunnum, J., Baker, R. & Peters, C. 2002. The ecology and evolutionary history of an emergent disease: Hantavirus pulmonary syndrome. *Bioscience*. 52(11): 989–998.

Youssef, R., Safi, N., Hemeed, H., Sediqi, W., Naser, J.A. & Butt, W. 2008. National malaria indicators assessment. *Afghan. Ann. Malaria J*. 1(1): 37–49.

Youtao, S., Jaroensutasinee, M. & Jaroensutasinee, K. 2008. Analysis of influenza cases and seasonal index in Thailand. *Int. J. Biolog. & Life Sci*. 4(2): 113–116.

Zimmerman, T. 1992. Ecology of malaria vectors in the Americas and future direction. *Mem. Inst. Os waldo Cruz* 87: 371–383.

Chapter 3

Water, water quality and health

S.I. Zeeman[1] & P. Weinstein[2] (Auth./eds.) with; E. Fearnley[3], C. Skelly[4], E.N. Naumova[5], J.S. Jagai[6], D. Castronovo[7], J. McEntee[5], M. Koch[8], S. Hamner[9] & T. Ford[10]

[1] *University of New England, Biddeford, ME, US*
[2] *University of South Australia, Adelaide SA, AU*
[3] *South Australian Department of Health, Adelaide, SA, AU*
[4] *University of Queensland, Brisbane, Qld, AU*
[5] *Tufts University, Medford, MA, US*
[6] *US Environmental Protection Agency, Chapel Hill, NC, US*
[7] *Mapping Sustainability, LLC, Palm City, FL, US*
[8] *Boston University, Boston, MA, US*
[9] *Department of Microbiology, Montana State University, Bozeman, MT, US*
[10] *College of Graduate Studies, University of New England, Portland, ME, US*

ABSTRACT: This chapter identifies the role environmental tracking plays in identifying public health water hazard and water quality issues. It outlines public health issues to be examined and provides an integrated overview of water and diseases by combining knowledge of the hydrologic cycle, which describes how water serves as a significant component of exposure pathways, with the tenants of public health surveillance.

Disclaimer: The views expressed by Doctor Jagai in this chapter do not necessarily reflect the views or policies of the US Environmental Protection Agency.

1 INTRODUCTION

Water is life. This is true in an evolutionary sense for all life on Earth and it is also true in terms of the essential role water plays in the sustenance of health and well being. Water is essential for basic hydration, sanitation, and hygiene; and, to more complicated roles in food preparation and production processes. Access to a safe and reliable water supply has been described as a basic human right (WHO 2003a), and cultural practices throughout the world are defined by its availability or absence in local environments. Human societies are not just dependent upon water and its quality for our health and well being. Water also shapes our social-cultural practices, our environmental heritage, and our economic systems.

However, water can also bring illness and injury. The most direct health impacts are felt through its periodic absence from environments that result in personal dehydration, degraded hygiene and sanitation. Although beyond the scope of this chapter, it is important also to remember that one-third of the world's seven billion people live under some degree of water stress (physically limited supply) and suffer adverse socio economic effects as a consequence: time spent on water haulage, often many hours each day, erodes opportunities for education and community development. Crop failure, through drought, impacts nutrition and can ultimately result in starvation. Even when water is available, its quality may be compromised directly by pathogen or chemical contamination, and it can expose populations to water vectored diseases like schistosomiasis and legionellosis. Drowning and physical injury from exposure to water in recreational settings and during flood events (including drownings within and around the home) are significant public health and safety

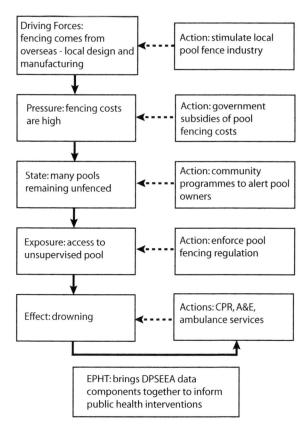

Figure 1. DPSEEA framework applied to swimming pool environments as an environmental public health tracking approach to assess risk of drowning (Modified from Briggs *et al.*, 1966).

concerns. Water also plays complex roles in population health through ecosystem linkages that are fundamental to the existence of diseases such as malaria, dengue and lymphatic filariasis, which are transmitted by mosquito vectors that require access to water for reproduction. Water is essential to life, but the relationships between water, water quality and human health are complex.

This Section identifies the role that environmental tracking can play in identifying public health water hazards and water quality issues; and, to provide an integrated overview of exposure pathways through water by combining knowledge of the hydrologic cycle with practices for public health surveillance. Primary emphasis is on e*xposure pathways* that can be identified through environmental tracking and by linking public health surveillance systems directly to public health outcomes. By monitoring directly and intervening at the level of environmental drivers of exposure, rather than basing public health interventions solely on the surveillance of disease outcomes, one can obtain longer lasting and potentially more sustainable public health interventions. This approach forms the basis of the WHO Headlamp project and is conveyed conceptually through the driving force, pressure, state, exposure, effects and action (DPSEEA) model (Figure 1) (Briggs *et al.*, 1996).

Traditional environmental public health tracking is more likely to provide earlier warning of impending health effects, thereby maximising the opportunity for prevention. When traditional methods are combined with DPSEEA exposure and effect surveillance for a population, the limitations of traditional health outcome surveillance are improved. However, to better understand the water related causal pathways that environmental public health tracking seeks to capture, one must first understand fundamental components of the hydrologic cycle.

Table 1. Sample linkages between exposure pathways, potential health hazards, health effects, and opportunities for environmental health tracking and health indicators.

Exposure pathway	Health hazards	Health effects	Environmental health and tracking indicators
Drinking water	Physical injury; infection; toxin	Dehydration; cholera; arsenicosis	Water supply level; treatment process quality; water quality
Recreational water	Physical injury; infection; toxin	Drowning; schistosomiasis; algal blooms	Compliance with pool fence legislation or beach patrol directions; open water quality
Sanitation water (lack of)	Infection; socio-economic deprivation	Trachoma; time spent in haulage	Proportion of population with running water; mean population distance to external water sources
Water storage/ overflow	Indirect infection	Malaria; a few others	Hydrological monitoring of water courses, wetlands and other water bodies
Water engineering	Aerosolisation; heating	Legionellosis; amoebic meningitis	Prevalence of water tower cooling systems; survey of hot water tank settings in domestic settings

The hydrologic cycle defines natural flows of water through the environment. As a planetary cycle, it includes environmental circumstances that pose health risks to its quality, environments where it is likely to be a physical hazard, and opportunities for extracting it for use. Although, water in the atmosphere can be contaminated directly (e.g. acid rain in heavily industrialised regions), the greatest risks for contamination occur when it moves into and across land surfaces toward the oceans via rivers, lakes and/or ground water. When water in and on the land is extracted for any human use, it carries with it the potential for exposure to diseases and other human health outcomes, or quality of life consequences.

Water in the landscape is channelled through catchments (or drainage basins) that define natural geographic divisions for using and managing water use. Modelling water flow through catchments allows it to be audited into two major components: (1) surface flows over land and into streams, rivers, and freshwater lakes or into reservoirs, and thence to the sea; and, (2) subsurface flows into ground water aquifers. These natural boundaries are the focus of interest to target water related health issues geographically (e.g. flood and drowning risk, disease risk, agricultural chemical risk, etc.). However, in many regions of the world, exposure to water and potential negative health effects of water contamination are increased through human diversions from the natural cycle through canals and other water reticulation systems. These water related health effects also have important environmental drivers that can be monitored to provide early warning indicators of potentially adverse health outcomes (Table 1).

In the material that follows, the introductory section is provided by Doctors Zeeman and Weinstein Section 2 was contributed by Doctors Weinstein, Fearnley and Skelly; Section 3 is provided by Doctors Naumova, Jagai, Castronovo, McEntee and Koch; and Section 4 is material proved by Doctors Hamner and Ford.

Traditional environmental public health tracking is more likely to provide earlier warning of impending health effects, thereby maximising the opportunity for prevention. When traditional methods are combined with DPSEEA exposure and effect surveillance for a population, the limitations of traditional health outcome surveillance are improved. However, to better understand the water related causal pathways that environmental public health tracking seeks to capture, one must first understand fundamental components of the hydrologic cycle.

The hydrologic cycle defines natural flows of water through the environment. As a planetary cycle, it includes environmental circumstances that pose health risks to its quality, environments where it is likely to be a physical hazard, and opportunities for extracting it for use. Although, water in the atmosphere can be contaminated directly (e.g. acid rain in heavily industrialised regions), the greatest risks for contamination occur when it moves into and across land surfaces toward the oceans via rivers, lakes and/or ground water. When water in and on the land is extracted for any human use, it carries with it the potential for exposure to diseases and other human health outcomes, or quality of life consequences.

2   ENVIRONMENTAL TRACKING AND PUBLIC HEALTH SURVEILLANCE

The principles of modern public health surveillance were established in the mid-20th Century to collect health outcomes data systematically (Teutsch & Churchill 2000). While these principles remain sound, the variety of data and the ways they are managed continue to develop (cf. O'Carroll *et al.*, 2003). The increasing concern over bio-terrorism, the growing complexity of global food systems, and the size and speed of reticulated water outbreaks have driven the desire for new and more sophisticated health surveillance techniques, including environmental health indicators, syndromic surveillance, and highly automated forms of real-time, data-based aberration detection methods (Wagner *et al.*, 2006).

2.1   *Public health actions and reactions to water-borne diseases*

Environmental public health tracking is distinct from other tracking systems in that its emphasis is on data integration across human health effects, exposures and environmental hazards. The focus is on collecting data across these three health issues irrespective of data source, and interpreting them based on relationships between the three components. The new effort here is in tracking the changes seen in one component against changes seen in the other two, with a view to building a more rigorous and more directly relevant argument for local public health action.

Data integration across several sources has been rare in public health surveillance, and is perhaps most easily illustrated by the public health intervention of pool fencing. By carrying out traditional health surveillance (notification of drowned toddlers with case investigation), it was determined that domestic swimming pools pose a significant drowning hazard. Legislation was therefore introduced in Australia to have pools fenced, (i.e. a public health intervention to reduce exposure to the hazard), and to permit local authorities to routinely inspect pool areas to ensure that fencing is maintained and that self-closing gates operate properly. In effect, these are hazard avoidance measures that reduce the more unpleasant task of recording drowned toddlers. *Integration* of data from pool fencing and gate closing records with toddler drowning records is an example of *environmental monitoring in the service of public health.* Enforcement of the environmental component is made easier with modern aerial and satellite observations.

Monitoring exposure pathways and comparing changes in each component is effective environmental public health tracking. It has been defined as *the on-going collection, integration, analysis and dissemination of data from environmental hazard monitoring, human exposure tracking, and health effect surveillance.* This integrated system combines rigorous and validated scientific information about environmental conditions and exposures with adverse health conditions to identify spatial or temporal relationships between the exposures and outcomes (McGeehin *et al.*, 2004).

2.2   *Microbial aspects*

Consumption of poor quality water is a significant and important transmission pathway for infectious diarrhoeal diseases, with an estimated ninety-four per cent of the global burden of diarrhoeal disease attributable to the modifiable environmental factors of unsafe drinking water supplies and poor sanitation and hygiene. The majority of the burden of diarrhoeal disease is in developing countries, and particularly among children (Prüss-Üstün & Corvalan 2006). Both endemic and epidemic acute-gastroenteritis is caused by consuming water contaminated with intestinal bacterial, viral or protozoan (Ford 1999). Common water-borne pathogens include *Vibrio cholerae, Shigella,* toxigenic *Escherichia coli, Campylobacter, Giardia* and *Cryptosporidium.*

While water-borne disease incidence has been reduced greatly in developed countries by implementing water treatment and sanitation procedures, people in lesser developed areas are still exposed to water at risk of microbial contamination from agricultural and municipal wastes, deterioration and malfunction of treatment and distribution systems, mismatches between water supply and demand, and the presence of organisms resistant to disinfection. The largest recorded water-borne outbreak in a developed country occurred in Milwaukee, US when the public water supply was contaminated with *Cryptosporidium*, leading to approximately 403,000 cases of gastroenteritis and fifty four deaths (Hoxie *et al.*, 1997). The public water supply was sourced from Lake Michigan, which was contaminated from nearby cattle faeces and other waste sources after heavy spring rains and snowmelt (MacKenzie *et al.*, 1994).

Another notable water-borne outbreak occurred in Walkerton, Canada, in May 2000, with an outbreak of *E. coli* O157:H7 and *Campylobacter jejuni* propagated through the reticulated public water supply. It resulted in 2300 cases of gastroenteritis, twenty seven cases of haemolytic uraemic syndrome (HUS) and seven deaths in a town of only 4800 residents (Hrudey *et al.*, 2003). This outbreak was subject to an intensive inquiry and report (O'Connor 2002) in which numerous contributing factors were identified. These included: systemic reporting failures; falsifications by cognizant water supply authorities; an extraordinarily high rainfall before the outbreak; the poor design and location of the well where the contamination occurred; and, poor chlorination practices (Hrudey *et al.*, 2003). Analysis of the well design and location concluded that it was very shallow, proximal to agricultural land, and located in an area where the surrounding geology allowed incursions of surface water into the well. The Walkerton outbreak serves as a reminder that reliable environmental monitoring of chlorine residuals and microbiological indicators, along with adequate operator training and use of weather monitoring and forecast systems are necessary (Auld *et al.*, 2004; Hrudey *et al.*, 2003). The association of many other water-borne disease outbreaks in developed countries that have had significant environmental and meteorological contributing factors, highlights the roles environmental monitoring can play in preventing disastrous outbreaks (Curriero *et al.*, 2001; Hrudey *et al.*, 2003; Auld *et al.*, 2004; Patz & Olson 2006; Nichols *et al.*, 2009).

In context of environmental monitoring for water-borne diseases, it is important to remember that broader ecological drivers are responsible for disease production in the majority of the world's population. Hydrological cycles and human disruptions of the natural cycle lead to flooding and high rainfall events that can contaminate drinking water sources with run-off from animal grazing areas, as well as inundate and cause overflow of sewage systems. Climate change is expected to increase both the severity and frequency of such events, and monitoring therefore needs to go well beyond the traditional microbes per ml approach. Upstream approaches to monitoring and eventually tracking would, therefore, ultimately include climate and environmental change monitoring and monitoring of intervention and remediation efforts.

There are numerous examples of multiple-barrier approaches to managing public health risks, for example regarding integrated pest management, hazard analysis critical control point (HACCP), and modern drinking water treatment and reticulation. The critical lesson in all approaches is not to rely on any single intervention. In terms of water treatment and reticulation the first defence is securing the safety of the drinking water source by preventing or limiting contamination. The second defence is to maintain a secure reticulation system that prevents contaminants from entering the system between the source and the delivery point. The third defence is to ensure that the water treatment facility, itself, consists of multiple barriers. Rigorous monitoring systems provide multiple checks to ensure that the existing barriers function correctly.

Relying only on surveillance of certain diseases in human populations underestimates greatly the true burden of water-borne disease due to under-reporting inherent in disease surveillance systems; but surveillance can also be a source of over-reporting, as many common water-borne pathogens are transmissible also by person-to-person contact, the faecal oral route, and via food contamination. Outbreaks rather than sporadic cases are more likely to be attributed to water sources within human public health surveillance systems. Therefore, there is a need for environmental monitoring and surveillance to promote health and environmental surveillance systems integration, and to provide more information regarding disease causation and ecology, both of which can lead to better public health interventions and preventative actions.

Water-borne disease can occur also via inhalation, ingestion and physical contact with recreational waters. Recreational waters are subject to microbial contamination from numerous sources including storm water, sewage overflows, industrial and agricultural run-off, bird faeces and bathers themselves. Respiratory and gastrointestinal diseases have been linked epidemiologically to microbial organism exposure, namely the indicator organisms of coliforms and enterococci, in recreational waters, and these studies form the basis of microbial limits set in recreational water quality guidelines (Cabelli *et al.*, 1982; Dufour, 1984; Calderon *et al.*, 1991; Kay *et al.*, 1994; Fleisher *et al.*, 1996). However, recreational water quality guidelines have developed beyond just microbial counts in environmental water bodies with the introduction of a risk assessment and risk management approach to recreational water guidelines, introduced with the *Guidelines for Safe Recreational Water Environments* (WHO 2003b). The risk assessment process facilitates integrating long-term microbiological data with an inspection-based assessment of the susceptibility of an area to direct influence via human faecal contamination. This allows a sanitary inspection category, or summary of environmental risk factors, to be derived (NHMRC 2004). The risk assessment approach for recreational waters is a positive example of integrated environmental monitoring data to assess recreational water quality and predict times or events that would be considered a high risk to public health.

2.2.1 *Legionnaires' disease*

Legionella pneumophilia is a ubiquitous environmental pathogen found in rivers, lakes and soils, and can also survive and grow in hot and warm water systems. It is a common risk to human health via colonisation in cooling towers, air conditioning systems, hospital water distribution systems and spa pools (Ford 1999; Fields *et al.*, 2002). Inhalation of aerosolised water containing *L. pneumophilia* can cause a severe, and sometimes fatal, pneumonic disease, called Legionnaires' disease, in susceptible individuals. Legionellosis is considered largely to be a preventable illness, as control of the organism in water systems and other artificial reservoirs will reduce the chance of human infections. As a consequence environmental hazard monitoring systems, guidelines, and legislation already exist in many countries. Routine cleaning, general maintenance, microbiological testing, and cooling tower registries form the basis of most environmental monitoring systems for cooling towers. Lessons have been learned from a number of large community outbreaks of Legionnaires disease (more than 100 cases) linked to cooling towers, including changes to guidelines and legislation regarding registration and maintenance of cooling towers and risk management plans (Garcia-Fulgueiras *et al.*, 2003; Greig *et al.*, 2004).

Outbreaks of Legionnaires' disease have educated public health officials about risk factors for the disease; however, most reported cases in human surveillance systems (up to eighty per cent) are sporadic cases. Risk factors for sporadic illness are less well defined (Marston *et al.*, 1994; Stout & Yu 1997). Two recent studies assessed relationships between meteorological factors and sporadic Legionnaires' disease in efforts to better describe environmental risk factors. Hicks (*et al.*, 2007) identified higher temperatures and rainfall as risk factors for sporadic cases in five mid-Atlantic States in the US, while Fisman (*et al.*, 2005) found increased humidity and rainfall to be important factors. These examples further reinforce the value of monitoring systems in so far as the ultimate ecological drivers of legionellosis lie beyond the cooling towers that produce the infective droplets. The very need for cooling towers arises in part perhaps because of unsustainable population densities in urban areas and in hotter climates that are not conducive to such densities, as well as in part to climate warming exacerbated by the emissions resulting from these dense populations. Only by monitoring and integrating a spectrum of ecological, environmental, and human indicators concurrently would it be possible to make evidence-based recommendations for public health interventions that extend to sustainable environmental management.

Hospitals are potential locations for Legionnaires' disease transmission because the exposed population is often immuno-compromised and at higher risk, or because water systems are sometimes old, and because water temperatures are favourable for growth of *Legionella* spp. Recent studies in a number of countries that assess environmental monitoring for legionella in hospital water systems and hospital-acquired cases of legionnaires' disease have had mixed results. Some studies do not recommend environmental surveillance (Tablan *et al.*, 2004; Leoni *et al.*, 2007); others highlight links between environmental surveillance and human cases (Stout *et al.*, 2007); and still others suggest actions in high risk patient areas only (Boccia *et al.*, 2006). It has been noted that cases of hospital-acquired Legionnaires' disease have not been reported in hospitals without legionella in their water supply (Stout *et al.*, 2007).

2.2.2 *Primary amoebic meningoencephalitis*

Naegleria fowleri is a thermophilic (warmth-loving) amoeba found in freshwaters. It is the causative agent of a rare, but often fatal illness known as primary amoebic meningoencephalitis (PAM), first identified and reported in South Australia in 1965 (Fowler & Carter 1965; Visvesvara *et al.*, 2007). The thermophilic organism grows well in waters of 30°C, with survival in waters up to 45°C. Adequate chlorination eliminates the organism from swimming pools, but naturally warm water bodies where the organism is commonly found, cannot be chlorinated. Infection occurs when contaminated water enters the nose, rather than via direct consumption, allowing it to migrate to and infect the brain, with rapid progression to death in the absence of prompt treatment. PAM is not a notifiable disease in most countries because it is rare. Its incidence is generally estimated to be 200 cases worldwide per year; but it is most likely underestimated significantly. Cases are reported in the US, Europe, and Australia (Esterman *et al.*, 1984; Cabanes *et al.*, 2001; Cogo *et al.*, 2004).

An epidemiological review of all 111 known PAM cases in the US (1962–2008) is described by Yoder (*et al.*, 2010). Most fit with previous descriptions of case reports and reviews where cases were mainly children (sixty two per cent), males (seventy nine per cent), those involved in vigorous recreational activity in natural warm water bodies rather than swimming pools, most cases occurred in the warmer summer months (eighty seven per cent), and all but one were fatal (Yoder *et al.*, 2010).

On-going routine monitoring of environmental waters for *N. fowleri* is not necessarily feasible or cost effective in all countries. However, in areas of known environmental risk (warm water conditions, high nutrients), environmental monitoring followed by specific warnings may be considered as an effective public health measure, rather than monitoring for the fatal illness (Visvesvara *et al.*, 2007). Environmental monitoring does not occur routinely in the two US states where the majority of US PAM cases have been identified (Yoder *et al.*, 2010). Environmental monitoring has been implemented in France for thermally polluted waters (from nearby power plants) where a specific hazard and risk has been identified (Cabanes *et al.*, 2001). In Australia, where water and environmental temperatures are both high and drinking water is largely sourced from surface environmental catchments, with long pipelines, the monitoring of environmental waters and the publicly supplied drinking water for the hazard of *N. fowleri* is recommended in the Australian drinking water guidelines (NHMRC 2004). Subsequent public health alerts are provided to the public when the organism is detected. Although infection does not occur via oral consumption of contaminated water, public water supplies are often used for swimming pools and other recreational areas. Therefore, the importance of adequate chlorination in swimming pools is emphasised.

An analysis of two years of environmental water monitoring data for *N. fowleri* was conducted in South Australia (Esterman *et al.*, 1984). The presence of the organism was associated with high water and air temperatures and low chlorination levels. The availability and analysis of further long-term environmental data would contribute to a better understanding of the organism's ecology and inform the need and structure for future surveillance systems.

2.3 *Chemical contaminants in water*

2.3.1 *Trihalomethanes*

Trihalomethanes (TTHMs) are of increasing interest in developing countries. The treatment of drinking water with disinfectants, such as chlorine and chloramine, is undoubtedly an important public health intervention, significantly reducing the risks of microbial water-borne diseases. However, a drawback of these treatments is the generation of disinfectant by-products (DBPs), some of which are potentially harmful to human health. Various forms of DBPs are found in drinking water that are created by the reaction of chemical disinfectants with natural organic matter in water; but, it is total TTHMs that have been the primary focus of past research. Adverse human health outcomes associated with TTHMs include bladder cancer, and adverse reproductive outcomes such as low birth weight, prematurity, pregnancy loss and birth defects (Bove *et al.*, 1995; ILSI 1995; Dodds *et al.*, 1999; Hwang & Jaakkola 2003; Dodds *et al.*, 2004; Chisholm *et al.*, 2008; Luben *et al.*, 2008; Nieuwenhuijsen *et al.*, 2008; Nieuwenhuijsen *et al.*, 2009a). Recent reviews have noted that the association between DBPs and birth defects is based on limited evidence (Hrudey 2009; Nieuwenhuijsen *et al.*, 2009b), with numerous studies inadequately measuring maternal exposure to DBPs and limited environmental sampling. Further research is needed to understand the mechanisms of disease and examine different combinations of DBP exposure (Nieuwenhuijsen *et al.*, 2009a).

The need for integrated surveillance with environmental monitoring and public health outcomes is essential to determine the extent and nature of DBPs on adverse birth outcomes, because the structure and quantity of natural organic matter in water sources influences the quantity and type of DBPs formed upon disinfection; and these, in turn vary with geographic location and season (Rodriguez *et al.*, 2004). With pressure mounting for high-quality water sources worldwide, with changing weather patterns, and with on-going droughts and population growth, drinking water from lower quality sources and alternatives such as direct or indirect use of recycled water must be considered. These waters are more likely to contain higher levels of DBPs, and this increased exposure could lead to an increased public health burden. The benefits of monitoring water sources for precursors to DBPs, and the implementation of treatment processes to reduce DBPs concentrations would have greater public health significance than relying on surveillance of adverse birth outcomes.

2.3.2 *Arsenic*
Arsenic is the main constituent of over 200 minerals. It is a ubiquitous element in natural environment (Ng 2005; Rahman *et al.*, 2009). Mining wastes contain large amounts of arsenic. These contribute elevated levels of arsenic to ground water supplies and to drinking water derived from them, posing risks to human health (Ng 2005). Natural groundwater contamination has been recorded in Bangladesh, India, China, Taiwan, and Chile all of which have high levels of contamination and limited alternative drinking water sources (Kapaj *et al.*, 2006; Mazumder 2007; Smedley & Kinniburgh 2002). Chronic exposure to arsenic-rich drinking water has been linked to skin lesions and hyperpigmentation, respiratory problems, and both bladder and lung cancer (Kapaj *et al.*, 2006). Similar to many other health risks, children are at greater risk than adults due to their larger consumption of water on a body weight basis. Children are also at risk of neuro-behavioural and intellectual and memory impairment, particularly in association with malnutrition (Calderon *et al.*, 2001; Tsai *et al.*, 2003). Interventions and prevention of adverse public health effects from arsenic contamination are greatly enhanced by environmental monitoring of drinking water sources. As is true for many illnesses where only a small percentage of the population affected show clinical symptoms at any particular time, longer term monitoring is a beneficial and positive public health action for timely public health education messages. Improving scientific understanding of environmental contamination and arsenic cycling is also necessary (Kapaj *et al.*, 2006).

2.3.3 *Fluoride*
Fluoride is a common chemical constituent of water from natural geological sources. The positive benefit of low level fluoridation of public water supplies to reduce dental cavities is well known. Levels as low as 0.1 mg/L have positive effects; but levels ranging between 0.5 mg/L – 1 mg/L are recommended for broader health benefits (McDonagh *et al.*, 2000; Jones *et al.*, 2005; Kirkeskov *et al.*, 2010). Of more importance for environmental monitoring and surveillance are the higher concentrations of natural fluoride contamination and leaching into groundwater from geogenic origin. These constitute the majority of drinking water supplies in India, China, South America and

Africa – all countries with large and growing populations (Fawell & Nieuwenhuijsen 2003; Ayoob & Gupta 2006). The health effects of chronic exposure to fluoridated drinking water with concentrations above 1.5 mg/L are dental fluorosis or discolouration of teeth; at concentrations higher than 6 mg/L, skeletal fluorosis can occur that leads bone deformities and fractures (WHO 2008d).

Skeletal fluorosis is exacerbated by poor nutritional status, and can cause severe pain and bone malformations that render people severely physically disadvantaged. The need for environmental monitoring of fluoride in drinking water is therefore an essential public health measure in endemically exposed regions, and can be done in conjunction with compliance to international drinking water guideline of 1.5mg/L (WHO 2008). The integration of *environmental chemical monitoring* of drinking water with observed health outcomes would provide further evidence of the causal relationship and the effectiveness of interventions in place to protect public health.

2.3.4   *Harmful algal blooms*

"Red tides" or harmful algal blooms (HABS) represent local overgrowth of single-celled planktons that include algae and dinoflagellates (see Chapter 9). They often contain toxins, and when large coastal blooms occur, they pose a hazard to recreational water users, local seafood supplies, and downwind coastal residents (Weinstein 2013).

The toxins produced include extremely potent neurotoxins (e.g. brevetoxin and saxitoxin, commonly from species of *Gymnodinium* and *Alexandrium,* respectively), and various gastrointestinal irritants (e.g. dinophysistoxin and domoic acid, commonly from species of *Dinophysis* and *Pseudonitzschia*, respectively) (Heymann 2004). The effect on humans depends on the dose and exposure pathway. Common symptoms include tingling and paralysis after consuming toxic shellfish having high amounts of the neurotoxins that get absorbed into the intestine. However, in windy conditions, toxins can also be aerosolised as sea spray, with eye and respiratory irritations occurring in beachgoers and coastal residents.

The mechanisms that drive red tides are complex, and include both natural and anthropogenic phenomena like coastal upwelling (deep, nutrient-rich waters rising), nutrient runoff from agricultural lands, and wind and sea surface temperature changes. The interactions of these factors are not yet well enough understood to allow public health officials to predict blooms based on environmental monitoring, and they are therefore not able to issue proactive warnings. The hazard can therefore only be managed reactively, generally by public warnings not to engage in recreational activities in or on the water – always as a response to people already having been adversely affected by the toxins. Because of the complexities of bloom ecology as well as human behaviour, both disease and disease intervention can be highly variable regionally. To plan hazard mitigation strategies, it is therefore important to regionalise recommendations based on the relevant scientific literature (Kirkpatrick *et al.*, 2004).

The hazard posed by red tides and other HABs is likely to increase substantially as a result of increasing nutrient runoff from expanding agriculture; increasing sea surface temperatures; and increasing human contact with coastal waters as populations expand (Weinstein 2013). It will therefore become increasingly important to develop predictive models based on environmental monitoring and ecological understanding, rather than continuing to rely on disease surveillance to identify and protect populations at risk.

2.4   *Hydrological disasters*

2.4.1   *Floods*

Floods account for approximately forty per cent of all natural disasters affecting large parts of the world's population. They can occur as independent hydrological events, or as an outcome linked to other events, including earthquakes and tsunamis (Guha-Sapir *et al.*, 2004; Jones 2006). Floods can cause significant injury, disease and death (mainly drowning and head trauma) via the physical inundation of large volumes of rapidly flowing water, and the associated destruction of infrastructure. A recent preliminary analysis of 771 of the 1118 deaths recorded in the state of Louisiana after Hurricane Katrina, and the associated flooding of New Orleans in 2005, identified that approximately two thirds of the deaths were most probably related to direct physical impacts of the flood. Drowning was the most common cause of death (Jonkman *et al.*, 2009). Analysis by area indicated that most of the deaths occurred in areas with the deepest water. One-third of the deaths were related to issues associated with poor public health facilities and general environmental conditions after the flood, including strokes and heart attacks. An environmental monitoring or flood warning system that would reduce the number of drownings would be to proactively move people out of high risk zones before an expected event. The combination of flood risk maps and meteorological warning systems are important aspects to such environmental monitoring systems for public health.

Some recent flooding events and other natural disaster emergencies within Australia have high-lighted the importance of good environmental warning systems and good communication networks for delivering the warnings. The need for formal emergency management agreements with local and national radio broadcasting services are critical for alerting the general public, as for example, in the wake of the Hunter Valley flood in New South Wales in 2007 (Cretikos *et al.*, 2007). Since the time lapse between environmental monitoring information signifying the potential for a flood event is often minimal, an essential component of environmental monitoring for disaster management is the rapid distribution of information. Currently in Australia, automated emergency messages are sent to telephones at residential addresses in an area of predicted emergency via a system called *Emergency alert*. Further testing is being conducted to expand alerts to include mobile telephones based on handset location at the time of emergency, rather than just the registered customer address (Australian Government 2010).

2.4.2 *Tsunamis*

Tsunamis occur typically as a series of large ocean waves and are another example of a natural event caused by oceanic disturbances caused by earthquakes, underwater landslides, and volcanic eruptions. They often cause serious physical health effects largely in the form of injuries and deaths due to drowning. The Indian Ocean tsunami of 26 December 2004 was triggered by an earthquake in the Indian Ocean and caused over 200,000 deaths in Indonesia, Sri Lanka, India, Malaysia, Maldives, Seychelles, Somalia and Thailand (Thailand 2005). Since this tragic event, planning efforts have been undertaken to develop and implement tsunami early warning systems. The Inter-governmental Oceanographic Commission (IOC) Tsunami programme supports member states in assessing tsunami risk, implementing tsunami early warning systems and educating communities at risk about preparedness (IOC 2010). An example of a currently-operating early warning system is the Australian Tsunami Warning System. It is a collaborative project between the Bureau of Meteorology, Geoscience Australia, and Emergency Management Australia. The system consists of monitoring sea levels changes resulting from seismic events and from deep-ocean buoys, and tsunami modelling systems that estimate size, impact area, and arrival time of tsunamis. These data are used to distribute warnings to State & Territory emergency management services, media and the public (Bureau of Meteorology 2010). In February 2010 the Australian Tsunami Warning System detected an earthquake and distributed warning messages that a Tsunami would arrive at Sydney, however, despite the warnings to stay away from beaches, people deliberately went to beaches to watch the event, and others went swimming (Dominey-Howes & Goff 2010). Although physical effects to the Sydney beaches were minimal, the account highlights the effectiveness of the environmental monitoring system and the need to better convey and enforce warnings to the public; and possibly, to improve actions from emergency services.

2.5 *Water-borne disease outbreaks – coming full circle*

There are many environmental and management/intervention activities related to water and public health issues. They can be monitored directly and tracked together. The hope is that by modifying environmental drivers of exposure, rather than basing public health intervention solely on the surveillance of disease outcomes, sustainable disease prevention becomes more likely. Further, from an integrated ecological perspective, it is equally important to appreciate that societies can use environmentally driven disease outcomes to successfully monitor states of the environment. Jardine (*et al.*, 2004) contend that *human disease incidence is in fact one of the most useful and practical bio-indicators of the health of an ecosystem and [can be used to] assist in guiding rapid and appropriate ecosystem interventions.*

A major advantage in using disease outbreaks as bio-indicators of even subtle ecosystem disruptions is that the health of human populations is generally subject to more widespread and accurate surveillance than is ecosystem health (Spiegel & Yassi 1997). Many sources of data, such as those obtained from disease registries, infectious disease notification systems, and hospitalizations, provide on-going measurement and monitoring of human communities. In this context disease surveillance is used as a surrogate indicator of ecosystem health, contributing to environmental monitoring in the up-stream sense of pool fence monitoring. Thus, human disease surveillance data can be used both to identify water-related health problems and to better target environmental interventions to alleviate them. Public health practitioners recognise this pattern as a move from tertiary and secondary prevention up-stream to primary prevention, where the hazard is removed in such a way as to prevent the disease from ever occurring. Environmental monitoring, including the use of water-related disease surveillance data, is therefore a facilitator of primary prevention – the best form of public health practice.

3 ASSESSING CRYPTOSPORIDIUM AND GIARDIA ENTERIC INFECTIONS

3.1 *Introduction*

Worldwide, water-borne diseases add substantially to global morbidity and mortality. Diarrheal diseases continue to be one of leading cause of mortality worldwide. Annually there are 4 billion cases of diarrhoea, 2.2 million resulted in deaths, eighty per cent of them occur in the first two years of life and eighteen per cent of deaths are in children under five years of age (WHO 2008b). In resource-poor settings eighty-eight per cent diarrheal infections are thought to be due to unsafe water, inadequate sanitation, and poor hygiene. Despite great efforts to provide clean drinking water and to regulate both the quality and use of recreational water, the burden caused by water-borne infection in developed countries should not be neglected. In the US, the annual rate of hospitalizations due to protozoan infection among the growing population of aging baby boomers is steadily increasing (Mor *et al.*, 2009). Although awareness of emerging enteric infections, including cryptosporidiosis, has been increased, under-diagnosis and under-reporting of cases remains a major barrier to accurate surveillance in many countries. Enteric infections are thought to be driven by numerous social and environmental factors, drinking water quality, hygiene and sanitation. However, the relative contribution of demographic, behavioural and environmental drivers with respect to their causal pathways is poorly understood. This gap in knowledge jeopardizes the development of sustainable preventive strategies that will be effective under potentially critical conditions, including the response to an increase of extreme weather events worldwide due to climate change.

Seasonal fluctuations in infectious diseases are well known and a well-documented phenomenon and many prominent theories attempt to explain the nature of such fluctuations. The core of seasonality is thought to be related to temporal oscillations in the governing transmission cycles of pathogenic agents and host susceptibility. It has been shown for example, that cryptosporidiosis rates increases in seasons when precipitation and temperature reach their extremes. In tropical climates the incidence of cryptosporidiosis increases during the warm, rainy season (Adegbola *et al.*, 1994; Newman *et al.*, 1999; Perch *et al.*, 2001) and in temperate climates it is shown to be high in the spring and fall (Shepherd *et al.*, 1988; Naumova *et al.*, 2000; Naumova *et al.*, 2005). However, the consistency of such behaviour on a global scale has not been investigated due to: a) limited data on laboratory-confirmed infections collected over a long time period with reasonable temporal resolution; and b), the difficulty in assessing local meteorological parameters in remote areas. Despite growing attention to disease seasonality, especially in enteric infections, a solid theoretical underpinning and relevant analytical tools are limited.

Seasonal factors operate at many levels, creating temporal and geographic shifts in human behaviour, animal husbandry, exposure concentration, and vector diversity and abundance. The extent to which human, animal and environmental factors contribute to the spread of infectious agents remains a mystery. Water-borne diseases, particularly cryptosporidiosis and giardiasis, are associated with zoonotic transmission (Hunter & Thompson 2005). Domestic livestock, especially cattle, are major reservoirs for these protozoa. Both protozoa have prolonged survival times in water and are resistance to chemical and physical agents (Brandonisio 2006). Outbreaks of infections with zoonotic nature have been shown to be associated with drinking water contaminated by livestock (Meinhardt 1996; Clark 1999; Hunter & Thompson 2005). Large scale animal production facilities, such as confined animal feeding operations (AFOs) and concentrated animal feeding operations (CAFOs) are increasing in number and size in the US (EPA 2006) and their potential for water contamination is a primary concern for large scale outbreaks. Waste from cattle production contains protozoa, such as *Cryptosporidium spp.* and *Giardia*, which can be transmitted to humans, therefore people residing in areas of high cattle density may be at increased risk for protozoan infections. In fact, county-level rates of hospitalization due to gastrointestinal infections in the US elderly were linked to high cattle density (Jagai *et al.*, 2010). Controlling the progression of a water-borne epidemic involves understanding and predicting the fate and transport of pathogens through the water distribution network (e.g. rivers, groundwater, wells, pipes and sewers) and the spread of infection from animal to human and in human population. Although conceptual models explain how diseases and organisms move through both systems, little work has been done to merge the two perspectives into a single unified framework.

Surveillance systems are a powerful tool for characterizing the spread of these diseases, identifying the aetiological agents, establishing timing and location of water-borne outbreaks, guiding regulatory agencies responsible for water quality and sanitation, developing vaccination programs, training public health practitioners on how to investigate disease outbreaks and for collaborating with local, state, federal and international agencies to control and prevent diseases. Surveillance systems are useful to evaluate technologies for providing safe drinking and recreational waters and guide research to improve water-quality regulations. The nature of water-borne diseases and their intricate connections with man-made and natural environments pose serious challenges for building effective surveillance systems. The difficulties in identification and specification of transmission of enteric infections via water routes require novel approaches to outbreak detection and disease reporting, tracking of pathogens in drinking water sources and evaluating the effects of meteorological factors. The outlined challenges suggest a need for integrated surveillance systems that utilize novel sources of environmental data, a wide implementation of state-of-art environmental and epidemiological testing, and an understanding of the intertwining connections between humans, animals and the environment.

Global climatic change combined with industrialization of food production results in modification of risk profiles for exposure to water and food-borne diseases. Predicting and measuring these changes as well as the associated occurrence of disease outbreaks has traditionally been an exercise in reactionary response with data and knowledge occurring after the fact. With increased availability of novel information technologies and developments, such as GEOSS, forecasting seasonal increases in enteric diseases becomes a reality. By assuming intrinsic relationships between meteorological conditions and disease outcomes we can use remote sensing as a proxy for the combined effects of temperature and precipitation. Remote sensing can be advantageous especially for geographic areas where consistent meteorological monitoring is lacking. Remote sensing data can also provide characteristics related to vegetation cover, landscape structure, and water content. For example, a vegetation index (VI) calculated from multispectral remote sensing data has been used as a measure of density and vigour of plant growth. There is growing interest in using VIs in epidemiological studies (Cringoli *et al.*, 2005); VIs have demonstrated good predictive properties for onchocerciasis in Ethiopia (Gebre-Michael *et al.*, 2005), schistosomiasis in Brazil (Bavia *et al.*, 2005), West Nile Virus in the US (Brownstein *et al.*, 2002), and cryptosporidiosis in Africa (Jagai *et al.*, 2009).

In this section a meta-analysis approach is described to assessment the possible seasonality of two common enteric infections with respect to temporal oscillations in remote sensing data. The presented findings suggest that satellite borne imagery can be useful for predicting seasonal increases in enteric infections on a global scale.

3.2 *Meta-analysis approach to study disease seasonality*

The systematic approach to seasonality characterization form a foundation for an extensive analysis of seasonal patterns in multiple geographical locations. One of the most intriguing questions in temporality of emerging pathogens is to determine factors governing the synchronization of seasonal peaks in various locations and/or subpopulations. To investigate seasonality of infection on a large geographical scale we developed a methodology which borrows ideas from meta-analysis research on systematic literature review coupled with mixed effect modeling techniques. This approach is illustrated using the transmission dynamics of two common enteric infections common in resource-poor areas: cryptosporidium infections (CI) and rotavirus infections (RI). Both infections are leading causes of diarrhoea in developing countries; both are resistant to traditional disinfectants, can survive for long periods in water and soil, have strong zoonotic transmission potential and are often implicated in water-borne outbreaks. Both infections are sensitive to environmental conditions, including ambient temperature and humidity, and exhibit well-defined seasonal patterns, however their seasonal patterns differ even within homogeneous climate zones and the causes for the striking seasonality are unknown.

The overall methodology consists of four steps: 1) literature search and data abstraction; 2) spatio-temporal linking and normalization; 3) data analysis and model diagnosis, and 4) result interpretation. For abstracting published data in a systematic manner we developed a set of criteria: 1) Studies must involve observational data on infection in humans that are not immunocompromised; 2) Studies must include at least a full year of data to cover annual seasons; and 3) Studies must provide records on a daily, weekly, monthly or quarterly basis. Based on each study site's longitude and latitude, we supplemented monthly relative prevalence data with time specific ambient temperature and precipitation, obtained from the National Climatic Data Center databases (NCDC 2012). When time specific temperature and precipitation data were unavailable, monthly averages were used from NCDC's global historical climatology network (GHCN).

Each study site was classified with respect to climate conditions. The Köppen scheme for classifying climates divided geographic regions into thirty categories based on ambient temperature and precipitation characteristics. Using the study sites' latitude and longitude information, each study site was supplemented with remote sensing data. Remote sensing data were used to classify geographical areas with respect to climatic conditions and to provide information of temporal oscillations as a proxy for the combined effects of temperature and precipitation in a time series fashion. Typically, VI provides a range of values, so low values of VI correspond to barren areas of rock, sand, or snow. Moderate values represent shrub and grassland, while high values indicate temperate and tropical rainforests. Thus, by assigning study sites an annual average VI value we characterize its climatic profile and by providing monthly VI values we characterize seasonal oscillations. This dual use allows us to conduct the analysis of association with the seasonal patterns of diseases controlling for climatic zone. In the presented case studies we have utilized one measure of vegetation based on two bands of multispectral data. Schematically the data abstraction process is shown in Figure 2.

Health outcomes provided in the published literature can be expressed in different forms such as death records, counts of hospitalizations, numbers of diarrheal episodes, positive tests, etc. These records are often adjusted for population at risk, and presented as morbidity or mortality rates. In order to standardize these different health outcome measures, the raw values extracted from the published literature were normalized into z-scores on a study-by-study basis as an internal reference level. Monthly z-scores were also calculated for temperature and precipitation data in order to assess relative associations.

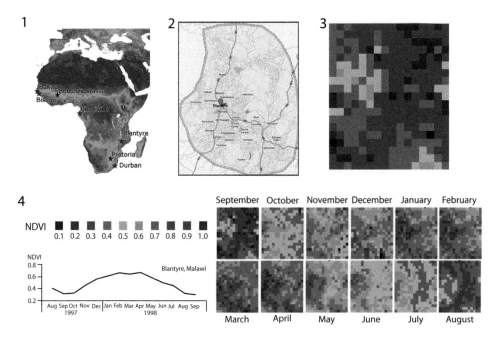

Figure 2. Schematic representation of steps in the data abstraction process for monthly vegetation index data from remote sensing imagery. Large numbers represent sequential steps in the process. (see colour plate 19)

A linear mixed effects model was used to link the z-score of monthly disease values with z-score of temperature and z-score of precipitation individually. The model also included fixed effects to control for absolute level of exposure at various locations; for example, a rainy month in a tropical location such as Bangladesh will have different implications compared to a rainy month in a dry arid location such as Kuwait. The mixed effect models were also examined utilizing the lagged exposure values to assess whether increased disease rates may be due to temperature and precipitation exposure from the previous month. The models were also run adjusting for latitude and distance of the study site from the equator.

A set of criteria was developed for abstracting published data, such as, in brief: 1) Studies must involve observational data on infection in humans that are not immuno-compromised; 2) Studies must include at least a full year of data to cover annual seasons; and 3) Studies must provide records on a daily, weekly, monthly or quarterly basis. Based on each study site's longitude and latitude, we supplemented monthly relative prevalence data with time specific ambient temperature and precipitation, obtained from the National Climatic Data Center (NCDC) databases. When time specific data were not available, temperature and precipitation averages were used from the Global Historical Climatology Network (GHCN). Each study site can be classified with respect to climate conditions.

Using the study sites' latitude and longitude information each was supplemented with information obtained from remote sensing. Remote sensing data can be used as a proxy for the combined effects of temperature and precipitation, and can be advantageous for studies conducted in areas where consistent meteorological monitoring data may not be available. Remote sensing data are primarily used to derive characteristics related to vegetation cover, landscape structure, and water content. The VI is a measure of density of plant growth which is calculated from remote sensing data. Very low values of VI (0.1 and below) correspond to barren areas of rock, sand, or snow. Moderate values represent shrub and grassland (0.2 to 0.3), while high values indicate temperate and tropical rainforests (0.6 to 0.8).

An analysis of global seasonal patterns for two water-borne diseases caused by a protozoan (cryptosporidiosis) and a rotavirus infection are presented. A meta-analysis approach was developed to assess seasonal patterns on a global scale. The data on health were collected from studies published on each disease from 1966 to 2011 and satisfy specific criteria: 1) studies must involve observational data on the disease in humans that are not immuno-compromised (HIV/AIDS patients); 2) studies must include at least a full year of data to cover all seasons; and 3), studies must provide data on a daily, weekly, monthly or quarterly basis. Based on the study site's longitude and latitude, monthly disease prevalence data were supplemented with time specific ambient temperature and precipitation, obtained from the NCDC databases. When time specific data were unavailable, temperature and precipitation normals were used from GHCN.

Health outcomes provided in the published literature can be expressed in different forms, including death records, counts of hospitalizations, numbers of diarrheal episodes, positive tests, etc. These records are often adjusted for population at risk, and presented as morbidity or mortality rates. In order to standardize these different outcome measures, the raw values extracted from the published literature were normalized into z-scores on a study-by-study basis, as an internal reference level. Monthly z-scores were also calculated for temperature and precipitation data in order to assess relative associations.

A linear mixed effects model was used to link the z-score of monthly CPI values with z-score of temperature and z-score of precipitation individually. The model also includes fixed effects to control for absolute level of exposure at various locations; for example, a rainy month in a tropical location such as Bangladesh will have different implications compared to a rainy month in a dry arid location such as Kuwait. The mixed effect models were also examined utilizing the lagged exposure values to assess whether increased CPI rates may be due to temperature and precipitation exposure from the previous month. The models were also adjusted for latitude and distance of the study site from the equator.

3.2.1 Cryptosporidiosis infections

The objective of this study was to improve understanding of seasonality of cryptosporidiosis infection and associations with climatic characteristics in sub-Saharan Africa. A time series dataset was assembled representing monthly incidence of cryptosporidiosis from studies conducted in sub-Saharan Africa. This data set was then examined for relationships between disease incidence and NDVI, a proxy for the combined effects of temperature and precipitation derived from remote sensing data.

3.2.1.1 Data abstraction

A meta-analysis was conducted of studies published on cryptosporidiosis infection (CI) from 1966 to May 2008. Literature was gathered using the Ovid MEDLINE search engine and specific keywords. A total of 138 studies met the inclusion criteria of which forty-nine studies presented data in a published journal article. After abstracting values of infection directly from publications or from authors, we compiled monthly time series for sixty-one sites. To assess the use of NDVI to predict the disease seasonal pattern, a sub-analysis was conducted for twelve sites located in sub-Saharan Africa (Table 2).

Since chlorophyll in plant leaves absorbs visible light and the cell structure of leaves reflects near infrared, these wavelengths are affected based on the density of vegetation. The Normalized Difference Vegetation Index (NDVI) measures vegetation cover, and over time can be used to monitor plant growth. NDVI is calculated as follows (Rouse et al., 1974):

$$NDVI = \frac{\rho_{nir} - \rho_{red}}{\rho_{nir} + \rho_{red}}, \tag{1}$$

where ρ_{nir} is the reflectance in the near-infrared (NIR) portion of the electromagnetic spectrum and ρ_{red} is the reflectance in the red region (RED). In the red region of the electromagnetic spectrum, sunlight is absorbed by chlorophyll, causing low red-light reflectance. In the near-infrared portion of the spectrum, the leaf's spongy mesophyll causes high reflectance. NDVI is a nonlinear function which varies between -1 and $+1$ (undefined when NIR and RED are zero). Low values of NDVI (0.1 and below) correspond to barren areas of rock, soil, sand, snow, or ice (USGS 2012). Healthy vegetation has increasingly positive values. Moderate values represent shrub and grassland (0.2 to

Table 2. Location, climate category, latitude, longitude, and period of NDVI data for studies used in analysis.

Reference	Location	Climate	Latitude	Longitude	Years of study
Bogaerts et al., 1987	Kigali, Rwanda	A	−1°57'	30°03'	1984–1985
Miller & van den Ende 1986	Durban, S. Africa	C	−29°51'	31°01'	1986
Fripp et al., 1991	Pretoria, S. Africa	C	−25°44'	28°11'	1986–1989
Molbak et al., 1993	Bissau, Guinea-Bissau	A	11°50'	−15°34'	1987–1989
Steele et al., 1989	Pretoria, S. Africa	C	−25°44'	28°11'	1987
Molbak et al., 1990	Bissau, Guinea-Bissau	A	11°50'	−15°34'	1987
Moodley et al., 1991	Durban, S. Africa	C	−29°53'	30°53;	1987
Duong et al., 1991	Libreville, Gabon	A	0°22'	9°26'	1990
Perch et al., 2001	Bissau, Guinea-Bissau	A	11°50'	−15°34'	1991–1997
Adegbola et al., 1994	Bakau, Gambia	A	13°28'	−16°39'	1991
Peng et al., 2003	Blantyre, Malawi	A	−15°46'	35°0'	1998
Tumwine et al., 2003	Kampala, Uganda	A	0°19'	32°35'	2000

0.3), while high values indicate temperate and tropical rainforests (0.6 to 0.8). NDVI is commonly used because it emphasizes spectral differences in surface materials while minimizing the effect of variable illumination conditions caused by changing topography and/or atmospheric conditions. This latter point is very important when generating global composite data sets where numerous image scenes are stitched together to create one seamless global coverage.

NDVI data were obtained from the famine early warning system (FEWS), Africa data dissemination service (ADDS) (FEWSnet 2008). The FEWS-NET NDVI data are a product from the NASA's Global Inventory Modeling and Mapping Studies (GIMMS) group, which provides eight kilometre resolution NDVI for every ten days (Pinzon et al., 2004; Tucker et al., 2005). ESRI's ArcMap 9.1 software and Python scripts were used to extract NDVI statistics for each city's study area. A study area was defined by determining the bounding coordinates for the study location using the aerial imagery within Google Earth. In defining the study area, we avoided large water bodies such as rivers, lakes, or oceans since these can skew NDVI measurements. The NDVI data were imported into ArcMap and the data range was divided by 250 to recover the original NDVI range of −1 to +1. ArcToolbox's Zonal Statistics function was used to calculate summary statistics for the minimum, maximum, mean, and standard deviation of all the NDVI pixel values within each defined study area. These summary statistics were then aggregated on a monthly basis.

3.2.1.2 Data normalization

The studies selected for this analysis used different measures for cryptosporidiosis outcome, number of cases, per cent positive stools, or prevalence. In order to standardize the different outcome measures used in each study, the raw values were normalized into z-scores on a study-by-study basis. The z-score was calculated using the mean and standard deviation for the complete duration of each study as follows:

$$Z_{ij} = \frac{x_{ij} - \bar{x}_i}{s_i}, \tag{2}$$

where Z_{ij} is the z-score for x_{ij}, the actual value for outcome (CI cases, prevalence or per cent positive stools) for study i in month j and $\bar{x}_i$ and s_i are the mean and standard deviation for each study, respectively.

The studies conducted in sub-Saharan Africa fell into two climate categories, Climate A, arid and semiarid climates, and Climate C, humid mid-latitude areas. A linear mixed effects model was used to link the z-score of monthly CPI values with NDVI individually (Model 1). The relationships were examined for all the study sites as well as for each climate category. The mixed effect regression

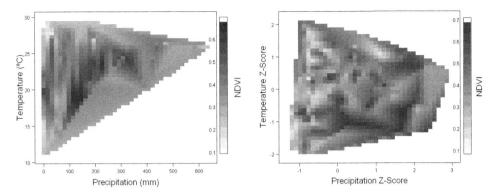

Figure 3. Left: Interpolation of NDVI in relation to precipitation and temperature; Right: precipitation z- score and temperature z-score (below).

Table 3. Descriptive statistics for all sites by climate category.

	All sites	Climate A Moist tropical	Climate C Humid mid-latitude
Number of sites	12	8	4
Number of months	302	218	84
NDVI mean/deviation	0.36/±0.12	0.36/±0.13	0.34/±0.10
Regression results Model 1[a]	0.214	−0.988	3.284*
Regression results Model 2[b]	−0.938	−3.363*	4.549*

*Significant regression parameters for fixed effects (p < 0.05)

model was defined as follows:

$$Z_{ij} = \beta_0 + \beta_1 x_{ij} + b_{0i} + b_{1i} x_i, \qquad (3)$$

where Z_{ij} is the cryptosporidiosis z-score, and x_{ij} is NDVI, the fixed effects: β_0 is the population intercept, β_1 is the population slope, and random effects: b_0 is the study intercept and b_1 is the study slope. The mixed effect models were also examined utilizing the lagged NDVI values to assess whether increased CI rates may be due to NDVI exposure from the previous month (Model 2).

3.2.1.3 Analysis
The twelve studies conducted in sub-Saharan Africa provided 302 months of data; on average each study site had twenty-three months of data. The sites were resided in the tropical and humid mid-latitude climates and the average NDVI values were about equal in both climate categories. Figure 3 demonstrates the relationship between NDVI, temperature, and precipitation for study sites in sub-Saharan Africa. The left panel shows the actual temperature and precipitation values and the right panel shows the z scores for temperature and precipitation. Higher NDVI values are associated with higher level of precipitation. Initial descriptive analysis illustrates the average values of NDVI for study sites in sub-Saharan Africa and also by climate subcategory (Table 3).

3.2.1.4 Results and interpretation
Based on the results of the regression models we found that in the humid mid-latitude areas (Climate C) an increase in NDVI predicts an increase in cryptosporidiosis for the current (Model 1) and the following (Model 2) months. In contrast, in tropical climates an increase in NDVI predicts a

103

decrease in cryptosporidiosis for the following month. When assessing synchronized (Model 1) and lagged (Model 2) relationships, NDVI with the infections were not significant (Table 3).

The results indicated that NDVI can be useful for a local forecast as a proxy for environmental exposure to the protozoa Cryptosporidium. The FEWS-NET NDVI values were used because they have been calibrated to take into account several issues including, volcanic ash from an eruption in South Africa in 1991, intra-sensor degradation, and inter-sensor degradation. The FEWS-NET data are also spatially and temporally smoothed to remove clouds and sub-pixel cloud contamination and are inter-calibrated with another high-resolution multispectral satellite. A more detailed analysis demonstrated that the NDVI values extracted from various sources may not be fully comparable (Jagai et al., 2007). Therefore, the results need to be further validated in a larger sample and in longer, more refined time series for various climate categories. Since NDVI was strongly associated with precipitation of a study location, it would be important to investigate how well NDVI correlates with water contamination. Continued interest in the use of remote sensing for public health research (Patz 2005) may lead to the development of new products which supersede NDVI and may be more appropriate for this type of analysis.

3.2.2 Rotavirus infections

The objective of this case study was to improve understanding of seasonality of RI and associations with climatic characteristics in South Asia. A time series data set was assembled representing monthly incidence of rotavirus from studies conducted in South Asia in a manner described above. They were then examined for relationships between disease incidence and the Enhanced Vegetation Index – another remote sensing measure of the combined effects of temperature and precipitation.

3.2.2.1 Data abstraction

A meta-analysis was conducted of studies published on rotavirus infection from 1966 to 2010 in South Asia (defined to include the countries of India, Pakistan, Bangladesh, Sri Lanka, Nepal, and Bhutan). As in the meta-analysis of cryptosporidiosis cases, a literature search was conducted and selected studies were evaluated based on the same criteria. A total of forty studies met the study criteria of which twenty-six studies presented data in the published journal article. We contacted the authors of eleven studies and were able to supplement data for six sites in the proper format. The final analysis incorporates data from thirty-nine studies. Four studies provided data for more than one location or period of time. Records of RI were aggregated into forty-seven monthly time series, reflecting discrete time periods and geographic regions (Table 4).

Although MODIS Terra provides both NDVI and enhanced vegetation index (EVI) products, it was decided to analyse the EVI product for several important reasons. EVI is a modification on NDVI that adds adjustment factors for soil and aerosol scattering. EVI is shown to better perform at correcting for distortions caused by reflected light from particles in the air and from the ground cover below the vegetation. In areas with heavy vegetation, EVI is more sensitive to small changes and does not become saturated as easily as NDVI. NDVI responds mostly to variations in the red band which occur from chlorophyll absorption, whereas EVI depends more on the near-infrared band that is responsive to structural variations in the canopy (NASA 2012). EVI is defined as follows (Liu & Huete 1995):

$$EVI = G \times \frac{\rho_{nir} - \rho_{red}}{\rho_{nir} + (C_1 \times \rho_{red} - C_2 \times \rho_{blue}) + L}, \tag{4}$$

where L is a soil adjustment factor. C_1 and C_2 are coefficients that correct the red band for aerosol scattering by factoring in the blue band. Reflectance values for the near-infrared, red, and blue wavelengths are designated by ρ_{nir}, ρ_{red}, and ρ_{blue}, respectively. Usually, $G = 2.5$, $C_1 = 6.0$, $C_2 = 7.5$, and $L = 1$.

EVI data were obtained from two sources: GIMMS data for locations with study dates from 1981 to 2000 and NASA's MODIS Terra satellite for locations with study dates from 2000 to 2007. GIMMS data are hosted by the Global Land Cover Facility at the University of Maryland (GLCF

Table 4. Study reference, location of study, climate, latitude, longitude, start and end date period of data used for each time series in analysis.

Reference	Location	Climate*	N Lat	E Long	Start	End	RS Data
Singh et al., 1989	Chandigarh, India	C	30°44'	76°47'	1Mar82	31Dec85	GIMMS
Ram et al., 1990	Chandigarh, India	C	30°44'	76°47'	1Dec84	31May97	GIMMS
Yachha et al., 1994	Chandigarh, India	C	28°36'	76°47'	1Oct88	28Feb91	GIMMS
Broor et al., 1993	New Delhi, India	C	28°36'	77°12'	1May88	31May90	GIMMS
Patwari et al., 1994	New Delhi, India	C	28°36'	77°12'	1Jun89	31May	GIMMS
Bahl et al., 2005	New Delhi, India	C	28°36'	77°12'	1Aug00	31Jul01	MODIS
Sharma et al., 2008	New Delhi, India	C	28°36'	77°12'	1Aug00	31Jul07	MODIS
Chakravarti et al., 2010	New Delhi, India	C	28°36'	77°12'	1Feb05	31Mar07	MODIS
Kang et al., 2009	New Delhi, India	C	28°36'	77°12'	1Nov05	30Nov07	MODIS
Sherchand et al., 2009	Kathmandu, Nepal	C	27°42'	85°12	1Nov05	31Oct07	MODIS
Phukan et al., 2003	Dibrugarh, India	C	27°30'	95°0'	1Apr99	31Mar00	GIMMS
Mishra et al., 2010	Lucknow, India	C	25°19'	80°59'	1Sep04	30Apr08	MODIS
Nath et al., 1992	Varanasi, India	C	24°52'	83°0'	1Aug88	31Jul89	GIMMS
Nishio et al., 2000	Karachi, Pakistan	B	24°52'	67°02'	1Jan90	31Dec97	GIMMS
Qazi et al., 2009	Karachi, Pakistan	B	24°52'	67°02'	1Jun05	30Jun07	MODIS
Stoll et al., 1982	Dhaka, Bangladesh	A	23°45'	90°15'	1Dec79	30Nov80	NA
Khan et al., 1988	Dhaka, Bangladesh	A	23°45'	90°15'	1Jan83	31Dec84	GIMMS
Tabassum et al., 1994	Dhaka, Bangladesh	A	23°45'	90°15'	1Jan89	31Dec89	GIMMS
Unicomb et al., 1993	Dhaka, Bangladesh	A	23°45'	90°15'	1Jun89	31Jul90	GIMMS
Unicomb et al., 1997	Dhaka, Bangladesh	A	23°45'	90°15'	1Jan90	31Dec93	GIMMS
Tanaka et al., 2007	Dhaka, Bangladesh	A	23°45'	90°15'	1Jan93	31Dec04	MODIS
Rahman et al., 2007	Dhaka, Bangladesh	A	23°45'	90°15'	1Jan01	31May06	MODIS
Qadri et al.,2007	Dhaka, Bangladesh	A	23°45'	90°15'	1Apr02	31Oct04	MODIS
Ahmed et al., 2010a	Dhaka, Bangladesh	A	23°45'	90°15'	1Jul05	30Jun06	MODIS
Black et al. 1982	Matlab, Bangladesh	A	23°21'	90°45'	1Apr78	31Mar79	NA
Bingnan et al., 1991	Matlab, Bangladesh	A	23°21'	90°45'	1Jun87	31May89	GIMMS
Unicomb et al., 1993	Matlab, Bangladesh	A	23°21'	90°45'	1Jun89	31Jul90	GIMMS
Zaman et al., 2009	Matlab, Bangladesh	A	23°21'	90°45'	1Jan00	31Dec06	MODIS
Rahman et al., 2007	Matlab, Bangladesh	A	23°21'	90°45'	1Jan01	31May06	MODIS
Saha et al., 1984	Kolkata, India	A	22°31'	88°25'	1Jul79	30Jun81	NA
Kang et al., 2009	Kolkata, India	A	22°31'	88°25'	1Nov05	30Nov07	MODIS
Nair et al., 2010	Kolkata, India	A	22°31'	88°25'	1Nov07	31Oct09	MODIS
Kang et al., 2009	Mumbai, India	A	18°54'	72°49'	1Nov05	30Nov07	MODIS
Purohit et al., 1998	Pune, India	A	18°45'	73°45'	1Jul92	30Jun96	GIMMS
Tatte et al., 2010	Pune, India	A	18°45'	73°45'	1Jan93	31Dec96	GIMMS
Tatte et al., 2010	Pune, India	A	18°45'	73°45'	1Jan04	31Dec07	MODIS
Kang et al., 2009	Pune, India	A	18°45'	73°45'	1Nov05	30Nov07	MODIS
Moe et al., 2005	Yangon, Myanmar	A	16°46'	96°10'	1Jan023	31Dec03	MODIS
Ananthan & Saravanan 2000	Chennai, India	A	13°04'	80°15'	1Dec97	31Mar99	GIMMS
Bhat et al., 1985	Bangalore, India	A	12°58'	77°34'	1Jan83	31Dec83	GIMMS
Brown et al., 1988	Vellore, India	A	12°55'	79°11'	1Aug83	31Jul85	GIMMS
Banerjee et al., 2006	Vellore, India	A	12°55'	79°11'	1Jan02	31Dec03	MODIS
Kang et al., 2009	Vellore, India	A	12°55'	79°11'	1Nov05	30Nov07	MODIS
Paniker et al., 1982	Calicut, India	A	11°15'	75°46'	1Sep76	28Feb78	NA
Kang et al., 2009	Trichy, India	A	10°48'	78°41'	1Nov05	30Nov07	MODIS
Amed et al., 2010b	Colombo, Sri Lanka	A	06°54'	79°52'	1Apr05	31Oct06	MODIS
Nyambat et al., 2009	Colombo, Sri Lanka	A	06°54'	79°52'	1Nov05	30Nov07	MODIS

*Climate categories: A – moist tropical climates, B – arid and semiarid climates, C – humid mid-latitude areas, D – colder temperate areas

**Remote sensing data source: Global Inventory Monitoring and Mapping Studies (GIMMS), MODIS

2012) and are derived from AVHRR sensors on board NOAA-7, NOAA-9, NOAA-11, NOAA-14 and NOAA-16. GIMMS processing aims to create a consistent time series of NDVI from the AVHRR satellite series by correcting for differences in solar illumination angles and sensor view angles over the period of record, caused by satellite overpass time drift. GIMMS processing also reduces NDVI variations that arise from sensor band calibration, volcanic aerosols, cloud cover and water vapour. GIMMS NDVI data were acquired on a bimonthly basis with one kilometre spatial resolution (GLCF 2012). MODIS Terra data are hosted by NASA's Reverb ECHO website and are available from 2000 to present at temporal resolutions of 16-days and 1-month, and at spatial resolutions of 250 m, 500 m, 1 km and 0.25°lat./lon. From the MODIS Terra satellite, monthly composite EVI data were acquired at one kilometre resolution. A one kilometre spatial resolution was chosen to remain consistent with the GIMMS data. Although MODIS data are available on a bimonthly basis, we acquired monthly composite data to help remove errors caused by clouds, sun glare, and the satellite instrument itself and to match the monthly rotavirus time series (NASA 2012). The MODIS sensor has many advantages over the AVHRR satellite series for calculating vegetation indices. MODIS offers a higher spatial resolution and also has narrower bandwidths in the red and near infrared that improve its sensitivity to chlorophyll. The MODIS satellite also produces less distortion from atmospheric water vapour (NASA 2012).

Vegetation indices were not available for four studies prior to 1981. The only instrument acquiring daily global imagery was the AVHRR sensor on board the NOAA-6 satellite, which was launched in 1979. NOAA-6 has a daylight overpass time of 0730 hours whereas NOAA-7, launched in 1981, has a daylight overpass time of 1430 hours. NDVI products are not created from the NOAA satellites with early morning daylight overpass times (NOAA-6 and NOAA-8) because the lower solar zenith angle produces inconsistent and less intense radiance measurements than the afternoon solar illumination conditions from the other NOAA satellites (Goward et al., 1985). Landsat multispectral scanner MSS offers an alternative to the AVHRR sensor for vegetation studies prior to 1981 (Deering et al., 1975). However, due to its eighteen day repeat cycle Landsat MSS does not produce enough cloud-free images to create a monthly time series for measuring seasonal vegetation dynamics (Goward et al., 1985).

Each study area was defined by locating the city on aerial imagery within Google Earth and creating a polygon around the city measuring forty square kilometres that avoided large water bodies. Esri's ArcMap 9.2 software and the Arc2Earth extension converted the polygon KML files to ArcGIS format. Bimonthly GIMMS data were acquired as georeferenced TIF images in a geographic coordinate system, which are fully compatible with ArcGIS 9.2. Duke University's marine geospatial ecology toolset (MGET) converted the monthly MODIS data from the HDF file format to ArcGIS rasters. For quality control, each study location polygon was overlaid onto the satellite imagery in conjunction with global country data to ensure proper alignment. Python scripts and ArcToolbox's Zonal Statistics function were used to create a time series of the mean vegetation index for each study area. For the GIMMS data, bimonthly statistics were averaged to create one monthly value.

3.2.2.2 Data normalization

As in the previous case study, studies selected for this analysis used different measures for rotavirus outcome. Of the thirty-nine studies, twenty-five (64.1%) presented outcome data as *number-of-cases*, thirteen (33.3%) presented outcome as *per cent positive stools*, and one study (2.6%) presented the outcome as *incidence*. To standardize these different outcome measures, the raw values were normalized into z-scores on a time series-by-time series basis. EVI data was also normalized to z-scores on a time series-by-time series basis.

3.2.2.3 Analysis

The relationship between the normalized outcome and EVI was examined using a regression model adapted to time series data. Similar to the analysis for cryptosporidiosis, a linear mixed effects model was used to link the z-score of monthly rotavirus values with z-score of EVI individually and control for the length of the time series at each study location. The overall relationships for all the study

Table 5. Descriptive characteristics for all sites and by climate category. Mean and standard deviation for temperature and precipitation, vegetation index, number of time series, and number of months covered.

Exposure variable	All sites	Climate A Moist tropical	Climate B Arid & semi-arid	Climate C Humid mid-latitude
Temperature (°C)	25.55 ± 4.81	26.24 ± 3.33	26.70 ± 4.22	24.22 ± 6.53
Precipitation (mm)	90.13 ± 129.47	111.28 ± 147.88	24.20 ± 36.30	58.82 ± 83.37
Number of time series	47	30	2	15
Months	1046	649	36	361
Vegetation index	0.31 ± 0.11	0.33 ± 0.90	0.13 ± 0.02	0.30 ± 0.11
Number of time series	43	26	2	15
Months	972	575	36	361

Table 6. Regression parameters and confidence intervals for models 1 and 2 for all studies and by climate category. Analysis was not conducted for Climate B because there were only two locations in this category.

	All studies		Climate A Moist tropical		Climate C Humid mid-latitude	
Exposure variable	Estimate	95% CI	Estimate	95% CI	Estimate	95% CI
Model 1						
Temperature	−0.379*	−0.435, −0.323	−0.444*	−0.513, −0.75	−0.310*	−0.408, −0.211
Precipitation	−0.246*	−0.305, −0.188	−0.259*	−0.334, −0.185	−0.271*	−0.370, −0.171
Vegetation index	−0.253*	−0.314, −0.193	−0.240*	−0.320, −0.161	−0.281*	−0.380, −0.182
Model 2						
Temperature	−0.288*	−0.347, −0.229	−0.363*	−0.434, −0.291	−0.209*	−0.314, −0.105
Precipitation	−0.186*	−0.242, −0.130	−0.209*	−0.284, −0.134	−0.173*	−0.261, −0.084
Vegetation index	−0.154*	−0.217, −00.91	−0.161*	−0.242, −0.081	−0.135*	−0.241, −0.029

*Significant parameters (p < 0.05)

sites were examined as well as for tropical climates (Climate A) and humid mid-latitude climates (Climate C). Relationships for the arid/semi-arid (Climate B) category were not examined because there were only two studies for this climate type.

A total of forty-seven time series were used in this analysis to assess the relationship between monthly rotavirus incidence in South Asia and EVI values. All study locations provided a total of 1046 months of data; on average, each study location had twenty-two months of data. Descriptive analysis illustrates the average values of temperature, cumulative precipitation and vegetation index for all study sites and also for each climate subcategory (Table 5). As all studies are from South Asia, there is little difference in the average temperature range by climate categories. The primary variation by climate category is seen in precipitation. The vegetation index is similar for moist tropical areas (A) and humid mid-latitude areas (C); but averaged much smaller values in the arid and semiarid region (B).

The highest levels of rotavirus are seen in the colder, drier months of the year (roughly December to March). When assessing synchronized relationships (Model 1) EVI z-score was a significant negative predictor of rotavirus z-score and this relationship held for both type A and C climates (Table 6). For all studies, a one unit increase in EVI z-score resulted in a −0.253 (95% CI: −0.314, −0.193) unit decrease in the rotavirus z-score. A decrease of 0.1 units in vegetation index for a seasonal change compared to the annual norm of 0.3 is associated with a 3.8 per cent increase above the annual rotavirus level.

The primary reasons for seasonal fluctuation in rotavirus incidence are still unknown. Generally, it is not expected that the seasonal pattern would be primarily driven by a favourable environment for the virus' survival and transmission. Temporal and geographic trends in the rotaviral infection have been better recognized in temperate than tropical climates. It has been suggested the seasonal pattern seen in rotavirus may be driven by air-borne transmission of the disease (Ansari *et al.*, 1991; Parashar *et al.*, 1998; Torok *et al.*, 1997). The drop in humidity and rainfall dries the soils which may increase the aerial transport of the contaminated faecal matter. In the US the seasonal pattern in rotavirus shifts by geographic location with peak activity occurring first in the Southwest from October through December and last in the Northeast in April or May (Torok *et al.*, 1997). This pattern is similar to that seen in for other respiratory viruses such as influenza and measles (Wenger & Naumova 2008). A recent paper by Pitzer (*et al.*, 2009) demonstrates that the seasonal pattern seen in rotavirus in the US may be influenced by spatiotemporal patterns in birth rate. Given high birth rates in most tropical countries, it may be expected that disease would start in tropical regions and spread to areas with lower birth rates, but these data may be difficult to obtain, given the relative lack of reporting of rotavirus disease in these countries. Preliminary evidence suggests that the seasonal pattern for rotavirus may vary by strain type (Sarkar *et al.*, 2008). Recent publication by Levy (*et al.*, 2008), indicates that high incidence in rotaviral infection is associated with relatively cold and dry weather in South Asia. Our findings agree with the published results. Furthermore, the use of remote sensing offered a reliable proxy for meteorological characteristics for areas in which ground measures of temperature and precipitation are not available.

In the presented case studies, we capitalized on previously published methodology for assessing the relationships between meteorological characteristics and disease outcomes (Jagai *et al.*, 2009; Naumova *et al.*, 2007). Continual land use changes and man-made disturbances pose the practical challenge of effectively monitoring changing environmental variables that could be used to reliably predict outbreaks. By focusing on pathogen sensitivity to climate and weather variations and by studying disease transmission and manifestation in a changing environment triggered by extreme natural conditions, we will better understand potential scenarios in non-intentional pathogen release. Extreme meteorological events like downpours, floods, droughts, heat waves, and rapid snowmelts affect both the environment and human society. They also affect natural habitats, livelihood and infrastructure. On the other hand, extreme weather events create a unique natural laboratory for scientists to better understand our biological world and our ability to detect, control, and prevent the spread of infection triggered by an event to become better trained and prepared to respond.

Land use and meteorological characteristics can be effectively measured using satellite imagery and readily utilized to monitor and forecast disease outbreaks. Earth monitoring satellites collect large-scale and time-sensitive data globally for relatively low cost. Furthermore, data obtained in near real-time might enable rapid responses. Satellite borne imagery can provide a whole range of new measures tailored for specific locations and purposes by accounting for different sources of variations. The proposed methodology capitalizes on the secondary data analysis of ready-for-use health records (surveillance records, medical claims, and vital statistics) and continuous environmental monitoring. By better understanding the intrinsic relationships between meteorological conditions and disease outcomes, this methodology can be applied to a variety of infectious pathogens (Naumova *et al.*, 2007). The strong predictive properties of remote sensing data offer new approaches for innovations in disease forecasting, early disease warning, and the design of preventive programs at the global and regional levels. With improved disease surveillance, testing and diagnosis, weather monitoring and the use of remote sensing global disease forecasting on a very refined spatial scale is a feasible goal.

4 CLEAN WATER AND SANITATION

Recognizing the importance of sanitation as a corner stone of public health efforts, the United Nations General Assembly named 2008 as the *International Year of Sanitation*. This is not a new item of global, political concern – over sixty years ago, during India's struggle for sovereignty, Mahatma Gandhi noted that, *sanitation is more important than independence*. The year 2008 also marks the centennial anniversary of the first use of chlorine to routinely disinfect a public drinking water supply in the US (EPA 2000). Although the importance of safe drinking water may be the first thought that comes to mind during discussion of how to address the problem of water-borne disease, sanitation is an equally, if not a more, important factor in preventing disease. Decades after Gandhi's declaration, the WHO estimates that over one billion people still do not have access to clean water, and 2.6 billion do not have access to basic sanitation (WHO/UNICEF 2006). Clearly, modern societies have failed to address the basic needs of its most disadvantaged citizens in not providing these essential elements of public health and wellbeing.

The WHO estimates that diarrheal diseases cause the deaths of some 1.7 million people, most of them children, annually, and that eighty-eight per cent of this diarrheal disease burden is related to unsafe water and the associated factors of poor sanitation and hygiene (Prüss-Üstün et al., 2008). These estimates are only an educated guess, however, and the true extent of disease incidence linked to water cannot be known. The poorest and most vulnerable members of society bear the greatest burden of water-borne and diarrheal disease. With limited access to health care, many of the poor fail to have their diseases diagnosed or treated. Many illnesses affecting the poor are simply never seen and recorded at health care centres, or reported to surveillance agencies.

Water-borne disease can be strictly defined as disease caused directly by consumption of pathogenic microorganisms present in contaminated water. More realistically, however, water-borne diseases are intimately and inextricably linked both to diseases transmitted through the oral-faecal route and to water-related diseases, and should be considered in this context. Contaminated water may be used to wash and prepare food, resulting in food borne transmission of disease, or may be aerosolized, allowing for air-borne transmission. Pathogenic microbes are also passed back and forth between water and soil, for example, through flooding events and agricultural irrigation. For these and related reasons, microbial agents that cause water-borne disease can quickly have their route of transmission altered to cause diseases interchangeably designated as food-borne, soil-borne or otherwise. Additionally, many pathogens that may not cause disease directly through a water-borne route nonetheless rely on a host that depends on water for at least a part of its life cycle and are clearly water-related. All of these water-borne/food borne/water-related disease threats can be effectively addressed by targeting the issues of unsafe water, poor sanitation and hygiene.

4.1 *How and why water-borne diseases are quantified*

A critical element of preventing and managing infectious disease outbreaks, including those of water-borne diseases, is surveillance. Surveillance involves reporting and gathering information on occurrences of infectious disease, followed by analysis and sharing of information by public health authorities. Identification of outbreaks and trends in turn informs decision-making and control measures taken by public health agencies. In many resource-rich, developed countries, public health systems have well established infectious disease surveillance programs. Great importance is placed both on preventing disease outbreaks as well as on being able to identify water-borne disease outbreaks at an early stage so that control measures can quickly be implemented. Both of these aims benefit greatly from well-designed and well-implemented surveillance efforts.

Surveillance programs may be mandated both by regional and national law. Sweden and the US were some of the earliest of the developed countries to establish programs for the surveillance of infectious diseases. In both countries, emphasis is placed on selected diseases, with doctors informing government agencies of notifiable diseases (Andersson & Bohan 2001). In the US, health care providers report notifiable diseases to local and state public health authorities, which in turn forward these data to national agencies like the CDC. Additionally, other federal agencies in the US may be assigned responsibility for specific aspects of surveillance; for example, the US Food and Drug Administration (FDA) is tasked with examining occurrences of food-borne disease outbreaks for foods falling under FDA regulation (GAO 2004).

Water-borne diseases that are selected for surveillance are chosen in part on the basis of disease severity and frequency of outbreaks. In Sweden, notifiable diseases that can be related to water-borne outbreaks include amoebiasis, camplylobacteriosis, cholera, enterohaemorrhagic

Escherichia coli infections, giardiasis, hepatitis A, paratyphoid and typhoid fevers, salmonellosis, shigellosis, and yersiniosis (Andersson & Bohan, 2001). In the US, the CDC's listing of water-related notifiable diseases essentially mirrors the Swedish listing, but also includes additional diseases such as cryptosporidiosis, a protozoan disease that has had greater impact in the US (CDC 2008). Cryptosporidiosis only relatively recently came to the forefront of public health awareness in the US when an outbreak affecting more than 400,000 people occurred in Milwaukee in 1993. Such a large outbreak of water-borne disease reveals shortcomings of a country's public health system. Lessons can be learned from such outbreaks, however, leading to improvements in surveillance and disease prevention. One consequence of the Milwaukee outbreak was an implementation of stricter standards for drinking water quality, specifically with regard to turbidity.

The WHO and many developed countries have provided assistance and expertise to less experienced countries in establishing and improving infectious disease surveillance efforts. Complementing the efforts of individual countries, the WHO also coordinates international, cooperative efforts of surveillance and fosters information gathering and exchange. In an important development to international surveillance efforts in 2005, over 190 nations signed a revised panel of international health regulations (IHR) developed under the auspices of the World Health Assembly. The IHR guides the activities of the WHO and cooperating member states in identifying, responding to, and sharing information on disease outbreaks and other public health-related events in order to prevent and limit the spread of disease while minimizing interference with international trade and travel (WHO 2005).

4.2 *How can surveillance be improved for diseases not being quantified?*

Water-borne diseases are obviously not confined to national boundaries, but individual countries place different emphases onto which diseases they focus. A country's focus on monitoring and addressing particular diseases may be based on a variety of factors, including but not limited to: the historic occurrence and importance of a disease; the flexibility of government institutions in responding to changing public health conditions, especially to newly emerging or re-emerging disease threats; and constraints imposed by limited resources including poverty and a government's ability and willingness to commit resources. This last issue of limited finances and resources is related to issues of poverty and disparities in access to health care and sanitation alluded to in the Introduction. To examine this issue in greater detail, two case studies are reviewed in which our research group has been involved in two very different areas of the world: India, and Native American Reservations in Montana, US.

4.2.1 *India*
In areas such as the Indian subcontinent, surveillance efforts have included monitoring endemic and seasonally recurring old world water-borne diseases such as cholera. Significant efforts have been made to address cholera and other water-borne diseases through the operation of dedicated research, surveillance, and treatment programs, including those operated by the National Institute for Cholera and Enteric Diseases (NICED 2005) in Kolkata, India, and the International Centre for Diarrhoeal Disease Research in Dhaka, Bangladesh (ICDDRB 2008). In spite of these meaningful efforts, due to persistent conditions of poverty and poor sanitation, India continues to experience a high incidence of water-borne and diarrheal disease. The combination of a large burden of disease and the poor's limited access to health care contribute to a situation in which newly recognized pathogens and associated emerging disease syndromes are not well screened for and are not well recognized as public health issues. For example, until recently, little recognition has been given in India to emerging diarrheal disease agents such as Shiga-toxin producing *Escherichia coli* (STEC), and more specifically, enterohaemorrhagic *Escherichia coli* (EHEC) that have come under close scrutiny in more developed nations. As recently as 2002, it was noted that STEC was not a potential cause of human diarrhoea in India, and did not portray the same threat as in other Western countries (Khan *et al.*, 2002). Reports identifying *E. coli* O157:H7 and STEC bacteria in the Ganges River, and EHEC in potable water distribution systems in the major city of Lucknow, suggest that these bacteria

have a well-established presence in India (Hamner *et al.*, 2007; Ram *et al.*, 2007; Ram *et al.*, 2008). These findings raise the question of how large a role these bacteria might be playing in causing diarrhoea in a country having a large burden of diarrheal disease and in which hundreds of millions of people living in poverty have little access to health care. The O157:H7 strains isolated from the Ganges in Varanasi were all found to have a sorbitol-positive phenotype (Hamner *et al.*, 2007), unlike most strains described in the US and other countries that are sorbitol-negative. A government of India website guide mirrors US recommendations that suspected infections be screened for O157:H7 using sorbitol-MacKonkey medium selective for the sorbitol-negative phenotype (MHFW 2008). It may well be that India's strains of O157:H7 bacteria are predominantly sorbitol-positive and that they are simply not being detected using current clinical screening techniques. Such a scenario, involving the issues of poverty, disparities in access to health care and inadequate consideration of genetic variation among related pathogenic bacteria, may partially explain the lack of recognition and surveillance for O157:H7 in India.

4.2.2 *Native American Montana*
More developed countries, on the other hand, have been able to devote resources to monitor and address recently recognized, emerging pathogens such as *Cryptospiridum parvum* and *E. coli* O157:H7. Like India, however, similar issues of poverty, inadequate sanitation facilities, and other disparities in health intervention also plague certain communities in the US. Native American tribal communities living in rural reservation settings are representative of such groups. In recent years, our research group has been collaborating with local Native American tribal college faculty and students in Montana. In consultation with a tribal community's Health Steering Committee, a pilot research project was initiated in 2008 to screen for the presence of EHEC bacteria in local tribal rivers. The rationale for this study was three fold: 1) the absence of a sewage treatment plant at the town in question; 2) the fact that there was only an antiquated sewage lagoon that leaked into the adjacent river; and 3) the presence of a large cattle ranching operation upstream from the town. Despite having only one dedicated tribal college student collecting water samples and being trained in bacterial isolation and molecular biology techniques during an eight-week period, the team was able to identify an isolate of *E. coli* that tested positive for Shiga toxin-I and intimin genes, genetic elements indicative of virulent EHEC. Follow-up testing has revealed that the isolate is not O157:H7, but is instead from the related serotype O111:H8, also associated with bloody diarrhoea and haemolytic uremic syndrome. At the time of this writing, the largest outbreak of previously rare O111-associated illness in the US had just occurred in Oklahoma, affecting over 300 people and causing one death (OSDH 2008). With the discovery of a potentially pathogenic O111 isolate in river water used for swimming as well as for drinking during traditional Native American religious practices, the lab group and the tribal health committee are now expanding monitoring and testing of tribal rivers to assess how prevalent these and related diarrhoea-genic bacteria may be. Water monitoring and discovery of EHEC bacteria cannot predict and may not prevent a future outbreak of disease. However, these findings may prove useful to the local health steering committee in its attempts to secure funding to upgrade their town's failing sewage collection system and build a sewage treatment plant, thereby improving public health infrastructure. Such findings also underscore the need to more adequately measure the occurrence of related diarrheal and water-borne disease in an underserved community which has not enjoyed the benefits of public health interventions found in more developed communities of the mainstream US.

Of course, public health disparities are not the only reason why water-borne diseases are under-reported and under-quantified. A variety of factors and constraints that limit water-borne disease surveillance could be identified and discussed. For example, specific diseases and their aetiological agents may be masked during co-infection with other disease agents. At a national level, governments might at times deliberately choose not to monitor or report on certain waterborne/infectious diseases for fear of unjustified travel and trade-related sanctions imposed by other countries (WHO 2007). The list goes on. Perhaps the most cogent summary of why infectious diseases are not adequately detected and addressed is offered in this summary statement from a 2000 US government report: *Development of an effective global surveillance and response system probably is at least a decade or more away, owing to inadequate coordination and funding at the international level and lack of capacity, funds, and commitment in many developing and former communist states. Although overall global health care capacity has improved substantially in recent decades, the gap between rich and poorer countries in the availability and quality of health care...is widening* (National Intelligence Council 2000).

4.3 Developments in infectious disease surveillance

Disease outbreaks are difficult to identify immediately. During a process that may take days if not weeks, doctors treating a patient with a notifiable disease must contact local health departments, which then forward information to a regional or national agency such as the CDC. The sooner an outbreak is identified, the more rapidly the cause can be determined and interventions taken. New applications of computer and information technology are showing promise in identifying outbreaks more quickly than is possible using traditional public health surveillance practices. One such application is HealthMap, developed in 2006 by medical and bioinformatics personnel at the Children's Hospital in Boston. With an understanding that the internet has much of the information necessary for early detection and surveillance of disease outbreaks, HealthMap has been developed as a data-mining tool to query Web sources in real-time for information about disease outbreaks. Once data are gathered and evaluated, graphic summaries are presented on a Google map with information on disease alerts, locations, and links to information sources (Brownstein *et al.*, 2008; Freifeld *et al.*, 2008). Internet sources include media reports and news sites, as well as blog sites, chat room discussions, and email listings related to public health. Given the possibility of false positives from such a disparate array of sources, HealthMap also incorporates official information such as WHO announcements, and takes into account expert sources including the program for monitoring emerging diseases (ProMED). ProMED is operated by the International Society for Infectious Diseases (ISID), and releases email announcements and alerts of disease outbreaks. Reports are compiled based on information from traditional sources, the internet, and from the ProMED subscriber base. Compiled information is screened by experts and then emailed to a subscriber list of over 30,000 participants worldwide (Madoff 2004).

A precursor to HealthMap, the global public health intelligence network (GPHIN) was developed by the Public Health Agency of Canada (Mykhalovskiy & Weir 2004). Like HealthMap, GPHIN gathers its information primarily from internet sites. GPHIN is a partner of the WHO's global outbreak alert and response network (GOARN) set up in 1997. ProMED, acting in parallel with GPHIN, provided the outside world with some of the first news of the outbreak of severe acute respiratory syndrome (SARS) in China in 2003, before any official government announcements were forthcoming. The SARS outbreak provided GOARN with its first serious challenge of detecting and responding to a major disease outbreak (Heymann & Rodier 2004).

4.4 Burden of disease and DALYs

In 1995, the WHO began publishing their world health reports in an effort to provide an annual assessment of global health. Statistical information on the global burden of communicable and non-communicable diseases is presented in these reports as a result of the global burden of disease study that was a collaboration between the WHO and Harvard University in the US in 1990 (Murray & Lopez 1996). Today's WHO global burden of disease project is described as a response to the need for comprehensive, consistent and comparable information on diseases and injuries at global, regional, and national levels (WHO 2008a). Although each edition is characterized by a general theme, most world health reports have included estimates of morbidity and mortality for selected infectious and parasitic diseases, including the category of diarrhoea. In general, this information has been and continues to be taken as the authoritative source for burden of disease. For the first three years that data are provided for all ages, 1995, 1996 and 1997, estimates for mortality from diarrheal disease decreased from about three million in 1995 to about 2.5 million in 1996 and 1997.

Table 7.   Data on mortality, incidence and DALYs from world health reports (WHO 2008b).

Year	Mortality	Rank	Incidence	Rank	DALYs	Rank
2002	1,797,972	7	4,512,989,000	1	61,966,183	5
2001	2,001,193	7	4,440,192,000	1	62,450,782	5
2000	1,797,073	6	4,402,571,000	1	62,287,598	5
1999	2,213,000	6			72,063,000	4
1998	2,219,000	6			73,100,000	3
1997	2,455,000	6	4,000,000,000	1		
1996	2,473,000	6	4,002,000,000	1		
1995	3,115,000	NA	4,002,000,000			
1993*	3,010,000	4	1,821,000,000			

*Data on diarrhoeal disease episodes provided for 1993 are only for children under five years of age

However, estimates of morbidity remained constant at around four billion episodes (increasing to 4.5 billion in 2002) (Table 7). All world health reports are available through the WHO website (WHO 2008b).

What do these numbers really mean in terms of burden of water-borne disease? A review paper published in 1999 (Ford 1999) discusses these numbers and, in particular focuses on the problem of underreporting of water-borne diseases. For example, in a study of water-borne disease incidence in Hyderabad, India, we found that self-reported incidence of gastrointestinal disease was on average 200 fold higher than officially reported numbers (Mohanty *et al.*, 2002). This underreporting is not surprising given the fact that diarrheal disease is not a reportable disease, even in developed countries, unless defined as an outbreak. In addition, most diarrheal episodes do not result in hospitalization or even a visit to a clinic. Infections may also be asymptomatic or not be reported by a clinician, even if sufficiently severe to promote a visit to a hospital or clinic. There is also the discussion of how much diarrheal disease is attributable to water? Amongst communities with limited access to sanitation, the water route of exposure is likely to be extremely significant. The World Health Report for 1996 suggested that up to seventy per cent of diarrheal episodes could be caused by contaminated food, and of course person-to-person transmission (secondary transmission) is also a major route of exposure to pathogens that cause diarrhoea and other disease outcomes. Pathogens can multiply rapidly in food materials, and it is likely that food preparation with contaminated water is a major cause of food-borne diarrheal disease. It would be impossible for the clinician to distinguish between water-borne, food-borne and person-to-person transmission of diarrheal diseases, and only epidemiological investigations of outbreaks can provide an indication of source of exposure to the causative agent.

Recognizing that reporting incidence of disease episodes provided a false impression of true disease burden, the global burden of disease study developed the concept of DALYs to more accurately reflect the burden of disease on human health. For example, if we compare incidence of disease tables for 2002 (WHO 2008c), annual incidence of diarrheal disease episodes was estimated at 4.5 billion, and annual incidence of HIV/AIDS was estimated at 8,350,000. From these data, diarrheal disease episodes occur 540 times as often as HIV/AIDS cases, yet no-one would compare an episode of diarrhoea with a case of HIV/AIDS. However, prior to 1998, diarrheal disease was ranked as the top cause of morbidity, based on incidence alone. DALYs allow an assessment of years lived with disability to be included in the calculation of disease burden. In the case of diarrheal disease and HIV/AIDS, the DALYs calculation for diarrheal disease episodes is 61,966,183 and for HIV/AIDS it is 84,457,784, numbers that give a far heavier weighting to a single case of HIV/AIDS than to an episode of diarrheal disease, considered far more appropriate for public health and policy debate.

Although generally the accepted parameter for measuring burden of disease, DALYs are not without controversy due to the number of assumptions that are made in their calculations and the manner in which specific weights for different diseases are calculated. The WHO website

113

remains the best source for information on calculating burden of disease. Quoting from Mathers (*et al.*, 2001), *The DALY is a health gap measure that combines both time lost due to premature mortality and non-fatal conditions. The DALY extends the concept of potential years of life lost due to premature death (PYLL) to include equivalent years of 'healthy' life lost by virtue of being in states other than good health.*

A DALY is therefore the sum of years of life lost (YLL) and years of life living with disability (YLD). Each of these parameters, however, contains both assumptions and weights that are subject to criticism and continued discussion. A calculation of a DALY requires social value choices that include: 1) How long 'should' people in good health expect to live?; 2) Is a year of healthy life gained now worth more to society than a year of healthy life gained sometime in the future, for instance in 20 years' time?; 3) How should we compare years of life lost through death with years lived with poor health or disability of various levels of severity?; 4) Are lost years of healthy life valued more at some ages than others?; and 5), Are all people equal? Do all people lose the same amount of health through death at a given age, even if there are variations in current life expectancies between population groups?

Some of these value choices raise serious ethical considerations that prompted considerable debate, particularly in the 1990s. Calculating DALYs is not trivial but fortunately the WHO provides a template to perform these calculations. To calculate YLL requires knowledge of the number of deaths from a specific disease, the age of death and the sex of each affected individual. Data for each disease is clustered into age groups and sex, and life expectancy (LE) if death had not occurred is then calculated from standard life tables. To address social value choice #1, WHO uses life tables that assume the average male lives to be eighty years old and the average female lives to be over eighty-two years of age.

4.4.1   Calculation of DALYs

One DALY is designed to represent the loss of one year of full health. The metric is calculated from years of life lost from premature mortality (YLL), plus years of life living with disability (YLD) (Mathers *et al.*, 2001). The simple calculation for YLL in each age and sex category is a function of numbers of deaths (N) and the standard life expectancy for that particular age and sex (L), $YLL = N \times L$. YLD can be simply calculated as a function of the number of incident cases (I) in each age and sex category, the average duration of the disability (L) and a disability weighting (DW) which is an assessment of the severity of a specific disease based on both empirical evidence and subjective evaluation, $YLD = I \times DW \times L$. These equations become far more complex when social values are considered in the development of these estimations. For example, YLL and YLD are calculated using the following equations when non-uniform age-weights and discounting for future years lost are applied:

$$YLL = NCe^{(ra)}/(\beta+r)^2[e^{-(\beta+r)(L+a)}[-(\beta+r)(L+a)-1]-e^{-(\beta+r)a}[-(\beta+r)a-1]] \quad (5)$$

$$YLD = IDWCe^{(ra)}/(\beta+r)^2[e^{-(\beta+r)(L+a)}[-(\beta+r)(L+a)-1]-e^{-(\beta+r)a}[-(\beta+r)a-1]], \quad (6)$$

where r is the discount rate (GBD standard value is 0.03), C is the age-weighting correction constant (GBD standard value is 0.1658), β is the parameter from the age-weighting function (GBD standard value is 0.04), a is the age of onset, and L is the duration of disability or time lost due to premature mortality.

In the WHO calculations, a standard discount rate of three per cent and an age weighting is then applied to calculate YLL. The discount rate of three per cent is applied to the calculation to give greater weighting to a year of healthy life gained today than in the future and addresses social choice #2. Age weighting is applied to address social choice #4, and assumes that a year lived by a young adult is more valuable than a year lived by a young child or older adult. Both discounting and age weighting are controversial because of the values placed on human life at different ages. However, they are relatively minor assumptions in relation to other aspects of data uncertainty.

Calculation of YLD requires incidence of a specific disease, disaggregated by age of onset, sex and duration. A disability weighting is then applied which is specific to each disease and YLD then calculated using discounting and age weighting, as above. Disability weighting is a somewhat subjective scale between zero and one, where zero represents perfect health and one is equivalent to death. Tables of disability weights for specific disease states are provided on the WHO website and have been estimated for different age groups. For example, the disability weighting for a diarrheal episode for a child under five is 0.119, but 0.086 for ages between fifteen and fifty-nine, reflecting the increased severity of diarrhoea in young children. In contrast, the disability weighting for HIV is 0.505 for all ages. The disability weighting, duration of disability and years of life lost are the real drivers of the DALY metric that provide a far better measure of disease burden than incidence alone.

The WHO and Harvard group initiated the global burden of disease (GBD) study, conducted an enormous amount of work in collating data and determining appropriate weights for different disease states. What they have produced, while not perfect, is a useful mechanism to quantify the impact of disease. To some, this may seem only to be putting numbers to something that is obvious; there is a vast burden of diarrheal disease, but an episode of diarrhoea is far less severe than a case of HIV. A key question is how do we define an episode of diarrheal disease? There are people that have never had a solid stool movement, and who are constantly re-inoculated by water-borne pathogens. Yet for DALY calculations, the WHO recommends modelling the duration of a diarrheal episode as short as a few days for uncomplicated episodes, to more or less two weeks for complicated episodes.

Quantification of disease burden is both the strength of the DALY approach and its weakness. Numbers are useful for decision makers who are faced with limited resources and who need to prioritize public health interventions. However, using global data can at times be misleading. In some areas of the world, death from diarrheal disease in children younger than five years of age exceeds all other causes of mortality, not including a category referred to as neonatal and other ill-defined causes. To put this in perspective, Morris (*et al.*, 2003) conducted a meta-analysis of studies reporting under-five mortality in developing countries without vital registration systems. Their analysis of observed proportion of deaths from thirty-eight separate studies indicated that twenty-two per cent of deaths in children under five were attributable to diarrheal disease, 20.5 per cent to pneumonia, 8.2 per cent to malaria, 2.7 per cent to measles and 46.6 per cent to neonatal, other, and undetermined. For this age group at least, diarrheal diseases are a major killer and should be a primary target for public health intervention, particularly as they are largely preventable.

Using GBD data, Prüss-Üstün (*et al.*, 2002) provided one of the first exposure-based estimates of disease burden from water sanitation and hygiene. Their analysis concluded that four per cent of all deaths and 5.7 per cent of DALYs were attributable to water, sanitation and hygiene. They focused on diarrheal diseases, schistosomiasis, trachoma, ascariasis, trichuriasis and hookworm disease, acknowledging that many potential water-borne diseases were as yet unquantifiable. In their later analysis, the authors include diarrhoea, malnutrition, intestinal nematode infections (e.g. ascariasis, trichuriasis and hookworm), lymphatic filariasis, trachoma, schistosomiasis, malaria, drowning and a category labelled *other* that included dengue, Japanese encephalitis, and onchocerciasis. In this updated analysis, they suggest that global improvements in water, sanitation and hygiene could potentially prevent at least 9.1 per cent of disease burden in DALYs or 6.3 per cent of all deaths (Prüss-Üstün *et al.*, 2006). These percentages are, of course, proportionally much higher in children. Figure 4 shows the percentage of total global burden of disease represented by the water-borne diseases discussed paper and calculated from the 2002 data (Prüss-Üstün *et al.*, 2008; WHO 2008c). HIV/AIDS is included for comparison. This undoubtedly remains an underestimate for water-borne diseases as there are potentially a large number of diseases that remain unquantifiable by the water route of exposure.

Other diseases, viral such as Hepatitis A and E, bacterial such as melioidosis, and protozoal, such as toxoplasmosis, cause an enormous burden of infectious disease, both in certain areas of the world and in specific populations, yet are not quantified within the Global Burden of Disease project. Hepatitis E is common in many parts of the world with outbreaks affecting up to 100,000 people reported (Schwartz *et al.*, 2006). In addition mortality can be as high as twenty per cent

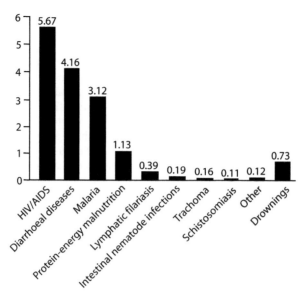

Figure 4.   Water-borne diseases as a percentage of total global burden of DALYs 2002: All causes (1.5×10^9); diarrhoeal causes (61.9×10^6).

in pregnant women. Melioidosis, caused by *Burkholdaria pseudomallei,* is endemic in Southeast Asia and northern Australia, and sporadic cases occur in many other parts of the developing world. Mortality from melioidosis ranges from ten to ninety per cent, if disseminated septicemia is present (Rega *et al.*, 2007). The primary route of transmission of hepatitis A and E is through faecally-contaminated water, and melioidosis is transmitted through contaminated soil and water (Ford & Hamner 2009). Toxoplasmosis is common throughout the world, with up to eighty per cent of people in parts of Europe sero-positive for subclinical infection with *T. gondii* (Singh & Sinert 2007). Clinical disease can be fatal for the foetus or for the immune compromised and it is likely the burden of disease from toxoplasmosis on a global scale is far higher than reported. The primary source of toxoplasmosis is cats, but there is accruing evidence that water may also be an important pathway of transmission. Ford & Hamner (2009) provide a review of many potentially water-borne diseases and approaches for their control.

Can we predict the future burden of diseases from water, sanitation and hygiene? One of the Global Burden of Disease Project's original goals was to provide future prediction of disease burden. Murray and Lopez used 1990 data to extrapolate burden of disease to 2020 (Murray & Lopez 1996). This work was subsequently updated to predict disease burden for 2030 based on 2002 data (Mathers & Loncar 2006). Both studies used similar approaches to modelling future disease burden, with baseline, pessimistic and optimistic projections based on socioeconomic variables, namely average income per capita, average number of years of schooling in adults, and time as a proxy measure for the impact of technological change (Mathers & Loncar 2006). In the case of the latter study, variable smoking was also added for projections of cancers, cardiovascular disease and respiratory disorders.

Obviously, major assumptions have to be made, but assuming economic improvements in less developed parts of the world, then the burden from infectious diseases decreases with the exception of HIV/AIDS, which increases in rank from fourth to third place (Table 8). The other big killers, malaria, tuberculosis and diarrheal diseases all decrease in ranking on the assumption of improved public health interventions. Predictions of this nature can of course dramatically change. For example, through a cure for HIV/AIDS or any breakdown in social structure through conflict or natural disasters on a scale that results in breakdown of the basic public health infrastructure,

Table 8. Projections of rankings in burden of disease measured in DALYs from 2002 to 2030 Baseline projections (Mathers & Loncar 2006).

Disease or injury	Rank (2002)	Disease or injury	Rank (2030)
Ischaemic heart disease	1	Ischaemic heart disease	1
Cerebrovascular disease	2	Cerebrovascular disease	2
Lower respiratory infections	3	HIV/AIDS	3
HIV/AIDS	4	COPD	4
COPD	5	Lower respiratory infections	5
Perinatal conditions	6	Trachea, bronchus, lung cancers	6
Diarrhoeal diseases	7	Diabetes mellitus	7
Tuberculosis	8	Road traffic accidents	8
Trachea, bronchus, lung cancers	9	Perinatal conditions	9
Road traffic accidents	10	Stomach cancer	10
Diabetes mellitus	11	Hypertensive heart disease	11
Malaria	12	Self-inflicted injuries	12
Hypertensive heart disease	13	Nephritis and nephrosis	13
Self-inflicted injuries	14	Liver cancers	14
Stomach cancer	15	Colon and rectum cancers	15
		Diarrhoeal diseases	16
		Malaria	22
		Tuberculosis	23

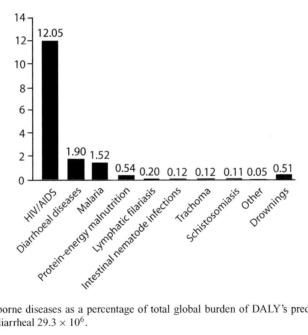

Figure 5. Water-borne diseases as a percentage of total global burden of DALY's predicted for 2030: All causes 1.5×10^9; diarrheal 29.3×10^6.

Figure 5 presents predicted water-borne diseases as a percentage of total global burden of disease for 2030.

Projections of increasing mortality and disability rates from HIV/AIDS raise an interesting question. To what extent is HIV/AIDS mortality caused by water-borne viruses, bacteria, protozoa or fungi? We are already trying to model disease burden from a majority of countries that lack even a most basic vital registration system. Reliable vital registration systems that can supply accurate data on cause of death are only available in countries with comparatively low disease burdens. It seems at this stage in the development of global society that the likelihood of assigning causative agents to mortality from HIV/AIDS, and indeed many other infectious diseases is far in the future and perhaps an unattainable goal. Teasing out the effects of malnutrition on both infectious and non-infectious diseases is also a daunting task (Prüss-Üstün et al., 2008). These authors estimate that 70,000 deaths per year from malnutrition are directly attributable to diarrhoea or nematode infections from contaminated water, and that the total number of malnutrition-related deaths associated both directly and indirectly with poor water quality, lack of sanitation or hygiene is actually about 860,000 in children less than five years old. This is about 1.5 per cent of the total global burden of mortality based on 2002 data.

Prediction involves considerable uncertainty as the world is constantly changing. Given that predicted burden of known disease is only a crude estimate, we must also consider the unknown diseases, both the future pandemic diseases that have yet to emerge and the burden of chronic disease that may actually have an infectious and potentially water-borne agent as their source. Unanticipated sequellae from potentially water-borne diseases have been reviewed previously (Ford & Colwell 1996; Ford 1999). They include associations between *Salmonella, Campylobacter, and Yersinia* and reactive arthritis;*Helicobacter pylori* and stomach cancer; *Campylobacter* and Guillain-Barr and Miller-Fisher Syndromes; E. coli 0157:H7 and kidney failure and brain damage; enteric viruses and miscarriages, diabetes and heart disease; and multiple episodes of diarrheal disease that have serious nutritional consequences that cause stunted growth and impair intellectual development. At this point in time, it is impossible to quantify the burden of chronic diseases from poor water quality, sanitation and hygiene, but improvements in this basic infrastructure on a global scale are likely to have enormous future implications for human health.

In addition to infectious disease, what is the burden of chemical contamination from water and how does it affect susceptibility to disease? Potentially hundreds of millions of people are exposed to elevated nitrates, arsenic, fluoride and a host of other chemicals in their drinking water. As a result, arsenocosis is epidemic in Bangladesh and West Bengal and skeletal fluorosis is at epidemic proportions in India. Just as nutritional deficiencies affect the immune system, exposure to chemicals may also reduce the body's ability to fight infectious disease. To date, we have no reliable method to quantify this additional burden of disease through exposure to contaminated water.

A provocative article proclaims that water-borne diseases may be history (Fenwick 2006). Yes, with global economic development we may be able to reduce the burden of the major water-borne diseases such as cholera, typhoid, and the parasitic killers such as malaria. However, the lessons learned from more developed countries suggest that protozoal (e.g. *Cryptosporidium parvum*), bacterial (e.g. *Legionella pneumophila, E. coli* 0157, and others), viral (e.g. norovirus), and fungal (e.g. *Aspergillus* spp.) infections will continue to occur through exposure to contaminated water and aerosols.

5 CONCLUSION

In terms of water related health issues, environmental public health tracking seems currently to be a distant dream. Even the technically less demanding monitoring of environmental health drivers is not yet an entrenched component of public health intelligence. The public health surveillance paradigm is still overwhelmingly focused on counting health outcomes: deaths, hospitalisations, positive laboratory reports, outbreak numbers and frequencies. There is a systemic structural problem within most public health systems that locate public health intelligence activities within a particular discipline, often in communicable diseases, where it can be difficult to shift or obtain new surveillance and monitoring resources for environmental driver monitoring, let alone environmental public health tracking.

Nonetheless, it does seem obvious that if monitoring environmental drivers makes sense, then environmental public health tracking makes more sense, because in essence it uses the same data to greater effect. It is fundamental to most environmental driver monitoring programs that the data already exist outside of the health system. The challenge is to be able to obtain those data, manage them, and use them effectively. The development of syndromic surveillance systems over the last decade, offers much hope to those interested in linking non-health data sets with the more traditional health outcomes surveillance.

The hydrologic cycle provides a blueprint for designing environmental public health tracking systems for water and water quality related health issues, and DPSEEA provides a framework for the structuring of tracking information. Once such a framework for tracking information is in place, the challenge in implementing environmental public health tracking becomes one of developing the inter-agency cooperation required to obtain and share information. Fortunately, it seems possible that at least all water-related health issues, if not all public health issues, could use a single tracking framework to connect health outcomes with environmental drivers and the management/intervention activities being employed that would significantly reduce costs. However, there will need to be a shift away from the current surveillance paradigm and a much greater demonstration of the intelligence value to organisations in designing, building, and employing environmental public health tracking systems.

REFERENCES

Adegbola, R.A., Demba, E., De Veer, G. & Todd, J. 1994. Cryptosporidium infection in Gambian children less than 5 years of age. *J. Trop. Med. Hyg.* 97(2): 103–107.

Ahmed, K., Ahmed, S., Mitui, M.T., Rahman, A., Kabir, L., Hannan, A., Nishizono, A. & Nakagomi, O. 2010a. Molecular characterization of VP7 gene of human rotaviruses from Bangladesh. *Virus Genes* 40(3): 347–356.

Ahmed, K., Batuwanthudawe, R., Chandrasena, T.G., Mitui, M.T., Rajindrajith, S., Galagoda, G., Pun, S.B., Uchida, R., Kunii, O., Moji, K., Abeysinghe, N., Nishizono, A. & Nakagomi, O. 2010b. Rotavirus infections with multiple emerging genotypes in Sri Lanka. *Arch. Virol.* 155(1): 71–75.

Ananthan S. & Saravanan P. 2000. Genomic diversity of group A rotavirus RNA from children with acute diarrhoea in Chennai, south India. *Indian J. Med. Res.* 111: 50–56.

Andersson, Y. & Bohan, P. 2001. Disease surveillance and waterborne outbreaks. Available from: http://www.who.int/water_sanitation_health/dwq/iwachap6.pdf [Accessed 12th February 2012].

Ansari, S.A., Springthorpe, V.S. & Sattar, S.A. 1991. Survival and vehicular spread of human rotaviruses: possible relation to seasonality of outbreaks. *Revs. Infect. Dis.* 13(3): 448–461.

Auld, H., MacIver, D. & Klaassen, J. 2004. Heavy rainfall and waterborne disease outbreaks: The Walkerton example. *J. Toxicol. & Environ. Health* 67(20–22): 1879–1887.

Australian Government. 2010. Emergency management in Australia. Available from: http://www.ema.gov.au/ [Accessed 12th February 2012].

Ayoob, S. & Gupta, A. K. 2006. Fluoride in drinking water: A review on the status and stress effects. *Crit. Rev. Environ. Sci. & Tech.* 36(6): 433–487.

Bahl, R., Ray, P., Subodh, S., Shambharkar, P., Saxena, M., Parashar, U., Gentsch, J., Glass, R., Bhan, M.K. & the Delhi Rotavirus Study Group. 2005. Incidence of severe rotavirus diarrhea in New Delhi, India, and G and P types of the infecting rotavirus strains. *J. Infect. Dis.* 192 (Suppl): 114–119.

Banerjee, I., Ramani, S., Primrose, B., Moses, P., Iturriza-Gomara, M., Gray, J.J., Jaffar, S., Monica, B., Muliyil, J.P., Brown, D.W., Estes, M.K. & Kang, G. 2006. Comparative study of the epidemiology of rotavirus in children from a community-based birth cohort and a hospital in South India. *J. Clin. Microbiol.* 44(7): 2468–2474.

Bavia, M.E., Carneiro, D.D.M.T., Gurgel, Hd.C., Madureira, F.C & Barbosa, M.G.R. 2005. Remote sensing and geographic information systems and risk of American visceral leishmaniasis in Bahia, Brazil. *Parassitol.* 47(1). 165–169.

Bhat, P., Macaden, R., Unnykrishnan, P. & Rao, H.G. 1985. Rotavirus & bacterial enteropathogens in acute diarrhoeas of young children in Bangalore. *Indian J. Med. Res.* 82: 105–109.

Bingnan, F., Unicomb, L., Rahim, Z., Banu, N.N., Podder, G., Clemens, J., Loon, P.L.V., Rafhava, M., Rao, M.R., Malek, A. & Tzipori, S. 1991. Rotavirus-associated diarrhea in rural Bangladesh: Two-year study of incidence and serotype distribution. *J. Clin. Microbiol.* 29(7): 1359–1363.

Black, R.E., Brown, K.H., Becker, S., Alim, A.R. & Huq, I. 1982. Longitudinal studies of infectious diseases and physical growth of children in rural Bangladesh. II. Incidence of diarrhea and association with known pathogens. *Amer. J. Epidemiol.* 115(3): 315–324.

Boccia, S., Laurenti, P., Borella, P., Moscato, U., Capalbo, G., Cambieri, A., Amore, R., Quaranta, G., Boninti, F., Orsini, M., Branca, G., Fadda, G., Spica, V.R. & Ricciardi, G. 2006. Prospective 3-year surveillance for nosocomial and environmental Legionella pneumophila: Implications for infection control. *Infect. Cont. & Hosp. Epidem.* 27(5): 459–465.

Bogaerts, J., Lepage, P., Rouvroy, D., Van Goethem, C., Nsengumuremye, F., Mohamed, O., Habyalimana, J.B. & Vandepitte, J. 1987. Cryptosporidiosis in Rwanda. Clinical and epidemiological features. *Ann. Soc. Belg. Med. Trop.* 67(2): 157–165.

Bove, F., Fulcomer, M., Klotz, J., Esmart, J., Dufficy E. & Je, S. 1995. Public drinking water contamination and birth outcomes. *Amer. J. Epidem.* 141(9): 850–862.

Brandonisio, O. 2006. Waterborne transmission of Giardia and Cryptosporidium. *Parassitol.* 48(1–2): 91–94.

Briggs, D., Corvalán, C. & Nurminen, M. 1996. *Linkage methods of environment and health analysis: General guidelines. A report of the Health and Environment Analysis for Decision-making (HEADLAMP) project.* Geneva: WHO.

Broor, S., Husain, M., Chatterjee, B., Chakraborty, A. & Seth, P. 1993. Temporal variation in the distribution of rotavirus electropherotypes in Delhi, India. *J. Diarrhoeal Dis. Res.* 11(1): 14–18.

Brown, D.W., Mathan, M.M., Mathew, M., Martin, R., Beards, G.M. & Mathan, V.I. 1988. Rotavirus epidemiology in Vellore, south India: group, subgroup, serotype, and electrophoretype. *J. Clin. Microbiol.* 26(11): 2410–2414.

Brownstein, J.S., Rosen, H., Purdy, D., Miller, J.R., Merlino, M., Mostashari, F. & Fish, D. 2002. Spatial analysis of West Nile virus: rapid risk assessment of an introduced vector-borne zoonosis. *Vect-Borne & Zoon. Dis.* 2(3): 157–164. [erratum in *Vect.-borne Zoon. Dis.* 2003. 3(3): 155].

Brownstein, J.S., Freifeld, C.C., Reis, B.Y. & Mandl, K.D. 2008. Surveillance sans frontières: Internet-based emerging disease intelligence and the HealthMap project. PLoS Med. 5: 1019–1024.

Bureau of Meteorology. 2010. Australian tsunami warning system. Available from: http://www.bom.gov.au/tsunami/about/atws.shtml. [Accessed 12th February 2012].

Cabanes, P.A., Wallet, F., Pringuez, E. & Pernin, P. 2001. Assessing the risk of primary amoebic meningoencephalitis from swimming in the presence of environmental *Naegleria fowleri. Appl. & Environ. Microbiol.* 67(7): 2927–2931.

Cabelli, V.J., Dufour, A., McCabe, L.J. & Levin, M.A. 1982. Swimming associated gastroenteritis and water quality. *Amer. J. Epidem.* 115(4): 606–616.

Calderon, J., Navarro, M.E., Jiminez-Capdeville, M.E., Santos-Diaz, M.A., Golden, A., Rodriguez-Leyva, I., Borja-Aburto, V. & Diaz-Barriga, F. 2001. Exposure to arsenic and lead and neuropsychological development in Mexican children. *Environ. Res. Sect. A* 85(2): 69–76.

Calderon, R.L., Mood, E.W. & Dufour, A. 1991. Health effects of swimmers and nonpoint sources of contaminated water. *Int. J. Environ. Health* 1: 21–31.

CDC. 2008. Morbidity and mortality weekly report: Summary of notifiable diseases – United States, 2006. Available from: http://www.cdc.gov/mmwr//PDF/wk/mm5553.pdf [Accessed 12th February 2012].

Chakravarti, A., Chauhan, M.S., Sharma, A. & Verma, V. 2010. Distribution of human rotavirus G and P genotypes in a hospital setting from Northern India. *SE Asian J. Trop. Med. & Pub. Health* 41(5): 1145–1152.

Chisholm, K., Cook, A., Bower, C. & Weinstein, P. 2008. Risk of birth defects in Australian communities with high levels of brominated disinfection by-products. *Environ. Health Perspect.* 116(9): 1267–1273.

Cogo, P.E., Scaglia, M., Gatti, S., Rossetti, F., Alaggio, R., Laverda, A.M., Zhou, L., Xiao, L. & Visvesvara, G.S. 2004. Fatal *Naelgeria fowleri* meningoencephalitis, Italy. *Emerg. Infect. Dis.* 10(10): 1835–1837.

Cretikos, M.A., Merritt, T.D., Main, K., Eastwood, K., Winn, L., Moran, L. & Durrheim, D.N. 2007. Mitigating the health impacts of a natural disaster: The June 2007 long-weekend storm in the Hunter region of New South Wales. *Med. J. Austrl.* 187(11–12): 670–673.

Cringoli, G., Ippolito, A. & Taddei, R. 2005. Advances in satellite remote sensing of pheno-climatic features for epidemiological applications *Parassitol.* 47(1): 51–62. [see comment].

Curriero, F.C., Patz, J.A., Rose J.B. & Lele, S. 2001. The association between extreme precipitation and waterborne disease outbreaks in the United States, 1948–1994. *Amer. J. Pub. Health* 91(8): 1194–1199.

Deering, D.W., Rouse, J.W., Haas, R.H. & Schell, J.S. 1975. Measuring forage production of grazing units from Landsat MSS data. In: *Proceedings of the 10th International Symposium on Remote Sensing of Environment,* Ann Arbor, MI. 1169–1178.

Dodds, L., King, W., Woolcott, C. & Pole, J. 1999. Trihalomethanes in public water supplies and adverse birth outcomes. *Epidem.* 10(3): 233–237.

Dodds, L., King, W., Allen, A.C., Armson, B.A., Fell, D.B. & Nimrod, C. 2004. Trihalomethanes in public water supplies and risk of stillbirth. *Epidem.* 15(2): 179–186.

Dominey-Howes, D. & Goff, J. 2010. Tsunami: Unexpected blow foils flawless warning system. *Nature* 350. DOI: 10.1038/464350a.

Dufour, A. 1984. Bacterial indicators of recreational water quality. *Can. J. Pub. Health* 75: 49–56.

Duong, T.H., Kombila, M., Dufillot, D., Richard-Lenoble, D., Owono Medang, M., Martz, M., Gendrel, D., Engohan, E. & Moreno, J.L. 1991. Role of cryptosporidiosis in infants in Gabon: Results of two prospective studies. *Bull. Soc. Pathol. Exot.* 84(5Pt5): 635–644.

EPA 2000. The history of drinking water treatment. Available from: http://www.epa.gov/safewater/consumer/pdf/hist.pdf [Accessed 12th February 2012].

EPA 2006. Fact Sheet: Concentrated Animal Feeding Operations Proposed Rulemaking. Available from: http://www.epa.gov/npdes/regulations/cafo_revisedrule_factsheet. pdf [Accessed 16th March 2012].

Esterman, A., Dorsch, M., Cameron, S., Roder, D., Robinson, B., Lake, J. & Christy, P. 1984. The association of *Naegleria fowleri* with the chemical, microbiological and physical characteristics of South Australian water supplies. *Water Res.* 18(5): 549–553.

Fawell, J. & Nieuwenhuijsen, M.J. 2003. Contaminants in drinking water. *Brit. Med. Bull.* 68: 199–208.

Fenwick, A. 2006. Waterborne infectious diseases: Could they be consigned to history? *Science* 313: 1077–1081.

FEWSnet. 2008. Available from: http://www.fews.net/Pages/default.aspxAfrica [Accessed 15th March 2012].

Fields, B.S., Benson, R.F. & Besser, R.E. 2002. *Legionella* and Legionnaires' disease: 25 years of investigation. *Clin. Microbio. Rev.* 15(3): 506–526.

Fisman, D.N., Lim, S., Wellenius, G.A., Johnson, C., Britz, P., Gaskins, M., Maher, J., Mittleman, M.A., Spain, C.V., Haas C.N. & Newbern, C. 2005. It's not the heat, it's the humidity: Wet weather increases Legionellosis risk in the greater Philadelphia metropolitan area. *J. Infect. Dis.* 192(12): 2066–2073.

Fleisher, J.M., Kay, D., Salmon, R.L., Jones, F., Wyer M.D. & Godfree, A.F. 1996. Marine waters contaminated with domestic sewage: Nonenteric illnesses associated with bather exposure in the United Kingdom. *Amer. J. Pub. Health* 86(9): 1228–1234.

Ford, T.E. 1999. Microbiological safety of drinking water: United States and global perspectives. *Environ. Health Perspect.* 107(Supp1): 191–206.

Ford, T.E. & Colwell, R.R. 1996. *A Global Decline in Microbiological Safety of Water: A Call for Action.* Washington DC: American Academy of Microbiology.

Ford, T.E. & Hamner, S. 2009. Control of water-borne pathogens in developing countries. In: R. Mitchell & J-D. Gu, (eds.), *Environmental Microbiology.* 2nd edition. Hoboken: Wiley-Blackwell.

Fowler, M. & Carter, R.F. 1965. Acute pyogenic meningitis probably due to *Acanthamoeba* sp: a preliminary report. *Brit. Med. J.* 2: 740–742.

Freifeld, C.C., Mandl, K.D., Reis, B.Y. & Brownstein, J.S. 2008. HealthMap: Global infectious disease monitoring through automated classification and visualization of internet media reports. *J. Am. Med. Inform. Assoc.* 15: 150–157.

Fripp, P.J., Bothma, M.T. & Crewe-Brown, H.H. 1991. Four years of cryptosporidiosis at GaRankuwa Hospital. *J. Infect.* 23(1): 93–100.

Garcia-Fulgueiras, A., Navarro, C., Fenoll, D., Garcia, J., Gonzalez-Diego, P., Jimenez-Bunuales, T., Rodriguez, M., Lopez, R., Pacheco, F., Ruiz, J., Segovia, M., Baladron, B. & Pelaz, C. 2003. Legionnaires' disease outbreak in Murcia, Spain. *Emerg. Infect. Dis.* 9(8): 915–921.

Gebre-Michael, T., Malone, J.B. & McNally, K. 2005. Use of geographic information systems in the development of prediction models for onchocerciasis control in Ethiopia. *Parassitol.* 47(1): 135–144.

GLCF 2012. Available from: http://glcf.umiacs.umd.edu/ [Accessed 15th March 2012].

Goward, S.N., Tucker, C.J., Dye & D.G. 1985. North American vegetation patterns observed with the NOAA-7 Advanced Very High Resolution Radiometer. *Vegetatio* 64: 3–14.

Greig, J.E., Carnie, J.A., Tallis, G.I., Ryan, N.J., Tan, A.G., Gordon, I.R., Zwolak, B., Leydon, J.A., Guest, C.S. & Hart, W.G. 2004. An outbreak of Legionnaires' disease at the Melbourne aquarium, April 2000: Investigation and case-control studies. *Med. J. Austrl* 180(11): 566–572.

Guha-Sapir, D., Hargitt, D. & Hoyois, P. 2004. Thirty years of natural disasters 1974–2003: Available from: http://www.cred.be/ [Accessed 15th March 2012].

Hamner, S., Broadaway, S.C., Mishra, V.B., Tripathi, A., Mishra, R.K., Pulcini, E., Pyle, B.H. & Ford, T.E. 2007. Isolation of potentially pathogenic *Escherichia coli* O157:H7 from the Ganges River. *Appl. Environ. Microbiol.* 73: 2369–2372.

Heymann, D.L. 2004. Control of Communicable Diseases Manual. *Int. J. Epidemiol.* 34 (6): 1446–1447. doi: 10.1093/ije/dyi210.

121

Heymann, D.L. & Rodier, G. 2004. Global surveillance, national surveillance, and SARS. *Emerg. Infect. Dis.* 10: 173–175.

Hicks, L.A., Rose, C.E., Fields, B.S., Drees, M.L., Engel, J.P., Jenkins, P.R., Rouse, B.S., Blythe, D., Khalifah, A.P., Feikin, D.R. & Whitney, C.G. 2007. Increased rainfall is associated with increased risk for legionellosis. *Epidem. & Infec.* 135(5): 811–817.

Hoxie, N.J., Davis, J.P., Vergeront, J.M., Nashold, R.D. & Blair, K.A. 1997. Cryptosporidiosis associated mortality following a massive waterborne outbreak in Milwaukee, Wisconsin. *Amer. J. Pub. Health* 87(12): 2032–2035.

Hrudey, S.E. 2009. Chlorination disinfection by-products, public health risk tradeoffs and me. *Water Research* 43(8): 2057–2092.

Hrudey, S.E., Payment, P., Huck, P.M., Gillham, R.W. & Hrudey, E.J. 2003. A fatal waterborne disease epidemic in Walkerton, Ontario: Comparison with other waterborne outbreaks in the developed world. *Water Sci. & Techn.* 47(3): 7–14.

Hunter, P.R. & Thompson, R.C.A. 2005. The zoonotic transmission of Giardia and Cryptosporidium. *Int J Parasitol.* 35(11–12): 1181–1190.

Hwang, B.F. & Jaakkola, J.J.K. 2003. Water chlorination and birth defects: A systematic review and meta-analysis. *Arch. Environ. Health* 58(2): 83–91.

ICDDR 2008. Available from: http://www.icddrb.org [Accessed 12th February 2012].

ILSI 1995. Report of epidemiological workshop for disinfection by-products and reproductive effects. Washington DC: International Life Science Institute.

IOC 2010. Available from: http://www.ioc-tsunami.org/index.php?option=com_content&view=article&id=1&Itemid=2&lang=en [Accessed 12th February 2012].

Jagai, J.S., Monchak, J., McEntee, J.C., Castronovo, D.A. & Naumova, E.N. 2007. The use of remote sensing to assess global trends in seasonality of Cryptosporidiosis. In: *Proceedings of the International Symposium on Remote Sensing of the Environment*, San Jose, Costa Rica, Vol. 32.

Jagai, J.S., Castronovo, D.A., Monchak, J, & Naumova, E.N. 2009. Seasonality of cryptosporidiosis: A meta-analysis approach. *Environ. Res.* 109(4): 465–478.

Jagai, J.S., Griffiths, J.K., Kirshen, P.H., Webb, P. & Naumova, E.N. 2010. Patterns of protozoan infections: Spatiotemporal associations with cattle density. *EcoHealth* 7(1): 33–46.

Jardine, C.N., Boardman, B., Osman, A., Vowles, J. & Palmer, J. 2004. *Methane UK*. Research report 30, Environmental Change Institute. Oxford: Oxford UP.

Jones, J. 2006. Mother Nature's disasters and their health effects: A literature review. *Nursing Forum* 41(2): 78–87.

Jones, S., Burt, B.A., Petersen, P.E. & Lennon, M.A. 2005. The effective use of fluorides in public health. *Bull. WHO* 83(9): 670–676.

Jonkman, S.N., Maaskant, B., Boyd, E. & Levitan, M.L. 2009. Loss of life caused by the flooding of New Orleans after Hurricane Katrina: Analysis of the relationship between flood characteristics and mortality. *Risk Anal.* 29(5): 676–698.

Kang, G., Arora, R., Chitambar, S.D., Deshpande, J., Gupte, M.D., Kulkarni, M., Naik, T.N., Mukherji, D., Venkatasubramanium, S., Gentsch, J.R., Glass, R.I. & Parashar, U.D. 2009. Multicenter, hospital-based surveillance of rotavirus disease and strains among Indian children aged <5 years. *J. Infect. Dis.* 200(Suppl): 147–153.

Khan, M.U., Eeckels, R., Alam, A.N. & Rahman, N. 1988. Cholera, rotavirus and ETEC diarrhoea: Some clinico-epidemiological features. *Trans. Roy. Soc. Trop. Med. & Hyg.* 82(3): 485–488.

Kapaj, S., Peterson, H., Liber, K. & Bhattacharya, P. 2006. Human health effects from chronic arsenic poisoning: A review. *J. Environ. Sci. & Health, Part a-Toxic/Hazardous Substances & Environ. Engin.* 41(10): 2399–2428.

Kay, D., Fleisher, J.M., Salmon, R.L., Jones, F., Wyer, M.D., Godfree, A.F., Zelenauch-Jacquotte, Z. & Shore, R. 1994. Predicting likelihood of gastroenteritis from sea bathing: Results from randomised exposure. *Lancet* 344: 905–909.

Khan, A., Das, S.C., Ramamurthy, T., Sikdar, A., Khanam, J., Yamasaki, S., Takeda, Y. & Nair, G.B. 2002. Antibiotic resistance, virulence gene, and molecular profiles of shiga toxin-producing *Escherichia coli* isolates from diverse sources in Calcutta, India. *J. Clin. Microbiol.* 40: 2009–2015.

Kirkeskov, L., Kristiansen, E., Boggild, H., Platen-Hallermund, F. von, Sckerl, H., Carlsen, A., Larsen, M.J. & Poulsen, S. 2010. The association between fluoride in drinking water and dental caries in Danish children: Linking data from health registers, environmental registers and administrative registers. *Comm. Dent. & Oral Epidem.* 38(3): 206–212.

Kirkpatrick, B., Fleming, L., Squicciarini, D., Backer, L.C., Clark, R., Abraham, W., Benson, J., Cheng, Y.S., Johnson, D., Pierce, R., Zaias, J., Bossart, G. & Baden, D.G. 2004. Literature review of Florida red tide: Implications for human health. *Harmful Algae.* 3(2): 99–115.

Leoni, E., Sacchetti, R., Aporti, M., Lazzari, C., Donati, M., Zanetti, F., De Luca, G., Finzi, G.F. & Legnani, P.P. 2007. Active surveillance of legionnaires disease during a prospective observational study of community and hospital-acquired pneumonia. *Infec. Cont. & Hosp. Epidem.* 28(9): 1085–1088.

Levy, K., Hubbard, A.E. & Eisenberg, J.N. 2008. Seasonality of rotavirus disease in the tropics: a systematic review and meta-analysis. *Int. J. Epidemiol.* 38(6): 1487–1496.

Liu, H.Q. & Huete, A. 1995. A feedback based modification of the NDVI to minimize canopy background and atmospheric noise. *IEEE Trans. Geosci. & Rem. Sens.* 33(2): 457–465.

Luben, T.J., Nuckols, J.R., Mosley, B.S., Hobbs, C. & Reif, J.S. 2008. Maternal exposure to water disinfection by-products during gestation and risk of hypospadias. *Occup. & Environ. Med.* 65(6): 420–427.

MacKenzie, W.R., Hoxie, N.J., Proctor, M.E., Gradus, M.S., Blair, K.A., Peterson, D.E., Kazmierczak, J.J., Addiss, D.G., Fox, K.R., Rose, J.B. & Davis, J.P. 1994. A massive outbreak in Milwaukee of Cryptosporidium infection transmitted through the public water supply. *New Engl. Journ. Med.* 331(3): 161–167.

Madoff, L.C. 2004. ProMED-mail: An early warning system for emerging diseases. *Clin. Infect. Dis.* 39: 227–232.

Marston, B.J., Lipman, H.B. & Breiman, R.F. 1994. Surveillance for Legionnaires' disease: Risk factors for morbidity and mortality. *Arch. Internal Med.* 154: 2417–2422.

Mathers, C.D., Vos, T, Lopez, A.D., Salomon, J. & Ezzati, M. 2001. *National Burden of Disease Studies: A Practical Guide.* Edition 2.0. Global program on evidence for health policy. Geneva: WHO.

Mathers, C.D. & Loncar, D. 2006. Projections of global mortality and burden of disease from 2002 to 2030. *PLoS Med.* 3(11): e442 doi:10.1371/journal.pmed.0030442.

Mazumder, D.N.G. 2007. Effect of drinking arsenic contaminated water in children. *Indian Pediatrics* 44(12): 925–927.

McDonagh, M.S., Whiting, P.F., Wilson, P.M., Sutton, A.J., Chestnutt, I., Cooper, J., Misso, K., Bradley, M., Treasure, E. & Kleijnen, J. 2000. Systematic review of water fluoridation. *Brit. Med. J.* 321(7265): 855–859.

McGeehin, M.A., Qualters, J.R. & Niskar, A.S. 2004. National environmental public health tracking program: Bridging the information gap. *Environ. Health Perspect.* 112(14): 1409–1413.

Meinhardt, P.L., Casemore, D.P., Miller, K.B. 1996. Epidemiologic aspects of human cryptosporidiosis and the role of waterborne transmission. *Epidemiol. Rev.* 18(2): 118–136.

MHFW 2008. Food safety India: *Escherichia coli* O157:H7. Available from: http://foodsafetyindia.nic.in/ecolifaq.htm [Accessed 15th March 2012].

Miller, N.M., van den Ende, J. 1986. Seasonal prevalence of Cryptosporidium associated diarrhoea in young children. *So. Afr. Med. J.* S70(10): 636–637.

Mishra, V., Awasthi, S., Nag, V.L. & Tandon, R. 2010. Genomic diversity of group A rotavirus strains in patients aged 1-36 months admitted for acute watery diarrhoea in northern India: A hospital-based study. *Clin. Microbiol. Infect.* 16(1): 45–50.

Moe, K., Hummelman, E.G., Lwin, T. & Htwe, T.T. 2005. Hospital-based surveillance for rotavirus diarrhea in children in Yangon, Myanmar. *J. Infect. Dis.* 192 (Suppl): 111–113.

Mohanty, J.C., Ford, T.E., Harrington, J.J. & Lakshmipathy, V. 2002. A cross-sectional study of enteric disease risks associated with water quality and sanitation in Hyderabad City. *J. Water Supply: Res. & Techn. (AQUA)* 51(5): 239–251.

Molbak, K., Hojlyng, N., Ingholt, L., Da Silva AP, Jepsen, S. & Aaby, P. 1990. An epidemic outbreak of cryptosporidiosis: a prospective community study from Guinea Bissau. *Pediatr. Infect. Dis. J.* 9(8): 566–570.

Molbak, K., Hojlyng, N., Gottschau, A., Sa, J.C., Ingholt, L. & da Silva, A.P. 1993. Cryptosporidiosis in infancy and childhood mortality in Guinea Bissau, west Africa. *BMJ.* 307(6901): 417–420.

Moodley, D., Jackson, T.F., Gathiram, V. & van den Ende, J. 1991. Cryptosporidium infections in children in Durban. Seasonal variation, age distribution and disease status. *So. Afr. Med. J.* S79(6): 295–297.

Mor, S.M., DeMaria, A. Jr., Griffiths, J.K., Naumova, E.N. 2009. Cryptosporidiosis in the elderly population of the United States. *Clin. Infect. Dis.* 48(6): 698–705.

Morris, S.S., Black, R.E. & Tomaskovic, L. 2003. Predicting the distribution of under-five deaths by cause in countries without adequate vital registration systems. *Int. J. Epidemiol.* 32: 1041–1051.

Murray, C.J.L. & Lopez, A.D. 1996. *The global burden of disease.* Cambridge: Harvard UP.

Mykhalovskiy, E. & Weir, L. 2004.The global public health intelligence network and early warning outbreak detection: A Canadian contribution to global public health. *Can. J. Pub. Health* 97: 42–44.

Nair, G.B., Ramamurthy, T., Bhattacharya, M.K., Krishnan, T., Ganguly, S., Saha, D.R., Manna, B., Ghosh, M., Okamoto, K. & Takeda, Y. 2010. Emerging trends in the etiology of enteric pathogens as evidenced from an active surveillance of hospitalized diarrhoeal patients in Kolkata, India. *Gut Pathog.* 2(1): 4. doi:10.1186/1757-4749-2-4.

NASA. 2012. Available from: http://earthobservatory.nasa.gov/Features/MeasuringVegetation/measuring_vegetation_2.php [Accessed 17th March 2012].

Nath, G., Singh, S.P. & Sanyal, S.C. 1992. Childhood diarrhoea due to rotavirus in a community. *Indian J. Med. Res.* 95: 259–262.

National Intelligence Council. 2000. The global infectious disease threat and its implications for the United States. Available from: http://www.dni.gov/nic/special_globalinfectious.html [Accessed 12th February 2012].

Naumova, E.N., Chen, J.T., Griffiths, J.K., Matyas, B.T., Estes-Smargiassi, S.A. & Morris, R.D. 2000. Use of passive surveillance data to study temporal and spatial variation in the incidence of giardiasis and cryptosporidiosis. *Pub. Health Rep.* 115(5): 436–447.

Naumova, E.N., Christodouleas, J., Hunter, P.R. & Syed, Q. 2005. Effect of precipitation on seasonal variability in cryptosporidiosis recorded by the North West England surveillance system in 1990–1999. *J. Water Health* 3(2): 185–196.

Naumova, E.N., Jagai, J.S., Matyas, B., DeMaria, A, Jr., MacNeill, I.B. & Griffiths, J.K. 2007. Seasonality in six enterically transmitted diseases and ambient temperature. *Epidemio. & Infect.* 135(2): 281–292.

NCDC 2012. Available from: http://www.ncdc.noaa.gov/oa/ncdc.html [Accessed 17th March 2012].

Newman, R.D., Sears, C.L., Moore, S.R., Nataro, J.P., Wuhib, T., Agnew, D.A., Guerrant, R.L. & Lima, A.A.M. 1999. Longitudinal study of Cryptosporidium infection in children in northeastern Brazil. *J. Infect. Dis.* 180(1): 167–175.

Ng, J.C. 2005. Environmental contamination of arsenic and its toxicological impact on humans. *Environ. Chem.* 2(3): 146–160.

NHMRC 2004. *Australian Drinking Water Guidelines.* Available from: http://www.nhmrc.gov.au/_files_nhmrc/ publications/attachments/adwg_11_06.pdf [Accessed 10th April 2012].

NICED 2005. Available from: http://www.niced.org/ about_niced.htm [Accessed 12th February 2012].

Nichols, G., Lane, C., Asgari, N., Verlander, N.Q. & Charlett, A. 2009. Rainfall and outbreaks of drinking water related disease in England and Wales. *J. Water & Health* 7(1): 1–8.

Nieuwenhuijsen, M.J., Toledano, M.B., Bennett, J., Best, N., Hambly, P., Hoogh, C. de, Wellesley, D., Boyd, P.A., Abramsky, L., Dattani, N., Fawell, J., Briggs, D., Jarup L. & Elliott, P. 2008. Chlorination disinfection by-products and risk of congenital anomalies in England and Wales. *Environ. Health Perspect.* 116(2): 216–222.

Nieuwenhuijsen, M.J., Martinez, D., Grellier, J., Bennett, J., Best, N., Iszatt, N., Vrijheid, M. & Toledano, M.B. 2009a. Chlorination disinfection by-products in drinking water and congenital anomalies: Review and meta-analyses. *Environ. Health Perspect.* 117(10): 1486–1493.

Nieuwenhuijsen, M.J., Smith, R., Golfinopoulos, S., Best, N., Bennett, J., Aggazzotti, G., Righi, E., Fantuzzi, G., Bucchini, L., Cordier, S., Villanueva, C.M., Moreno, V., La Vecchia, C., Bosetti, C., Vartiainen, T., Rautiu, R., Toledano, M., Iszatt, N., Grazuleviciene, R. & Kogevinas, M. 2009b. Health impacts of long-term exposure to disinfection by-products in drinking water in Europe: HIWATE. *J. Water & Health* 7(2): 185–207.

Nishio, O., Matsui, K., Oka, T., Ushijima, H., Mubina, A., Dure-Samin, A. & Isomura, S. 2000. Rotavirus infection among infants with diarrhea in Pakistan. *Pediatr. Int.* 42(4): 425–427.

Nyambat, B., Gantuya, S., Batuwanthudawe, R., Wijesinghe, P.R., Abeysinghe, N., Galagoda, G., Kirkwood, C., Bogdanovic–Sakran, N., Kang, J.O. & Kilgore, P.E. 2009. Epidemiology of rotavirus diarrhea in mongolia and sri lanka, march 2005-february 2007. *J. Infect. Dis.* 200(Suppl1): 160–166.

O'Carroll, P.W., Yasnoff, W.A., Ward, M.E., Ripp, L.H. & Martin E.L. 2003. *Public health informatics and information systems.* Health Informatics Series. New York: Springer.

O'Connor, D.R. 2002. Report of the Walkerton Inquiry: Part 1 – the events of May 2000 and related issues. Toronto: The Walkerton Inquiry.

OSDH. 2008. Available from: http://www.ok.gov/health/ [Accessed 12th February 2012].

Paniker, C.K., Mathew, S. & Mathan, M. 1982. Rotavirus and acute diarrhoeal disease in children in a southern Indian coastal town. *Bull. WHO* 60(1): 123–127.

Parashar, U.D., Bresee, J.S., Gentsch, J.R. & Glass, R.I. 1998. Rotavirus. *Emerg. Infect. Dis.* 4(4): 561–570.

Patwari, A.K., Srinivasan, A., Diwan, N., Aneja, S., Anand, V.K. & Peshin, S. 1994. Rotavirus as an aetiological organism in acute watery diarrhoea in Delhi children: Reappraisal of clinical and epidemiological characteristics. *J. Trop. Pediatr.* 40(4): 214–218.

Patz, J. 2005. Satellite remote sensing can improve chances of achieving sustainable health. *Environ. Health Perspect.* 113(2): A84–85.

Patz, J.A. & Olson, S.H. 2006. Climate change and health: Global to local influences on disease risk. *Ann. Trop. Med. & Parasit.* 100(5–6): 535–549.

Peng, M.M., Meshnick, S.R., Cunliffe, N.A., Thindwa, B.D.M., Hart, C.A., Broadhead, R.L. & Xiao, L. 2003. Molecular epidemiology of cryptosporidiosis in children in Malawi. *J. Eukaryot. Microbiol.* 50(Suppl): 557–559.

Perch, M., Sodemann, M., Jakobsen, M.S., Valentiner-Branth, P., Steinsland, H. & Fischer, T.K. 2001. Seven years' experience with Cryptosporidium parvum in Guinea-Bissau, West Africa. *Ann. Trop. Paediatr.* 21(4): 313–318.

Phukan, A.C., Patgiri, D.K. & Mahanta, J. 2003. Rotavirus associated acute diarrhoea in hospitalized children in Dibrugarh, north-east India. *Indian J. Pathol. Microbiol.* 46(2): 274–278.

Pinzon, J., Brown, M.E. & Tucker, C.J. 2004. Satellite time series correction of orbital drift artifacts using empirical mode decomposition. In: N. Huang & S.S.P. Shen (eds.), *Hilbert-Huang Transform: Introduction and Applications*: 167–186. Hackensack: World Scientific.

Pitzer, V.E., Viboud, C., Simonsen, L., Steiner, C., Panozzo, C.A., Alonso, W.J., Alonso, W.J., Miller, M.A., Glass, R.I., Glasser, J.W., Parashar, U.D. & Grenfell, B.T. 2009. Demographic variability, vaccination and the spatiotemporal dynamics of rotavirus epidemics. *Science* 325(5938): 290–294.

Prüss-Üstün, A., Kay, D., Fewtrell, L. & Bartram, J. 2002. Estimating the burden of disease from water, sanitation and hygiene at a global level. *Environ. Health Perspec.* 110: 537–542.

Pruss-Üstün, A. & Corvalan, C. 2006. *Preventing disease through healthy environments: Towards an estimate of the environmental burden of disease.* Geneva: WHO.

Prüss-Üstün, A., Bos, R., Gore, F. & Bartram, J. 2008. *Safer water, better health: Costs, benefits and sustainability of interventions to protect and promote health.* Geneva: WHO. Available from: http://whqlibdoc.who.int/publications/2008/9789241596435_eng.pdf [Accessed 12th February 2012].

Purohit, S.G., Kelkar, S.D. & Simha, V. 1998. Time series analysis of patients with rotavirus diarrhoea in Pune, India. *J. Diarrhoeal Dis. Res.* 16(2): 74–83.

Qadri, F., Saha, A., Ahmed, T., Al Tarique, A., Begum, Y.A. & Svennerholm, A-M. 2007. Disease burden due to enterotoxigenic Escherichia coli in the first 2 years of life in an urban community in Bangladesh. Infect Immun 75(8): 3961–3968.

Qazi, R., Sultana, S., Sundar, S., Warraich, H., un-Nisa, T., Rais, A. & Zaidi, A.K. 2009. Population-based surveillance for severe rotavirus gastroenteritis in children in Karachi, Pakistan. *Vaccine* 27(Supp5): F25–F30.

Rahman, M., Sultana, R., Ahmed, G., Nahar, S., Hassan, Z.M., Saiada, F., Podder, G., Faruque, A.S., Siddique, A.K., Sack, D.A., Matthijnssens, J., Van Ranst, M. & Azim, T. 2007. Prevalence of G2P[4] and G12P[6] rotavirus, Bangladesh. *Emerg. Infect. Dis.* 13(1): 18–24.

Rahman, M.M., Ng, J.C. & Naidu, R. 2009. Chronic exposure of arsenic via drinking water and its adverse health impacts on humans. *Environ. Geochem. & Health* 31: 189–200.

Ram, S., Khurana, S., Khurana, S.B., Sharma, S., Vadehra, D.V. & Broor, S. 1990. Bioecological factors & rotavirus diarrhoea. *Indian J. Med. Res.* 91: 167–170.

Ram, S., Vajpayee, P. & Shanker, R. 2007. Prevalence of multi-antimicrobial-agent resistant, shigatoxin and enterotoxin-producing *Escherichia coli* in surface waters of river Ganga. *Environ. Sci. Technol.* 41: 7393–7399.

Ram, S., Vajpayee, P. & Shanker, R. 2008. Contamination of potable water distribution systems by multiantimicrobial-resistant enterohemorrhagic *Escherichia coli*. *Environ. Health Perspec.* 116: 448–452.

Rega, P.P., Mothershead, J.L., Talavera, F., Kulkarni, R., Halamka, J. & Darling, R.G. 2007. Glanders and melioidosis. Available from: http://www.emedicine.com/emerg/topic884.htm [Accessed 12th February 2012].

Rodriguez, M., Serodes, J. & Levallois, P. 2004. Behaviour of trihalomethanes and haloacetic acids in a drinking water distribution system. *Water Res.* 38: 4367 82.

Rouse, J.W., Haas, R.H., Schell, J.A. & Deering, D.W. 1974. Monitoring vegetation systems in the Great Plains with ERTS. In: *Proceedings of the Third Earth Resources Technology Satellite-1 Symposium* 3010–3017, Greenbelt: NASA SP-351.

Saha, M.R., Sen, D., Datta, P., Datta, D. & Pal, S.C. 1984. Role of rotavirus as the cause of acute paediatric diarrhoea in Calcutta. *Trans. R. Soc. Trop. Med. & Hyg.* 78(6): 818–820.

Sarkar, R., Gladstone, B.P., Ajjampur, S.S.R., Kang, G., Jagai, J.S., Ward, H. & Naumova, E.N. 2008. Seasonality of pediatric enteric infections in tropical climates: Time-series analysis of data from a birth cohort on diarrheal disease In: *Proc. Int. Soc. Environ. Engin.* 19(Epidemiology): 1471.

125

Schwartz, J.M., Ingram, K. & Flora, K.D. 2006. Hepatitis E. Available from: http://www.emedicine.com/MED/topic995.htm [Accessed 12th February 2012].

Sharma, S., Ray, P., Gentsch, J.R., Glass, R.I., Kalra, V. & Bhan, M.K. 2008. Emergence of G12 rotavirus strains in Delhi, India in 2000 to 2007. *J. Clin. Microbiol.* 46(4): 1343–1348.

Shepherd, R.C., Sinha, G.P., Reed, C.L. & Russell, F.E. 1988. Cryptosporidiosis in the West of Scotland. *Scot. Med. J.* 33(6): 365–368.

Sherchand, J.B., Nakagomi, O., Dove, W., Nakagomi, T., Yokoo, M., Pandey, B.D. 2009. Molecular epidemiology of rotavirus diarrhea among children aged <5 years in Nepal: Predominance of emergent G12 strains during 2 years. *J. Infect. Dis.* 200(Suppl): 182–187.

Singh, V., Broor, S., Mehta, S. & Mehta, S.K. 1989. Clinical and epidemiological features of acute gastroenteritis associated with human rotavirus subgroups 1 and 2 in northern India. *Indian J. Gastroent.* 8(1): 23–25.

Singh, D. & Sinert, R. 2007. Toxoplasmosis. Available from: http://www.emedicine.com/emerg/ topic601.htm [Accessed 12th February 2012].

Smedley, P.L. & Kinniburgh, D.G. 2002. A review of the source, behaviour and distribution of arsenic in natural waters. *Appl. Geochem.* 17(5): 517–568.

Spiegel, J. & Yassi, A. 1997. The use of health indicators in environmental assessment. *J. Med. Syst.* 21: 275–89.

Steele, A.D., Gove, E. & Meewes, P.J. 1989. Cryptosporidiosis in white patients in South Africa. *J. Infect.* 19(3): 281–285.

Stoll, B.J., Glass, R.I., Huq, M.I., Khan, M.U., Holt, J.E. & Banu, H. 1982. Surveillance of patients attending a diarrhoeal disease hospital in Bangladesh. *Br. Med. J.* 285(6349): 1185–1188.

Stout, J.E. & Yu, V.L. 1997. Legionellosis. *New Engl. Journ. Med.* 337: 682–687.

Stout, J.E., Muder, R.R., Mietzner, S., Wagener, M.M., Perri, M.B., DeRoos, K., Goodrich, D., Arnold, W., Williamson, T., Ruark, O., Treadway, C., Eckstein, E.C., Marshall, D., Rafferty, M.E., Sarro, K., Page, J., Jenkins, R., Oda, G., Shimoda, K.J., Zervos, M.J., Bittner, M., Camhi, S.L., Panwalker, A.P., Donskey, C.J., Nguyen, M.H., Holodniy, M. & Yu, V.L. 2007. Role of environmental surveillance in determining the risk of hospital-acquired legionellosis: A national surveillance study with clinical correlations. *Infect. Contr. & Hosp. Epidem.* 28(7): 818–824.

Tabassum, S., Shears, P. & Hart, C.A. 1994. Genomic characterization of rotavirus strains obtained from hospitalized children with diarrhoea in Bangladesh. *J. Med. Virol.* 43(1): 50–56.

Tablan, O.C., Anderson, L.J., Besser, R., Bridges, C. & Hajjeh, R. 2004. Guidelines for preventing health care associated pneumonia: Recommendations of CDC and Healthcare Infection Control Practices Advisory Committee. *MMWR* 53(RR-3): 1–37.

Thailand. 2005. Rapid health response, assessment and surveillance after a tsunami – Thailand, 2004–2005. *MMWR* 54(3): 61–64.

Tanaka, G., Faruque, A.S., Luby, S.P., Malek, M.A., Glass, R.I. & Parashar, U.D. 2007. Deaths from rotavirus disease in Bangladeshi children: Estimates from hospital-based surveillance. *Pediatr. Infect. Dis. J.* 26(11): 1014–1018.

Tatte, V.S., Gentsch, J.R. & Chitambar, S.D. 2010. Characterization of group A rotavirus infections in adolescents and adults from Pune, India: 1993–1996 and 2004–2007. *J. Med. Virol.* 82(3): 519–527.

Teutsch, S.M. & Churchill, R.E. 2000. *Principles and Practice of Public Health Surveillance*, 2nd ed., New York: Oxford UP.

Torok, T.J., Kilgore, P.E., Clarke, M.J., Holman, R.C., Bresee, J.S. & Glass, R.I. 1997. Visualizing geographic and temporal trends in rotavirus activity in the United States, 1991 to 1996. National Respiratory and Enteric Virus Surveillance System Collaborating Laboratories. *Pediat. Infect. Dis. J.* 16(10): 941–946.

Tsai, S.Y., Chou, H.Y., The, H.W., Chen C.M. & Chen C.J. 2003. The effects of chronic arsenic exposure from drinking water on the neurobehavioural development in adolescence. *Neurotoxic.* 24: 747–753.

Tucker, C.J., Pinzon, J., Brown, M.E., Slayback, D., Pak, E., Mahoney, R., Vermote, E. & Saleous, N. 2005. An extended AVHRR 8km NDVI dataset compatible with MODIS and SPOT vegetation NDVI data. *Int. J. Rem. Sens.* 26(20): 4485–4498.

Tumwine, J.K., Kekitiinwa, A., Nabukeera, N., Akiyoshi, D.E., Rich, S.M., Widmer, G., Feng, X. & Tziporo, S. 2003. Cryptosporidium parvum in children with diarrhea in Mulago Hospital, Kampala, Uganda. *Amer. J. Trop. Med. Hyg.* 68(6): 710–715.

UMD 2012. Available from: http://glcf.umiacs.umd.edu/data/modis [Accessed 26th February 2012].

Unicomb, L.E., Bingnan, F., Rahim, Z., Banu, N.N., Gomes, J.G., Podder, G., Munshi, M.H. & Tzipori, S.R. 1993. A one-year survey of rotavirus strains from three locations in Bangladesh. *Arch. Virol.* 132(1–2): 201–208.

Unicomb, L.E., Kilgore, P.E., Faruque, S.G., Hamadani, J.D., Fuchs, G.J., Albert, M.J. & Glass, R.I. 1997. Anticipating rotavirus vaccines: hospital-based surveillance for rotavirus diarrhea and estimates of disease burden in Bangladesh. *Pediatr. Infect. Dis. J.* 16(10): 947–951.

GAO 2004. Emerging infectious diseases: review of state and federal disease surveillance efforts. Available from: http://www.gao.gov/new.items/d04877.pdf [Accessed 12th February 2012].

USGS 2012. Available from: http://ivm.cr.usgs.gov/whatndvi.php [Accessed 17th March 2012].

Visvesvara, G.S., Moura, H. & Schuster. F.L. 2007. Pathogenic and opportunistic free-living amoebae: *Acantamoeba* spp., *Balamuthia mandrillaris, Naegleria fowleri* and *Sappinia diploidea. FEMS Immunol. Med. Microbio.* 50(1): 1–26.

Vu Wein. 2012. Available from: http://koeppen-geiger.vu-wien.ac.at/ [Accessed 24th February 2012].

Wagner, M.M, Moore, A.W. & Aryel, R.M. 2006. *Handbook of Biosurveillance.* Burlington: Elsevier Academic Press.

Weinstein, P. 2012. Red Tides. In: P. Bobrowsky (ed.), *Encyclopaedia of Natural Hazards.* Heidelberg: Springer Verlag.

Wenger, J.B. & Naumova, E.N. 2008. What Happens in Vegas, Doesn't Stay in Vegas: Traveling Waves of Influenza in the US Elderly Population, 1991–2004. In: *Proceedings of the Syndromic Surveillance Conference:* 197. Raleigh: Advances in Disease Surveillance

WHO 2003. Available from: http://www.who.int/water_sanitation_health/diseases/diarrhoea/en/ [Accessed 13th March 2012].

WHO 2003b. *Guidelines for safe recreational water environments, Volume 1: Coastal and fresh waters.* Geneva: WHO.

WHO 2003a. *The right to water.* Geneva: WHO.

WHO 2007. Weekly Epidemiological Record. Cholera, 2006. Available from: http://www.who.int/wer/2007/wer8231.pdf [Accessed 12th February 2012].

WHO 2008a. About the Global Burden of Disease Project. Available from: http://www.who.int/ healthinfo/bodabout/en/index.html [Accessed 12th February 2012].

WHO 2008b. The world health report. Available from: http://www.who.int/whr/en/ [Accessed 12th February 2012].

WHO 2008c. Global Burden of Disease Estimates. Available from: http://www.who.int/healthinfo/ bodestimates/en/index.html [Accessed 12th February 2012].

WHO 2008d. Guidelines for drinking-water quality: incorporating 1st and 2nd addenda, Vol.1, Recommendations, 3rd ed. Geneva: WHO.

WHO 2012. Available from: http://www.who.int/ihr/en/ [Accessed 17th March 2012].

WHO/UNICEF 2006. Meeting the MDG drinking water and sanitation target: the urban and rural challenge of the decade. Available from: http://www.who.int/water_sanitation_health/ monitoring/ jmpfinal.pdf [Accessed 12th February 2012].

Yachha, S.K., Singh, V., Kanwar, S.S. & Mehta, S. 1994. Epidemiology, subgroups and serotypes of rotavirus diarrhea in north Indian communities. *Indian Pediatr.* 31(1): 27–33.

Yoder, J.S., Eddy, B.A., Visvesvara, G.S., Capewell, L. & Beach, M.J. 2010. The epidemiology of primary amoebic meningoencephalitis in the USA, 1962–2008. *Epidemiol. & Infect.* 138(7): 968–975.

Zaman, K., Yunus, M., Faruque, A.S., El Arifeen, S., Hossain, I., Azim, T., Rahman, M., Podder, G., Roy, E., Luby, S. & Sack, D.A. 2009. Surveillance of rotavirus in a rural diarrhoea treatment centre in Bangladesh, 2000–2006. *Vaccine* 27(Supp 5): F31–34.

Environmental Tracking for Public Health Surveillance – Morain & Budge (eds)
© 2013 Taylor & Francis Group, London, ISBN 978-0-415-58471-5

Chapter 4

Air quality and human health

D.W. Griffin[1] & E.N. Naumova[2] (Auth./eds.) with; J.C. McEntee[3], D. Castronovo[4],
J.L. Durant[5], M.B. Lyles[6], F.S. Faruque[7] & D.J. Lary[8]
[1] *US Geological Survey, Tallahassee, FL, US*
[2] *Tufts University School of Medicine, Boston, MA, US*
[3] *Cardiff University, Cardiff, UK*
[4] *Mapping Sustainability LLC, Walpole, MA, US*
[5] *Tufts University, Medford, MA, US*
[6] *US Naval War College, Newport, RI, US*
[7] *University of Mississippi Medical Center, Jackson, MS.; US*
[8] *W.B. Hanson Center for Space Science, University of Texas, Richardson, TX, US*

ABSTRACT: This chapter addresses how atmospheric constituents and events impact human health, and how environmental monitoring has aided our understanding of them. The principal focus is on planetary processes (volcanoes and desert dust storms); a secondary theme addresses surface emissions originating from anthropogenic sources.

1 INTRODUCTION

Earth's atmosphere is composed of chemically and physically distinct layers known as the troposphere, stratosphere, mesosphere, thermosphere, and exosphere (Figure 1). Between these layers are zones of transition known as the thermopause (where temperature no longer decreases with altitude), stratopause, mesopause, with each of these zones marking a change in temperature trend with increasing altitude above the Earth's surface. Environmental monitoring satellites orbit the Earth above the thermosphere and beyond from about 600 km to about 40,000 km. The lower atmospheric layer, the troposphere, ranges from an upper-level altitude of about eight kilometres over the Arctic and Antarctic to about seventeen kilometres over the equator. Approximately seventy-five per cent of Earth's atmospheric mass exists in the troposphere in addition to about ninety-nine per cent of its aerosols (particles and gas). The majority of all weather phenomena occur within this layer. Within the troposphere, humans exist in the planetary boundary, or atmospheric boundary layer, defined as the lowest part of the troposphere where winds are under the influence of surface topography. The planetary boundary layer typically reaches altitudes into the troposphere of about fifty metres over the Arctic and Antarctic to about two kilometres around the equator.

Humans and other animals have evolved to extract oxygen from the atmosphere through respiration. Consequently, air quality chemistry is an up-front and personal health risk for every person. Broad-scale weather events also impact health in context of disasters and hazards that if not managed properly lead to disease outbreaks or exacerbate well being issues within a population. A few sample data for weather-related mortality are presented in Table 1. Although lightning is a most intimidating hazard, the US Centers for Disease Morbidity and Mortality Weekly Report documented only 1318 deaths in the US between 1980 and 1995 (MMWR 2003a). The data also illustrate that weather-related mortality in the United States is linked to large storms and extreme temperature events. Globally, mortalities due to extreme-weather events have dropped about ninety-five per cent since the early 19th Century due to higher awareness and preparedness (Goklany 2007). Between

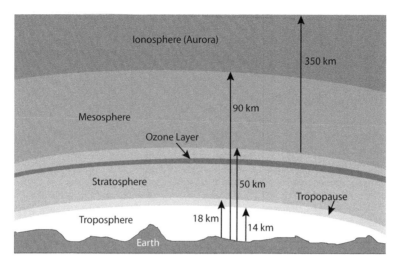

Figure 1.   Earth's atmospheric layers, from the troposphere to the ionosphere.

Table 1.   Weather-related deaths in the US, 2000–2008. Data courtesy NOAA, National Weather Service.

Year	Lightning	Tornadoes	Storms*	Extreme temp	Flood	Other**
2008	27	126	76	115	82	3
2007	45	81	28	152	43	2
2006	48	67	38	255	72	49
2005	38	38	1072	182	43	31
2004	32	35	108	33	82	35
2003	43	54	75	56	86	74
2002	51	55	96	178	49	53
2001	44	40	112	170	48	35
2000	51	41	95	184	38	49
Average	42	60	189	147	65	37

*Thunderstorm winds, tropical cyclones, winter
**Rain, dust storms, dust devils, mudslides, volcanic ash, high wind, misc

2000 and 2006, annual weather related deaths accounted for only about 0.03 per cent of total global deaths. These data pale in comparison to mortality caused by non-communicable illnesses (about fifty-nine per cent) and communicable diseases (about thirty-two per cent) These, include flu viruses that move from host to host through the atmosphere) (Goklany 2007). Economic health risk can be just as devastating. It has been estimated that ninety-six severe weather-events that occurred between 1980 and 2009 caused in excess of $700 billion USD in damages (NCDC 2010).

Although inclement weather poses obvious risks to human health, there are many respiratory health threats that are invisible to the unaided eye. The most indefensible challenge to the respiratory tract is caused by natural releases of toxic gases, such as carbon dioxide. In August 1986 approximately 1700 people were asphyxiated by carbon dioxide released from the depths of Lake Nyos in Cameroon, West Africa (Kling *et al.*, 1987; Giggenbach 1990). A similar event occurred several years earlier alongside another regional lake that asphyxiated thirty-seven people (Giggenbach *et al.*, 1991). Several degassing studies have been conducted in Italy to examine event fatalities of small numbers of humans and other life forms, and to develop strategies to limit future risks (Tassi *et al.*, 2005; Nadkarni *et al.*, 2008). Degassing events are not uncommon in tectonically active areas and highly concentrated clouds of carbon dioxide may be emitted from water and soil that cause fatalities in exposed populations (Rogie *et al.*, 2001; Carapezza *et al.*, 2003). It is clear that changes in atmospheric gas composition can be fatal and are perennial risks to respiratory systems.

Respiratory tracts have evolved to protect life from inhaling particulates. The basic aetiology of lung inflammation begins with deposition of particles into the lung. Breathing patterns, the branched morphology of the airways, and particle size and shape can influence the location of particle deposition (Cullen *et al.*, 2002). The first line of defence in the human system lies with nasal hairs and mucus glands that provide moisture to the hairs to facilitate particle capture. For particles larger than five to ten micrometres in size, nasal filtering efficiency or entrapment is near 100 per cent compared to only about twenty-five per cent for particles in the 0.25 μm range (Brown *et al.*, 1950). The greatest deposition rate beyond the nasal chamber, from the trachea to the alveoli, occurs with particles between one to two micrometres in size, and continues to decrease as particle size approaches 0.25 μm. It then increases with further decrease in size. Among ultrafine particles (median diameter of twenty-six nanometres) deposition in the respiratory tract increases with a decrease in particle size (smallest particles, median diameter of 8.7nm) and deposition rate increases over 4.5-fold with exercise (Daigle *et al.*, 2003). Particles that penetrate and are deposited beyond the nasal cavity are cleared via secretion, mucociliary transport, and cough or ingestion. Optimal mucociliary transport occurs at 100 per cent humidity; and, decreases as lower humidity levels influence transport/clearance efficiency negatively humidity influence transport/clearance efficiency negatively (Corbett *et al.*, 1999). Research has demonstrated that carbonaceous ultrafine (100 nm) isotope-labelled particles remain in the lung system up to 3 days or more, but no evidence of movement across the interstitial barrier of the alveoli into the bloodstream was been noted (Wiebert *et al.*, 2006). Other investigations also reported that they were not able to detect movement through the interstitial barrier (Brown & Hovmoeller 2002; Mills *et al.*, 2006). In contrast, several studies have reported movement of ultrafine particles from the lung environment into the circulation system of humans and rats (Nemmar *et al.*, 2002; Takenaka *et al.*, 2006). Particles that become lodged in the main airways are cleared by the mucociliary escalator (Lehnert 1993). If a particle penetrates into the non-ciliated alveolar level, it may be cleared by phagocytic alveolar macrophages. However, if the particles impair macrophage-mediated removal they are not cleared. Some particles may directly penetrate the alveolar epithelium and reach the lung's deep lymph-node environment. Studies have demonstrated different immune responses based on particle size (Donaldson *et al.*, 2000; Samuelsen *et al.*, 2009). The latter has demonstrated that ultrafine particles in the range of 64nm elicited *pronounced inflammatory response* versus challenge with fine scale particles larger than 200 nm.

Studies addressing the accumulation of silica (most common mineral in Earth's crust) in the lung have demonstrated allergic response, asthmatic stress, and silicosis/pulmonary fibrosis risk (Norboo *et al.*, 1991; Saiyed *et al.*, 1991; Kwon *et al.*, 2002; Park *et al.*, 2005; Chang *et al.*, 2006). Macrophages have been shown to become inundated with inert particles, which subsequently halt the clearance of all particles at the alveolar level (Morrow 1992). Macrophages release toxins that destroy particles; however, the released IL-1, IL-6, tumour-necrosis factor (TNF), fibroblast-growth factor, and the affluence of polymorphonuclear neutrophils can damage lung tissue (Cullen *et al.*, 2002). The short-term effect of the above processes is lung inflammation, which manifests itself as difficulty in breathing and increased susceptibility to respiratory infections. Long-term prognoses include the possibility of fibrosis and carcinogenesis (Driscoll 1996).

Desert-dust storms are the primary naturally occurring means of both short- and long-term exposure to silica. Volcanic events are another primary source of exposure to atmospherically suspended soils. With both dust storms and volcanic events, significant quantities of mineral material can be entrained into the atmosphere with larger events capable of global dispersion (Simkin & Siebert 1994; Griffin 2007). Exposure to significant concentrations of soil particles is common over extended periods of time from dust storms and soil particles, or ash from volcanic eruptions. Earthquakes and landslides also can generate dust clouds that adversely impact human health (Schneider *et al.*, 1997; Cook *et al.*, 2005). Another source of heavy loads of airborne particulates consists of clouds of smoke from forest fires that can transport ash, soil, microorganisms and other organic constituents thousands of kilometres from their point of origin (Morawska & Zhang 2002; Zhang & Morawska 2002; Mims & Mims 2004). A review of literature reveals that scientists have sought to characterize these heavy-particle-load events, to elucidate loading mechanisms, transport range, particulate composition, and their impact on ecosystems and human health. Earth observing satellites have contributed greatly to our understanding of transcontinental and transoceanic dispersal by providing data and imagery of atmospheric events as they develop. Assimilation of these data and/or fusion of data with imagery into numerical models are beginning to reveal associated risks from these processes.

Traditionally used to observe land-cover information and biogeophysical process phenomena on the Earth's surface, remote sensing has been used increasingly by scientists, engineers, public health communities, and epidemiologists to measure environmental variables that impact human health. A growing global network of atmospheric sensors is utilized for monitoring clouds, precipitation, chemistry, aerosols, oceanic winds, and changing environmental events (SMD 2012). Information on individual satellites can be found at (NASA 2012a). Satellites that provide full-colour, bird's eye views of Earth have produced incredible imagery of dust storms, volcanic eruptions, fire plumes, and other aerosols that have fostered a more comprehensive view of the nature of these events on a global scale, as well as the potential implications to human and ecosystem health in downwind environments. Many of these images can be accessed from websites hosted by national space agencies such as NASA's Earth Observatory (EO 2012). There are numerous other national and international databases from which to view images or to download spectral data for use in models.

Material in Section 2, volcanic ash, is provided by Doctors Naumova, McEntee, Castronova and Durant. Section 3, desert dust, is contributed by Doctor Griffin; Section 4, anthropogenic contaminants, is provided by Doctor Lyles; and Section 5, emerging Earth observing sensors and data, is contributed by Doctors Lary and Faruque.

2 VOLCANIC EMMISSIONS AND HEALTH

2.1 *Volcanic emissions, air quality and human health*

A 1985 earthquake in Mexico killed some 30,000 people, leading the United Nations General Assembly to designate the 1990s as the International Decade for Natural Disaster Reduction (IDNDR). One of the greatest risks is the danger of living near active volcanoes, as exemplified by the eruption of Nevado del Ruiz in the Columbian Andes, which resulted in 23,000 deaths (Villegas 2004). Prior to the IDNDR, most sensor data relating to volcanoes were limited to aerial photographs before and after the event (Villegas 2004). Today, Earth observing technology is used for a variety of increasingly complex volcano-related tasks, including pre-disaster loss estimations, integrated measurements, animations, and pre-event 3D modelled simulations. Remote-sensing data facilitate accurate and timely data collection to estimate human exposure to volcanic ashes in both highly accessible and hard-to-reach locations.

In the US, the Cascade Range of the Pacific Northwest has thirteen potentially active volcanoes extending across 1600 kilometres in highly populated areas of Washington, Oregon, and northern California. This range is the most volcanically hazardous area in the US with over 100 eruptions in the past 4000 years. On average, it has two volcanic eruptions per century including: Mount St. Helens from 1980–1986 and Lassen Peak, California from 1914–1917. The Mount St. Helens eruption 1980 caused fifty-seven deaths and over one billion US dollars in damages. There are more than one million residents at risk of a volcanic eruption from Mount Rainier, one of the Nation's most dangerous volcanoes, in the Seattle-Tacoma, Washington area (Smith *et al.*, 2005). Alaskan volcanoes also pose significant risk not only to local populations, but also to communities hundreds and even thousands of miles away. For instance, the ash cloud from the Augustine volcano drifted as far as Colorado, Arizona, and even Virginia (Kienle & Shaw 1979; Waythomas & Miller 1998). During the 1989–90 eruption of the Redoubt volcano, ash clouds were recorded as far as West Texas (Casadevall 1994; Waythomas & Miller 1998). While local communities typically experience the most immediate effects of volcanic eruptions, distant areas can also be adversely affected. The long-distance transport of ash is not only possible, but probable if recent activity of the Alaskan and Cascadian volcanoes is any indication (USGS 2006; AVO 2010).

The hazards caused by volcanic eruptions are detrimental to human health. Recent developments in Earth observations make it possible to monitor these events from afar. Availability of these data combined with public health exposure information presents an opportunity to determine whether remotely acquired data can serve as a relative proxy for ground-based measurements. This Section explores the possibility by describing what is needed to establish an observational link to ground-based information. This is followed by a discussion of the health impacts of volcanic ash, and results from an actual application.

2.1.1 Satellite monitoring of volcanoes

Volcanoes emit gases and ash into the atmosphere that impact the human respiratory system (Delmelle et al., 2001), natural environments (Robock 2000; Tank et al., 2008), structures, and both ground-based and air-borne equipment, especially commercial aviation (Guffanti et al., 2005). Traditionally, sensor data from volcanoes consisted of ground-based measurements collected at the source of degassing. Volcanologists are primarily interested in chemical species, such as H_2O, CO_2, SO_2, HCl, HF, and H_2S (McGonigle 2005), because fluctuations in chemical composition can indicate whether a future eruption is imminent. The direct impact of certain chemicals on human and natural environments is the primary concern here.

The standard procedure for ground-based measurements involves sample collection with subsequent laboratory analysis (Symonds et al., 1994). However, this approach has a number of limitations including: (a) samples collected only where temperatures and safety permit, because such low-temp vents may not be representative of the whole volcano; (b) the potential for the chemical composition of the sample to be altered during storage; and (c) delays in analysis, making real-time applicability difficult (McGonigle 2005). Sensing technologies have improved continuously upon these limitations with advanced in situ and Earth observing equipment that permits relatively safe, accurate, and frequent measurement.

Advances in satellite-based observation systems have introduced an entirely new toolbox to researchers. Spectroscopic information of debris flows obtained from the SRTM (Stevens et al., 2003) or the Air-borne Visible/Infrared Imaging Spectrometer (AVIRIS) (Crowley et al., 2003) combined with digital elevation models (Glaze & Baloga 2003) are increasingly successful in predicting lava flow and lahar paths for risk zonation (Crowley et al., 2003; Tralli et al., 2005). Establishing the link between high ozone signals and ultraviolet (UV) radiation absorption made it possible to use satellite sensors for volcanic-plume monitoring (McGonigle 2005; Bluth et al., 1993).

Both spectrally opaque and translucent clouds can be analysed using satellite acquired data (Webley & Mastin 2009; Webley et al., 2009). To study opaque clouds, thermal infrared data ($\lambda = 10$–$12\,\mu m$) are used (Dean et al., 2004), whereas translucent clouds are studied using the split-window-brightness temperature method (also known as the reverse absorption method) (Prata 1989; Pergola et al., 2004). Once ash-clouds are detected, particle size, mass loadings, ash cloud volume, and even plume height estimations can be obtained (Webley & Mastin 2009; Holasek et al., 1996). A number of satellite sensors are available for volcanic ash observation, including MODIS (Watson et al., 2004) and ASTER (Pieri & Abrams 2004). In addition, techniques continue to be developed to improve the accuracy and sensitivity of sensors (Gangale et al., 2009).

2.1.2 Impacts of volcanic ash
2.1.2.1 Societal impacts

Impacts of volcanic eruptions stretch far beyond geophysical damage. Evacuations, whether voluntary or mandatory, can have social, political, economic, emotional, and physical impacts on evacuees and surrounding residents. The response to eruption warnings is closely related to the perception of risk (Inhorn & Brown 2000). Middle and higher socioeconomic classes are able to call on family and friends to assist with moving belongings, while lower economic classes are at a distinct disadvantage. Day labourers often do not have access to telephones and may not learn about an evacuation until the work day is over. Evacuation may be further complicated by cessation of public transportation and roads clogged with personal vehicles. Forced military evacuations often target the poor unfairly causing fear and resentment among this group. Wealthier families relocate with greater ease than their poorer counterparts, who lack automobile access and are forced to evacuate with less notice. Shelters are typically over-crowded with unsanitary communal facilities and few health-related resources. These conditions lead to nutritional deficiencies and weakened resistance to illness, resulting in the spread of infectious diseases such as chickenpox and measles (Inhorn & Brown 2000). Emotional conditions worsen when families are separated during the evacuation process. Increased stress and psychological disorders are common among resettled people. Evacuees, especially children, suffer increased health problems due to unfamiliar climates and disruption of normal nutrition (Whiteford & Tobin 2004). Strife between evacuees and locals is amplified when resettled people are seen as getting *unfair* amounts of aid. The resettled people may have sold their homes and land or have lost their jobs or agricultural assets. Most likely they have no source of income while they are displaced. In their newly settled areas, resettled people are excluded from politics and feel no sense of community, history, or future. Consequently, despite the health risks, many displaced people return to their own towns (Whiteford & Tobin 2004). All of these well being issues exacerbate direct respiratory health effects in individuals.

2.1.2.2 Human health impacts

Volcanic eruptions are associated typically with toxic emissions; pyroclastic flows (the mixture of rock fragments and superheated gases that can achieve speeds over 100 km/hour and temperatures over 300°C); the release of ash; volcanic gases containing a host of harmful compounds such as carbon dioxide, water vapour, sulphur dioxide, hydrogen chloride, hydrogen sulphide, hydrogen fluoride, among others; and polycyclic aromatic hydrocarbons produced whenever any complex organic material is burned by hot pyroclastic flows (Cullen *et al.*, 2002; Hansell *et al.*, 2006). Despite the fact that almost ten per cent of the global population of seven billion people live within the potential exposure range of an active volcano, information on the acute and chronic respiratory health effects of volcanic emissions is relatively sparse. A few studies have begun to document these impacts (Baxter *et al.*, 1981; Small & Naumann 2001; Horwell & Baxter 2006; Horwell 2007). The acute effects of volcanic ash falls and gases on respiratory conditions vary from undetected to well-defined (Hansell *et al.*, 2006; Horwell & Baxter 2006). During and shortly after eruptions with heavy ash fall, the transient, acute irritant effects of volcanic ash and gases on mucus membranes in the upper-respiratory tract and the exacerbation of chronic lung diseases have been documented (Nania & Bruya 1982; Baxter *et al.*, 1983). These effects have been found following eruptions in Cerro Negro, Nicaragua, in 1992 (Malilay *et al.*, 1997); Mt. Sakura-jima in Japan (Wakisaka *et al.*, 1988); and Mt. Tungurahua in Ecuador (Tobin & Whiteford 2001; Tobin & Whiteford 2002a,b).

Volcanic ash is capable of inducing acute respiratory problems in susceptible people, especially children and those with a history of respiratory illness such as asthma or bronchitis (Johnson *et al.*, 1982; Baxter *et al.*, 1983; Horwell & Baxter 2006). The precise mechanism causing these problems has not been well defined; however, its complexity has been outlined (Baxter *et al.*, 1983; Horwell & Baxter 2006). The irritant effects of volcanic ash, on human airways depend on the physical and chemical properties of the ash including particle size (Baxter *et al.*, 1983; Martin *et al.*, 1986), the concentration of respirable ash particles, mineralogical composition, and duration of exposure (Horwell & Baxter 2006). They also depend on how well one's respiratory tract is functioning, such as ventilation rate, nasal filtration efficiency, and the mucociliary clearance rate (Martin *et al.*, 1986). Several in-vitro and in-vivo experiments on different lung cells and animals suggest that inhaled volcanic ash is less toxic to the lung than other compounds like fine and coarse ambient particulate matter, quartz, or free crystalline silica.

Some observations support the notion that volcanic ash is not a potent stimulus to lung inflammation since it does not stimulate the release of IL-8, a quimiotactic factor for neutrophils, nor does it depress γ-interferon and TNF-α secretion from either human alveolar macrophages or normal human bronchial epithelial cells (Martin *et al.*, 1984; Becker *et al.*, 2005). However, it has been shown that physical immunologic barriers such as ciliary beating frequency and mucus lining can be

134

altered after a short exposure to volcanic ash (Schiff *et al.*, 1981). It has been reported that humoral immunologic parameters can also be affected by volcanic ash; for instance, workers exposed to volcanic ash had significantly lower C3 and C4 levels as well as a marked decrease in serum immunoglobulin-G (IgG) levels when compared to unexposed controls (Olenchock *et al.*, 1983). Immunological assays have shown that IgA, IgG, and albumin, airway proteins, play a protective role in mediating the effect of inhaled dust in human lungs (Martin *et al.*, 1986). Therefore, young and malnourished children with pre-existing low levels of immunoglobulin are at a higher risk for respiratory problems than healthy children. Moreover, some rodent models indicate that exposure to different particulate air pollutants, including volcanic ash, stimulates immuno-competent cells (e.g. monocytes) more strongly than alveolar macrophages to produce oxidative response (Becker *et al.*, 2002). Exposure to volcanic ash was also found to be related to an impairment of stimulated superoxide production while resting superoxide anion production is normal, indicating that volcanic ash might pose a risk for infection by compromising phagocyte antibacterial functions (Castranova *et al.*, 1982). Such mechanisms may explain why individuals with underlying lung impairment, including chronic inflammation, are more susceptible to the harmful effect of air contaminants than healthy individuals.

At high altitudes where oxygen demand is high, even a small increase in CO, SO_2, CO_2, and similar compounds that might affect haematocrit level may trigger a substantial increase in emergency room (ER) visits for asthma exacerbation, especially during fumarole activity. Very little is known about the effects of fresh fractured silica particles on developing lung tissue in young children. One may speculate that rough edges of silica particles due to micro-abrasion may damage the epithelial lining and promote pathogen colonization. Due to a child's high susceptibility for viral infections, exposure to silica particles may explain an increase in acute lower-respiratory infections of bacterial origin in young children.

2.1.3 *Monitoring volcanoes using air-borne and satellite sensors*
Distance imaging and chemical monitoring provide quick, accurate, and timely methods for monitoring exposure to health hazard outcomes, and improving communication of health risks to the public. Such data and imagery also can provide advanced situational information for volcanoes that are under-monitored. The ability to forecast volcanic activity and to mitigate hazard exposure depends upon knowledge of previous activity at a site and at similar sites worldwide. Sensing technology has been used increasingly by geographers and public health professionals to predict, monitor, and quantify the impacts of natural disasters.

2.1.3.1 Estimating hazardous human health exposure from ash & sulphur dioxide
In a four-year longitudinal study of children aged six to thirty-six months in Quito, Ecuador, 4450 cases of acute upper-respiratory infection and 518 physician-diagnosed cases of pneumonia were observed. Among these six (1.2 per cent) required hospitalization (F. Sempertegui, personal communication). Among all the acute upper-respiratory infection episodes, 102 (2.3 per cent) progressed into acute lower-respiratory infection. Therefore, it is anticipated that for every case of hospitalization due to pneumonia, there are approximately eighty-six cases of mild pneumonia. If true, this observation testifies to the severity of respiratory conditions. In fact, the observed effect was very similar with a reported increase in acute upper-respiratory infection (2.6 times), acute lower-respiratory infection (2.5 times), and asthma (2.1 times), after eruptions of Tungurahua in 1999 (ReliefWeb 2010).

Quito was affected by several episodes of ash emission between October and December, 1999; April, 2000; and July, 2000. Figure 2 is a composite of ash advisories available from the volcano ash advisory archive (VAAA) (NOAA 2012). It is possible that estimates of the ER rates before and after the eruptions in April may have been elevated by these prior eruptions, causing the effects to be underestimated. It is also plausible that the effects of volcanic activity were exacerbated due to little or no rain, which can clean the air between eruptions. Additional studies are needed to better quantify the effect of ash fallouts and related pollutants.

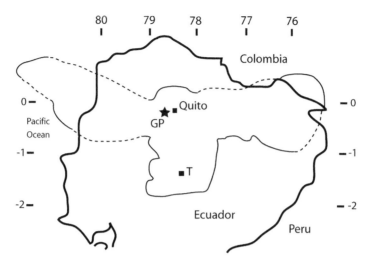

Figure 2. Extent of Guagua Pichincha ash deposition (thin line) compiled from NOAA ash advisories October 1999–July 2000 (Courtesy Geophysical Institute of Ecuador).

2.1.3.2 Significance

Recent research has emphasized the spectrum of health problems associated with direct and indirect effects of volcanic ash and gasses, including displacement, disruptions in infrastructure (Tobin *et al.*, 2002; Whiteford *et al.*, 2002), and soil, air, and water contamination (Delmelle *et al.*, 2001; Loehr *et al.*, 2005). Studies stress the need for implementing efficient surveillance systems to monitor the health effects associated with various environmental and socioeconomic factors, especially in the most vulnerable populations (Malilay *et al.*, 1997). In the absence of such surveillance systems, respiratory health outcomes that often reflect the most severe conditions requiring ER visits and hospitalization cannot be characterized. The results of these efforts provide important information for environmental public-health policy and a better understanding of the short-term health effects of environmental exposure to volcanic ashes. Observed effects in Quito's children most likely reflect an increase in severity of acute respiratory infections and exacerbation of pre-existing asthma-related conditions. Because no surveillance systems exist, health outcomes from urgent ER visits and hospitalizations were used, reflecting the most severe conditions. It is expected that young children with pneumonia and asthma-related conditions are more likely to be hospitalized than older children or those with acute upper-respiratory infection.

2.1.3.3 Limitations

Observed increases in respiratory infections in April, 2000 were most likely triggered by aerosols and ash falls produced by Pichincha eruptions. However, it is important to consider that the main limitation in this analysis; that is, the absence of direct measures of exposure, must be considered. It is also plausible that ash particles deposited during 1999 ash falls could have been re-suspended in the air without elevated volcanic activity. Unfortunately, systematic monitoring of air quality in Quito for critical air pollution was initiated after 2000, so only indirect proxies such as satellite imagery for ash emissions and daily records on seismic activity of Pichincha can be used to establish the timing of high exposure (NOAA 2012).

In this four-year longitudinal study, the correlation between ER visits and timing of high exposure was examined. However, limitations to this approach exist. First, health effects of volcanic aerosols and ambient traffic-related pollution are difficult to separate. Secondly, volcanic activity is a continuous process in which the delayed health effects of one event may overlap with others. Thirdly, a seasonal elevation in incidence of respiratory infections and pneumonia could coincide with the eruption and therefore bias the results. Finally, temporal variations in ER visits could be prompted by a number of social factors (e.g. weekends, holidays, strikes) as well as by changes in perception associated with volcano alerts. Although potential confounding factors from air pollution, continuous volcanic activity, seasonal infection rates, and reporting bias could have affected this study, their role most likely is not substantial.

2.1.4 *Need for air-borne and satellite sensors: A case for Quito and Guagua Pichincha*

Volcanic ashes worsened the already poor air quality in Quito (Estrella *et al.*, 2005). Typically, the chemical compositions of these ashes have been distinguished from by-products of anthropogenic air pollutants in urban settings. However, one of the most predominant components of particulate matter (PM_{10}), are found in automobile emissions as well as volcanic ash. In fact, a few samples of PM_{10} concentrations collected in urban Quito in the late fall of 1999 demonstrated extremely high levels in both the northern and southern parts of the city. Data reveal that during the 1999 Pichincha eruption, PM_{10} increased progressively from 58 $\mu g/m^3$ on September 30, to 407 $\mu g/m^3$ on October 6–7, to 1487 $\mu g/m^3$ on November 26, 1999. All of these exceeded the allowed level by eight to twenty-eight times (OPS 2005). Unfortunately, specific information on ash composition, transport, and deposition is limited, especially for eruptions prior to 2000 (Diaz *et al.*, 2006). Reported ash composition had substantial respirable material and elevated cristobalite content (Baxter *et al.*, 1999; Baxter 2003; Garcia-Aristizabal *et al.*, 2007). For three consecutive eruptions in November 1999, July 2000, and February 2002 percentages of PM_{10} in the Guagua Pichincha ash ranged from 7.9 to 13.6; not dissimilar to the Mount St. Helens 1980 ash composition (Baxter *et al.*, 1983, 2003). Data from the 2001 eruptions of Tungurahua and El Reventador volcanoes in Ecuador indicated that the ash layer in the Quito area was mainly composed of medium-to-fine ash particles with about twenty per cent free crystals (Le Pennec *et al.*, 2006). That study indicated that the ash was re-suspended during the windy afternoons and caused ocular irritation, respiratory troubles, and other health problems.

Although higher levels of respiratory disease in Quito's children after volcanic activity have been documented, the studies acknowledge that the absence of direct measures of exposure was a considerable limitation. The authors suggest that ash composition and deposition can be monitored using air-borne and satellite sensors, thereby enabling detailed tracking and exposure assessments of events that may be detrimental to human health.

2.1.5 *A novel methodology to estimate hazardous human health exposure*

For epidemiology, use of imagery and data collected by remote systems is a relatively new and emerging field with most studies surfacing in the past two decades. Several techniques based on remote surveillance have been developed to assess the ability of detectors to monitor volcanic plume composition (Prata 1989; Yu, *et al.*, 2002; Pieri & Abrams 2004; Watson *et al.*, 2004). It is essential to investigate whether acquired data can serve as a proxy for ground-based air-quality assessment (Andronico *et al.*, 2009; Carter & Ramsey 2009; Tupper & Wunderman 2009; Lyons *et al.*, 2010). Reliable methods and algorithms are needed for linking remotely sensed data with outcomes observed through public health surveillance.

2.1.6 *Detecting and measuring ash dispersal using existing sensor methodologies*

Volcanoes emit a mixture of acid gases, water, and solid-phase ash particles. The fates of these species are typically complex, involving physical (i.e. aggregation, absorption, and fallout) and chemical processes (i.e. the oxidation of sulphur dioxide [SO_2] to H_2SO_4). They pose a hazard to air traffic safety as well as ground conditions in the local environment; and, in sufficient volume, the global climate system. In almost every case they pose hazardous environments for human health. A variety of techniques has arisen to monitor each of these species using satellite-based instruments (Prata 1989; Krueger *et al.*, 1995; Seftor *et al.*, 1997; Krotkov *et al.*, 1999; Schneider *et al.*, 1999; Wright *et al.*, 2002; Ellrod 2003; Ellrod *et al.*, 2003; Tupper *et al.*, 2004; Chrysoulakis *et al.*, 2007; Filizzola *et al.*, 2007). However, measurements are still commonly taken on the ground (Prata &

Table 2. Sensors and associated bandwidths.

Sensor	Bandwidths (μm)	Target species	Spatial resolution (km)	Repeat times (days)
OMI	0.3–0.34	SO_2, aerosols	~14	1–2
GOES	11–12	ash, ice	>14	~0.01
MODIS	7.3, 8.6, 11–12	SO_2, ash, ice	1	0.5
ASTER	8.6, 11–12	SO_2, ash, ice	0.09	16

Bernardo 2009), such as on the shield volcanoes of Hawaii (Dockery 2006; Sutton et al., 2006) where accessibility is not a major obstacle.

Today, algorithms utilize a wide range of wavelengths to detect, track, and model volcanic plume components, including silicate ash and sulphur dioxide (SO_2) (Watson et al., 2004). Each technique is dependent on a series of sensor-based observational issues relating to temporal, spatial, and spectral resolution (Pieri & Abrams 2004). Satellite data can quantify the horizontal spreading velocity of a volcanic cloud, allowing for prediction of future movement (Oppenheimer 1998; Tralli et al., 2005). MODIS AVHRR, the Total Ozone Mapping Spectrometer (TOMS), and GOES sensors supply data for the majority of studies attempting to identify and quantify volcanic-ash- and gas-containing clouds (Prata 1989; Wen & Rose 1994; Krueger, Walter et al., 1995; Seftor et al., 1997; Krotkov et al., 1999; Schneider et al., 1999; Wright et al., 2002; Ellrod 2003; Ellrod et al., 2003; Tupper et al., 2004; Watson et al., 2004; Gu et al., 2005; Chrysoulakis et al., 2007; Filizzola et al., 2007; Novak et al., 2008). Sensors on both geostationary and polar-orbiting satellites are acquiring data. Geostationary satellites about 36,000 km above Earth orbit at the same velocity as Earth's rotation, maintaining a constant observation window over a particular region of the globe. Polar-orbiting satellites collect data at 700 km to collect nearly global coverage on a predictable revisit cycle. A trade-off exists between these two orbit types: geostationary satellites re-acquire images every few minutes but have a spatial resolution of several kilometres, whereas polar-orbiting satellites make no more than one or two measurements of the same location per day with resolutions at the meter scale. Each sensor has a different suite of attributes, making it useful for particular phenomena. Table 2 provides details on commonly used sensors and Figure 3 shows the footprints.

2.1.7 Remote data acquisition
Periods of heightened volcanic activity can be identified by examining imagery from the Smithsonian Institute global volcanism program archives and the MODVOC global-hotspot archive (MODVOC 2009; Smithsonian 2009). For each study period, instances when substantial ash was dispersed are identified and EO data for these instances are collected. Satellite imagery can be retrieved through file transfer protocol (FTP) servers for GOES-8 East (hereinafter GOES-8), MODIS, ASTER, and OMI archives. GOES-8 data are archived at the University of Wisconsin-Madison Space Science and Engineering Center (SSEC). Data are delivered in the man computer interactive data access system (McIDAS) format and converted into GeoTiff format using image processing software (SSEC 2005). MODIS sensor data are distributed through NASA's Level 1 atmosphere archive distribution system (LAADS) (NASA 2009). ASTER data are available from the NASA data gateway (NASA 2009). After data are acquired, they are analysed using digital image-processing software.

2.1.8 Ash detection
Ash presence can be estimated using GOES-8, MODIS and ASTER data. Prata developed the brightness temperature difference (BTD) technique using two spectral channels at 10.3–11.3 μm and 11.5–12.5 μm, which are atmospheric spectral windows where the effects of other atmospheric gases are minimized (Prata 1989). These channels are used to discriminate volcanic from meteorological clouds containing ice, water vapour, and water droplets. The BTD method involves

138

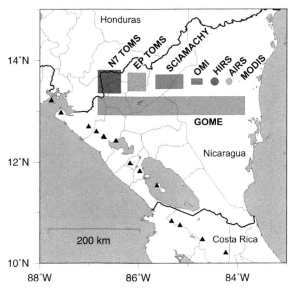

Figure 3. The nadir footprint of SO_2 sensors, excluding ASTER, whose footprint would not register at this scale. GOES has a slightly larger footprint than MODIS (Courtesy Carn, University of Maryland, Baltimore County). (see colour plate 20)

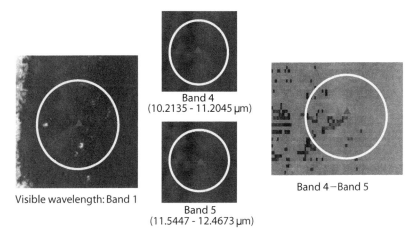

Figure 4. GOES-8 imagery, Tungurahua Volcano, Ecuador 14:45 UTC 16 September 2001. (see colour plate 21)

subtracting GOES channel-5 image data from channel-4 data (T4–T5). T4 greater than T5 indicates the presence of hydro meteoric clouds, whereas T4 less than T5 indicates the presence of silicate-rich ash with stronger negative numbers representing greater ash richness (Wen & Rose 1994; Watson *et al.*, 2004). The BTD method acquires data both day and night because the sensors operate in the thermal infrared region. The same strategy can be used for GOES-8 channels 4 (10.2135–11.2045 μm) and 5 (11.5447–12.4673 μm) (Figure 4) and the MODIS channels 31 (10.780–11.280 μm) and 32 (11.770–12.270 μm) (Figure 5).

Both GOES and MODIS data are used because they offer different temporal and spatial resolutions. The GOES platform is advantageous when monitoring smaller eruptions with a lifespan of no

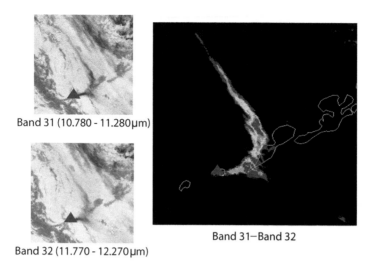

Band 31 (10.780 - 11.280 µm)

Band 32 (11.770 - 12.270 µm)

Band 31–Band 32

Figure 5. MODIS Imagery, Mount Cleveland Volcano, Alaska 23:10 UTC 19 February 2001. (see colour plate 22)

longer than a few hours, because the sensor updates every thirty minutes and has a four kilometre spatial resolution. Currently, MODIS imagery for many locations on the Earth is acquired once a day, but has a one kilometre spatial resolution. Whereas the temporal resolution for the MODIS instrument is less frequent, its spatial resolution is finer. By integrating both data types into the analysis, volcanic ash-plume trajectories are identified, tracked, and illustrated using a GIS. Volcanic ash plumes can be confirmed using advisories from NOAA's Volcanic Ash Advisory Centre. These advisories are disseminated as both text and graphic messages and are updated on a daily basis. Findings can be confirmed by using a dispersion model.

2.1.9 *Dispersion model*

Although remote sensing data are a powerful tool for observing and analysing volcanic-plume dispersal (including plume footprint, horizontal size and location), satellite data cannot forecast ash trajectories. Volcanic ash transport- and dispersion-models are used for predicting ash trajectories. Dispersion models can obtain data relating to ash trajectory; ash transport, concentration, size distribution, vertical dimensions, and age of cloud (Searcy *et al.*, 1998; Draxler & Hess Pudykiewicz 1989). Dispersion models and sensor data are used for reciprocal validation by comparing observed data with modelled data. Modelling the spatial extent of ash clouds, including details such as particle size and future trajectories, is particularly informative in complimenting and verifying sensor data, especially when imagery is obscured by meteorological conditions (e.g. clouds) (Peterson *et al.*, 2009).

Data can be used to create rolling thirty minute GOES and daily MODIS volcanic plume maps for air-borne ash particles within a GIS. Volcanic plumes can be compared with ground-based air-quality monitoring data from the same location and dispersion-model results from the same time period. By having frequent ground-data measurements and dispersion model results to compare with sensor data from the same time period, one can determine whether there is a correlation between what is observed in the air and what has fallen to the ground. This helps to establish whether sensor data can serve as a relative proxy for ground-based measurements. It is probable that higher amounts of ash particles in the air correspond with higher amounts on the ground, although the magnitude of the relation is unknown. Determining what proportion of ash does end up on the ground warrants further investigation. Some studies have explored this issue (Andronico

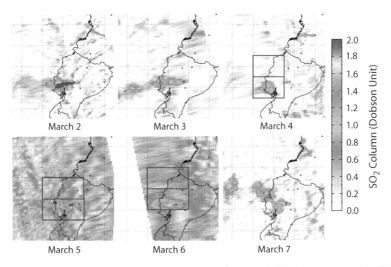

March 2   March 3   March 4

March 5   March 6   March 7

Figure 6.   OMI SO_2 imagery for Tungurahua volcano, Ecuador, March 2007. (see colour plate 23)

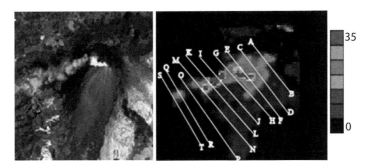

Figure 7.   Emissions from Pacaya volcano, Guatemala are similar to emissions from Tungurahua (Courtesy Watson, University of Bristol). (see colour plate 24)

et al., 2009; Carter & Ramsey 2009; Tupper & Wunderman 2009; Lyons *et al.*, 2010) but none has undertaken these efforts with explicit focus on public health consequences.

2.1.10   *SO_2 detection*

Multiple sensors are used for SO_2 data acquisition to optimize the temporal and spatial resolution. For SO_2 detection, data are acquired by the Ozone Monitoring Instrument (OMI), MODIS, and ASTER. Having ninety metre pixels, ASTER can provide the finest detail. Advanced SO_2 detection uses a tool known as interface description language (IDL) developed at the Jet Propulsion Laboratory (JPL).

OMI imagery was used to estimate SO_2 dispersal from the Tungurahua volcano. Figure 6 shows SO_2 emissions from this volcano moving westward. Dynamic series of these maps can provide a visual-analytical method for viewing plume-dispersion, such as those shown in Figure 7 (Castronova *et al.*, 2009).

Sulphur dioxide data also are acquired by the OMI on board NASA's Earth Observing System (EOS) Aura Satellite. The OMI sulphur dioxide group (SDG) produces daily composited SO_2 images for select areas. Archived images are available from May 2006 and are available within two to three days after each acquisition. Archived data dating back to 2004 are also available from the SDG website (OMISDG 2007).

The MODIS sensor on board EOS Terra and Aqua platforms use channels 28 (7.2–7.5 μm) and 29 (8.400–8.700 μm) to measure SO_2. Each band uses a different measurement method (Realmuto et al., 1994; Realmuto et al., 1997; Realmuto 2000; Watson et al., 2004). Channel-28 measures differences in radiance between SO_2-filled atmosphere and SO_2 free atmosphere. Channel-29 identifies SO_2 by measuring the difference in radiance between clear ground measurements and ground measurements with SO_2 cloud cover (Novak et al., 2008).

Channels-10 to -12 of ASTER are sensitive to SO_2 and capable of detecting small-scale degassing (Watson et al., 2004). Because of its high spatial resolution, ASTER is the only instrument in orbit that detects plumes less than 1.5 km wide (Pieri & Abrams 2004).

2.1.11 *Developing volcanic ash and SO_2-exposure maps*
Remotely collected SO_2 data can be fused or assimilated with geospatial data and relevant dispersion models to simulate exposure to SO_2. Simulations can be created every half hour from the GOES sensor and augmented with daily data from MODIS. They can also be overlaid with demographic data to assess how many people might have been affected. Dynamic maps of these time-stamped outputs can be created to provide an initial visual assessment for how well areas of elevated exposure correspond with health surveillance data obtained from ER records (Castronovo et al., 2009).

2.1.12 *Linking sensor data with health outcome data*
Working with data from a variety of sources requires a sound research hypothesis, developing a conceptual framework, selecting potential candidates for measures of exposure, and outlining properties of health outcomes clearly (Jagai et al., 2007). Also, there needs to be compatibility among health outcome measures and exposure measures derived from sensor data. This is achieved analytically by considering normalization schemes, inclusion of a proper indicator in the modelled procedure, stratification, and other techniques. Though it is rarely possible, there should be complete uninterrupted temporal and spatial overlap in health and exposure data. Temporal completeness expands the array of applicable analytical techniques because many statistical methods for time-series data have serious limitations if there are gaps in the series. Poor overlap may lead to a substantial reduction in statistical power to detect an effect, since the sample size available for the analysis will be smaller in a joint time series. Typically, time-referenced health outcomes are recorded as daily, weekly, monthly, quarterly, or annual counts of health-related events. The temporal period of availability of satellite-acquired data depends on their orbital designs and parameters (e.g. every thirty minutes or daily; sun-synchronous or geostationary). Therefore, to properly link sensor data with health outcome data, it is important to ensure convertibility of time units and perform a basic time unit alignment. Finally, the spatial alignment has to be carefully addressed. Health outcomes are typically derived from specific geographic areas and may differ by specificity, sensitivity, and population coverage. Moreover, these data must have geographic coordinates to ensure proper selection of the target or catchments area for abstracting sensor data.

Detection of a meaningful increase in number of health-related events by using statistical scanning in a given location should then be linked to an elevated exposure, if a causal relationship exists. The ability to detect such a link depends on characteristics of geographical area, spatial adjacency, and a spatial distribution of event occurrences. In practice, rarely do uniform or normal distributions correctly reflect population density. Similarly, a simple regular shape like a concentric circle (a primary choice in many spatial-clustering algorithms) rarely represents a geographic area of interest in a proper manner. For example, a geographical area representing communities effected directly by a chemical plume is likely to resemble a funnel with respect to factors contributing to contamination; that is, wind speed, wind direction, diffusion, and photochemical reactions (Naumova et al., 1997). Similarly, a pattern of communities effected by a hurricane is likely to resemble a storm trajectory (Chui et al., 2006). In some situations, assumptions related to cluster properties are reasonable as, for example, when a point-source exposure is well defined; when everyone is affected in a similar way; or, when the population within a study area is relatively stable. Examples include diseases that can be effectively scanned for clustering are exacerbations of chronic disorders (e.g. asthma) or injuries. But even in well-defined cases, spatial methods have to take into consideration

142

Figure 8. Aerial photograph of Quito, Ecuador. Pichincha volcano lies west of Quito in the central background. A narrow zone of haze stretches from left to right across the city (Geophysical Institute of Ecuador). (see colour plate 25)

complex non-Euclidean topologies of an outbreak signature. Such topologies are very likely to be observed in spatio-temporal patterns of diseases when transmission is amplified by re-suspended ash deposits, or distorted by multiple population relocations, when tracking of exposure location is unwarranted. Complex spatial patterns of georeferenced exposure-transmission-detection pathways have to be better understood. An example illustrates the potential for using remote sensing data to better understand health effects of volcanic eruptions.

2.2 *Potential applications: Guagua Pichincha*

2.2.1 *Guagua Pichincha in April 2000*

Quito, capital of Ecuador, is located 2800 m above sea level and is surrounded by four active volcanoes: Guagua Pichincha, Cotopaxi, Antisana and Tungurahua (de la Cadena 2006). The strato-volcano Guagua Pichincha is located thirteen kilometres west of Quito and became active in 1998 with emissions of vapour, ashes, and fumes after 340 years of dormancy. In spring, 2000, volcanic ash containing silica, sulphurs and particulate matter were again reported, and yellow alerts were issued for Quito. Nestled in a long, narrow valley between the base of the mountain range to the west and the precipitous canyon of the Machángara River to the east, Quito possesses an unmatched setting: a dangerous chamber created by nature and enhanced by humans (Figure 8).

Over the last two decades, rapid urbanization and sprawl have resulted in hazardously high levels of air pollutants (Jurado & Southgate 2001; Cifuentes *et al.*, 2005). The unique topographical location of Quito and the pattern of prevailing wind, traps stagnant, contaminated air. At that altitude, where the demand for oxygen is high, even a small increase in air pollution may trigger a substantial increase in ER visits for respiratory conditions.

Studies indicate strong effects of air pollution on respiratory health in Quito children (Estrella *et al.*, 2005). In this study, elevated rates of ER visits for acute and lower-respiratory infections and asthma-related conditions were documented in the younger children of Quito in association with the volcanic activity of Guagua Pichincha in April 2000. Before, during, and after ash falls in Quito, fluctuations in paediatric ER visits represented the negative impact volcanic activity has on human health. In all 5169 ER records were abstracted for patients treated between 1 and 10 December 2000, and were subsequently classified into three non-overlapping categories: acute upper-respiratory infection (2392); acute lower-respiratory infection (2319); and, asthma-related hospitalizations (431). Volcanic activity was documented from a variety of governmental and media sources and was compared with a time series of daily counts of respiratory infections in the studied population. Applying a Poisson regression model to examine the relations between the timing of the eruption and ER visits, it was found that increases in daily cases relative to the annual average were significant for all ER visits and for each disease group.

After geocoding 4786 records (93.4 per cent), categorizing them into the city of Quito, the city suburbs, or the area outside the city's metropolitan boundaries, the spatial distribution of cases and changes during eruption was examined. The daily means of ER visits in the youngest children were two times higher during the four periods of volcanic activity than pre- and post-eruption, after adjusting for baseline differences. In addition to the level of six to eight cases per day typically observed in girls and boys treated in the ER, respectively, a 2.25-fold increase in daily ER visits was observed in all parts of the city. Overall, during twenty-eight days of volcanic activity and ash releases, on average, 138 girls (CI-95 per cent = 104, 207) and 206 boys (CI-95 per cent = 137, 252) were treated in ER in contrast to the typically observed level of 6.17 and 8.18 cases per day in girls and boys, respectively. Not only did this study demonstrate an increase in ER visits to the Baca Ortiz Hospital after volcanic activity, it emphasized the usefulness of hospital-based surveillance for monitoring the health effects associated with volcanic activity.

Rapid deployments of disease surveillance systems that incorporate novel data collection technologies demonstrate the potential for increasing information exchange. The list of prospective candidates for disease monitoring is growing and includes not only traditional hospitalization records and laboratory-confirmed tests, but also the data streams driven by new information technologies. Personal digital assistant (PDA)-based records collected directly from outbreak investigations, hits and queries targeting specific websites, and searches for media and news are examples of these new data streams. These novel data sources are the primary candidates for linkages with sensor data.

2.2.2 *Eyjafjallajökull volcano, Iceland April 2010*
A volcanic eruption of Eyjafjallajökull in Iceland reminded everyone of the need to better understand risks associated with volcanic activity (Gudmundsson 2011). Eleven centuries of settlement in Iceland were marked by an explosive eruption of Eyjafjallajökull that started on 14 April 2010. Eyjafjallajökull is a glacier volcano that rises to 1666 m above sea level on the southern shore of Iceland about 150 km from Reykjavik. The explosive interaction of magma at over 1000°C with ice and water generated large volumes of finely comminuted ash. Erupting from beneath the ice cap, the plume of ash and vapor reached heights of almost ten kilometers on the first day. Alerts were issued to people with respiratory illnesses. The next three days of sustained tephra production resulted in ash dispersal in the southeast area populated by several hundred people. Fortunately, the Reykjavík metropolitan area of some 300,000 inhabitants was located just outside the margins of the active volcanic zone. Eruptions halted by the end of May 2010. Since then, on windy days, released ash is re-suspended into the air and, as a result, particulate matter concentration levels have been unusually high in southern and southwestern Iceland, including Reykjavik. Ash-fall deposits can remain in the environment, sometimes for decades and can be redistributed by wind and by human activities.

The up-to-date studies suggest that the acute and chronic health effects of volcanic ash depend on: a) particle size and its respirable fraction; b) mineralogical composition, especially the crystalline silica content; and, c) physico-chemical properties of the surfaces of ash particles. The effects vary between volcanoes and even between eruptions, making the comparison of specific effects difficult. The fresh ash from Eyjafjallajökull was very well characterized: it contained twenty-five per-cent respirable particles (i.e. less than 10 μm). The main constituents were sixty per cent SiO_2, sixteen per cent AlO_3 and ten per cent FeO with the negligible content of crystalline silica content. The plume was carefully monitored via remote sensing. Such large-scale events offer opportunities for detailed investigation of the long-term effects of ash deposition on human and animal health and the health of ecosystems.

3   DESERT DUST STORMS AND HEALTH

Deserts account for approximately thirty-three per cent of Earth's terrestrial environments. The two largest deserts are the cold deserts of the Arctic and Antarctica, with each covering approximately $13 \times 10^6 \, km^2$. Although they are the largest, they are in general covered by snow and ice for most of the year and their surface soils contribute little to Earth's annual atmospheric desert-dust load. Current global estimates for the quantity of desert dust moving some distance in Earth's atmosphere ranges between one-half to five billion metric tons per year, and some believe that the upper limit is an underestimate (Goudie & Middleton 2001).

Dust particles are loaded into the atmosphere by storm fronts that generate high winds across the surface. These winds cause saltation, which is the movement and bounce of soil particles along the surface (Grini & Zender 2004). Saltation in conjunction with sandblasting of downwind surfaces by particles results in dust-cloud formation characterized by fine grain sizes, typically less than $10\,\mu m$ (Cahill et al., 1998; Gillette et al., 2004; Grini & Zender 2004). These less than one micrometre, ultrafine particles are capable of long-range global dispersion, although grain sizes as large as $75\,\mu m$ have been identified at distances up to 10,000 km from their source (Betzer 1988). Atmospheric particulate load for a 1995 African dust storm that impacted air quality in Tampa Bay, Florida more than 6500 km from the source, was over ten times that observed during normal atmospheric conditions; and, over ninety-nine per cent of the particles were less than one micrometre in diameter. This small size is capable of penetrating into the deep lung environment (Griffin 2007).

Primary sources of dust entrained into Earth's atmospheric boundary layer are the Saharan and Sahelian zone of North Africa. These regions are currently believed to contribute approximately fifty to seventy-five per cent of the total desert dust budget (Moulin et al., 1997; Goudie & Middleton 2001; Prospero & Lamb 2003). The second greatest source is the desert region of Asia that includes the Gobi, Takla Makan and Badain Jaran deserts. Dust storms also are common from other world regions. The major ones include North America's Great Basin and Sonoran deserts, Mexico's Chihuahuan desert, South America's Patagonia and Atacama deserts, South Africa's Etosha and Mkgadikgadi deserts, Australia's Sandy and Great Victorian deserts, and the Syrian and Arabian desert regions of the Middle East. Due to periodic short-term fluxes in climate that influence regional rates of precipitation and anthropogenic influences (i.e. deforestation), these smaller desert regions can contribute significantly to atmospheric-particle loading. This is best demonstrated by the American Dust Bowl of the 1930s and in the recent large dust storms originating on the continent of Australia (Griffin et al., 2002; De Deckker et al., 2008).

3.1.1 North African dust storms
Large dust storms capable of transoceanic movement occur year round in the Saharan and Sahelian zones. Common transport routes from the northern fringe of the Sahara are into and over the northern Atlantic, Europe and Middle East during the spring months (Graham & Duce 1979). Less frequently, large dust storms moving out of North Africa have been tracked eastward over the Middle East, across Asia, over the Pacific, and into North America (McKendry et al., 2007). Yearly dust storms generated on the mid- and southern Sahara and Sahel move west across the Atlantic Ocean and impact air quality in the Caribbean and Americas. Typically, between the Northern Hemisphere summer months of May through October, dust storms move across the Atlantic dust corridor between 15N–25N latitude and impact air quality in the Caribbean and south eastern US (Graham & Duce 1979). A cloud of dust arising off the west coast of North Africa typically takes three to five days to reach the Caribbean, and in cases of prolonged transmission can form atmospheric bridges of dust across the Atlantic (Figure 9).

The equatorial trade winds drive and carry dust and hurricanes across the Atlantic. Between the Northern Hemisphere's winter months (November–April), transatlantic transport is farther south due to equatorial shifts of Hadley cells and dust moves to the southern Caribbean and South America (Swap et al., 1996; Giraudi 2005). The primary source regions of transatlantic dust are the Bodele Depression and Lake Chad regions of the Sahara (Goudie & Middleton 2001; Koren et al., 2006). Fine textured sediments from lakebeds are more easily mobilized into the atmosphere than coarse textured terrestrial soils.

Interestingly, Lake Chad is only five per cent of its 1960's area due to a combination of climate change and anthropogenic diversion of its source waters for cultivation. Similarly, the dried lakebed of Owens Lake in southern California and the exposed sediments of the Aral Sea are prime sources

Figure 9.   An atmospheric bridge of dust observed on 8 August 2001 extends along the arrow from its source in North Africa to the Caribbean (Courtesy NASA & ORBIMAGE). (see colour plate 26)

of entrained dust that is due primarily to engineering diversions of source waters for consumption and agricultural purposes (Griffin *et al.*, 2001). Another influence on dust transport is natural and anthropogenic desertification along border regions. Although desertification has been documented on annual scales, the overall size of the Sahara has not changed significantly in recent geologic time (Tucker & Nicholson 1999).

Naturally occurring climate cycles that are driven by large-scale planetary processes like the North Atlantic Oscillation (NAO) and El Niño influence long-range dust transport out of Africa significantly (Prospero & Nees 1986; Moulin *et al.*, 1997; Giannini *et al.*, 2003). The NAO is a duel pressure system that moves annually in a northerly and southerly flux over the equatorial and North Atlantic Ocean. When this system is in a more southerly position, there is more annual rainfall over North Africa and thus less dust transport off the continent. Likewise, when the NAO is in a more northerly location, there is less rainfall over North Africa and greater dust transport out of the region. Since the late 1960s the NAO has predominantly held a more northerly location. This started the current long-running North African drought that has resulted in a noted increase of North African dust transport across the Atlantic to the Caribbean and Americas (Prospero & Nees 1986; Moulin *et al.*, 1997). Looking at annual dust-transport records collected on Barbados, this increase in transport is apparent. A majority of peak-transport years between the early 1960s and the late 1990s occurred during El Niño years (Prospero & Nees 1986). Approximately forty million tons of Saharan and Sahelian dust are deposited in the Amazon Basin each year (Swap *et al.*, 1992; Koren *et al*., 2006). Nutrients such as iron, phosphorus, and nitrogen carried in desert dust impact primary production in oceans on both long- and short-term temporal scales (Graham & Duce 1979; Ridgwell 2003). Aeolian iron influx in oceans has been estimated between about one to ten billion mol/year using dust-iron solubilities of one and ten per cent, respectively (Fung *et al.*, 2000). Ice-core analyses of dust deposition on geological time scales has demonstrated that during glacial periods of elevated dust transport, oceanic productivity increased and atmospheric carbon concentrations decreased (Broecker 2000; Ridgwell 2003; Lambert *et al.*, 2008). Although some organisms may benefit from this nutrient rich dust, the same dust may be harmful to others. Correlation has been made between African desert-dust deposition in Florida Bay and algal blooms of *Karinia brevis*, the red tide agent. Blooms of this algae routinely impacts marine ecosystem and human health negatively (Lenes *et al.*, 2001; Walsh & Steidinger 2001).

3.1.2   *Asian dust storms*
Clouds of dust moving out of Asia's deserts occur primarily in the spring between February and May. Like the NAO's influence on transatlantic transport of North African dust, the Pacific decadal oscillation (PDO) influences dust transport from Asia to the Americas, and during larger scale dust events, around the globe (Gong *et al.*, 2006; Hara *et al.*, 2006). The PDO spawns El Niño/La Niña-like events that are characterized by sea surface temperature anomalies above the latitude of 20°N. They typically occur on time scales of two to three decades but there is great variability in their frequency and duration. During positive-phase events there is typically less dust transport from Asia

146

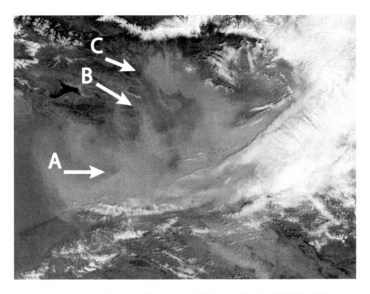

Figure 10. A large dust storm originating in western China on 24 April 2010. This storm resulted from dust converging from three locations. Arrows indicate direction of dust movement from each source region (Courtesy MODIS image, NASA). (see colour plate 27)

across the Pacific to North America than during negative-phase periods, when the eastern Pacific cools. Positive-phase PDO reduces the movement of northern fronts into and across the deserts of Asia restricting the transpacific transport window. The increase in large clouds of Asian dust moving eastward to North America has coincided with recent negative-phase years. El Niño and La Niña events influence dust transport out of Asia with transport across the Pacific occurring at roughly 45°N latitude during El Niño years (greater dust transport) and 40°N during La Niña years (less dust transport) (Gao *et al.*, 2003; Hara *et al.*, 2006). The desertification rate along Asia's desert borderland averaged approximately 2100 km^2 per year between 1975 and 1987 (Zhenda 1993). Anthropogenic activities contributing to this rate are deforestation, overgrazing and population growth. A report from the US Embassy in China in April 1998 cited range wars in the Ningxia Hiu Autonomous Region between herders grazing the border region grasslands and farmers harvesting them (Griffin *et al.*, 2001). A dust storm that impacted air quality in China's Gansu and Xinjiang Provinces in April 2010 (Figure 10) was reported to be one of the largest in since the early 1990s, with visibility in some areas approaching zero and reports of three deaths and one missing person (Tan 2010). Although early web-based reports of this storm indicated that it was restricted to the Asian continent, large dust events such as that one are capable of global dispersion. The chemical fingerprint of a 1990 Asian dust storm moving eastward off the Asian coast crossed the Pacific, North America, and the Atlantic and was detected via isotopic analysis in samples collected in the French Alps (Grousset *et al.*, 2003). Similarly, a large Asian dust storm that impacted North American air quality in 1998 reduced solar-radiation levels by as much as forty per cent and left a chemical fingerprint inland to the state of Minnesota (Husar *et al.*, 2001).

3.2 *Desert dust and human health*

3.2.1 *Exposure to desert dust*
Silicosis is a disease caused by chronic inhalation of silica. This disease is typically occupational in nature and is a health risk to those most frequently exposed to mineral dusts, such as miners and cement workers. It is caused by silica penetrating into the lungs where it causes nodular lesions. The disease causes impaired lung function and can be fatal. One understudied field regarding silica

and human health is the occurrence of this disease in populations that are frequently exposed to desert dust-storms. These include populations close to dust sources as well as those who reside within downwind dust-transport corridors.

Studies focused on the incidence of silicosis have demonstrated potential risk. For example researchers detected silicosis in fourteen of sixteen screened Himalayans in a group residing in a common region of dust, compared to ten of twenty-four in a region where dust exposure was less common (Norboo et al., 1991). In a related study, where dust storm silica concentrations were as high as seventy per cent, the average prevalence of silicosis in 449 individuals from three different regional villages was 16.5 per cent, and in one village as high as 45.3 per cent (Saiyed et al., 1991). This study further noted many cases of massive fibrosis; that is, nodules larger than one centimetre.

In a study conducted in China, where populations are commonly exposed to dust, the silica content in regional soils ranged from about fifteen to twenty-six per cent, the soil particle size range of two to five micrometres was twelve to twenty-five per cent and the prevalence rate of silicosis was seven per cent for 395 screened individuals, or about twenty-one per cent for those over forty years of age (Xu et al., 1993). Those authors also reported the occurrence of silicosis in a village camel and that no human cases of silicosis were detected in a control group (Xu et al., 1993).

Another concern of dust exposure is the development of asthma and the risk to those susceptible to respiratory stress. High asthma rates and the risk to children developing the disease in populations frequently exposed to desert dust have been documented (Bener et al., 1996; Al Frayh et al., 2001). Interestingly, the asthma rates on Barbados increased seventeen-fold between 1973 and 1996, the period corresponding to a Saharan drought and an increase in transatlantic dust transport (Howitt et al., 1998 Prospero 1999). On Trinidad, researchers have demonstrated a link between African desert-dust and paediatric respiratory stress (Gyan et al., 2005). The link between Asian desert dust exposure and human morbidity and mortality has been established by numerous research teams (Kwon et al., 2000; Kwon et al., 2002; Park et al., 2005; Chang et al., 2006; Yang 2006; Kan et al., 2007; Meng & Lu 2007; Belser et al., 2008; Chan et al., 2008; Chiu et al., 2008).

A widely reported but poorly understood illness associated with exposure to desert dust is a disease known by a number of names: desert-dust pneumonia; Al Eskan disease; Gulf War syndrome; and Persian Gulf War syndrome. Cases of desert-dust pneumonia were widely reported during the 1930's American Dust Bowl. Rates then were a hundred per cent higher than those reported during periods of lower dust storm frequency; but, the causative agent was not identified (Brown et al., 1935). It was rumoured that the disease was caused by breathing through wet cloth, a method used for protection during dust events (Egan 2006).

A similar dust-related disease has been widely reported in military personnel deployed in the Middle East during the first and second Gulf Wars (Korenyi-Both et al., 1992; Korenyi-Both 2000; MMWR 2003b; Shoor et al., 2004; Aronson et al., 2006). A survey of over 15,000 deployed personnel between 2003 and 2004 reported that over sixty-nine per cent experienced some form of respiratory illness (Sanders et al., 2005). In addition to respiratory stress, deployed personnel have reported neurological illnesses, fever, and fatigue; and, although some epidemiological studies have shown a link between deployed personnel and disease relative to non-deployed control groups, others have shown no differences between deployed and control groups (Haley et al., 1997; Rook & Zumla 1997; Fukuda et al., 1998; Ismail et al., 1999; Doebbeling et al., 2000; Knoke et al., 2000; Hyams et al., 2001; Gray et al., 2002; Kang et al., 2002; Lee et al., 2002; Hooper et al., 2008; Smith et al., 2009). In recognition of the health risk to personnel from dust-storm exposure, the US military currently provides dust-storm warnings in Korea via its web-based Yellow Sand Activity Warning System (YSAWS) (KMA 2008). This system provides a real-time, colour-coded flag system to provide health advisories to personnel. Ground-based monitoring and satellite sensors are being utilized by China, Korea, and Japan to monitor KOSA (yellow sand) activity to protect public and economic health, address mitigation strategies, and understand the influence of climate change (Yamamoto 2007).

The Dust Regional Atmospheric Model (DREAM) has been utilized to forecast dust entrainment, transport, and deposition in south western US and in China (Nickovic et al., 2001; Papayannis et al., 2007; Yin et al., 2007; Sprigg et al., 2009). In the south-western US, this model is nested within the

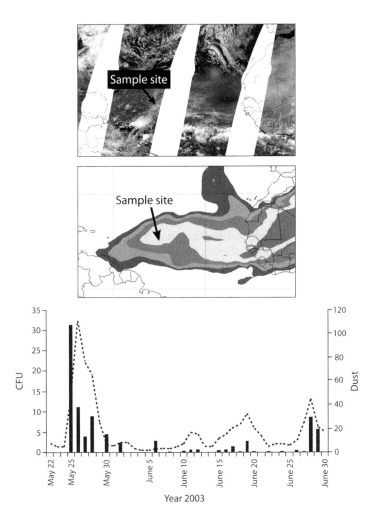

Figure 11.   African dust-borne microorganisms transported across the Atlantic Ocean between 22 May and 30 June 2003: Ocean Drilling Program, Leg 209. The black bars are CFUs and the Y-axis on the left shows their values in μg/litre^3 of air; the dotted line shows modelled NAPPS dust concentrations on the right side Y-axis in μg/litre^3 of air. (see colour plate 28)

National Weather Service NCEP/eta weather forecast model to assimilate satellite digital surface terrain observations that enable forecasting of dust cloud evolution, movement, and deposition over periods of 24–72 hours (Morain *et al.*, 2007; Morain *et al.*, 2009; Morain & Budge 2010). Similarly, the Naval Aerosol Analysis and Prediction System (NAAPS) Global Aerosol Model uses digital satellite and surface measurement data to predict the distribution of dust aerosols at various altitudes on a global scale (Johnson *et al.*, 2003; Schollaert *et al.*, 2003; Honrath *et al.*, 2004; Reid *et al.*, 2004). This model accounts for mobilization, spatial mixing, advection, wet removal and dry deposition, and has been utilized successfully to measure desert-dust-associated microbial concentrations in the tropical mid-Atlantic (Figure 11) (Griffin *et al.*, 2006). Image A shows a series of satellite images of African dust crossing the Atlantic, and the location of a marked research site, May 26, 2003; image B is a contoured unit-less optical depth global aerosol model output for May 26, 2003 with the same marked research site in A. Image C shows total bacterial- and fungal-colony-forming unit (CFU) concentrations obtained from atmospheric samples normalized to per cubic meter for the study period 22 May to 30 June 2003. The blue line shows the trend of predicted

NAAPS dust concentrations per cubic meter at ten metres above sea level for those dates that were acquired several months after the cruise. The majority of CFU was detected during periods of elevated dust (Griffin et al., 2006). Both the NAAPS and DREAM/eta models have great potential for public health dust advisories and monitoring.

3.2.2 Dust-storm microorganisms & their transport

The number of bacteria found in a gram of forested, arid, and other types of soils ranges from 10^7 to 10^9, respectively (Whitman et al., 1998). Within these samples there are an estimated 10,000 bacterial types (Torsvik et al., 1996; Torsvik et al., 2002). The number of fungi found in soil has been reported at 10^6/gram (Tate 2000). Viral concentrations have been reported as high as 10^8/gram (Reynolds & Pepper 2000; Williamson et al., 2003). Factors that influence microbial survival and diversity in soils include pH, temperature, elemental composition, nutrient content, and moisture concentration (Yates & Yates 1988; Gans et al., 2005; Janssen 2006). Growth of microorganisms in laboratory cultures has been the historical approach for their identification in soil samples. However non-culture techniques, such as direct-count assays in which cell or nucleic stains allow accurate counts per weight or volume of soil and water, have demonstrated that less than one to up to nineteen per cent of bacteria are culturable in any given sample (Torsvik et al., 1990; Amann et al., 1995; Sait et al., 2002). Clearly, the limitations of culture-based methods prevent the determination of viability for many microorganisms; but, modern molecular-based assays such as the polymerase chain reaction enable one to determine presence/absence in samples (Griffin 2007). Studies of desert microorganisms have reported concentrations of bacteria and fungi per cubic meter of air during normal atmospheric conditions to be 0–1000 CFU and 0–291 CFU, respectively. In these same environments during periods when dust storms were present, concentrations ranged from 0–15,700 CFU and 0–703 CFU, respectively (Griffin 2007). On an international scale, dust storm microbiology studies have demonstrated that very diverse communities of both bacteria and fungi are capable of surviving long-range atmospheric transport (Griffin 2007). Atmospheric sources of stress to dust-borne microorganisms include exposure to UV, desiccation, temperature, and both high- and low-humidity levels (Mohr 1997). Several factors influence the ability of many microorganisms to withstand the rigors of long range transport. These include: (a) attenuation of UV light by up to fifty per cent depending on particle load; (b) the interactions among particles like partial UV shielding and particle tumbling during transport; and (c), variable humidity levels as dust clouds move over aquatic environments (Gregory 1961; Mohr 1997; Herman et al., 1999; Dowd & Maier 2000; Griffin 2007). Various species of both bacteria and fungi are more capable of surviving long-distance transport or suspension at extreme altitudes because their cell pigmentation provides some UV shielding, tolerance to extreme UV doses, or the ability to form spores that provide stress protection (Griffin 2004; Griffin 2007; Griffin 2008; Smith et al., 2010).

With regard to bacteria, one of the most well-known pathogens associated with dust storms is *Neisseria meningitides*. Outbreaks of meningitis are reported frequently following dust events in the meningitis belt of North Africa. These outbreaks typically occur between February and May of each year and effect on the order of 200,000 individuals (Molesworth et al., 2003; Sultan et al., 2005). The current belief is that inhaled dust particles cause abrasions in the nasopharyngeal mucosa allowing virulent strains residing in the nose access to its host's bloodstream, and thus the ability to cause infection (Molesworth et al., 2002). Recently, *N. meningitides* was cultured from settled dust samples collected in Kuwait, demonstrating that dust storms may serve as a vector for this pathogen (Lyles et al., 2005). Other species of bacteria known to cause disease in humans and isolated in a number of different desert-dust storm studies include *Bacillus licheniformis* (peritonitis), *Ralstonia paucula* (speticemia, tenosynovitis), *Acinetobacter calcoaceticus* (respiratory infections), *Kocuria rosea* (bacteremia), *Brevibacterium casei* (sepsis), *Staphylococcus epidermidis* (endocarditis), and *Pseudomonas aeruginosa* (can cause fatal infections in burn victims) (Griffin 2007). Dust-storm exposure of personnel to ubiquitous pathogens such as *Acinetobacter baumannii* may present significantly unique human health risk in desert environments (Holt 2007).

The most well documented human pathogen known to be transported in dust clouds is the fungus *Coccidioides immitis* (Fiese 1958). This pathogen is only found in the Americas, and outbreaks

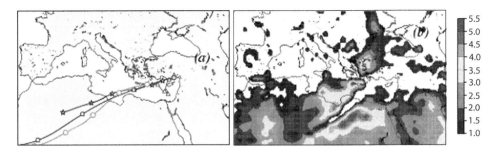

Figure 12. (Left): Air-mass back-trajectory from Erdemli, Turkey; (Right): Earth Probe TOMS satellite data (trajectory data courtesy ECMWFTOMS; TOMS image courtesy NASA). (see colour plate 29)

of the disease known as coccidioidomycosis are reported annually following exposure to clouds of desert dust (Jinadu 1995; Hector & Laniado-Laborin 2005). Groups at highest risk include, African-Americans, Asians, and people younger than twenty years old (Williams *et al.*, 1979; Durry *et al.*, 1997). Infections by this pathogen can be fatal (Schneider *et al.*, 1997). Other genera of fungi that have been identified in desert-dust studies, and whose spores are known to be human allergens include *Acremonium, Alternaira, Aspergillus, Cladosporium, Curvularia, Emericella, Fusarium, Nigrospora, Paecilomyces, Pithomyces, Phoma, Penicillium, Trichophyton* and *Ulocladium*) (Griffin 2007). In a desert-dust study conducted at a coastal Mediterranean research site in Erdemli, Turkey, fungal CFU were more commonly recovered from atmospheric samples during dust events than bacterial CFU. This differed from what was observed in the Caribbean where bacterial CFU were equal to Bacterial CFU, or in Mali, Africa where bacterial CFU were more frequently recovered (Griffin *et al.*, 2003; Kellogg *et al.*, 2004; Griffin *et al.*, 2007). These data demonstrate different microbial dust-storm fingerprints that were influenced by distance from source, distance of transport, and difference in storm source regions. In the Turkish dust samples, the dominant fungal CFU were species of *Alternaria*. Spores of this fungus are known to be potent human allergens. Childhood exposure to spores of *Alternaria* in semi-arid environments has been associated with the onset of asthma (Halonen *et al.*, 1997). Figure 12 illustrates the combined use of satellite data to visualize dust aerosol transport into Turkey from Africa and air mass back-trajectory data. Back-trajectory is used to determine the origin of an air mass to determine impact regions as related to observed microbial CFU. These data clearly illustrate an increase in microbial CFU as the dust cloud passed over the research site, and the air mass/dust-cloud source region as the deserts of North Africa (Griffin *et al.*, 2007). The top image shows a seventy-two hour air mass back trajectory identifying Africa as the source. The red line is 1000 hPa; the grey line is 850 hPa; the green line is 700 hPa; and the blue line is 500 hPa. One thousand hPa is about one per cent of atmospheric pressure, or one millibar. The TOMS image on the bottom shows African dust impacting the research site in Erdemli, Turkey.

Table 3 lists bacterial and fungal CFU aero microbiology data for April 14–16, 2002. CFU = colony-forming units. Blank # = negative controls. CFU data show few fungal CFU and no bacteria CFU at the research site on 14 April 2002, followed by an increase and peak (at 1733 hrs.) in CFU on the 15th of April as the dust storm impacted the site, followed by a decrease in CFU on the 16th as the storm passed and the atmosphere started to clear of particulates.

One of the most neglected fields of desert-dust microbiology is virology, which is reflected in the limited number of papers cited in scientific literature. An atmospheric study of African desert dust in the Caribbean reported viral concentrations of 1.8×10^4/cubic metre during normal atmospheric conditions and 2.13×10^5/cubic metre during a dust event (Griffin *et al.*, 2001). A report demonstrated an increase in concentrations of air-borne influenza viruses during dust versus non-dust events in Asia (Chen *et al.*, 2010). This report is particularly noteworthy, given that transpacific atmospheric transport of infectious influenza viruses from Asia to North America was previously hypothesized (Hammond *et al.*, 1989).

Table 3. Aero-microbiology for Erdemli, Turkey (courtesy Griffin *et al.*, 2005).

Sample	Date	Time	Bacterial CFU	Fungal CFU
24	4/14/02	09:00	0	4
25	4/14/02	14:54	0	2
Blank 5	4/14/02	15:16	0	0
26	4/15/02	08:27	5	38
27	4/15/02	13:42	3	35
28	4/15/02	17:33	6	121
29a	4/16/02	15:03	1	33
29b	4/16/02	15:03	0	34
Blank 6	4/16/02	15:47	0	0

With regard to dust-storm microbiology, it is still unclear how microbial ecology in various desert soils differs; and, more importantly, how the ecology of air-borne dust-associated communities change with distance travelled. Nor is it clear how susceptible community members succumb to stress or how new organisms are added as particles sand-blast downwind environments and new members are added when air masses from different sources mix. Smoke transport of fungi from the Yucatan to Texas across the Gulf of Mexico has been reported (Mims & Mims, 2004). This potential vector of long-range atmospheric transport warrants investigation. Obviously, more investigations are needed of dust-borne viral transport given the recent influenza report by Chen (*et al.* 2010). In addition to the direct transport of human viral pathogens in dust-storms, there is the possibility that bacteriophages, known to move virulence genes from pathogenic strains of bacteria to non-pathogenic strains, may be capable of surviving long-range dust transport and thus are shuttling these genes between distant populations. Many of the molecular tools that will advance these fields have only recently become available; but, it seems clear that these methods will be augmented by aerial and satellite technology.

4   ANTHROPOGENIC POLLUTANTS

Loads of chemicals and hazardous wastes have been introduced into the atmosphere that didn't even exist in 1948. The environmental condition of the planet is far worse than it was 42 years ago (Gaylord Nelson, Earth Day 1990).

4.1   Air-borne dust and chemicals

Another desert-dust related health risk is exposure to anthropogenic chemicals that are utilized in dust-cloud source regions and are present in surface soils or are scavenged by the dust clouds. Scavenging occurs when emitted chemicals adhere to particles and are entrained in the interstitial spaces of the clouds as the clouds pass over emission areas (agriculture and industrial). In the late 1960s an air-quality research project conducted on samples from Barbados found that atmospheric-dust samples were of European and/or African origin and contained pesticide residues. Based on pesticide concentrations in these samples, the study concluded that atmospheric delivery of these chemicals to oceans was greater than from river sources (Risebrough *et al.*, 1968).

Dusts entrained around the Aral Sea have been shown to carry agricultural pesticides. A pesticide known as phosalone has been detected in regional air-borne dust at a concentration of 126 mg/kg and hospitalizations due to phosalone exposure have been reported (O'Malley & McCurdy 1990; O'Hara *et al.*, 2000). Worse, the surface area of this Sea has decreased by half since 1960, primarily due to the diversion of source waters for agricultural irrigation (Micklin 1988; Nandalal & Hipel 2007). This has left a dry shoreline area over 27,000 km^2 of exposed sediments that, in addition to

152

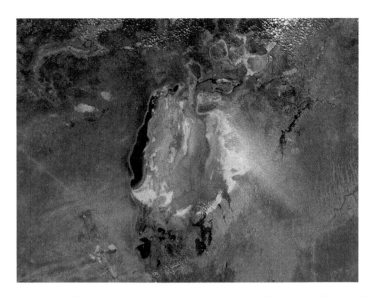

Figure 13.   MODIS image for 26 March 2010 showing a dust event originating in the Aral Sea region and moving to the southeast, along the Kazakhstan/Uzbekistan border (Courtesy NASA Earth-observatory). (see colour plate 30)

regional agricultural lands, are a prime source for atmospheric dust (Micklin 2007). Research has detected pesticides such as DDT in the breast milk of women and in the blood of children at some of the highest concentrations ever reported (O'Malley & McCurdy 1990; Hooper *et al.*, 1998). The extent of this environmental disaster has yet to reach its pinnacle. The Aral Sea region is a prime example of how humans can contribute significantly to an increase in the quantity of dust moving in Earth's atmosphere.

The Aral Sea disaster has been widely monitored by Earth-observing satellites. The USGS has a website (http://earthshots.usgs.gov/Aral/Aral) that contains a time series of satellite images from 1964 that dramatically illustrate the demise of the sea. NASA's Earth-observing satellites routinely capture images of dust storms in the region and this can be seen in Figure 13.

Long-range transport of industrial pollutants from Asia to North America, with and without associated desert dust, was identified during six events that occurred between 1993 and 2001 (Jaffe *et al.*, 2003). Industrial pollutants, which included carbon monoxide, peroxyacetyl nitrate, and non-methane hydrocarbons, were detected in samples collected during these episodes at various mountain-top locations and by aircraft. Other studies have documented the movement of industrial dust and smoke aerosols out of Asia that impact air quality throughout the Pacific and in North America (Jaffe *et al.*, 1999; Liang *et al.*, 2004; Jaffe *et al.*, 2005; Weiss-Penzias *et al.*, 2006; Weiss-Penzias *et al.*, 2007; Zhang *et al.*, 2008a). In these studies, Earth-observing sensor data and models were employed to identify pollutant sources and regions of impact. Atmospheric samples were used to identify sources but satellite observations provided the technology for tracking downwind plume dispersion. In 2003, export of atmospheric pollutants from China across the Pacific was estimated at twenty million tons, of which 5.6 million tons were estimated to have reached the west coast of North America (Yu *et al.*, 2008).

Aerosolized dust on smaller scales in non-arid regions also pose a risk to human health (Griffin 2007). Clean-up following Hurricane Katrina in New Orleans revealed that contaminated dry muds aerosolized into dust containing toxic metals, wastewater pollutants and pathogenic microorganisms. Other sources of hazardous toxic metals from dust followed the Aznalcollar mine tailing accident and the World Trade Centre attacks of September 2011 (Grimalt *et al.*, 1999; Plumlee & Ziegler 2005; Plumlee *et al.*, 2006; Plumlee *et al.*, 2007; Griffin *et al.*, 2009).

4.2 Chemical warfare

Chemicals weapons in warfare are intended to expose combatant forces to fatal or disabling doses of air-borne contaminants. In World War I, there were approximately 100,000 fatalities caused by releases of chlorine and mustard gas, and another 1.2 million non-fatal injuries. Approximately eighty-five per cent of the fatalities were due to mustard gas. The French were the first to deploy a toxic chemical weapon (tear gas), but the attack had no impact on the German Army. Similarly, the German military attempted to gas the Russian Army in January, 1915 near Warsaw, but the chemical quickly froze and had no effect. A prime example of the danger of using such weapons occurred in September 1915, when the British released a chemical attack at the Battle of Loos. A change in wind direction first stalled the gas cloud over no-man's-land, then pushed the cloud back over British positions (Heller 1984).

During the eight-year Iran-Iraq war (1980–1988), mustard gas and nerve agents were used against Iran and its own civilian population, resulting in an estimated 100,000 casualties and approximately 20,000 deaths. In contrast biological weapons have so far been relatively ineffective. This is due primarily to flawed delivery mechanisms and issues in dose response with various pathogens. Releasing harmful chemicals into the atmosphere as a military strategy seems to be not only indiscriminate, but highly risky. Their use in the early 20th Century proved only that chemical and biological warfare is sinister. More problematical in today's world are exposures resulting from unintentional industrial accidents.

4.2.1 Agent Orange

One of the best known cases of herbicide use impacting human health was Agent Orange during the Vietnam War. This program utilized several different herbicides in addition to Agent Orange to clear foliage around bases, roads, and trails, and to kill crops to limit food production. Agent Orange was a mixture of two chemicals, 2,4-dichlorophenoxyacetic acid and 2,4,5-trichlorophenoxyacetic acid that contained the dioxin 2,3,7,8-tetrachlorodibenzodioxin (aka TCDD). It is a by-product of 2,4,5-trichlorophenoxyacetic-acid production and is a known human carcinogen (Schecter et al., 2001). Approximately 50 million litres were sprayed over approximately 2.6 million acres on multiple occasions (Ngo et al., 2006). TCDD exposure has been shown to induce physiological responses such as shape changes, cause membrane damage in red blood cells, and has been linked to an increased prevalence of diabetes and insulin resistance (Suwalsky et al., 1996; Duchnowicz et al., 2005). There has been a number of studies that have suggested or linked TCDD exposure to birth defects; but also, there have been a number that challenged these findings and concluded that the link is questionable (Erickson et al., 1984; Friedman 2005; Ngo et al., 2006; Schecter & Constable 2006). Obviously, many military personnel and civilians were exposed to TCDD during the Vietnam War. In 1970, one human breast-milk sample was reported to contain 1832ppt. Subsequent analyses in 1973–1995 showed a decline in concentrations, but persistence in exposed populations over time. Tissue-sample concentrations collected in 1984 from an area that was not sprayed were less than three parts per thousand. Recent studies have shown that TCDD is still present in soil and water samples and in current residential tissue/fluid samples in areas where this chemical was utilized (Schecter et al., 2001; Dwernychuk et al., 2002).

Two metre high spatial resolution, declassified satellite photography collected between 1960 and 1972 provide data that can be utilized to map 'use areas'. One reason for doing this is to identify contaminated regions that might serve as sites for long-term human health exposure studies, and to design remediation projects (Hatfield-Consultants 2000). A GIS database has been developed for agent orange and other pesticides used in Vietnam as a tool to supplement epidemiological and sensor-based studies (Stellman et al., 2003).

4.3 Industrial accidents

In the US between 1994 and 2001, there were 696 ammonia, 534 chlorine, 100 flammable mixtures, 98 hydrogen fluoride/hydrofluoric acid, 59 chlorine dioxide, and 447 other reported accidents with the twenty-five most commonly involved chemicals and mixtures (Kleindorfer *et al.*, 2003). In July, 2003 a cloud of chlorine gas was released from a plant in Baton Rouge, Louisiana, that resulted in eight employees requiring medical attention and causing residents within a half-mile radius to seek in-door shelter. Morbidity and mortality in incidents such as these within the United States have occurred primarily within manufacturing facilities and have not caused significant risk to local populations. This is not the case everywhere.

4.3.1 *Bhopal industrial leak*
Two devastating incidents marked the 1980's: Bhopal, India in December, 1984 and Chernobyl, Ukraine in April, 1986. The Bhopal incident involved a leak of almost twenty-seven tons of methyl isocyanate and other chemical gases from a pesticide production facility that ultimately exposed hundreds of thousands of people. Immediate deaths were estimated at 3800 individuals with total fatalities climbing to an estimated 6000 due to post-exposure medical complications (Dhara *et al.*, 2002; Broughton 2005). Approximately 50,000 injured still suffer from exposure injuries. For a review of chemical accidents in industrializing countries (de Souza Porto & Freitas 1996). The limited availability of satellites capable of data acquisition from industrial accidents following the Bhopal disaster was recognized (Chrysoulakis & Cartalis 2003). Currently, this need is driving the development of software capable of utilizing satellite data to track and monitor the release of toxic plumes in the atmosphere (Chrysoulakis *et al,* 2005).

4.3.2 *Three Mile Island, Chernobyl, and Fukushima radiation leaks*
Accidental release of radioactive particles into the atmosphere is another source of human health risk. Two of the most familiar events are the nuclear power plant accidents at Three Mile Island in the US and Chernobyl, in the Ukraine. The Three Mile Island accident occurred in March, 1979 near Middletown, Pennsylvania. That accident was caused by a loss of reactor coolant due to a pilot-operated relief valve that stuck open, and a failure of reactor personnel to recognize the loss of coolant when system alarms sounded (Walker 2004). The event released approximately thirteen million curies of radioactive gas. Of the thirteen million curies, only about fifteen curies (carcinogenic iodine-131) were believed to be harmful (Walker 2004). Approximately 140,000 residents within a five-mile radius of the power plant temporarily evacuated the area in the following days. Exposure from the event was reported to be less than that adsorbed from a chest X-ray. Although there were later reports of elevated cases of infant mortality and effects on wildlife in exposed populations, epidemiological investigations found no evidence of event-related morbidity or mortality in exposed populations (USNRC 2009). Fortunately, there were no identified cases of morbidity or mortality in plant personnel or residents who lived close to the plant. This was not the case with the Chernobyl event.

The Chernobyl incident released several forms of particulate and gaseous radioisotopes into the atmosphere. It is the most significant unintentional release of radiation into the environment to date. It remains a problem because radioactive particulates deposited far downwind persist in soils subject to atmospheric entrainment. It took ten days to contain the event during which fifty million curies of noble gases and fifty million curies of non-noble gases were released into the atmosphere (Ginsburg & Reis 1991). Dispersion of these materials occurred throughout the Northern Hemisphere. During the first three months following the event, thirty-one individuals (most were first responders) died out of 237 individuals who acquired acute radiation sickness (Hallenbeck 1994). One hundred and thirty-five thousand residents within thirty kilometres of the immediate site of the power plant were permanently evacuated (Chesser & Baker 2006). The International Atomic Energy Agency (IAEA) has estimated that cancer deaths may reach 4,000 out of 600,000 who received the greatest exposure. Thyroid cancer is the primary long-term concern, but fortunately this disease is very treatable. Outside of the exclusion zone, fatal cancer risk from the event is believed to be minor. Within the exclusion zone, surface and ground waters were heavily contaminated. Forests within this zone are now known as the red forests due to the discoloured and dead pine trees (Arkhipov *et al.*, 1994). Initial radiation levels were fatal to indigenous animals and abandoned livestock. Interestingly, wildlife currently flourishes within the exclusion zone, although bone and tissue samples show elevated radioactivity (current radiation levels are only about three per cent of those directly following the event). Genetic analyses of voles captured within the exclusion zone showed no evidence of mutation and there is currently no evidence that elevated rates of morbidity or mortality occur in animal populations within the exclusion zone. Although human health risk outside of the exclusion-zone is believed to be minimal, there is concern about atmospheric remobilization of radioactive material from this event in both the exclusion zone and in down-wind regions of deposition where arid zone surface soils are susceptible to atmospheric entrainment. Remobilization of cesium-137 and dispersion to downwind environments have been noted (Hao *et al.*, 2008). Scientists in Greece detected elevated cesium-137 during a Saharan desert dust event that demonstrated remobilization and long-range dispersion of previously deposited Chernobyl radio nucleotides (Papastefanou *et al.*, 2001). Although availability of Earth-observing satellites was limited at the time of that accident, the Landsat Program provided images of the region, which highlight the change in cropland cover from one month to six years post-accident.

The more recent Fukushima Daiichi nuclear power plant incident started on 11 March 2011 following the Tohoku 9.0 magnitude earthquake and succeeding tsunami. While the operational reactors shutdown following the earthquake, the fourteen meter tsunami (This coastal plant was designed to withstand a 5.7 meter tsunami) that hit the plant knocked out backup generator systems used for emergency cooling operations, causing the subsequent event. Failures in three of the six reactors were rated Level 7, the highest level on the International Nuclear Event Scale (Three Mile Island was rated Level 5 and Chernobyl Level 7). Between the onslaught of the incident and 12 April 2011, an estimated ten million curies of radiation were released into surrounding and downwind environments. While this incident fortunately did not cause any direct casualties, due in part to a much more effective response than was documented with the Chernobyl incident, it has been estimated that total radiation leakage could eventually surpass Chernobyl levels. Approximately 200,000 people were displaced due to mandatory (twenty kilometre exclusion zone) and recommended evacuations (Dauer *et al.*, 2011). Early economic cost estimates including remediation and compensation range from roughly $100 to $250 billion US dollars. Historical and current information on this event can be found at the International Atomic Energy Agency Fukushima Nuclear Accident (IAEA 2012).

4.3.3 *Sverdlovsk biological release*

An accidental release of a virulent strain of *Bacillus anthracis* (causative agent of anthrax) from a bioweapon production facility occurred near Sverdlovsk in 1979. It caused sixty-four human deaths and outbreaks of the disease in local livestock populations (Meselson *et al.*, 1994). In the early 20th Century ground observation was limited to planes, balloons, and blimps; and images of a number of the World War-I gas attacks show smoke-like clouds of vapour being carried downwind from their points of origin. Modern Earth observing platforms are capable of imaging these types of clouds, as can be seen in numerous archived 'fire-smoke' images. LiDAR (light detection and ranging) is an optical sensing (light scattering) technology being used on satellites to study atmospheric aerosols. They are being used also to investigate and to monitor the release and movement of aerosolized biological weapons (Immler *et al.*, 2005; Glennon *et al.*, 2009).

4.4 *Pesticides and herbicides*

The historical public health concern related to pesticide exposure, which includes herbicides has been restricted to individuals involved in their application, and to those in close proximity to application sites. An 1990 estimate for the quantity of pesticides used globally was reported to be about 2.5 million tons (Pimentel *et al.*, 1992). Such widespread use of these chemicals is due to annual crop-loss estimates of thirteen per cent from insects, twelve per cent from plant pathogens, and ten per cent from weeds (Cramer 1967). The primary means of large-scale application of either group of chemicals is through hand sprayers, automobiles, or aircraft. Risks from these classes of chemicals are obvious. Just look at a package at your local store and read the application and exposure warning labels. While residential use of pesticides and herbicides present health risks, it is the more potent variants of these that are utilized by government organizations for agricultural purposes that are the main health concern in downwind environments. The governments of Canada and the US (like many other national, state and regional governments) provide guidelines or laws for using these chemicals, and numerous documents addressing their acquisition, use, and storage can be found at their respective websites. Guidelines for hand and vehicle application include limiting application to wind-speed conditions of less than 15 km/hour to protect those in downwind environments and ten metre buffer zones around wells and surface waters to protect potential drinking water supplies and aquatic ecology. Agency guidelines on aerial application (crop dusting) include limiting application to wind speeds of less than 4.4 m/second and an altitude of five metres or less if near a no-spray zone. Health risk from exposure to these chemicals was recognized early, and scientific data addressing this issue have been reviewed periodically (Hallenbeck & Cunningham-Burns 1985; Oehme 1991; Ecobichon & Joy 1994; Safe 1995; Margni et al., 2002).

Civilian uses of pesticides pose significant health risks. In South Africa there are approximately 100 to 200 cases of pesticide poisoning annually, and approximately ten per cent of these cases are fatal (London & Bailie 2000). In India, injuries from sprayer-applied chemical exposure is about 4.5 per 1,000 sprayers per year (Nag & Nag 2004). In Sri Lanka, there were approximately 10,000 to 20,000 pesticide poisonings per year with a fatality rate of nearly ten per cent between 1986 and 2000 (Roberts et al., 2003). Sri Lanka banned class-I organophosphate pesticides in 1995 resulting in a drop in fatalities from this form of pesticide, but there was a marked increase in fatalities from toxic replacements. The annual incident rate of acute pesticide poisoning in developing countries is approximately 18.2 per 100,000 agricultural workers and 7.4 per million school children (Thundiyil et al., 2008). In 1993, twenty-seven Provinces in China reported approximately 52,000 cases of acute pesticide poisoning, of which some 6,000 were fatal (He et al., 1999). Between 1985 and 1990, US poison centres reported about 338,000 cases and ninety-seven deaths from pesticide, herbicide, fungicide, and rodenticide exposure (Klein-Schwartz & Smith 1997). The data reported here from the various countries include cases of self-poisoning, but most are related to handling and application.

Health risk from handling (mixing, loading sprayers) and to the applicators can be minimized with protective breathing devices but risk to individuals in downwind environments is not as easy to mitigate. It has been estimated that less than 0.1 per cent of pesticides reach their target organisms and thus present risk through penetration into ground-waters, precipitation run-off to surface waters, spray-drift, and remobilization into the atmosphere via field dust as crops are harvested and/or the fields are later prepared for their next crop (Pimentel 1995). Applicator unit nozzle selection (that is, the size of the spray droplets) is the main factor influencing drift. Aerosol droplets larger than $100\,\mu m$ are less susceptible to drift, regardless of the application technique used. Low-volume ground sprayers have been shown to be the most susceptible to drift (Salyani & Cromwell 1992). Small-spray droplets in the size range of 10–$50\,\mu m$ can drift distances exceeding thirteen kilometres (Akesson & Yates 1964). Computer programs have been developed as tools to calculate spray-drift in different geographical and atmospheric conditions (Zhu et al., 1995; Teske et al., 2002). As with most diseases, the young, old, and immune-compromised are at greatest health-risk. The US/EPA and the University of Washington's Center for Child Environmental Health Risks Research conduct studies on pesticide exposure pathways, and health outcomes. They provide a database and related publications on exposure and prevention (UW 2012). Researchers have utilized a combination of geospectral and geospatial techniques to successfully demonstrate the feasibility of looking at historical and current pesticide use to identify potentially exposed populations for epidemiological studies (Ward et al., 2000). A similar combined approach was utilized that included land-use, atmospheric conditions, and pesticide-use data to determine exposure for cancer research (Maxwell et al., 2009). This same publication provides a review of the current state of knowledge for this genre of pesticide exposure assessments, including the pros and cons of the methodologies.

4.5 Urban air contaminants (persistent organic pollutants)

As planet Earth becomes increasingly urbanized, critical research is needed that addresses links between global atmospheric processes and air pollutant exposures to these growing populations. The health effects of vehicle emissions on urban life are well investigated and are being supported by sensor data from ground, air, and satellite platforms. Satellites deployed over the past few decades carry a very sophisticated array of measurement instruments to monitor atmospheric trace gases (e.g. O_3, NO, NO_2, HCHO, CO_2, and SO_2) and particulates generated at scales relevant to urban ecosystems (one to ten kilometre). These measurements are useful for characterizing pollution source areas and detecting transient events, as well as for delineating and tracking of pollutant plumes. There is also considerable interest in using sensor observations to estimate ground level concentrations of toxic air pollutants such as nitrogen dioxide (NO_2) and respirable aerosols and particulates. This is done by comparing satellite observations of pollutant levels in tropospheric air columns with ground-level (boundary layer) concentrations obtained from direct measurements, and modelling the results. NO_2 is particularly amenable to satellite observation because it is present at high levels in the surface boundary layer relative to the free troposphere; and it, therefore, produces a strong boundary layer signal (Martin 2008). Satellite observation of tropospheric NO_2 columns began in 1995 with the Global Ozone Monitoring Experiment (GOME-1), and has continued since then with the SCanning Imaging Absorption spectroMetre for Atmospheric CHartographY (SCIAMACHY), OMI, and GOME-2 (Lamsal et al., 2008). Observations from GOME and SCIAMACHY have been used to demonstrate the correlation between ground-level NO_x emissions and tropospheric column measurements of NO_2 (Leue et al., 2001; Martin et al., 2003; Jaeglé et al., 2005; Richter et al., 2005; Zhang et al., 2007). Petritoli (et al., 2004) and Ordóñez (et al., 2006) found a significant correlation (R2 ≥ 0.78) between ground-level NO_2 measurements and GOME tropospheric NO_2 columns. Figure 14 shows a one-year average of tropospheric NO_2 columns over Western Europe retrieved from OMI. Pronounced enhancements are evident over major urban and industrial areas. From the high degree of spatial heterogeneity, it is evident that the dominant signals are from the boundary layer. Martin (et al., 2004) estimate that seventy five per cent of the NO_2 column occurs in the lowest 1500 m of the atmosphere. Panel (a) shows the region with an overlay of the main roads and regional topography. The Rhine, Rhone, and Seine river valleys are also enumerated as 1, 2, & 3, respectively. Details of Randstad in the Netherlands (panel b), the Ruhr area in Germany (panel c), and Paris (panel d) and also shown. The colour scales have been adjusted for each region, as compared to the colour scale of Panel (a).

Satellite observation of AOD, a measure of light scattering by atmospheric particles, began in 1999 with MISR, and has continued with MODIS, the Cloud and Aerosol LiDAR with Orthogonal Polarization (CALIOP, pronounced calîopç), and OMI (Martin 2008; Hoff & Christopher 2009). A growing body of literature suggests that AOD can be used as a predictor of ground-based $PM_{2.5}$ mass concentrations. For example, Gupta et al., (2006) compared MODIS AOD and $PM_{2.5}$ mass concentration in several locations including Sydney, Delhi, Hong Kong, and New York City and obtained a linear correlation coefficient ($R^2 = 0.96$) between daily mean AOD and ground-based $PM_{2.5}$. However, they found that the $PM_{2.5}$–AOD relationship strongly depended on aerosol concentrations, RH, cloud cover and mixing height. The highest correlation was found for clear sky conditions with RH less than 40–50 per cent and mixing height from 100 m to 200 m. Engle-Cox (et al., 2004) compared MODIS/AOD with ground-based $PM_{2.5}$ and PM_{10} mass concentration measurements from across the US, and found relatively poor correlations for western parts of the US compared to the east and Midwest. This was attributed to differences between MODIS and ground-based sensor measurements, regression artefacts, and terrain variations as well as the MODIS cloud mask and aerosol optical depth algorithms. In a comprehensive review of the subject, Hoff & (Carpenter 2009) estimated that the precision in measuring AOD is about twenty per cent, and the relationship to $PM_{2.5}$ is at best thirty per cent in controlled measurements. They concluded that, although this precision is insufficient to meet regulatory needs in terms of air quality prediction, other information provided by orbiting sensors fill gaps in areas where there are few ground-based monitors, or for delineating plumes, and for assessing plume dispersion.

Studies by the National Research Council (2007), Martin (2008), and references therein have identified needs for improving sensor design parameters for use at the urban scale. Some of the most pressing needs include: (1) developing instruments and data processing algorithms to achieve higher spatial and temporal resolution for urban areas; (2) improving validation processes,

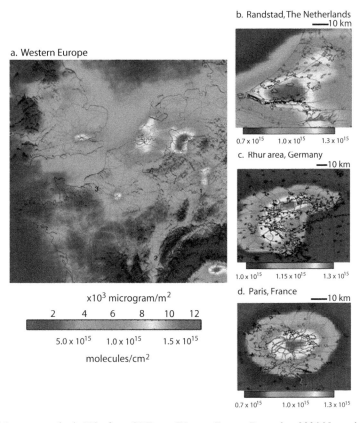

b. Randstad, The Netherlands
——10 km

a. Western Europe

0.7×10^{15} 1.0×10^{15} 1.3×10^{15}

c. Rhur area, Germany
——10 km

1.0×10^{15} 1.15×10^{15} 1.3×10^{15}

d. Paris, France
——10 km

$\times 10^3$ microgram/m^2

2   4   6   8   10   12

5.0×10^{15} 1.0×10^{15} 1.5×10^{15}

molecules/cm^2

0.7×10^{15} 1.0×10^{15} 1.3×10^{15}

Figure 14.   Mean tropospheric NO$_2$ from OMI over Western Europe December 2004-November 2005. Data from Boersma *et al.*, 2007 (Courtesy J.P. Veefkind). (see colour plate 31)

especially for tropospheric NO$_2$ (e.g. different terrain, seasons, and boundary layer concentrations); and (3), improving characterization of geophysical fields (e.g. cloud contamination, surface reflectivity).

Finally, effective utilization of novel sources of information calls for inter-, intra-, and multi-disciplinary research; and, policy development in data manipulation, archiving, analysis, sharing, and dissemination. Of necessity, emerging global and local health surveillance systems will be built upon analytical approaches and techniques that enable data processing and their more timely use for public health practice and research (Jajai *et al.*, 2009; McEntee *et al.*, 2010; Fefferman & Naumova 2010).

5   EMERGING EARTH OBSERVING SENSORS AND DATA

Key factors for many public health incidents are the environmental conditions leading to the specific public health issue under consideration. The information on those environmental conditions typically come from non-medical sources and involve estimating parameters such as the abundance of air-borne particulates, water inundation etc. Trying to address these factors comprehensively over a large area is a challenge. Addressing the challenge involves EO data, numerical modeling, and machine learning. Machine learning is a subfield of artificial intelligence that is concerned with the design and development of algorithms that allow computers to learn the behaviour of data sets empirically. Since environmental conditions change across all temporal scales of reference,

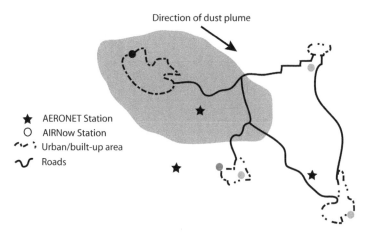

Figure 15. Relationships between anthropogenic (fugitive) dust/aerosols and atmospheric dust. Courtesy, Earth Data Analysis Center, University of New Mexico.

it should be possible to provide timely alerts of health issues through environmental forecasting systems. Hence, remote sensing, numerical modelling and machine learning are the focus of this Section.

Satellite and aerial observations contribute a wide variety of data to better understanding how natural environments relate to various disease categories. Key to the effective use of these data is identifying the appropriate sensors and the properties of their acquired data sets that can be used to improve models and to enhance surveillance systems, decision support tools, and early warning systems. To illustrate the roles these data sets can play, an example is provided for particle matter less than 2.5 μm in aerodynamic diameter ($PM_{2.5}$). Data on $PM_{2.5}$ are useful for studying asthma and respiratory conditions. Frequent, if not continuous, monitoring of fine particulate concentrations removes a critical barrier in studying health impacts, especially in the vast majority of rural and remote areas where there are no *in situ* monitors. Results from repetitive, synoptic measurements can assist public policy decisions for coping with $PM_{2.5}$ events and episodes.

5.1 $PM_{2.5}$ information for asthma studies

Today there are twenty-two million people in the US alone with asthma, and one of the environmental triggers for asthma is the concentration of air-borne particulates such as $PM_{2.5}$. Various networks of ground-based sensors provide routine measurements of $PM_{2.5}$, but their spatial coverage is sparse, especially in the developing world. Moreover, ground observation sites do not necessarily represent their broader surrounding areas. Many of the EPA ground-based monitors in the US are located near roads where air quality is influenced by local dust entrainment and vehicle exhaust that is not representative of surrounding neighbourhoods. Multiple space-borne sensors provide an array of data products on atmospheric aerosols. There is growing interest in using these data for air quality and health purposes, but there is also a growing realization that currently available products do not provide what is often needed, the near surface $PM_{2.5}$ abundance.

Simulating both windblown dust and anthropogenic air pollution events is intensive computationally in dusty regions. Figure 15 is a schematic illustrating this complexity. The diagram is a hypothetical region within a model domain that has a low level of ambient dust and aerosols. In this region there are five hypothetical AIRNow monitors (circles) and three AERONET Stations (stars). Each of these sites has a shade of grey indicating its observed concentration of dust and aerosol measured at a specific time, as contributed by its respective local population and economic activities. Each of the population centres (irregular shapes) has a shade of grey to indicate its ambient level of anthropogenic dust and aerosols at the moment of measurement. A dust cloud (pale brown)

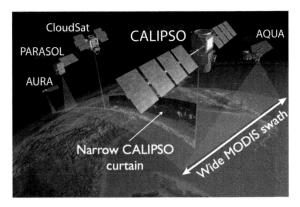

Figure 16. Visualization of the two complimentary types of aerosol information obtained by NASA's A-TRAIN constellation, detailed vertical information from CALIPSO curtains and global coverage of the total optical depth from MODIS. (see colour plate 32)

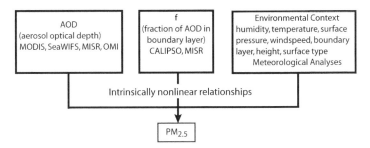

Figure 17. Schematic of the complexity surrounding multi-variate, non-linear relationships between satellite platforms and sensors for measuring AOD, humidity, temperature, surface pressure, wind speed, boundary layer height, and surface type to measure $PM_{2.5}$.

is shown moving toward the northeast that will add an atmospheric dust component to the ambient anthropogenic signal. The population centre in the upper left of the domain shows a particularly heavy concentration of atmospheric dust mixed with the ambient anthropogenic dust/aerosol load. One can imagine that levels of air contaminants will rise in the population centre near the centre of the image as the dust cloud passes through. Modelled dust concentrations can therefore consist of three different phenomena: atmospheric dust only; the ambient level of urban dust only; or a combination of both. To separate the atmospheric from the ambient concentrations, one must know the dust loads of each.

An approach to overcoming this limitation is to bring together data from multiple sensors, models of their meteorological context, and advanced machine learning. The suites of observations used include LiDAR data from CALIOP on-board the CALYPSO platform. CALIOP provides global vertical profiles of the atmosphere and fuses them into their appropriate meteorological context to provide three-dimensional vertical curtains of aerosol distribution (Figure 16).

Figure 17 is another schematic showing how multiple data sets converge on $PM_{2.5}$ measurements. This prototype system involves several components. The approach takes into account the cardinally multi-variate, non-linear relationship between currently available EO data products and the parameters needed to estimate $PM_{2.5}$ routinely. The absence of such an approach has been a stumbling block in using satellite observations to infer $PM_{2.5}$ for health studies.

Science is close to delivering global $PM_{2.5}$ analyses on a $0.1° \times 0.1°$ lat/lon grid. These analyses can then be used in several ways: 1) to display a daily global visualization of the $PM_{2.5}$ abundance in Google Earth, so that for any day and location for which there are data, an estimate of the breathable $PM_{2.5}$ abundance with an associated uncertainty can be provided; 2) the analyses could be used to provide personalized $PM_{2.5}$ accumulated dosage if a timeline of locations is provided; and 3), people with respiratory issues could check the history of $PM_{2.5}$ for any location. This would be of use if they are considering moving to a new location. This is just the first of many examples envisioned for an integrated environmental health system. The goal is to provide people with optimized personal alerts relevant to their health to enable smart lifestyle choices based on timely environmental information.

5.1.1 *Realizing the goal*

Realizing the goal requires two components. The first uses the appropriate temporally and spatially varying meteorological context of the latest version of each satellite product, as well as *in-situ* ground truth observations of $PM_{2.5}$ abundance. The precise context of observations is critical, as there is significant temporal and spatial variability in the abundance of $PM_{2.5}$. Careful attention must be paid to ingesting/fusing satellite observations in both time and place. The second component uses nonlinear, nonparametric, multivariate machine learning to address the issues for which there is not yet a complete theoretical description. Ideally, science will eventually have a complete theoretical understanding of the multivariate, nonlinear relationships between $PM_{2.5}$ and AOD, in which case machine learning would not be required. Currently, the array of tools for multivariate, nonlinear, nonparametric machine learning has proven to be valuable for a variety of applications. For estimating $PM_{2.5}$, the three nonlinear, multivariate issues that must be addressed are: 1) the inter-instrument bias between satellite AOD products and ground truth (e.g. from AERONET); 2) the inter-instrument bias between the suite of satellite AOD products (MODIS Terra, MODIS Aqua, MISR, SeaWiFS and OMI); and 3) the multivariate, nonlinear dependence of the abundance of $PM_{2.5}$ in the atmospheric boundary layer on humidity, temperature, boundary-layer height, surface pressure, wind speed, and surface type.

A major focus of machine-learning research is to produce (induce) empirical models from data automatically. This approach is usually used because of the absence of adequate and complete theoretical models that are more desirable conceptually. Currently there is not an adequate and complete theoretical model describing the complex relationships between AOD, as retrieved by satellite instruments, and the near-surface abundance of $PM_{2.5}$ in $\mu g/cm^3$. Neither is there an adequate and complete theoretical explanation for the inter-instrument biases that are present between the suite of AOD satellite products and ground truth data such as AERONET. Both of these factors have hindered use of EO data to estimate near-surface abundance of $PM_{2.5}$ in $\mu g/cm^3$.

Machine learning assists construction of a more complete theoretical model and offers two important advantages relative to more traditional statistical prediction models. These advantages are: 1) they use traditional statistical prediction models; and 2) they specify *a priori* that the relation between the predictors and the outcome is linear. In many cases, however, the form of the relationship is unknown or poorly understood. Machine-learning models offer a nonparametric alternative to parametric modelling. In addition, they are capable of learning relations in the data that may not be evident in *a priori* model specification. Machine-learning models include the capability for identifying and simulating nonlinear relations among variables. Support vector machines (SVM) are a type of machine learning found to be useful in this discussion (Lary *et al.*, 2009).

SVMs were used initially for classification. They are based on a concept of decision planes that define decision boundaries (Vapnik 1995; 1998). SVMs were subsequently extended to include support vector regression (SVR) (Scholkopf *et al.*, 2000; Smola & Scholkopf 2004). SVM and SVR use kernel functions to map data into a different space. The concept of a kernel mapping function is powerful. The SVM model algorithmic process utilizes a higher dimensional space to achieve excellent predictive power and offers an important advantage compared with neural network approaches. Specifically, neural networks can suffer from multiple local minima; in contrast, the solution to an SVM is global and unique.

5.1.2 *Technical approach and methodology*

NASA has a constellation of satellites flying in close formation called the A-Train (see Figure 16). Several of these satellites host instruments that make aerosol observations. These include Terra

MODIS and MISR launched in 1999, and Aqua MODIS launched in 2002, Aura OMI launched in 2004, and CALIOP launched in 2006 (Kahn *et al.*, 2005; Remer *et al.*, 2005; Mcgill *et al.*, 2007; Torres *et al.*, 2007; *Winker et al.*, 2007). Aerosol observations from SeaWiFS are also available from GeoEye OrbView-2 satellite launched in 1997 (Hooker & Mcclain 2000).

Several of these instruments provide a daily global picture of AOD. It is a measure of the total light extinction by atmospheric aerosols in a vertical column at a given wavelength from Earth's surface to the top of the atmosphere. For example, MODIS provides AOD across its swath at a spatial resolution of ten kilometres; SeaWiFS' resolution is 1.1 kilometre. A new MODIS product at three kilometre spatial resolution should soon be available. MODIS, OMI and SeaWiFS provide the total global aerosol burden, but not how it is vertically distributed. Other instruments provide detailed vertical aerosol structure while not providing the contiguous global coverage of MODIS, OMI and SeaWiFS. CALIPSO provides corrected backscatter and extinction profiles at 120 metre vertical resolution at altitudes below twenty kilometres, but does not provide contiguous horizontal coverage. MISR provides some vertical information for cases with higher optical depths and distinct plume boundaries but at a coarser resolution than CALIPSO. The CALIPSO observations provide a set of high vertical resolution, narrow swath curtains underneath the satellite overpass. These curtains span the globe daily, but there are substantial gaps between them. Since CALIPSO completes fewer than fifteen orbits per day, there is a longitudinal separation of 24.7° between successive curtains at the equator.

The horizontal resolution of the raw CALIPSO data depends on the altitude. From the surface to eight kilometres, it is 0.33 km; from eight to twenty kilometres, it is one kilometre; from twenty to thirty kilometres, it is 1.66 kilometres; and above thirty kilometres, it is five kilometres. However, the resolution of the corrected CALIPSO backscatter and extinction is a minimum five kilometres and as much as twenty to forty kilometres, if the aerosol loading is light. In Figure 16, one can imagine two complimentary types of aerosol information: a wide swath of Aqua MODIS at the head of the A-Train; followed immediately by a narrow curtain of vertical information collected by CALIPSO.

5.2 *Relating aerosol extinction to PM$_{2.5}$ abundance*

The relationship between PM$_{2.5}$ abundance at Earth's surface and the boundary layer optical depth or aerosol extinction depends on a variety of factors that change both seasonally and geographically. These factors include humidity, temperature, boundary layer height, surface pressure, wind speed and surface type (Liu *et al.*, 2004a; Liu *et al.*, 2004b; Hutchison *et al.*, 2005; Gupta *et al.*, 2006; Koelemeijer *et al.*, 2006; Liu *et al.*, 2007a; Liu *et al.*, 2007b; *Liu et al.*, 2007c; Pelletier *et al.*, 2007; Gupta & Christopher 2008; Hutchison *et al.*, 2008; Zhang *et al.*, 2009).

If the relationship between the AOD observed by MODIS and PM$_{2.5}$ are analysed, it is found that the best correlations are mostly observed over the eastern US in summer and fall (Zhang *et al.*, 2009). The south eastern US has the highest correlation coefficients of more than 0.6. The south western US has the lowest correlation coefficient of approximately 0.2. Several factors are at work. One is that the entire aerosol load does not usually reside within the boundary layer, hence using AOD alone as a proxy for PM$_{2.5}$ introduces a significant error. For example, on the Pacific coast a, a significant fraction of the AOD is due to smoke events in which substantial amounts of aerosol are above the boundary layer. This issue can be addressed by estimating what fraction of the total AOD comes from the boundary layer for each satellite data pixel as described later in Section 5.4.1. In addition, the correlation depends on the version of the satellite retrieval, for example, MODIS v5.2.6 AOD retrievals demonstrate better correlation with PM$_{2.5}$ than v4.0.1 retrievals; but they have much less coverage because of the differences in the cloud-screening algorithm (Zhang *et al.*, 2009). The correlation between AOD and PM$_{2.5}$ is also related to the surface pressure and wind-speed (Smirnov *et al.*, 1995; Lyamani *et al.*, 2006; Choi *et al.*, 2008; Rajeev *et al.*, 2008). This issue can be addressed by using the surface pressure and wind-speed contemporary with each observation used. The surface pressure and wind-speed comes from the meteorological analyses.

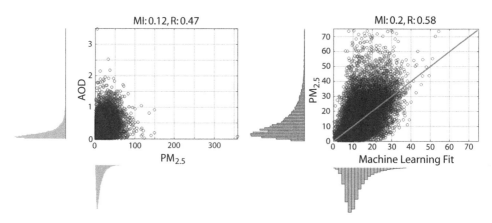

Figure 18.   (Left) scatter diagram comparison between MODIS AOD and EPA $PM_{2.5}$; (Right) scatter diagram obtained by machine learning and inferring $PM_{2.5}$ to be a function of the AOD, surface pressure, surface temperature, surface humidity, and planetary boundary layer height.

It has been found that the correlation between AOD and $PM_{2.5}$ increases as the mixing layer height decreases (Gupta *et al.*, 2006). Also higher wind speeds can induce higher mixing layer height, which changes the correlation. Relative humidity (RH) can affect the AOD-$PM_{2.5}$ correlation by changing the optical properties of the aerosols. The higher the relative humidity, the larger the proportion of light scattered, and the larger AOD (Hoff & Christopher 2009). This issue can be addressed by using the humidity and boundary layer height together with each observation used. The humidity comes from the meteorological analyses. The boundary layer height comes from the meteorological analyses and can be verified from the available LiDAR data. The meteorological analyses are available from the modern era retrospective analysis for research and applications (MERRA) system housed at the Goddard Space Flight Centre (GSFC) global modelling and assimilation office (GMAO).

As can be seen from Figure 18, a machine learning multivariate estimate of $PM_{2.5}$ as a function of the AOD, surface pressure, surface temperature, surface humidity, and planetary boundary layer height is more successful at estimating $PM_{2.5}$ than using the AOD alone. The correlation coefficient for Figure 18 (left) is 0.47 and the mutual information 0.12. The correlation coefficient for Figure 18 (right) is 0.58 and the mutual information 0.2. A prototype study conducted by the authors indicates that accounting for the shape of the aerosol vertical profile and estimating the fraction of the total AOD in the boundary layer, f, is important.

Determination of the fraction of total aerosol extinction due to aerosols in the boundary layer is an important missing parameter needed to improve accuracy in the estimate of the $PM_{2.5}$ abundance. To determine the fraction of the total AOD close to the surface for each grid point it is necessary to combine the CALIOP extinction profile curtains in their appropriate meteorological context. There are typically two or three CALIPSO curtains obtained over the US every day. These can be modelled to extend the aerosol height information to the entire grid. The fraction can be estimated in two ways. The first is to use the NAAPS transport model in a reanalysis mode that includes operational MODIS AOD assimilation as well as CALIOP 3D-variational data assimilation (Zhang *et al.*, 2008b; Campbell *et al.*, 2009; Reid *et al.*, 2009; Xian *et al.*, 2009). Several groups have used three-dimensional models to study long-range aerosol transport (Karyampudi *et al.*, 1999; Colarco *et al.*, 2003a; Colarco *et al.*, 2003b; Colarco *et al.*, 2004; Hoff *et al.*, 2005; Hoff *et al.*, 2006; Matichuk *et al.*, 2007; 2008). The fraction can also be estimated by using the Lagrangian/trajectory approach adopted by a variety of NOAA operational forecasts; for example, the smoke forecasting system, the volcanic ash transport and dispersion system, and the emergency assistance activities system. Each of these systems uses a trajectory/dispersion approach to track the motion of air parcels utilizing the HYbrid Single-Particle Lagrangian Integrated Trajectory (HYSPLIT) model.

Several groups have been using trajectory models and CALIPSO data to study long-range aerosol transport (Escudero *et al.*, 2006; Kovacs 2006; Charles *et al.*, 2007; Huang *et al.*, 2007; Feldman *et al.*, 2008; Huang *et al.*, 2008; Hutchison *et al.*, 2008; Liu *et al.*, 2008).

5.2.1 *Data gaps in the AOD products*

While the primary physical challenge in relating $PM_{2.5}$ to AOD observations concerns partitioning total column measurements into vertical components, observational and retrieval challenges of the satellite sensors remain and must be addressed. AOD information provided by an individual sensor might be spatially incomplete, or in some other way insufficient to generate daily global or regional maps. In such cases, judicious merging of co-located data from MODIS, OMI, MISR or SeaWiFS, can fill in many of the data gaps created by missing data, and can create a consistent, reliable, and complete global or regional map. AOD data from these sensors can be optimally combined with error estimates to produce improved AOD estimates. Missing data from one sensor can be provided by available co-located data from another sensor, thus increasing the total data coverage spatially.

The GES-DISC has studied data merging methods for daily AOD from MODIS Terra and Aqua (Leptoukh *et al.*, 2007; Levy *et al.*, 2009; Zubko *et al.*, 2009). They experimented with three methods for pure merging (no interpolation): simple arithmetic averaging (SIM), maximum likelihood estimate (MLE), and weighting by pixel counts (WPC). They applied them to both the global maps and regional subsets and found that the methods produce comparable AODs. The MLE is slightly better with respect to the AOD standard deviation, and the SIM is slightly better with respect to the AOD mean. Other techniques could include data interpolation with empirical orthogonal functions (DINEOF) and independent component analysis (ICA).

5.2.2 *Data biases between AOD products*

Before merging the data as just described one needs to ensure that the various merged outputs agree within the measurement uncertainties and are not biased relative to each other. As part of the joint ESA ENVISAT/NASA Aura validation, a framework has been developed to detect and correct inter-instrument bias effectively as a multi-variate non-linear function of many explanatory factors. What the explanatory factors are depends on the data product being considered.

While progress has been made in understanding the biases between MODIS and AERONET, there is still an imperfect understanding of all the root causes of bias. Biases in a variety of EO data sets have been corrected by using an empirical machine learning approach (Lary *et al.*, 2007; Brown *et al.*, 2008; Lary & Aulov 2008; Lary *et al.*, 2009). As shown in Figure 18 machine-learning algorithms are able to adjust the AOD bias effectively between the MODIS instruments and AERONET (Lary *et al.*, 2009). SVMs performed the best, improving the correlation coefficient between the AERONET AOD and the MODIS AOD from 0.86 to 0.99 for MODIS Aqua and from 0.84 to 0.99 for MODIS Terra. Key in allowing the machine-learning algorithms to correct the MODIS bias was provision of the surface type and other ancillary variables that explain the variance between MODIS and AERONET AOD.

The same techniques can be applied to adjust and merge different EO data sets. Different sensors provide different measurements and make different sets of assumptions. Some data sets have smaller uncertainties under specific conditions than others, and these differences can be using machine learning techniques to optimize a merged product. A partial optimization using this approach was achieved by combining MODIS and MISR AOD, taking into account MISR's superiority in the North American west and MODIS's superior spatial coverage (Van Donkelaar *et al.*, 2006).

5.3 *Public health implications*

The public health implication from this discussion is that a global data product of boundary layer $PM_{2.5}$ may one day be possible. Such a product is useful for a variety of public health applications from long-term epidemiological studies of respiratory conditions, to pro-active real time alerts for people with asthma.

5.3.1 *Classification*

There are numerous occasions in public health when classification based on the information available must be made. In optimal situations there is *a priori* information that makes classification possible. Most often, however, situations arise that require a classification but there is no *a priori* basis for making it. In such situations unsupervised machine learning proves very useful. In the first part of this Section a global $PM_{2.5}$ data product was described based on observations. A similar forecast product could be derived also based on model simulations. A key factor to providing a good $PM_{2.5}$ prediction would be to delineate surface dust and aerosol sources accurately. Providing such a classification is non-trivial. Machine learning takes a radically different approach to provide an unsupervised multi-variate and non-linear classification of surface types using multi-spectral satellite data. The process would allow automatic identification of dust sources that, in turn, lead to air-borne particulates and their associated health impacts.

5.3.2 *Self-organizing maps*

Self-organizing maps (SOMs) are created by a data visualization and unsupervised classification technique that reduce the dimensions of data through self-organizing neural networks (Kohonen 1982; 1990). They help address the issue that humans simply cannot visualize high dimensional data. The way SOMs reduce dimensionality is by producing a map, usually with two dimensions, that plots the similarities of the data objectively by grouping similar data items together. SOMs learn to classify input vectors according to how they are grouped in the input space. They differ from competitive layers in that neighbouring neurons in the self-organizing map learn to recognize neighbouring sections of the input space. Thus, self-organizing maps learn both the distribution and topology of the input vectors they are trained on. This approach allows SOMs to accomplish two things, reduce dimensions and display similarities.

A SOM attempts to replicate the computational power of biological neural networks, with the following important features: 1) non-linearity – the ability to represent non-linear functions or mappings; 2) non-parametric – they do not presume *a priori* a functional form for the data being analysed; 3) adaptability – the ability to change the internal representation (connection weights, network structure) if new data or information are available; 4) robustness – ability to handle missing, noisy or otherwise confusing data; and 5), power/speed – the ability to handle large data volumes in acceptable time due to inherent parallelism (Kohonen 1995; 1997; 2001).

5.4 *EO data for dust source classification*

For many environmental applications, satellite spectral imaging has become a tool of preference for collecting information about Earth's surface. These multi-wavelength data provide signatures relevant for a wide variety of applications. However, classifying these intricate and high-dimensional data sets is far from trivial. Discriminating among many surface cover classes and discovering spatially small but interesting spectral categories has proved to be an insurmountable challenge to many traditional clustering and classification methods (Villmann *et al.*, 2003). By customary measures, such as principal component analysis, the intrinsic spectral dimensionality of hyperspectral images appears to be surprisingly low. Yet, dimensionality reduction has not always been successful in terms of preservation of important surface class distinctions. The spectral bands, many of which are highly correlated, may lie on a low-dimensional but non-linear manifold, which is a scenario that eludes many classical approaches. In addition, these data comprise enormous volumes and are frequently noisy. This motivates research into advanced and novel approaches for image analysis and, in particular, machine learning using neural network self-organizing maps (Merenyi *et al.*, 1996).

Data from the two MODIS instruments aboard the Terra (EOS AM) and Aqua (EOS PM) satellites are the basis for this case study (NASA 2012b). Terra and Aqua MODIS are viewing the entire Earth's surface every one to two days, acquiring data in thirty-six spectral bands. These data are used to provide an albedo product of both directional hemispherical reflectance (black-sky albedo, BSA) and bi-hemispherical reflectance (white-sky albedo, WSA). The output product is based on

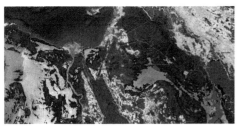

Figure 19. (Left) shows a region centred on Iraq. All the land pixels that could contain dust sources were classified by a SOM into 1000 classes. (Right) shows the small subset of SOM classes (shown in red) that have the largest overlap with regions identified as dust sources by NRL. (see colour plate 33)

the level-3 gridded data product containing sixteen days of data projected onto a 0.05° (5.6 km) lat/lon grid.

5.4.1 *Methodology*

The goal was to identify all the surface locations on the planet that are dust sources. A SOM was used to classify all land surface locations into a set of categories, a subset of which encompasses regions that are dust sources. There are many types of dust sources ranging in size, physiographic position, elevation and land use practices that were delineated. For example in Figure 19 the Nile delta is in a different class from mountainous regions. The SOM did an acceptable job of distinguishing the surface types, even placing Chad's Bodele in a class all of its own.

To achieve a comprehensive classification it was necessary to consider surface conditions throughout the year. An entire year of daily data at 0.05° (roughly 7.5 km^2) was therefore created. A massive data set was generated from which to calculate the mean, μ, for each grid point. Attention was first restricted to those broad MODIS surface categories that could include dust sources. Those are barren or sparsely vegetated surfaces, croplands, savannah, grasslands, and open and closed shrub lands. For each of these cover types an input vector was constructed that contained seven values, one for each of the seven bands provided in the MODIS product the mean, μ, of the WSA. When training the SOM, the Euclidean distance was used to compare the input vectors.

To provide a fine gradation of categories the SOM grouped the surface locations into 1000 classes, only a small subset of which actually correspond to regions that are dust sources (Figure 19). Once the classes corresponding to potential dust sources were identified, an automated method that could be executed routinely was used to identify actual dust sources. Figure 19 (left) shows all the land surface pixels over a region centred on Africa classified by a SOM into 1000 classes. Figure 19 (right) shows a subset of those 1000 classes that have the largest overlap with regions that were selected laboriously by an analyst as dust sources. There is a high degree of overlap between a subset of the SOM classes shown in blue and cyan, and those selected by the analyst shown in red. The regions are hard to identify manually so the red regions have an associated uncertainty.

Now that the signatures for potential dust sources have been identified using the SOM, areas having the same signature can be selected automatically for other land areas of the globe. An example of this is shown in Figure 20. The regions identified are very plausible dust sources, typically around the periphery of desert regions. These selections were regarded as very impressive by the same analyst that manually created the red regions in Figure 19 (right). From these data, there seems to be a method for automatically identifying global dust sources, and monitoring how they change with time. These sources can then be used in the forecast models used to produce PM$_{2.5}$ forecasts, with obvious public health applications (e.g. for personalized alerts for the twenty-two million plus asthma sufferers in the US).

167

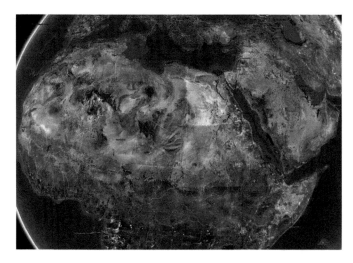

Figure 20. The shaded pixels (blue and cyan) show all the land surface pixels in the SOM classes identified as dust sources. These classes were the same SOM classes displayed in Figure 19 (right). (see colour plate 34)

6 FUTURE DIRECTIONS

Current and future Earth-observing satellite systems will enable scientists to more accurately track and study atmospheric events (Haynes 2009). Several countries have used data from these systems to formulate actions for improving data collection, access, and analysis capabilities (Williamson & Obermann 2002). As a result of growing evidence and concern for global air quality impacts on health, the Group on Earth Observations was created in 2003. The developing world suffers most from severe consequences of natural disasters and also faces inadequacies in accessing satellite sensor data that can aid in disaster warning, relief mobilization, and telemedical support (Jayaraman et al., 1997). The volcanic eruption of Eyjafjallajökull in Iceland was a sharp reminder of the far reaching economic, social, and health-related impacts of volcanoes (Laursen 2010). Based on recent estimates, the Eyjafjallajökull eruptions have cost airlines upward of two billion $USD (Batty 2010). Currently, WMO is developing a sand and dust storm warning advisory and assessment system (SDSWAS) which, when fully implemented, will provide global data accessible through regional processing centres in Spain, China, and Japan. The WMO website provides links that report current global weather events as they might impact human and societal health. There are numerous websites that provide very specific, current and archived environmental data from Earth-observing platforms, as well as a growing toolbox of models designed for use by health communities of practice. What is needed regarding public health are studies on the presence of pathogenic microorganisms, naturally occurring toxic metals, and pollutants such as pesticides, herbicides, and industrial emissions that utilize model-derived data. Validation of models is needed to advance their use as public health tools that may one day provide widespread real-time data to healthcare and public-health protection personnel, and the public in general.

A growing number of research groups utilize model and satellite data to illustrate relationships between atmospheric constituents, their sources, impacts on humans, and transmission pathways. More of these types of projects are needed to advance aerosol monitoring for public health (Jaffe et al., 2003; Griffin 2007). Limitations of current methodologies will diminish in time. Advances will continue to be made in computing, molecular biology, chemistry, and sensor system technologies. Existing capabilities are already sufficiently robust to advance public-health studies at all geographic scales from local to global. Although there are toxicology programs, a few aerobiology programs, and numerous meteorology programs, most are discipline-specific and do not offer the opportunity to investigate global scale questions that require integrated approaches. A broadening of research areas in meteorology programs is needed. The need for research to address the movement of particles in the atmosphere was recognized in 1935 by Meier & Lindbergh in which they state…*The potentialities of world-wide distribution of spores of fungi and other organisms caught up and carried abroad by trans-continental winds may be of tremendous economic consequence.*

REFERENCES

Akesson, N.B. & Yates, W.E. 1964. Problems relating to application of agricultural chemicals and resulting drift residues. *Ann. Rev. Entom.* 9(1): 285–318.

Al Frayh, A.R., Shakoor, Z., Gad El Rab, M.O. & Hasnain, S.M. 2001. Increased prevalence of asthma in Saudi Arabia. *Ann. Allerg. Asth. & Immun.* 86(3): 292–296.

Amann, R.I., Ludwig, W. & Schleifer, K.H. 1995. Phylogenetic identification and in situ detection of individual microbial cells without cultivation. *Microbio. Rev.* 59(1): 143–169.

Andronico, D., Spinetti, C., Cristaldi, A. & Buongiorno, M.F. 2009. Observations of Mt. Etna volcanic ash plumes in 2006: An integrated approach from ground-based and polar satellite NOAA-AVHRR monitoring system. *J. Volcan. & Geoth. Res.* 180(2–4): 135–147.

Arkhipov, N.P., Kuchma, N.D. Askbrant, S., Pasternak, P.S. & Musica, V.V. 1994. Acute and long-term effects of irradiation on pine (*Pinus silvestris*) stands post-Chernobyl. *Sci. Total Environ.* 157: 383–386.

Aronson, N.E., Sanders, J.W. & Moran, K.A. 2006. In harm's way: Infection in deployed American military forces. *Clin. Infect. Dis.* 43: 1045–1051.

AVO. 2010. Alaska Volcano Observatory. Available from: http://www.avo.alaska.edu [Accessed 20th March 2012].

Batty, D. 2010. Icelandic volcano now appears to be dormant say scientists. *Guardian Weekly 23May2010*

Baxter, P.J. 2003. The eruption of El Reventador volcano 2002: Health hazards and the implications for volcano risk management in Ecuador. Washington DC: PAHO

Baxter, P.J., Ing, R., Falk, H., French, J., Stein, G.F., Bernstein, R.S., Merchant, J.A. & Allard, J. 1981. Mount St. Helens eruptions, May 18 to June 12, 1980. An overview of the acute health impact. *J. Amer. Med. Assoc.* 246(22): 2585–2589.

Baxter, P.J., Ing, R., Falk, H. & Plikaytis, B. 1983. Mount St. Helens eruptions: The acute respiratory effects of volcanic ash in a North American community. *Arch. Environ. Health* 38(3): 138–143.

Baxter, P.J., Bonadonna, Dupree, C.R., Hards, V.L., Kohn, S.C., Murphy, M.D., Nichols, A., Nicholson, R.A., Norton, G., Searl, A., Sparks, R.S. & Vickers, B.P. 1999. Cristobalite in volcanic ash of the Soufriere Hills volcano, Montserrat, British West Indies. *Science* 283(5405): 1142–1145.

Becker, S., Soukup, J.M. & Gallagher, J.E. 2002. Differential particulate air pollution induced oxidant stress in human granulocytes, monocytes and alveolar macrophages. *Toxicol. in Vitro* 16(3): 209–218.

Becker, S., Mundandhara, S., Devlin, R.B. & Madden, M. 2005. Regulation of cytokine production in human alveolar macrophages and airway epithelial cells in response to ambient air pollution particles: Further mechanistic studies. *Toxicol. & Appl. Pharmacol.* 207(2): 269–275.

Belser, J.A., Blixt, O., Chen, L.M., Pappas, C., Maines, T.R., Van Hoeven, N., Donis, R., Busch, J., Mcbride, R., Paulson, J.C., Katz, J.M. & Tumpey, T.M. 2008. Contemporary North American influenza H7 viruses possess human receptor specificity: Implications for virus transmissibility. *Proc. Nat. Acad. Sci.* 105(21): 7558–7563.

Bener, A., Abdulrazzaq, Y.M., Al-Mutawwa, J. & Debuse, P. 1996. Genetic and environmental factors associated with asthma. *Human Biology* 68(3): 405–414.

Betzer, P.R., Carder, K.L., Duce, R.A., Merrill, J.T., Tindale, N.W., Uematsu, M., Costello, D.K., Young, R.W., Feely, R.A., Freland, J.A., Bernstein, R.E. & Greco, A.M. 1988. Long-range transport of giant mineral aerosol particles. *Nature* 336(6199): 568–571.

Bluth, G.J.S., Schnetzler, C.C. Krueger, A.J. & Walter, L.S. 1993. The contribution of explosive volcanism to global atmospheric sulphur dioxide concentrations. *Nature* 366(6453): 327–329.

Broecker, W.S. 2000. Abrupt climate change: Causal constraints provided by the paleoclimate record. *Earth-Sci. Rev.* 51(1–4): 137–154.

Broughton, E. 2005. The Bhopal disaster and its aftermath: A review. *Environ. Health* 4: 6.

Brown, E.G., Gottlieb, S. & Laybourn, R.L. 1935. Dust storms and their possible effect on health. *Public Health Repts.* 50(40): 1369–1383.

Brown, J.H., Cook, K.M., Ney, F.G. & Hatch, T. 1950. Influence of particle size upon the retention of particulate matter in the human lung. *Amer. J. Pub. Health* 40(4): 450–458.

Brown, J.S., Zeman, K.L. & Bennett, W.D. 2002. Ultrafine particle deposition and clearance in the healthy and obstructed lung. *Amer. J. Respir. & Criti. Care Med.* 166 (9):1240–1247.

Brown, M.E., Lary, D.J., Vrieling, A., Stathakis D. & Mussa, H. 2008. Neural networks as a tool for constructing continuous NDVI time series from AVHRR and MODIS. *Int. J. Rem. Sens* 29(24): 7141–7158.

Cahill, T.A., Gill, T.E., Reid, J.S., Gearhart, E.A. & Gillette, D.A. 1998. Saltating particles, playa crusts and dust aerosols at Owens (dry) Lake, California. *Earth Surf. Proc. & Landforms* 21(7): 621–639.

Campbell, J.R., Reid, J.S. Westphal, D.L., Zhang, J., Hyer, E.J. & Welton, E.J. 2010. CALIOP aerosol subset processing for global aerosol transport model data assimilation. *J. Sel. Topics in Appl. Earth Obs. & Rem. Sens.* 3(2): 203–214.

Carapezza, M.L., Badalamenti, B., Cavarra, L. & Scalzo, A. 2003. Gas hazard assessment in a densely inhabited area of Colli Albani volcano (Cava dei Selci, Roma). *J. Volcan. & Geoth. Res.* 123(1–2): 81–94.

Carter, A.J. & Ramsey, M.S. 2009. ASTER- and field-based observations at Bezymianny Volcano: Focus on the 11 May 2007 pyroclastic flow deposit. *Rem. Sens. Environ.* 113(10): 2142–2151.

Casadevall, T.J. 1994. The 1989–1990 eruption of Redoubt volcano, Alaska: Impacts on aircraft operations. *J. Volcan. & Geoth. Res.* 62(1–4): 301–316.

Castranova, V., Bowman, L. Shreve, J.M., Jones, G.S. & Miles, P.R. 1982. Volcanic ash: Toxicity to isolated lung cells. *J. Toxicol. & Environ. Health,* 9(2Part A): 317–325.

Castronovo, D., Chui, K. & Naumova, E. 2009. Dynamic maps: A visual-analytic methodology for exploring spatio-temporal disease patterns. *Environ. Health* 8(1): 61–70.

Chan, C.C., Chuang, K.J., Chen, W.J., Chang, W.T., Lee, C.T. & Peng, C.M. 2008. Increasing cardiopulmonary emergency visits by long-range transported Asian dust storms in Taiwan. *Environ. Res.* 106(3): 393–400.

Chang, C.C., Lee, I.M., Tsai, S.S. & Yang, C.Y. 2006. Correlation of Asian dust storm events with daily clinic visits for allergic rhinitis in Taipei, Taiwan. *J. Toxicol. & Environ. Health* A 69(3): 229–235.

Charles, L., Gross, B., Moshary, F., Wu, Y., Vladutescu, V. & Ahmed, S. 2007. Atmospheric transport of smoke and dust particulates and their interaction with the planetary boundary layer as observed by multi-wavelength lidar and supporting instrumentation. *Lidar Rem Sens. Environ. Monitor.* VIII. UN Singh. 6681: U157–U167.

Chen, P.S., Tsai, F.T., Lin, C.K., Yang, C.Y., Chan, C.C., Young, C.Y. & Lee, C.H. 2010. Ambient influenza and avian influenza virus during sandstorm days and background days. *Environ. Health Perspec.* 118(9): 1211–1216.

Chesser, R.K. & Baker, R.J. 2006. Growing up with Chernobyl. *Amer. Sci.* 94(6): 542–549.

Chiu, H.F., Tiao, M.M., Ho, S.C., Kuo, H.W., Wu, T.N. & Yang, C.Y. 2008. Effects of Asian dust storm events on hospital admissions for chronic obstructive pulmonary disease in Taipei, Taiwan. *Inhal. Toxicol.* 20(9): 777–781.

Choi, Y.S., Ho, C.H., Chen Ho, D., Noh,Y.H. & Song, C.K. 2008. Spectral analysis of weekly variation in PM10 mass concentration and meteorological conditions over China. *Atmosph. Environ.* 42(4): 655–666.

Chrysoulakis, N. & Cartalis, C. 2003. A new algorithm for the detection of plumes caused by industrial accidents, based on NOAA/AVHRR imagery. *Int. J. Rem. Sens.* 24(17): 3353–3368.

Chrysoulakis, N., Adaktylou, N. & Cartalis, C. 2005. Detecting and monitoring plumes caused by major industrial accidents with JPLUME, a new software tool for low-resolution image analysis. *Environ. Model. & Software* 20(12): 1486–1494.

Chrysoulakis, N., Herlin, I., Prastacos, P., Yahia, H., Grazzini, J. & Cartalis, C. 2007. An improved algorithm for the detection of plumes caused by natural or technological hazards using AVHRR imagery. *Rem. Sens. Environ.* 108(4): 393–406.

Chui, K.H., Castronovo, D.A., Jagai, J.S., Kosheleva, A.A. & Naumova, E.N. 2006. Gastroenteritis infections in the U.S. elderly and extreme weather events: Exposures to Atlantic tropical storms of 1998–2002. *Epidemiology* 17(6): S477.

Cifuentes, L.A., Krupnick, A.J. 2005. Urban air quality and human health in Latin America and the Caribbean. Inter Amer. Dev. Bank. Available from: http://www.iadb.org/sds/env [Accessed 20th March 2012].

Colarco, P.R., Toon, O.B. & Holben, B.N. 2003a. Saharan dust transport to the Caribbean during PRIDE: 1. Influence of dust sources and removal mechanisms on the timing and magnitude of downwind aerosol optical depth events from simulations of in situ and remote sensing observations. *J. Geophys. Res. Atmosph.* 108(D19): 8589.

Colarco, P.R., Toon, O.B., Reid, J.S., Livingston, J.M., Russell, P.B., Redemann, J., Schmid, B., Maring, H.B., Savoie, D., Welton, E.J., Campbell, J.R., Holben, B.N. & Levy, R. 2003b. Saharan dust transport to the Caribbean during PRIDE: 2. Transport, vertical profiles, and deposition in simulations of in situ and remote sensing observations. *J. Geophys. Res. Atmosph. 108(D19):*8590.

Cook, A.G., Weinstein, P. & Centeno, J.A. 2005. Health effects of natural dust – Role of trace elements and compounds. *Biol. Trace Element Res.* 103(1): 1–15.

Corbett, E.L., Churchyard, G.J., Clayton, T., Herselman, P., Williams, B., Hayes, R., Mulder, D. & De Cock, K.M. 1999. Risk factors for pulmonary mycobacterial disease in South African gold miners. A case-control study. *Amer. J. Respir. Criti. Care Med.* 159(1): 94–99.

Cramer, H.H. 1967. Plant protection and world crop production. *Pflanzenschutz-Nachrichten 'Bayer'* 20: 1–524.

Crowley, J.K., Hubbard, B.E. & Mars, J.C. 2003. Analysis of potential debris flow source areas on Mount Shasta, California, by using airborne and satellite remote sensing data. *Rem. Sens. Environ.* 87(2–3): 345–358.

Cullen, R.T., Jones, A.D., Miller, B.G., Tran, C.L., Davis, J.M.G., Donaldson, K., Wilson, M., Stone, V. & Morgan, A. 2002. Toxicity of volcanic ash from Montserrat. *Edinburgh. Inst. Occupat. Med.* Research Report TM/02/01.

Daigle, C.C., Chalupa, D.C., Gibb, F.R., Morrow, P.E., Oberdorster, G., Utell, M.J. & Frapton, M.W. 2003. Ultrafine particle deposition in humans during rest and exercise. *Inhal. Toxicol.* 15(6): 539–552.

Dalton, M.P., Bluth, G.J.S., Prata, A.J., Watson, I.M. & Carn, S.A. 2006. Ash, SO_2, and aerosol analysis of the November 2002 eruption of Reventador volcano, Equador, using TOMS, HIRS, and MODIS satellite sensors. *Cities on Volcanoes, Fourth Conference.* Quito, Ecuador.

Dauer, L.T., Zanzonico, P., Tuttle, R.M., Quinn, D.M. & Strauss, H.W. 2011. The Japanese tsunami and resulting nuclear emergency at the Fukushima Daiichi power facility: Technical, radiologic, and response perspectives. *J. Nucl. Med.* 52:1423–1432.

de Deckker, P., Abed, R.M.M., de Beer, D., Hinrichs, K.U., O'loingsigh, T., Schefufl, E., Stuut, J.B.W., Tapper, N.J. & van der Kaars, S. 2008. Geochemical and microbiological fingerprinting of airborne dust that fell in Canberra, Australia in October 2002. *Geochem. Geophy. Geosyst.* 9, Q12Q10, doi:10.1029/2008GC002091.

de la Cadena, M. 2006. Risk management of volcanic crises in Quito's metropolitan area. *Cities on Volcanoes, Fourth Conference.* Quito, Ecuador.

de Souza Porto, M.F. & Freitas, C.M. 1996. Major chemical accidents in industrializing countries: The social-political amplification of risk. *Risk Anal.* 16(1): 19–29.

Dean, K.G., Dehn, J., Papp, K.R., Smith, S., Izbekov, P., Peterson, R., Kearney, C. & Steffke, A. 2004. Integrated satellite observations of the 2001 eruption of Mt. Cleveland, Alaska. *J. Volcan. & Geoth. Res.* 135(1–2): 51–73.

Delmelle, P., Stix, J., Bourque, C.P., Baxter, P.J., Garcia-Alvarez, J. & Barquero, J. 2001. Dry deposition and heavy acid loading in the vicinity of Masaya volcano, a major sulfur and chlorine source in Nicaragua. *Environ. Sci. & Tech.* 35(7): 1289–1293.

Dhara, V.R., Dhara, R., Acquilla, S.D. & Cullinan, P. 2002. Personal exposure and long-term health effects in survivors of the Union Carbide disaster at Bhopal. *Environ. Health Perspec.* 110(5): 487–500.

Diaz, V., Parra, R., Ibarra, B. & Paez, C. 2006. Monitoring of ash from volanic eruptions in Quito, Ecuador. *Cities on Volcanoes, Fourth Conference.* Quito, Ecuador.

Dockery, D. 2006. Concentration and acidity of airborne particulate matter in communities of the big island of Hawaii. *Cities on Volcanoes, Fourth Conference.* Quito, Ecuador.

Doebbeling, B.N., Clarke, W.R., Watson, D., Torner, J.C., Woolson, R.G., Boelker, M.D., Barrett, D.H. & Schwartz., D.A. 2000. Is there a Persian Gulf War syndrome? Evidence from a large population-based survey of veterans and nondeployed controls. *Amer. J. Med.* 108(9): 695–704.

Donaldson, K., Stone, V., Gilmour, P.S., Brown, D.M. & Macnee, W. 2000. Ultrafine particles: Mechanisms of lung injury. *Phil. Trans. Roy. Soc. Lond. Series A: Math., Phys. & Engin. Sci.* 358(1775): 2741.

Dowd, S.E. & Maier, R.M. 2000. Aeromicrobiology. In: R.M. Maier, I.L Pepper. & C.P. Gerba, (eds.), *Environmental microbiology.* San Diego: Academic Press.

Draxler, R.R. & Hess, G.D. 1998. An overview of the HYSPLIT_4 modelled system for trajectories, dispersion, and deposition. *Austral. Meteorol. Mag.* 47(4): 295–308.

Driscoll, K.E. 1996. Role of inflammation in the development of rat lung tumors in response to chronic particle exposure. In J.J.L. Mauderly & R.J. McCunney (eds), *Particle overload in the rat lung and lung cancer: Implications for human risk assessment*: 139–153. Philadelphia: Taylor & Francis.

Duchnowicz, P., Szczepaniak, P. & Koter, M. 2005. Erythrocyte membrane protein damage by phenoxyacetic herbicides and their metabolites. *Pestic. Biochem. & Physiol.* 82(1): 59–65.

Durry, E., Pappagianis, D., Werner, S.B., Hutwagner, L., Sun, R.K., Maurer, M., Mcneil, M.M. & Pinner, R.W. 1997. Coccidioidomycosis in Tulare County, California, 1991: Reemergence of an endemic disease. *Med. Mycol.* 35(5): 321–326.

Dwernychuk, L.W., Cau, H.D., Hatfield, C.T., Boivin, T.G., Hung, T.M., Dung, P.T. & Thai, N.D. 2002. Dioxin reservoirs in southern Vietnam–A legacy of Agent Orange. *Chemosphere* 47(2): 117–137.

Ecobichon, D.J. & Joy, R.M. 1994. *Pesticides and Neurological Diseases*. Boca Raton: CRC Press.

Egan, T. 2006. *The worst hard time*. New York: Houghton Mifflin.

Ellrod, G.P. 2003. Development of volcanic ash image products using MODIS multi-spectral data. *Amer. Met. Soc. Ann. Mtg*. Long Beach, CA.

Ellrod, G.P., Connell, B.H. & Hillger, D.W. 2003. Improved detection of airborne volcanic ash using multispectral infrared satellite data. *J. Geophys. Res.* 108(D12): 4356.

EO. 2012. Available from: http://earthobservatory.nasa.gov [Accessed 4th February 2012].

Erickson, J.D., Mulinare, J., Mcclain, P.W., Fitch, T.G., James, L.M., Mcclearn, A.B. & Adams Jr., M.J. 1984. Vietnam veterans' risks for fathering babies with birth defects. *J. Amer. Med. Assoc.* 252(7): 903–912.

Escudero, M., Stein, A. Draxler, R.R., Querol, X., Alastuey, A., Castillo, S. & Avila, A. 2006. Determination of the contribution of northern Africa dust source areas to PM10 concentrations over the central Iberian penin-sula using the hybrid single-particle Lagrangian integrated trajectory model (HYSPLIT) model. *J. Geophys. Res. Atmosph.* 111(D6): 210.

Estrella, B., Estrella, R., Oviedo, J., Narvez, X., Reyes, M.T., Gutierrez, M. & Naumova, E.N. 2005. Acute respiratory diseases and carboxyhemoglobin status in school children of Quito, Ecuador. *Environ. Health Perspec.* 113(5): 607–611.

Fiese, M.J. 1958. *Coccidioidomycosis*. Springfield: Charles C. Thomas.

Filizzola, C., Lacava, T., Marchese, F., Pergola, N., Scaffidi, I. & Tramutoli, V. 2007. Assessing RAT (Robust AVHRR Techniques) performances for volcanic ash cloud detection and monitoring in near real-time: The 2002 eruption of Mt. Etna (Italy). *Rem. Sens. Environ.* 107(3): 440–454.

Friedman, J.M. 2005. Does Agent Orange cause birth defects? *Teratology* 29(2): 193–221.

Fukuda, K., Nisenbaum, R., Stewart, B., Thompson, W.W., Robin, L., Washko, R.M., Noah, D.L., Barrett, D.H., Randall, B., Herwaldt, B.L., Mawle, A.C. & Reeves, W.C. 1998. Chronic multisymptom illness affecting Air Force veterans of the Gulf War. *J. Amer. Med. Assoc.* 280(11): 981–988.

Fung, I.Y., Meyn, S.K., Tegen, I., Doney, S.C., John, J.G. & Bishop, J.K.B. 2000. Iron supply and demand in the upper ocean. *Global Biogeochem. Cycles* 14(1): 281–295.

Gangale, G., Prata, A.J. & Clarisse, L. 2009. The infrared spectral signature of volcanic ash determined from high-spectral resolution satellite measurements. *Rem. Sens. Environ.* 114(2): 414–425.

Gans, J., Wolinsky, M. & Dunbar, J. 2005. Computational improvements reveal great bacterial diversity and high metal toxicity in soil. *Science* 309: 1387–1390.

Gao, T., Yu, X., Ma, Q., Li, H., Li, X. & Si, Y. 2003. Climatology and trends of the temporal and spatial distribution of sandstorms in Inner Mongolia. *Water, Air, & Soil Pollu.* 3(2): 51–60.

Garcia-Aristizabal, A., Kumagai, H., Samaniego, P., Mothes, P., Yepes, H. & Monzier, M. 2007. Seismic, petro-logic, and geodetic analyses of the 1999 dome-forming eruption of Guagua Pichincha volcano, Ecuador. *J. Volcan. Geoth. Res.* 161(4): 333–351.

Giannini, A., Saravanan, R. & Chang, P. 2003. Oceanic forcing of Sahel rainfall on interannual to interdecadal time scales. *Science* 302: 1027–1030.

Giggenbach, W.F. 1990. Water and gas chemistry of Lake Nyos and its bearing on the eruptive process. *J. Volcan. & Geoth. Res.* 42(4): 337–362.

Giggenbach, W.F., Sano, Y. & Schmincke, H.U. 1991. CO_2-rich gases from Lakes Nyos and Monoun, Cameroon; Laacher See, Germany; Dieng, Indonesia, and Mt. Gambier, Australia–Variations on a common theme. *J. Volcan. & Geoth. Res.* 45(3–4): 311–323.

Gillette, D., Ono, D. & Richmond, K. 2004. A combined modeling and measurement technique for estimating windblown dust emissions at Owens (dry) Lake, California. *J. Geophys. Res. – Earth* 109(F1): 337–362.

Ginzburg, H.M. & Reis, E. 1991. Consequences of the nuclear power plant accident at Chernobyl. *Pub. Health Repts.* 106(1): 32.

Giraudi, C. 2005. Eolian sand in peridesert northwestern Libya and implications for Late Pleistocene and Holocene Sahara expansions. *Palaeogeo., Palaeoclim., Palaeoecol.* 218(1–2): 161–173.

Glaze, L.S. & Baloga, S.M. 2003. DEM flow path prediction algorithm for geologic mass movements. *Environ. & Engin. Geosci.* 9(3): 225–240.

Glennon, J.J., Nichols, T., Gatt, P., Baynard, T., Marquardt, J.H. & Vanderbeek, R.G. 2009. *System perfor-mance and modeling of a bioaerosol detection lidar sensor utilizing polarization diversity*. Laser Radar Technology & Applications XIV, Orlando, FL.

Goklany, I.M. 2007. Death and death rates due to extreme weather events: Global and U.S. trends, 1900–2004. In: *The Civil Society Report on Climate Change*:47–59. London: International Policy Press.

Gong, S.L., Zhang, X.Y., Zhao, T.L., Zhang, X.B., Barrie, L.A., Mckendry, I.G. & Zhao, C.S. 2006. A sim-ulated climatology of Asian dust aerosol and its trans-Pacific transport. Part II: Interannual variability and climate connections. *J. Climate* 19(1): 104–122.

172

Goudie, A.S. & Middleton, N.J. 2001. Saharan dust storms: Nature and consequences. *Earth Sci. Rev.* 56: 179–204.

Graham, W.F. & Duce, R.A. 1979. Atmospheric pathways of the phosphorus cycle. *Geochim. et Cosmochim. Acta* 43: 1195–1208.

Gray, G.C., Reed, R.J., Kaiser, K.S., Smith, T.C. & Gastanaga, V.M. 2002. Self-reported symptoms and medical conditions among 11,868 Gulf War-era veterans: The Seabee health study. *Amer. J. Epidem.* 155(11): 1033.

Gregory, P.H. 1961. *The Microbiology of the Atmosphere*. London: Leonard Hill Books Ltd.

Griffin, D.W. 2004. Terrestrial microorganisms at an altitude of 20,000 m in Earth's atmosphere. *Aerobiologia* 20: 135–140.

Griffin, D.W. 2007. Atmospheric movement of microorganisms in clouds of desert dust and implications for human health. *Clin. Microbio. Rev.* 20(3): 459–477.

Griffin, D.W. 2008. Non-spore forming eubacteria isolated at an altitude of 20,000 m in Earth's atmosphere: extended incubation periods needed for culture-based assays. *Aerobiologia* 24(1): 19–25.

Griffin, D.W., Garrison, V.H., Herman, J.R. & Shinn, E.A. 2001. African desert dust in the Caribbean atmosphere: Microbiology and public health. *Aerobiologia* 17(3): 203–213.

Griffin, D.W., Kellogg, C.A. & Shinn, E.A. 2001. Dust in the wind: Long-range transport of dust in the atmosphere and its implications for global public and ecosystem health. *Global Change & Human Health* 2(1): 20–33.

Griffin, D.W., Kellogg, C.A., Garrison, V.H. & Shinn, E.A. 2002. The global transport of dust. *Amer. Sci.* 90(3): 228–235.

Griffin, D.W., Kellogg, C.A., Garrison, V.H., Lisle, J.T., Borden, T.C. & Shinn, E.A. 2003. African dust in the Caribbean atmosphere. *Aerobiologia* 19(3–4): 143–157.

Griffin, D.W., Westphal, D.L. & Gray, M.A. 2006. Airborne microorganisms in the African desert dust corridor over the mid-Atlantic ridge, Ocean Drilling Program, Leg 209. *Aerobiologia* 22(3): 211–226.

Griffin, D.W., Kubilay, N., Kocak, M., Gray, M.A., Borden, T.C. & Shinn, E.A. 2007. Airborne desert dust and aeromicrobiology over the Turkish Mediterranean coastline. *Atmosph. Environ.* 41(19): 4050–4062.

Griffin, D.W., Petrosky, T., Morman, S.A. & Luna, V.A. 2009. A survey of the occurrence of *Bacillus anthracis* in North American soils over two long-range transects and within post-Katrina New Orleans. *Appl. Geochem.* 24(8): 1464–1471.

Grimalt, J.O., Ferrer, M. & Macpherson, E. 1999. The mine tailing accident in Aznalcollar. *Sci. Total Environ.* 242(1–3): 3–11.

Grini, A. & Zender C.S. 2004. Roles of saltation, sandblasting, and wind speed variability on mineral dust aerosol size distribution during the Puerto Rican Dust Experiment (PRIDE). *J. Geophys. Res.* 109: D07202.

Grousset, F.E., Ginoux, P., Bory, A. & Biscaye, P.E. 2003. Case study of a Chinese dust plume reaching the French Alps. *Geophys. Res. Lett.* 30(6): 1277–1277.

Gu, Y., Rose, W.I., Schneider, D.J., Bluth, G.J.S. & Watson, I.M. 2005. Advantageous GOES IR results for ash mapping at high latitudes: Cleveland eruptions 2001. *Geophys. Res. Lett.* 32(2): L02305.

Gudmundsson, G. 2011. Respiratory health effects of volcanic ash with special reference to Iceland: A review. *Clin. Respir. J.* 5: 2–9.

Guffanti, M., Ewert, J.W., Gallina, G.M., Bluth, G.J.S. & Swanson, G.L. 2005. Volcanic-ash hazard to aviation during the 2003–2004 eruptive activity of Anatahan volcano, Commonwealth of the Northern Mariana Islands. *J. Volcan. & Geoth. Res.* 146(1–3): 241–255.

Gupta, P. & Christopher, S.A. 2008. Seven year particulate matter air quality assessment from surface and satellite measurements. *Atmosph. Chem. & Phys.* 8(12): 3311–3324.

Gupta, P., Christopher, S.A., Wang, J., Gehrig, R., Lee, Y. & Kumar, N. 2006. Satellite remote sensing of particulate matter and air quality assessment over global cities. *Atmosph. Environ.* 40(30): 5880–5892.

Gyan, K., Henry, W., Lacaille, S., Laloo, A., Lamesee-Eubanks, C., Mckay, S., Antoine, R.M. & Monteil, M.A. 2005. African dust clouds are associated with increased paediatric asthma accident & emergency admissions on the Caribbean island of Trinidad. *Int. J. Biomet.* 49(6): 371–376.

Haley, R.W., Kurt, T.L. & Hom, J. 1997. Is there a Gulf War syndrome? Searching for syndromes by factor analysis of symptoms. *J. Amer. Med. Assoc.* 277(3): 215–222.

Hallenbeck, W.H. 1994. *Radiation Protection*. Boca Raton: CRC.

Hallenbeck, W.H. & Cunningham-Burns, K.M. 1985. *Pesticides and Human Health*. New York: Springer-Verlag.

Halonen, M., Stern, D.A., Wright, A.L., Taussig, L.M. & Martinez, F.D. 1997. *Alternaria* as a major allergen for asthma in children raised in a desert environment. *Amer. J. Respir. Criti. Care Med.* 155(4): 1356–1361.

Hammond, G.W., Raddatz, R.L. & Gelskey, D.E. 1989. Impact of atmospheric dispersion and transport of viral aerosols on the epidemiology of influenza. *Revs. Infect. Dis.* 11(3): 494–497.

Hansell, A.L., Horwell, C.J. & Oppenheimer, C. 2006. The health hazards of volcanoes and geothermal areas. *Brit. Med. J.* 63(2): 149–156.

Hao, W.M., Bondarenko, O.O., Zibtsev, S. & Hutton, D. 2008. Vegetation fires, smoke emissions, and dispersion of radionuclides in the Chernobyl exclusion zone. *Dev. Environ. Sci.*: 8: 265–275.

Hara, Y., Uno, I. & Wang, Z. 2006. Long-term variation of Asian dust and related climate factors. *Atmos. Environ.* 40(35): 6730–6740.

Hatfield Consultants. 2000. Development of Impact Mitigation Strategies Related to the Use of Agent Orange Herbicide in the Aluoi Valley, VietNam. Volume 1: Report; Volume 2: Appendices. Hatfield Consultants Ltd., West Vancouver, BC, Canada; 10-80 Committee, Ha Noi, Viet Nam.

Haynes, J.A. 2009. *NASA satellite observations for climate research and applications for public health:* 407–414. Int. Sem. Nuclear War & Planetary Emergencies, 42th Session, Erice, Italy. Singapore: World Scientific.

He, F., Chen, S. & China, P.R. 1999. Health impacts of pesticide exposure and approaches to prevention. *Asian Pac. Newsletter Occup. Health Safety* 6: 60–63.

Hector, R.F. & Laniado-Laborin, R. 2005. Coccidioidomycosis – A fungal disease of the Americas. *PLoS Med.* 2(1): e2.

Herman, J.R., Krotkov, N., Celarier, E., Larko, D. & Labow, G. 1999. The distribution of UV radiation at the Earth's surface from TOMS measured UV-backscattered radiances. *J. Geophys. Res.* 104: 12059–12076.

Hoff, R.M. & Christopher, S.A. 2009. Remote sensing of particulate pollution from space: Have we reached the promised land? *J. Air & Waste Manage. Assoc.* 59(6): 645–675.

Hoff, R.M., Palm, S.P., Engel-Cox, J.A. & Spinhirne, J. 2005. GLAS: A long-range transport observation of the 2003 California forest fire plumes to the northeastern US. *Geophys. Res. Letts.* 32(22): 08.

Hoff, R.M., Huff, A.K. & Szykman, J.J. 2006. Three-dimensional air quality system: 3D-AQS. Available from: http://appliedsciences.nasa.gov/pdf/AQ-3D- AQS_Final_BenchmarkReport.pdf [Accessed 18th March 2012].

Holasek, R.E., Self, S. & Woods, A.W. 1996. Satellite observations and interpretation of the 1991 Mount Pinatubo eruption plumes. *J. Geophys. Res.* 101(27): 27635–27655.

Holt, G. 2007. Repair of facial fractures in the Iraq war combat theater. *J. Amer. Med. Assoc.* 298(24): 2905.

Honrath, R.E., Owen, R.C., Val Martin, M., Reid, J.S., Lampina, K., Fialho, P., Dziobak, M.P., Kleissl, J. & Westphal, D.L. 2004. Regional and hemispheric impacts of anthropogenic and biomass burning emissions on summertime CO_2 and O_3 in the North Atlantic lower free troposphere. *J. Geophys. Res.* 109(D24310): doi:10.1029/2004JD005147.

Hooker, S.B. & Mcclain, C.R. 2000. The calibration and validation of SeaWiFS data. *Prog. Oceanogra.* 45(3–4): 427–465.

Hooper, K., Petreas, M.X., Chuvakova, T., Kazbekova, G., Durz, N., Seminova, G., Sharmanov, T., Hayward, D., She, J.W., Visita, P., Windler, J., Mckinnery, M., Wade, T.J., Grassman, J. & Stephens, R.D. 1998. Analysis of breast milk to assess exposure to chlorinated contaminants in Kazakstan: High levels of 2,3,7,8-tetrachlorodibenzo-p-dioxin (TCDD) in agricultural villages of southern Kazakstan. *Environ. Health Perspec.* 106(12): 797–806.

Hooper, T.I., DeBakey, S.F., Nagaraj, B.E., Bellis, K.S., Smith, T.C. & Gackstetter, G.D. 2008. The long-term hospitalization experience following military service in the 1991 Gulf War among veterans remaining on active duty, 1994–2004. *BMC Public Health* 8: 60.

Horwell, C.J. 2007. Grain-size analysis of volcanic ash for the rapid assessment of respiratory health hazard. *J. Environ. Monitor.* 9(10): 1107–1115.

Horwell, C.J. & Baxter, P.J. 2006. The respiratory health hazards of volcanic ash: A review for volcanic risk mitigation. *Bull. Volcan.* 69(1): 1–24.

Howitt, M.E., Naibu, R. & Roach, T.C. 1998. The prevalence of childhood asthma and allergy. In Barbados. The Barbados National Asthma and Allergy Study. *Amer. J. Respir. & Criti. Care Med.* 157: A624.

Hsu, N.C., Tsay, S.C., King, M.D. & Herman, J.R. 2006. Deep blue retrievals of Asian aerosol properties during ace-asia. *IEEE Trans. Geos. & Rem. Sens.* 44(11): 3180–3195.

Huang, J.P., Minnis, P., Yi, Y., Tang, Q., Wang, X., Hu, Y., Liu, Z., Ayers, K., Trepte, C. & Winker, D. 2007. Summer dust aerosols detected from CALIPSO over the Tibetan Plateau. *Geophys. Res. Letts.* 34(18): 805.

Huang, J., Minnis, P., Chen, B., Huang, Z., Liu, Z., Zhao, Q., Yi, Y. & Ayers, J.K. 2008. Long-range transport and vertical structure of Asian dust from CALIPSO and surface measurements during PACDEX. *J. Geophys. Res. Atmosph.* 113(D23): 212.

Husar, R.B., Tratt, D.M., Schichtel, B.A., Falke, S.R., Li, F., Jaffe, D., Gasso, S., Gill, T., Laulainen, N.S., Lu, F., Reheis, M.C., Chun, Y., Westphal, D., Holben, B.N., Gueymard, C., Mckendry, I., Kuring, N., Feldman, G.C., Mcclain, C., Frouin, R.J., Merrill, J., Dubois, D., Vignola, F., Murayama, T., Nickovic, S.,

Wilson, W.W., Sassen, K., Sugimoto, N. & Malm, W.C. 2001. Asian dust events of April 1998. *J. Geophys. Res.-Atmosph.* 106(D16): 18317–18330.

Hutchison, K.D., Smith, S. & Faruqui, S.J. 2005. Correlating MODIS aerosol optical thickness data with ground-based PM2.5 observations across Texas for use in a real-time air quality prediction system. *Atmosph. Environ.* 39(37): 7190–7203.

Hyams, K.C., Riddle, J., Trump, D.H. & Graham, J.T. 2001. Endemic infectious diseases and biological warfare during the Gulf War: A decade of analysis and final concerns. *Amer. J. Trop. Med. & Hyg.* 65(5): 664–670.

IAEA 2012. Available from: www.iaea.org/About/japan-infosheet.html [Accessed 18th March 2012].

Immler, F., Engelbart, D. & Schrems, O. 2005. Fluorescence from atmospheric aerosol detected by a lidar indicates biogenic particles in the lowermost stratosphere. *Atmosph. Chem. & Phys.* 5: 345–355.

Inhorn, M.C. & Brown, P.J. 2000. The anthropology of infectious disease. *Ann. Rev. Anthro.* 19(1): 89–117.

Ismail, K., Everitt, B., Blatchley, N., Hull, L., Unwin, C., David, A. & Wessely, S. 1999. Is there a Gulf War syndrome? *The Lancet* 353(9148): 179–182.

Jaffe, D., Anderson, T., Covert, D., Kotchenruther, R., Trost, B., Danielson, J., Simpson, W., Berntsen, T., Karlsdottir, S. & Blake, D. 1999. Transport of Asian air pollution to North America. *Geophys. Res. Lett.* 26(6): 711–714.

Jaffe, D., McKendry, I., Adnerson, T. & Price, H. 2003. Six 'new' episodes of trans-Pacific transport of air pollutants. *Atmos. Environ.* 37(3): 391–404.

Jaffe, D., Prestbo, E., Swartzendruber, P., Weiss-Penzias, P., Kato, S., Takami, A., Hatakeyama, S. & Kajii, Y. 2005. Export of atmospheric mercury from Asia. *Atmos. Environ.* 39(17): 3029–3038.

Jagai, J.S., Monchakb, J., Mcentee, J.C., Castronovo, D.A. & Naumova, E.N. 2007. The use of remote sensing to assess global trends in seasonality of cryptosporidiosis. *32nd Int. Symp. Rem. Sens. Environ.*, San Jose, Costa Rica.

Janssen, P.H. 2006. Identifying the dominant soil bacterial taxa in libraries of 16S rRNA and 16S rRNA genes. *Appl. & Environ. Microb.* 72(3): 1719–1728.

Jayaraman, V., Chandrasekhar, M.G. & Rao, U.R. 1997. Managing the natural disasters from space technology inputs. *Acta Astronaut.* 40(2–8): 291–325.

Jensen, J.R. 2000. *Remote Sensing of the Environment.* Upper Saddle River, NJ: Prentice Hall.

Jinadu, B.A. 1995. Valley Fever Task Force Report on the control of *Coccidioides immitis.* Bakersfield: California: Kern County Health Department.

Johnson, K.G., Loftsgaarden, D.O. & Gideon, R.A. 1982. The effects of Mount St. Helens volcanic ash on the pulmonary function of 120 elementary school children. *Amer. Rev. Respir. Dis.* 126(6): 1066–1089.

Johnson, K.S., Elrod, V.A., Fitzwater, S.E., Plant, J.N., Chavez, F.P., Tanner, S.J., Gordon, R.M., Westphal, D.L., Perry, K.D., Wu, J. & Karl, D.M. 2003. Surface ocean-lower atmosphere interactions in the Northeast Pacific Ocean Gyre: Aerosols, iron and the ecosystem response. *Global Biogeochem. Cycles* 17(2): 1063. doi:1010.1029/2000JC000555.

Jurado, J. & Southgate, D. 2001. Dealing with air pollution in Latin America: The case of Quito, Ecuador. *Environ. & Devel. Econ.* 4(03): 375–388.

Kahn, R.A., Gaitley, B.J. Martonchik, J.V., Diner, D.J., Crean, K.A. & Holben, B. 2005. Multi-angle imaging spectroradiometer (MISR) global aerosol optical depth validation based on 2 years of coincident aerosol robotic network (AERONET) observations. *J. Geophys. Res. Atmosph.* 110(D10): S04.

Kan, H., London, S.J., Chen, G., Zhang, Y., Song, G., Zhao, N., Jiang, L. & Chen, B. 2007. Differentiating the effects of fine and coarse particles on daily mortality in Shanghai, China. *Environ. Int.* 33(3): 376–384.

Kang, H.K., Mahan, C.M., Lee, K.Y., Murphy, F.M., Simmens, S.J., Young, H.A. & Levine, P.H. 2002. Evidence for a deployment-related Gulf War syndrome by factor analysis. *Arch. Environ. Health* 57(1): 61–68.

Karyampudi, V.M., Palm, S.P., Reagen, J.A., Fang, H., Grant, W.B., Hoff, R.M., Moulin, C., Pierce, H.F., Torres, O., Browell, E.V. & Melfi, S.H. 1999. Validation of the Saharan dust plume conceptual model using Lidar, Meteosat, and ECMWF data. *Bull. Amer. Meteorol. Soc.* 80(6): 1045–1075.

Kellogg, C.A., Griffin, D.W., Garrison, V.H., Peak, K.K., Royall, N., Smith, R.R. & Shinn, E.A. 2004. Characterization of aerosolized bacteria and fungi from desert dust events in Mali, West Africa. *Aerobiol.* 20(2): 99–110.

Kienle, J. & Shaw, G.E. 1979. Plume dynamics, thermal energy and long-distance transport of vulcanian eruption clouds from Augustine volcano, Alaska. *J. Volcan. & Geoth. Res.* 6(1–2): 139–164.

Klein-Schwartz, W. & Smith, G.S. 1997. Agricultural and horticultural chemical poisonings: Mortality and morbidity in the United States. *Ann. Emerg. Med.* 29(2): 232–238.

Kleindorfer, P.R., Belke, J.C., Elliott, M.R., Lee, K., Lowe, R.A. & Feldman, H.I. 2003. Accident epidemiology and the U.S. chemical industry: Accident history and worst-case data from RMP*Info. *Risk Anal.* 23(5): 865–881.

175

Kling, G.W., Clark, M.A., Wagner, G.N., Compton, H.R., Humphrey, A.M., Devine, J.D., Evans, W.C., Lockwood, J.P., Tuttle, M.L. & Koenigsberg, E.J. 1987. The 1986 Lake Nyos gas disaster in Cameroon, West Africa. *Science* 236(4798): 169.

KMA 2008. Available from: http://www.korea.amedd.army.mil/webapp/yellowSand/Default.asp [Accessed 4th February 2012].

Knoke, J.D., Smith, T.C., Gray, G.C., Kaiser, K.S. & Hawksworth, A.W. 2000. Factor analysis of self-reported symptoms: Does it identify a Gulf War syndrome? *Amer. J. Epidem.* 152(4): 379–388.

Koelemeijer, R.B.A., Homan, C.D. & Matthijsen, J. 2006. Comparison of spatial and temporal variations of aerosol optical thickness and particulate matter over Europe. *Atmosph. Environ.* 40(27): 5304–5315.

Koren, I., Kaufman, Y.J., Washington, R., Todd, M.C., Rudich, Y., Martins, J.V. & Rosenfeld, D. 2006. The Bodele Depression: A single spot in the Sahara that provides most of the mineral dust to the Amazon forest. *Environ. Res. Lett.* 1: 014005.

Korenyi-Both, A.L. 2000. Al Eskan disease and no gaming please. *Military Medicine* 165(11): 3–4.

Korenyi-Both, A.L., Molnar, A.C. & Fidelus-Gort, R. 1992. Al Eskan disease: Desert Storm pneumonitis. *Military Med.* 157(9): 452–462.

Kovacs, T. 2006. Comparing MODIS and AERONET aerosol optical depth at varying separation distances to assess ground-based validation strategies for spaceborne Lidar. *J. Geophys. Res. Atmosph.* 111(D24): 203.

Krotkov, N.A., Torres, O., Seftor, C., Krueger, A.J., Kostinski, A., Rose, W.I., Bluth, G.J.S., Schneider, D.J. & Schaifer, S.J. 1999. Comparison of TOMS and AVHRR volcanic ash retrievals from the August 1992 eruption of Mt. Spurr. *Geophys. Res. Lett.* 26(4): 455–458.

Krueger, A.J., Walter, L.S., Bhartia, P.K., Schnetzler, C.C., Krotkov, N.A., Sprod, I. & Bluth, G.J.S. 1995. Volcanic sulfur dioxide measurements from the total ozone mapping spectrometer instruments. *J. Geophys. Res.* 100(14): 14057–14076.

Kwon, H.J., Cho, S.H. & Chun, Y.S. 2000. The effects of the Asian dust events on daily mortality in Seoul, Korea. *Epidem.* 11(4): 243.

Kwon, H.J., Cho, S.H., Chun, Y., Lagarde, F. & Pershagen, G. 2002. Effects of the Asian dust events on daily mortality in Seoul, Korea. *Environ. Res.* 90 (Sect. A): 1–5.

Lambert, F., Delmonte, B., Petit, J.R., Bigler, M., Kaufmann, P.R., Hutterli, M.A., Stocker, T.J., Ruth, U., Steffensen, J.P. & Maggi, V. 2008. Dust – climate couplings over the past 800,000 years from the EPICA Dome C ice core. *Nature* 452: 616–619.

Lary, D.J. 2010. Artificial intelligence in aerospace. In: T.A. Thawar (ed.), *Aerospace Technologies Advancements*: 492–516. Vukovar, Croatia: INTECH.

Lary, D.J. 2010. Artificial Intelligence in geoscience and remote sensing. In: P. Imperatore & D. Riccio (eds.), *Geoscience & Remote Sensing. New Achievements*: 105–128. Vukovar, Croatia: INTECH.

Lary, D.J. & Aulov, O. 2008. Space-based measurements of HCl: Inter-comparison and historical context. *J. Geophys. Res. Atmosph.* 113(D15): S04.

Lary, D.J., Müller, M.D. & Mussa, H.Y. 2004. Using neural networks to describe tracer correlations. *Atmosph. Chem. & Phys.* 4: 143–146.

Lary, D.J., Waugh, D.W., Douglass, A.R., Stolarski, R.S., Newman, P.A. & Mussa, H. 2007. Variations in stratospheric inorganic chlorine between 1991 and 2006. *Geophys. Res. Letts* 34(21): 811.

Lary, D.J., Remer, L.A., Macneill, D., Roscoe, B. & Paradise, S. 2009. Machine learning and bias correction of MODIS aerosol optical depth. *IEEE Geos. & Rem. Sens. Letts.* 6(4): 694–698.

Laursen, L. 2010. Iceland eruptions fuel interest in volcanic gas monitoring. *Science* 328: 410–411.

Le Pennec, J.L., Ruiz, A.G., Hidalgo, S., Samaniego, P., Ramon, P., Eissen, J.P., Yepes, H., Hall, M.L., Mothes, P., Vallee, A. & Vennat, J. 2006. Characteristics and impacts of recent ash falls produced by Tungurahua and El Reventador volcanoes, Ecuador. *Cities on Volcanoes, Fourth Conference*. Quito, Ecuador.

Lee, H.A., Gabriel, R., Bolton, J.P.G., Bale, A.J. & Jackson, M. 2002. Health status and clinical diagnoses of 3000 UK Gulf War veterans. *J. Roy. Soc. Med.* 95: 491–497.

Lehnert, B.E. 1993. Defense mechanisms against inhaled particles and associated particle-cell interactions. *Revs. Mineral. & Geochem.* 28(1): 427–469.

Lenes, J.M., Darrow, B.P., Cattrall, C., Heil, C.A., Callahan, M., Vargo, G.A., Byrne, R.H., Prospero, J.M., Bates, D.E., Fanning, K.A. & Walsh, L.J. 2001. Iron fertilization and the *Trichodesmium* response on the West Florida shelf. *Limnol. & Oceanogra.* 46(6): 1261–1277.

Leptoukh, G., Zubko, V. & Gopolan, A. 2007. Spatial aspects of multi-sensor data fusion: Aerosol optical thickness. *IGARSS: Int. Geos. & Rem. Sens. Symp. vols 1–12: Sensing & Understanding Our Planet*: 3119–3122.

Levy, R.C., Leptoukh, G.G., Kahn, R., Zubko, V., Gopalan, A. & Remer, L.A. 2009. A critical look at deriving monthly aerosol optical depth from satellite data. *IEEE Trans. Geos. Rem. Sens.* 47(8): 2942–2956.

Liang, Q., Jaegle, L., Jaffe, D.A., Weiss-Penzias, P., Heckman, A. & Snow, J.A. 2004. Long-range transport of Asian pollution to the northeast Pacific: Seasonal variations and transport pathways of carbon monoxide. *J. Geophys. Res.* 109: D23S07.

Liu, Y., Franklin, M., Kahn, R. & Koutrakis, P. 2007a. Using aerosol optical thickness to predict ground-level PM2.5 concentrations in the St. Louis area: A comparison between MISR and MODIS. *Rem. Sens. Environ.* 107(1–2): 33–44.

Liu, Y., Koutrakis, P. & Kahn, R. 2007b. Estimating fine particulate matter component concentrations and size distributions using satellite-retrieved fractional aerosol optical depth: Part 1 – method development. *J. Air & Waste Manag. Assoc.* 57(11): 1351–1359.

Liu, Y., Koutrakis, P., Kahn, R., Turquety, S. & Yantosca, R.M. 2007c. Estimating fine particulate matter component concentrations and size distributions using satellite-retrieved fractional aerosol optical depth: Part 2 – a case study. *J. Air & Waste Manag. Assoc.* 57(11): 1360–1369.

Liu, Y., Park, R.J., Jacob, D.J.Q., Li, B., Kilaru, V. & Sarnat, J.A. 2004a. Mapping annual mean ground-level PM2.5 concentrations using multiangle imaging spectroradiometer aerosol optical thickness over the contiguous United States. *J. Geophys. Res. Atmosph.* 109(D22): 10.

Liu, Y., Sarnat, J.A., Coull, B.A., Koutrakis, P. & Jacob, D.J. 2004b. Validation of multiangle imaging spec-troradiometer (MISR) aerosol optical thickness measurements using aerosol robotic network (AERONET) observations over the contiguous United States. *J. Geophys. Res. Atmosph.* 109(D6): 9.

Liu, Z., Liu, D. Huang, J., Vaughan, M., Uno, I., Sugimoto, N., Kittaka, C., Trepte, C., Wang, Z., Hostetler, C. & Winker, D. 2008. Airborne dust distributions over the Tibetan Plateau and surrounding areas derived from the first year of CALIPSO Lidar observations. *Atmosph. Chem. & Phys.* 8(16): 5045–5060.

Loehr, A., Bogaard, T., Heikens, A., Hendriks, M., Suarti, S., Bergen, M., van Gestel, K.C.A.M., Straalen, N., Vroon, P. & Widianarko, B. 2005. Natural pollution caused by the extremely acid crater lake Kawah Ijen, East Java, Indonesia. *Environ. Sci. & Poll. Res.* 12(2): 89–95.

London, L. & Bailie, R. 2001. Challenges for improving surveillance for pesticide poisoning: Policy implications for developing countries. *Int. J. Epidem.* 30(3): 564.

Lyles, M.B., Fredrickson, H.L., Bednar, A.J., Fannin, H.B. & Sobecki, T.M. 2005. The chemical, biological, and mechanical characterization of airborne micro-particulates from Kuwait. *8th Annual Force Health Protection Conference.* Louisville, Kentucky. Session #2586. Available from: http://www.apgea.army.mil/fhp/Archives/FHP2005/Users/ConferenceAgenda_Pop.aspx [Accessed 20th March 2012].

Lyamani, H., Olmo, F.J., Alcántara, A. & Alados-Arboledas, L. 2006. Atmospheric aerosols during the 2003 heat wave in southeastern Spain-I: Spectral optical depth. *Atmosph. Environ.* 40(33): 6453–6464.

Lyons, J.J., Waite, G.P., Rose, W.I. & Chigna, G. 2010. Patterns in open vent, strombolian behavior at Fuego volcano, Guatemala, 2005–2007. *Bull. Volcan.* 72(1): 1–15.

Malilay, J., Real, M.G., Vanegas, A.R., Noji, E. & Sinks, T. 1997. Public health surveillance after a volcanic eruption: Lessons from Cerro Negro, Nicaragua, 1992. *Rev. Panam. Sal. Publ.* 1: 213–219.

Margni, M., Rossier, D., Crettaz, P. & Jolliet, O. 2002. Life cycle impact assessment of pesticides on human health and ecosystems. *Agric. Ecosys. & Environ.* 93(1–3): 379–392.

Martin, T.R., Ayars, G., Butler, J. & Altman, L.C. 1984. The comparative toxicity of volcanic ash and quartz. Effects on cells derived from the human lung. *Amer. Rev. Respir. Dis.* 130(5): 778–782.

Martin, T.R., Wehner, A.P. & Butler, J. 1986. Evaluation of physical health effects due to volcanic hazards: The use of experimental systems to estimate the pulmonary toxicity of volcanic ash. *Amer. J. Publ. Health* 76(Suppl): 59–65.

Matichuk, R.I., Colarco, P.R., Smith, J.A. & Toon, O.B. 2007. Modeling the transport and optical properties of smoke aerosols from African savanna fires during the southern African regional science initiative campaign (Safari 2000). *J. Geophys. Res. Atmosph.* 112(D8): 203.

Matichuk, R.I., Colarco, P.R. Smith, J.A. & Toon, O.B. 2008. Modeling the transport and optical properties of smoke plumes from South American biomass burning. *J. Geophys. Res. Atmosph.* 113(D7): 208.

Maxwell, S.K., Meliker, J.R. & Goovaerts, P. 2009. Use of land surface remotely sensed satellite and airborne data for environmental exposure assessment in cancer research. *J. Exp. Sci. & Environ. Epidem.* 20: 176–185.

Mcgill, M.J., Vaughan, M.A., Trepte, C.R., Hart, W.D., Hlavka, D.L., Winker, D.M., & Kuehn, R. 2007. Airborne validation of spatial properties measured by the CALIPSO Lidar. *J. Geophys. Res. Atmosph.* 112(D20): 201.

McGonigle, A.J.S. 2005. Volcano remote sensing with ground-based spectroscopy. *Philos. Trans. Roy. Soc. Lond. – Series A: Mathematical, Physical & Engineering Sciences* 363(1837): 2915–2929.

McKendry, I.G., Strawbridge, K.B., Oneill, N.T., Macdonald, A.M., Liu, P.S.K., Leaitch, W.R., Anlauf, K.G., Jaegle, L., Gairlie, T.D. & Westphal, D.L. 2007. Trans-Pacific transport of Saharan dust to western North America: A case study. *J. Geophys. Res.* 112: D01103.

Meier, F.C. & Lindbergh, C.A. 1935. Collecting micro-organisms from the Arctic atmosphere. *Sci. Monthly* 40: 5–20.

Meng, Z. & Lu, B. 2007. Dust events as a risk factor for daily hospitalization for respiratory and cardiovascular diseases in Minqin, China. *Atmosph. Environ.* 41(33): 7048–7058.

Meselson, M., Guillemin, J., Hugh-Jones, M., Langmuir, A., Popova, I., Shelokov, A. & Yampolskaya, O. 1994. The Sverdlovsk anthrax outbreak of 1979. *Science.* 266(5188): 1202.

Micklin, P.P. 1988. Desiccation of the Aral Sea: A water management disaster in the Soviet Union. *Science* 241: 1170–1176.

Micklin, P.P. 2007. The Aral Sea disaster. *Ann. Rev. Earth & Planetary Sci.* 35: 47–72.

Mills, N.L., Amin, N., Robinson, S.D., Anand, A., Davies, J., Patel, D., De La Fuente, J.M., Cassee, F.R., Boon, N.A. & Macnee, W. 2006. Do inhaled carbon nanoparticles translocate directly into the circulation in humans? *Amer. J. Respir. & Criti. Care Med.* 173(4): 426.

Mims, S.A. & Mims, F.M. 2004. Fungal spores are transported long distances in smoke from biomass fires. *Atmosph. Environ.* 38(5): 651–655.

MMWR 2003a. Increase in coccidioidomycosis – Arizona, 1998–2001. *Morb. & Mort. Wkly. Rept.* 52(6): 109–112.

MMWR 2003b. Severe acute pneumonitis among deployed U.S. military personnel – southwest Asia, March – August 2003. *Morb. & Mort. Wkly. Rept.* 290(14): 1845–1846.

MODVOC. 2009. Near-real-time thermal monitoring of global hot-spots. Available from: http://modis.higp. hawaii.edu [Accessed 19th March 2012].

Mohr, A.J. 1997. Fate and transport of microorganisms in air. In C.J. Hurst, G.R. Knudsen, M.J. McInerney, L.D. Stetzenbach & M.V. Walter (eds.), *Manual of Environmental Microbiology*: 641–650. Washington: ASM Press.

Molesworth, A.M., Cuevas, L.E., Morse, A.P., Herman, J.R. & Thomson, M.C. 2002. Dust clouds and spread of infection. *Lancet.* 359(9300): 81–82.

Molesworth, A.M., Cuevas, L.E., Conner, S.J., Morse, A.P. & Thomson, M.C. 2003. Environmental risk and meningitis epidemics in Africa. *Emerg. Infect. Dis.* 9(10): 1287–1293.

Morain, S.A. & Budge A.M. 2010. Suggested practices for forecasting dust storms and intervening their health effects. In O. Altan, R. Backhaus, P. Boccardo & S. Zlatanova (eds.), *Geoinformation for Disaster and Risk Management*: 45–50. Copenhagen: Joint Board of Geospatial Information Societies & United Nations Office of Outer Space Affairs.

Morain, S.A., Sprigg, W.A., Benedict, K., Budge, A., Budge, T., Hudspeth, W., Barbaris, B., Yin, D. & Shaw, P. 2007. Public Health Applications in Remote Sensing: Verification and Validation Report. NASA Cooperative agreement NNS04AA19A.

Morain, S.A., Sprigg, W.A., Benedict, K., Budge, A., Budge, T., Hudspeth, W., Sanchez, G., Barbaris, B., Catrall, C., Chandy, B., Mahler, A.B., Shaw, P., Thome, K., Nickovic, S., Yin, D., Holland, D., Spear, J., Simpson, G. & Zelicoff, A. 2009. Public Health Applications in Remote Sensing: Final Benchmark Report. NASA Cooperative agreement. NNS04AA19A.

Morawska, L. & Zhang J.F. 2002. Combustion sources of particles. 1. Health relevance and source signatures. *Chemosphere* 49(9): 1045–1058.

Morrow, P.E. 1992. Dust overloading of the lungs: Update and appraisal. *Toxicol. & Appl. Pharmacol.* 113(1): 1–12.

Moulin, C., Lambert, C.E., Dulac, F. & Dayan, U. 1997. Control of atmospheric export of dust from North Africa by the North Atlantic Oscillation. *Nature.* 387: 691–694.

Nadkarni, N.M., McIntosh, A.C.S. & Cushing, J.B. 2008. A framework to categorize forest structure concepts. *For. Ecol. & Mangmnt.* 256(5): 872–882.

Nag, P.K. & Nag, A. 2004. Drudgery, accidents and injuries in Indian agriculture. *Indust. Health* 42(2): 149–162.

Nandalal, K.D.W. & Hipel, K.W. 2007. Strategic decision support for resolving conflict over water sharing among countries along the Syr Darya River in the Aral Sea Basin. *J. Water Res. Plan. & Mangmnt.* 133: 289–299.

Nania, J. & Bruya, T.E. 1982. In the wake of Mount St. Helens. *Annals Emerg. Med.* 11(4): 184–191.

NASA 2012a. Available from: http://www.nasascience.nasa.gov/earth-science/mission_list [Accessed 23rd February 2012].

NASA 2012b. Available from: http://lpdaac.usgs.gov/lpdaac/products/modis_products_table/albedo/16_day_l3_global_0_05deg_cmg/mcd43c3 [Accessed 18th March 2012].

Naumova, E.M., Haas, T.C. & Morris, R.D. 1997. *Estimation of Individual Exposure Following a Chemical Spill in Superior, Wisconsin.* New York: Springer-Verlag.

NCDC 2010. Available from: http://www.ncdc.noaa.gov/oa/reports/billionz.html [Accessed 19th March 2012].

Nemmar, A., Hoet, P.H.M., Vanquickenborne, B., Dinsdale, D., Thomeer, M., Hoylaerts, M.F., Vanbilloen, H., Mortelmans, L. & Nemery, B. 2002. Passage of inhaled particles into the blood circulation in humans. *Circulation* 105(4): 411.

Ngo, A.D., Taylor, R., Roberts, C.L. & Nguyen, T.V. 2006. Association between Agent Orange and birth defects: Systematic review and meta-analysis. *Int. J. Epidem.* 35(5): 1220–1230.

Nickovic, S., Kallos, G., Papadopoulos, A. & Kakaliagou, O. 2001. A model for prediction of desert dust cycle in the atmosphere. *J. Geophys. Res.* 106: 18113–18130.

NOAA 2012. Available from: http://www.ssd.noaa.gov/VAAC/ARCH11/archive.html. [Acessed 4th February 2012].

Norboo, T., Angchuk, P.T., Yahya, M., Kamat, S.R., Pooley, F.D., Corrin, B., Kerr, I.H., Bruce, N. & Ball, K.P. 1991. Silicosis in a Himalayan village population: Role of environmental dust. *Thorax* 46: 861–863.

Novak, M.A., Watson, I.M., Delgado-Granados, H., Rose, W.I., Cardenas-Gonzalez, L. & Realmuto, V. 2008. Volcanic emissions from Popocatepetl volcano, Mexico, quantified using Moderate Resolution Imaging Spectroradiometer (MODIS) infrared data: A case study of the December 2000–January 2001 emissions. *J. Volcan. & Geoth. Res.* 170(1–2): 76–85.

Oehme, M. 1991. Dispersion and transport paths of toxic persistent organochlorides to the Arctic: Levels and consequences. *Sci. Total Environ.* 106(1–2): 43–53.

O'Hara, S.L., Wiggs, G.F.S., Mamedov, B., Davidson, G. & Hubbard, R.B. 2000. Exposure to airborne dust contaminated with pesticide in the Aral Sea region. *The Lancet* 355(9204): 627.

Olenchock, S.A., Mull, J.C., Mentnech, M.S., Lewis, D.M. & Bernstein, R.S. 1983. Changes in humoral immunologic parameters after exposure to volcanic ash. *J. Toxicol. & Environ. Health.* Part A 11(3): 395–404.

O'Malley, M.A. & McCurdy, S.A. 1990. Subacute poisoning with phosalone, an organophosphate insecticide. *Western J. Med.* 153(6): 619–624.

OMISDG 2007. Ozone Monitoring Instrument Sulfur Dioxide Group. Available from: http://so2.gsfc. nasa.gov [Accessed 4th February 2012].

Oppenheimer, C. 1998. Review article: Volcanological applications of meteorological satellites. *Int. J. Rem. Sens.* 19(15): 2829–2864.

OPS 2005. Eruciones volcanicas y proteccion de la salud. Quito, Ecuador, Organization Panamercana de la Salud. Available from: http://www.ops-oms.org/Spanish/DD/PED/gv_modulo2-1.pdf?vm=r [Accessed 19th March 2012].

Papastefanou, C., Manolopoulou, M., Stoulos, S., Ioannidou, A. & Gerasopoulos, E. 2001. Coloured rain dust from Sahara Desert is still radioactive. *J. Environ. Radioact.* 55: 109–112.

Papayannis, A., Zhang, H.Q., Amiridis, V., Ju, H.B., Chourdakis, G., Georgoussis, G., Perez, C., Chen, H.B., Goloub, P. & Mamouri, R.E. 2007. Extraordinary dust event over Beijing, China, during April 2006: Lidar, Sun photometric, satellite observations and model validation. *Geophys. Res. Lett.* 34, L07806. doi: 10.1029/2006GL029125.

Park, J.W., Lim, Y.H., Kyung, S.Y., An, C.H., Lee, S.P., Jeong, S.H. & Ju, Y.S. 2005. Effects of ambient particulate matter on peak expiratory flow rates and respiratory symptoms of asthmatics during Asian dust periods in Korea. *Respirology* 10: 470–476.

Pelletier, B., Santer, R. & Vidot, J. 2007. Retrieving of particulate matter from optical measurements: A semiparametric approach. *J. Geophys. Res. Atmosph.* 112(D6): 18.

Pergola, N., Tramutoli, V., Marchese, F., Scaffidi, I. & Lacava, T. 2004. Improving volcanic ash cloud detection by a robust satellite technique. *Rem. Sens. Environ.* 90(1): 1–22.

Peterson, R., Webley, P.W., D'Amours, R., Servranckx, R., Stunder, R. & Papp, K. 2012. Volcanic ash cloud dispersion models. In K. Dean & J. Dehn (eds.) *Volcanoes of the North Pacific: Observations from Space*: Chapter 7. London: Praxis.

Pieri, D. & Abrams, M. 2004. ASTER watches the world's volcanoes: A new paradigm for volcanological observations from orbit. *J.Volcan. & Geoth. Res.* 135(1–2): 13–28.

Pimentel, D. 1995. Amounts of pesticides reaching target pests: Environmental impacts and ethics. *J. Agric. & Environ. Ethics* 8(1): 17–29.

Pimentel, D., Acquay, H., Biltonen, M., Rice, P., Silva, M., Nelson, J., Lipner, V., giordano, S., Horowitz, A. & D'amore, M. 1992. Environmental and economic costs of pesticide use. *BioScience* 42(10): 750–760.

Plumlee, G.S. & Ziegler, T.L. 2005. The medical geochemistry of dusts, soils, and other earth materials. In B.S. Lollar (ed.), *Environmental Geochemistry*: 263. Oxford: Elsevier.

179

Plumlee, G.S., Morman, S.A. & Zeigler, T.L. 2006. The toxicological geochemistry of Earth materials: An overview of processes and the interdisciplinary methods used to understand them. *Revs. Mineral. & Geochem.* 64(1): 5–57.

Plumlee, G.S., Foreman, W.T., Meeker, G.P., Demas, C.R., Lovelace, J.K., Hageman, P.L., Morman, S.A., Lamothe, P.J., Breit, G.N., Furlong, E.T. & Goldstein, H. 2007. Sources, mineralogy, chemistry, environmental reactivity, and metal bioaccessibility of flood sediments deposited in the New Orleans area by Hurricanes Katrina and Rita. USGS Open-File Report 2006–1023.

Prata, A.J. 1989. Observations of volcanic ash clouds in the 10–12 μm window using AVHRR/2 data. *Int. J. Rem. Sens.* 10(4): 751–761.

Prata, A.J. & Bernardo, C. 2009. Retrieval of volcanic ash particle size, mass and optical depth from a ground-based thermal infrared camera. *J. Volcan. & Geoth. Res.* 186(1–2): 91–107.

Prospero, J.M. & Nees, R.T. 1986. Impact of the North African drought and El Niño on mineral dust in the Barbados trade winds. *Nature* 320(6064): 735–738.

Prospero, J.M. 1999. Long-range transport of mineral dust in the global atmosphere: Impact of African dust on the environment of the southeastern United States. *Proc. Nat. Acad. Sci.* 96: 3396–3403.

Prospero, J.M. & Lamb, P.J. 2003. African droughts and dust transport to the Caribbean: Climate change implications. *Science* 302(5647): 1024–1027.

Pudykiewicz, L. 1989. Simulation of the Chernobyl dispersion with a 3-D hemispheric tracer model. *Tellus* 41(B): 391–412.

Pugnaghi, S., Gangale, G., Corradini, S. & Buongiorno, M.F. 2006. Mt. Etna sulfur dioxide flux monitoring using ASTER-TIR data and atmospheric observations. *J. Volcan. & Geoth. Res.* 152(1–2): 74–90.

Rajeev, K., Parameswaran, K., Nair, S.K. & Meenu, S. 2008. Observational evidence for the radiative impact of Indonesian smoke in modulating the sea surface temperature of the equatorial Indian Ocean. *J. Geophys. Res. Atmosph.* 113(D17): 201.

Realmuto, V.J. 2000. The potential use of Earth observing system data to monitor the passive emission of sulfur dioxide from volcanoes. In P.J. Mouginis-Mark, J.A. Crisp & J.H. Fink (eds.), *Remote Sensing of Active Volcanism*: 101–115. Washington DC: American Geophysical Union

Realmuto, V.J., Abrams, M.J., Fabrizia Buongiorno, M. & Pieri, D.C. 1994. The use of multispectral thermal infrared image data to estimate the sulfur dioxide flux from volcanoes: A case study from Mount Etna, Sicily *J. Geophys. Res.* 99(1): 481–488.

Realmuto, V.J., Sutton, A.J. & Elias, T. 1997. Multispectral thermal infrared mapping of sulfur dioxide plumes: A case study from the East Rift Zone of Kilauea volcano, Hawaii. *J. Geophys. Res.* 102(B7): 15057–15072.

Reid, J.S., Prins, E.M., Westphal, D.L., Schmidt, C.C., Richardson, K.A., Christopher, S.A., Eck, T.F., Reid, E.A., Curtis, C.A. & Hoffman, J.P. 2004. Real-time monitoring of South American smoke particle emissions and transport using a coupled remote sensing/box-model approach. *Geophys. Res. Lett.* 31(L06107). doi.10.1029/2203GL018845.

Reid, J.S., Hyer, E.J. Prins, E.M., Westphal, D.L., Zhang, J., Wang, J., Christopher, S.A., Curtis, C.A., Schmidt, C.C., Eleuterio, D.P., Richardson, K.A. & Hoffman, J.P. 2009. Global monitoring and forecasting of biomass-burning smoke: Description of and lessons from the Fire Locating and Modeling of Burning Emissions (FLAMBE) program. *IEEE J. Sel. Topics Appl. Earth Obs. & Rem. Sens.* 2(3): 144–162.

ReliefWeb. 2010. Efectos en la salud por les erupciones del Tungurahua. Available from: http://www.reliefweb.int/rw/dbc.nsf/doc108?OpenForm&rc=2&emid=VO-1999-0584-ECU [Accessed 4th February 2012.

Remer, L.A., Kaufman, Y.J. Tanre, D., Mattoo, S., Chu, D.A., Martins, J.V., Li, R.R., Ichoku, C., Levy, R.C., Kleidman, R.G., Eck, T.F., Vermote, E. & Holben, B.N. 2005. The MODIS aerosol algorithm, products, and validation. *J. Atmosph. Sci.* 62(4): 947–973.

Reynolds, K.A. & Pepper, I. L. 2000. Microorganisms in the environment. In R.M. Maier, I.L. Pepper & C.P. Gerba (eds.), *Environmental Microbiology*: 585. San Diego: Academic Press

Ridgwell, A.J. 2003. Implications of the glacial CO_2 iron hypothesis for Quarternary climate change. *Geochem. Geophys. Geosyst.* 4(9): 1076.

Risebrough, R.W., Huggett, R.J., Griffin, J.J. & Goldberg, E.D. 1968. Pesticides: Transatlantic movements in the Northeast Trades. *Science* 159(3820): 1233–1236.

Roberts, D.M., Karunarathna, A., Buckley, N.A., Manuweera, G., Sheriff, M.H.R. & Eddleston, M. 2003. Influence of pesticide regulation on acute poisoning deaths in Sri Lanka. *Bull. WHO* 81(11): 789–798.

Robock, A. 2000. Volcanic eruptions and climate. *Revs. Geophys.* 38(2): 191–219.

Rogie, J.D., Kerrick, D.M., Sorey, M.L., Chiodini, G. & Galloway, D.L. 2001. Dynamics of carbon dioxide emission at Mammoth Mountain, California. *Earth & Planetary Sci. Lett.* 188(3–4): 535–541.

Rook, G.A. & Zumla, A. 1997. Gulf War syndrome: Is it due to a systemic shift in cytokine balance towards a Th2 profile? *Lancet* 349(9068): 1831–1833.

Safe, S.H. 1995. Environmental and dietary estrogens and human health: Is there a problem? *Environ. Health Perspec.* 103(4): 346–351.

Sait, M., Hugenholtz, P. & Janssen, P.H. 2002. Cultivation of globally distributed soil bacteria from phylogenetic lineages previously only detected in cultivation-independent surveys. *Environ. Microbio.* 4(11): 654–666.

Saiyed, H.N., Sharma, Y.K., Sadhu, H.G., Norboo, T., Patel, P.D., Patel, T.S., Vendaiah, K. & Kashyap, S.K. 1991. Non-occupational pneumoconiosis at high altitude villages in central Ladakh. *Brit. J. Indian Med.* 48: 825–829.

Salyani, M. & Cromwell, R.P. 1992. Spray drift from ground and aerial applications. *Trans. Amer. Soc. Agric. & Biolog. Engin.* 35(4): 1113–1120.

Samuelsen, M., Nygaard, U.C. & Lovik, M. 2009. Particle size determines activation of the innate immune system in the lung. *Scandin. J. Immunol.* 69(5): 421–428.

Sanders, J.W., Putnam, S.D., Frankart, C., Frenck, R.W., Monteville, M.R., Riddle, M.S., Rockabrand, D.M., Sharp, T.W. & Tribble, D.R. 2005. Impact of illness and non-combat injury during operations Iraqi Freedom and Enduring Freedom (Afghanistan). *Amer. J. Trop. Med. & Hyg.* 73(4): 713–719.

Schecter, A., Dai, C., Papke, O., Prange, J., Constable, J.D., Matuda, M., Duc Thao, V. & Piskac, A. 2001. Recent dioxin contamination from Agent Orange in residents of a southern Vietnam city. *J. Occup. & Environ. Med.* 43(5): 435–443.

Schecter, A. & Constable, J.D. 2006. Commentary: Agent Orange and birth defects in Vietnam. *Int. J. Epidem.* 35(5): 1230–1232.

Schiff, L.J., Byrne, M.M., Elliott, S.F., Moore, S.J., Ketels, K.V. & Graham, J.A. 1981. Response of hamster trachea in organ culture to Mount St. Helens volcano ash. *Scan. elect. microsc.* (Pt 2): 169–178.

Schneider, D.J., Rose, W.I., Coke, L.R., Bluth, G.J.S., Sprod, U.E. & Krueger, A.J. 1999. Early evolution of a stratospheric volcanic eruption cloud as observed with TOMS and AVHRR. *J. Geophys. Res.* 104: 4037–4050.

Schneider, E., Hajjeh, R.A., Spiegel, R.A., Jibson, R.W., Harp, E.L., Marshall, G.A., Gunn, R.A., McNeil, M.M., Pinner, R.W., Baron, R.C., Hutwagner, L.C., Crump, C., Kaufman, L., Reef, S.E., Feldman, G.M., Pappagianis, D. & Werner, S.B. 1997. A coccidioidomycosis outbreak following the Northridge, California earthquake. *J. Amer. Med. Assoc.* 277(11): 904–908.

Schölkopf, B., Smola, A.J., Williamson, R.C. & Bartlett, P.L. 2000. New support vector algorithms. *Neural Computation* 12(5): 1207–1245.

Schollaert, S.E., Yoder, J.A., Westphal, D.L. & O'reilly, J.E. 2003. The influence of dust and sulfate aerosols on ocean color spectra and chlorophyll-*a* concentrations derived from SeaWiFS of the U.S. Coast. *J. Geophy. Res.* 108(C6): 3191. doi:3110.1029/2000JC000555.

Searcy, C., Dean, K., Stringer, W. 1998. PUFF: A high-resolution volcanic ash tracking model. *J. Volcan. & Geoth. Res.* 80(1–2): 1–16.

Seftor, J. & Hsu, N., 1997. Detection of volcanic ash clouds from Nimbus 7/total ozone mapping spectrometer. *J. Geophys. Res.* 102(D14): 16,749–16,759.

Shoor, A.F., Scoville, S.L., Cersovsky, S.B., Shanks, D., Ockenhouse, C.F., Smoak, B.L., Carr, W.W. & Petruccelli, B.P. 2004. Acute eosinophilic pneumonia among U.S. military personnel deployed in or near Iraq. *J. Amer. Med. Assoc.* 292(24): 2997–3005.

Simkin, T. & Siebert, L. 1994. *Volcanoes of the World.* Tucson: Geoscience Press & Smithsonian Institution Global Volcanism Program.

Simpson, J.J., Hufford, G., Pieri, D. & Berg, J. 2000. Failures in detecting volcanic ash from a satellite-based technique. *Rem. Sens. Environ.* 72(2): 191–217.

Sinigalliano, C.D., Gidley, M.L., Shibata, T., Whitman, D., Dixon, T.H., Laws, E., Hou, A., Bachoon, D., Brand, L., Amaral-Zettler, L., Gast, R.J., Steward, G.F., Nigro, O.D., Fujioka, R., Betancourt, W.Q., Vithanage, G., Mathews, J., Fleming, L.E. & Solo-Gabriele, H.M. 2007. Impacts of Hurricanes Katrina and Rita on the microbial landscape of the New Orleans area. *Proc. Nat. Acad. Sci. USA* 104(21): 9029–9034.

Small, C. & Naumann, T. 2001. The global distribution of human population and recent volcanism. *Global Environmental Change Part B: Environmental Hazards* 3(3–4): 93–109.

SMD 2012. Available from: www.nasascience.nasa.gov/earth-science/mission_list [Accessed 4th February 2012].

Smirnov, A., Villevalde, Y., O'Neill, N.T., Royer, A. & Tarussov, A. 1995. Aerosol optical depth over the oceans – analysis in terms of synoptic air-mass types. *J. Geophys. Res. Atmosph.* 100(D8): 16639–16650.

181

Smith, B., Wong, C.A., Smith, T.C., Boyko, E.J., Gackstetter, G.D. & Margaret, A.K. 2009. Newly reported respiratory symptoms and conditions among military personnel deployed to Iraq and Afghanistan: A prospective population-based study. *Amer. J. Epidem.* 170(11): 1433–1442.

Smith, D.B., Cannon, W.F., Woodruff, L.G., Garrett, R.G., Klassen, R., Kilburn, J.E., Horton, J.D., King, J.D., Goldhaber, M.B. & Morrison, J.M. 2005. Major-and trace-element concentrations in soils from two continental-scale transects of the United States and Canada. USGS Open File Report 2005-1253: 1–20.

Smith, D.J., Griffin, D.W. & Schuerger, A.C. 2010. Stratospheric microbiology at 20 km over the Pacific Ocean. *Aerobiologia* 26(1): 35–46.

Smithsonian. 2009. Smithsonian Institute's Global Volcanism Program Archives of Natural History. Available from: http://www.volcano.si.edu/world/ [Accessed 4th February 2012].

Smola, A.J. & Scholkopf, B. 2004. A tutorial on support vector regression. *Stats. & Comput.* 14(3): 199–222.

Sprigg, W., Barbaris, B., Morain, S.A., Budge, A.M., Hudspeth, W. & Pejanovic, G. 2009. Public-health applications in remote sensing. Available from: http://www.spie.org/x33688.xml?highlight=x2416& ArticleID=x33688 [Accessed 4th February 2012].

SSEC. 2005. Space Science and Engineering Center. Available from: http://www.ssec.wisc.edu/mcidas [Accessed 4th February 2012].

Stellman, J.M., Stellman, S.D., Weber, T., Tomasallo, C., Stellman, A.B. & Christian Jr., R. 2003. A geographic information system for characterizing exposure to Agent Orange and other herbicides in Vietnam. *Environ. Health Perspec.* 111(3): 321.

Stevens, N.F., Manville, V. & Heron, D.W. 2003. The sensitivity of a volcanic flow model to digital eleva-tion model accuracy: Experiments with digitised map contours and interferometric SAR at Ruapehu and Taranaki volcanoes, New Zealand. *J. Volcan. & Geoth. Res.* 119(1–4): 89–105.

Sultan, B., Labadi, K., Guegan, J.F. & Janicot, S. 2005. Climate drives the meningitis epidemics onset in West Africa. *PLoS Med.* 2(1): e6.

Sutton, A.J., Elias, T., Kauahikaua, J.P. & Moniz-Nakamura, J. 2006. Increased CO_2 emissions from Kilauea: A newly appreciated health hazard at Hawaii Volcanoes National Park, USA. *Cities on Volcanoes, Fourth Conference.* Quito, Ecuador.

Suwalsky, M., Benites, M., Villena, F., Aguilar, F. & Sotomayor, C.P. 1996. Interaction of 2,4-dichlorophenoxyacetic acid (2,4-D) with cell and model membranes. *Biochim. et Biophys. Acta (BBA)-Biomembranes* 1285(2): 267–276.

Swap, R., Garstang, M., Greco, S., Talbot, R. & Kallberg, P. 1992. Saharan dust in the Amazon Basin. *Tellus* 44(B): 133–149.

Swap, R., Ulanski, S., Cobbertt, M. & Garstang, M. 1996. Temporal and spatial characteristics of Saharan dust outbreaks. *J. Geophys. Res.* 101(D2): 4205–4220.

Symonds, R.B., Rose, W.I., Bluth, G.J.S. & Gerlach, T.M. 1994. Volcanic-gas studies: Methods, results, and applications. *Revs. Mineral. & Geochem.* 30(1): 1–66.

Takenaka, S., Karg, E., Kreyling, W.G., Lentner, B., Moller, W., Behnke-Semmler, M., Jennen, L., Walch, A., Michalke, B. & Schramel, P. 2006. Distribution pattern of inhaled ultrafine gold particles in the rat lung. *Inhal. Toxicol.* 18(10): 733–740.

Tan, K. 2010. Available from: http://shanghaiist.com/2010/04/26/gansu-sandstorm.php [Accessed 22nd February 2012].

Tank, V., Pfanz, H. & Kick, H. 2008. New remote sensing techniques for the detection and quantification of earth surface CO2 degassing. *J. Volcan. & Geoth. Res.* 177(2): 515–524.

Tassi, F., Vaselli, O., Capaccioni, B., Giolito, C., Duarte, E., Fernandez, E., Minissale, A. & Magro, G. 2005. The hydrothermal-volcanic system of Rincon de la Vieja volcano (Costa Rica): A combined (inorganic and organic) geochemical approach to understanding the origin of the fluid discharges and its possible application to volcanic surveillance. *J. Volcan. & Geoth. Res.* 148(3-4): 315–333.

Tate, R.L. III. 2000. *Soil Microbiology*. New York: John Wiley.

Teske, M.E., Bird, S.L., Esterly, D.M., Curbishley, T.B., Ray, S.L. & Perry, S.G. 2002. AgDRIFT®: A model for estimating near-field spray drift from aerial applications. *Environ. Toxicol. & Chem.* 21(3): 659–671.

Thundiyil, J.G., Stober, J., Besbelli, N. & Pronczuk, J. 2008. Acute pesticide poisoning: A proposed classification tool. *Bull. WHO* 86(3): 205–209.

Tobin, G.A. & Whiteford L.M. 2002a. Community resilience and volcano hazard: The eruption of Tungurahua and evacuation of the faldas in Ecuador. *Disasters* 26(1): 28–48.

Tobin, G.A. & Whiteford, L.M. 2002b. Economic ramifications of disaster: Experiences of displaced persons on the slopes of Mount Tungurahua, Ecuador. In *25th Appl. Geogr. Conf.* 316–324. Binghamton, NY.

Tobin, G.A. & Whiteford, L.M. 2001. Children's health characteristics under different evacuation strategies: The eruption of Tungurahua volcano, Ecuador. *24th Appl. Geogr. Conf.* Fort Worth, TX.

Tobin, G.A., Whiteford, L.M. & Connor, C.B. 2002. Modeling volcanic ash dispersion and its impact on human health and community resilience. Tampa: University of South Florida.

Torres, O., Tanskanen, A. Veihelmann, B., Ahn, C., Braak, R., Bhartia, P.K., Veefkind, P. & Levelt, P. 2007. Aerosols and surface UV products from ozone monitoring instrument observations: An overview. *J. Geophys. Res. Atmosph.* 112(D24): 1–14.

Torsvik, V., Salte, K., Sorheim, R. & Goksoyr, J. 1990. Comparison of phenotypic diversity and DNA heterogeneity in a population of soil bacteria. *Appl. & Environ. Microbio.* 56(3): 776–781.

Torsvik, V., Sorheim, R. & Goksoyr, J. 1996. Total bacterial diversity in soil and sediment communities: A review. *J. Indust. Microbio. & Biotech.* 17(3): 170–178.

Torsvik, V., Ovreas, L., & Thingstad, T.F. 2002. Prokaryotic diversity: Magnitude dynamics, and controlling factors. *Science* 296: 1064–1066.

Tralli, D.M., Blom, R.G., Zlotnicki, V., Donnellan, A. & Evans, D.L. 2005. Satellite remote sensing of earthquake, volcano, flood, landslide and coastal inundation hazards. *Int. J. Photogram. & Rem. Sens.* 59(4): 185–198.

Tucker, C.J. & Nicholson, S.E. 1999. Variations in the size of the Sahara Desert from 1980 to 1997. *Ambio.* 28(7): 587–591.

Tupper, A. & Wunderman, R. 2009. Reducing discrepancies in ground and satellite-observed eruption heights. *J. Volcan. & Geoth. Res.* 186(1–2): 22–31.

Tupper, A., Carn, S., Davey, J., Kamada, Y., Potts, R., Prata, F. & Tokuno, M. 2004. An evaluation of volcanic cloud detection techniques during recent significant eruptions in the western 'Ring of Fire'. *Rem. Sens. Environ.* 91(1): 27–46.

USEPA 2009. Global Earth Observation System of Systems Tools. Available from: http://www.epa.gov/GEOSS/eos/text_tool.html [Accessed 4th February 2012].

USGS 2006. USGS Cascade Range Update. Available from: http://volcano.wr.usgs.gov/cvo/current_updates.php [Accessed 4th February 2012].

USGS 2012. Available from: https://lpdaac.usgs.gov/lpdaac/products/modis_products_table/albedo/16_day_l3_global_0_05deg_cmg/mcd43c3 [Accessed 23rd February 2012].

USNRC 2009. Backgrounder on the Three Mile Island accident, United States Nuclear Regulatory Commission. Available from: http://www.nrc.gov/reading-rm/doc-collections/fact-sheets/3mile-isle.html [Accessed 19th March 2012].

UW 2012. Available from: http://depts.washington.edu/chc/ [Accessed 19th March 2012].

Van Donkelaar, A., Martin, R.V. & Park, R.J. 2006. Estimating ground-level PM2.5 using aerosol optical depth determined from satellite remote sensing. *J. Geophys. Res. Atmosph.* 111(D21): 201.

Vapnik, V.N. 1995. *The nature of statistical learning theory.* New York: Springer.

Vapnik, V.N. 1998. *Statistical learning theory.* New York: Wiley.

Villegas, H. 2004. Volcanic disaster seen from space: Understanding eruptions using Landsat-TM. *Geospat. Info. Managem. Int.* 18(5): 68–71.

Wakisaka, I., Yanagihashi, T., Tomari, T. & Ando, T. 1988. Effects of volcanic activity on the mortality figures for respiratory diseases. *Japanese J. Hyg.* 42(6): 1101–1110.

Walker, J.S. 2004. *Three Mile Island: A nuclear crisis in historical perspective.* Berkeley: UC Press.

Walsh, J.J. & Steidinger, K.A. 2001. Saharan dust and Florida red tides: The cyanophyte connection. *J. Geophys. Res. - Oceans* 106(C6): 11,597–11,612.

Ward, M.H., Nuckols, J.R., Weigel, S.J., Maxwell, S.K., Cantor, K.P. & Miller, R.S. 2000. Identifying populations potentially exposed to agricultural pesticides using remote sensing and a Geographic Information System. *Environ. Health Perspec.* 108(1): 5–12.

Watson, I.M., Realmuto, V.J., Rose, W.I., Prata, A.J., Bluth, G.J.S., Gu, Y., Bader, C.E. & Yu, T. 2004. Thermal infrared remote sensing of volcanic emissions using the moderate resolution imaging spectroradiometer. *J. Volcan. & Geoth. Res.* 135(1–2): 75–89.

Waythomas, C.F. & Miller, T.P. 1998. Preliminary volcano-hazard assessment for Iliamna Volcano, Alaska. USGS Open-File Report 99–373.

Webley, P. & Mastin, L. 2009. Improved prediction and tracking of volcanic ash clouds. *J. Volcan. & Geoth. Res.* 186(1–2): 1–9.

Webley, P.W., Dehn, J., Lovick, J., Dean, K.G., Bailey, J.E. & Valcic, L. 2009. Near-real-time volcanic ash cloud detection: Experiences from the Alaska Volcano Observatory. *J. Volcan. & Geoth. Res.* 186(1–2): 79–90.

Weiss-Penzias, P., Jaffe, D., Swartzendruber, P., Hafner, W., Chand, D. & Prestbo, E. 2007. Quantifying Asian and biomass burning sources of mercury using the Hg/CO ratio in pollution plumes observed at the Mount Bachelor Observatory. *Atmosph. Environ.* 41(21): 4366–4379.

Weiss-Penzias, P., Jaffe, D.A., Swartzendruber, P., Dennison, J.B., Chand, D., Hafner, W. & Prestbo, E. 2006. Observations of Asian air pollution in the free troposphere at Mount Bachelor Observatory during the spring of 2004. *J. Geophys. Res.* 111: D10304.

Wen, S. & Rose, W.I. 1994. Retrieval of sizes and total masses of particles in volcanic clouds using AVHRR bands 4 and 5. *J. Geophys. Res.* 99(D3): 5421–5431.

Whiteford, L.M. & Tobin, G.A. 2002. In the shadow of the volcano: Human health and community resilience following forced evacuation. Technical Report, Center for Disaster Management and Humanitarian Assistance. Tampa: University of South Florida.

Whiteford, L.M. & Tobin, G.A. 2004. Saving lives, destroying livelihoods: Emergency evacuation and resettlement policies. In: A. Castro & M. Springer (eds.), *Unhealthy Health Policies: A Critical Anthropological Examination*: 189–202. Walnut Creek: AltaMira Press.

Whitman, W.B., Coleman, D.C. & Wiebe, W.J. 1998. Prokaryotes: The unseen majority. *Proc. Nat. Acad. Sci.* 95: 6578–6583.

Wiebert, P., Sanchez-Crespo, A., Seitz, J., Falk, R., Philipson, K., Kreyling, W.G., Moller, W., Sommerer, K., Larsson, S. & Svartengren, M. 2006. Negligible clearance of ultrafine particles retained in healthy and affected human lungs. *Europ. Respir. J.* 28(2): 286–290.

Williams, P.L., Sable, D.L., Mendez, P. & Smyth, L.T. 1979. Symptomatic coccidioidomycosis following a severe natural dust storm. An outbreak at the Naval Air Station, Lemoore, California. *Chest* 76(5): 566–570.

Williamson, K.E., Wommack, K.E. & Radosevich, M. 2003. Sampling natural viral communities from soil for culture-independent analyses. *Appl. & Environ. Microbio.* 69(11): 6628–6633.

Williamson, R.A. & Obermann, R.M. 2002. Continuity of moderate resolution Earth observation data: Is an international solution the answer? *Space Policy* 18(2): 129–133.

Winker, D.M., Hunt, W.H. & McGill, M.J. 2007. Initial performance assessment of CALIOP. *Geophys. Res. Letts* 34(19): 803.

Wright, R., Flynn, L., Garbeil, H., Harris, A. & Pilger, E. 2002. Automated volcanic eruption detection using MODIS. *Rem. Sens. Environ.* 82(1): 135–155.

Xian, P., Reid, J.S., Turk, J.F., Hyer, E.J. & Westphal, D.L. 2009. Impact of modeled versus satellite measured tropical precipitation on regional smoke optical thickness in an aerosol transport model. *Geophys. Res. Letts* 36(16): 805.

Xu, X.Z., Cai, X.G., Men, X.S., Yang, P.Y., Yang, J.F., Jing, S.L., He, J.H. & Si, W.Y. 1993. A study of siliceous pneumoconiosis in the desert area of Sunan County, Gansu Province, China. *Biomed. Environ. Sci.* 6(3): 217–222.

Yamamoto, Y. 2007. Recent moves to address the KOSA (yellow sand) phenomenon. *Quart. Rev.* 22: 45–61.

Yang, C.Y. 2006. Effects of Asian dust storm events on daily clinical visits for conjunctivitis in Taipei, Taiwan. *J. Toxicol. & Environ. Health*, 69(18 Part A): 1673–1680.

Yates, M.V. & Yates, S.R. 1988. Modeling microbial fate in the subsurface environment. *CRC Criti Revs Environ. Cont.* 17(4): 307–344.

Yin, D., Nickovic, S. & Sprigg, W. 2007. The impact of using different land cover data on wind-blown desert dust modeling results in the southwestern United States. *Atmosph. Environ.* 41(10): 2214–2224.

Yu, H., Remer, L.A., Chin, M., Bian, H., Kleidman, R.G. & Diehl, T. 2008. A satellite-based assessment of transpacific transport of pollution aerosol. *J. Geophys. Res.* 113: D14S12.

Yu, T., Rose, W.I. & Prata, A.J. 2002. Atmospheric correction for satellite-based volcanic ash mapping and retrievals using 'split window' IR data from GOES and AVHRR. *J. Geophys. Res.* 107(D16): 4311.

Zhang, J.F. & Morawska, L. 2002. Combustion sources of particles: 2. Emission factors and measurement methods. *Chemosphere* 49(9): 1059–1074.

Zhang, L., Jacob, D.J., Boersma, K.F., Jaffe, D.A., Olson, J.R., Bowman, K.W., Worden, J.R., Thompson, A.M., Avery, M.A. & Cohen, R.C. 2008a. Transpacific transport of ozone pollution and the effect of recent Asian emission increases on air quality in North America: An integrated analysis using satellite, aircraft, ozonesonde, and surface observations. *Atmosph. Chem. & Phys.* 8(20): 6117–6136.

Zhang, J.L., Reid, J.S., Westphal, D.L., Baker, N.L. & Hyer, E.J. 2008b. A system for operational aerosol optical depth data assimilation over global oceans. *J. Geophys. Res. Atmosph.* 113(D10): 208.

Zhang, H., Hoff, R.M. & Engel-Cox, J.A. 2009. The relation between moderate resolution imaging spectroradiometer (MODIS) aerosol optical depth and PM2.5 over the United States: A geographical comparison by US Environmental Protection Agency regions. *J. Air & Waste Manag. Assoc.* 59(11): 1358–1369.

Zhenda, Z. & Tao, W. 1993. The trends of desertification and its rehabilitation in China. *Desertifi. Cont. Bull.* 22: 27–29.

Zhu, H., Reichard, D.L., Fox, R.D., Ozkan, H.E. & Brazee, R.D. 1995. DRIFTSIM, a program to estimate drift distances of spray droplets. *Appl. Engin. Agric.* 11(3): 365–369.

Zubko, V., Leptoukh, G. & Gopalan, A. 2010. Study of data merging and interpolation methods for use in an interactive online analysis system: MODIS terra and aqua daily aerosol case. *IEEE Trans. Geosci. & Rem. Sens.* 48(12): 4219–4235.

Chapter 5

Emerging and re-emerging diseases

C.J. Witt[1] (Auth./ed.) with; J.E. Pinzon[2], P.A. Manibusan[3], J.A. Pavlin[4], R.V. Gibbons[4],
T.E. Myers[5], A.L. Richards[5], P. Daszak[6], S.P. Luby[7], J.H. Epstein[6], M. Jahangir Hossain[7],
E.S. Gurley[7], J.R.C. Pulliam[8], A. Zayed[9], M.S. Abdel-Dayem[9] &
M.G. Collacicco-Mayhugh[10]

[1]*Armed Forces Health Surveillance Center, Silver Spring, MD, US*
[2]*Science Systems & Applications Inc., Greenbelt, MD, US*
[3]*Tripler Army Medical Center, Honolulu, HI, US*
[4]*Armed Forces Research Institute of Medical Sciences, Bangkok, TH*
[5]*Naval Medical Research Center, Silver Spring, MD, US*
[6]*EcoHealth Alliance New York, NY, US*
[7]*International Centre for Diarrhoeal Disease Research, Dhaka, BD*
[8]*University of Florida, Gainesville, FL, US*
[9]*Cairo University & U.S. Naval Medical Research Unit, Cairo, Egypt*
[10]*Walter Reed Army Institute of Research, Silver Spring, MD, US*

ABSTRACT: This chapter examines emerging and re-emerging diseases in context of the environmental and ecological conditions, trends and events that together drive a microbe's transition pathway to that emergence. The study of these naturally occurring phenomena provides insights into possible future disease outbreak events. It also suggests possible ways to monitor for such events through targeted, proactive surveillance. Disease emergence should not come as a surprise. It should be expected, and societies should be prepared for it.

Disclaimer: The opinions in this paper are those of the authors and do not represent the official position of the US Department of Defense.

1 INTRODUCTION

There are approximately 1400 recognized human pathogens in the world, of which about eighty seven are recognized as novel or having emerged since 1980 (Woolhouse & Gaunt 2007). These emerging pathogens share some basic biotic characteristics. They are primarily, but not exclusively, RNA viruses, and they possess an innate capability to adapt opportunistically to new hosts and environments for survival. They are mostly associated with animal reservoirs or arthropod vectors. They usually enter human awareness acutely as dangerous pathogens (Morse 2009). However, some emerging infections in humans are not yet known to be associated with a specific disease (e.g. Tioman virus) (Tee *et al.*, 2009).

Emergence is not new (Morens *et al.*, 2008; Holmes 2011). The process has been occurring for millennia, and it is reasonable to expect that microbes' biotic characteristics will continue to evolve as yet unknown and novel emerging infections in the future. What is new, however, is the unprecedented rate and frequency of emergence over the last thirty years (Gubler 2008; Daszak 2009).

An explanation offered for this trend is that emergence is not solely the result of an individual microbe's existence in nature. Rather, emergence is rooted in a complex interrelationship of biotic and abiotic drivers or facilitators. From an emerging zoonotic or vector-borne disease perspective, non-microbial biotic drivers can include those animals and arthropod vectors that serve as ecological

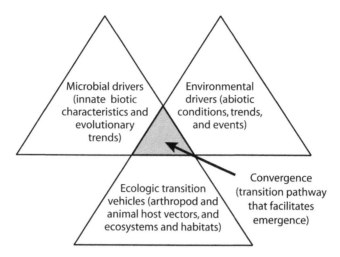

Figure 1. Transition pathway towards disease emergence: a nonlinear interaction between multiple microbiological and environmental drivers, and ecological vehicles that, together, facilitate an opportunity for disease emergence. The opportunity for pathogen emergence in humans is greatest at the convergence of the triad.

transition vehicles. Through their biology and behaviours such as presence, abundance, habitat and nutrition preferences, social behaviours, and other factors, they facilitate a microbe's ability and opportunity to interface with humans.

Abiotic drivers include environmental and ecological phenomena such as climate change, weather anomalies, and natural disasters like flooding or forest fires. Abiotic ecological drivers can also be thought of as transition vehicles, and include anthropogenic drivers such as habitat disruption derived from economic development, travel, trade, political and social instability, natural resource harvesting, unplanned urbanization, and agricultural practices (Smolinski et al., 2003). These natural and anthropogenic drivers can bring humans into direct exposure with new microbial agents. They can lead to the initiation of new transition pathways that enable a microbe to emerge as an infectious human pathogen.

Furthermore, the biotic drivers within the microbe can be impacted by both environmental and ecological drivers. A microbe can evolve under selection pressures that promote its compatibility with a new transition vehicle (arthropod vector, animal host, or ecologic habitat). Finally, if a microbe establishes itself in the vehicle, it can evolve to permit infection, proliferation and expression of disease in the human. If transmission, establishment and proliferation go beyond the initial human host, a new microbe enters human awareness as an emerging infectious disease with outbreak potential (Morse 2009).

Emergence, therefore, results from a convergence of opportunities: the expression of a microbe's biotic drivers (innate characteristics and evolutionary trends), abiotic environmental conditions, trends and events, and ecological transition vehicles (arthropod vectors, animal hosts, and evolving habitats and ecosystems) that together become a transition pathway for emergence (Figure 1). Science may not yet know exactly how the overall transition pathway gives rise to a specific disease's opportunity to emerge, or to what extent each member of the triad influences every instance; but clear, intuitive associations do exist (Patz & Reisen 2001; Patz & Olson 2006; Epstein 2007; Randolf & Rogers 2007; Woolhouse & Gaunt 2007; Gubler 2008; Morens 2008; Daszak 2009; Morse 2009; Tee et al., 2009).

If emergence is derived from opportunity, it follows that if one microbe can take advantage of a given opportunity to explore new territory (i.e. a new ecosystem or habitat vector or host), and it finds itself in a suitable position for survival and propagation, then other microbes with similar biotic characteristics might emerge also. This chapter examines what is currently understood about how some microbes are emerging as human pathogens. The authors focus on selected pathogens and explore what transition-pathway drivers may have facilitated their emergence. By using what is known about a specific emerging infection, assumptions can be drawn about it and other related, but yet to be discovered pathogens that might emerge under similar, perhaps recurring circumstances. The authors suggest how the identification of the convergence of the different types of drivers might be used proactively to initiate efficient emerging infection surveillance by targeting geographic areas or environmental conditions that possess characteristics conducive of the putative transition pathway. Such targeting could result in an ability to predict a pathogen's emergence before it occurs.

If one can discern the environmental driver for a pathogen to emerge, then it should be possible to prepare for emergence by either using a transition pathway's expected critical control point(s) to interrupt convergence and emergence; or, if that is not feasible, to prepare public health actions that mitigate the emerging pathogen's outbreak at its early stages. For example, one could use this predictive ability to design and advocate for the adoption of basic habitat control or human behaviour protocols based on what is known of the behaviour and requirements of the prototype microbe (Gubler 2001; IOM 2003; Smolinski 2003; Morens 2008; Nasci 2008). This would help communities cope with the initial surprise of an emerging infectious outbreak before a full understanding of the new pathogen is achieved. This approach was used during the initial stages of the H5N1 influenza outbreak in 2004. Authorities used surveillance and human infection control templates for a respiratory syndrome of unknown aetiology developed for SARS the previous year. Where practiced, the SARS template of initiating basic infection control and community public health practices slowed and prevented exposure of animals and humans to this new respiratory pathogen until a more specific understanding of the H5N1 outbreak dynamic was gained (Abdullah *et al.*, 2004; Webby & Webster 2004; WHO 2004).

Therefore, this chapter explores different emerging infectious diseases as prototypes for putatively associating similar, but yet-to-emerge pathogens, with identifiable environmental conditions, trends and events, and their ecological transition vehicles. The authors then suggest how those associations might be used to search for plausible signals that can mark the potential for a successful transition pathway, and use them as the foundation for emergence surveillance; to look for as yet unknown pathogens; and prepare for these surprises before they occur.

The emergence of different filoviruses across central Africa over the last thirty years underscores their importance as uniquely virulent emerging human diseases. In Section 2, Doctor. Pinzon explores what is known about Ebola virus (EBOV) and Marburg virus (MARV) as prototypic filoviruses whose human disease outbreaks are expressed through a convergence of human and animal transition pathways culminating in what is ultimately an anthropogenic transmission pathway. Doctor Pinzon presents what is known about the biology and behaviour of filovirus outbreaks. He then presents a framework for thinking about the biotic and abiotic drivers that may be predictably associated with the beginnings of those outbreaks to lay a foundation for future understanding about how; under what conditions; and when, human filovirus outbreaks emerge. From there he explores how one might use this framework to design a remote sensing monitoring system for detecting what appears to be a very complex spatial and temporal expression pattern resulting from a multi-component environmental dynamic.

In Section 3, Doctors Myers and Richards explore the emergence of rickettsiae, a group of obligate intracellular bacteria that constantly surprise us with their apparently ubiquitous presence. As a group, they have been known to exist for many years. They are not new, but are considered newly discovered and emerging in many parts of the world. Sixteen new rickettsiae have been discovered over the last twenty years. Climate change may be a driver in the apparent expansion of their tick hosts' geographic range. Anthropogenic activities like the suburbanization of the northeastern US that brings humans into more frequent contact with ticks, also serve as a transition vehicle that influences the opportunity for rickettsiae's emergence (Fish 1995; Randolph 2000). Just as possible, the detection of new rickettsial species like *Rickettsia parkeri*, the causative agent of Tidewater spotted fever, could be either a reflection of microbial evolution, changing biotic characteristics, or be the result of an environmental driver's facilitation of the tick host's geographic species range. Alternatively, it could be a reflection of improved diagnostic technology (e.g. abiotic anthropomorphic transition vehicle) that enables the distinction between different rickettsial subgroups. If the latter is the case, then improved diagnostic technology will drive the design of surveillance for new and emerging rickettsiae. The application of better diagnostics, targeted to geographic areas where favourable ecological and environmental drivers converge to form a transition pathway for rickettsiae emergence will facilitate the ability to anticipate rickettsial disease threats. This could be in response to changes in environmental factors at the eco-climatic level, or at the biologic level of the arthropod vector's internal environment. The use of advanced diagnostics can be the key to demonstrating that the convergence that is rickettsial emergence is probably occurring more frequently than previously known. Rickettsiae are probably present if one looks for them as Doctors Myers and Richards suggest in their contribution.

In Section 4, Doctors Manibusan, Pavlin and Gibbons discuss the relative roles of rapid unplanned urbanization and potentially, climate change in permitting the mosquito vectors, particularly *Aedes aegypti*, to introduce and sustain themselves, and to transmit dengue. Urbanization, exacerbated by the influence of poverty and its associated human activities of overcrowding, use of open-air, standing water supplies, and peri-domestic garbage build-up, and perhaps in combination with the environment becoming warmer, is permitting this anthropophilic mosquito to efficiently replicate and provide a favourable transition pathway for dengue viruses to flourish. The authors explore how climate change, while of great concern globally, is likely less important as a driver of dengue emergence in this Himalayan country than the rapid, unplanned urbanization over the last decade. Nepal may represent a case where the transition pathway for dengue emergence is tied to an inadequate infrastructure for delivering safe water to the poor than the more typically associated environmental driver – drought. The authors explore this scenario as a starting point for understanding the expected future emergence of dengue infections in other areas of the world that have similar infrastructures and ecosystems.

In Section 5, Doctors Daszak, Luby, Epstein, Hossain, Gurley and Pulliam present what is known about the complex ecologic drivers of the single Nipah virus outbreak in Malaysia, and compare it with the repeated outbreaks that have occurred in Bangladesh starting in 2001. Both scenarios involve bats as virus reservoirs. Both rely on bat feeding habits to bring Nipah into close proximity to humans or their domestic animals to effect transmission and infection. When examined together, these scenarios shed light on the role of intensified agricultural practices in driving this disease's emergence; and the breadth of medical and public health investigation needed to be prepared for and combat it.

Leishmaniasis is not really an emerging disease; but it is re-emerging in many parts of the world. In Section 6, Doctors Colacicco-Mayhugh, Zayed and Abdel-Dayem explore the major factors governing its re-emergence: environmental change, ecological disruptions, human emigration and immigration, and changes in vector and or reservoir distribution. For example, an outbreak of visceral leishmaniasis (VL) in the Upper Nile Province of southern Sudan began in the late 1980s and was attributed to the importation of leishmaniasis by the movement of people due to the war in Eritrea (Perea 1991; Zijlstra 1991). A probable ecological driver also promoting leishmaniasis' establishment in Sudan was the coincidental reforestation in the same geographic areas providing a favourable habitat of VL's arthropod vector *Phlebotomus orientalis* (Ashford & Thompson 1991; Schorscher & Goris 1992; Elnaiem *et al.*, 1997). In addition, there has been a general resurgence of leishmaniasis in both people and dogs in other areas of Africa, Asia, Europe, and the Americas over the past few decades. While this trend can be related to a variety of anthropogenic drivers such as the introduction through increases in travel from endemic regions or from a growing number of immune suppressed human hosts, they also suggest that a leishmaniasis transition pathway exists where specific environmental and ecological vehicles coincide and successfully accommodate the opportunistic behaviours of leishmaniasis' different biotic drivers (Ready 2010). Given what is currently known about leishmaniasis, case surveillance can be used to monitor at-risk areas for leishmaniasis establishment or resurgence; and if possible, contain it. However, if the public health goal is leishmaniasis elimination, remote sensing technology could aid in detecting the signals of environmental and ecological drivers of its re-emergence. These drivers could then be targeted for intensive intervention to block the transition pathway.

2 FILOVIRUS: AN INTEGRATED APPROACH IN TIME & SPACE

2.1 *Filoviruses and cross species transmission*

In recent decades, several newly emerging diseases have resulted in major threats to both affected communities and global public health. Viruses from wildlife hosts in particular, have exhibited a capacity for cross-species transmission (CST), and have caused high-impact human diseases. Ebola and Marburg haemorrhagic fevers are among these. It has been estimated that approximately sixty per cent of human infectious diseases have animal origins (zoonoses), including some important viral diseases that are traditionally considered to be of human origin (Jones *et al.*, 2008). For example, measles and smallpox may well have their prehistoric origins in wildlife (Wolfe *et al.*, 2007). It may be logical and prudent therefore, to anticipate that there are other filoviruses with a capacity to cross into human populations. To anticipate that these cross-overs will occur, and to prepare for their mitigation, will require an understanding of filoviruses as a biologic system in the environment. Scientists will need to know how the ecological dynamic of CST interacts with a 'new' virus's evolutionary factors to overcome environmental, demographic, and host-specific barriers that enable them to transmit and infect humans.

In the specific case of Ebola and Marburg, there is still limited knowledge about the true animal reservoir. Recent studies point to frugivorous bats as the likeliest hosts, but uncertainty remains (Leroy *et al.*, 2005; Swanepoel *et al.*, 2007; Towner *et al.*, 2007; Pourrut *et al.*, 2009). Knowing the reservoir (or reservoirs) is critical to inform *smart surveillance*; that is, targeting those regions believed to be high-risk hotspots for filovirus emergence and human risk (Pinzon *et al.*, 2004). Finally, to grasp the full complexity of a successful filovirus CST, one needs to consider other factors including the type and intensity of reservoir, intermediary stages, and final host contacts. These are affected by social, behavioural, seasonal, and ecological traits. To deal with the emergence of filoviruses, the whole dynamic of filovirus transmission and emergence as an integrated physical and biological system needs research. This assumes that emergence is a natural system, and that it possesses the essential characteristics of a complex adaptive system (CAS) (Holland 1995). These characteristics include heterogeneity of components (species, agents, vectors, hosts, humans, environment, and climate) which provides variability through nonlinear interactions among those components. Interactions among the parts determine and reinforce a self-organizing hierarchical structure. CAS defines the rules that govern which plants and animals can be expected in an environment, and how they have adapted there for successful living. By using the deductive perspective of CAS, together with the hierarchical organization as a starting point, it should be possible to extract the filoviruses' potential transmission enablers, and to explore their properties through simplified models that enable filovirus transmission and emergence into humans (Levin 1999). For example, it is known that some forms of contact between donor and recipient hosts are a precondition for virus transfer. It is therefore affected negatively by a hierarchical structure that separates the donor and recipient; the barriers of geography, ecology, and behaviour. In theory, factors that disrupt the geographical distribution of host species, or that decrease behavioural separation tend to promote new avenues of interaction between reservoirs and hosts. An example of the first is wildlife trade and migration due to eco-climatic and environmental shifts like drought or habitat destruction. An example of the second is bush-meat harvesting. Both provide opportunities for virus sharing and the emergence of infections in new host species.

In the case of filoviruses, these barrier disruptions can explain the spatial pattern of sporadic or occasional human outbreaks. In addition, the seasonality of climate and the corresponding uneven temporal distribution of resources needed for existence place hardships on both wildlife and humans. These also tend to decrease barriers to transmission, but in a more cyclical pattern. Thus, the CAS for filoviruses in general may comprise both an eco-climatic set of conditions or drivers that trend on a given temporal scale, and also a geographical and/or behavioural set of factors that trend on totally different scales. For example, seasonality is tracked by eco-climatic information, which in turn is used to characterize places by their patterns of temperature and precipitation. This tracking allows estimation of the distribution and patterns of vegetation growth, which in turn allows one to create schemes for relating vegetation to on-the-ground effects of climate. It is postulated here that one of those effects is disease emergence.

A wide variety of environmental monitoring data from satellite and ground-based systems are accessible and can be used to probe the overall associations of specific climate variables with disease incidence (Colwell 1996; Hay et al., 1998; Linthicum et al., 1999; Pinzon et al., 2004; Randolph & Rogers 2007; Anyamba et al., 2009).

Stepping back, science as a disciplined thought process begins from observation and interpretation. It then seeks explanation and mechanisms (Levin 1999). It formulates a strategic concept for examining apparently unpredictable natural events by averaging recorded observations over time and space. It is a starting point; but, more is needed. The average also needs to be applied at a proper temporal or spatial domain to relate its findings in a meaningful way. For example, natural occurrences tend to come into focus on particular domains (temporal, spatial), and not on others. Thus, specificity of domains matters when relating disease emergence characteristics and environmental conditions. Different degrees of aggregation show different views of system dynamics. On broad spatial scales and long temporal scales, ecological systems in their entirety exhibit well-established patterns upon which one can derive reliable generalizations relevant to modelling and prediction. Variation among observations however, does occur. Observations may be apparent events and non-consequential to a system's dynamics, or expressions of impactful trends expressed at different or smaller scales, and noticed periodically because they individually cross the magnitude barrier of the larger scale as it is in play; or because they converge with and augment the expression of the larger scale. So where does this lead in filovirus emergence? What is known about the relationship of these viruses to various domains? How do domains relate to CAS? Can one use scale or CAS to design an appropriate tool for conducting surveillance for filovirus emergence?

2.2 Tracking Ebola and Marburg

Marburg virus (MARV) and Ebola viruses (EBOV) comprise the known main aetiological agents in central Africa that cause filovirus-associated severe haemorrhagic fevers (HF) in both human and non-human primates (Sanchez et al., 2007). The high pathogenicity of these 'prototypic' filoviruses has been apparent since the discovery of MARV HF among vaccine plant workers in Marburg, Germany and the former Yugoslavia in 1967 (Kissling et al., 1968). Almost ten years later in 1976, a second filovirus, and the first found in a natural setting, emerged in two nearly simultaneous outbreaks about 800 km apart in southern Sudan and the present day Democratic Republic of Congo. These occurrences led to the discovery of two EBOV subtypes, Sudan Ebola virus (SEBOV) and Zaire Ebola virus (ZEBOV) (Johnson 1978; Smith 1978). The ZEBOV epicentre was in Yambuku, a rural village with the Ebola River running through it in northern Democratic Republic of Congo. The mortality rate for ZEBOV was about eighty per cent and fifty per cent for SEBOV. In 1989, a third strain/subtype of EBOV was identified in infected macaques in a quarantine facility in Reston, Virginia, US (hence, REBOV). The monkeys were imported from the Philippines (Jahrling et al., 1990). REBOV is lethal to monkeys but appears to be non-pathogenic in humans.

EBOV HF was not reported again until the end of 1994, when three outbreaks started almost simultaneously. In November of that year, ethnologists studying chimpanzees in the Tai National Park, Côte d'Ivoire, found dead chimpanzees and noticed the absence of others. A female researcher became infected during the necropsy of one of the dead chimpanzees, and a fourth novel strain of EBOV, the Cote d'Ivoire (CIEBOV), was isolated from her blood (Formenty et al., 1999). That December, multiple human ZEBOV HF cases were reported in gold panning camps in Gabon, and a large human outbreak began in the Kikwit District in Democratic Republic of Congo. The Kikwit outbreak resembled the 1976 outbreaks in that secondary transmission of the virus occurred through close personal contact between family members and among hospital workers.

A fifth EBOV subtype, Bundibugyo Ebola virus (BEBOV), was identified in November 2007 from cases in an outbreak in Bundibugyo district in Uganda. Unlike the other EBOV outbreaks, the new strain had a case fatality rate of twenty-five per cent (Towner et al., 2008). Since the EBOV outbreaks of the mid 1990's, there have been sporadic reports of human and non-human primate cases (Walsh *et al.*, 2003; Pourrut *et al.*, 2005; Towner *et al.*, 2008; Cardenas 2010).

Overall, the reported outbreaks have included: ZEBOV: one isolated case in 1996 from South Africa (imported from Gabon), and ten outbreaks with high near-uniform lethality of eighty per cent in Democratic Republic of Congo in 1995, 2007, and 2008; in Gabon in 1994/1995, 1996, 1997, and 2001/2002; and in the Congo in 2001, 2003, and 2005. There have been two SEBOV outbreaks: one in Uganda in 2000/2001, and one in Sudan in 2004. Both had a case fatality rate near fifty per cent. There has also been one isolated human case of CIEBOV in Côte d'Ivoire in 1994, and one reported BEBOV outbreak in Uganda in 2007/2008. MARV has also proven to be a recurring threat in central Africa with its severe morbidity and high fatality rate in the recent outbreaks in 1998 1999 and in Angola in 2004–2005.

The diversity in magnitude and location of these EBOV and MARV outbreaks, combined with their similar presentation and severity of disease, suggest that these filoviruses have similar pathogenic characteristics, host tropism, and possibly human-exposure dynamics. It also suggests that something in their pathways to outbreak causes variation in their observed temporal cycles and spatial patterns, which are not immediately evident through a linear analysis. Attempts to unravel the known and relevant details of filovirus emergence has caused epidemiologists to start with inferences from circumstantial evidence and correlations; not very satisfying scientifically for modelling and prediction, but inferences from such information does provide an adequate place to begin formulating hypotheses. Once formed, these heuristic models must be probed for verification or rejection (see also Chapter 7). Using the CAS rubric and approaching the interpretation of observations using multiple domains, one may tease apart the biological, environmental, and temporal connections that describe a plausible filovirus emergence model. Use of such a model to probe for scientifically verifiable explanations and to better understand the systems' complexity may give clues to how filoviruses in general emerge. This probing will require incorporating recent advances in a number of disciplines relevant to modelling and prediction; namely, collaborative initiatives between virology, veterinary, human medicine, climate and ecology (Colwell 1996; Hay *et al.*, 1998; Linthicum *et al.*, 1999; Pinzon *et al.*, 2004; Randolph & Rogers 2007; Anyamba *et al.*, 2009).

It is known that during observed outbreaks, the spread of Ebola virus among gorillas and chimpanzees along the border region between Gabon and Democratic Republic of Congo has caused both a high number of deaths among these primates, and a series of human cases with histories of contact with infected animals in the wild (Walsh *et al.*, 2003; Rouquet *et al.*, 2005). Specifically, there is strong evidence obtained during the ZEBOV HF epidemics in Gabon, that hunters acquired infection from scavenging gorilla, chimpanzee and duiker carcasses (Walsh *et al.*, 2003; Leroy *et al.*, 2004; Rouquet *et al.*, 2005; Lahm *et al.*, 2007). Leroy (*et al.*, 2004) found that great apes can be infected with ZEBOV under different ecological conditions through independent transmission events. This suggests that there is a wide distribution range for these viruses. It is further known that because of their high case fatality rates, great apes are presumed to be dead-end hosts; and thus, when combined with the discordance of filovirus event locations and the great apes' natural geographic distributions, that apes are unlikely reservoir hosts for this virus. Current evidence from virus isolation studies and the known geographic range of some species of frugivorous bats suggests strongly that these animals are the filoviruses' natural reservoir (Leroy *et al.*, 2005; Swanepoel *et al.*, 2007; Towner *et al.*, 2007; Pourrut *et al.*, 2009).

From what is known, these bats appear to bear their young during the wet season, when fruit is most abundant. These fruits are also eaten by great apes. Therefore, a presumption is growing that the behaviour around the process of birthing might be the mechanism for filovirus shedding, and the plausible biological driver for primate exposure to the virus. Recent analyses of environmental and climatic factors associated with EBOV support this idea. In the tropical forest areas of Gabon and Democratic Republic of Congo multiple incidents of ZEBOV have been reported at approximately two to four months after the onset of markedly drier environmental conditions (i.e. lower than 1.5 standard deviations from the mean NDVI value) during the longer and first of two yearly wet seasons (Tucker *et al.*, 2002; Peterson *et al.*, 2004; Pinzon *et al.*, 2004).

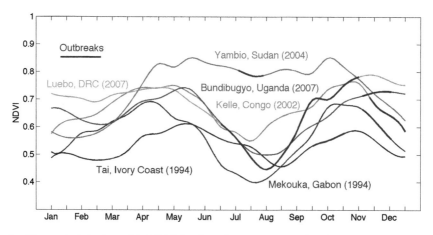

Figure 2.   Time series behaviour of the NDVI data from the documented outbreak sites of EBOV HF. Outbreak periods are denoted by the thick brown sections of the timelines. Note that the outbreaks tend to occur toward the middle of the second dry season. (see colour plate 35)

2.3   Hypothesis for future filovirus outbreaks

Using the arguments so far presented, and given what is known about filovirus emergence, a team of researchers at the Goddard Space Flight Center has formulated a strategy to model proxy indicators of EBOV HF risk, and to characterize areas endemic to this infectious disease. NDVI was used to characterize the vegetation type(s) associated with EBOV HF outbreaks and the temporal and spatial vegetation patterns in the wet and dry season during those events. Landsat data reveal that all reported Ebola haemorrhagic fever outbreaks occur in either tropical moist forests or gallery tropical forests in a savannah matrix (Tucker *et al.*, 2002). Riverine forest can be detected through classification of satellite imaging (DEM, NDVI, temperature), and one can track their wet/dry conditions to predict increase of contact, assuming that this determines high risk. In other words, environmental conditions affect population dynamics, but more importantly they affect behaviour that, in turn, triggers contact (possibly violent) when food is scarce. So in this case, one can develop an eco-climate-driven early warning system for filovirus by monitoring dry conditions that predict increased contact (risk) indirectly with the pathogen.

If one assumes that bats are the link to follow for modelling and predicting ZEBOV, the question becomes, what is known about the ecology and diet of the three frugivorous bat species suspected of being the reservoir hosts? They are widely distributed in equatorial Africa and found in riverine forests at elevations less than 1800 meters (where ZEBOV outbreaks have been reported). Figs are their preferred food, but they, along with the great apes, also eat mangos and bananas. According to Langevin & Barclay (1990), while male bats may forage over distances of up to fifteen kilometres to locate good quality food, female bats rely on established feeding routes that offer a constant supply of lower quality food.

The information gained from this categorization enabled the team to use extraction methods, such as SVD to identify specific environmental features (or constellations of features) associated with outbreak sites (Figure 2). This restricted the eco-climatic analysis to only tropical moist forest or gallery tropical forest areas in Africa.

The main result of this first step was the identification of the Gabon-Central African region as an important epicentre of ZEBOV transmission in Africa (Figure 3). A first step towards *smart surveillance* is more targeted detection and analysis of what is causing filovirus emergence in these specific areas (Figure 4). Several questions attend the analysis: Why is there a difference between dry seasons and animal behaviour? What is the difference in transmission opportunity between those two dry seasons in a year, and across different years? What is the nature of interaction between

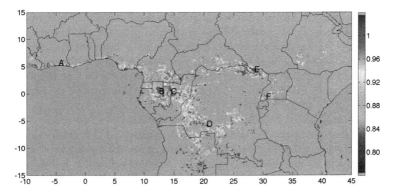

Figure 3. Regions where EBOV HF is endemic. The colour bar at right shows levels of ecological similarity between known Ebola outbreak sites and other areas in central Africa. Red (1) = high ecological similarity. Blue (0.1) = low ecological similarity. Sites of known Ebola outbreaks are denoted as: A = Tai, Côte d'Ivoire; B = Mekuoka, Gabon; C = Kelle, Congo; D = Luebo, Dem. Rep. Congo; E = Nzara, Sudan; and F = Bundibugyo, Uganda. Grey areas: areas with low ecologic similarity (<0.75). Red and orange areas outside the known outbreak sites are possible future *Ebola hot zones* (Updated from Pinzon *et al.*, 2004). (see colour plate 36)

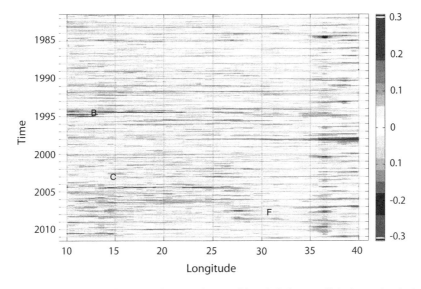

Figure 4. Longitudinal Hovmoller anomaly NDVI image of the *Ebola hot zone* linked to outbreak sites B, C and F from Figure 3. Colour bar at right denotes NDVI readings. Green denotes above average NDVI levels, or rain. Brown denotes below average NDVI levels, or no rain. The linear coloured markings over a geographic area (width) and time (height) show markedly drier environmental conditions in 1991, 1994, 2000–2001 and 2004. (see colour plate 37)

the biology and the behaviour of the reservoirs and their associated eco-climate interactions; or, is there a combination of drivers of the observed temporal scales causing both the longer twenty year cycle of outbreaks and a shorter cycle of sporadic cases?

Figure 3 confirms the drier conditions between 10° and 15° South latitude in Africa. Figure 4 shows that EBOV HF was dormant for about twenty years until the end of 1994, when three outbreaks started almost simultaneously. Since CST has been related to changes in fruit availability and dietary composition during environmental hardship, one wonders if virulence is governed solely by the reservoir, or possibly also by seasonal variations in the fruit attractant or its nutritional quality that affect exposure risks or susceptibility of primates to viral infection? This question needs to be addressed by future research into reservoir hosts and infected animals. Another intriguing question is whether the reservoir hosts are in fact similar in filovirus transmission capability; and if so, do all outbreaks follow a similar temporal pattern driven primarily by host factors in association with eco-climatic trends and conditions? The complexity of this question resides in the different ecological and temporal domains involved. If there is a common thread running through this biotic complexity, is there an underlying abiotic or environmental condition or trend, or convergence of conditions and trends that better explain what has been observed to date: filovirus outbreaks or sporadic cases? This variation suggests more than one *environmental temporal frequency*. Could the favourable conditions for outbreaks occur at different temporal scales, at different frequencies, from local through regional to multi-regional variability? Environmental processes occurring at the local level could affect the rates of virus exposure and transmission, including human and animal behaviour (e.g. hunting patterns). Human exposure to the risk also increases under these conditions, since bush meat preferences of village-hunters change to a wider array of animals (Walsh *et al.*, 2003; Leroy *et al.*, 2004; Rouquet *et al.*, 2005; Lahm *et al.*, 2007).

Environmental processes can affect regional-scale processes such as habitat conditions (e.g. vegetation, precipitation, and landscape patterns). Multiregional effects could also occur when climatic signals like ENSO and NAO indices influence cycles and patterns. The NAO index is typically measured through variations in the normal pattern of lower atmospheric pressure over Iceland and higher pressure near the Azores and the Iberian Peninsula (Hurrel 1995; Jones *et al.*, 1997). The ENSO has two cycles: strong negative (El Niño events) and strong positive (La Niña events). Several indices have been developed to indicate the onset of ENSO cycles (Glantz 2001) such as sea-level pressure and SOI, along with changes in Pacific SST. SOI is measured by the difference in pressure between Tahiti & Darwin at the opposite ends of the oscillation's seesaw. Wolter & Timlin (1998) combined SOI and SST along with four more observed variables over the tropical Pacific to create a more robust index called the MEI. These processes influence the risk of contact with the pathogen and occur at different spatial scales and temporal frequencies.

Recasting the hypothesis more specifically, seasonal forecasts of filovirus epidemics based on climate and vegetation patterns can be available for specific regions at different levels of accuracy. Their forecast lead-times vary according to inherently different scales of spatial, temporal, and biological vegetation parameters. Implicit in this assumption are the essential characteristics of the CAS system: heterogeneity, nonlinearity and hierarchical structure. Even though a linear approach is the traditional method to identify trends and oscillations in the structure from noise in a dataset (i.e. Fourier analysis-based filtering in frequency space), the Fourier method is not applicable here. This is because natural variations (CAS systems, in general) involve nonlinear processes and thus violate the underlying Fourier analysis assumptions of stationarity and linearity. As an alternative, the newly developed adaptive time domain EMD is proposed to study the nonlinear structure in filovirus epidemics (Huang *et al.*, 1998, Huang *et al.*, 2009; Wu & Huang 2009). EMD is a hierarchical approach that uses nonlinear methods to explore the observable interactions among the integrated physical and biological systems involved at different (temporal) scales. EMD is designed to seek these different intrinsic modes of oscillation within the data, based on the principle of local scale separation. Decomposing the NAO signal and Gabon-Democratic Republic of Congo NDVI-based EBOV HF risk into their different intrinsic EMD modes of oscillation, reveals a three year oscillation component of the NDVI-derived EBOV HF risk swings. One also sees a fifteen to twenty month lag with a three year NAO component that itself rides over a ten year mode which appears to be based on a simultaneous twenty years cycle (Figure 5). These components could be part of the mechanism for the observed twenty year cycle; perhaps explaining why filovirus outbreaks have been so erratic in their timing and location. If one analyses these compatible, observable swings, one could discern with a lead-time of several seasons, the physical mechanisms that presage the inter-annual temperature, precipitation and vegetation pattern that promotes filovirus transmission potential. Spatial uncertainty (scale dependent) can be improved while keeping robustness in the risk prediction by analysing shorter temporal cycles at higher spatial resolutions (Figure 4). This

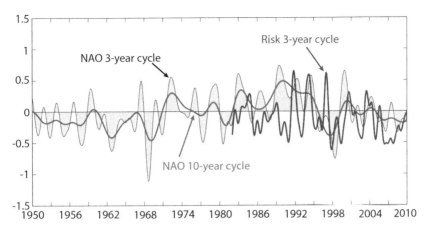

Figure 5. Decomposition of NAO signal and *Ebola hot-zone* NDVI-based EBOV HF risk into their different intrinsic EMD modes of oscillation. A 3-year oscillation component of the NDVI-derived EBOV HF risk swings in a 15–20 month lag with a 3-year NAO component that rides over a 10-years mode modulated simultaneously by a 20-year cycle.

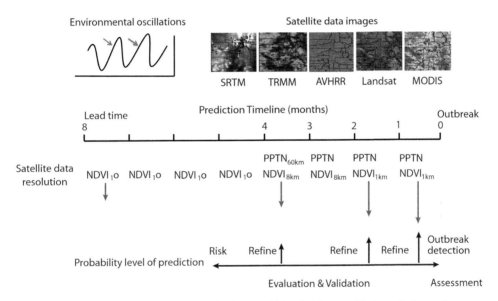

Figure 6. Model prediction framework: Developing a multi-level risk map with dynamic decreasing uncertainty and increasing temporal and spatial accuracy that reinforce expandability and accessibility. (see colour plate 38)

would require the integration of field observations data (epidemiological, animal and vector data collected from active or passive surveillance, or field surveys) into modelling the (local) explanatory mechanisms as well as validation tools (landscape epidemiology) that guarantee the robustness.

Interpreting the scientific content of satellite data and images, and predicting infectious diseases is challenging since both processes have accumulated uncertainties. One uncertainty is risk assessment, which depends on the level of understanding of the relevant science; another is the nature of associated, local risk factors; and, a third affects the types of decisions the risk assessment might influence. Nevertheless, once the model template is organized (Figure 6), a more concise context for decision making to explore scenario assumptions is attainable for risk management.

The concept of risk adopted in the prediction is the likelihood that some event will occur; when it is used in decision making, it translates into the magnitude of the consequences of that event when specific scenario assumptions are considered. In other words, there is a need for a versatile way to report this scientific and technological information to foster a dialogue between scientists and non-specialists (decision makers, citizens, consumers) regarding risks, decision costs, and decision-tool improvements.

In summary, monitoring climate factors based on the association of disease inter-annual cycles (local interactions) and linking them to climate variables at the regional and multiregional scales could be among the easily implemented tools to best manage and control filoviruses. Moreover, monitoring wildlife population dynamics and their associated environmental factors may help identify sentinels for infection threats and help in the investigation of natural reservoirs for EBOV and MARV. Such studies would allow a better understanding of some of the CST mechanisms of filoviruses, perhaps shedding light on their natural history. They may also serve as a future testing ground for educating local communities about filovirus risk factors, improving case detection, and developing a rapid, timely, and effective outbreak response for organizations like the WHO.

3 RICKETTSIA: THE UBIQUITOUS EMERGER

Rickettsiae are Gram-negative, obligate intracellular bacteria that cause some of the oldest vector-borne diseases known to humankind. Throughout history their associated diseases have been lumped together as 'typhus', and under this rubric have wreaked havoc on the world's population. It is estimated that since ancient times, epidemic typhus has probably caused more deaths than all injuries from all the wars in history (Kelly *et al.*, 2002). We now know that as aetiological agents, rickettsiae can be classified into two main groups based upon serologic and genetic typing: the typhus group that includes *Rickettsia prowazekii* (epidemic typhus) and *Rickettsia typhi* (murine typhus); and the larger, more diverse, spotted fever group. The spotted fever group comprises over twenty known species including the pathogens *Rickettsia rickettsii* (Rocky Mountain spotted fever) and *Rickettsia conorii* (Mediterranean spotted fever).

Rickettsiae comprise a heterogeneous family of bacteria. Their unifying characteristic, however, is that they all must live in eukaryotic cells. Since it is known that forms of eukaryotic cells inhabit almost every environment on Earth, it follows that rickettsiae probably inhabit most of the world's environments. If sought, they should be found, and evidence shows that this has been the case. Sixteen new tick-borne spotted fever rickettsiae (including subspecies) have been described since 1984 (Figure 7). Prior to 1984 only four spotted fever group rickettsiae were known (Parola *et al.*, 2005; Socolovschi *et al.*, 2009).

3.1 *Emergence and/or increased recognition of rickettsiae*

Increased recognition of rickettsiae as an emerging constellation of bacteria has several reasons. First and most likely, is that since 2000, molecular tools such as polymerase chain reaction (PCR), restriction fragment length polymorphism (RFLP), and genetic sequencing technologies have enabled discovery of genomic distinctions between different rickettsial species and strains. Prior to these technologies, most rickettsioses could only be identified by using non-species specific sero-logic tests such as indirect immunofluorescence assays (IFA) and enzyme-linked immunosorbent assays (ELISA). These were useful in showing that a disease's aetiological agent belonged to either the spotted fever group or typhus group of rickettsioses, but they could not distinguish between species or subspecies. In addition to serologies, rickettsioses could be diagnosed by isolating and growing the rickettsiae from clinical samples in cell cultures, an extremely time consuming prac-tice taking weeks and prone to contaminations and failures. Successfully cultured rickettsiae were often classified according to their staining properties and their behaviour in *in-vivo* virulence and pathogenicity challenges; again, time consuming and imprecise. The eukaryotic requirement for rickettsial growth, and the stringent extra-cellular conditions necessary for rickettsiae to remain

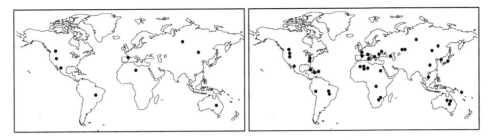

Figure 7. Regional geographic locations of known spotted fever group Rickettsiae species (subspecies are not shown). Left: Before 1984; Right: After 1984 (T. Myers, personal communication).

viable outside living cells, make it nearly impossible to use classical biochemical analyses like glucose and sucrose metabolism characteristics (normal techniques to identify bacteria) to identify and distinguish rickettsial species and strains. These conditions inhibited investigators' ability to identify or look for an unknown rickettsia's presence, or to discover its identity beyond a general, group-level classification. Modern molecular tools such as PCR and sequencing now allow investigators to clearly differentiate between rickettsial species.

A second putative driver of rickettsiae emergence is a growing awareness that they can establish themselves in essentially any environment using humans and animals as hosts. These vertebrate-host transition vectors transport rickettsia either directly as blood-borne infections, or indirectly by harbouring rickettsia-infected arthropod parasites. Once brought into an area, they become established and spread locally to indigenous arthropods when bitten by the introduced, infected vertebrate-host. Rickettsiae can also be transmitted and achieve establishment through the biting of indigenous vertebrate-hosts by the imported arthropod vector. If the rickettsia's transmission and establishment are sustainable, disease outbreaks emerge – the expression of successful convergence. A good case in point is the discovery of *Rickettsia africae* in the Caribbean (Kelly 2006). *Rickettsia africae* is carried by the tick *Amblyomma variegatum*. This tick was introduced from Senegal onto Guadeloupe in the early 1800s via imported livestock. From there it spread widely across the Caribbean, especially during the last forty years with increasing livestock trade between the islands (Barre *et al.*, 1996). This trade-associated spread probably led to the identified emergence of human spotted fever cases around the West Indies (Kelly 2006). Besides intentional transportation, rickettsial emergence can result from changes in arthropod vector/host location as vertebrate hosts migrate into new habitats looking for food or safer havens associated with land use/land cover changes. These dynamics introduce rickettsial species into new areas, and can eventually change the environment's regulatory processes (suppression or promotion) so that rickettsia can establish itself and emerge as a new endemic pathogen.

Finally, and most relevant for this discussion, the emergence of diverse rickettsiae species can be a reflection of the complex relationship between the rickettsiae themselves, their vectors, hosts, and global ecological niches. Rickettsiae probably have diversified evolutionarily through time in response to changes in hosts and environmental conditions. Environmental factors both within and outside the arthropod affect the likelihood of rickettsial survival and propagation. They also influence rickettsiae's virulence and ability to cause disease outbreaks in humans and animal hosts. Rickettsiae within ticks are subjected to temperature shifts, starvation, hemolymph osmotic pressure, pH changes, and varying oxygen and carbon dioxide levels (Needham & Teel 1986; Munderloh & Kurtti 1995; Lighton *et al.*, 1993; Socolovschi *et al.*, 2009). Studies have suggested that temperature differences within the tick affect greatly the activation and virulence of rickettsia. *R. rickettsii* organisms are less viable in *Dermacentor andersoni* held at 4°C than when the ticks are maintained at 21°C (Niebylski *et al.*, 1999). Since ticks have no internal temperature regulation, their body temperature fluctuates when ambient temperature changes (Niebylski *et al.*, 1999). It is also demonstrated that rickettsiae tend to be dormant when found in unfed ticks. The rickettsiae can reactivate and proliferate however, if those same ticks ingest a blood meal (Munerloh *et al.*, 2005; Socolovschi *et al.*, 2009).

Recognition that some rickettsiae, like *R. rickettsii*, are quite promiscuous with regard to their tick hosts complicates scientific understanding of their emergence and transition pathways. They can be found in several ticks, such as *Dermacentor variabilis, D. andersoni, Rhipicephalus sanguineus*, and *Amblyomma cajennense*; whereas other rickettsiae species, such as *R. conorii*, are quite host specific (tick host – *Rhipicephalus sanguineus*). Conversely, many tick species have different host preferences. This influences the distribution of their associated rickettsiae. For example, *Amblyomma hebraeum* ticks, the principle vector for *R. africae* (African tick-bite fever), have a high affinity for humans. They flourish mainly in the rural peri-domestic areas of Africa and the Caribbean where opportunities to feed on humans and their livestock exist (Raoult *et al.*, 2001). *R. rickettsii's* dispersion potential is greater due to the diversity of its arthropod vectors' vertebrate hosts. At the other end of the spectrum, *R. conorii* seems to restrict itself to *Rhipicephalus sanguineus* ticks, which have a strong host preference for dogs. Dogs in turn, often bring these ticks and therefore *R. conorii* rickettsiae into urban areas. Because of its strong preference for dogs, *R. sanguineus* will only attach itself to humans in the absence of the dog host. Once introduced into an urban area, if dogs are not numerous enough to sustain these ticks over time, *R. sanguineus* can, and will, feed on humans instead. This can result in the emergence and spread of human *R. conorii* (Mediterranean spotted fever) infections (Parola *et al.*, 2005).

Finally, the rickettsiae, via their arthropod vectors and vertebrate hosts, are influenced by climate and other ecological factors in their environments. For example, rickettsiae and their vector and biological host activities and environmental responses vary seasonally and as a result of abrupt or evolutionary changes in land cover, moisture and sunlight levels. As these parameters change individually or consequentially, they influence the direct rickettsiae-vector relationship and the indirect rickettsiae-host relationship. Ultimately, changes in both relationships correlate to a direct impact on the rickettsiae's presence, abundance and biologic activities. If both relationships are protective, because of its intracellular requirements, the rickettsiae can be brought into, become established, and survive wherever eukaryotic cells are found. As scientists discover these cells in new, even if extreme, environments using enhanced rickettsiae detection and identification techniques, health professionals should expect to find these ubiquitous emergents as well. The ubiquity of rickettsiae is therefore explicable by the breadth of its transition pathway; rickettsiae's plastic ability to utilize different transition vehicles opportunistically, and adapt to different and changing environmental conditions and trends because of its intracellular habitat and biotic characteristics.

4   DENGUE: URBANIZATION AND ROLE OF CLIMATE IN TRANSMISSION

4.1   *Emergence and spread of Dengue*

Transmission of dengue is dependent on the presence of mosquitoes (*Aedes spp*), especially *Ae. aegypti*, and an adequate host population to sustain it. Urbanization, tempered by the influence of socioeconomic status, is the quintessential driver of dengue transmission in climates that support *Aedes spp*. Urbanization provides a large, dense host population and breeding sites for the vector. Climates that support survival of *Aedes spp* are essential and may impact dengue transmission due to their effects on the virus, vector and vector replication, and human behaviour. Humans may provide peridomestic climates suitable for *Aedes*. The effects of climate change, while of great concern, are probably far less important than urbanization and other processes resulting from human behaviour. Climate and urbanization issues are discussed for Nepal in the context of how they enable the spread of endemic dengue.

Infection with a dengue virus may be: asymptomatic; cause a nonspecific viral syndrome, a febrile illness known as dengue fever (DF); or less commonly, cause dengue haemorrhagic fever (DHF), distinguished not by haemorrhage but by a vascular leak syndrome. There are four serotypes of dengue virus (DENV1-4) belonging to the genus *Flavivirus* in the family Flaviviridae. DENV is transmitted primarily by *Ae. aegypti* but also *Ae. albopictus*. The disease is especially prominent in tropical and subtropical areas where *Ae. aegypti* has been successful in an ecological niche with humans. There are an estimated thirty-six million cases of DF, 2.1 million cases of DHF, and 21,000 deaths annually. Case numbers are increasing in severity, and the geographic range is spreading. Approximately 3.6 billion people (fifty-five per cent of the world's population) in 124 countries are at risk for dengue transmission. The paediatric Dengue vaccine initiative (PVDI) has been inaugurated (PVDI 2012).

Because dengue is dependent on vector transmission, the feeding patterns, longevity, and reproductive biology of *Ae. aegypti* almost certainly influence the force of infection and the rate at which susceptible individuals become infected (i.e. rate of transmission). *Ae. aegypti* feed preferentially on humans, live in or near residential areas, and lay eggs in artificial water containers in and around homes (Jansen & Beebe 2010). The flight range of *Ae. aegypti* is only 100–200 meters, which emphasizes the importance of population density and movements in dengue transmission. After the mosquito takes a blood meal that contains DENV, the extrinsic incubation period of the virus in the mosquito is typically eight to ten days. During this period, the virus replicates in the gut of the mosquito and moves to the salivary gland, after which it can transmit the virus to another person while feeding or probing. It is unclear if the virus can be transmitted mechanically by a mosquito directly by feeding or probing on a viremic person and then transferring the virus directly to another person without incubating the virus – akin to a needle stick – but this would not be a major means of transmission. *Aedes* mosquitoes have been described as cautious feeders. The slightest disturbance will upset its feeding such that it may feed on three to five persons for one gonotrophic cycle and virus can be transmitted to the host with each probing (Gubler & Rosen 1976). This process, combined with potentially multiple gonotrophic cycles, can increase greatly the number of people exposed to the virus.

The range of dengue endemic and epidemic regions of the world has expanded in recent decades, making the virus the most common arboviral infection in the tropics and subtropics; and it has become an enormous public health threat (WHO 2009) (Figures 8 & 9) This growth depends on the expansion of the vector's habitat enabled by population growth and urbanization, lack of effective vector control programs, poor public health infrastructure, increased international travel, and perhaps gradual climate change (Figure 9).

By 2008 there were more people living in urban, rather than rural environments. Between 2000 and 2030 the world's urban population is expected to increase by seventy-two per cent and the geographical areas of cities of more than 100,000 people could increase by 175 per cent (Martine 2007). Urbanization in permissive climates represents a major driver of dengue transmission; however, its role is moderated by socioeconomic status and human behaviour. From 1906 to 2005, global average temperature increased by 0.74°C, and the rate of change seems to be accelerating (Patz *et al.*, 2008). However, the possible influence of climate warming on expanding the areas of dengue transmission via the expansion of *Ae. aegypti* is not straightforward. The present distribution is considerably smaller than historically, when it included southern Europe and the eastern US, northward; and larger areas of Australia and South America, southward (Reiter 2001; Jansen & Beebe 2010).

4.2 *Effects of temperature and precipitation*

Temperature has been shown to have direct effects on both *Ae. aegypti* and dengue viruses (Focks *et al.*, 1993; Patz *et al.*, 1998; Patz & Reisen 2001; Malla *et al.*, 2008; Enrique Morales Vargas *et al.*, 2010). Throughout the life cycle of the mosquito, temperature impacts development, longevity and reproductive rate (Yang *et al.*, 2009). The aquatic stages of the mosquito (egg, larval, and pupal) have lowest mortality between 15°C and 35°C. The transition rate is optimal at 25°C, with no progression from larvae to pupae above 40°C. The female adult mosquito mortality rate is lowest between 15°C and 30°C, with the highest oviposit rate at 26°C and the largest amount of offspring at 29°C (Yang *et al.*, 2009). Warmer temperatures reduce larval size and adult size, resulting in adults needing to feed more often to develop an egg batch (Patz *et al.*, 1996; Enrique Morales

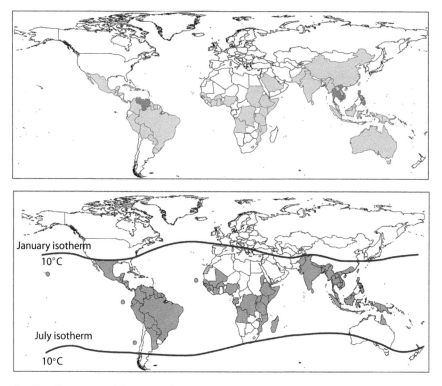

Figure 8.   Top: Emergence of Dengue and Dengue Haemorrhagic Fever. Darker grey is prior to 1960; light grey is after 1960. Bottom: Grey areas between the January and July isotherms represent countries or areas where dengue has been reported; The January and July 10°C isotherms define the geographical limits for year-round survival of *A. aegypti* as of 2006 (Courtesy WHO 2009).

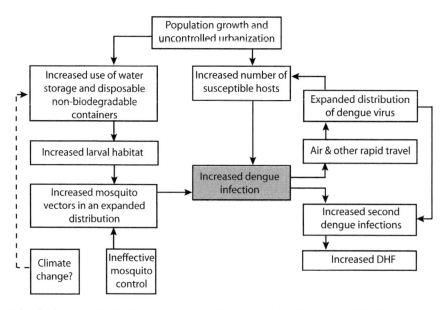

Figure 9.   Environmental and anthropogenic mechanisms potentially affect dengue infection rates.

Vargas *et al.*, 2010). As temperatures increase, the mosquito population can increase substantially (Focks *et al.*, 1993; Focks *et al.*, 2000; Malla *et al.*, 2008). *Ae. aegypti* mosquitoes have been documented to increase biting habits during warmer temperatures (Pant & Yasuno 1973; Focks *et al.*, 1993). The ratio of female to male mosquitoes has also been noted to increase at warmer temperatures (Tun-Lin *et al.*, 2000). As demonstrated in the laboratory, temperatures above 40°C would theoretically reduce mosquito survival; but, this might be offset by the ability of *Ae. aegypti* to exploit cooler microenvironments such as indoors, underground, in cement water tanks, coolers, and under vegetation that are common habitats in the built environment (Tyagi & Hiriyan 2004; Sharma *et al.*, 2008; Jansen & Beebe 2010).

Dengue viruses are also directly affected by temperature changes. The extrinsic incubation period in the mosquito for DEN-2 is shortened from twelve days at 30°C to seven days as temperature increases to 32–35°C. A shorter extrinsic incubation period means more mosquitoes reach the capacity to infect humans (Focks *et al.*, 1993; Watts *et al.*, 1987; Malla *et al.*, 2008). In addition, as temperatures and humidity increase, relative humidity has been shown to increase virus propagation (Patz *et al.*, 1998; Thu *et al.*, 1998).

Increases in temperature and humidity have been correlated with increased dengue transmission. In Taiwan, average temperatures higher than 18°C were associated with increased DF incidence, with every rise of one degree centigrade increasing the risk of DF transmission by 1.95 times (Wu *et al.*, 2009). In Puerto Rico, increases in temperature and rainfall correlated significantly with increases in dengue transmission over a twenty year period (Johansson *et al.*, 2009a); and in Thailand, ENSO contributed a fifteen to twenty-two per cent increase in dengue cases following those seasons (Tipayamongkholgul *et al.*, 2009). In these studies it is difficult to know whether the increased transmission is due to temperature and humidity alone, or to other unmeasured variables. Another study shows that while multiyear climate variability may play a role in endemic inter-annual dengue dynamics, there is no evidence of a strong, consistent relationship (Johansson *et al.*, 2009b). The role of ENSO may be confounded by local climate heterogeneity, insufficient data, random outbreaks, and other potentially stronger, intrinsic factors affecting transmission.

4.3 *Urbanization*

The mosquito vector of dengue has long exploited human migration to urban centres. In the 18th and 19th Centuries when global shipping increased and port cities expanded, both *Ae. aegypti* and DENV expanded into new geographical areas (Smith 1956; Gubler 2006). The worldwide expansion of dengue increased dramatically during and after World War II commensurate with large troop movements and the dislocation of resident populations into make-shift, temporary cities (Halsted 1984; UN 2005).

Unplanned and rapid urbanization can lead to dramatic changes in dengue transmission (Watts *et al.*, 1987; Nakhapakorn & Tripathi 2005; Petersen & Marfin 2005; Cox *et al.*, 2007; Kyle & Harris 2008; Rohani *et al.*, 2009; Wu *et al.*, 2009). Urbanization can lead to increased exposure to *Ae. aegypti* throughout the human population, especially when sanitation services are overwhelmed and breeding sites emerge in garbage-infested areas and water storage vessels. The *Aedes* mosquito is anthropophilic and thrives in built environments with a high human density and multiple accessible breeding sites. Urbanized areas provide a wide variety of potential breeding habitats for *Ae. aegypti*, one being potable water tanks (Garelli *et al.*, 2009). In poorer neighbourhoods, there is an increased need for water storage in the form of cisterns and rain barrels to guarantee an adequate water supply. In all neighbourhoods there are numerous small reservoirs of stagnant water, such as flowerpots, gutters, and fountains. Studies have documented that the lack of water supply and unreliable removal of solid waste contribute to an increase in dengue transmission (Barrera *et al.*, 1995; Caprara *et al.*, 2009).

Increasing numbers of dengue outbreaks, and circulating serotypes are occurring even in semi-arid and arid regions like Saudi Arabia, India, and Australia (Fakeeh & Zaki 2001; Tyagi & Hiriyan 2004; Sharma *et al.*, 2008; Beebe *et al.*, 2009). Ironically, in Australia there is concern that the reintroduction of *Ae. aegypti* will occur during droughts as a result of installing large domestic water storage tanks. Habitat modelling in south eastern Australia suggests that these tanks could result in an expansion of dengue and an increase in risk of transmission. It appears, therefore, that any location where there are vectors and dense human populations are susceptible to dengue transmission.

4.4   *Causes of Dengue Emergence in Nepal*

The first case of dengue fever in Nepal was diagnosed by serology and reported in 2004 from the Terrai region bordering India (Pandey *et al.*, 2004). Nepal's borders are open to India where dengue is well established. In 2006, the Nepal National Public Health Laboratory and the Armed Forces Research Institute of Medical Sciences (AFRIMS) in Thailand identified all four serotypes of dengue in Nepal, with twenty per cent of cases coming from Kathmandu (Malla *et al.*, 2008). These patients had not travelled to the Terrai region, and it is therefore assumed that the vector and DENV expanded its geographic range to Kathmandu. Investigators found *Ae. aegypti* mosquitoes in the transmission region, which previously had not been documented. Another study testing paired sera from febrile patients in Kathmandu in 2006 also found eight per cent positive for dengue immunoglobulin-M (IgM) antibodies (Blacksell *et al.*, 2007). What factors may have contributed to the emergence of dengue in this country? Certainly, capability for dengue laboratory testing is not available at most hospitals, and a majority of febrile illnesses go undiagnosed, so there could have been circulating cases for many years prior to the detection of this first case. However, a serosurvey testing for dengue IgM antibodies in 876 patients suffering from a fever in Kathmandu in 2001 found no evidence of dengue (Murdoch *et al.*, 2004).

In Nepal, the concern over urbanization and climate change are supported by data. Between 1980 and 1993, Nepal's urbanization rate of 7.7 per cent was the highest in South Asia (UN 2002). At the 2001 census, Nepal had an estimated annual population increase of 2.3 per cent, which will almost certainly increase through time. It has also suffered from the consequences of rapid, unplanned urbanization; namely poor sanitation and waste disposal (Pantha & Sharma 2001; UN 2002). Increased precipitation and gradually rising average temperatures in these areas produce rich breeding grounds for mosquitoes. The Nepal Department of Hydrology and Meteorology reported that average temperatures in Nepal are increasing approximately 0.06°C per year (Rai 2007). These warmer temperatures have been linked to increased melt water from glacier systems in the Himalayas. In terms of DENV ecology, these environmental changes have led to increased water, flooding, alteration of precipitation, sedimentation, and drought (Focks *et al.*, 1993; Karki 2007; Chhetri & Pandey 2009). Warmer temperatures have led to an increase in the habitat range for *Ae. aegypti* in many regions of Nepal that affect Nepalese living in poorer districts where piped water supply is inadequate (Hales *et al.*, 2002).

While global climate warming trends have been implicated in the spread of DENV and other mosquito vectors like West Nile virus, Japanese encephalitis virus, Rift Valley Fever virus, and the malaria parasite, there are other possible causes for the geographic expansion of these diseases (Brower 2001; Weaver & Reisen 2010). Interactions between the vector, the pathogen, and the environment are complex; and, it is difficult to determine which factor, if any, is primarily responsible. Climate warming allows vectors to live, and to live longer, in regions where they could not survive previously. Although temperature increases have been documented in Nepal, the Terrai area where dengue was first diagnosed has always had a suitable climate to sustain the *Aedes* mosquito. Seasonal transmission cycles in tropical areas almost always occur during the cooler, rainy season, demonstrating that increasing temperature alone will not cause an increase in dengue transmission.

An increase in precipitation may contribute to dengue fever emergence; but in drier areas the human response to drought or shortages of potable water is to build water storage facilities (Patz *et al.*, 1998; Caprara *et al.*, 2009). Thus, the urban environment itself seems to impact dengue transmission (Barclay 2008; Wilder-Smith & Gubler 2008). Another important consideration is that during the ten year civil war, large numbers of internally displaced people moved to urban centres like Kathmandu. Many of these remain displaced and are living in slums without running water, waste management, or vector control.

While urbanization may play the most significant role in dengue expansion, international travel has been thought to contribute most to the introduction of the virus to new locations. India has had a long history of dengue transmission, starting in the south and moving northward. The border between India and Nepal is open, with frequent commercial and personal travel. It is likely that the dengue virus and the *Ae. aegypti* mosquito were introduced into Nepal through travel from India. The *Ae. albopictus* mosquito has long been present in Nepal (Gaunt *et al.*, 2004), and can also contribute to the spread of the newly introduced dengue virus.

Dengue intensity, severity, and distribution are increasing alarmingly. It is tempting to attribute this to global warming, but the anthropophilic nature of *Ae. aegypti* and its ability to exploit microenvironments make urbanization a more important factor than climate. It should also be remembered that the historical distribution of dengue has included areas without current endemicity or recent epidemics. There are no simple explanations. Urbanization will continue without adequate public health infrastructure for water and sanitation. Vector control programs are expensive, often difficult to implement, and particularly difficult to sustain. No available licensed vaccine exists to prevent dengue fever, though several are in development.

While prevention strategies continue, including vector control, education, personal protective measures, and vaccine development, surveillance for potential cases should be implemented in newly endemic areas to track the spread and impact of the disease. Unfortunately, active case findings with laboratory confirmation will likely exceed the financial and workforce capacity of many endemic countries. In areas that are heavily endemic for dengue, simply tracking cases of fever during the transmission season can provide a way to detect potential dengue outbreaks (Nakhapakorn & Tripathi 2005). In a newly endemic country, dengue cases may be overshadowed by other febrile diseases. However, utilizing a syndromic system which incorporates symptoms more likely seen with dengue, such as headache, myalgias (muscle pains), arthralgias (joint pains), retro-orbital pain (headaches, migraines), and signs of vascular compromise, may help to target potential dengue cases and warn of significant increases, especially during normal transmission seasons (Jefferson *et al.*, 2008; Meynard *et al.*, 2008; Randrianasolo *et al.*, 2010). A percentage of those with otherwise undiagnosed febrile illnesses that meet a case definition could then be tested to track the burden of dengue infection and circulating serotypes. As a population is exposed to more than one dengue strain, hospital sentinel systems should also be in place to monitor for severe disease and death, with immediate investigation of all haemorrhagic fever.

5 NIPAH VIRUS IN MALAYSIA: THE ROLE OF AGRICULTURAL PRACTICES

5.1 *Two scenarios in the transition pathway to emergence*

Nipah virus (NiV) is a lethal zoonotic pathogen that has caused repeated outbreaks in South Asia since 2001. The reservoir hosts for this paramyxovirus are *Pteropus spp.*, fruit bats (also called flying foxes), which are abundant throughout the tropical Asian-Pacific region. Neutralizing antibodies to NiV have been found in *Pteropus vampyrus* and *P. hypomelanus* across Malaysia (Johara, 2001; Daszak *et al.*, 2006). NiV has been isolated from both bat species (Chua *et al.*, 2002a; Rahman *et al.*, 2010), and infection appears to be common throughout their range, including Cambodia (Olson *et al.*, 2002; Reynes *et al.*, 2005), Thailand (Wacharapluesadee *et al.*, 2005; Wacharapluesadee *et al.*, 2010), Indonesia (Sendow *et al.*, 2010; Breed *et al.*, 2011), India (Epstein *et al.*, 2008) and Bangladesh (Hsu *et al.*, 2004; Inst. Epid. 2004).

Nipah virus was first identified in peninsular Malaysia and Singapore during the 1997–99 outbreak in pigs and humans (MMWR 1999; Chua 2003). There were 268 human encephalitis cases reported and a case fatality rate of thirty-nine per cent (Goh *et al.*, 2000; Chua 2003). In Malaysia, NiV was recovered from the oropharynx, respiratory secretions, and urine of infected patients (Chua *et al.*, 2001). Contact with sick pigs was the primary risk factor (Parashar *et al.*, 2000), with the initial spill-over traced to a large commercial pig farm, on which NiV-infected pigs

Table 1. Time and location of Nipah outbreaks in Bangladesh: Number of cases and mortality.

Year	Location	Cases	Deaths
2001	Meherpur	13	9
2002	(no cases)	0	0
2003	Naogaon	12	8
2004	Rajbari, Faridpur	67	50
2005	Tangail	12	11
2006	(no cases)	0	0
2007	Thakurgaon, Kushlia, Pabna, Natore, Naogaon	18	9
2008	Manikgonj, Rajbari	11	9
2009	Rangpur, Gaibandha, Rajbari, Niphamari	4	1
2010	Faridpur, Rajbari, Gopalgonj, Kurigram	19	17
2011	Lalmonirhat, Dinajpur, Rangpur, Nilpahmari, Rajbari, Comilla	28	28
Total		184	142

developed fever, neurological and respiratory signs, including a loud, barking cough. There was only one case of likely person-to-person transmission. It was a nurse who cared for hospitalized NiV infected patients. The nurse had serologic evidence and MRI findings characteristic of NiV infection (Tan & Tan 2001). The human outbreak of Nipah infection ceased when pig farms in the region were de-populated (Chua *et al.*, 2000). The origins of the Malaysian outbreak were attributed to increased indirect contact between fruit bats and pigs after the planting of fruiting trees that encouraged bats to congregate directly over pig sties (Field *et al.*, 2001; Johara *et al.*, 2001; Chua *et al.*, 2002).

Several characteristics distinguish Nipah virus outbreaks in Bangladesh from that seen in Malaysia. In Bangladesh since 2001, over ten outbreaks and numerous sporadic cases of NiV have been recognized, all occurring between December and May (Chua *et al.*,2004; Inst. Epid. 2004; ICDDRB 2004; ICDDRB 2005; Luby *et al.*, 2006; Gurley *et al.*, 2007) (Table 1 & Figure 10). These have involved 184 human cases, with a primary presentation of fever and central nervous system pathology. Many cases also had respiratory involvement. Of these 184 cases, 142 have died (case fatality rate seventy-seven per cent). The recurring outbreaks have all been smaller than the Malaysian event, and seem to involve direct transmission from bats to humans, without the involvement of livestock (Luby *et al.*, 2006). In Bangladesh, drinking fresh date palm sap, which is collected in open pots from tapped trees at night, appears to be the principle source of human infections (Luby *et al.*, 2006; Nahar *et al.*, 2010).

Bats of multiple species, including *Pteropus giganteus* have been photographed licking sap and urinating in taps placed above pot openings (Khan *et al.*, 2008) (Figure 11). Some epidemiological data suggest potential livestock involvement in spill over events, however, neither NiV antibody nor the virus have actually been detected in intermediate livestock hosts (Luby *et al.*, 2009a). This may be due to the rarity of large herds of livestock in Bangladesh (which would be necessary for long-term transmission of NiV), so it seems domestic animals do not serve as amplifier hosts for NiV in that country.

In addition, in Bangladesh, most NiV outbreaks and sporadic human cases have been geographically distant from one another. The phylogeny of viral isolates suggest that multiple strains are circulating in bats, and taken together with the epidemiological evidence, suggest that each outbreak is caused by a distinct exposure of humans to the virus. Epidemiological evidence also suggests that person-to-person transmission of NiV occurs (Chadha *et al.*, 2006; Gurley *et al.*, 2007; Luby *et al.*, 2009b). From 2001 through 2007 there were ten persons known to transmit NiV to sixty-six other people, most likely through infectious respiratory secretions (Luby *et al.*, 2009b). This contrasts sharply with NiV outbreaks in Malaysia and Singapore, where intensive epidemiological investigations found little evidence of human-to-human transmission.

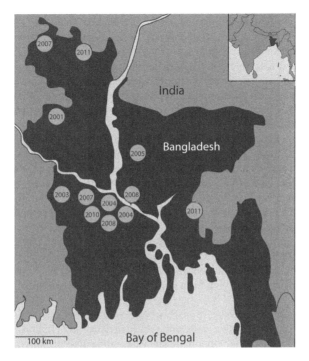

Figure 10. Map of the reported outbreaks of Nipah virus in Bangladesh 2001–2011. Of the 184 reported cases, there were 142 (77%) deaths. Note that some years involve multiple spill-over events.

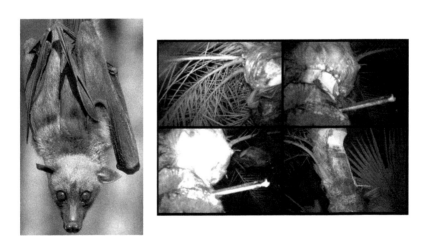

Figure 11. (Left) *Pteropus giganteus*, a major reservoir for Nipah virus in Bangladesh. They are common throughout the Indian subcontinent, Courtesy J.H. Epstein, EcoHealth Alliance; (Right) Infrared images of fruit bats, including *P. giganteus* feeding on date palm sap, Courtesy S.U. Khan, ICDDRB.

5.2 *Scenario 1: The Malaysia experience*

Considerable speculation exists on the origins of NiV in Malaysia. An early hypothesis posited that climatic anomalies during the late 1990s led to atypical movement of infected fruit bats, driving them into new areas, including the region near the index farm (Chua *et al.*, 2002b). These climatic anomalies included drought, resulting from an ENSO event, combined with a haze event, or elevated particulate matter in the atmosphere, from forest fires in Kalimantan, Indonesia, and Sumatra (Chua et al., 2002b). Haze events are a regular occurrence in peninsular Malaysia during dry seasons and have been linked to at least one other significant biological phenomenon, coral die-offs (Abram et al., 2003). Why did NiV emerge at this time? Following the initial NiV outbreak, it was proposed that an ENSO-exacerbated haze event was the principal driver prompting the movement of infected bats into the outbreak area. However, this is refuted by the retrospective diagnosis of human NiV cases on the index farm eight months prior to the elevation in particulate matter (Pulliam et al., 2012).

Subsequent efforts to identify the factors of NiV emergence in Malaysia broadened to include detailed analysis of livestock production records from the index farm, broad-scale data on Malaysian agricultural production, and a five-year serological survey of the main bat reservoirs (Pulliam et al., 2012). This study found NiV antibodies widely distributed in fruit bat colonies across peninsular Malaysia. Seropositive bats were found at all identified roosts and sampling sites. This, when combined with evidence that the P. vampyrus population and number of roost sites have been shrinking since the mid-1990s (Mohd-Azlan et al., 2001; Epstein et al., 2009), and that these bats regularly move between Indonesia, Malaysia, and Thailand in the absence of atmospheric haze, suggest that it is unlikely that emergence occurred due to a novel NiV introduction into the bat population, or expansion of the flying fox population into the area surrounding the affected or index farm in 1997.

Data from FAO and the Malaysian Ministry of Agriculture show that the district where the index farm was located was one of the few areas with very high densities of both commercial pig and fruit (particularly mango) production. On a national scale, production of both commodities increased dramatically from approximately 1970 until the time of the Nipah outbreak. In fact, the closure of piggeries during the 1998–1999 Nipah outbreak not only reduced Malaysian pig production by a third, it also reduced mango production by a third, as mangoes planted around piggeries were left un-harvested (Pulliam et al., 2012). Evidence further suggests that the early human cases associated with the index farm were most likely the result of virus' introduction into a completely susceptible pig population. Mathematical models, based on the size and production parameters of the index farm clearly suggest that initial introduction would have led to a large, rapid epizootic on the farm (and a tight cluster of human cases) increased pig mortality, herd immunity in the survivors, and extinction of the virus within one to two months. The history of the first five human cases in 1997 is consistent with this evidence. However, computational models show that it is extremely unlikely that a single introduction could lead to persistent circulation of the virus on the farm. Therefore, re-introduction of the virus is the most likely explanation for subsequent cases associated with the index farm. Such re-introductions are plausible given that field surveys of fruit bats in Malaysia identified a fruit bat colony within ten kilometres of the index farm, and that the property also had extensive mango, durian, and jackfruit orchards. These orchards attract bats for feeding. Within the eighteen months following the index case, long-term within-farm persistence of small numbers of sporadic NiV cases, both pig and human, occurred and were followed by outbreaks on other farms within the district, then later in the region. In computational models, the re-introduction of NiV supports the field data. These models show that the initial introduction of NiV into a pig population creates a priming effect that allows a secondary introduction to persist in what is now a partially-immune population.

As pigs born during the initial event gradually lose their maternal antibodies, the virus is able to circulate and persist for extended periods similar to those observed at the index farm (Pulliam et al., 2012). Thus, the emergence of NiV in Malaysia was the product of two drivers. Firstly, agricultural intensification, in the form of increased commercial pig production and patterns of dual-use agriculture, created a pathway for the repeated transmission of NiV from fruit bat reservoirs to pigs. Secondly, the initial spill-over primed the pig population for persistence of the pathogen upon subsequent spill-over events, in turn leading to increased transmission among pigs and to humans. Once infected pigs were sold outside the region, the opportunities for greater human exposure, infection, and disease followed.

Figure 12.   *Pteropus giganteus* (fruit bats) over a village in Faridpur, Bangladesh. They are often found close to or within villages and share mangoes, guava and date palm sap, among other foods, with humans.

5.3   *Scenario 2: The Bangladesh experience*

The mechanism of NiV emergence in Bangladesh is less clear. Outbreaks have been small and confined to remote rural areas making investigation difficult. However, investigations of individual outbreaks have identified three probable routes of transmission from NiV bat reservoirs into humans. The most frequently implicated is the ingestion of fresh date palm sap. Date palm sap is harvested from December through March, particularly in west central Bangladesh. A tap is cut into the tree trunk and sap flows slowly overnight into an open clay pot. Infrared camera studies confirm that the fruit bat species found in Bangladesh, *P. giganteus* and other bats frequently visit date palm sap trees and lick the sap during collection (Khan *et al.*, 2008; Nahar *et al.*, 2010). Under laboratory conditions NiV can survive for days in sugar rich solutions such as fruit pulp (Fogarty *et al.*, 2008). Most date palm sap is boiled to make molasses, a process which kills NiV. However, some sap is consumed as a fresh beverage, drunk raw within a few hours of collection. In the 2005 Nipah outbreak in Tangail District, Bangladesh, and in the 2008 Nipah outbreak in Manikgonj and Rajbari Districts, the drinking of raw date palm sap was the primary, significant factor associated with illness (Luby *et al.*, 2006; Rahman *et al.*, 2012). Twenty-one of the twenty-three recognized index NiV cases in Bangladesh developed their initial symptoms during the December through March date palm sap collection season (Luby *et al.*, 2009b).

A second possible route of NiV infection in Bangladesh is exposure of domestic animals to partially-eaten fruit contaminated with bat saliva. Domestic animals forage for such food and also eat date palm sap that is grossly contaminated with bat faeces. As in Malaysia, domestic animals can become infected with NiV and transmit the virus to other animals, including humans. Contact with a sick cow was strongly associated with NiV infection in Meherpur, Bangladesh in 2001 (Hsu *et al.*, 2004). A similar outbreak in 2003 in Naogaon was associated with contact with a pig herd brought into that community (ICDDRB 2003).

Third, some people may come into direct contact with NiV infected bat secretions. In the Goalando outbreak in 2004, NiV cases were more likely to have climbed trees than uninfected controls (Montgomery *et al.*, 2008). Fruit bats, including *P. giganteus* are gregarious, and roost in large colonies in trees close to human settlements (Figure 12). The size of these colonies, some up to several thousand bats, can make human exposure to excreta likely.

Although there is no documented evidence, it is likely that NiV was transmitted from wildlife to people in Bangladesh long before it was first confirmed in 2001. Phylogenetic analyses suggest that the virus co-evolved with their *Pteropus* hosts over thousands of years (Rahman *et al.*, 2010),

209

and these bats have been seen in the region for centuries (Sterndale 1884), probably millennia. Likewise, people living historically in the in the Bangladesh region have been harvesting date palm sap for centuries using the same basic techniques as today (Kamaluddin *et al.*, 1998). However, even if NiV outbreaks have occurred for this long, it is unlikely, given limited outbreak surveillance, that even lethal outbreaks would have been noticed beyond the immediate locality. Therefore, three recent changes are probably contributing to the recognition of NiV outbreaks in Bangladesh: 1) changes to bat populations in recent years, perhaps as a response to increased cultivation of food in the last forty years (Kahn 2001); 2) improved public health surveillance for this type of outbreak, coupled with increased media interest; and, 3) the 1998 identification of NiV in Malaysia and subsequent availability of laboratory diagnostic techniques for testing in Bangladesh.

5.4 *Designing a surveillance system for Nipah virus*

The single NiV emergence event in Malaysia contrasts with the repeated outbreaks in Bangladesh. This difference in outbreak pattern suggests that tailoring surveillance not just to the characteristics of the microbe itself, but also to the different pathways by which it emerged, would be beneficial. In Malaysia, there appears to be little or no consumption of raw date palm sap, although palm wine and molasses are produced. Fruit bat colonies are often found in heavily forested or swampy areas, rather than near human settlement, so direct human contact with fruit bats and their excreta is probably not very frequent. Indigenous Malaysians (aka Orang Asli) occasionally hunt fruit bats for food, and the licensed shooting of fruit bats for sport is a fairly common practice among urban populations (Pulliam *et al.*, 2012). The latter appears not to involve frequent contact with bat carcasses, which are often left without being retrieved. When carcasses are butchered, it is done within restaurants rather than in the field. So, it is likely that direct, anthropogenic spill-over is not the principle transmission pathway in Malaysia (Chong *et al.*, 2003; Epstein *et al.*, 2006). Apart from relapsed cases from the original outbreak there have been no further reports of human NiV in Malaysia, nor cases of pig infection.

Therefore, in Malaysia, the biggest risk for future NiV outbreaks likely remains spill-over from bats to pigs in intensive, but not bio-secure farms. Surveillance for these outbreaks could centre upon the detection of unusual clinical findings including high rates of respiratory, neurological, abortion events, etc. in intensively farmed pigs in close proximity to fruit production areas. Tracking bat hunting licenses could provide a crude assessment of changes in the level of bat hunting, and therefore increased risk of direct NiV transmission to hunters (Pulliam *et al*, 2012). However, it is important to note that in the original outbreak, the highest number of human NiV cases occurred long after the putative initial spill over from bats to pigs, and in regions where bat distribution does not overlap with pig farms. Therefore, effective surveillance will need to include areas with bat colonies, and separately, those with a high degree of overlap between fruit production and commercial pig farming (Pulliam *et al.*, 2012). The Malaysian Government has recently produced guidelines for biosecurity around pig farms, and this is hoped to provide a safeguard against a repeat of the conditions around the index farm prior to the 1998–99 outbreak.

In Bangladesh, NiV undergoes repeated spill over from bats to humans. This suggests that analysis of the complex relationship of geographical and seasonal variation in bat viral incidence, human-bat contact, date palm sap production and other risk factors together could provide valuable insights into the dynamic of NiV emergence. This, in turn, could be used to develop a targeted surveillance and prevention system. Targeted surveillance for pre-spill-over indicator events has not been attempted yet, but might include consideration of: 1) the routes of NiV infection and transmission, as well as the dynamics of viral shedding in bat colonies; 2) the meta-population structure and inter-colony connectivity and migrations of bats in Bangladesh (it has recently been proposed that changes in bat population connectivity is responsible for sporadic outbreaks of Hendra virus in Australia (Plowright *et al.*, 2011); 3) the contribution of specific human social behaviours to NiV outbreak expression including an evaluation of food-borne and person-to-person transmission risks; and an assessment of genetic heterogeneity in human NiV shedding through respiratory secretions, and contact patterns to produce *super-spreaders*; and 4), the potential impact of human travel on the spreading of NiV from village to village, village to cities, and from Bangladesh to other countries.

A foundation for this prospective surveillance already exists in Bangladesh. Civil surgeons located at each of the sixty-four districts throughout the country actively report clusters of unusual illnesses, including encephalitis, to authorities in the capital city, Dhaka. In addition, a commercial firm reviews all English and Bengali language newspapers published in Bangladesh every day and forwards news stories suggestive of potential outbreaks to the disease surveillance team at the Institute of Epidemiology Disease Control and Research (IEDCR 2004) within the Ministry of Health and Family Welfare of the Government of Bangladesh. Surveillance team members review media reports and follow-up as appropriate with local government authorities. Efforts to document the type and frequency of date palm sap consumption are being conducted. Studies of the cultural significance and importance of date palm sap might also inform efforts to educate the public against its consumption. Finally, the Government of Bangladesh ICDDRB monitors for hospitalized encephalitis cases at six government hospitals located in the region where Nipah outbreaks have repeatedly occurred. This surveillance includes patient interviews and laboratory diagnostic testing if cases are present during the at-risk season of January through March.

Proactive NiV surveillance in bats is also being conducted to facilitate the geographic refinement of NiV surveillance to those areas most at risk. Fruit bats are screened for Nipah virus antibodies as well as viral nucleic acid using PCR. Estimates of the rate of exposure in juvenile bat populations (under 2 years old) provide insights into recent viral circulation in a particular bat colony. This in turn provides a useful tool for assessing the risk of viral spill over to people.

Refinements to NiV surveillance in Bangladesh are on-going. When NiV outbreaks are suspected, a collaborative team of national and international experts is available for rapid mobilization. When an outbreak is confirmed, a multidisciplinary team of physicians, epidemiologists, anthropologists and veterinarians promptly investigates the outbreak to determine its particular pathways to transmission. Bat surveillance is integral to this process. This includes the identification of local bat colonies and the collection of samples for detecting viral shedding. Livestock in the vicinity are also examined for evidence of NiV infection. This multi-disciplinary surveillance and detection approach provides further insights into how NiV outbreaks emerge. It helps guide improvements in proactive surveillance as well as outbreak prevention and mitigation initiatives.

6 LEISHMANIASIS: THE COMPLEXITY OF SAND FLY VECTORS

Leishmaniasis causes morbidity and mortality throughout the subtropics and tropics. As discussed in Chapter 2 on zoonotic and vector-borne diseases, the different forms of leishmaniasis, (cutaneous – CL, visceral -VL), are vectored by species of Phlebotomine sand flies in the genera *Phlebotomus* and *Lutzomyia*. Over the past several decades, these sand flies have emerged in new locations as a result of environmental and ecologic phenomena like climate change; and anthropogenic drivers such as habitat modifications, travel, and political instabilities that lead to human population migrations. All these factors can influence the ability of a sand fly to thrive and transmit disease. Leishmaniasis has also re-emerged in areas where it was once eliminated through the concomitant use of DDT in indoor residual spraying for malaria's mosquito vectors.

Thus, the dynamic of leishmaniasis emergence or re-emergence, is a multi-factoral process dependent on a variety of conditions and events that lead to the opportunity and efficiency of sand flies as leishmaniasis transmission vehicles. Through this perspective, satellite-based remote sensing technology and ENM can be used to detect and characterize leishmaniasis emergence, no matter which form of the parasite, by focusing on the environmental drivers, and ecosystem and habitat requirements of their sand fly vectors, wherever in the world they are found. By using these tools to ultimately detect, characterize and track the process of leishmaniasis emergence or re-emergence, we can implement early warning systems that forewarn public health officials of leishmaniasis risk and promote the initiation of early responses to either prevent or mitigate the impact of emergence events.

6.1 Environmental and ecological phenomena in leishmaniasis emergence

Long-term climate change, recurring eco-climatic events, and habitat modification influence leishmaniasis emergence through their direct effect on the sand fly vector's biology. Just as it is true for arthropod-borne diseases in general, change in ambient temperature and precipitation in a given area affects the distribution and abundance of vectors and their efficiency in transmitting disease. In north-eastern India, air temperature is a crucial factor in the incidence of VL (Bhunia et al., 2010). The sand fly's ability to transmit leishmaniasis is most efficient in the 25.0–27.5°C temperature range. Temperatures outside that range are either too cold or too warm and trigger a sand fly species' biological mechanism for suspending development (diapause or quiescence). If this occurs, it ultimately delays the transition of sand flies from their immature to adult stages until ambient temperatures are suitable again for development, feeding, flight, mating, and oviposition. A prolonged period of diapause or quiescence in a sand fly population can reduce or eliminate the presence of adult sand flies in an area for extended periods, reducing the likelihood of disease transmission. Without efficient transmission, disease emergence does not proceed. In addition, long-term temperature change in a given area affects habitat suitability for sand flies. If the habitat is not right, sand flies may be unable to survive or may become less efficient vectors. A generalized climate warming trend then also shifts the distribution of the vectors, and with them the diseases they carry. González (et al., 2010) predicted that climate change could exacerbate the ecological risk of human exposure to leishmaniasis in areas far removed from current endemic areas, such as southern Canada. For this to happen however, suitable vectors and the pathogen would have to be present in the new area. This could occur by introductions that are gradual expansions of infected sand fly populations as they search for suitable habitats, or quickly by a man-made introduction through multiple individual travellers and their luggage, or shipments of infected sand fly-containing commercial cargoes. It could also occur through a combination of separate sand fly and pathogen introductions. The constant would be the presence of leishmaniasis-susceptible humans in the new area.

In addition to the potential for long-term climate change to ultimately cause the emergence of leishmaniasis in new areas, the effects of short-term cyclic climate patterns could also lead to epidemics of leishmaniasis. In the relationship between CL incidence and climate in Costa Rica, Chaves & Pascual (2006) found that the incidence of disease has a three year cycle, corresponding to temperature fluctuations associated with the ENSO. By studying the relationship between these short-term, predictable climate events and sand fly vector capability, it is possible to identify when conditions become appropriate for increased leishmaniasis transmission and human disease production.

Habitat modifications also affect leishmaniasis emergence by creating new sand fly niches and bringing people closer to leishmaniasis vectors (and reservoirs in a zoonotic transmission cycle). Sand fly fauna can be greatly affected by deforestation, dam building, and urban sprawl into rural areas (Ashford & Thompson 1991; Baneth et al., 1998; Al-Jawabreh et al., 2004; Chaves et al., 2008).

Through a greater understanding of the effect of environmental and eco-climatic changes on the sand flies, we could better predict potential human disease emergence events. RS tools can help signal or predict those events. ENM could be particularly useful in predicting the effect of long-term habitat or climate change on the distribution of various sand fly species. Other modelling techniques that employ RS technology to look at such factors as vegetation indices could be useful in signalling outbreaks due to short-term, repetitive eco-climatic changes, such as those caused by the ENSO.

6.2 Socioeconomic facilitators as anthropogenic drivers of leishmaniasis

Leishmaniasis is often considered a disease of the poor, with socioeconomic status correlated to disease incidence (Ahluwalia 2003; Alvar et al., 2006; Anoopa Sharma 2006). In general, socioeconomic conditions influence exposure to sand flies and thereby leishmaniasis risk by affecting a wide range of factors including population density, household location, sleeping patterns, mosquito net usage, and nutritional status (Cref 1987; Kolaczinski *et al.*, 2008). Leishmaniasis outbreaks in the poorer urban areas of Brazil, India, Afghanistan, and other countries have been linked to under-privileged socioeconomic and suboptimal health status in human populations (Desjeux 2004; Sharma *et al.*, 2007; Boelaert *et al.*, 2009). Historically, VL associated with *L. chagasi* was considered a rural disease in Brazil and other countries in South America. However, since the early 1980s, VL has increasingly become an urban disease as peri-domestic sand fly vector and zoonotic reservoir habitats have increased. The urban emergence of VL in Brazil first occurred in Teresina, the capital of Piauí State. This epidemic started in rural areas, but eventually spread into the peripheral areas of the city. The highest incidence rates in Teresina occurred in the north-eastern and southern suburbs, which were undergoing rapid rural/urban expansion. This brought residual ecological habitats (forests and pastures) conducive to sand fly and animal reservoir presence, poor sanitation infrastructure, and susceptible people into a convergence ripe for leishmaniasis emergence (Arias 1996; Gratz 1999; Werneck 2002; Costa 2005; Neto 2009). Similar patterns have been reported in the Brazilian states of Bahia, Minas Gerais and Rio Grande do Norte (Jeronimo 1994; Sherlock 1996; Silva 2001; Franke 2002). This pattern of VL emergence, driven by urbanization, translates into another model for predictive surveillance based on the focused targeting of peri-urban areas for the *L. chagasi* emergence risk. A combination of vector and pathogen surveillance field work and predictive modelling incorporating RS and ENM vegetation and land cover layers can indicate the most likely at-risk areas. An ability to do this can help target vector control and disease prevention efforts in these types of locations.

6.3 *Human travel, displacements and relocations as anthropogenic drivers of leishmaniasis*

Human relocations and displacements facilitate leishmaniasis emergence, either through leishmaniasis infected people moving into an area, or naïve individuals becoming infected while traveling to leishmaniasis endemic regions then returning to their previously non-endemic homes. If *Leishmania* parasites are transported into new areas where there are suitable sand fly vectors, the opportunity arises for sand fly exposure to the parasite. If other conditions are right, this can start the process of leishmaniasis establishment and emergence as a new disease threat.

The risk of this happening appears to be most acute when facilitated by air travel to remote areas of the world for business or eco-tourism. Imported cases of leishmaniasis in northern Europe have been associated with travellers returning from holidays in endemic areas of the Mediterranean region (Scope *et al.*, 2003; Aspöck *et al.*, 2008). While isolated human cases will not lead to emergence by itself, the presence of competent sand fly vectors due to a concomitant expansion of vector range in Europe associated with a climate warming trend could trigger an emergence event (Ready 2010). Competent sand fly vectors are currently present in many 'leishmaniasis free' areas of Europe (Aransay *et al.*, 2004; Naucke & Schmitt 2004). At present however, the sand flies' temperature-dependent developmental time is too long in these areas to permit leishmaniasis establishment. For instance, while *P. ariasi is* a competent vector of *L. infantum*, its temperature-dependent inter-generational development time is long, with just over one generation of these sand flies produced per year (Dye *et al.*, 1987). This makes leishmaniasis establishment and maintenance inefficient and hence unlikely through that vector. However, if the number of sand fly generations per year increases as a result of consistent elevations in ambient temperatures, and leishmaniasis continues to be repetitively introduced by travellers, the combination of the two drivers could lead to establishment of the disease; initially in southern France, then subsequently in more northern Europe.

Armed conflicts and the inevitable social and community destabilization they produce are known to have negative effects on socio-economic conditions, health care availability, and the ability to conduct vector control in affected communities. These factors exacerbate the potential for many types of disease epidemics to occur. Unlike leisure or business travel between endemic and uninfected areas, conditions during war can lead to massive, disorganized movements of people under suboptimal disease prevention, detection and containment conditions. Again, this may be either through mass movement of infected (overt disease or occult and not suspected) people from an endemic area into a non-endemic region where a competent vector already resides, or through movement of naïve people into endemic areas where they become at risk for initial infection.

Leishmaniasis outbreaks have occurred in naïve soldiers entering endemic areas. For example, Brazilian soldiers were infected after being sent to the Amazon for training. (Brandão-Filho *et al.*, 1998; de Silveira *et al.*, 2002; Oliveira Guerra *et al.*, 2003). Similarly, cases of CL have occurred among troops in the US Army's jungle training centre in Panama (Takafuji *et al.*, 1980; Sanchez *et al.*, 1992). During the 1991 Gulf War, there were cases of *Leishmania tropica* with visceral symptoms among US soldiers (Magill *et al.*, 1994). In 2003, there was an outbreak of CL caused by *Leishmania major* among US service members in Iraq (Aronson *et al.*, 2006). In Afghanistan, there have been periodic epidemics of *Leishmania major* among NATO service members in the northern regions of Afghanistan, a known leishmaniasis *hot spot* (Faulde 2006; 2008).

Combatants returning to their non-endemic garrisons or home communities can lead to disease spread if competent, previously uninfected vectors are indigenous to those places. The civil war in Sudan led to the emergence of VL when infected soldiers moved into a non-endemic area of that country where suitable sand fly vectors had recently become established. Entomological surveys indicated that *Phlebotomus orientalis* was the primary vector in this instance (Schorscher 1992). This species is most commonly associated with woodland areas composed of *Acacia* spp and *Balamites* spp trees (Schorscher 1992). The area at the epicentre of the VL epidemic in the Upper Nile province experienced heavy flooding from 1960–1964, which destroyed the existing forests (Ashford 1991). Twenty years after the flooding, the forests were regenerated as were well-established populations of these sand flies.

In 1984, an epidemic of VL began in what is now the Upper Nile State of South Sudan (Perea 1991). Prior to this epidemic, this area was considered free of VL. From 1961 to 1971, this area experienced civil war. Twelve years later, in 1983, another civil war broke out. In the fall of 1988, some sick members of the Nuer tribe travelled to Khartoum for treatment and were diagnosed with VL (DeBeer 1990). Perea (*et al.*, 1991) characterized the epidemic and determined that the high proportion of adult clinical cases and high seroprevalence among individuals over the age of fifteen indicated that VL had been recently introduced into this region. Subsequent epidemiological investigations revealed that this outbreak began in 1984, shortly after the start of the second conflict (Seaman 1996).

The primary inhabitants of this region are members of the Nuer and Dinka tribes. During the wet season, individuals cultivate sorghum and maize near their villages (Seaman 1996). In the dry season, they move with their cattle to find water (Seaman 1996). The peak VL transmission season in this area is during the dry season when the Nuer and Dinka move during the night when temperatures are cooler, through the forests on their way to known water camps (Schorscher 1992). With the start of the civil war in 1983, *Leishmania donovani* was introduced into these forests, probably by soldiers coming from training camps on the Ethiopian border where VL is endemic (Perea 1991). Once human movements and the earlier reforestation efforts brought the vector, pathogen, and susceptible humans together, VL spread rapidly, causing high levels of disease among the Nuer and Dinka adult population. In the first decade of this outbreak, there were an estimated 100,000 cases of VL in the area of the Upper Nile province (Ritmeijer 2003). Since the introduction of VL, the disease has become established, with approximately 23,000 patients being treated at specialty clinics from 1989 through 2002 (Ritmeijer 2003).

The emergence of VL in Sudan illustrates the potential utility of, and opportunity for RS-based surveillance to prevent or reduce the impact of sand fly vector and thus leishmaniasis emergence events. Had the technology for RS-based predictive modelling been available prior to this emergence event, a model incorporating vegetation, temperature, and rainfall data could have predicted the risk for *P. orientalis* vector expansion from the endemic areas near Ethiopia. This sort of model, coupled with data on human movements from VL endemic areas could have signalled a heightened risk of the VL emergence among the Nuer and Dinka tribes. While such a model would not have been able to directly prevent that emergence, it could have been used by those working in public health to gain a more rapid understanding of the scope of the outbreak's origins and help focus interventions to mitigate the crisis earlier and more efficiently.

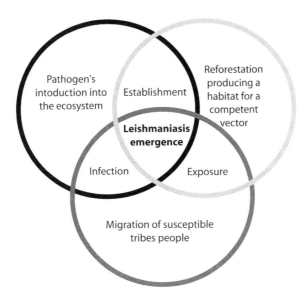

Figure 13.   Transition pathway of leishmaniasis emergence in Sudan.

In the context of this chapter's framework, Figure 13 shows the transition pathway of leishmaniasis emergence in Sudan; its three anthropomorphic drivers: 1) migration of susceptible Nuer and Dinka tribes' people; 2) reforestation to habitats permissive of the competent sand fly vector; and 3) the pathogen's introduction into the forest ecosystem; and its emergence dynamics: *establishment,* the ability of the pathogen to be present and transmissible in the ecosystem; *exposure,* the opportunity of the sand fly vector to bite susceptible people; and *infection,* the entrance of the pathogen into a new host and cause disease. Leishmaniasis emergence itself is the *convergence* of the interacting drivers and dynamics.

6.4   *Applying RS technology to sand fly epidemiology and predicting leishmaniasis risk*

A wide variety of satellite-based remote sensing tools are available to science and public health practice for potential use in developing a predictive surveillance system for detecting leishmaniasis emergence. Harnessing that potential and transforming it for practical application however, requires knowledge-driven selectivity in what tools should be used and how they can be used to avoid being overwhelmed by the biotic and abiotic complexity that is leishmaniasis. Different leishmaniasis vectors can display different behavioural patterns, such as when and where they feed. They may also have different ecological and habitat requirements, such as the ideal temperature and humidity requirements for larval development. The differences between the anthroponotic and zoonotic transmission cycles must be accounted for in any generated models to avoid confounding the identification of constancy in the sand flies' requirements across many different geographic conditions. That said, however, modelling a non-zoonotic epidemiological cycle is simpler than modeling a cycle where there is a non-human animal reservoir, such as rodents or dogs. The two systems are very different.

Scientists and public health practitioners must remain cautious to not over-generalize interpretations of RS data across dissimilar vectors, and where the various geographic areas possess underlying ecological systems that are too site-specific or unique for only one model to adequately contain. Generalizations are useful, but must be weighed in balance with the critical site-specific or unique vector characteristics that accommodate for and permit diverse leishmaniases to exist and be transmitted by different species of sand flies. With that said, while the epidemiology of the leishmaniases can be very complex, there are a few fundamental remotely-sensed parameters that

215

are universal elements in leishmaniasis modelling. These include elevation, slope, temperature, and precipitation since the distribution of sand flies and leishmaniases are impacted by these factors to one extent or another throughout the global range of sand flies. While these variables have been well correlated with sand fly and leishmaniasis distributions, they are not the whole story. For example, remote-sensing generated vegetation indices do help in predicting habitat suitability for sand fly vectors. However, their applicability to all sand fly species vectoring leishmaniases is questionable. Sand flies are often associated with vegetation because of their biological requirement of taking plant-based sugar meals, their need for larval development sites in soil with adequate moisture and organic matter, and their use of vegetation for sand fly resting sites (Schlein & Yuval 1987; Schlein & Jacobsen 1999). The NDVI have been used in several studies to show a positive correlation between sand fly density and vegetation levels (Bavia et al., 2005). However, the strength of the correlation is not consistent across all geographic habitats and sand fly species. In the case of a purely anthroponotic transmission, in some urban centres in Afghanistan, sand flies may have alternate sources of sugar meals and may have breeding and resting sites in animal pens, mud walls, and other areas closely associated with their human blood source; not vegetation levels. Thus there may not always be a clear, direct relationship between sand fly abundance and vegetation indices. Vegetation indexes reflect the presence of some other more subtle environmental variable that is the actual driver of sand fly population presence and abundance, and/or leishmaniasis transmission in Afghanistan. For instance, vegetation indices may indirectly reflect soil moisture, precipitation and ambient temperature as a constellation, and only when present in proper sequence together do they serve as the main factor governing sand fly development. In addition, temperature also affects vector competence independently. Therefore, while the putative link between vegetation indices and sand flies is not necessarily direct, these indices can serve as indirect indicators of an area's ability to support sand fly population presence and ability to transmit disease. The influence of these indirect factors, or other constellations, need to be very precisely interpreted for different levels of predictive correlation. Analyses must account for the ecologies of the different geographic areas, the species of vectors, and the species of leishmaniasis parasites themselves for meaningful interpretation.

Given these nuances, how might RS technology be applied? It is known that larval sand flies develop in soil with high moisture and rich organic content. Either too much or too little can adversely affect sand fly populations. Remotely sensed soil moisture data are available; but are not always reliable, and its spatial resolution is low (Schneider et al., 2008; Parajka et al., 2009). Once fully developed and refined, remotely sensed soil moisture data could help identify soil moisture levels in different endemic areas, and be analysed for key similarities, which if built into spectral indexes, could signal predictable habitat suitability for different sand fly species.

LST is another promising tool for predicting sand fly abundance and leishmaniasis transmission potential. Sand flies are primarily active at night and typically fly within two meters of the ground surface. Their activity strongly relates to air temperature close to the surface soil (Sawalha et al., 2003; Boussa et al., 2005). This could be used to develop predictive measures of sand fly activity. In addition, LST could also serve as an indicator for an area's suitability for sand fly larval development since temperature drives larval development time. Consistently higher LST may indicate faster larval development times and could be used to estimate the timing of sand fly population blooms, an indicator of heightened leishmaniasis transmissibility risk.

As mentioned earlier, the ecological impact of land use can also be analysed using RS technology. Leishmaniasis emergence has been linked to habitat change, such as deforestation and urbanization (Costa 2005; Chaves et al., 2008). To take advantage of this, satellite imagery of expanding urban areas led to the identification of likely areas of leishmaniasis emergence at the rural/urban interface. Models using land use data should focus on well-defined areas with high resolution imagery.

6.5 Use of ENM in understanding the potential for leishmaniasis emergence

Ecological niche modelling is another approach that helps highlight the current distribution of sand fly and leishmaniasis risks. Peterson & Shaw (2003) used ENM to describe the niches of three *Lutzomyia* species that vector leishmaniasis in Brazil. They incorporated climate change scenarios into their model, but low spatial resolution (10 km) was a limitation of their model and analysis (Peterson & Shaw 2003). Colacicco-Mayhugh *(et al.,* 2010) modelled the distribution of *P. papatasi* and *P. alexandri* across portions of their known ranges in the Middle East using a one kilometre resolution. This more granular approach may be better suited toward modelling the potential for sand fly range expansion due to global warming or other climate change influences on the biotic and abiotic drivers that lead to leishmaniasis emergence. While niche models alone are not the best tool to indicate near-term leishmaniasis emergence, in conjunction with sensor data processing technology they are central to informing more sophisticated, multi-perspective models that represent the complexity of sand fly biology and diversity, and thus possible leishmaniasis emergence scenarios under different geographic conditions.

Ultimately, RS-based predictive platforms will need to incorporate the vector, pathogen, and environmental variables to present a comprehensive model that has the power to accurately predict short term leishmaniasis outbreak events as well as look at the long-term potential for disease spread. One example of a dynamic approach to modelling this shorter term outbreak risk comes from work by Nieto (*et al.,* 2006) in Bahia, Brazil. As described in Chapter 2, this approach allows for a synergistic look at the potential for a leishmaniasis outbreak based on a variety of key factors for vector development and disease. This sort of approach is likely to be successful in modelling short-term emergence or outbreak events based on the effects of repetitive, predictable climate changes such as El Niño and La Niña. This two-pronged modelling approach, incorporating infection and real time sand fly data, is an example of the comprehensive effort most likely to be effective in signalling outbreak events in the short-term.

6.6 *Summary*

Worldwide, the incidence of leishmaniasis is increasing. Cases of CL and VL have increased dramatically in parts of Brazil, Afghanistan, and Africa in recent decades (Desjeux 2001). In addition, the disease has emerged in new foci in the tropics, subtropics, and temperate regions. This emergence and increased incidence are mainly attributable to eco-climatic and environmental conditions and trends, and anthropogenic drivers on the sand fly vector's success as a group of species and disease transmitters.

Like other arthropod-borne diseases, the distribution of the leishmaniases is very closely linked to many abiotic and biotic drivers, direct and indirect, that influence the development and survival of the vector. By examining the known events of leishmaniasis emergence through the prism of a joint RS and ENM modelling approach, one should be able to distinguish between the factors common to all leishmaniasis outbreaks and those which are site- or vector species-specific in diverse scenarios. With the resulting understanding of what leishmaniasis emergence entails in different parts of the world, a surveillance system that incorporates many diverse models could help predict both long-term and short-term emergence events.

REFERENCES

Abdullah, A.S.M., Tomlinson, B., Thomas, G.N. & Cockram, C.S. 2004. Impacts of SARS on health care systems and strategies for combating future outbreaks of emerging infectious diseases. In: S. Knobler, A. Mahmoud, S. Lemon, A. Mack, L. Sivitz & K. Oberholtzler (eds.), *Learning from SARS: Preparing for the next disease outbreak:* 83–86. Washington DC: Institute of Medicine.
Abram, N.J., Gagan, M.K., McCulloch, M.T., Chappell, J. & Hantoro, W.S. 2003. Coral reef death during the 1997 Indian Ocean dipole linked to Indonesian wildfires. *Science* 301: 952–955.
Ahluwalia, I.B., Bern, C., Costa, C., Akter, T., Chowdhury, R., Ali, M., Alam, D., Kenah, E., Amann, J., Islam, M., Wagatsume, Y., Haque, R., Breiman, R.F. & Maguire. J.H. 2003 Visceral leishmaniasis: Consequences of a neglected tropical disease in a Bangladeshi community. *Amer. J. Trop. Med. & Hyg.* 69(6): 624–628.

Al-Jawabreh, A., Schnur, L.F., Nasereddin, A., Schwenkenbecher, J.M., Abdeen, Z., Barghuthy, F., Khanfar, H., Presber, W. & Schonian, G. 2004. The recent emergence of *Leishmania tropica* in Jericho (A'riha) and its environs, a classical focus of *L. major*. *Trop. Med. & Int. Health.* 9(7): 812–816.

Alvar, J., Yactayo, S. & Bern, C. 2006. Leishmaniasis and poverty. *Trends in Parasit.* 22(12): 552–557.

Anyamba, A., Chretien, J., Small, J., Tucker, C.J., Formenty, P., Richardson, J.H., Britch, S.C., Schnabel, D.C., Erickson, R.L. & Linthicum, K.J. 2009. Prediction of a Rift Valley Fever outbreak. *Proc. Natl. Acad. Sci.* 106: 955–959.

Aransay, A.M., Testa, J.M., Morillas-Marquez, F. Lucientes, J. & Ready, P.D. 2004. Distribution of sand-fly species in relation to canine leishmaniasis from the Ebro Valley to Valencia, northeastern Spain. *Parasit. Res.* 94: 416–420.

Arias, J.R., Monteiro, P.S. & Zicker, F. 1996. The re-emergence of Viscceral Leishmaniasis in Brazil. *Emerg. Infec. Dis.* 2(2): 145–146.

Aronson, N.E., Sanders, J.W. & Moran, K.A. 2006. In harm's way: Infections in deployed American military forces. *Clinic. Infect. Dis.* 43(8): 1045–1051.

Ashford, R.W. & Thompson, M.C. 1991. Visceral leishmaniasis in Sudan. A delayed development disaster. *Ann. Trop. Med. & Parasit.* 85(5): 571–572.

Aspöck, H., Gerersdorfer, T., Formayer, H. & Walochnik, J. 2008. Sandflies and sand fly-borne infections of humans in Central Europe in the light of climate change. *Mid. Europe. J. Med.* 120(Supp4): 24–29.

Baneth, G., Dank, G., Keren-Kornblatt, E., Sekeles, E., Adini, I., Eisenberger, C.L., Schnur, L.F., King, R. & Jaffe, C.L. 1998. Emergence of Visceral Leishmaniasis in Central Israel. *Amer. J. Trop. Med. & Hyg.* 59(5): 722–725.

Barclay, E. 2008. Is climate change affecting dengue in the Americas? *Lancet* 371(9617): 973–974.

Barre, N., Camus, E., Fifi, J., Fourgeaud, P., Numa, G., Rose-Rosette, F. & Borel, H. 1996. Tropical bont tick eradication campaign in the French Antilles: Current status. *Ann. N.Y. Acad. Sci.* 791: 412–20.

Barrera, R., Navarro, J.C., Mora, J.D., Dominguez, D. & Gonzalez, J. 1995. Public service deficiencies and *Aedes aegypti* breeding sites in Venezuela. *Bull. Pan. Am. Health Organ.* 29(3): 193–205.

Bavia, M.E., Carneiro, D.D., Gurgel, H.C., Madureira, F.C & Barbosa, M.G. 2005 Remote sensing and geographic information systems and risk of American visceral leishmaniasis in Bahia, Brazil. *Parasit.* 47(1): 165–9.

Beebe, N.W., Cooper, R.D., Mottram, P. & Sweeney, A.W. 2009. Australia's dengue risk driven by human adaptation to climate change. *PLoS Negl. Trop. Dis.* 3(5): e429.

Beer, P. de, El Harith, A., van Grootheest, M. & Winkler, A. 1990. Outbreak of kala-azar in the Sudan. *Lancet* 27: 224.

Bhunia, G.S., Kesari, S., Jeyaram, A., Kumar, V. & Das, P. 2010. Influence of topography on the endemicity of Kala-azar: A study based on remote sensing and geographical information system. *Geospat. Health* 4(2): 155–165.

Blacksell, S.D., Sharma, N.P., Phumratanaprapin, W., Jenjaroen, K., Peacock, S.J. & White, N.J. 2007. Serological and blood culture investigations of Nepalese fever patients. *Trans. Roy. Soc. Trop. Med. Hyg.* 101(7): 686–690.

Boelaert, M., Meheus, F., Sanchez, A., Singh, S.P., Vanlerberghe, A., Picado, A., Meesen, B. & Sundar, S. 2009. The poorest of the poor: A poverty appraisal of households affected by visceral leishmaniasis in Bihar, India. *Trop. Med. & Int. Health* 14(6): 639–644.

Boussa, S., Guernaoui, S., Pesson, B. & Boumezzough. A. 2005. Seasonal fluctuations of phlebotomine sand fly populations (*Diptera: Psychodidae*) in the urban area of Marrakech, Morocco. *Acta Tropica* 2: 85–91.

Brandão-Filho, S.P., Felinto de Brito, M.E., Martins, C.A.P., Sommer, I.B., Valença, H.F., Almeida, F.A. & Gomes. J. 1998. American cutaneous leishmaniasis in military training unit localized in Zona da Mata of Pernambuco State, Brazil. *Rev. Soc. Brasil. de Med. Trop.* 31(6): 575–578.

Breed, A., Field, H., Wacharapluesadee, S., Sendow, I. & Meers, J. 2011. On the distribution of henipaviruses in the Australasian region: Does Nipah virus occur east of the Wallace Line? *EcoHealth* 7: S37.

Brower, V. 2001. Vector-borne diseases and global warming: Are both on an upward swing? *EMBO Rep.* 2(9): 755–757.

Brown, N.G. 1999. *Challenge of climate change.* New York: Routledge.

Caprara, A., Lima, J.W., Marinho, A.C., Calvasina, P.G., Landim, L.P. & Sommerfeld, J. 2009. Irregular water supply, household usage and dengue: A bio-social study in the Brazilian Northeast. *Cad Saude Publica,* 25(Suppl): 125–136.

Cardenas, W.B. 2010. Review: Evasion of the interferon-mediated antiviral response by filoviruses. *Viruses* 2: 262–282.

Chadha, M.S., Comer, J.A. Lowe, L., Rota, P.A., Rollin, P.E., Bellini, W.J., Ksiazek, T.G. & Mishra, A. 2006. Nipah virus-associated encephalitis outbreak, Siliguri, India. *Emerg. Infect. Dis.* 12: 235–240.

Chaves, L.F. & Pascual, M. 2006. Climate Cycles and forecasts of Cutaneous Leishmaniasis, a nonstationary vector-borne disease. *PLoS Med.* 3(8): e295.

Chaves, L.F., Cohen, J.M., Pascual, M. & Wilson, M.L. 2008. Social exclusion modifies climate and deforestation impacts on a vector-borne disease. *PLoS Neglected Trop. Dis.* 2(2): e176.

Chhetri, M.P. & Pandey, R. 2009. Effects of climate change in Nepal: Time for policy to action. Available from: http://iopscience.iop.org/1755-1315/6/41/412037/pdf/1755-1315_6_41_412037.pdf. [Accessed 12th February 2012].

Chong, H.T., Tan, C.T., Goh, K.J., Lam, S.K. & Chua, K.B. 2003. The risk of human Nipah virus infection directly from bats (*Pteropus hypomelanus*) is low. *Neurol. J. SE Asia* 8: 31–34.

Chua, K.B. Bellini, W.J., Rota, P.A., Harcourt, B.H., Tamin, A., Lam, S.K., Ksiazek, T.G., Rollin, P.E., Zaki, S.R., Shieh, W., Goldsmith, C.S., Gubler, D.J., Roehrig, J.T., Eaton, B., Gould, A.R., Olson, J., Field, H., Daniels, P., Ling, A.E., Peters, C.J., Anderson, L.J. & Mahy, B.W. 2000. Nipah virus: A recently emergent deadly paramyxovirus. *Science* 288: 1432–1435.

Chua, K.B., Goh, K.J., Hooi, P.S., Ksiazek, T.G., Kamarulzaman, A., Olson, J. & Tan, C.T. 2001. The presence of Nipah virus in respiratory secretions and urine of patients during an outbreak of Nipah virus encephalitis in Malaysia. *J. infect.* 42: 40–43.

Chua, K.B., Koh, C.L., Hooi, P.S., Wee, K.F., Khong, J.H., Chua, B.H., Chan, Y.P., Lim, M.E. & Lam, S.K. 2002a. Isolation of Nipah virus from Malaysian Island flying-foxes. *Microbes & Infect.* 4: 145–151.

Chua, K.B., Chua, B.H. & Wang, C.W. 2002b. Anthropogenic deforestation, El Nino and the emergence of Nipah virus in Malaysia. *Malaysian J. Pathol.* 24: 15–21.

Chua, K.B. 2003. Nipah virus outbreak in Malaysia. *J. Clin. Virol.* 26: 265–275.

Colacicco-Mayhugh, M.G., Masuoka, P. & Grieco. J.P. 2010. Ecological niche model of *Phlebotomus alexandri & P. papatasi* (Diptera: Psychodidae) in the Middle East. *Int. J. Health Geogra.* 9:2. Available from: http://www.ij-healthgeographics.com/content/9/1/2 [Accessed 12th February 2012].

Colwell, R.R. 1996. Global climate and infectious disease: The cholera paradigm. *Science* 274: 2025–2031.

Costa, C.H.N, Werneck, G.L., Rodrigues Jr., L., Santos, M.V., Arau'jo, I.B., Moura, L.S., Moreira, S., Gomes, R.B.B. & Lima, S.S. 2005. Household structure and urban services: neglected targets in the control of visceral leishmaniasis. *Ann. Trop. Med. & Parasit.* 99(3): 229–236.

Cox, J., Grillet, M.E., Ramos, O.M., Amador, M. & Barrera, R. 2007. Habitat segregation of dengue vectors along an urban environmental gradient. *Am. J. Trop. Med. & Hyg.* 76(5): 820–826.

Cref, B.J., Jones, T.C., Badar, R., Sampaio, D., Teixeira, R. & Johnson, W.D.J. 1987. Malnutrition as a risk factor for severe Visceral Leishmaniasis. *J. Infec. Dis.* 156: 1030–1033.

Daszak, P. 2009. Can we predict future trends in disease emergence? In D.A. Relman, M.A. Hamburg, E.R. Choffnes & A. Mack (eds.), *Microbial evolution and co-adaptation: Attribute to the life and scientific legacies of Joshua Lederberg:* 252–266. Washington DC: Institute of Medicine.

Daszak, P., Plowright, R., Epstein, J.H., Pulliam, J., Abdul Rahman, S., Field, H.E., Smith, C.S., Olival, K.J., Luby, S., Halpin, K., Hyatt, A.D. & the Henipavirus Ecology Research Group. 2006. The emergence of Nipah and Hendra virus: Pathogen dynamics across a wildlife-livestock-human continuum. In R.S Collinge (ed.), *Disease Ecology: Community structure and pathogen dynamics:* 186–201. Oxford: Oxford UP.

Desjeux, P. 2001. The increase in risk factors for leishmaniasis worldwide. *Trans. Roy. Soc. Trop. Med. & Hyg.* 95(3): 239–243.

Desjeux. P. 2004. Leishmaniasis: current situation and new perspectives. *Comparat. Immunol., Microbiol. & Infect. Dis.* 27(5): 305–318.

Dye, C., Guy, M.W., Elkins, D.B., Wilkes, T.J. & Killick-Kendrick. R. 1987. The life expectancy of phlebotomine sandflies: First field estimates from southern France. *Med. & Veterin. Entomol.* 1: 417–425.

Elnaiem, D., Hassan, H. & Ward, R. 1997. Phlebotomine sandflies in a focus of visceral leishmaniasis in a border area of eastern Sudan. *Ann. Trop. Med. Parasit.* 91(3): 307–318.

Enrique Morales Vargas, R., Ya-Umphan, P., Phumala-Morales, N., Komalamisra, N. & Dujardin, J.P. 2010. Climate associated size and shape changes in *Aedes aegypti* (Diptera: Culicidae) populations from Thailand. *Infect. Genet. Evol.* 10(4): 580–585.

Epstein, P.R. 2007. Chikungunya fever resurgence and global warming. *Am. J. Trop. Med. & Hyg.* 76(3): 403–404.

Epstein, J.H., Rahman, S.A., Zambriski, J.A., Halpin, K., Meehan, G., Jamaluddin, A.A., Hassan, S.S., Field, H.E., Hyatt, A.D., Daszak, P. & Henipavirus Ecology Research Group. 2006. Feral cats and risk for Nipah virus transmission. *Emerg. Infect. Dis.* 12: 1178–1179.

Epstein, J.H., Prakash, V., Smith, C.S., Daszak, P., McLaughlin, A.B., Meehan, G., Field, H.E., & Cunningham, A.A. 2008. Henipavirus infection in fruit bats (*Pteropus giganteus*), India. *Emerg. Infect. Dis.* 14: 1309–1311.

Epstein, J., Olival, K.J., Pulliam, J.R.C., Smith, C., Westrum, J., Hughes, T., Dobson, A.P., Zubaid, A., Rahman, S.A., Basir, M.M., Field, H.E. & Daszak, P. 2009. *Pteropus vampyrus*, a hunted migratory species with a multinational home-range and a need for regional management. *J. Appl. Ecol.* 46: 991–1002.

Fakeeh, M. & Zaki, A.M. 2001. Virologic and serologic surveillance for dengue fever in Jeddah, Saudi Arabia, 1994–1999. *Am. J. Trop. Med. & Hyg.* 65(6): 764–767.

Faulde, M., Heyl, G. & Amirih, M. 2006. Zoonotic cutaneous leishmaniasis outbreak in Mazar-e Sharif, Afghanistan. *Emerg. Infect. Dis.* 12: 1623–1624.

Faulde, M., Schrader, J., Heyl, G., Amirih, M. & Hoerauf. A. 2008. Zoonotic cutaneous leishmaniasis outbreak in Mazar-e Sharif, northern Afghanistan: An epidemiological evaluation. *Int. J. Med. Microbio.* 298(5–6): 543–50.

Field, H., Young, P., Yob, J.M., Mills, J., Hall, L. & Mackenzie, J. 2001. The natural history of Hendra and Nipah viruses. *Microbes & Infec.* 3: 307–314.

Fish, D. 1995. Environmental risk and prevention of Lyme Disease. *Am. J. Med.* 98(S4A): 2–9.

Focks, D.A., Haile, D.G., Daniels, E. & Mount, G.A. 1993. Dynamic life table model for *Aedes aegypti* (Diptera: Culicidae): simulation results and validation. *J. Med. Entomol.* 30(6): 1018–1028.

Focks, D.A., Brenner, R.J., Hayes, J. & Daniels, E. 2000. Transmission thresholds for dengue in terms of *Aedes aegypti* pupae per person with discussion of their utility in source reduction efforts. *Am. J. Trop. Med. & Hyg.* 62(1): 11–18.

Fogarty, R., Halpin, K., Hyatt, A.D., Daszak, P. & Mungall, B.A. 2008. Henipavirus susceptibility to environmental variables. *Virus Res.* 132: 140–144.

Formenty, P., Hatz, C., LeGuenno, B., Stoll, A., Rogenmoser, P. & Widmer, A. 1999. Human infection due to Ebola virus, subtype Côte d'Ivoire: Clinical and biologic presentation. *J. Infect. Dis.* 179: 48–53.

Franke, C.R., Staubach, C., Ziller, M. & Schluter, H. 2002. Trends in the temporal and spatial distribution of visceral and cutaneous leishmaniasis in the state of Bahia, Brazil, from 1985 to 1999. *Trans. Roy. Soc. Trop. Med. & Hyg.* 96: 236–241.

Garelli, F.M., Espinosa, M.O., Weinberg, D., Coto, H.D., Gaspe, M.S. & Gurtler, R.E. 2009. Patterns of *Aedes aegypti* (Diptera: Culicidae) infestation and container productivity measured using pupal and Stegomyia indices in northern Argentina. *J. Med. Entomol.* 46(5): 1176–1186.

Gaunt, C.M., Mutebi, J.P. & Munstermann, L.E. 2004. Biochemical taxonomy and enzyme electrophoretic profiles during development, for three morphologically similar *Aedes* species (Diptera: Culicidae) of the subgenus Stegomyia. *J. Med. Entomol.* 41(1): 23–32.

Glantz, M.H. 2001. *Current of Change: Impacts of El Niño and la Niña on climate and society.* New York: Cambridge UP.

Goh, K.J., Tan, C.T., Chew, N.K., Tan, P.S.K., Kamarulzaman, A., Sarji, S.A., Wong, K.T., Abdullah, B.J.J., Chua, K.B. & Lam, S.K. 2000. Clinical features of Nipah virus encephalitis among pig farmers in Malaysia. *N Engl. J. Med.* 342: 1229–1235.

González, C., Wang, O., Strutz, S.E., González-Salazar, C., Sánchez-Cordero, V. & Sarkar, S. 2010. Climate change and risk of leishmaniasis in North America: Predictions from ecological niche models of vector and reservoir species. *PLoS Negl. Trop. Dis.* 4(1): e585.

Gratz, N.G. 1999. Emerging and resurging vector-borne diseases. *Ann. Rev. Entomol.* 44: 51–75.

Gubler, D.J. 2001. Prevention and control of tropical diseases in the 21st century: Back to the field. *Am. J. Trop. Med. & Hyg.* 65(1): 5–11.

Gubler, D.J. 2006. Dengue hemorrhagic fever: History & current status. *Novartis Found. Symp.* 277: 3–16; discussion 16–22, 71–73, 251–253.

Gubler, D.J. 2008. The global threat of emergent/reemergent vector-borne diseases. In S.M. Lemon, P.F. Sparling, M.A. Hamburg, D.A. Relman, E.R. Choffnes & A. Mack (eds.), *Vector-borne diseases: Understanding the environmental, human health, and ecological connections:* 43–64. Washington DC: Institute of Medicine.

Gubler, D.J. & Rosen, L. 1976. A simple technique for demonstrating transmission of dengue virus by mosquitoes without the use of vertebrate hosts. *Am. J. Trop. Med. & Hyg.* 25(1): 146–150.

Guerra, J.A. de O., Talhari, S., Paes, M.G, Garrido, M. & Talhari, J.M. 2003. Clinical and diagnostic aspects of American tegumentary leishmaniais in soldiers simultaneously exposed to the infection in the Amazon Region. *Rev. Soc. Brasil. Med. Trop.* 36(5): 587–590.

Gurley, E.S., Montgomery, J.M., Hossain, M.J., Bell, M., Azad, A.K., Islam, M.R., Molla, M.A.R., Carroll, D.S., Ksiazek, T.G., Rota, P.A., Lowe, L., Comer, J.A., Rollin, P., Czub, M., Grolla, A.,

Feldmann, H., Luby, S.P., Woodward, J.L. & Breiman, R.F. 2007. Person-to-person transmission of Nipah virus in a Bangladeshi community. *Emerg. Infect. Dis.*13: 1031–1037.

Hales, S., de Wet, N., Maindonald, J. & Woodward, A. 2002. Potential effect of population and climate changes on global distribution of dengue fever: An empirical model. *Lancet* 360(9336): 830–834.

Halstead, S.B. 1984. Selective primary health care: Strategies for control of disease in the developing world. XI. Dengue. *Rev. Infect. Dis.* 6(2): 251–264.

Hay, S.I, Snow, R.W. & Rogers, D.J. 1998. From predicting mosquito habitat to malaria seasons using remotely sensed data: Practice, problems and perspectives. *Parasit. Today* 14(8): 306–313.

Holland, J. 1995. *Hidden order: How adaptation builds complexity*. Reading: Addison Wesley.

Holmes, E.C. 2011. What does virus evolution tell us about virus origins? *J. Virol.* 85(11): 5247–5251.

Hsu, V.P., Hossain, M.J., Parashar, U.D., Ali, M.M., Ksiazek, T.G., Kuzmin, I., Niezgoda, M., Rupprecht, C., Bresee, J. & Breiman, R.F. 2004. Nipah virus encephalitis recmergence, Bangladesh. *Emerg. Infect. Dis.*10: 2082–2087.

Huang, N.E, Shen, Z, Long, S.R., Wu, M.C., Shih, E.H., Zheng, Q., Tung, C.C. & Liu, H.H. 1998. The empirical mode decomposition method and the Hilbert spectrum for non-stationary time series analysis, *Proc. Roy. Soc. Lond.* 454A: 903–995.

Huang, N.E., Wu, Z., Pinzon, J.E., Parkinson, C.L., Long, S.R., Blank, K., Gloersen, P. & Chen, X. 2009. Reductions of noise and uncertainty in annual global surface temperature anomaly data. *Adv. Adapt. Data Anal.* 1(3): 447–460.

Hurrell, J.W. 1995. Decadal trends in the North Atlantic Oscillation: Regional temperatures and precipitation. *Science* 269: 676–679.

ICDDRB 2003. Outbreaks of encephalitis due to Nipah/Hendra-like Viruses, Western Bangladesh. *Health & Sci. Bull.*1: 1–6 Available from: https://centre.icddrb.org/pub/publication.jsp?classificationID=56& pubID=4673 [Accessed 12th February 2012].

ICDDRB 2004. Person-to-person transmission of Nipah virus during outbreak in Faridpur District, 2004. *Health & Sci. Bull.* 2: 5–9.

ICDDRB 2005. Nipah virus outbreak from date palm juice. *Health & Sci. Bull.* 3: 1–5.

IEDCR 2004. Nipah encephalitis outbreak over wide area of western Bangladesh. *Health & Sci. Bull.* 2: 7–11.

IOM 2003. *Microbial threats to health: Emergence, detection, and response*. Washington DC: National Academy Press.

Jahrling, P.B., Geisbert, T.W., Dalgard, D.W., Johnson, E.D., Ksiazek, T.G., Hall, W.C. & Peters, C.J. 1990. Preliminary report: Isolation of ebola virus from a monkey imported to USA. *Lancet* 335: 502–505.

Jansen, C.C. & Beebe, N.W. 2010. The dengue vector *Aedes aegypti*: What comes next? *Microbes. Infect.* 12(4): 272–279.

Jefferson, H., Dupuy, B., Chaudet, H., Texier, G., Green, A., Barnish, G., Green, A., Barnish, G, Boutin, J.P. & Meynard, J.B. 2008. Evaluation of a syndromic surveillance for the early detection of outbreaks among military personnel in a tropical country. *J. Pub. Health* 30(4): 375–383.

Johansson, M.A., Dominici, F. & Glass, G.E. 2009a. Local and global effects of climate on dengue transmission in Puerto Rico. *PLoS Negl Trop Dis* 3(2): e382.

Johansson, M.A., Cummings, D.A. & Glass, G.E. 2009b. Multiyear climate variability and dengue: El Niño Southern Oscillation, weather and dengue incidence in Puerto Rico, Mexico, and Thailand: A longitudinal data analysis. *PLoS. Med.* 6(11): e1000168.

Johara, M.Y., Y., Field, H., Rashdi, A.M., Morrissy, C., van der Heide, B., Rota, P., bin Adzhar, A., White, J., Daniels, P., Jamaluddin, A. & Ksiazek, T.G. 2001. Nipah virus infection in bats (order Chiroptera) in peninsular Malaysia. *Emerg. Infect. Dis.*7: 439–441.

Johnson, K.M. 1978. Ebola haemorrhagic fever in Zaire, 1976. *Bull. WHO* 56: 271–293.

Jones, K.E., Patel, N.G., Levy, M.A., Storeygard, A., Balk, D. & Gittleman, J.L. 2008. Global trends in emerging infectious diseases. *Nature* 451: 990–994.

Jones, P.D., Jónsson, T. & Wheeler, D. 1997. Extension to the North Atlantic Oscillation using early instrumental pressure observations from Gibraltar and South-West Iccland. *Int. J. Clim.* 17: 1433–1450.

Kamaluddin, M., Nath, T.K. & Jashimuddin, M. 1998. Indigenous practice of Khejur palm (*Phoenix Sylvestris*) husbandry in rural Bangladesh. *J. Trop. Forest Sci.* 10: 357–366.

Karki, K.B. 2007. Greenhouse gases, global warming and glacier ice melt in Nepal. *J. Agric. & Environ.* 8: 1–7.

Kelly, D.J., Richards, A.L., Temenak, J.J., Strickman, D. & Dasch, GA. 2002. The past and present threat of rickettsial diseases to military medicine and international public health. *Clin. Infect. Dis.* 34(Supp4): 145–169.

Kelly, P.J. 2006. *Rickettsia africae* in the West Indies. *Emerg. Infect. Dis.* 12: 224–226.

Khan, A.R.K. 2001. Status and distribution of bats in Bangladesh with notes on their ecology. *Zoos' Print J.* 16: 479–483.

Khan, M.S.U., Nahar, N., Sultana, R., Hossain, M.J., Gurley, E.S. & Luby, S.P. 2008. Understanding bats access to date palm sap: Identifying preventative techniques for Nipah virus transmission *Amer. J. Trop. Med. & Hyg.* 79: 1131.

Kissling, R.E., Robinson, R.Q., Murphy, F.A. & Whitfield, S.G. 1968. Agent of disease contracted from green monkeys. *Science* 160: 888–890.

Kolaczinski, J.H., Hope, A., Ruiz, J.A., Rumunu, J., Richer, M. & Seaman, J. 2008. Kala-azar Epidemiology and Control, Southern Sudan. *Emerg. Infect. Dis.* 14(4): 664.

Kyle, J.L. & Harris, E. 2008. Global spread and persistence of dengue. *Ann. Rev. Microbio.* 62: 71–92.

Lahm, S.A., Kombila, M., Swanepoel, R. & Barnes, R.F.W. 2007. Morbidity and mortality of wild animals in relation to outbreaks of Ebola hemorrhagic fever in Gabon, 1994–2003. *Trans. Royal. Soc. Trop. Med. & Hyg.* 101(1): 64–78.

Langevin, P. & Barclay, M.R.R. 1990. *Hyspignathus monstrosus. Mamm. Species* 357: 1–4.

Leroy, E.M., Rouquet, P., Formenty, P., Souquière, S., Kilbourne, A., Froment, J.M., Bermejo, M., Smit, S., Karesh, W., Swanepoel, R., Zaki S.R. & Rollin, P.E. 2004. Multiple Ebola virus transmission events and rapid decline of central African wildlife. *Science* 303: 387–390.

Leroy, E.M., Kumulungui, B., Pourrut, X., Rouquet, P., Hassanin, A., Yaba, P., Delicat, A., Paweska, J. Gonzalez, J.P. & Swanepoel, R. 2005. Fruit bats as reservoirs of Ebola virus. *Nature* 438: 575–576.

Levin, S.A. 1999. *Fragile dominion: Complexity and the commons.* Reading: Perseus Books.

Lighton, J.R.B., Fielden, L.J. & Rechav, Y., 1993. Discontinuous ventilation in a non-insect, the tick Amblyomma marmoreum (Acari: Ixodidae): Characterization and metabolic modulation. *J. Exp. Zool.* 180: 229–245.

Linthicum, K.J., Anyamba, A., Tucker, C.J., Kelley, P.W., Myers, M.F. & Peters, C.J. 1999. Climate and satellite indicators to forecast Rift Valley Fever epidemics in Kenya. *Science* 285: 397–400.

Luby, S.P., Rahman, M., Hossain, M.J., Blum, L.S., Hussain, N.M., Gurley, E., Khan, R., Rahmin, S., Nahar, N., Kenah, E., Comer, J.A. & Ksiazek, T.G. 2006. Foodborne transmission of Nipah virus, Bangladesh. *Emerg. Infect. Dis.* 12: 1888–1894.

Luby, S.P., Gurley, E.S. & Hossain, M.J. 2009a. Transmission of human infection with Nipah virus. *Clinic. Infect. Dis.* 49: 1743–1748.

Luby, S.P., Hossain, M.J., Gurley, E.S., Ahmed, B-N, Banu, S., Khan, S.U., Homaira, N. Rota, P.A. Rollin, P.E., Comer, J.A., Kenah, E., Ksiazek, T.G. & Rahman, M. 2009b. Recurrent zoonotic transmission of Nipah virus into humans, Bangladesh, 2001–2007. *Emerg. Infect. Dis.* 15(8): 1229–1235.

Magill, A.J., Grogl, M., Gasser, R.A., Sun, W. & Oster, C.N. 1993. Visceral infection caused by *Leishmania tropica* in veterans of Operation Desert Storm. *New Engl. J. Med.* 328: 1383–1387.

Malla, S., Thakur, G.D., Shrestha, S.K., Banjeree, M.K., Thapa, L.B. & Gongal, G. 2008. Identification of all dengue serotypes in Nepal. *Emerg. Infect. Dis.* 14(10): 1669–1670.

Martine, G. 2007. State of world population 2007: Unleashing the potential of urban growth. Available from: http://www.unfpa. org/webdav/site/global/shared/documents/publications/2007/695_filenam e _sowp2007_eng.pdf [Accessed 13th February 2012].

Meynard, J.B., Chaudet, H., Green, A.D., Jefferson, H.L., Texier, G., Webber, D., Dupuy, B. & Boutin, J-P. 2008. Proposal of a framework for evaluating military surveillance systems for early detection of outbreaks on duty areas. *BMC Pub. Health* 8: 146.

MMWR 1999. Update: Outbreak of Nipah virus-Malaysia and Singapore. *MMWR* 48(16): 32–348.

Mohd-Azlan, J., Zubaid, A. & Kunz, T.H. 2001. Distribution, relative abundance, and conservation status of the large flying fox, *Pteropus vampyrus*, in peninsular Malaysia: A preliminary assessment. *Acta Chiropterol.* 3: 149–162.

Montgomery, J.M., Hossain, M., Gurley, E., Carroll, D.S., Croisier, A., Bertherat, E., Asgari, N., Formenty, P., Keeler, N., Comer, J., Bell, M.R., Akram, K., Molla, A.R., Zaman, K., Islam, M.R., Wagoner, K., Mills, J.N., Rollin, P.E., Ksiazek, T.G. & Breiman, R.F. 2008. Risk factors for Nipah virus encephalitis in Bangladesh. *Emerg. Infect. Dis.* 14: 1526–1532.

Morens, D.M. 2008. Confronting vector-borne diseases in an age of ecologic change. In S.M. Lemon, P.F. Sparling, M.A. Hamburg, D.A. Relman, E.R. Choffnes & A. Mack (eds.), *Vector-borne diseases: Understanding the environmental, human health, and ecological connections:* 274–287. Washington DC: Institute of Medicine.

Morens, D.M., Kolkers, G.K. & Fauci, A.S. 2008. Emerging infections: A perpetual challenge. *Lancet Infect. Dis.* 8: 710–719.

Morse, S.S. 2009. Emerging infections: Condemned to repeat? In D.A. Relman, M.A. Hamburg, E.R. Choffnes & A. Mack (rapporteurs), *Microbial evolution and co-adaptation: A tribute to the life and scientific legacies of Joshua Lederberg*: 195–208. Washington DC: Institute of Medicine.

Munderloh, U.G. & Kurtti, T.J. 1995. Cellular and molecular interrelationships between ticks and prokaryotic tick-borne pathogens. *Annu. Rev. Entomol.* 40: 221–243.

Munderloh, U.G., Jauron, S.D. & Kurtti, T.J. 2005. The tick: A different kind of host for human pathogens. In J.L. Goodman, D. Dennis & D.E. Sonenshine (eds.), *Tick-borne diseases of humans:* 37–64 Washington DC: ASM Press.

Murdoch, D.R., Woods, C.W., Zimmerman, M.D., Dull, P.M., Belbase, R.H. & Keenan, A.J. 2004. The etiology of febrile illness in adults presenting to Patan hospital in Kathmandu, Nepal. *Am. J. Trop. Med. & Hyg.* 70(6): 670–675.

Nahar, N., Sultana, R., Gurley, E.S., Hossain, M.J. & Luby, S.P. 2010. Date palm sap collection: Exploring opportunities to prevent Nipah transmission. *Ecohealth* 7: 196–203.

Nakhapakorn, K. & Tripathi, N.K. 2005. An information value-based analysis of physical and climatic factors affecting dengue fever and dengue haemorrhagic fever incidence. *Int. J. Health Geogr.* 4: 13.

Nasci, R.S. 2008. Integration of strategies: Surveillance, diagnosis, and response. In S.M. Lemon, P.F. Sparling, M.A. Hamburg, D.A. Relman, E.R. Choffnes & A. Mack (eds.), *Vector-borne diseases: Understanding the environmental, human health, and ecological connections:* 263–267. Washington DC: Institute of Medicine.

Naucke, T.J. & Schmitt, C. 2004. Is leishmaniasis becoming endemic in Germany? *Int. J. Med. Microbio.* 293(Supp37): 179–181.

Needham, G.R. & Teel, P.D. 1986. Water balance by ticks between bloodmeals. In J.R. Sauer & J.A. Hair (eds.), *Morphology, physiology, and behavioural biology of ticks:* 100–151. Chichester: Ellis Horwood.

Neto, J.C., Werneck, G.L. & Costa, C.H.N. 2009. Factors associated with the incidence of urban visceral leishmaniasis: An ecological study in Teresina, Piauí State, Brazil. *Cadernos de Saúde Pública.* 25(7): 1543–1551.

Niebylski, M.L., Peacock, M.G. & Schwan, T.G. 1999. Lethal effect of *Rickettsia rickettsii* on its tick vector (*Dermacentor andersoni*). *Appl. Environ. Microbiol.* 65: 773–778.

Nieto, P., Malone, J.B. & Bavia, M.E. 2006. Ecological niche modeling for visceral leishmaniasis in the state of Bahia, Brazil, using genetic algorithm for rule-set prediction and growing degree day-water budget analysis. *Geospatial Health* 1: 115–126.

Olson, J., Rupprecht, C., Rollin, P., An, U., Niezgoda, M., Clemins, T., Walston, J. & Ksiazek, T.G. 2002. Antibodies to Nipah-like virus in bats (*Pteropus lylei*), Cambodia. *Emerg. Infect. Dis.* 8: 987–988.

Pandey, B.D., Rai, S.K., Morita, K. & Kurane, I. 2004. First case of Dengue virus infection in Nepal. *Nepal Med. Coll. J.* 6(2): 157–159.

Pant, C.P. & Yasuno, M. 1973. Field studies on the gonotrophic cycle of *Aedes aegypti* in Bangkok, Thailand. *J. Med. Entomol.* 10(2): 219–223.

Pantha, R., Sharma, B.J. 2001. Population size, growth and distribution. *Nepal Popul. J.* 11(10): 67–78.

Parajka, J., Naeimi, V., Bloschl, G. & Komma, J. 2009. Matching ERS scatterometre based soil moisture patterns with simulations of a conceptual dual layer hydrologic model over Austria. *Hydrol. & Earth Syst. Sci.* 13: 259–271.

Parashar, U.D., Sunn, L.M., Ong, F., Mounts, A.W., Arif, M.T., Ksiazek, T.G., Kamaluddin, M.A., Mustafa, A.N., Kaur, H., Ding, L.M., Othman, G., Radzi, H.M., Kitsutani, P.T., Stockton, P.C., Arokiasamy, J., Gary, H.E. & Anderson, L.J. 2000. Case-control study of risk factors for human infection with a new zoonotic paramyxovirus, Nipah virus, during a 1998–1999 outbreak of severe encephalitis in Malaysia. *J. Infect. Dis.* 181: 1755–1759.

Parola, P., Paddock, C.D. & Raoult, D. 2005. Tick-borne rickettsioses around the world: Emerging diseases challenging old concepts. *Clin. Microbiol. Rev.* 18(4): 719–718.

Patz, J.A. & Olson, S.H. 2006. Climate change and health: Global to local influences on disease risk. *Ann. Trop. Med. Parasitol.* 100(5–6): 535–549.

Patz, J.A. & Reisen, W.K. 2001. Immunology, climate change and vector-borne diseases. *Trends Immunol.* 22(4): 171–172.

Patz, J.A., Epstein, P.R., Burke, T.A. & Balbus, J.M. 1996. Global climate change and emerging infectious diseases. *JAMA* 275(3): 217–223.

Patz, J.A., Martens, W.J., Focks, D.A. & Jetten, T.H. 1998. Dengue fever epidemic potential as projected by general circulation models of global climate change. *Environ. Health Perspect.* 106(3): 147–153.

Patz, J.A., Olson, S.H., Uejio, C.K. & Gibbs, H.K. 2008. Disease emergence from global climate and land use change. *Med. Clin. North Am.* 92(6): 1473–1491, xii.

PVDI 2012. http://www.pasteur-international.org/ip/resource/filecenter/document/01s-000042-0m0/dengue-en-basse-def.pdf [Accessed 22nd March 2012].

Perea, W.A., Ancelle, T. Moren, A., Nagelkerke, M. & Sondorp, E. 1991. Visceral leishmaniasis in southern Sudan. *Trans. Roy. Soc. Trop. Med. & Hyg.* 85: 48–53.

Petersen, L.R. & Marfin, A.A. 2005. Shifting epidemiology of Flaviviridae. *J. Travel Med.* 12(Suppl): S3–S11.

Peterson, A.T.D. & Shaw, J. 2003. *Lutzomyia* vectors for cutaneous leishmaniasis in Southern Brazil: Ecological niche models, predicted geographic distributions, and climate change effects. *Int. J. Parasit.* 33: 919–931.

Peterson, A.T.D., Bauer, J.T., Mills, J.N. 2004. Ecologic and geographic distribution of filovirus disease. *Emerg. Infect. Dis.* 10: 40–47.

Pinzon, J.E., Wilson, J.M., Tucker, C.J., Arthur, R., Jahrling, P.B. & Formenty, P. 2004. Trigger events: Enviroclimatic coupling of Ebola haemorrhagic fever outbreaks. *Am. J. Trop. Med. & Hyg.* 71: 664–674.

Pinzon, J.E, Brown, M.E. & Tucker C.J. 2005. Empirical mode decomposition correction of orbital drift artifacts in satellite data stream. In N.E. Huang & S.S. Shen (eds.), *Applications in Hilbert-Huang Transform and Applications*: 167–186. Singapore: World Scientific.

Plowright, R.K., Foley, P., Field, H.E., Dobson, A.P., Foley, J.E., Eby, P. & Daszak, P. 2011. Urban habituation, ecological connectivity and epidemic dampening: The emergence of Hendra virus from flying foxes (*Pteropus* species). *Proc. Roy. Soc. B-Biolog. Sci.* 278: 3703–3712.

Pourrut, X., Kumulungui, B., Wittmann, T., Moussavou, G., Delicat, A., Yaba, P., Nkoghe, D., Gonzalez, J.P. & Leroy, E.M. 2005. The natural history of Ebola virus in Africa. *Microbes Infect.* 7: 1005–1014.

Pourrut, X., Souris, M., Towner, J.S., Rollin, P.E., Nichol, S.T., Gonzalez, J.P. & Leroy, E.M. 2009. Large serological survey showing cocirculation of Ebola and Marburg viruses in Gabonese bat populations, and a high seroprevalence of both viruses in *Rousettus aegyptiacus*. *BMC Infect. Dis.* 9: 159.

Pulliam, J.R.C., Epstein, J.H., Dushoff, J., Rahman, S.A., Bunning, M., Jamaluddin, A.A., Hyatt, A.D., Field, H.E., Dobson, A.P., Daszak, P. & the Henipavirus Ecology Research Group. 2012. Agricultural intensification, priming for persistence, and the emergence of Nipah virus, a lethal bat-borne zoonosis. *J. Roy. Soc. Interface* 9(66): 89–101. doi: 10.1098/rsif.2011.0223.

Rai, M. 2007. Climate change and agriculture: A Nepalese case. *J. Agric & Environ.* 8: 92–95.

Rahman, S.A., Hassan, S.S., Olival, K.J., Mohamed, M., Chang, L.Y., Hassan, L., Suri, A.S., Saad, N.M., Shohaimi, S.A., Mamat, Z.C., Naim, M.S., Epstein, J.H., Field, H.E., Daszak, P. & the Henipavirus Ecology Research Group. 2010. Characterization of Nipah Virus from Naturally Infected *Pteropus vampyrus* Bats, Malaysia. *Emerg. Infect. Dis.* 16: 1990–1993.

Rahman, M.E., Hossain, M.J., Sultana, S. Homaira, N., Khan, S.U., Rahman, M., Gurley, E.S., Rollin, P.E., Lo, M.K., Comer, J.A., Lowe, L., Rota, P.A., Ksiazek, T.G., Kenah, E., Sharker, Y. & Luby, S.P. 2012. Date palm sap linked to Nipah virus outbreak in Bangladesh. *Vector-borne & Zoonot. Dis.* 12(1): 65–72.

Randolph, S.E. 2000. Ticks and tick-borne disease systems in space and from space. *Adv Parasit* 47: 218–243.

Randolph, S.E & Rogers, D.J. 2007. Ecology of tick-borne disease and the role of climate. In F.O. Ergonul & C.A. Whitehouse (eds.), *Crimean-Congo hemorrhagic fever*: 167–186. Dordrecht: Springer.

Randrianasolo, L., Raoelina, Y., Ratsitorahina, M., Ravolomanana, L., Andriamandimby, S., Heraud, J.M., Rakotomanana, F., Ramanjato, R., Randrianarivo- Solofoniaina, A. E. & Richard, V. 2010. Sentinel surveillance system for early outbreak detection in Madagascar. *BMC Pub. Health* 10: 31.

Raoult, D., Founier, P.E., Fenollar, F., Jensenius, M., Price, T., de Pina, J.J., Caruso, G., Jones, N., Laferi, H., Rosenblatt, J.E. & Marrie, T.J. 2001. *Rickettsia africae*, a tick-borne pathogen in travelers to sub-Saharan Africa. *N. Engl. J. Med.* 344(20): 1504–1510.

Ready, P.D. 2010. Leishmaniasis emergence in Europe. *Euro. Surveill.* 15(10): 19505.

Reiter, P. 2001. Climate change and mosquito-borne disease. *Environ. Health Perspect.* 109(Suppl): 141–161.

Reynes, J.M., Counor, D., Ong, S., Faure, C., Seng, V., Molia, S., Walston, J., Georges-Courbot, M.C., Deubel, V. & Sarthou, J.L. 2005. Nipah virus in Lyle's flying foxes, Cambodia. *Emerg. Infect. Dis.* 11: 1042–1047.

Ritmeijer, K. & Davidson, R.N. 2003. Royal Society of Tropical Medicine & Hygiene Joint Meeting with Medecins Sans Frontieres @ Manson House, London, 20 March 2003. *Trans. Roy. Soc. Trop. Med. & Hyg.* 97: 609–613.

Rohani, A., Wong, Y.C., Zamre, I., Lee, H.L. & Zurainee, M.N. 2009. The effect of extrinsic incubation temperature on development of dengue serotype 2 and 4 viruses in *Aedes aegypti* (L.). *SE-Asian J. Trop. Med. & Pub. Health* 40(5): 942–950.

Rouquet, P., Froment, J.M., Bermejo, M., Kilbourne, A., Karesh, W., Reed, P., Kumulungui, G., Yaba, P., Delicat, A., Rollin, P.E. & Leroy, E.M. 2005. Wild animal mortality monitoring and human Ebola outbreaks, Gabon and Republic of Congo, 2001–2003. *Emerg. Infect. Dis.* 11: 283–290.

Sanchez, J.L., Diniega, B.M., Small, J.W., Miller, R.N., Andujar, J.M., Weina, P.J., Lawyer, P.G., Ballou, R. & Loveplace, J.K. 1992. Epidemiologic investigation of an outbreak of cutaneous leishmaniasis in a defined geographic focus of transmission. *Amer. J. Trop. Med. & Hyg.* 47(1): 47–54.

Sanchez, A., Geisbert, T.W. & Feldmann, H. 2007. Filoviridae: Marburg and Ebola viruses. In D.M. Knipe & P.M. Howley (eds.), *Fields Virology:* 1409–1448. Philadelphia: Lippincott Williams & Wilkins.

Sawalha, S.S., Shtayeh, M.S., Khanfar, H.M., Warburg, A. & Abdeen, Z.A. 2003. Phlebotomine sand flies (Diptera: Psychodidae) of the Palestinian West Bank: Potential vectors of leishmaniasis. *J. Med. Entomol.* 40: 321–328.

Schlein, Y & Yuval, B. 1987. Leishmaniasis in the Jordan Valley. IV. Attraction of *Phlebotomus papatasi* (Diptera: Psychodidae) plants in the field. *J. Med. Entomol.* 24: 87–90.

Schlein, Y. & Jacobson, R.L. 1999. Sugar meals and longevity of the sand fly *Phlebotomus papatasi* in an arid focus of Leishmania major in the Jordan Valley. *Med. & Veterin. Entomol.* 13(1): 65–71.

Schneider, K., Huisman, J.A., Breuer, L., Zhao, Y. & Frede, H.G. 2008.Temporal stability of soil moisture in various semi-arid steppe ecosystems and its application in remote sensing. *J. Hydrol.* 359(1–2): 16–29.

Schorscher, J.A. & Goris, M. 1992. Incrimination of *Phlebotomus (Larroussius) orientalis* as a vector of visceral leishmaniasis in western Upper Nile Province, southern Sudan. *Trans. Roy. Soc. Trop. Med. & Hyg.* 86: 622–623.

Scope, A., Trau, H., Anders, G., Barzilai, A., Confino, Y. & Schwartz, E. 2003. Experience with New World cutaneous leishmaniasis in travelers. *J. Amer. Acad. Derm.* 49(4): 672–678.

Seaman, J., Mercer, A.J., Sondrop. 1996. The epidemic of Visceral Leishmaniasis in Western Upper Nile, Southern Sudan: Course and impact from 1984 to 1994. *Int. J. Epidemiol.* 25(4): 862.

Sendow, I., Field, H.E., Adjid, A., Ratnawati, A., Breed, A.C., Morrissy, C.D. & Daniels, P. 2010. Screening for Nipah Virus Infection in West Kalimantan Province, Indonesia. *Zoonoses & Pub. Health* 57: 499–503.

Sharma, U., Redhu, N.S., Mathur P. & Singh. S. 2007. Re-emergence of visceral leishmaniasis in Gujarat, India. *J. Vector-borne Dis.* 44: 230.

Sharma, K., Angel, B., Singh, H., Purohit, A. & Joshi, V. 2008. Entomological studies for surveillance and prevention of dengue in arid and semi-arid districts of Rajasthan, India. *J. Vector-borne Dis.* 45(2): 124–132.

Sherlock, I.A. 1996. Ecological interactions of Visceral Leishmaniasis in the State of Bahia, Brazil. *Memorias do Inst. Oswaldo Cruz* 91(6): 671–683.

Silva, E.S., Gontijo, C.M.F., Pacheco, R.S., Fiuza, V.O.P. & Brazil, R.P. 2001. Visceral Leishmaniasis in the metropolitan region of Belom Horizonte, State of Minas Gerais, Brazil. *Memorias do Instituto Oswaldo Cruz.* 96(3): 285–291.

Silveira, F.T., Ishikawa, E.A.Y., de Souza, A.A.A. & Lainson, R. 2002. An outbreak of cutaneous leishmaniasis among soldiers in Belém, Para State, Brazil, caused by *Leishmania (Viannia) lindenbergi* n. sp.: A new leishmanial parasite of man in the Amazon Region. *Parasite* 9(1): 43–50.

Smith, C.E. 1956. The history of dengue in tropical Asia and its probable relationship to the mosquito *Aedes aegypti. J. Trop. Med. & Hyg.* 59(10): 243–251.

Smith, D.I.H. 1978. Ebola haemorrhagic fever in Sudan, 1976. *Bull. WHO* 56: 247–270.

Smolinski, M.S., Hamburg, M.A. & Lederberg, J. 2003 *Microbial threats to health: Emergence, detection, and response.* Washington DC: National Academies Press.

Socolovschi, C., Mediannikov, O., Raoult, D. & Parola, P. 2009. Update on tick-borne diseases in Europe. *Parasite* 16(4): 259–273.

Sterndale, R.A. 1884. *Natural History of Mammalia of India and Ceylon.* Calcutta: Thacker, Spink & Co.

Swanepoel, R., Smit, S.B., Rollin, P.E., Formenty, P., Leman, P.A., Kemp, A., Burt, F.J., Grobbelaar, A.A., Croft, J, & Bausch, D.G. 2007. Studies of reservoir hosts for Marburg virus. *Emerg. Infect. Dis.* 13(12): 1847–1851.

Takafuji, E.T., Hendricks, L.D., Daubek, J.L., McNeil, K.M., Scagliola, H.M. & Diggs, C.L. 1980. Cutaneous leishmaniasis associated with jungle training. *Amer. J. Trop. Med. & Hyg.* 29(4): 516–520.

Tan, C.T. & Tan, K.S. 2001. Nosocomial transmissibility of Nipah virus. *J. Infect. Dis.* 184: 1367–1367.

Tee, K.K., Takebe, Y. & Kamarulzaman, A. 2009. Emerging and re-emerging viruses in Malaysia, 1997–2007. *Int. J. Infect. Dis.* 13: 307–318.

Thu, H.M., Aye, K.M. & Thein, S. 1998. The effect of temperature and humidity on dengue virus propagation in *Aedes aegypti* mosquitoes. *SE Asian J. Trop. Med. Pub. Health* 29(2): 280–284.

Tipayamongkholgul, M., Fang, C.T., Klinchan, S., Liu, C.M. & King, C.C. 2009. Effects of the El Niño-Southern Oscillation on dengue epidemics in Thailand, 1996–2005. *BMC Pub. Health* 9: 422.

Towner, J.S., Pourrut, X., Albarino, C.G., Nkogue, C.N., Bird, B.H., Girard, G., Ksiazek, T.G., Gonzalez, J.-P., Nichol, S.T. & Leroy, E.M. 2007. Marburg virus infection detected in a common African bat. *PLoS One* 2: e764.

Towner, J.S., Sealy, T.K., Khristova, M.L., Albarino, C.G., Conlan, S., Reeder, S.A., Quan, P.L., Lipkin, W.I., Downing, R., Tappero, J.W., Okware, S., Lutwama, J., Bakamutumaho, B., Kayiwa, J., Comer J.A., Rollin, P.E., Ksiazek, T.G. & Nichol, S.T. 2008. Newly discovered Ebola virus associated with hemorrhagic fever outbreak in Uganda. *PLoS Pathog.* 4(11): e1000212.

Tucker, C.J., Wilson, J.M., Mahoney, R., Anyamba, A., Linthicum, K. & Myers, M.F. 2002. Climatic and ecological context of the 1994–1996 Ebola outbreaks. *Photogr. Engin. Rem. Sens.* 2: 147–152.

Tun-Lin, W., Burkot, T.R. & Kay, B.H. 2000. Effects of temperature and larval diet on development rates and survival of the dengue vector *Aedes aegypti* in north Queensland, Australia. *Med. Vet. Entomol.* 14(1): 31–37.

Tyagi, B.K. & Hiriyan, J. 2004. Breeding of dengue vector *Aedes aegypti* (Linnaeus) in rural Thar desert, northwestern Rajasthan, India. *Dengue Bull.* 28: 220–222.

UN 2002. *Country Profile Nepal.* Available from: http://www.un.org/esa/earthsummit/nepal-cp.htm. [Accessed 12th February 2012].

UN 2005. World Urbanization Prospects: Available from: http://www.un.org/esa/population/publications/WUP2005/2005wup.htm [Accessed 12th February 2012].

Wacharapluesadee, S., Lumlertdacha, B., Boongird, K., Wanghongsa, S., Chanhome, L., Rollin, P., Stockton, P., Rupprecht, C.E., Ksiazek, T.G. & Hemachudha, T. 2005. Bat Nipah Virus, Thailand. *Emerg. Infect. Dis.* 11: 1949–1951.

Wacharapluesadee, S., Boongird, K., Wanghongsa, S., Ratanasetyuth, N., Supavonwong, P., Saengsen, D., Gongal, G.N. & Hemachudha, T. 2010. A longitudinal study of the prevalence of Nipah Virus in *Pteropus lylei* Bats in Thailand: Evidence for seasonal preference in disease transmission. *Vect.-borne & Zoon. Dis.* 10: 183–190.

Walsh, P.D., Abernethy, K.A., Bermejo, M., Beyers, R., DeWachter, P., Akou, E.M., Huijbregts, B.I. Mambounga, D., Toham, K.A. & Kilbourn, M, 2003. Catastrophic ape decline in western equatorial Africa. *Nature* 422: 611–614.

Watts, D.M., Burke, D.S., Harrison, B.A., Whitmire, R.E. & Nisalak, A. 1987. Effect of temperature on the vector efficiency of *Aedes aegypti* for dengue 2 virus. *Am. J. Trop. Med. & Hyg.* 36(1): 143–152.

Weaver, S.C. & Reisen, W.K. 2010. Present and future arboviral threats. *Antiviral Res.* 85(2): 328–345.

Webby, R.J. & Webster, R.G. 2004. Are we ready for pandemic influenza? In S. Knobler, A. Mahmoud, S. Lemon, A. Mack, L. Sivitz, & K. Oberholtzler (eds.), *Learning from SARS: Preparing for the next disease outbreak:* 208–217. Washington DC: Institute of Medicine.

WHO 2004. WHO consultation on priority public health interventions before and during an influenza pandemic. Geneva, Switzerland 16–18 March 2004 and Response 2004. Available from: http://afro.who.int/index.php?option=com_docman&task+doc_download&gid=5116 [Accessed 12th February 2012].

WHO 2009. Dengue and dengue haemorrhagic fever. Available from: http://www.who.int/mediacentre/factsheets/fs117en [Accessed February 12th 2012].

Wilder-Smith, A. & Gubler, D.J. 2008. Geographic expansion of dengue: The impact of international travel. *Med. Clin. North Am.* 92(6): 1377–1390, x.

Wolfe, N.D., Dunavan, C.P. & Diamond, J. 2007. Origins of major human infectious diseases. *Nature* 447: 279–283.

Wolter, K. & Timlin, M.S. 1998. Measuring the strength of ENSO: How does 1997/98 rank? *Weather* 53: 315–324.

Woolhouse, M. & Gaunt, E. 2007 Ecological origins of novel human pathogens. *Criti. Revs. Microbiol.* 33: 231–242.

Wu, P.C., Lay, J.G., Guo, H.R., Lin, C.Y., Lung, S.C. & Su, H.J. 2009. Higher temperature and urbanization affect the spatial patterns of dengue fever transmission in subtropical Taiwan. *Sci. Total Environ.* 407(7): 2224–2233.

Wu, Z. & Huang, N.E. 2009. Ensemble empirical mode decomposition: A noise-assisted data analysis method. *Adv. Adapt. Data Anal.* 1: 1–41.

Yang, H.M., Macoris, M.L., Galvani, K.C., Andrighetti, M.T. & Wanderley, D.M. 2009. Assessing the effects of temperature on the population of *Aedes aegypti*, the vector of dengue. *Epidemiol. Infect.* 137(8): 1188–1202.

Zijlstra, E.E., Ali, A.S., El-Hassan, A.M., El-Toum, I.A., Satti, M., Ghaalib, H.W. & Kager, P.A. 1991. Lala-azar in displaced people from southern Sudan: Epidemiological, clinical and therapeutic findings. *Trans. Roy. Soc. Trop. Med. Hyg.* 85: 365–9.

Data, modelling, and information systems

Chapter 6

Data discovery, access and retrieval

S. Kempler (Auth./ed.)[1] with; G.G. Leptoukh[1], R.K. Kiang[1], R.P. Soebiyanto[1], D.Q. Tong[2],
P. Ceccato[3], S. Maxwell[4], R.G. Rommel[4], G.M. Jacquez[4], K.K. Benedict[5], S.A. Morain[5],
P. Yang[6], Q. Huang[6], M.L. Golden[7], R.S. Chen[7], J.E. Pinzon[8], B. Zaitchik[9], D. Irwin[10],
S. Estes[10], J. Luvall[10], M. Wimberly[11], X. Xiao[12], K.M. Charland[13], R.P. Stumpf[14],
Z. Deng[15], C.E. Tilburg[16], Y. Liu[17], L. McClure[18], & A. Huff[19]

[1] *NASA GSFC, Greenbelt, MD, US*
[2] *SISS, George Mason University, Fairfax, VA, US*
[3] *IRI, Columbia University, Palisades, NY, US*
[4] *BioMedware, Ann Arbor, MI, US*
[5] *EDAC, University of New Mexico, Albuquerque, NM, US*
[6] *George Mason University, Fairfax, VA, US*
[7] *SEDAC, Columbia University, Palisades, NY, US*
[8] *SSAI, Lanham, MD, US*
[9] *Johns Hopkins University, Baltimore, MD, US*
[10] *Marshall Space Flight Center, Huntsville, AL, US*
[11] *South Dakota State University, Brookings, South Dakota, US*
[12] *University of Oklahoma, Norman, OK, US*
[13] *Children's Hospital, Boston, MA, US*
[14] *NOAA National Ocean Service, Silver Spring MD, US*
[15] *Civil & Environmental Engineering, Louisiana State University, Baton Rouge, LA, US*
[16] *University of New England, Biddeford, ME, US*
[17] *School of Public Health, Emory University, Atlanta, GA, US*
[18] *University of Alabama, Birmingham, AL, US*
[19] *Battelle Memorial Institute, Columbus, OH, US*

ABSTRACT: This chapter explores the complex, and sometimes frustrating, world of data discovery, access, delivery and use by reference to the US National Aeronautics & Space Administration's (NASA's) public health applications portfolio in 2011. It also provides examples of global information system applications in health.

1 INTRODUCTION

NASA's Applied Sciences Division (ASD) promotes innovation by public and private sector organizations applying satellite data, models, products, and scientific findings for air quality management and policy activities benefitting human health and safety. The program focuses on themes of air quality planning, forecasting and compliance and the two crosscutting themes of climate and emissions inventories. It also focuses on areas of infectious disease, emergency preparedness and response, and environmental impacts.

Although satellites obviously cannot monitor swarms of malaria-carrying mosquitoes or other zoonotics, they carry sensors that record environmental attributes that control mosquito populations and distributions, and are excellent tools for gathering and transmitting data from remote regions. This makes it possible to record and monitor geographic or meteorological factors favouring mosquito habitats and the possible onset of diseases before they occur (CNES 2011).

In general, international space assets, infrastructure and expertise aim to improve human understanding of the solar system and beyond. Earth observations form a subset of this understanding

Table 1. URLs referenced in section 2.0.

Product/service	URL (all accessed 18th January 2012)
A. Resilience human populations	http://www.enotes.com/public-health-encyclopedia/vector-borne diseases
B. SERVIR	http://www.servirglobal.net/en/AboutSERVIR.aspx
C. EASTWeb	http://globalmonitoring.sdstate.edu/eastweb
D. Hong Kong observatory	http://www.hko.gov.hk/wxinfo/pastwx/extract.htm
E. Climate and health effects	http://www.cdc.gov/climatechange/effects/default.htm
F. Healthy water	http://www.cdc.gov/healthywater/disease
G. Tracking algal blooms	http://ccma.nos.noaa.gov/stressors/extremeevents/hab/RSFieldOps.aspx

as scientists probe for Earth-like planets. The knowledge gained by looking back to Earth helps ground-positioning, communication satellites and associated applications, becoming key tools towards strengthening preparedness, improving surveillance, and providing effective early-warning (ESA 2011).

The Japanese Space Agency (JAXA) also has a concept on space initiatives for health (Igarashi 2010). It is founded on a similar principal that space based remote sensing data can be used to monitor and predict air quality conditions that enable decision support tools and early warning systems for specific diseases. In addition, the international GEO recognizes that collaboration is essential for exploiting the growing potential of Earth observations to support decision making in an increasingly complex and environmentally stressed world (GEO 2011a). In regards to health, GEO defined two tasks in its 2012–2015 Work Plan: HE-01 *Tools and Information for Health Decision Making*, and HE-09-02 *Tracking Pollutants* (GEO 2011b).

These are all testaments to the awareness and willingness of the international community of space agencies and Earth observing organizations to invest resources that bridge scientific discoveries and practical applications that benefit society. These initiatives form the backdrop stimulating remote sensing aided public health research and applications projects. The keystones for facilitating the success of these projects is both the ability to transform long-term environmental data into relevant information useful for public health studies, and the research community's understanding of how emerging and evolving sources of information can be utilized to enhance their work. Using satellite remote sensing data provides the following benefits to environmental data collection: 1) enables continuous data acquisition; 2) helps to refresh environmental data sets with each satellite overpass, conditions permitting; 3) offers synoptic coverage and good spectral resolution to augment point source ground measurements 4) offers accurate data for information and analysis; and 5) serves as a large archive of historical data (ClearLead 2011). The purpose of this Chapter is to provide insights on the availability of EO data and information services applicable to public health interests, and how this information is currently being applied in public health research and development.

Sections 2, 3, and 4 address different aspects and perspectives for how convergence of EO data and health can occur. Section 2 provides synopses of ASD projects to give readers a quick introduction to the variety of topics being developed. They have been contributed by their respective project Principal Investigators as indicated in Tables 2, 3, and 4, noting that project titles have been edited for length. Section 3 is an in-depth look at three initiatives showing how sensor data often are used for public health, and how remote sensing data enhance research. Section 4 focuses on three case studies on how to access and use appropriate data sets.

2   EARTH SCIENCE SENSOR DATA FOR PUBLIC HEALTH APPLICATIONS

Section 2 is a sample of project synopses from ASD's health and air quality program 2011 portfolio of health projects (NASA 2012). The purpose for profiling them is to show how they link environment and health communities; how they advance the roles of spectral and spatial science; and how they lead to social and economic benefits. It is clear that both communities need to form collaborations to synergize their distinct skill sets, and to fuel ideas for further data and methods

Table 2.   Selected NASA-funded research applications on vector-borne diseases.

Principal investigator	Brief title of project
J.E. Pinzon, GSFC/SSAI	Predicting zoonotic haemorrhagic fever in sub-Saharan Africa
B. Zaitchik, Johns Hopkins Univ.	Detection and early warning for malaria risk in the Amazon
D. Irwin & J. Kessler, MSFC	SERVIR Africa
S. Estes, MSFC/USRA	Potential range expansion of *Aedes aegypti* in Mexico
M. Wimberly, SDSU	Forecasting mosquito-borne disease outbreaks using AMSR-E
R. Kiang, GSFC	Avian influenza risk prediction in Southeast Asia
X. Xiao, Univ. of OK	Sensor imagery and satellite telemetry for avian influenza
K. Charland, Children's Hospital, Boston	Data for an influenza forecasting system

sharing amongst new and experienced users. Since 2003, ASD has funded some thirty projects related to health. They describe uses of EO data sets, collateral data sets; methods for using data; techniques for modelling and transforming data into public health information; and, anticipated project outputs and benefits to appropriate user communities. Projects are presented in three broad disease categories: vector-borne, water-borne, and air-borne. For ease of presentation, all URLs for this section are listed in Table 1. Other citations for URLs are given in the reference section to this Chapter in standard reference format. Table 1 is a list of URLs referenced in Section 2, and Table 2 lists a selection of NASA-funded projects focused on for vector-borne diseases.

2.1   Data sets for vector-borne disease studies

Vectors are the transmitters of disease-causing organisms that carry pathogens from one host to another. By common usage, vectors are considered to be invertebrate animals, usually arthropods. Furthermore, key components that trigger vector-borne diseases include: 1) abundance of vectors and reservoir hosts; 2) prevalence of disease-causing pathogens suitably adapted to the vectors and the human or animal host; 3) local environmental conditions, especially temperature and humidity; and 4) resilience behaviour and immune status of the human population (Table 1, A). It is the third component, environmental conditions that most projects address. The significance of these project synopses is realized when considering the fragile balance of our physical and biological environment. Abrupt or long-term changes in abiotic factors may lead to an alteration in ecosystem equilibrium, resulting in more or less favourable vector habitats. Anticipated changes in temperature and precipitation resulting from global warming will affect vector reproduction and longevity, rate of development of the parasites, and pathogens in the vector, as well as the geographic extent of invasions (WHO 1990). Abiotic environmental factors also affect vegetation patterns that determine distributions of disease-causing vectors. Some of these are temperature, precipitation, relative humidity, wind, solar radiation, topography, and fresh water rivers, ponds, and lakes (Kay *et al.*, 1989). Research continues in understanding and modelling environmental factors used to predict the magnitude and direction of an infected vector, translate these predictions to early warning systems, and aid policy decision makers.

2.1.1   Predicting zoonotic haemorrhagic fever in sub-Saharan Africa
Knowledge generated about infectious diseases through basic research on associations, causes and effects of climate has been an important source of information for public health decision makers. The processes involved and their dynamics show a need for synergistic interpretations with field observation that include monitoring and analysis of remotely sensed data. This project involves a multidisciplinary team of earth scientists, epidemiologists and public health experts using satellite data to map the risk of outbreaks of deadly viruses like EBOV, MARV and RVFV.

Emerging infectious diseases are a global and regional security issue with the capacity to have serious human health and economic impacts, worldwide. This project is integrating EO data into a global emerging infectious surveillance and response system (GEIS) to complement it with a systematic method for monitoring and forecasting environmental and climatic risk factors associated with emerging infectious diseases. NDVI temperature data from the MODIS instruments on Terra and Aqua; TRMM/GPCP (precipitation) data, SRTM (elevation data) and simulated products from both the future NPOESS preparatory project (NPP) and the Global Precipitation Mission (GPM) are projected. These data sets will aid in migrating a reactive surveillance system to one that is proactive. The goal of this synergy is to facilitate decision making with early warning tools.

The end-of-project goal is an eco-climate monitoring algorithm that assesses environmental and climatic risk factors quantitatively, and that could lead to outbreaks of vector-borne diseases. It will provide risk maps that highlight areas where targeted surveillance should be implemented. Monthly environmental risk maps for zoonoses focusing on EBOV, MARV filoviruses and RVF in Africa will be provided. Environmental risk maps support GEIS efforts to improve surveillance systems crucial for preventing, detecting and containing emerging infections that threaten military personnel, their families, and national security, and enhance GEIS overseas military research units with their service to host country counterparts, WHO, and FAO to improve local epidemiological capabilities.

2.1.2 Detection & early warning system for malaria risk in the Amazon

Malaria is a leading cause of morbidity in the Amazon basin. Major challenges remain in targeting, intervention, and control strategies; in particular the distribution of health resources (treatments, diagnostics, and long-lasting impregnated nets) due to eco-social dynamics that result in differences between where people are infected and where they are diagnosed. An effective malaria risk monitoring system begins with reliable estimates of the eco-physiological factors that drive *Anopheline* mosquito populations. On the Amazon frontier, as in many other malaria-prone regions, *Anopheline* density is known to be associated with meteorological factors, including precipitation and air temperature, with land surface characteristics, including land cover and local topography, and with surface hydrological conditions, including soil moisture and surface water ponding. Thus, this project's first objective is to merge multiple EO data sets through a land data assimilation system (LDAS) that can be used to drive spatially explicit ecological models of anopheline mosquito distribution. Assimilations will be driven by observational data. LDAS simulations will make use of satellite-derived meteorological forcing data, parameter data sets, and assimilation observations, including: 1) precipitation from TRMM and GPM; 2) land cover from MODIS, Landsat, ASTER, and LDCM; 3) soil moisture from AMSR-E (where applicable) and Soil Moisture Active-Passive (SMAP); 4) terrestrial water storage from the Gravity Recovery and Climate Experiment (GRACE and GRACE-II); 5) surface temperature from MODIS, Landsat, ASTER, LDCM; 6) Vegetation Fraction/Leaf Area Index from MODIS, Landsat, ASTER, LDCM; and 7) topography from SRTM. Earth science data will be important drivers of models for mosquito population dynamics. The LDAS will drive ecologically based models of mosquito dynamics, developed using data from larva and adult mosquito collection sites maintained in the study region.

The second objective is to develop a human activity and settlements map that uses a spatially explicit model of human settlement locations derived from census and regional studies, areas of forest concessions, and indicators of forest disturbance to identify permanent and temporary sites of human activity. Sensor data will help generate permanent and temporary areas of human activity and augment maps of human settlements.

The third objective integrates ecological and human population models resulting from the companion objectives to create spatial risk maps of human malaria risk. Overall, the project builds on collaborations between investigators expert in land use, climate and ecological modelling, and epidemiologists who are expert in vector-borne diseases, biostatistics, and demography to inform health interventions.

2.1.3 SERVIR Africa

Efforts to implement public health early warning and decision support systems are not limited to small teams using new technologies and information sources to improve state-of-the-art health surveillance. For example, SERVIR (a Spanish acronym for *Regional Visualization and Monitoring System*) (Table 1, B) enables EO data sets and predictive models to be used for timely decisions through regional platforms in Mesoamerica, East Africa, and the Hindu-Kush Himalayas. The system was developed initially in 2004 through joint sponsorship of NASA, the US Agency for International Development (US/AID), the World Bank, and the Central American Commission for Environment and Development (CCAD). It is an initiative that applies earth observations and predictive models to support decision making by government officials, managers, scientists, researchers, students, and the public. In 2005, the Water Center for the Humid Tropics of Latin America and the Caribbean (CATHALAC) in Panama became the first regional SERVIR facility, serving Central America and the Dominican Republic. In late 2008, a SERVIR facility at the Regional Center for Mapping of Resources for Development (RCMRD) in Nairobi, Kenya, was dedicated to serve East Africa. A SERVIR facility for the Hindu-Kush Himalaya region in Asia was inaugurated in Kathmandu, Nepal in October, 2010. US/AID and NASA provide primary support for SERVIR, with the long term goal of transferring its capabilities to other host countries.

The objective of this project is to initiate a SERVIR-like system using NASA science research results to improve decision support in Africa. The initial focus is on flooding and RVF because they are both related to several of the environmental parameters necessary to monitor them: precipitation, topography, soil moisture and land cover. Data from multiple missions, sensors, and models are being used. These include SRTM, AMSR-E, TRMM, MODIS, the global hazard model-flood (GHMF), and the infectious disease eco-climatic link (IDEL) algorithm. The goal is to create a functioning SERVIR-Africa node at RCMRD and to improve current decision making processes of the Kenya flood response system. Additional benefits will be improved flood forecasting and monitoring capabilities at the operational SERVIR facility in Central America.

In addition, SERVIR has become a link to several relevant articles (Kalnay et al., 1996; Huffman et al., 2001; Uppala et al., 2008; Caminade et al., 2011). Caminade (et al., 2011), in particular, discusses using climatic indicators to map RVF and malaria over West Africa. In order to model and map both diseases, different climate data sets are used. Daily rainfall is estimated using mixed satellite and rain gauge observations from the global precipitation climatology project (GPCP) data set, and rainfall and temperature from both the National Centers for Environmental Prediction (NCEP) and the European Centre for Medium Range Weather Forecasts (ECMWF) reanalysis projects, as described in ECMWF Newsletter No 110 (ECMWF 2007). Dynamic models driven by daily rainfall and temperature simulate malaria incidence in the human population, and a methodology for tracking weather and climate events that enhance RVF risk over West Africa are being employed.

2.1.4 *Range expansion of Aedes aegypti in Mexico*

This project is an international collaboration between the Universities Space Research Association (USRA), NASA, The National Center for Atmospheric Research (NCAR), CDC, and the University of Veracruz in Mexico. By modelling the social, economic, environmental and epidemiological factors that influence the survival and abundance of *Ae. Aegypti*, it will be easier to control the primary transmitters of dengue viruses. The ultimate goal is to employ this integrated modelling approach to understand the potential of *Ae Aegypti* to expand its range into heavily populated, high elevation areas like Mexico City under various climate and socioeconomic scenarios.

The first objective is to employ EO data to augment the environmental monitoring and modelling component of the National Science Foundation project. Several remote sensing products have been proposed and their usefulness is being determined through ground verification. The data and their uses include: 1) MODIS NDVI to monitor vegetation and seasonality in a transect area from Veracruz to Mexico City. This 250 m resolution product is a sixteen-day average ratio of the reflectance of red and near infrared spectral bands. It is a gridded, product, masked for water, clouds, and cloud shadows; 2) MODIS LST is used as an indicator of Earths' surface energy balance often used to relate climate, hydrological, ecological and other environmental variables; 3) MODIS LCLU data are used to represent land cover derived from a year's input of Terra observations. The scheme identifies seventeen land cover classes defined by the International Geosphere Biosphere Programme (IGBP); 4) soil moisture data from the AMSR-E are used in regions of low to moderate vegetation and processed to provide soil moisture conditions to a depth of one to two centimetres (Njoku *et al.*, 2003); 5) digital elevation from SRTM are used to provide surface topography (Farr *et al.*, 2007); and 6) climate prediction centre morphing technique (CMORPH) will provide precipitation data based on microwave observations from multiple sensors. Products are available at several time intervals, the most appropriate for this purpose being thirty minute and three hour intervals. These data are critical for understanding the habitat requirements necessary for mosquito survival and spread.

2.1.5 *Forecasting mosquito-borne disease outbreaks using AMSR-E*

This research uses data from AMSR-E to develop improved environmental models of mosquito-borne disease risk. AMRS-E products provide several critical environmental variables that are directly relevant to mosquito ecology, including near-surface air temperature, soil moisture, and fractional water cover. These variables will be used to model and forecast inter-annual fluctuations in WNV cases and mosquito populations. A limitation of vegetation indices, land surface temperature, and many other products derived from existing remote sensing platforms is that they provide indirect measurements of the proximal environmental factors influencing mosquito populations and disease risk. The recent development of a new set of daily global land surface parameters derived from AMSR-E offers novel environmental metrics and expanded opportunities for mosquito-borne disease risk forecasting. The specific objectives of this research are to: 1) develop statistical models of WNV risk and mosquito population dynamics; 2) compare their performance with models based on MODIS and TRMM products to quantify potential improvements in forecasting WNV outbreaks; and 3) generate and disseminate early warning predictions from these models to gain qualitative feedback from vector control experts and public health practitioners.

These objectives are to use the archive of AMSR-E land surface observations commencing in 2002 and to summarize the data temporally to match eight-day MODIS composites. These will be spatially matched to human case data at the county level and mosquito data at the city boundary level. Appropriate statistical methods are being applied to account for spatial and temporal autocorrelation, seasonality, and the effects of other environmental variables such as land cover and land use. Data from the AMSR-2 sensor on the JAXA GCOM-W satellite will be used in future to produce seasonal forecasts of WNV risk for the South Dakota Department of Health. The expected benefits of this project are: 1) its prospect for incorporating novel information from a region of the electromagnetic spectrum currently underutilized in mosquito-borne disease research; and 2) its prospect for producing a more accurate and effective early-warning system.

The larger dimension of the effort is to develop early warning systems that can forecast areas of future disease risk based on environmental variability in space and time. The epidemiological applications of spatial technologies project (Table 1, C) is focused on WNV in the northern Great Plains of the US, and malaria in the Amhara region of Ethiopia (EASTWeb 2012). So far progress has been made applying geospatial technologies for mapping, risk analysis, and ecological forecasting of infectious diseases. An earlier project applied merged sensor data with a GIS to manage and process spatial statistics for analysing disease patterns and developing predictive models to explain how climate and land use patterns influence outbreaks of vector and host populations.

2.1.6 *Avian influenza risk prediction and early warning of pandemic influenza*

This research utilizes sensor data, models and analysis techniques to enhance decision support capabilities for avian influenza (AI) and pandemic influenza (PI) risks. The first objective focused on assessing AI risks for poultry farms and humans and on the potential for early detection of pandemic influenza. In particular, spatial and-temporal risks of H5N1 outbreaks for selected districts in Indonesia were generated, along with short- and medium-term influenza-like illness forecasts for selected regions. The second objective was to advance the established capabilities for modelling seasonal influenza created in the former project, by refining the models and extending their capabilities to a global scale. Toward this end, climate-based models to predict influenza risks in major cities around the world are being developed.

Another research focus uses weekly counts of laboratory-confirmed influenza viruses, and climatic and meteorological parameters collected from two primary sources: ground-based and satellite-derived measurements. Daily meteorological observations were retrieved from the Hong Kong Observatory (Table 1, D) including maximum, mean, minimum temperature, mean dew point temperature, mean relative humidity, global solar radiation and total evaporation. For the US study site in Maricopa County, Arizona, daily climatic observations were acquired from the local flood control district. Data were aggregated from thirty-two stations. These data include daily mean air temperature, minimum, mean and maximum dew point, minimum and maximum relative humidity, maximum wind speed, minimum and maximum air pressure, and maximum solar radiation. EO measurements of daily rainfall also were obtained for both Hong Kong and Maricopa County from instruments on TRMM. Daily LST data from MODIS also were extracted. Both Terra and Aqua missions carry this instrument. Temperature and precipitation are very important environmental determinants of infectious disease transmission (Kiang *et al.*, 2006; Xiao *et al.*, 2007; Soebiyanto *et al.*, 2010). Consequently, the multinational GPM is an important successor to TRMM. It is scheduled for launch in 2013. If successful, it should provide a key data set for continued influenza monitoring.

Understanding and controlling AI outbreaks brings substantially more benefits to the society than just the farms where the outbreaks occur. It spares extensive culling, preserves the livelihood of small farmers, and protects food security and biodiversity. The expansion of developed capabilities to the global scale requires collaboration with CDC's international epidemiology and response team, influenza division, to acquire data from targeted countries. Predictive capabilities for seasonal influenza in major population centres are being developed. The predictive and early warning capabilities developed by this project will promote influenza surveillance, prediction and control at CDC and key public health agencies around the world (Table 1, E).

Research on malaria surveillance modelling uses NASA data, model outputs, and analytical and modelling expertise to enhance decision support capabilities for malaria risk assessment and control. The capabilities that are developed concern detection, prediction and reduction of malaria risk. Since rainfall provides vector breeding sites and prolongs vector life span by increasing humidity, anomalies in precipitation are the attribute most frequently used for predicting malaria outbreaks. However, it has also been shown that rainfall, or the lack of it, has a complex effect on malaria transmission for various parts of the world (Kovats *et al.*, 2003). For example, although moderate rainfall may promote malaria transmission, intense and prolonged rainfall may flush away larval habitats and thus reduce transmissions. Similarly, lack of rainfall does not always reduce larval populations. On the contrary, lack of rainfall may create new habitats, such as pools and puddles, in some regions and therefore increase larval population. In addition, droughts may be deleterious to predator populations or may cause human populations with no immunity to move to areas endemic with malaria (Kovats *et al.*, 2003). These factors may indirectly increase overall malaria transmissions.

Another meteorological variable that is often used for predicting malaria transmission is temperature. Warmer temperature hastens larval and vector development and therefore increases the rate of vector production (Craig *et al.*, 1999). It also shortens the time for spores to reproduce via multiple fission, which thereby produce more vectors, and more time for them to transmit malaria. In addition, warmer air holds more moisture and therefore enhances mosquito survivorship.

Relative humidity is important for the survivorship of malaria vectors. While it is not a standard remote sensing data product, relative humidity can be computed from dew point and air temperatures, which are usually provided for from satellite instruments that measure atmosphere properties.

Vegetation is often associated with vector breeding, feeding, and resting locations. A number of vegetation indices have been used in remote sensing and Earth science disciplines. The most widely used is NDVI (Tucker 1979). It is simply defined as the difference between the red and the near infrared bands normalized by twice the mean of these two bands. For green vegetation, the reflectance in the red band is low because of chlorophyll absorption, and the reflectance in the near infra-red band is high because of the spongy mesophyll leave structure. The more vigorous and denser the vegetation is, therefore, the higher the NDVI becomes.

Remote sensing instruments and geophysical parameters used in these studies include: ASTER (ground cover, vegetation index, rainfall), AVHRR (ground cover, vegetation index, rainfall), MODIS (ground cover, vegetation index, surface temperature, humidity), TRMM (rainfall), IKONOS (ground cover, vegetation index, rainfall), and SRTM (DEM).

2.1.7 *Sensor data and satellite telemetry of wild birds for decision support of avian influenza*

This research also addresses avian influenza (H5N1 subtypes), but from a slightly different perspective. It examines the interplay between the local persistence of HPAI (an avian-adapted strain of H5N1) in poultry, and episodic long-distance dispersal by migratory wild birds, which pose serious threats to poultry production, wild birds, and human health. Geospatial technologies are needed to understand migration patterns among wild birds that spread diseases over long distances. Specifically, MODIS time series data are used to map cropping intensity (double and triple crop cultivation per year), crop calendar (planting dates and harvesting date), and paddy rice fields. PALSAR data are used to map water bodies and natural wetlands. In addition, satellite telemetry data are used to track wild waterfowl migrating from Southern China, to Siberia and Russia. This project provides updated and improved geospatial data sets for cropping intensity, paddy rice, wetlands and LST; and uses the emergency prevention system (EMPRES) model for priority animal, plant pests, and diseases to assess, forecast, and communicate risks of avian influenza.

Maps of avian influenza *hot-spots* and *hot times* provide timely support for disease surveillance, pandemic preparedness planning, and disease management and response. Another objective is to train young scientists in the era of one world/one health, including agro-ecology, wild bird biology, veterinary, epidemiology, public health, eco-informatics and geo-informatics.

2.1.8 *Develop an influenza forecasting system*

This research combines data on influenza activity from CDC with key environmental data sets on climate and phenology obtained by TOPS to understand environmental drivers affecting influenza transmission patterns, and the timing of peaks in influenza activity. Influenza is a contagious respiratory illness caused by viruses that affect between five and twenty per cent of the US population each year, resulting in more than 200,000 hospitalizations and 36,000 deaths on average. Influenza in temperate regions exhibits a pronounced seasonality that has been well described. A number of hypotheses have been proposed to explain this seasonality, and recent studies have demonstrated both a temperature threshold effect (in animal studies) and a significant correlation between latitude and the timing of annual influenza epidemics. These results suggest that environmental drivers play a role in influenza transmission, innate immunity, and/or virus-host interaction, and could be used to forecast the date of peak influenza activity. In particular, the average difference in peak week for southern *vs.* northern cities is approximately sixty days, which matches phenological patterns closely in the US and suggests a potential link between the two.

In recent years, absolute humidity, solar radiation, temperature and relative humidity have emerged as important determinants of the timing of seasonal influenza epidemics. To date, research results have shown that drops in absolute humidity tend to be followed by increases in influenza mortality (Shaman *et al.*, 2010). Other studies observed a relationship between latitude, solar radiation and the timing of peak influenza activity (Charland *et al.*, 2009; Finkelman *et al.*, 2007). Precipitation is also an important predictor of increased influenza incidence in tropical and sub-tropical regions. This research utilizes various statistical analyses to relate climatological information to health data. Daily estimates of average saturation vapour pressure deficit, solar radiation, maximum temperature and precipitation were obtained from TOPS. Interestingly, results show limited evidence that absolute humidity, solar radiation, precipitation and maximum temperature could add to successful forecasting of influenza. Time series models that accounted for seasonality performed as well as the models with meteorological inputs. Only peak timing appeared to be meaningfully associated with vapour pressure deficit, solar radiation, maximum temperature, but the forecast error suggested that forecasts of the peak timing could be miss-forecast by between two to three weeks on average.

Table 3. Selected NASA-funded research applications on water-borne diseases.

Principal investigator	Brief title of project
R. Stumpf & T. Wynne, NOAA	Monitoring and forecasting cyanobacterial blooms
Z. Deng, LSU	Satellite detection and forecasting of oyster norovirus outbreak
C. Tilburg, Univ. of New England	Land-use and precipitation for hydrology & public health

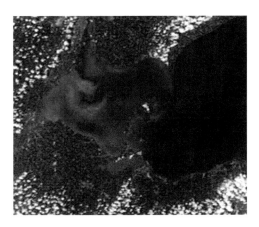

Figure 1. MODIS true colour image showing a bloom in the western basin of Lake Erie, August 2009. (see colour plate 39)

2.2 Data sets for water-borne disease studies

Many illnesses, contaminants, and injuries can be water-, sanitation-, or hygiene-related. Water-borne diseases are caused by organisms that are spread through water directly. Water-related illnesses can be acquired through an absence of water for good hygiene, poor sanitation, or increasing insect populations that breed in water and then spread disease (Table 1, F). The projects discussed in Table 3 describe how the addition of remote sensing data has aided in detecting and forecasting water-borne diseases.

2.2.1 Monitoring and forecasting cyanobacterial blooms

This project addresses cyanobacterial harmful algal blooms (CyanoHABs), a global problem affecting public and environmental health (see also Chapter 10). The project is developing a modelling and forecasting system (MFS) to identify, document, and forecast CyanoHABs in major water bodies to aid environmental and public health managers in planning responses. Satellite sensors are used to collect data on ocean colour to characterize the amount of live phytoplankton biomass and turbidity data, identify algal blooms, and track plumes (Table 1, G). This project combines environmental, meteorological, and disease surveillance data with colour and temperature satellite data from MODIS and VIIRS. Ocean colour sensors provide large synoptic sampling otherwise impossible to collect with standard *in situ* techniques that may be useful for identifying harmful CyanoHABs (Kahru 1997; Kutser 2004; Wynne *et al.*, 2008). Once an algal bloom has been identified, a method must be in place to track the bloom's position and to compensate for missing surface data caused by cloud cover (Figure 1). The nowcast is created using imagery that is generally older than one day. Wynne (*et al.*, 2011) discusses this further and provides detailed methodologies for sensor data, hydrodynamic models and particle tracking. The MFS capability includes: 1) routine bloom identification; 2) forecasting CyanoHAB events; and 3) transferring results to health and environmental management agencies. Successful implementation should reduce monitoring costs, improve water management practices, and reduce public health impacts.

2.2.2   *Satellite-assisted detection and forecasting of oyster norovirus outbreak*

This project utilises Terra and Aqua MODIS data to detect and forecast oyster norovirus outbreaks in coastal Louisiana. The strategy is to combine environmental data with *in-situ* bacteriological data from field samplers and laboratory analysis. The system consists of a series of retrieval algorithms or water quality models that link MODIS data to water quality indicators controlling norovirus disease outbreaks, an artificial neural network model for detecting and forecasting faecal coliform (norovirus indicator organism), and a hierarchical Bayesian model for detecting and forecasting norovirus disease outbreak risks.

The system enables shellfish managers to make two types of decisions: detection management decisions (open/close) and forecasting management decisions (classification and reclassification). The decision management capability makes it possible to reduce decision making time from its current time lag of two to four months, to one day. The forecast management function predicts oyster norovirus outbreaks in a probabilistic fashion. The system is essential to classify/reclassify oyster growing areas and for long-term planning and sustainable oyster management.

2.2.3   *Influence of land-use and precipitation on regional hydrology & public health*

The central objective of this project is to determine if a relatively simple regression model can be used to predict water quality along the coast of Maine. In the north eastern US, climate change scenarios typically indicate an increase in overall precipitation, which would lead to larger river discharge. Increased discharge would then lead to greater pathogen loading in rivers and coastal waters from anthropogenic sources due to runoff from contaminated sites. To achieve the project's objective, EO and *in-situ* data of the Gulf of Maine watersheds, estuaries, and coastal ocean are being used in simple regression models. Observations for the project were collected over a twelve year time period from TRMM's TMI sensor designed to measure rainfall over the swath under the TRMM satellite. TRMM is able to distinguish precipitation variations at a resolution of 0.25°C. In addition, imagery from Landsat is being used to quantify changes in LCLU in the modelled region. River data from USGS and other sources allow hydrology and nutrient discharges to be quantified. Water quality data obtained from state agencies allow models to be calibrated and enable predictions of water quality for these same state agencies. Information from the models is used to determine the feasibility of this approach to climate scenarios.

The most common method for testing water for pathogens is to measure the concentration of *E. coli* and total coliforms, bacteria that are normally present in the intestinal tract of humans and other animals and that are used as indicators for recent faecal contamination. Since the State of Maine does not currently have the ability to test for *E. coli* and total coliforms quickly, they use river discharge as a simple proxy for water quality. Maine's current method relates low water quality events to high measured discharge. High discharge events would trigger increased run-off and low water quality. Water resource managers are particularly interested in the ability of a method to predict low water quality events but also to minimize false alarms or times when a method predicts a low water quality event that does not materialize.

Bulk measurements of discharge are not able to differentiate between run-off from high pollution areas and run-off from less-developed, pristine areas. A method that identifies where run-off occurs would likely predict reduced water quality events more accurately. A precipitation-based method that uses simple linear regression of observed precipitation and observed water quality shows promise for calibrating the model. To compare the effectiveness of the precipitation-based method and Maine's current method, the percentage of actual reduced water quality events that were predicted by each method and the percentage of predicted reduced water quality events that preceded actual events have been examined. The precipitation-based method predicted low water quality events more accurately, and with fewer false alarms. Models developed by this project are determining that accurate measurements of precipitation, land-use, and river discharge can be

Table 4. Selected NASA-funded research applications on air-borne diseases.

Principal investigator	Brief title of project
Y. Liu, Emory Univ.	Tracking and modelling particle exposure for health/epidemiology
J. Luvall, MSFC	Air-borne pollen prediction for asthma alerts and health decisions
L. McClure, Univ. of Alabama	Environmental data linked to cohort data for health decisions
S. Morain, Univ. of NM	Adding EO data to EPHTN via the NM/EPHT system
A. Huff, Battelle Mem. Inst.	Using AOD data to create $PM_{2.5}$ fields for health and epidemiology

used to develop regression models to achieve accurate predictions of water quality and that can be expanded to larger temporal and spatial scales.

2.3 Data sets useful for air-borne diseases, air quality, and health

Air pollution affects health in many ways with both short-term and long-term effects. Examples of short-term effects include irritation to the eyes, nose, throat and upper respiratory infections such as bronchitis and pneumonia. Short-term air pollution can aggravate individuals suffering with asthma and emphysema. Long-term health effects can include COPD, lung cancer, heart disease, and even damage to the brain, nerves, liver, or kidneys. Chronic exposure to persistent organic pollutants (POPs) affects the lungs of growing children and may aggravate or complicate medical conditions in the elderly (LBL 2011). Earth observation data are advancing scientific understanding of the effects of elevated concentrations of POPs, and in tracking their movements for more informed policy decisions. Table 4 lists five projects in NASA's 2011 portfolio for air-borne diseases.

2.3.1 Environmental public health tracking using particle exposure models and epidemiology

As one of six ambient air pollutants, fine particulate matter (i.e. air-borne particles less than or equal to 2.5 micrometres in aerodynamic diameter-$PM_{2.5}$) has been linked to various acute and chronic health problems. Because satellite aerosol remote sensing expands the coverage of $PM_{2.5}$ monitoring to rural and suburban areas, it is important to examine how satellite aerosol remote sensing might extend CDC's coverage of fine particulates in the national environmental public health tracking network (EPHTN), and specifically, the utility of satellite-derived air quality estimates.

This project is developing spatial statistical models that integrate aerosol information from: 1) MODIS level-2 aerosol product; 2) the GOES aerosol and smoke product (GASP); 3) the MISR level-2 aerosol product; 4) the OMI aerosol index; 5) meteorological observations; 6) land use information; and, 7) EPA $PM_{2.5}$ measurements. The aim is to produce a spatially resolved daily $PM_{2.5}$ concentration surface. These data sets with different spatial resolutions are all reprojected to the twelve kilometre resolution community multi-scale air quality (CMAQ) model grid and compared to EPHTN's tracking network of existing CMAQ-simulated $PM_{2.5}$ concentrations. In addition, satellite predictions are validated prospectively with independent field sampling. Satellite predictions are also evaluated in epidemiological models that link $PM_{2.5}$ levels with cardiorespiratory morbidity outcomes collected in the Atlanta, Georgia metropolitan area. Outputs include detailed analyses of the spatial and temporal patterns of $PM_{2.5}$ pollution in the domain, a statistical evaluation of the advantages of estimating $PM_{2.5}$ concentrations, and an assessment of the potential benefit of ingesting satellite observations into EPHTN.

2.3.2 Air-borne dust prediction and vegetation phenology for tracking pollen episodes

This health application explores details of phenology and meteorology, and their dependencies to produce a first-generation deterministic model for predicting and simulating pollen emission and downwind pollen concentration. Pollen can be transported great distances. In fact, pollen from *Juniperus spp.* can be transported 200–600 km downwind (Van de Water *et al.*, 2003). Hence, local observations of plant phenology may not be consistent with the timing and sources of pollen collected by air sampling instruments. Based on the NCEP non-hydrostatic meteorological model, satellite sensor data, and an *in-situ* network of phenology cameras, a real-time and rapid response pollen release and transport system is being prototyped. This prototype is based on the rapid response MODIS direct broadcast system to acquire daily data with one hour lag time, similar to the MODIS rapid response system for fire detection. Research outputs will be used to support EPHTN, which includes the State of New Mexico's environmental public health tracking system, and the syndrome reporting information system (SYRIS®) for asthma and allergy alerts and decision support.

2.3.3 *Satellite data augment a national public health cohort study for better health decisions*
This project links three national data sets: EPA's *in-situ* network of PM$_{2.5}$ observations; solar insolation data from the North American regional reanalysis (NARR) network; and LST from satellite sensor observations. These data sets are merged with public health data from the reasons for geographic and racial differences in stroke (REGARDS) national cohort study, to determine whether these environmental risk factors are related to cognitive decline. Environmental data sets and public health linkage analyses will be disseminated to end-users for decision making, priority setting, program evaluation, public health research, and resource allocation through CDC's wide-ranging online data for epidemiological research (WONDER) system. For further information on this project, see Section 4.3 below.

2.3.4 *Adding air quality observations to state and national tracking systems*
This application addresses respiratory public health at both the individual and community levels by forecasting atmospheric ozone, dust, and other aerosols that trigger asthmatic responses or myocardial infarction; and, by enhancing the State of New Mexico's ability to prepare and provide early warning forecasts and alerts to populations at risk through its EPHTS. Between 2003 and 2008 a dust entrainment model based on DREAM (Nickovic 2001) was re-configured to assimilate EO data, and nested within the NCEP/eta national weather forecast model, to develop a numerical system to simulate dust entrainment, transport and deposition on hourly meteorological conditions. Several static DREAM model parameters for land surface conditions were replaced by assimilating more frequently updated measurements of land cover from MODIS, elevation from SRTM, and surface roughness length from a look-up table based also on MODIS. Replacing these parameters showed marked improvement in DREAM's performance for forecasting dust entrainment in the American southwest. The model simulates timely forecasts of hourly mineral dust-patterns, but is less successful forecasting particle concentrations (Morain & Budge 2010). Rolling hourly simulations can be converted into animated gifs to produce forty-eight to seventy-two hour forecasts of dust episodes. These animations represent an important step in verifying that satellite sensor data not only improve model performance, but also add value to daily weather forecasts. The NCEP/eta model alone does not forecast atmospheric dust; and therefore, cannot be used for intervening possible asthma interventions at the local, state, or regional levels.

   After 2008, capabilities from the NCEP/DREAM model system were extended to include hourly ozone and aerosol data collected by US/EPA. The new model system was improved by sharing hourly dust loadings from NCEP/DREAM with CMAQ to enhance its ability to distinguish between atmospheric dust entrainment and fugitive dust, and to add simulations for 107 species of POPs. Two other improvements were made: 1) by replacing the MOD12Q1 product generated in 2002 with the MCD12 product generated in 2008 (Friedl 2010); and 2) by refreshing the NCEP/DREAM model with sixteen-day NDVI rolling updates of changing dust source geographies across the model domain. NDVI up-dates of dust sources are attributed overwhelmingly to agricultural practices and seasonality. These changes were assimilated into the model regularly to better quantify barren land distributions in the Southwest. To improve CMAQ outputs, the model domain for ozone and aerosol observations was reduced to accommodate a finer resolution grid cell spacing of 7.5 km instead of the original 17 km spacing. This is important in context of health applications because epidemiologists need exposure data across large areas that are then aggregated to zip code or census tract levels, most of which have complex shapes and attendant edge-pixel problems, best managed statistically at high resolution (Morain & Budge 2010).

Creating the nested model and combining it with CMAQ ozone and aerosol data represents an engineering approach to provide data that support health tracking and decision making. The aim throughout the effort has been to verify and validate outputs, and to build archives of hourly model runs for $PM_{2.5}$, PM_{10}, and 107 anthropogenic EPA air quality contaminants. As the archives continue to grow, they support short-term episodic alerts for daily health and long term data sets for epidemiological studies of exposures and outcomes.

2.3.5 *Using aerosol optical depth to represent $PM_{2.5}$ fields*

This project supports environmental public health tracking systems by establishing that AOD data from satellite observations can generate $PM_{2.5}$ data sets. Such data sets reflect spatial and temporal variations in ambient concentrations occurring at local scales, and are representative of the true $PM_{2.5}$ field. Fine particulates are critical air pollutants, and their adverse impacts on human health are well established. Traditionally, studies that analyse health effects of exposure to $PM_{2.5}$ use data from broadly scattered ground-based monitoring stations. However, due to large spatial and temporal gaps between these point source monitors, daily synoptic satellite AOD data provide information about particulate concentrations in areas where monitors do not exist.

Using a hierarchical Bayesian model (HBM) to combine monitored data with estimates of $PM_{2.5}$ derived from AOD and CMAQ, this approach could represent a significant step toward creating accurate and representative $PM_{2.5}$ data sets that can be used to make informed public health decisions. The accuracy of the combined monitor/satellite/air quality model data sets are being determined in relation to monitor values, and their performance over data sets analogous to the current best estimates of $PM_{2.5}$ fields are being quantified. Environmental public health tracking programs associated with Maryland, the CDC, and EPA have expressed interest in using the results of the feasibility study to enhance their existing decision making activities.

3 CONVERTING ESR DATA INTO HEALTH RESEARCH AND APPLICATIONS

This Section provides data and information services for EO data that: 1) facilitate their use in health research and applications; 2) invoke new ways of thinking about environmental health tracking; and 3) stimulate new ideas about how these services advance health surveillance. Section 3.1 is an overview by Doctors Tong and Soebiyanto of atmospheric data used for health applications. This is followed in Sections 3.2 through 3.6 by contributions addressing specific uses of sensor data, and the tools necessary to facilitate their use. The first of these by Doctor Ceccato describes EO data for malaria surveillance. Doctor Maxwell then presents a tool for evaluating environmental measurements of, and responses to, extreme heat events. The third by Doctor Benedict describes an IT system for delivering dust forecasts, and the fourth by Doctors Golden and Chen describes data, tools, and services provided by the Socioeconomic Data and Applications Center (SEDAC). The last sub-Section by doctors Kempler, Lynnes, Vollmer, and Leptoukh describes health data and services from GES-DISC.

Key information requirements for epidemiology are the spatial and temporal distributions of environmental agents, their levels of concentration, and their proximity (exposure potential) to effected cohorts. Due to lack of exposure data, many epidemiological studies use measurements from stationary ground monitors as surrogates of exposures for individuals proximal to those sites (Ito *et al.*, 2005; Jerrett *et al.*, 2009). This central monitoring approach is usually oversimplified (Tong *et al.*, 2009). However, sensor data supply area-wide, synoptic values for monitoring time, space, and concentrations of environmental agents across all land, air, and water surfaces. Nevertheless, satellite sensor data have only recently entered the realm of either epidemiological studies that establish the linkage between exposure and risks, or health assessments that quantify health effects based on epidemiological knowledge. The underlying obstacles that inhibit the marriage of these two promising partners are discussed. If these obstacles can be overcome, EO data and their delivery systems will be of great service to health surveillance practitioners.

3.1 *Atmospheric data for health applications*

Since the launch of the first satellites, efforts have been made to measure atmospheric constituents and Earth's surface characteristics from space. The Nimbus satellites, which operated from 1963 to 1993, began the age of space-based Earth observations. The Nimbus program expanded human knowledge of the upper atmosphere significantly, including the discovery of the ozone hole. Like Nimbus, most of the early satellites focused on the upper part of the atmosphere. The TOMS ultraviolet instrument is the first to observe the total ozone column as well as the tropospheric ozone column (Krueger & Jaross 1999). The GOME observes not only ozone, but also ozone precursors, such as nitrogen dioxide (NO_2), formaldehyde (HCHO), and bromine oxide (BrO) that catalyse polar ozone destruction (Burrows *et al.*, 1999). SCIAMACHY sensors extended these efforts by providing observations of additional atmospheric constituents [methane (CH_4), carbon monoxide (CO), and carbon dioxide (CO_2)] at even finer resolution (Bovensmann *et al.*, 1999). Continuous, routine measurements of trace gas distributions are being provided by several satellite instruments, including the MOPITT/Terra (Drummond & Mand 1996), AIRS/Aqua (Aumann & Pagano 1994), TES/Aura (Beer *et al.*, 2001) and the Infrared Atmospheric Sounding Interferometer (IASI on MetOP) (Schlussel *et al.*, 2005). Meanwhile, utilizing higher spectral resolution in the limb geometry, vertical profiles of atmospheric constituents are now being observed with the Michelson interferometer for passive atmospheric sounding (MIPAS) on board ENVISAT and Atmospheric Chemistry Experiment (ACE) instruments (Bernath *et al.*, 2005; Clerbaux *et al.*, 2008; Fischer *et al.*, 2008). EO data sets expanded in both spatial and temporal resolution and the capacity for retrieving atmospheric chemistry have provided unprecedented opportunities for science data users to study Earth systems and the societal impacts of environmental changes from a space perspective.

This article reviews the current status, challenges and opportunities for applying satellite data to environmental monitoring and health surveillance. Environmental tracking is central to understanding levels of hazards imposed on society. Epidemiological studies around the world have associated pollutant exposure with adverse health effects, including both morbidity and mortality (Bell *et al.*, 2004; Jerrett *et al.*, 2009). The WHO estimated that 800,000 annual premature deaths, or 1.2 per cent of all deaths, are caused by exposure to urban fine particles (Cohen *et al.*, 2004). It has been estimated that the global health burden of air pollution is much larger than the WHO study, perhaps by as much as five-fold or six per cent of all deaths, which excluded ozone impacts and included only part of the urban population (Annenberg *et al.*, 2010). Given the profound impacts of air pollution on human health, it is critical to understand the sources, formation, and chemical characterization of air pollution, so that linkages and mechanisms from pollution sources to health endpoints can be revealed and understood before effective mitigation strategies can be designed to protect public health.

Several approaches have been developed to extend point measurements to broader areas so that larger populations can be included. Kriging, for example has been applied to the relationship between fine particle exposures and mortality risks within-city that result in considerably higher risk levels than those found in central monitoring approaches, and three dimensional air quality models such as the CMAQ are able to estimate population exposure to air pollutants at places where monitoring data are not available (Jerrett *et al.*, 2005; Bell 2006). Both methods can provide exposure estimates at additional locations, thus alleviating some limitations of the central monitoring method. These methods, however, are subject to inherent limitations; in particular, the need to validate the estimates to ensure appropriate representations of ambient concentrations (Bell 2006). The distribution of monitoring sites for surface ozone (O_3) from the two major networks, AQS and CASTNET, reveals that there are insufficient numbers of observations to verify gross national patterns in the air quality system (Tong *et al.*, 2007; Avnery *et al.*, 2011). In most cases, across the field of environmental health, one of the most pressing challenges is obtaining accurate exposure assessment data (Patz 2005). Data for statistical and sensor studies at monitoring sites are obtained from US/EPA. Human population data are county-level, all-age sums provided by the US Bureau of Census, and corn production data are compiled by the USDA, National Agricultural Statistics Service (NASS) (see section 3.1.1.3).

3.1.1 The range of air quality applications for health

Because of its broad spatial coverage, satellite sensor data have considerable promise for expanding air quality data from ground monitoring networks into broader urban, suburban and rural air quality contexts (Patz 2005; Liu *et al.*, 2009). A series of vignettes is provided to reprise these opportunities. While each makes a case for successful satellite data use, such data have not been used widely in either epidemiological studies that might establish linkages between exposure and risks, or in health assessments that quantify the levels of health damage based on epidemiological linkages. What are the obstacles that inhibit the marriage of two seemingly promising partners? How can science move forward to bridge the gap between satellite data and their health application?

3.1.1.1 Surface levels of air pollution

In general, satellite data provide column loading of atmospheric components, while health studies mostly focus on surface concentrations at the breathing level. To bridge the gap, several algorithms have been developed to derive surface concentration from existing satellite products. A simple approach was presented in which ground-level $PM_{2.5}$ concentrations were estimated by applying a localized vertical profile from a global chemical transport model to aerosol optical thickness retrieved from MISR (Liu *et al.*, 2004). They found the derived MISR $PM_{2.5}$ concentrations to be in good agreement with the ground measurements, with a correlation coefficient $r = 0.81$, an estimated slope of 1.0 and an insignificant intercept when three outliers are excluded. A similar approach was applied to another region using vertical profiles from the same global model to retrieve surface $PM_{2.5}$ concentrations from TOMS, MODIS and MISR measurements (Hu *et al.*, 2009). Regardless of its successful application, this approach is compromised by inherent uncertainties with several key inputs, including the cloud screening for satellite products, modelled vertical profile, and model and satellite spatial resolution. A two-stage generalized additive model (GAM) was subsequently used to estimate ground-level $PM_{2.5}$ concentrations in the north eastern US based on land use data, meteorological data and the GASP AOD product (Liu *et al.*, 2009). This approach, while immune to several weaknesses in the atmospheric model-based approach is able to achieve a higher model performance (model correlation coefficient $= 0.89$), but faces its own limitation in that the two-stage GAM cannot be applied in regions where ground $PM_{2.5}$ measurements are too sparse in space.

3.1.1.2 Long-range transport of air pollutants

The large areal coverage of satellite sensors provides a reliable means for monitoring long-distance transport of air pollutants and other exposure agents. The magnitude of transboundary air pollutant transport has long been at centre-stage in a global debate on how to effectively reduce air pollution; through domestic emission control, or background mitigation (i.e. blaming one's upwind neighbours?). For instance, a global chemical transport model was used to estimate the impact of transPacific transport of mineral dust on aerosol concentrations in North America. The authors found that transPacific sources are responsible for forty-one per cent of the worst dust days in the western US (Fairlie *et al.*, 2007). Within the borders of the US it has been estimated that for over eighty per cent of the conterminous states, interstate transport was more important than local emissions for summertime peak ozone concentrations (Tong & Mauzerall 2008). These results are all based on model calculations that cannot be verified easily or independently through traditional ground-based measurements.

Several studies have used EO data to quantify air quality and health effects of long-range pollutant transport. The data can detect transboundary transport because of their large temporal and spatial scales. Quantitative or semi-quantitative assessments of intercontinental and hemispherical transport of aerosols, for instance, have been conducted since 1970s (Fraser 1976; Lyons *et al.*,

243

1978; Fraser *et al.*, 1984; Herman *et al.*, 1997). However, these studies are limited by poor accuracy of earlier satellite measurements. Recent improvements in data quality and enhanced new capabilities of satellite sensors are enabling new opportunities to investigate this issue with a more robust data set (Yu *et al.*, 2008). In particular, multi-wavelength, multi-angle, and polarization measurements have provided additional information on the physical and chemical characteristics of air pollutants, such as particle size (fine *vs.* coarse), shape (spherical *vs.* non-spherical), and absorptiveness (absorptive and scattering) (Higurashi & Nakajima 2002; Tanre *et al.*, 2001; Holzer-Popp *et al.*, 2008). The new capabilities associated with passive sensors are further complemented by an increased data pool of active LiDAR data that shed light on the vertical structure of pollution distribution (Spinhirne *et al.*, 2005). Built on these recent advances, scientists at GSFC assessed the pollution flux from Asia to North America using a four-year (2002–2005) climatological archive of MODIS AOD, relative humidity and vertical distribution data from field campaigns and satellite data retrievals (Yu *et al.*, 2008). With uncertainties within a factor of two, they estimated an influx of aerosols into the western coasts of North America to be approximately 4.4Tg/yr.

Dust and aerosols not only impose adverse health effects directly, but act also as carriers for long-range transport of diseases (Prospero *et al.*, 2002). A recent study in Japan attempted to apply a satellite-based monitoring approach to measure the influx of Chinese pollutant loads and their association with bronchitis mortality (Goto *et al.*, 2010). They found no significant association by examining the relationship between the annual average amount of incoming pollutant load and annual average mortality from asthma. However, this is perhaps the first attempt to apply synoptic sensor data to epidemiological research investigating chronic mortality effects.

3.1.1.3 Crop exposure and ecosystem health

Environmental health and human health are inextricably intertwined. Ecosystem health is an emerging field in satellite remote sensing that aims to assess impacts of air pollution and climate change. Previous field experiments using the open top chamber (OTC) approach have shown that ambient ozone, alone or with acid rain gases, accounts for up to ninety per cent of crop losses from air pollution in the US. Crop yield loss from ozone exposure is of particular concern because surface ozone levels remain high in many major agricultural regions (Tong *et al.*, 2007; Avnery *et al.*, 2011). Quantitative assessment of ozone damage to crops is challenging because of difficulties in measuring reductions in crop yield that result from exposure to surface ozone. Ozone monitors are sparse or non-existent in major corn production areas in the US.

A number of approaches have been proposed to estimate crop exposure to ambient ozone. Most of these fall into two categories: those that use ground monitoring to determine ozone exposure through spatially extrapolated measurements at discrete locations (Adams & Croker 1989; Hertstein *et al.*, 1995; Felzer *et al.*, 2004); and those that use atmospheric chemical transport models (CTM) to calculate ozone exposure (Tong *et al.*, 2007; Avnery *et al.*, 2010). The major limitation of the monitoring approach is that all extrapolation techniques, such as the widely used kriging, cannot capture the actual spatial variability in ozone concentrations perturbed by chemical and physical processes. The modelling approach alleviates this limitation by accounting for all underlying processes that control ozone variations. Using this approach, annual approximate losses of ten per cent for soybean yields in the US due to ozone exposure were reported (Tong *et al.*, 2007). Globally, ozone exposure may be reducing soybean, wheat, and maize yields by as much as 8.5–14 per cent, 3.9–15 per cent, and 2.2–5.5 per cent, respectively (Avnery *et al.*, 2010). It should be pointed out, however, that modelling approaches have their own limitations, such as the uncertainties in parameterization and the need of independent verification of the implementations of these complex processes.

A recent study by Fishman and colleagues is the first to utilize both ground and satellite ozone measurements to verify crop yield losses from ozone exposure (Fishman *et al.*, 2010). By using a multiple linear regression model, they found that soybean crop yields during a five year period across the mid-section of the US were on the order of ten per cent less, consistent with earlier modelling by Tong (*et al.*, 2009). Interestingly, their research demonstrates that space-based measurements provide a means for quantifying crop losses that could be employed on a global basis. They also argue that satellite data may even be a better measure of ozone amounts outside of urban areas because of the general paucity of surface sites in predominately farmland regions (Fishman *et al.*, 2010).

3.1.1.4 Linking human ecosystems to infectious diseases

Human-induced land use changes have had complex impacts on human health, mainly by altering food supply, shelter, and sanitation that in-turn have unintended health consequences (Patz *et al.*, 2004; Patz, 2005). Rural and urban developments, including roads, dams, and dwellings, all modify transmission of infectious diseases. In tropical regions, irrigation increases schistosomiasis and malaria by supplying habitat and breeding sites. Similarly, construction of hydroelectric dams can lead to mosquito proliferation and subsequent infectious diseases like filariasis and elephantiasis (Thompson *et al.*, 1996). An important feature of human ecosystems is that they be designed to withstand severe natural hazards and mitigate injuries and causalities (Glantz & Jamieson 2000). Hurricane Katrina demonstrated the combined effects of urban infrastructure failures and extreme weather by claiming 1,836 lives and causing widespread property destruction (Knabb *et al.*, 2006).

In the 1990s NASA's Ames Research Center (ARC) executed several projects in collaboration with New York Medical College and Yale School of Medicine to develop sensor-based models for Lyme disease transmission risk in New England (Beck *et al.*, 2000). One project used canine seroprevalence rate (CSR) as a surrogate of human exposure risk and compared municipality level CSR with Landsat TM data to examine relationships between Lyme disease transmission rate and tick bites on dogs near their owners' property (Dister *et al.*, 1993). A similar approach was employed to map relative adult tick abundance on residential properties based on TM-derived indices of vegetation greenness and wetness, assuming a relationship between forest patch size and white-tailed deer populations that are the major host of the adult tick and its primary mode of transportation (Dister *et al.*, 1997; Beck *et al.*, 2000).

3.1.1.5 Linking weather and climate to diseases

The impact of weather on disease progression and spread has long been recognized since the beginning of medical science by Hippocrates (NRC 2001). However, the underlying mechanism in which weather and climate variability change the spatiotemporal dynamics of disease transmission remains poorly understood. Advances in technology and modelling techniques in the past few decades have enabled satellite meteorological data to be used to understand such phenomena in a more quantitative manner. In fact, a number of successful uses of EO data in disease spread have led to developing early warning systems that allow public health agencies to monitor, prevent and control diseases (Grover-Kopec *et al.*, 2005; Anyamba *et al.*, 2009; Witt *et al.*, 2011). In addition, with the surge of evidence on the impacts of climate variability and climate cycles on health, EO data have become even more popular in recent years. Elevated average temperature and frequent extreme heat events increase the risk for heat-related mortality and complicate chronic illnesses such as cardiovascular diseases. They also will expand the disease's geographic range, especially for vector-borne and zoonotic diseases such as dengue and WNV, where the vectors thrive under warm temperature and adequate rainfall. Satellite sensor observations can provide health organizations with decision tools for mitigating future risks and possibly preventing or controlling diseases.

In vector-borne and zoonotic diseases, environmental conditions play critical roles in propagating the vector populations. Vector abundance determines the number of new disease cases, and hence is important for controlling vector populations as a means for preventing disease (Hay *et al.*, 1997). Studies have indicated that the rate of mosquito egg development depends on temperature, and to a lesser extent relative humidity (Hoshen & Morse 2004). Rainfall also impacts mosquito breeding sites. Pools and puddles from rain can create breeding sites, but heavy, intense and prolonged rainfall may flush away larval habitats (Kovats 2003; Kiang *et al.*, 2011). Vegetation is another factor that often is correlated with mosquito breeding, feeding and resting sites. Remote sensing measurements of temperature, rainfall and vegetation are used frequently to study the spread of mosquito-borne diseases and to identify areas at risks, especially for malaria (Craig *et al.*, 1999; Brooker *et al.*, 2006; Rahman *et al.*, 2006; Gomez-Elipe *et al.*, 2007; Kelly-Hope *et al.*, 2009; Adimi *et al.*, 2010; Haque *et al.*, 2010). For RVF, Linthicum (*et al.*, 1999) coupled satellite NDVI data with Pacific and Indian Ocean SST anomalies to predict outbreaks in East Africa up to five months in advance. Surface temperature has been used successfully at one kilometre^2 resolution from TOPS, along with mosquito field data to assess WNV risk in California (Nemani *et al.*, 2009; Barker *et al.*, 2010). Currently there are fewer applications of EO data for studying dengue transmission (Harrington *et al.*, 2001). This is due to the intrinsic nature of the dengue vector mosquito that is capable of breeding within a house or in an urban area where environmental conditions are difficult to measure remotely. However, Hales (*et al.*, 2002) have shown that the current geographical limits of dengue can be modelled using vapour pressure data. Global circulation model projections from the IPCC were further used to estimate the world population at risk under climate change scenarios.

In addition to vector-borne diseases, EO data have been used to model the spread of seasonal and pandemic influenza. Influenza's spatiotemporal dynamics have been observed to vary historically with latitude, which further suggests the role of weather and environmental factors. Several experimental studies have in fact indicated that influenza virus survival depends on temperature, and that transmission effectiveness is subjected to variability in temperature and humidity. Soebiyanto (*et al.*, 2010) used satellite derived LST and rainfall, in combination with ground station data, to predict influenza cases in Hong Kong, New York City and Maricopa County in Arizona with reasonable accuracy (see Chapter 2). It has also been shown that satellite solar radiation estimates are related significantly to the timing of influenza epidemic in thirty-five cities in the US (Charland *et al.*, 2009).

Another disease application that has shown promise for EO applications is meningitis, which is epidemic in Africa. Since epidemics seem to start during the dry season and end at the onset of rainfall, factors such as low absolute humidity and dusty atmospheric conditions have been associated with it. NDVI, cold cloud duration and rainfall estimates were used along with aerosol index, soil and land cover types to predict the outbreaks (Thomson *et al.*, 2006).

3.1.2 *Remaining challenges*

Although there are several successful applications, the mainstream epidemiological community is just beginning to use EO data as a component of their research. In regulatory applications that estimate costs and benefits of pollutant control strategies, satellite data have not yet been used to cover a large population, as required by national scale assessments. Currently, computer models and ground monitoring data are often combined to measure regional or national scale exposure levels. So what are the challenges facing more routine use of satellite data in health applications where spatial coverage plays an essential role?

There seem to be three major obstacles: uncertainty, accessibility and data quality. Epidemiological studies of the health effects of air pollution are an example. From a human exposure perspective, the best exposure data are those that represent the dosage, or to a less rigorous extent, the pollutant level in the microenvironment immediately surrounding the studied cohorts. Because such data are difficult to obtain, most large scale epidemiological studies use the ambient concentration as the exposure surrogate. Although there are many complicating factors when such a surrogate is used, this approximation has been widely employed in epidemiological studies. Satellite data, however, cannot provide surface level concentrations of air pollutants directly. Current approaches have been to obtain surface level atmospheric aerosols (Liu *et al.*, 2004; Morain 2011), but the results are yet to be used in epidemiological or assessment studies, most possibly because of the large uncertainties in the input data and processing algorithms.

Another challenge is the accessibility of satellite data for environmental epidemiologists. With the proliferation of satellite platforms, the task of using a combination of EO data sets has become intimidating for users lacking the expertise for analysing them (Patz 2005). This includes understanding that different satellite platforms measure different properties and phenomena of Earth's surface and its atmosphere in varying temporal and spatial resolutions (Kaufman *et al.*, 2002). In most cases, satellite data need to be further processed, modelled, and interpreted to match health data, or be in GIS formats that are compatible with existing tools used routinely by epidemiologists. This challenge calls for developing data sets and tools that consider the special requirements of health studies and make the needed data accessible conveniently to users having little or no remote sensing background.

Beside accessibility, quality is also a major concern for EO data used in health studies. Health studies put high priorities on temporal resolution because of their fundamental research approaches. For instance, epidemiologists often use the time-series method to investigate the acute effect of environmental factors on a concerned health endpoint (Bell *et al.*, 2005). Satellite data, when used in such studies have to be associated with detailed data quality information, so that users can interpret their results properly. In other cases, such as cohort studies, long-term average data are used to derive the accumulated effect of environmental factors (Pope *et al.*, 2002; Jerrett *et al.*, 2009). Monthly averages are mostly aggregated from all valid data that extend from full temporal coverage to those having large data voids because of cloud cover or other quality issues. A careful study of the data quality section in metadata records is needed, so that proper caution can be exercised to avoid misinterpreting subsequent scientific analysis.

Finally, efforts are needed to reach out to the health community. Since 1985, NASA has hosted and participated in a series of workshops to solicit input from various health communities on potential applications of EO data in epidemiology, infectious diseases, and ecosystem health. These experiences and a review of literature (Beck *et al.*, 2000) have not lead to consensus regarding universal requirements for a remote sensing system in health applications. For example, environmental epidemiologists rely heavily on ground-based monitor data, which are considered ground verification. However, the narrow spatial and temporal coverage of monitoring sites imposes many limitations on these studies (Tong *et al.*, 2007; 2009). A system that is able to integrate information on pollutants, weather and other confounding factors will enhance risk assessment greatly (Patz, 2005).

3.1.3 *Conclusion*

It is well recognized that satellite-acquired environmental measurements hold great promise for epidemiologic applications. Limited by available measurements, many epidemiological studies rely on central monitor approaches to estimate exhort (i.e. force or impel in an indicated direction) exposures in the same area, which also bear their own inherent limitations. Because of their broad spatial coverage and continuous operations, satellite data hold great promise for alleviating some of the spatial data continuity limitations recognized by health communities. Numerous community efforts are being made to address these issues. It is expected that the future will see a further increase in health applications nurtured by more satellite products, reduced uncertainties, as well as user-oriented data services.

3.2 *Using EO data for malaria surveillance*

Major human diseases like malaria and dengue are sensitive to inter-seasonal and inter-decadal environmental and climatic changes (Thompson *et al.*, 1996). Monitoring variations in surface conditions such as temperature, rainfall and vegetation helps decision makers at Ministries of Health to assess the risk levels of malaria epidemics. The International Research Institute (IRI) for climate and society has developed products based on EO data to monitor these changes and provide information directly to decision makers. The mission of the IRI is to help societies to understand, anticipate and manage the impacts of seasonal climate fluctuations to improve human welfare and to better monitor environments, especially in developing countries. The URLs for this section are presented in Table 5.

3.2.1 *The role of climate*

Given its impact on populations and the gravity of its pathology, malaria remains one of the most significant infectious diseases. It is essentially an environmental disease since the vectors require specific habitats with surface water for reproduction and humidity for adult mosquito survival.

Table 5.   URLs referenced in section 3.2.

Product/service	URL (accessed 18th January 2012)
A. MEWS	http://iridl.ldeo.columbia.edu/maproom/Health/Regional/Africa/Malaria/MEWS/
B. IRI data library	http://iridl.ldeo.columbia.edu/SOURCES/USGS/LandDAAC/MODIS

The development rates of both the vector and parasite populations are influenced by temperature. In sub-Saharan Africa the pattern of malaria transmission varies markedly from region to region, depending on climate and biogeography. The association between rainfall and malaria epidemics has been recognized for many decades (Christophers 1911), but while increasing precipitation may increase vector populations in many circumstances by increasing available anopheles breeding sites, excessive rains may also have the opposite effect by flushing out small breeding sites, such as ditches or pools (Fox 1957), or by decreasing the temperature, which in regions of higher altitude can hinder malaria transmission.

Temperature also plays an important role in the variability of malaria transmission. The development rate of mosquito larvae and the malaria parasite within the mosquito host is regulated highly by temperature. It is also one of the factors that influence the survival rate of mosquitoes. Generally, mosquitoes develop faster and feed earlier in their life cycle and at a higher frequency in warmer conditions. The plasmodium parasite in the mosquito also multiplies more rapidly at higher temperatures (Gilles 1993; NRC 2001). Humidity impacts the survival rate of the mosquito as well. Mosquitoes will generally not live long enough to complete their transmission cycle where relative humidity is constantly less than sixty per cent (Pampana 1969; Gilles 1993; NRC 2011).

3.2.2 Monitoring rainfall
In the majority of countries in Africa, monitoring rainfall is a major problem because there are too few rain gauges and very sparse coverage. Therefore, it is necessary to use rainfall estimations derived from satellite measurements. Using rainfall estimate products updated approximately every ten days through the Africa data dissemination service (ADDS), IRI has developed a web-based Malaria early warning system (MEWS) interface that enables users to gain a contextual perspective of the current rainfall season by comparing data to previous seasons and recent short-term averages (Table 5, A) (see Chapter 9 for details). The interface is in the IRI data library (Table 5, B) and takes the form of an online interactive map. It displays the most recent decadal rainfall map (Figure 2) over which national and district administrative boundaries and the epidemic risk zone can be overlaid, in this case as a guide rather than an absolute mask that excludes districts of local interest).

These visual features can be toggled on or off and the user can zoom in to any region for more clarity. In addition, the map can be downloaded in different formats compatible with common image analysis and GIS software such as ArcView® developed by Esri or HealthMapper developed by WHO (WHO 2012).

Decadal rainfall can be averaged spatially over a variety of user-selected areas, including administrative districts at $11 \, km^2$, $33 \, km^2$, $55 \, km^2$ and $111 \, km^2$ grids. By selecting a sampling area and a specific location of interest, four time-series graphs are generated by clicking the map at the area selected. Graphs are generated to enable an analytical comparison of recent rainfall with respect to that of recent seasons and the long-term series. A description of the time-series figures, the data used and their source is also provided (Grover-Kopec et al., 2005). A newer version of this capability using the CPC morphing technique (CMORPH) developed by NOAA-CPC is scheduled to replace the current product with one that exhibits better agreement with field measurements (Joyce et al., 2004; Dinku et al., 2007).

3.2.3 Monitoring vegetation and water bodies
Vegetation type and growth stage play critical roles in determining mosquito abundance, irrespective of associated rainfall. The type of vegetation that surrounds breeding sites, and thereby provides

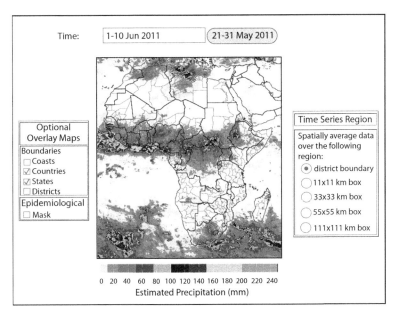

Figure 2. MEWS *clickable map* for rainfall monitoring 1–10 June 2011. (see colour plate 40)

potential resting and protection from solar radiation and desiccation, are important in determining mosquito abundance (Beck *et al.*, 1994). Surface water provides the habitat for the juvenile stages of malaria vectors (egg, larvae, and pupae). Monitoring the state of water bodies and wetlands is therefore important to identify sources of malaria vectors. To monitor vegetation and water bodies, high resolution MODIS data obtained at 250 m spatial resolution are optimal because they provide frequent observations and are available at no cost. Sixteen-day L3 Global 250M SIN GRID V004 vegetation indices can be downloaded from the USGS land processes DAAC and provided to users via the IRI data library. Users can download either the raw data as single spectral channels in blue, red, NIR and SWIR wavelengths, and in different formats compatible with common image analysis and GIS software, or as NDVI and EVI (Huete *et al.*, 2002). Using the online IRI data library, users can create a variety of secondary products. For instance, they can: 1) combine the different NIR-SWIR channels to generate tailored spectral indices for monitoring vegetation status in terms of moisture content (Ceccato *et al.*, 2002); 2) visualize a colour composite of the SWIR-NIR and Red channels (Red-Green-Blue) where vegetation appears in green, bare soil in brown and water in blue; 3) integrate the colour composite into GIS software with ancillary data such as roads and villages, as shown in Figure 3; 4) extract weighted averages of the different indices per GIS layers such as district boundaries or several other shape files; and 5) create long-term series of vegetation indices. Products such as these are already being used operationally by Ministries of Health for malaria control. They allow users to forecast the risk of malaria epidemics.

3.3 *Internet-based heat evaluation and assessment tool (I-HEAT)*

Over the past two decades heat waves have been responsible for more deaths in the US than any other natural hazard (Borden & Cutter 2008; NWS 2012). In 1995, at least 700 deaths were attributed to a heat wave in Chicago (Semenza *et al.*, 1996). Western Europe experienced an unusually intense heat wave in summer, 2003 that resulted in over 70,000 heat-related deaths (Robine *et al.*, 2008), and Paris alone attributed 4867 deaths to this heat wave (Dousset 2009). A report by the IPCC concluded that climate warming is unequivocal and most Earth scientists agree that global warming continues as the interglacial age deepens. Health professionals need to plan and prepare

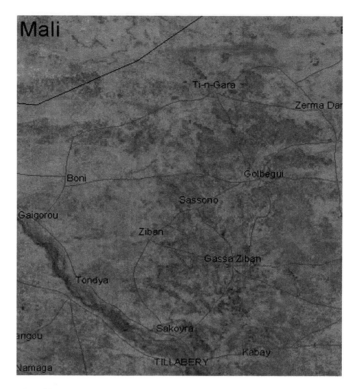

Figure 3. TERRA-MODIS colour composite RGB where the SWIR channel is displayed in red, the NIR channel in green and the visible red channel in blue. This image was acquired during the rainy season in north Niamey, Niger. It is roughly 200 km^2 in area. Small water bodies in blue are breeding sites for mosquitoes and locations where nomads water their cattle. (see colour plate 41)

for heat-related emergencies because future extreme events are expected to increase the frequency, duration and magnitude (Meehl & Tebaldi 2004; IPCC 2007).

The US global change research program (USGCRP) identified mapping and modelling tools for identifying populations vulnerable to climate changes as an important need. A report published by the Trust for America's Health Foundation concluded that special efforts should be made to address the impact of climate change on at-risk and vulnerable communities. The report recommends that all state and local health departments include public health considerations as part of their climate change preparedness. Yet, the report noted, only five States in the US developed a strategic climate change plan that included a public health response, and effective software tools are almost entirely lacking. Perhaps this is because there is less perceived health risk from gradual climate warming than from more immediate extreme weather events.

Impacts of climate variability on human health vary by region, demographics, duration of exposure and ability to adapt or cope with the changes (Ebi & Semenza 2008; English *et al.*, 2009). Elderly populations over sixty-five years, for example, are particularly vulnerable to the effects of extreme heat in part because of physiology, disabilities and medications (Semenza *et al.*, 1996; Ebi & Meehl 2007; Kovats & Hajat 2008; Schifano *et al.*, 2009). To identify populations vulnerable to heat waves, demographic data such as age and poverty status need to be linked with local-scale environmental data such as land cover and temperature (English *et al.*, 2009; Rinner *et al.*, 2010). An extensive review of heat-wave impacts on human health concluded that geospatial and remote sensing technologies provide a key capability for mitigating heat-related deaths. In particular, understanding the dynamics between urban heat islands and vulnerable populations is critical for reducing heat-related deaths (Wilhelmi *et al.*, 2004).

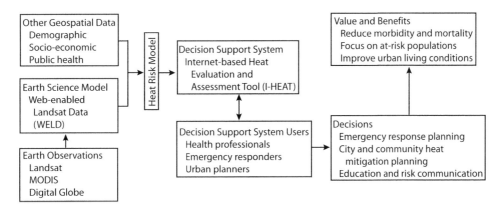

Figure 4. I-HEAT system diagram.

3.3.1 *I-HEAT system description*

The Internet-based heat evaluation and assessment tool (I-HEAT) is a new software system to provide health professionals and risk assessors with advanced geospatial web-based tools for preparing and responding to emergency heat events, developing mitigation strategies, and educating the public. The system couples demographic and environmental data obtained from Landsat data and imagery with browser-based software to provide health professionals with a tool to model and map heat-related morbidity and mortality risks at the neighbourhood level (Figure 4). Landsat data will be integrated with demographic, socio-economic, and health data in a heat-risk model that incorporates elements of the vulnerability mapping (Table 6).

Local surface temperature derived from EO digital data is a significant predictor of heat-related mortality (Vandentorren *et al.*, 2006; Smargiassi *et al.*, 2009). Landsat provides these data at 30–120 m spatial resolution and temporal repeat cycles of eight to sixteen days for modelling heat risk in urban environments (UHI 2010). Landsat data are used routinely by environmental scientists, yet their potential has not been fully realized by health practitioners, perhaps because of the difficulty and expense of integrating the data, either by fusing or assimilating them into applications by non-technical users. Only a few studies have demonstrated the utility of these data in heat risk modelling (Vandentorren *et al.*, 2006; Smargiassi *et al.*, 2009).

Although Landsat data are available at no cost through the USGS Earth Resources Observation and Science (EROS) Centre, the data are not provided in a format that can be imported and used easily without intermediate analysis by analysts with access to high-end computing facilities. Products from Web-Enabled Landsat Data (WELD) are needed to integrate them into user-friendly software applications. WELD provides a nationally consistent data source that has been temporally composited, mosaicked, and converted to geo-physical and bio-physical products (Roy *et al.*, 2010). One of the initial goals of the I-HEAT project is to determine the feasibility of using land surface temperature and vegetation measures from WELD for heat risk assessment.

Managing and processing geo-spatial and image data requires technical expertise that often is not available in many health organizations. I-HEAT will be designed for use by heath officials who require a user-friendly tool to access and analyse WELD and other geospatial data used in Table 6. Geospatial mapping and modelling can be intensive computationally, especially if satellite image data are incorporated. Web-based software tools enable users to access and process these large data sets without acquiring and maintaining computer resources in-house. They also stimulate better collaboration between, and among health organizations to ensure accessibility to the software and data security. In addition, developers can maintain and update software easily to meet various user requirements in a single secured environment, thus reducing the burden on health professionals to maintain desktop applications.

251

Table 6. Selection of variables in a heat risk model (adapted from Reid *et al.*, 2009).

Category/variable	Source
Land cover	
Vegetation greenness (NDVI)	Landsat (WELD)
Land surface temperature	Landsat (WELD)
Demographic	
Per cent population below the poverty line	Census
Per cent population with less than a high school diploma	Census
Per cent population of a race other than white	Census
Per cent population living alone	Census
Per cent population ≥ 65	Census
Per cent population ≥ 65 living alone	Census
Per cent population ≥ 65 single	Census
Diabetes prevalence	
Per cent population ever diagnosed with diabetes	Behavioural Risk Factor Surveillance System
Air conditioning	
Per cent households without central AC	American Housing Survey
Per cent households without any AC	American Housing Survey

I-HEAT is designed initially for Detroit, Michigan. Detroit lacks a heat wave health warning system and has significant racial/ethnic and socio economic disparities in heat exposure and heat-related health effects (O'Neill *et al* 2003; Schwartz 2005), and climate change scenarios suggest the area will experience high temperature increases (Meehl & Tebaldi 2004). Surveys to assess the efficacy of I-HEAT include vulnerable residents, local governmental officials and community leaders to quantify perceptions of heat risk, and to identify the prevention and intervention programs needed. Software for I-HEAT is designed for two user groups: those monitoring heat risks to develop response strategies; and those modifying the underlying risk model. Both groups receive displays of a web-delivered map that provides an intuitive representation of geographic patterns (Figure 5). Street maps and satellite images are provided for easy reference. The design allows users to switch easily between the model and results display. The primary data displayed are at urban residential scales so users can view underlying relationships among risk model variables: surface temperature, land cover, and demographic factors. The modelling interface allows user to adjust coefficient weightings to compare different outcomes heuristically.

3.3.2 *Summary*

Over the past two decades heat waves have resulted in many deaths in the US and globally. Extreme heat events are expected to rise in frequency and metropolitan and regional health authorities need to prepare for the needs of vulnerable populations. I-HEAT will enable these authorities to prepare and respond more effectively to heat emergencies by planning and targeting intervention programs. The system supports queries such as: show residential areas with low air conditioning availability, high proportion of elderly populations, and daytime temperatures during heat events in excess of 90°F (32°C); and, where should first responders be based during heat events to better serve vulnerable populations? It is anticipated that future I-HEAT developments will incorporate real-time heat warning capabilities.

3.4 *An IT system for interoperable multi-resolution dust modelling*

Environmental forecast products are valuable to public health professionals for developing alerts and planning resource allocations. Their value is in direct proportion to the following characteristics of the modelling system: 1) timeliness of the forecast products, particularly in terms of lead time to identify future events; 2) spatial resolution of forecasts relative to county, zip-code, census tract or

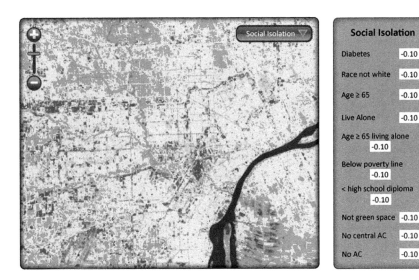

Figure 5.   The I-HEAT interface showing a heat-risk map of Detroit, Michigan. (see colour plate 42)

Table 7.   URLs referenced in section 3.4.

Product/ service	URL (all accessed 18th January 2012
A. PHAiRS	http://phairs.unm.edu
B. SOAP	http://www.w3.org/standards/techs.soap#w3c_all
C. OGC web coverage service	http://www.opengeospatial.org/standards/wcs
D. Global Forecast Service	http://www.nco.ncep.noaa.gov/pmb/products/gfs/
E. OGC web mapping service	http://www.opengeospatial.org/standards/wms

block group that are the preferred analytical and alert units for public health surveillance; 3) forms and formats of products that have maximum utility for public health system users; and 4) accuracy of the timing and location of events. This section describes a feasibility study for developing a dust forecasting system having improved system performance for items 1–3, and establishes a framework for streamlined evaluation of item 4, all while building upon dust forecasting model components developed over the past decade of research and application development projects. Two health application projects contributed components to this environmental forecast model: the public health applications in remote sensing (PHAiRS) project; and the interoperability and high-performance computing testbed project to expand the capabilities of PHAiRS model components. Table 7 is a list of URLs referenced in this sub-Section.

The PHAiRS project (Table 7, A) included the University of Arizona, the University of Malta, and the University of New Mexico, to produce an enhanced version of DREAM (Nickovic *et al.*, 2001) for the south western US. Resulting model outputs were made available to public health surveillance personnel and decision makers through a set of web services that deliver data via simple object access protocols (SOAP) (Table 7, B). The interoperability and high-performance computing (HPC) testbed advanced the products and services developed through PHAiRS by expanding the number of modelled particle size bins form four to eight, and by migrating from DREAM/eta to HPC's version of DREAM/nmm. The 8-bin DREAM/eta and DREAM/nmm models are the core systems for developing the interoperable, multi-resolution IT system described here.

The project's goal was to perform a test implementation of the modelling systems and to assess their interacting performance characteristics and source data through open standards-based

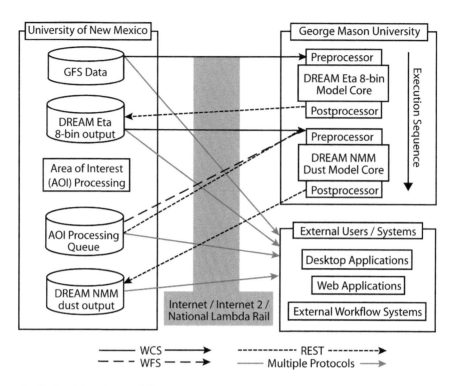

Figure 6. System integration workflow.

interoperable data services. The potential benefits of such a system are: 1) improved flexibility and an improved capability for integrating new data sources into the modelling system; 2) an ability to separate data management and publication physically from modelling; and 3), insert a degree of platform and file format independence between system components (i.e. only having to know the protocol for requesting data from a service, and not having to know a model's operating system or source data format for a particular data service.

The system integration approach is illustrated in Figure 6. The overall integration model is one in which all persistent data are stored and published via Open Standards at the Earth Data Analysis Center (EDAC) at the University of New Mexico, and all modelling is performed on separate systems at George Mason University (GMU). Data exchange between EDAC and GMU was accomplished through two primary OGC web service standards, and OGC web coverage services (Table 7, C) and the representational state transfer (REST) web service model, which in turn is based on the Internet standard hypertext transfer protocol (HTTP) (Fielding 1999; Fielding 2000; Whiteside & Evans 2006). These standards provide the connectivity between separate data management and processing components in the system.

Individual components performing the integrated modelling steps include the data storage and management systems represented by the cylindrical objects on the left in Figure 6. They provide persistent storage of the global forecast system (GFS) data (Table 7, D); low-resolution DREAM/eta model outputs; calculated areas of interest for which high-resolution model runs should be executed; and high-resolution DREAM/nmm model outputs. All three sets of model products (GFS, DREAM/eta, and DREAM/nmm) are published both within the system and for external users via WCS, while the calculated areas of interest are published via REST. The modelling and data processing components are represented by the rectangular boxes on the right in Figure 8. They perform model runs and analytical functions. For the modelling components hosted at GMU, model

Figure 7. Feasibility study model domain and areas of interest bounding boxes for high-resolution model runs.

pre-processors were modified to enable dynamic acquisition of initialization and boundary conditions from remote data servers via WCS and REST, and post-processors were modified to push model outputs to the remote data storage systems at EDAC. In both cases modification of the pre- and post-processors allow streamlined integration of the existing DREAM/eta and DREAM/nmm modelling cores within a broader services-oriented architecture.

The DREAM/eta and DREAM/nmm modelling cores remained largely unchanged in this project. The primary changes consisted of parameterising the models to allow modifications of the model domain without having to recompile the model. Figure 7 shows the full model domain. A complete set of model runs consists of full-domain low-resolution (50 km) runs of DREAM/eta, from which sub-domain regions of elevated dust concentrations are derived by the areas of interest processing component run on the servers at EDAC. The set of areas of interest for a specific low-resolution model run are then retrieved by the DREAM/nmm server for execution of individual areas of interest at higher spatial resolution. This targeted execution of a high-resolution model only for areas for which a more detailed analysis is required reduces the overall computational requirements significantly for generating a comprehensive picture of a large model domain.

3.4.1 *Data sources*

A variety of sensor data, model, and geospatial data were employed. MODIS land cover, from which potential dust production sites are identified, are obtained from both MOD12Q1 and MCD12Q1 products using the *Barren/Sparsely Vegetated* class. Digital elevation data were obtained from the global digital elevation and topography data set (GTOPO30); and aerodynamic surface roughness data were derived from a table look-up based on land cover developed by Stennis Space Center (SSC) (Sanchez 2007; Morain & Sprigg 2008). GFS data at one degree resolution are available every six hours. These data constitute the key, externally provided, initial and boundary condition parameters used by both modelling cores.

In previous NASA applications projects, data and products have been stored locally with the model and used as fixed data sources that are linked statically to the model through direct recon-figuration of the model code. This static approach was overcome in this project by using modified model pre-processors that use WCS for acquiring essential GFS data for a specific model run. This approach provides two reductions in data management and transfer costs. First, the current collection of over 1.3 terabytes of GFS data stored at EDAC does not need to be replicated at GMU to support local modelling. Second, WCS allows transfer of only the subset of needed meteorological parameters from each GFS product instead of transferring complete data files. Specifically,

Table 8.   Feasibility test data transfer times.

Activity	Time (HH:MM:SS)	Notes
Retrieval of GFS parameters from EDAC	00:01:32	Initialization and boundary conditions for DREAM/eta from WCS
Delivery of DREAM/eta results to EDAC	00:04:37	REST-based submission
Retrieval of DREAM/eta and AOI data from EDAC servers	00:03:50	Time required for the retrieval of the required initialization and boundary condition data for a single area of interst DREAM/nmm run
Delivery of DREAM/nmm results to EDAC servers	00:01:30	
Minimum transfer time	00:11:59	If each AOI was executed in parallel and transfers took place simultaneously
Maximum transfer time	01:42:09	If each AOI was executed sequentially- and transfers took place one after another

only five meteorological parameters are required to run DREAM/eta: geopotential height, relative humidity, temperature, u winds and v winds (Benedict *et al.*, 2010). These five parameters, when retrieved for a 24-hour model run, represent a file transfer of about 1.2 megabytes, compared to a transfer size of 360 megabytes to transfer the complete set of source GRIB1 files.

3.4.2 *Project outcomes*

The feasibility test was developed and executed for a specific set of dust events that took place on 1 July 2007 as an end-to-end test of the performance characteristics. The test included WCS services based upon the THREDDS data server and REST services hosted at EDAC for model submission to EDAC's servers and retrieval of areas of interest; and DREAM/eta and DREAM/nmm modelling systems hosted at GMU. The test measured data transfer times between EDAC and GMU over Internet-2, and model execution times for both DREAM/eta and DREAM/nmm on the servers maintained by GMU. The test was performed over the DREAM/eta domain shown in Figure 9. Eighteen areas of interest were identified in the test, each of which was used to execute a high-resolution (3 km) DREAM/nmm model run. Data transfer times are presented in Table 8 with the minimum and maximum data transfer times summarized at the bottom of the table. One can see that individual transfer times are relatively short over the Internet 2 connection between EDAC and GMU; and, given a scenario where data for individual areas of interest execution on separate systems takes place, the minimum transfer time for parallel transfer and execution is just under twelve minutes.

Model execution times are summarized in Table 9 for both the full-domain DREAM/eta and eighteen individual areas of interest for the DREAM/nmm model. As with the data transfer times discussed above, while the total execution time for all of the individual models could be quite long if executed sequentially over twelve hours, if the DREAM/nmm model runs were executed simultaneously on separate computers, the total execution time for the DREAM/eta and all eighteen DREAM/nmm areas of interest model runs is only 1.75 hours. This total execution time is significantly shorter than the estimated seventy-two hours that it would take to execute a twenty-four hour DREAM/nmm model run over the entire model domain.

These results suggest two conclusions: 1) that the interoperable nested modelling approach is feasible as an alternative to monolithic model integration frameworks in which specialized data transfer protocols and connectivity models are employed; and 2) significant gains in accelerated access to multi-resolution model outputs can be obtained through parallel execution of independent high-resolution areas of interest models when compared to the time required to execute a full-domain, high-resolution model.

Table 9.   Model execution times for DREAM/eta and DREAM/nmm models.

Model run type	Time (HH:MM)	Notes
DREAM/eta	0:20:00	Full 37 × 20 degree model domain at a ~50 km resolution
DREAM/nmm	Min: 00:27	Summary statistics for 18 NMM-dust runs (3 km × 3 km)
	Max: 01:25	for the AOIs identified following the 1 July 2011
	Mean: 00:39	DREAM/eta model run
	Median 00:32	
	Sum: 11:43 n = 18	
Minimum total execution time	01:45	Combined DREAM/eta and DREAM/nmm execution time if all DREAM/nmm AOI model runs are executed simultaneously on separate computers
Maximum total execution time	12:03	Combined DREAM/eta and DREAM/nmm execution time if all DREAM/nmm AOI model runs are executed sequentially on a single server

Overall, the outcome of this analysis is that this approach has potential for accelerating delivery of actionable dust forecasts to public health officials, while also establishing open standards based services for performing automated QA/QC on products generated by the system. Furthermore, the simplified parallel execution model has significant potential for deploying scalable modelling systems into commodity computing environments (that is, cloud computing), providing scalable modelling capabilities that do not require investment in fixed computing infrastructure, but instead take advantage of the expanding availability of on-demand computational resources.

3.5   Health data from SEDAC

The Socioeconomic Data and Applications Center (SEDAC) is operated by the Center for International Earth Science Information Network (CIESIN) at Columbia University. It provides data and services related to human interactions with the environment, and in particular demographic and socioeconomic data that can be integrated with EO data and imagery. A number of SEDAC data and information resources have been used extensively in public health research and surveillance, especially those related to the spatial distribution of population and associated demographic characteristics. SEDAC also provides a number of interactive tools and resources for visualization and analysis of interdisciplinary data useful to a wide range of users concerned with public health issues. The sub-sections below describe a variety of data sets and services for generating health-related information on-the-fly from numerous URLs. Table 10 is a list of these services and their access URLs.

3.5.1   Basic services
SEDAC's interactive map client provides visualization of various geospatial data layers available from SEDAC (Table 10, A). It is organized by data set collections, selected interdisciplinary topics and major world regions. It retrieves data from SEDAC map servers through open geospatial standards, and overlays them for visual analysis. Users can perform basic map operations by turning base layers and overlays on or off, and by zooming into areas of interest. Key layers of interest to the public health community include data on population distribution, land cover and land use, infant mortality, child malnutrition, air and water quality, and natural hazards. Planned developments include the ability to subset directly and download data to retrieve layers from external servers, and to integrate spatial queries and other analytic functions. SEDAC's data layers are also available to other map clients that can access data through OGC standards.

The population estimation service (PES) (Table 10, B) estimates population totals and related statistics within a user-defined region. It enables one to obtain estimates of the number of people

Table 10.   URLs referenced in section 3.5.

Products/services	URL (accessed 18th January 2012)
A. Interactive map client	http://sedac.ciesin.columbia.edu/maps/client
B. Population estimation service	http://sedac.ciesin.columbia.edu/gpw/wps.jsp
C. SEDAC TerraViva! Viewer	http://sedac.ciesin.columbia.edu/terraVivaUserWeb
D. Map gallery	http://sedac.ciesin.columtia.edu/maps/gallery/browse
E. Photostream	http://www.flickr.com/photos/54545503@N04/
F. Global rural urban mapping project	http://sedac.ciesin.columbia.edu/gpw
G. Global poverty mapping project	http://sedac.ciesin.columbia.edu/povmap/
H. United States census grid data	http://sedac.ciesin.columbia.edu/usgrid/
I. Population, landscape, and climate estimates	http://sedac.ciesin.columbia.edu/place
J. SeaWiFS chlorophyll concentrations	http://sedac.ciesin.columbia.edu/es/seawifs.html
K. Environmental performance index	http://sedac.ciesin.columbia.edu/es/epi/
L. Anthropogenic biomes	http://sedac.ciesin.columbia.edu/es/anthropogenicbiomes.html
M. Confidentiality in geospatial data applications	http://sedac.ciesin.columbia.edu/confidentiality

residing in specific areas quickly without having to download and analyse large amounts of spatial data. The service accepts polygons that define areas of interest and returns population totals, land area, quality measures, and basic parametric statistics for the requested polygons based on version-3 of SEDAC's gridded population of the world (GPWv3) data. The service can be used to estimate populations at risk of exposure to a public health threat or natural hazard event or episode.

The PES service is accessible through three standard protocols used by many online map tools and clients: the OGC WPS standard, a REST interface, and a SOAP interface. Standards-based clients such as uDig are able to submit requests using the OGC WPS. Users of ArcGIS software from Esri can submit requests through SOAP. The REST interface is intended for use with lightweight java script clients. The parametric statistics returned for each polygon include the count (number of grid cells used in the analysis), minimum population count, maximum population count, range of population counts, mean population counts, and standard deviation of population counts. Two measures of data quality are included in the service results. The first measure reflects the precision of the input data and the second indicates when the requested polygons were too small in area compared with the underlying input data to produce reliable population statistics. SEDAC provides a simple client interface and related interfaces that allow users to submit a single polygon for analysis. More complex queries can be submitted through a GIS software package that supports spatial queries through one of the three supported protocols.

Terra Viva! SEDAC Viewer (Table 10, C) is a Windows-based standalone software application that provides a powerful data-viewing engine and tools to visualize and integrate hundreds of different socioeconomic and environmental variables and layers, including a range of remote sensing data sets. The 2011 version has been updated with several new SEDAC data sets and includes fifty-one ready-made maps, ten GeoData indicator collections with hundreds of variables, and other features such as scatter plots, tabular data display, map image production, and Web-based download of additional data layers.

The SEDAC TerraViva! Viewer is unique for many reasons. It is easy to use and does not require an Internet connection. Its global data viewing engine enables users to open multiple windows to view multiple layers side-by-side, rather than as overlays. These windows may be linked to show different layers for the same area or unlinked to show the same layer at different spatial scales or using different geographic projections. The gazetteer helps to pinpoint cities, states, provinces, countries, water bodies, and other locations. The Viewer offers a library of maps by theme: population distribution, land cover, physical geography, and more. Users may also create customized maps and charts based on their data and areas of interest.

SEDAC's map gallery provides ready-made maps based on key SEDAC data resources addressing a wide range of topics relevant to the public health community (Table 10, D). Maps have been created for individual countries, selected regions, and for the globe as appropriate for each data set or data collection. Most maps are available both in jpeg and PDF formats, and are made available under a creative commons 3.0 attribution license as long as acknowledgement is given. The maps are also available directly via Flickr through the photo-stream service (Table 10, E). Maps relevant to public health applications include those developed for the GPW and global rural-urban mapping project (GRUMP) collections, the indicators of coastal water quality data set, the US Census Grids collection, the poverty mapping collection, and the environmental performance index data sets.

3.5.2 *Data and information for health applications*

GPWv3 and GRUMP collections represent complementary efforts to improve understanding of human spatial distributions across the globe (Table 10, F). GPWv3 provides estimates of population totals and densities on a latitude-longitude grid for 1990, 1995, and 2000, based on data available in 2000 from national population censuses and related sources. GPWv3 has a nominal resolution of 2.5 arc minutes. Grid cells are about $21\,km^2$ at the equator or $15\,km^2$ at 45°N or 45°S, and use the best available subnational boundary and population data for more than 125,000 subnational administrative units to determine population distribution within countries at this grid scale. GRUMP attempts to better delineate urban versus rural population distribution by incorporating data on urban extent from satellite observations of night lights and other supplementary data sources. GRUMP provides population counts and densities for 1990, 1995, and 2000 on a thirty arc-second grid (about one square kilometre). GRUMP also includes data on human settlements and urban extent.

Several complementary data sets have also been developed. GPW future estimates (GPWfe) projects population distribution grids for 2005, 2010, and 2015 using simple population growth rates. The purpose of these projections is to show a scenario of future spatial distribution for population at a subnational resolution, assuming a continuation of recent demographic patterns (Balk *et al.*, 2005). The low elevation coastal zone urban-rural estimates (LECZ-URE) combine GRUMP data with SRTM elevation data to produce population estimates on a thirty arc-second grid for populations resident on coastal lands less than ten meters in elevation above mean sea level (McGranahan *et al.*, 2007).

These population data sets have been used widely in public health research and surveillance, as they provide information on the population potentially exposed to infectious disease, environmental stresses and disease. Combining the GPW with satellite-derived ground-level values of fine particulate matter ($PM_{2.5}$), scientists have been able to estimate global long-term exposures to this air pollutant (van Donkelaar *et al.*, 2010). GPW data also have been used to determine the effect of population density and population growth on global temporal and spatial patterns of emerging infectious diseases (Jones *et al.*, 2008). GPW and GRUMP data were used in several studies on malaria incidence and the effectiveness of prevention programs (Teklehaimanot *et al.*, 2007; Hay *et al.*, 2010; Tatem *et al.*, 2010). Gridded population counts and population density from GRUMP enabled estimation of the number of children under age five at risk of malaria in a study on insecticide-treated net coverage in Africa (Noor *et al.*, 2009). In modelling the interaction of short-distance commuting and long-distance airline patterns and their impacts on global epidemics, both GPW and GRUMP were used to divide the world into a grid with cells that could be considered at different resolution levels (Balcan *et al.*, 2009). These are only a few of many health-related studies that have utilized these gridded population data sets as an integral part of their analyses.

3.5.2.1 Poverty mapping

The Global Poverty Mapping project seeks to enhance current understanding of the global distribution of poverty and the geographic and biophysical conditions of where the poor live (Table 10, G). The project aims to assist policy makers, development agencies, and the poor themselves in designing socioeconomic and health interventions to reduce poverty and achieve the Millennium Development Goals. Subnational, spatially explicit, poverty data sets are provided for selected

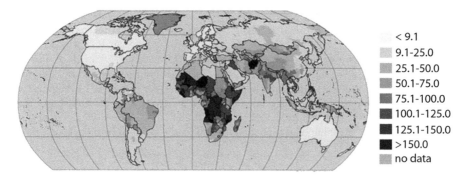

	< 9.1
	9.1-25.0
	25.1-50.0
	50.1-75.0
	75.1-100.0
	100.1-125.0
	125.1-150.0
	>150.0
	no data

Figure 8. Subnational infant mortality rate data serves as one proxy in mapping global poverty. (see colour plate 43)

proxy measures of poverty at global and national scales. The global data are of varying resolution, but primarily coarse; the national data sets are at considerably higher-resolution. Data catalogues for Poverty Mapping describe the variables available in each data set, and the underlying spatial, survey, and census data sets used to construct the integrated collection. Metadata records also provide details on source data and methods.

At a global scale, poverty is usually represented by national indicators such as gross domestic product or population living on less than one US dollar/day. These indicators are not available at a subnational level for most countries. The poverty mapping site provides global estimates of poverty based on subnational infant mortality rates and child malnutrition data, recognizing that both are proxies for poverty and welfare rather than direct measures (Figure 8). Data were drawn from demographic and health surveys, multiple indicator cluster surveys, national human development reports, and other sources. These data were first linked to boundary data for their reporting regions and then translated to a common grid (Storeygard et al., 2008).

Numerous methods have been used to construct estimates of poverty that are finely resolved. These methodologies utilize both indirect and direct estimation techniques. Indirect estimation includes small area estimates of poverty and inequality, poverty and food security country case studies, and combining surveys with remote sensing data. Direct estimation is based on the unsatisfied basic needs data set. Poverty and welfare are generally measured through proxy consumption variables like estimates of expenditures and consumer goods, or basic needs like sanitation, water, housing, and education. Data sets representing both approaches are available in this collection.

3.5.2.2 Populations near US superfund sites

In response to a request by the National Institute of Environmental Health Sciences (NIEHS), CIESIN conducted an assessment of populations living in proximity to superfund National Priorities List (NPL) sites for 2000 (Golden et al., 2008) (Figure 9). This assessment improved on earlier studies in several ways: 1) SEDAC's US census grids data set was used to determine population totals and demographic breakdowns in proximity to NPL sites (Table 10, H); 2) a methodology was applied to eliminate double-counting of populations in proximity of more than one NPL site; 3) the study used a majority rule to determine when population data should be included in the summation; 4) the total population in proximity to two or more Superfund sites was estimated; and 5) these analyses were conducted twice: once for populations living within a one-mile buffer, and again for those within four miles of the site.

SEDAC provides access to the assessment report and to the three key spatial data sets in shapefile format used for the analyses: 1) Agency for Toxic Substances and Disease Registry (ATSDR) hazardous waste site polygon data, 1996; 2) US/EPA's NPL point data with CIESIN modifications, 2008; and 3) ATSDR hazardous waste site polygon data with CIESIN modifications, 1996. In the future, an online interactive map service with information on vulnerable populations in proximity to Superfund sites across the US will also be available to the public.

Figure 9.   US population and NPL superfund sites with four mile buffers. (see colour plate 44)

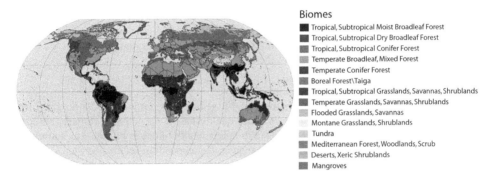

Figure 10.   PLACE II provides population and land area distribution for classes of several different demographic, physical, biological, and climatic variables, including Biomes. (see colour plate 45)

3.5.2.3   Population, landscape and climate estimates

The population, landscape and climate estimates (PLACE II) data set contains country-level measures of the spatial characteristics of 228 nations to researchers for whom national aggregates are more useful than GIS data (Table 10, I). PLACE II provides the numbers and percentages of people and the land area (square kilometres and percentages) represented within each class of a number of demographic, physical, biological, and climatic variables for each country around the world, for the years 1990 and 2000. These variables include biomes, climate zones, coastal proximity zones, elevation zones, and population density zones as shown in Figure 10. The full data array of nearly 300 variables, tabulated by country, is available for download in spread-sheet format, together with supporting documentation. The PLACE II map collection displays examples of input variables and country dynamics via more than forty maps at global, continental, and national scales.

The methodology used to develop PLACE II has been refined since the first version of PLACE. PLACE II employs population estimates from version 3 of the gridded population of the world data set (GPWv3) for both 1990 and 2000, in order to assess trends and population shifts through time. Thematic classes and land area estimates by country were held constant, to provide a uniform spatial geography across the decadal span. PLACE II also improves upon the spatial processing workflow to better account for updated coastlines and country boundaries used in GPWv3. Codebooks describing each variable used are provided along with a methodology paper (SEDAC 2007). PLACE II enables users to address such questions such as: 1) Which countries have the highest percentages of their population living within 100 or 200 km of a marine coast?; 2) What proportion of the

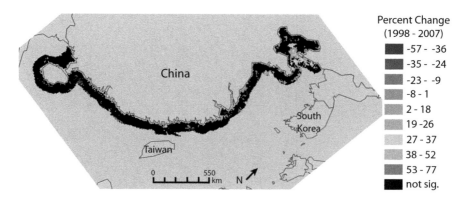

Figure 11.   Percentages of change in chlorophyll concentrations as an indicator of algal biomass along the China coast, 1998–2007. (see colour plate 46)

Kenyan population lives at an elevation lower than where malaria-carrying mosquitoes are found?; and 3), In which climate types do most of the US population live?

3.5.2.4   Indicators of coastal water quality

The flow of nutrients into coastal waters from land-based sources has increased around the world in recent decades. The resulting change in water quality has many potential impacts on coastal and marine ecosystems that in turn could lead to public health impacts. Phosphorus and nitrogen contribute to enhanced algae growth, and subsequent decomposition reduces oxygen availability to benthic life forms. Changes to nutrient loadings can also change the phytoplankton species composition and diversity. In extreme cases, eutrophication can lead to hypoxia, oxygen-depleted dead zones, and harmful algal blooms (see Chapter 9).

Measuring chlorophyll concentrations as an indicator of algae biomass may be a way to assess coastal water quality and changes over time. Figure 11 shows chlorophyll-α concentrations derived from SeaWiFS (Table 10, J). The data have been used to analyse trends over a ten year period (1998–2007), helping to identify near-coastal areas with improving, declining, and stable chlorophyll concentrations that can provide guidance for environmental management decisions. In Figure 11, black indicates no significant change; orange 53–77 per cent; light orange 38–52 per cent; yellow, 27–37 per cent; light green, 19–26 per cent; and green, 2–18 per cent. Areas of declining concentrations are shown in light blue, −1 to −8 per cent; medium blue, −9 to −23 per cent; blue, −24 to −35 per cent; and dark blue, −36 to −57 per cent. The website also provides a description of the global data set and how it was derived. Raster data are available for download in Esri GRID and GeoTiff format, and the indicators of change are provided in tabular format. Ancillary data are provided in Esri GRID and shape-file formats.

Yearly average concentrations of chlorophyll-α in nanograms/metre^3 are based on annual composites of SeaWiFS satellite data provided by GSFC and GeoEye at a resolution of nine km^2. Level-3 products were derived from true-colour images generated from sub-sampled, calibrated, Rayleigh-corrected level-2 data, which were derived from raw radiance counts by applying sensor calibration, atmospheric corrections, and bio-optical algorithms. HDF files were converted to Esri GRID format and coastline buffers were clipped by country and exclusive economic zone. For defining coastal zones, the first ten kilometres of coastal waters were excluded because of the potential for bottom-reflectance or suspended sediments affecting the satellite measurements in close proximity to the coast. The extent of coastal zones was limited to 100 km offshore as an arbitrary limit beyond which impacts from land-based sources on oceanic eutrophication are unlikely.

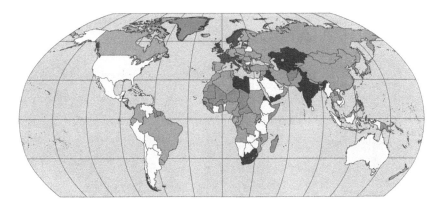

Figure 12. 2012 Environmental performance index scores. Dark green: strongest index; light green, strong; ivory, modest; orange, weaker; red, weakest; and grey, no index (Courtesy CIESIN & YCELP). (see colour plate 47)

3.5.2.5 Environmental performance index 2010

The environmental performance index (EPI) ranks 132 countries on their environmental performance based on twenty-two indicators in ten policy categories including environmental health, air quality, water resource management, biodiversity and habitat, forestry, fisheries, agriculture, and climate change (Table 10, K). Figure 12 shows EPI rankings for 2012. The website provides access to materials relating to the 2012 EPI, including a summary for policy makers, a detailed report, country profiles, access to the EPI component data and metadata, and an interactive visualization tool. The website also includes an archive of data and reports for the 2006 and 2008 EPIs. The 2012 EPI was developed by the Yale Center for Environmental Law and Policy (YCELP) and CIESIN, in collaboration with the World Economic Forum and the Joint Research Centre of the European Commission. The 2012 EPI centres on two broad environmental protection objectives: reducing environmental stresses on human health, and promoting ecosystem vitality and sound natural resource management. It utilizes a proximity-to-target methodology focused on a core set of environmental outcomes linked to policy goals (YCELP *et al.*, 2010).

Each nation is benchmarked against others that are similarly situated with groupings, based on geographic regions, level of development, trading blocs and demographic characteristics. Many of the EPI indicators are directly relevant to public health. These include: 1) *child mortality*: the probability of dying between one's first and fifth birthdays per 1000 children at age one; 2) *access to sanitation*: percentage of a country's population that has access to an improved source of sanitation; 3) *access to drinking water*: percentage of a country's population that has access to an improved source of drinking water, that is: piped water into dwelling, plot or yard; public tap/standpipe, tubewell/borehole, protected dug well, protected spring, and rainwater collection; 4) *indoor air pollution*: percentage of population using solid fuels in households; 5) *particulate matter*: population weighted exposure to $PM_{2.5}$ in micro grams per cubic meter; 6) *SO_2emissions per capita*: amount of SO_2 in kilograms per person; 7) *SO_2 emissions per \$ GDP*: amount of SO_2 per 2005 constant dollar USD; 8) *change in water quantity*: area weighted per cent reduction of mean annual river flow from "natural" state owing to water withdrawals and reservoirs; 9) *pesticide regulation*: based on a twenty-two point legislative scale on status of pesticide use and toxic chemical controls; and, 10) *renewable electricity*: renewable electricity production as a percentage of total electricity production.

3.5.2.6 Anthropogenic Biomes

Anthropogenic biomes (aka anthromes or human biomes) describe the terrestrial biosphere in its contemporary, human-altered form using global ecosystem units defined by patterns of sustained

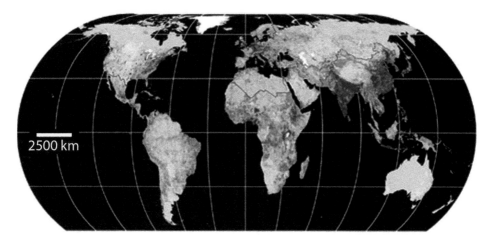

Figure 13.   Anthromes based on population density, land use and vegetation cover. Blue, purple and red colours define heavily cultivated regions in Asia and Europe; greens and yellows are semi-natural and croplands; pale to darker orange colours are defined as rangelands; and pale green to grey areas are "wild" lands (Courtesy Brandon, 2010). (see colour plate 48)

direct human interaction (Figure 13). It delineates twenty-one anthropogenic biomes based on population density, land use and vegetative cover. Anthropogenic biomes are grouped into six major categories: dense settlements, villages, croplands, rangeland, forested and wild lands (Ellis & Ramankutty 2008). A full description of these biomes is provided in their website (Table 10, L). Users may download the data as one global grid or in six separate grids for the populated continents. The data are available in GeoTif and GRID formats. Data may be useful to public health surveillance and research by providing contextual information important for understanding human-environment-disease interactions.

3.5.2.7   Confidentiality issues in geospatial data applications

The synthesis of geospatial data with socioeconomic and medical data could lead to many benefits for society, especially in terms of improving public health. Geographic information systems are powerful integrating technologies capable of bringing together information from a variety of sources, including remotely sensed data from instruments aboard aircraft and orbiting satellites and precise spatial coordinates from GPS instruments. The analytical potential of linking spatially explicit data with health surveys and other demographic and behavioural data is great. However, location-specific data at the household or even neighbourhood level may provide sufficient information so that the identity of study subjects can be determined either directly or indirectly.

In the field of public health, good science and successful policies depend on developing effective strategies to balance the rights of individuals with the needs of the community. In order to understand, diagnose, monitor, treat, and prevent diseases and injuries that harm sectors of the population, it is necessary to enlist the cooperation of those at risk. Only with detailed information on individual exposures, behaviour, and socio-demographic and health conditions can researchers begin to understand the aetiology of illnesses. Survey and study data combined with extensive georeferenced data from multiple projects across diverse disciplines can further reveal the dynamic interactions of environment, infrastructure, populations, and disease. However, such linkages also have the potential of disclosing the identities of individual study participants. Some approaches that have been implemented to protect confidentiality while still providing data access include: aggregation, masking, and suppression; research contracts with confidentiality clauses and disclosure penalties; safe houses for restricted data access by approved users; and, protected online data-sharing co-laboratories (Golden *et al.*, 2005).

Table 11. URLS referenced in section 3.6.

Product/service	URL (all accessed 18th January 2012)
A. Distributed Active Archive Centers	http://esdis.eosdis.nasa.gov/dataaccess/datacenters.html
B. DAACs and data archives	http://nasadaacs.eos.nasa.gov/bout.html
C. Mirador	http://mirador.gsfc.nasa.gov/
D. OPeNDAP	http://disc.sci.gsfc.nasa.gov/services/opendap/
E. IDV	http://www.unidata.ucar.edu/software/idv/
F. McIDAS-V	http://www.ssec.wisc.edu/mcidas/software/v/
G. Panoply	http://www.giss.nasa.gov/tools/panoply/
H. Ferret	http://ferret.wrc.noaa.gov/Ferret/
I. GrADS	http://www.iges.org/grads/
J. GrADS data server	http://disc.sci.gsfc.nasa.gov/services/grads-gds
K. Simple subset wizard	http://disc.gsfc.nasa.gov/SSW/
L. Giovanni	http://giovanni.gsfc.nasa.gov
M. Giovanni documentation	http://disc.sci.gsfc.nasa.gov/giovanni/additional/users-manual

The SEDAC website on confidentiality issues in geospatial data Applications (Table 10, M) promotes awareness of confidentiality and privacy issues related to the integration of geo-referenced data from the natural, social, and public health sciences. It provides access to resources that analyse these issues and that offer concrete tools, techniques, and policies for safeguarding confidential information. It is intended for natural, social, and public health scientists, educators, data managers, and decision makers who use and disseminate geospatial data, especially remote sensing data.

3.6 *Health data and information services at GES DISC*

As noted in Section 2.0, public health and environmental scientists have become increasingly resourceful in using satellite sensor data sets to facilitate their research. Although a core set of environmental parameters has emerged, many other parameters are utilized as a result of particular research interest and/or the need of specific projects. In addition there is an increasing number of choices about where to access and utilize specific geophysical measurements. As a result, project personnel often need to become students of the data products they employ to understand the assumptions embedded in data creation, data quality, and data validation. Maximizing the combination of the data user's knowledge of the product, and data services that more easily provide that knowledge will yield the most refined research results.

NASA, through its Earth Observing System Data and Information System (EOSDIS) program, has developed, and continues to evolve, information technologies that facilitate maximum return on its Earth science data investments. The GES DISC is one of twelve Earth science data centres, aka Distributed Active Archive Centres (DAACs). Each specializes in one or more specific Earth science disciplines. URLs associated with this section are listed in Table 11

GES DISC specializes in archiving, distributing, stewarding, and providing user data access and analysis services for remote sensing data associated with global atmospheric composition, atmospheric dynamics, hydrology, precipitation, and global modelling data. In addition, GES DISC has developed several tools and services that promote easier use and usability of Earth science data and information. As data and information management needs of science researchers have become more sophisticated, services have taken advantage of maturing information technologies to develop and implement tools and services that help researchers extract the information they seek from data.

Although this discussion focuses on the products and services residing at GES DISC, the same, or similar data and services, are available from other DAACs and data archives (Table 11, A & B). For example, the source of MODIS data products shown in Table 12 to support public health

Table 12. Types of environmental parameters and sensors that provide data.

Environmental parameter	Sensor measurements
Precipitation	TRMM; Aqua/AIRS (daily); Aqua/AIRS (monthly); GLDAS (monthy); MERRA (monthly 2D & 3D, chem., hourly 2D & 3D-3 hourly)
Relative humidity	MODIS (daily); Aqua/AIRS (daily); Aqua/AIRS (monthly); Aura MLS
Water runoff	GLDAS (monthly); MERRA (monthly 2D and 3D); MERRA (monthly chem.); MERRA (hourly 2D, 3D, and 3-hourly)
Vegetation indices	MERRA (monthly 2D and 3D); MERRA (hourly 2D, 3D, and 3-hourly)
Soil moisture	GLDAS (monthly); NEESPI (daily, monthly); MERRA (monthly 2D and 3D); MERRA (hourly 2D; 3D, and 3-hourly)
Surface temperature	Aqua/AIRS (daily, monthly); Aura MLS; MERRA (monthly 2D and 3D; hourly 2D, 3D, and 3-hourly); GLDAS (monthly)
Air quality	Aura TES
Aerosols	AEROSOL (daily, monthly); Aura (OMI L3, L2G); MODIS (daily, Monthly); MISR; TOMS; MERRA (monthly 2D, 3D; monthly chem); MERRA (hourly 2D, 3D, and 3-hourly)
Wind	MERRA (monthly 2D, 3D; monthly chem.); MERRA (hourly 2D, 3D and 3-hourly); GLDAS (monthly)

research can be found at several locations. The remainder of this section describes the data and information services offered by GES DISC that are relevant to public health research.

3.6.1 Health related Earth science data sets at GES DISC

Various remote sensing measurements identified by public health research projects reside in many NASA and non-NASA archives. In Table 12 previously used and potential data sets relevant to public health research available from GES DISC are listed by measurement and remote sensing instrument. In addition, relevant modelled data of global geophysical parameters are included in the modern era retrospective-analysis for research and applications (MERRA), and other data sets generated by GSFC's global modelling and assimilation office (GMAO), the North American land data assimilation system (NLDAS), and the global land data assimilation system (GLDAS). Data products generated by GSFC's Hydrological Sciences Branch are also listed. Modelled data provide additional parameters useful to environmental health studies.

Whereas, several GES DISC products listed in Table 13 have proven to be useful in relating environmental factors to public health conditions, others have potential use to complement or verify EO data and ground based measurements currently being captured, or may be useful depending on the temporal or spatial resolutions required of the research.

3.6.2 Data search, access, exploration, and discovery services

Acquiring NASA data and information has been expedited greatly with advances in information technology and implementation of promising technologies to perform pre-research functions quickly, such as searching for specific data, exploring terabytes of data, and visualizing chosen data of interest. The use and usability of data initially generated for science research has expanded to applications research where data users are typically less familiar with EO data than with their usual data and data sources. Information services have facilitated inclusion of sensor data into applications R&D, and in particular public health research. The following GES DISC services have been developed and continue to evolve to enhance the access of data and glean information from the data on the researcher's behalf.

3.6.2.1 Data search and access

Mirador is an Earth science data search tool developed at GES DISC (Table 11, C). It has a simplified, clean interface that combines a free text data set search with a relational database that

Table 13.   Satellite data sets, parameters measured, spatial resolutions, and lengths of record.

Data set*	Measurement	Spatial resolution	Temporal resolution
TRMM	Precipitation	0.25 – 1 deg	1997-present
GPM	Precipitation		Starting in 2003
AIRS	Precipitation; surface air temperature	50 × 50 km ; 1 × 1 deg	2002-present
GLDAS	Precipitation; water runoff; soil moisture; surface air; temperature; wind	1/4 × 1/4 deg; 1 × 1 deg	1979-present
NLDAS	Precipitation; water runoff; vegetation index; soil moisture; surface air temperature; wind	1/8 × 1/8 deg	1979-present North America
MERRA	Precipitation; water runoff; vegetation index; soil moisture; surface air temperature; wind; aerosols	1.25 × 1.25 deg; 2/3 × 1/2 deg	1979-present
NEESPI- AMSR-E	Soil moisture	1 × 1 deg	2002-present
LPRM using AMSR-E	Soil moisture	25 × 25 km	2002-present
TOVS	Surface air temperature	1 × 1 deg	1984–1995
OMI	Aerosols; solar irradiance	24 × 13 km; 1 × 1 deg; 1/4 × 1/4 deg	2004-present
TOMS	Aerosols	1.25 × 1 deg	2004-present
DUST-DISC	Aerosols	4 × 4 km; 1/2 × 1/2 deg	1997-present
GOCART	Aerosols	2.5 × 2 deg	2000–2007
GSSTF2b	Wind	1 x 1 deg	1998–2008
SORCE	Solar irradiance		2003-present

*For full names of data sets see Acronyms.

stores file information, as well as gazetteer locations for locations and geophysical events. Other features include quick response, spatial and variable sub-setting, data file hit estimator, gazetteer (geographic search by feature name capability), and an interactive shopping cart. The Mirador starting page has a keyword data search that requests three basic search entries: Keyword (e.g. instrument, parameter), date range for which data are desired, and geographic location for which data are desired. Location can be entered manually by pointing, clicking and dragging across a desired area on the map. All data sets residing at GES DISC can be located through Mirador and documentation for data and services is available.

Two other Mirador tabs are the *Projects* and *Science Areas* tabs. The former lists and describes all available data sets by sensor source or assimilation model. The latter provides data for Earth science categories and further breaks the categories into measurements that can be searched and accessed. The system continues to evolve as user communities request new services and data access.

3.6.2.2   OPeNDAP

The open source project for a network data access protocol (OPeNDAP, Table 11, D) provides alternative access to individual variables within data sets in a form usable by many tools, such as the integrated data viewer (IDV) (Table 11, E), McIDAS-V (Table 11, F), Panoply (Table 11, G), Ferret (Table 11, H) and GrADS (Table 11, I). This system provides software that makes local data accessible to remote locations regardless of local storage format. OPeNDAP software is available at no cost. GES DISC has installed OPeNDAP to provide users access to data that can be used easily in their local data analysis tools. Currently, health related data sets associated with the AIRS, TRMM, TOMS, and OMI instruments and the MERRA modelled data sets are accessible through OPeNDAP. Within the core of OPeNDAP is the GrADS data server GDS. This is a stable, secure data server that provides sub-setting and analysis services across the internet (Table 11, J). It is used for data networking to make local data accessible to remote locations. Currently TRMM, MERRA, GLDAS, and NLDAS data products are available through GDS.

3.6.2.3 Web mapping service

The OGC/WMS described in Section 3.4 is a service that provides map depictions for any spatial area over the network via a standard protocol, enabling users to build customized maps based on data coming from a variety of distributed sources. GES DISC offers WMS layers for AIRS surface temperature, TRMM rainfall, and OMI atmospheric parameters. WMS requests return an XML file that can be read by an OGC compliant client that includes many GIS tools. Subsequent WMS GetMap requests return image files to be rendered in the client. WMS-produced maps are generally rendered in a pictorial format such as PNG, GIF or JPEG. Occasionally they are rendered as vector-based graphical elements in scalable vector graphics (SVG) or web computer graphics metafile (WebCGM) formats (de la Beaujardiere 2006).

3.6.2.4 Simple Subset Wizard

GES DISC has teamed with nine of other EOSDIS Data Centres to develop the simple subset wizard (SSW) (Table 11, K). This feature provides a unified user interface for submitting spatial subset requests for data across EOSDIS data centres. Geophysical parameters (keywords) related to public health can be searched by data centre.

3.6.3 *Information exploration and discovery*
3.6.3.1 Giovanni

Giovanni is an online tool allowing application users to explore, visualize, inter-compare, and analyse EO data easily using only a Web browser. It has proven to be a valuable tool for locating and visualizing data, whether the user knows the specific data set of interest, or is seeking measurements with particular signatures. Starting with basic information: Location, date/time, parameter, and type of visualization, users can explore, view, and download specific data for further analysis. If the viewed data are not of interest, users can repeat the operation with different basic information. More advanced operations and visualizations are also available, such as multi-data comparisons, scatter plots, time series, animations, and more. Examples are shown in Figures 14–16. Currently, over forty project and discipline specific versions of Giovanni exist (Table 11, L).

Giovanni contains several hundred parameters from GES DISC data holdings, plus MISR AOD and data from the Tropospheric Emission Spectrometer (TES). The latter are distributed from the Langley Research Center (LaRC, Atmospheric Science Data Center (ASDC)). The MODIS aerosol and water vapour products are archived and distributed from LAADS at GSFC. As an example, precipitation data available through Giovanni are listed under measurements relevant to public health research. These include TRMM; Aqua/AIRS (daily and monthly); GLDAS (monthly); MERRA (hourly -2D, -3D; monthly-2D and -3D; monthly chemistry; and 3-hourly). Documentation is available (Table 11, M).

3.6.3.2 Data Quality Screening Service

The data quality screening service (DQSS) filters bad-quality data points from a data file by using quality variables included within the file (Lynnes 2010). This service is extremely useful in applications research because it uncovers and utilizes higher quality data efficiently to give more accurate results. DQSS is available through the Mirador search tool. It generates output files that have the same structure as the input, albeit with a few extra variables, and is thus usable in any tool made to work with the original product. The service replaces data arrays with the corresponding filtered arrays. Data values deemed unusable are set to *fill values*. The output file also includes the quality mask used for filtering and the original (unscreened) data values in additional data objects. The DQSS steps shown in Figure 17 are accessible through Mirador for an AIRS data set. MODIS and MLS data sets are planned.

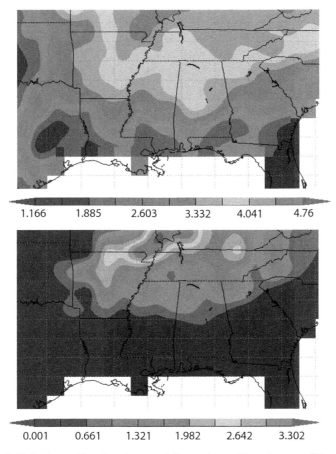

Figure 14. (Top) GLDAS monthly soil moisture and (Bottom) monthly surface runoff for the south eastern US between April and June 2011. Soil moisture ranges from <1.2 in purple and blue areas to more than 4.5 in the orange and red areas. Surface runoff values range from <0.6 to >2.6. (see colour plate 49)

4 ACCESSING NASA EARTH SCIENCE INFORMATION

Section 2 provided the breadth of public health research that takes advantage of the availability of remote sensing data. Section 3 described the tools and services available to access remote sensing data for this research. Section 4 provides an in depth look at three projects describing in greater detail how remote sensing data are used to enhance public health applications. The first study by Doctors Tilburg and Zeeman addresses precipitation events to predict coastal water quality. The second study by Doctor Zaitchik addresses the value of remotely acquired data for developing a malaria early warning system in the Peruvian Amazon; and the third, by Doctor McClure links environmental data with a national public health cohort study to enhance public health decision making.

4.1 *Precipitation events for predicting coastal water quality*

Rainfall and runoff are associated with the rise in water-borne diseases (Curriero *et al.*, 2001). Heavy rainfall events are more likely to lead to marked decline in microbiological quality of inland and marine recreational waters and drinking water supplies as a result of heavy runoff (Fayer *et al.*, 2004). Rain events result in movement of land-based pollutants to streams and rivers. These pollutants travel downstream with river flow to the coastal ocean and adjacent waters, which

269

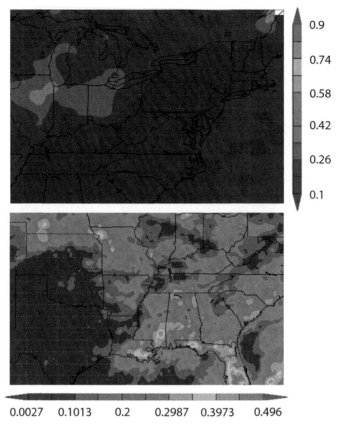

Figure 15. (Top) MODIS AOD values range from lowest in purple (<0.1) to highest in pale green (>0.3). (Bottom) Average TRMM precipitation rate at 00:00 UTC 22June to 22 August 2011. Purple, blue, turquois represent <.0.002 to 0.2; green, yellow represent >0.2 <0.39; and orange, red >0.39 <0.49 mm/hr. (see colour plate 50)

results in reduced water quality events (RWQEs). In the north eastern US, climate change scenarios typically indicate an increase in overall precipitation, which would lead to larger river discharge (IPCC 2007). Increased discharge would then lead to greater pathogen loading in rivers and coastal waters from anthropogenic sources due to runoff from contaminated sites. Consequently, a method for predicting RWQEs has become crucial for effective water resource management. The central objective of this study is to determine if a relatively simple model using easily accessible remotely-sensed data can be used to predict RWQEs along the coast of Maine. Information from this study can be used to enhance every day decision making at the state and county level or aid in long term plans for water treatment plants in the region.

The most common method for testing water for RWQEs due to the presence of pathogens is to measure the concentration of *Escherichia coli* and total coliforms, which are bacteria that are normally present in the intestinal tract of humans and other animals and are used as indicators for recent faecal contamination. Unfortunately, the method is costly and time-consuming. Since most states do not currently have the ability or resources to quickly test for *E. coli* and Total Coliforms (typically 18–24 hrs.), proxies that can predict RWQEs are needed. A common method in the state of Maine is river discharge since discharge is highly correlated with upstream precipitation and the associated runoff within watersheds (A. Bouravkovsky pers. comm.). However, bulk measurements of discharge are not able to differentiate between run-off from high pollution areas that would tend

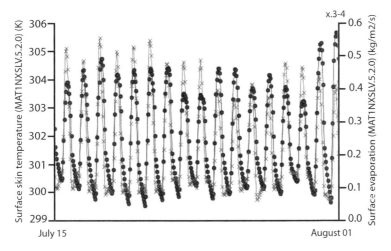

Figure 16. Hourly area-averaged temperature time series along the Texas/Louisiana coast 15July to 1August 2011. The blue line is surface skin temperature ranging from 299°K to 306°K, and the red line is surface evaporation ranging from 0.0 to 1.2 kg/m². (see colour plate 51)

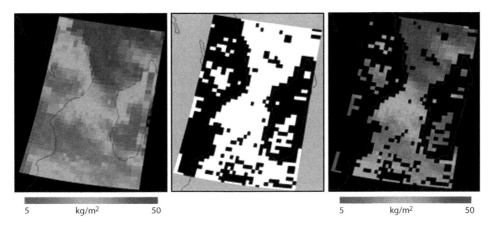

Figure 17. Total column precipitable water (kg/m²) from AIRS near Madagascar on 4 June 2009. (Left) raw data; (Center) quality mask; (right) post screen out-put. Red is ±5 km²; blue is ±50 kg/m². (see colour plate 52)

to result in RWQEs and run-off from less developed, pristine areas that would not affect water quality during high run off events. Additionally, not all rivers are equipped with river gauges to measure discharge. A method that can incorporate the source of run-off in its predictions of RWQEs would likely be more accurate and dependable.

Here, we describe the methods used to create a model that predicts RWQEs using remotely-sensed precipitation data and then test the accuracy and usefulness of that model using *in-situ* observations of water quality. The ease of access and the regular spacing of the remotely acquired precipitation data are likely to lead to timely and accurate predictions of RWQEs that were not available by earlier methods.

The model was used to examine water quality in three different watersheds: the Saco, Kennebec, and Androscoggin River watersheds located in Maine and New Hampshire (Figure 18). All empty into the Gulf of Maine, a marginal sea in the northwest Atlantic Ocean. Maine is an ideal test

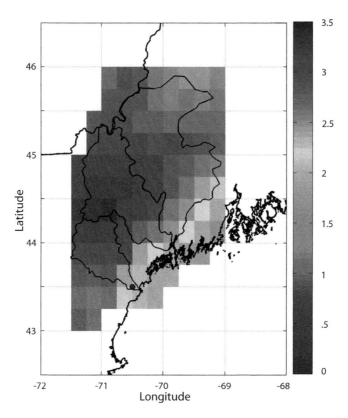

Figure 18.   Mean daily precipitation (mm/day) observed for TRMM grid cells between 1998 & 2009. Black lines represent boundaries of three different watersheds. Blue circle is a water sampling site in Saco River Watershed. (see colour plate 53)

hed for this project because of its large number of rivers, varied watersheds, and strong reliance on coastal waters for both recreation and seafood. In addition, state agencies have little access to biological testing and are in need of new methods for the detection of RWQEs. The relatively low computational cost of the model (it can run on any laptop computer) allows one to examine different scenarios and instances, and their effects on water quality along the Atlantic coast.

4.1.1   *Data sets*
To achieve the main objective of the project, both remote-sensing and *in-situ* data were used. Observations for the project were collected over a twelve year time period from 1998–2009 from TRMM's TMI sensor. It is able to distinguish precipitation variations at a grid cell resolution of 0.25°. Figure 18 shows the location of the different TRMM grid cells and outlines of the three watersheds. The colours within each cell represent the mean daily precipitation measured over the twelve year time period. Since the long-term goal of this project was to produce an easily accessible tool to users who might be unfamiliar with satellite data, no effort was made to process the downloaded data, but instead to determine how minimally manipulated data could be used to predict RWQEs. The TRMM data time series is available from Giovanni. The data were manually downloaded and processed in MATLAB. Eventually, the method will be exported to an Excel spread-sheet or some other easily accessed software tool. The large area of the watersheds (ranging from 4410 to 9100 km^2) allowed for ten to twenty-five TRMM grid cells to occupy each watershed, providing details of small-scale spatial patterns of variability across a watershed. Data for water quality were obtained from state agencies to both calibrate and validate model outputs.

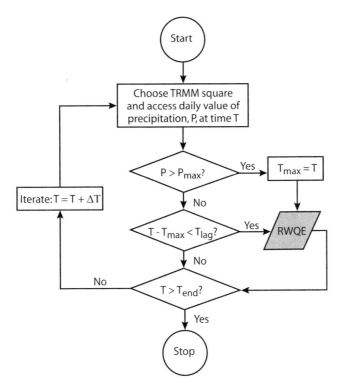

Figure 19.   Flow diagram used to predict RWQEs from precipitation data.

4.1.2 *Comparison methods*

Any method that predicts RWQEs must balance two competing goals: (1) accurate prediction of individual reduced water quality events and (2) reduction of false alarms (i.e. a predicted RWQE that did not materialize). Resource managers depend on knowledge of water quality in the coastal regions to make decisions in order to protect consumers of shellfish, users of recreational areas, and others from pollutants in the water. However, they are also under pressure to avoid costly shutdowns that remove revenue from beach-goers, consumers of shellfish, and others from local economies. Closing a healthy shellfish bed or beach due to an inaccurate prediction of reduced water quality (i.e. a false alarm) can have devastating consequences for local businesses. Consequently, resource managers need access to a reliable method that predicts as many RWQEs as possible with as few false alarms as possible.

To examine the link between remotely acquired precipitation data within a watershed to RWQEs downstream of the precipitation locations, the question was: *What amount of precipitation at some location within the watershed will result in RWQEs at a particular location downstream?* The general model used to predict the presence of RWQEs from precipitation data is represented in a flow diagram in Figure 19. Three variables were used to determine the optimum settings to predict RWQEs: (1) P_{max}, the amount of rainfall that defines an RWQE-producing precipitation event is measured in mm/day over a two to twenty day period; (2) the location of TRMM grid cell; and (3), T_{lag}, the time period over which a particular precipitation event can effect a region measured over a three to seven day period.

To measure the accuracy of the model and determine the optimum amount of precipitation, ideal location, and time lag of precipitation for each water quality site, an examination was performed on the ability of the model to predict known RWQEs with the current method (river discharge) and to avoid false alarms. Maine's current method of predicting reduced water quality events relies on measured discharge. Since higher river discharge is a direct result of increased run-off, there is a strong connection between measured discharge and coastal and riverine water quality. Each river is assigned a discharge that will result in a predicted reduced water quality event. When river discharge exceeds this value, shellfish beds and recreational areas are closed and testing of the region begins. For this study, we used a value of 1000 ft^3/s (28 m^3/s), which resulted in the most accurate prediction of RWQEs in the coastal ocean.

RWQEs can occur due to a number of different mechanisms, such as pollution events or wastewater treatment plant malfunctions. Although precipitation is likely to trigger an RWQE, there will be a number of precipitation events that do not result in RWQEs as well as RWQEs that occur even after no rainfall. To provide a quantitative measure of model skill that accounts for these additional events and to determine the level of significance of agreement between the model and observations, we used a simple randomization test, which compared the ability of the model to predict events and avoid false alarms with distributions from a large number of synthetic data sets. The synthetic data sets were constructed by determining the number of different precipitation events in the observed data set that exceeded P$_{max}$ (n events) and then randomly placing those throughout the time period. The random data set is then used to predict RWQEs. Once a large number of data sets have been created for each water quality measurement location (100–1000), a frequency distribution of the percentage of predicted RWQEs and false alarms can be constructed.

Comparison of the model predicted RWQEs with the distribution from the synthetic data set provides an estimate of the significance of the model skill (Bishop *et al.*, 2010). A model prediction that is greater than ninety-five per cent of the synthetic prediction percentage indicates that the model skill outperforms random chance at a significance of ninety-five per cent. Model false alarms that are less than ninety-five per cent of the synthetic data set indicate that the model outperforms random chance at a significance of ninety-five per cent. An example of this is shown in Figure 20 for the location in the Saco River watershed shown in Figure 18 for two different TRMM cells. Examination of Figure 21 reveals that the location of the precipitation events has a very strong effect on the accuracy of the method. Precipitation within TRMM square seven provides no better predictive skill than random guessing. The fraction of the number of observed RWQEs successfully predicted by precipitation within TRMM square seven (0.28) is less than all of the synthetic data sets in which the time of RWQEs were randomly chosen throughout the time period, while the percentage of false alarms (0.67) is higher than eighty-five per cent of the synthetic data sets. Precipitation within TRMM square eighty-five can be used to predict RWQE (0.67 prediction rate and 0.45 false alarm fraction) with much greater accuracy than random chance, providing a valuable tool for resource managers. The river discharge model using a discharge threshold of 28 m^3/s produces similar results, indicating that satellite recorded precipitation data can provide as accurate a predictive capability as more costly river discharge data. Note that P$_{max}$ = 8 mm/day and T$_{lag}$ = 3 days for both TRMM cells and random synthetic data sets. A perfect method would predict all of the RWQEs (Fraction = 1.0) and have no false alarms (Fraction = 0.0). Varying the amount of precipitation that can trigger an RWQE also results in quite different effects on RWQE prediction and false alarms (Figures 21 & 22). A low threshold (P$_{max}$) results in the method predicting almost all of the observed RWQEs, while a high P$_{max}$ results in the method missing a large number of observed RWQEs. Interestingly, an increase in P$_{max}$ does not result in a monotonic decrease in the number of false alarms, indicating that large precipitation events do not always result in RWQEs.

4.1.3 *Conclusions*

The use of readily available precipitation data at a spatial resolution of eighteen to twenty five kilometres may provide an ideal method for water resource managers to aid in their prediction of RWQEs. However, the method must be tested and the model calibrated before it can be used by resource managers. Examination of the method's statistical significance using a randomization test suggests that the method has promise. TRMM data provide for an easily accessible, effective method of predicting water quality, although the model still requires a degree of calibration for individual downstream sites.

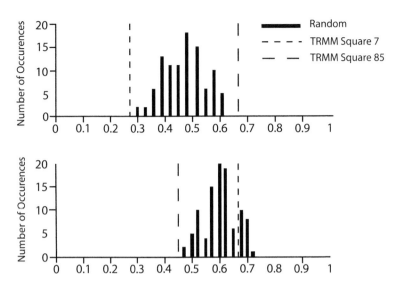

Figure 20. (Top): The fraction of RWQEs (x-axis) preceded by high precipitation events. The dashed vertical lines are for TRMM square 7 (left) and TRMM square 85 (right); (Bottom): The fraction of false alarms predicted by high precipitation events. The dashed vertical lines are for TRMM square 85 (left) and TRMM square 7 (right).

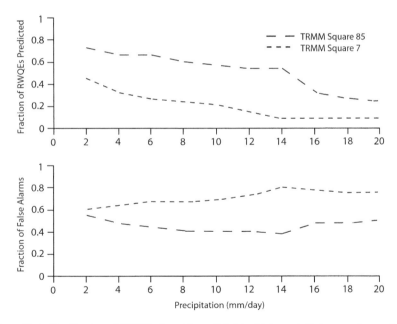

Figure 21. (Top): The fraction of RWQEs (y-axis) predicted by high precipitation events as a function of P_{max}.; (Bottom): The fraction of false alarms for high precipitation events as a function of P_{max}.

4.2 Developing a detection and early warning system for malaria risk in the Amazon

This application makes use of satellite-derived data in a number of ways. The first and most obvious motivation for satellite analysis is the fact that it is very difficult to perform land surveys

275

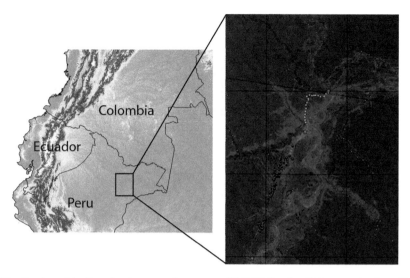

Figure 22. Portion of the Peruvian Amazon, shown on a 432-RGB Landsat TM composite from 22 August 2005. Red and yellow points show mosquito collection sites along the Nauta-Iquitos and Iquitos-Mazan roads, respectively. Green points are settlements. (see colour plate 54)

over thousands of square kilometres in the Peruvian Amazon (Figure 22). Accurate characterization of evolving land cover, of shifting rivers, and of topographic variations that can influence hydro-ecology and mosquito breeding sites, all require the application of remotely sensed data. Second, satellite platforms provide novel observations to inform spatially distributed hydrological and meteorological analysis over large areas. These observations are directly relevant to the characterization of mosquito breeding habitats, and they represent an unprecedented opportunity to integrate observation-based eco-hydrological analysis into disease risk monitoring and early warning systems.

In the context of this application, satellite observations of landscape, hydrology, and climate are used both in offline analysis of *in situ* mosquito and malaria case count data, and as inputs for the land surface models at the core of the Peruvian Amazon frontier land data assimilation system (PAF-LDAS) being implemented to integrate data streams relevant to estimates of transmission risk.

One example of offline satellite-based analysis is the use of Landsat images to characterize LCLUC around mosquito monitoring sites. Previous studies have indicated that counts of mosquitoes at monitoring sites in the Peruvian Amazon can be correlated with patterns in surrounding land cover (Vittor *et al.*, 2009). One proposed explanation for this relationship is that changes in land cover, including deforestation, change local hydrology in a way that favours mosquito breeding; indeed, the promise of this explanation serves as a primary motivation for the coupled weather/hydrology/land cover analysis at the heart of the current project. As a starting point, then, it is important to confirm that the relationship between land cover and mosquito density holds in the current study area for recently established mosquito collection sites. Next, the spatial scale of land cover influence on mosquito density must be further investigated to inform skilful extrapolation of the method to regions that lack active mosquito monitoring sites. To accomplish these analyses, Landsat-derived land cover characteristics within a radius of 250 m to 10 km of mosquito collection sites are being tested as predictive variables for mosquito counts, both as independent predictors and as inputs to a multivariate statistical model. Along this same line of questioning, the investigators are examining temporal relationships between land use change and mosquito presence: are

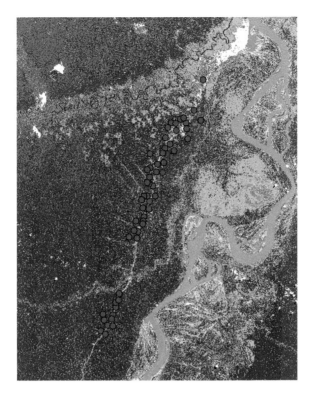

Figure 23.   Collection sites reported along the Nauta-Iquitos road, overlain on a supervised classification of a Landsat image. Iquitos is located in the northeast corner of the image. (see colour plate 55)

the mosquito counts largest in sites near recently cleared forest or in sites surrounded by long-established human settlements? The archived Landsat record offers an unparalleled opportunity to investigate these spatio-temporal dynamics, as shown in Figure 23.

Initial efforts with single-scene Landsat image classifications have proved to be quite promising. Nevertheless, given the known complications in detecting tropical forest disturbance from satellite over larger scales (Oliveira *et al.*, 2007), the research team is prepared to employ a number of more sophisticated land cover analysis techniques, including: (1) supervised and unsupervised classification of stacked seasonal Landsat images from a single year (Ichii *et al.*, 2003; Matricardi *et al.*, 2005), (2) unsupervised classification of Landsat image stacks from multiple years, in which clustering of like pixels can distinguish between pixels that have undergone change and those that have not (Steininger *et al.*, 2001), (3) spectral end-member un-mixing of single-scene Landsat images that accounts for green vegetation, senescent vegetation, and bare ground (Oliveira *et al.*, 2007), and (4) Fourier cycle similarity analysis of MODIS 16-day NDVI composites for each year of analysis, through which differences in phenology can be used to classify different LUC and end-member un-mixing is applied to determine per cent cover within 250m MODIS pixels (Geerken *et al.*, 2005; Evans & Geerken 2006). These advanced techniques are expected to be particularly important as the project scales up from well-known sites around Iquitos to the broader Peruvian Amazon region. Evaluation of land cover classifications is being performed through field survey with local experts as well as comparisons with selected high-resolution commercial imagery.

When integrated to PAFLDAS, these land cover classifications represent one of numerous satellite-derived products used to inform ecologically-based mosquito predictions. An LDAS is a numerical modelling scheme that integrates observations from various sources within such models, using data assimilation and other techniques to produce optimal maps of land surface states

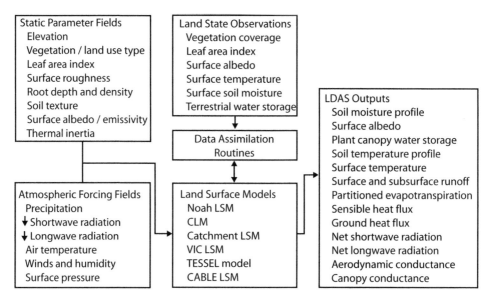

Static Parameter Fields	Land State Observations	LDAS Outputs
Elevation	Vegetation coverage	Soil moisture profile
Vegetation / land use type	Leaf area index	Surface albedo
Leaf area index	Surface albedo	Plant canopy water storage
Surface roughness	Surface temperature	Soil temperature profile
Root depth and density	Surface soil moisture	Surface temperature
Soil texture	Terrestrial water storage	Surface and subsurface runoff
Surface albedo / emissivity		Partitioned evapotranspiration
Thermal inertia	**Data Assimilation Routines**	Sensible heat flux
		Ground heat flux
Atmospheric Forcing Fields	**Land Surface Models**	Net shortwave radiation
Precipitation	Noah LSM	Net longwave radiation
Shortwave radiation	CLM	Aerodynamic conductance
Longwave radiation	Catchment LSM	Canopy conductance
Air temperature	VIC LSM	
Winds and humidity	TESSEL model	
Surface pressure	CABLE LSM	

Figure 24. Components of an LDAS. Some parameters (e.g. surface albedo) appear in multiple boxes and can be updated as an observational input and subsequently predicted as an LDAS output.

and fluxes (Figure 24). Data assimilation is a computational process by which two independent estimates of a variable can be combined to determine one best estimate, which is less uncertain than either of the two inputs. In this case, the variable is a land surface state such as soil moisture. One estimate comes from the land surface model and the other from a satellite observation. The error characteristics of each are used to weight their contributions to the final estimate. Other projects have demonstrated that these techniques can be implemented at regional and global scales (Mitchell *et al.*, 2004; Rodell *et al.*, 2004).

Land data analysis system models use physical equations to simulate the storage and movement of water and energy on and within the land surface. Net surface radiation, for example, is typically solved as:

$$R_{net} = (1 - \alpha) \cdot S_{\downarrow} + \varepsilon \cdot L_{\downarrow} - \varepsilon \sigma T^4, \tag{1}$$

where α and ε are surface properties albedo and emissivity, respectively, σ is the Stephan-Boltzmann constant, T is simulated surface temperature, and $S_{\downarrow}$ and $L_{\downarrow}$ are incoming shortwave and longwave radiation drawn from atmospheric data. Net radiation is then partitioned into surface energy fluxes and storages using additional physically equations.

PAFLDAS will include four initial land surface models in parallel analysis: 1) Noah (Chen *et al.*, 1996; Ek, *et al.*, 2003); 2) the variable infiltration capacity (VIC) model (Liang *et al.*, 1996); 3) Catchment (Koster *et al.*, 2000); and 4) the Common Land Model (CLM) (Dai *et al.*, 2003). Additional land surface models are being considered also. These include the tiled ECMWF scheme for surface exchanges over land (TESSEL) (Van den Hurk *et al.*, 2000) and CSIRO's atmosphere biosphere land exchange (CABLE) model (Kowalczyk *et al.*, 2006).

The basic implementation of PAFLDAS has been accomplished at one kilometre resolution (Figure 25), with plans to improve data inputs to achieve an ultimate spatial resolution of 500 metres, which is sufficient to capture most features of the Amazonian land use mosaic. LDAS techniques have been applied at sub-kilometre resolutions before, but modelling at this scale is not common. The low topographic relief of the study area allows for confidence in the application of one-dimensional land surface models, as the lateral transfer of moisture across grid cells is unlikely to be a significant influence on near-surface conditions. At the same time, land use transitions are quite dramatic, so there is reason to believe that high-resolution application of LDAS models

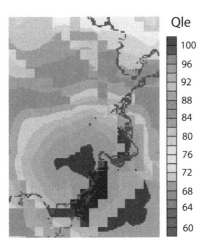

Qle

100
96
92
88
84
80
76
72
68
64
60

Figure 25.   Monthly averaged latent heat flux (W/m²) for February 2009 over a portion of the PAFLDAS domain. Data cells have different sizes due to differing resolution of input parameters and forcing data, but can be improved as LDAS incorporates additional satellite data sets and downscaling techniques. (see colour plate 56)

and assimilation techniques will add value to the analysis of conditions relevant to malaria risk. In implementing PAFLDAS, the research team is focused on the representation of temporal and spatial variability in near surface soil moisture, surface temperature and humidity, and the impact of land use on hydrologic states. Data assimilation efforts are challenged by the fact that available passive microwave soil moisture observations from AMSR-E have coarse spatial resolution and very limited reliability in tropical forest environments. For this reason, the team is particularly focused on the potential of the future SMAP mission to provide soil moisture estimates for PAFLDAS, as well as on innovative assimilation options such as GRACE (Zaitchik et al. 2008), which will become more relevant when the feasibility study is scaled to larger areas, and thermally-based soil moisture estimates (Anderson *et al.*, 2007; Meng *et al.*, 2009).

All remote sensing and data assimilation work in this project is motivated by the value these analyses have for the development of spatially-distributed malaria risk monitoring and early warning systems. Land cover analysis and PAFLDAS fields will drive spatial-temporal models of *Anopheles* mosquito ecology. These models, derived using mosquito databases collected across four areas around Iquitos over four different time periods, will use a multilevel Poisson modelling framework (Raudenbush & Bryk 2002). Since results will be used for development of risk maps, models will be fit using Bayesian analysis to allow more effective means of smoothing the map surface (Clayton & Kaldor 1987). This analysis will also involve examination of land fragmentation metrics to capture the importance of landscape structure and function on *Anopheles* distribution (Turner 1989). Once appropriate scales and fragmentation measures are identified, a multilevel simultaneous equation model will be used to estimate jointly the presence of larva and adults, particularly for the Iquitos-Nauta road where larva and adult data were collected concurrently over time. These models will be used to create predictive models of *Anopheles* presence for the entire study region.

Beyond ecological modelling, mosquito prediction models and environmental determinants need to be linked to malaria risk through accurate measures of population exposure. The research team will develop models to link environmental changes like forest conversions and changes in non-forest cover, to regions of permanent and temporary human occupation including urban expansion, agricultural expansion, logging, oil exploration activities, and mining. The goal will be to create plausible population risk sets that are defined not just where a person lives, but by their social and economic spaces that are directly and indirectly linked to land use and land use cover (LCLUC). The resulting output will be seasonal *Human Activity and Settlement Maps* that depict population density based on permanent human settlement, and human activities that show regions of recent and long-term anthropogenic LCLUC. These include forest conversion due to logging and agriculture, urban expansion, road construction, oil exploration and production, mining, and other identifiable forms of LCLUC. Three types of data sources are being used for: 1) census data of the regional population (1981, 1993, 2005, 2007) and agriculture (1994 & 2010); 2) demographic and economic survey data available from demographic and health surveys (DHS: 1991–92, 1996, 2000, 2004–08, 2009) and the World Bank's living standards measurement surveys (LSMS: 1985, 1991, 1994); and 3), data related to development projects such as road paving between 1995 and 2001, and forest and oil exploration concessions.

Annual population density will be estimated for all of the State of Loreto as well as at smaller geographic subdivisions (provinces and districts) by first estimating annual intercensal population for the Loreto area by age-sex composition using the cohort-component method (Shryock *et al.*, 1976). This method will be adjusted accordingly based on fertility, mortality and migration data obtained from DHS and LSMS surveys as well as national mortality data when available. Second, population will be disaggregated by province and district by computing population percentages for each sub-region in each census year and conducting a constrained interpolation to produce annual estimates of population distribution. Finally, population will be further disaggregated within districts by assigning population values to known cities and towns like Iquitos, Nauta, and Mazan, and by using the LCLUC output to identify permanent areas of human settlement. The *Human Activity and Settlements Map* will identify at least crudely those areas undergoing or potentially undergoing change. Non-forested areas that have not experienced change for a minimum of five years will be considered urban or agricultural based on their particular land cover. Forested areas that have not undergone change for five years will be identified as potential areas of human activity, while areas that have undergone change in the previous five years will be identified as current areas of human activity. The Map will take advantage of auxiliary data on road construction, infrastructure development, and logging and oil concession areas. With the benefit of the Landsat data archive and retrospective PAFLDAS simulations, the research team will examine several spatial scales to identify areas as well as examine the sensitivity of the five year cut-off.

Results of the Human Settlements and Activity analysis and the spatially explicit *Anopheles* ecological models will provide inputs to malaria transmission models that can be used predictively for risk mapping and early warning. These applications are of considerable importance to Ministries of Health operating in Iquitos and in other malaria-prone regions of the Amazon.

4.3   Linking environmental data and public health cohort data to enhance decision making

4.3.1   Project objectives and research design

There are seven objectives in this application effort: 1) produce daily gridded estimates of $PM_{2.5}$ for the conterminous US for 2003 to2008 using MODIS Aqua data; 2) produce daily gridded solar inso-lation maps for the same area and time frame using data from the NARR; 3) produce daily gridded LST maps from MODIS; 4) link the estimates of $PM_{2.5}$, insolation and LST with data from more than 30,000 participants in REGARDS; 5) determine whether exposure to $PM_{2.5}$ or solar insolation are related to the rate of cognitive decline among the participants in the REGARDS study, inde-pendent of other known risk factors for cognitive decline; 6) examine the relationships between the estimated $PM_{2.5}$ and insolation and other health-related conditions among REGARDS participants, including diminished kidney function, hypercholesterolemia, hypertension, and inflammation mea-sured by C-reactive protein; and 7) deliver daily gridded environmental data sets ($PM_{2.5}$, SI, and LST) to CDC-WONDER for the six year period.

A summary of our research methodology is provided in Figure 26. It shows the different data elements, how they will be linked, and the outputs generated from the project. In addition, it provides a description of the decision support system and identifies who potential end-users may be. Scientists associated with NASA's Marshall Space Flight Center (MSFC) downloaded and processed daily AQUA MODIS AOD data for the conterminous US for 2003–2008. The data processing included extracting and mosaicking different swaths of data into one national dataset.

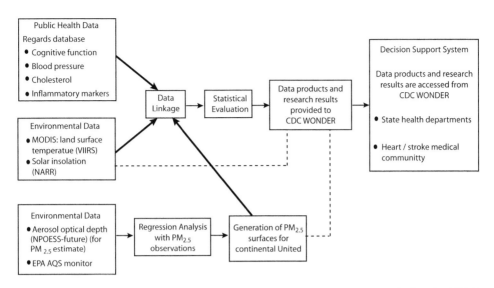

Figure 26.   Schematic describing the research methodology (Courtesy McClure, UAB and Alhamdan USRA).

Part of the processing was also to estimate ground-level fine particulates (PM$_{2.5}$) from MODIS AOD using regression equations per EPA region per season from (Zhang *et al.*, 2009). The MSFC team also obtained and processed EPA Air Quality System (AQS) PM$_{2.5}$ data for the whole conterminous US for 2003–2008. The regional surfacing algorithm of Al-Hamdan (*et al.*, 2009) was modified and used to generate continuous spatial surfaces of daily PM$_{2.5}$ for the conterminous US, which also involved a quality control procedure for the EPA AQS data and a bias adjustment procedure for the MODIS data (Figure 27).

PM$_{2.5}$ data processing is nearly complete. The surfacing algorithm has been modified and the MODIS and AQS data are processed; thus, preparation of daily national surfaces of PM$_{2.5}$ is underway. This procedure also includes a quality control element for examining the US/EPA data, as well as a bias adjustment for the MODIS data. Upon completion of these surfaces, the PM$_{2.5}$ data will be linked spatially and temporally to the REGARDS participants, and the data incorporated with the public health data.

In Figure 28 solar insolation (SI) and LST data sets have been linked to the REGARDS participants and analyses of the linked data have begun. The original aim of the project did not include an assessment of the association between SI, LST and stroke. However, the project assessed several periods of exposure to SI and LST to utilize the most robust measure in models describing the relationship between SI and stroke. The determination was that the one-year prior to baseline was most appropriate to include as the primary exposure. Table 14 provides the hazard ratios and ninety-five per cent confidence intervals for the associations between SI and stroke, and for maximum temperature averages and stroke, when exposure is defined as occurring the year prior to the in-home interview. The results for solar insolation indicate that for those participants exposed to SI at levels below the median, the risk of stroke is 1.44 times higher than for those exposed to SI at levels above the median (95% CI: 1.16, 1.79), but that after the addition of LST, the risk of death is actually higher for those exposed to lower levels of SI, compared to those exposed to higher (HR: 1.73, 95% CI: 1.25, 1.29). This association is not attenuated after multivariable adjustment for known stroke risk factors, such as demographics, behavioural and medical risk factors (HR: 1.61, 95% CI: 1.15, 2.26). Similarly, there is a U-shaped association between temperature and stroke, with those exposed to the lowest temperatures, and those exposed to the higher temperatures, at higher risk for stroke than those exposed to mid-temperatures. These patterns are detailed in Table 14, including after multivariable adjustment.

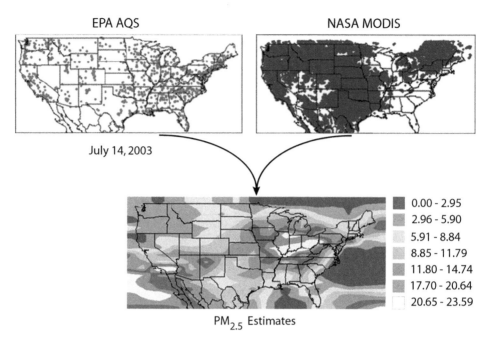

EPA AQS   NASA MODIS

July 14, 2003

■	0.00 - 2.95
	2.96 - 5.90
	5.91 - 8.84
	8.85 - 11.79
■	11.80 - 14.74
■	17.70 - 20.64
□	20.65 - 23.59

$PM_{2.5}$ Estimates

Figure 27. A schematic showing how EPA/AQS data are merged with NASA MODIS data to form a smoothed $PM_{2.5}$ surface (Courtesy McClure & Alhamdan, USRA). (see colour plate 57)

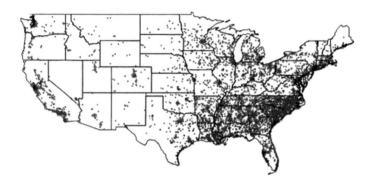

Figure 28. The REGARDS participants; those in red are whites, those in blue are African Americans. (see colour plate 58)

In addition to the assessment of stroke incidences, the project has assessed the relationship between SI and cognition. This assessment utilizes the *six-item screener* as the cognitive outcome, and includes LST as a potential confounder of the relationship between SI and cognition (Callahan et al., 2002). The analysis included an assessment of exposure periods to determine which were most robust to include. Again the 1-year prior to baseline was determined to be most suitable for modelling. The relationship between SI, LST and cognitive impairment was more complicated than for stroke. The interaction between SI and cognitive impairment exists but the relationship between them differs as a function of quartile of LST. Thus, it is not possible to provide results for each association separately. However after multivariable adjustment among those participants for whom LST exposures were in the 1st tertile; that is, those exposed to lower SI levels, were 1.26 times more likely to experience incident cognitive impairment than those exposed to higher levels (95%

Table 14. Hazards ratios and 95% confidence intervals for associations between insolation and maximum temperature averages for the year previous to in-home interview with stroke incidence (N = 16.529).*

Model	Below median insolation	Quartile of maximum temperature				
		1st	2nd	3rd		4th
Univariable**	*1.44 (1.16, 1.79)*	*1.86 (1.39, 2.52)*	Ref	1.22 (0.88, 1.69)		1.36 (0.99, 1.89)
Unadusted***	*1.73 (1.25, 2.29)*	*1.68 (1.25, 2.29)*	Ref	*1.51 (1.07, 2.14)*		*2.05 (1.36, 3.11)*
Adjusted model****	*1.59 (1.14, 2.23)*	*1.49 (1.05, 2.14)*	Ref	*1.73 (1.19, 2.51)*		*1.95 (1.29, 2.96)*
Mediation model*****	*1.61 (1.15, 2.26)*	1.41 (0.99, 2.03)	Ref	*1.69 (1.17, 2.46)*		*1.91 (1.27, 2.91)*

*Italic values indicate variables with chi-square p-values <0.05
**Univariate models contain only a measure of insolation or temperature with stroke of any subtype as outcome
***Unadjusted model contains both measures of insolation or temperature with stroke of any subtype as outcome
****Adjusted model adds all demographic, behavioural and medical confounders from Table 1 to the unadjusted model
*****Mediation model adds medical mediators to the adjusted model

CI: 0.94, 1.68); for those in the 2nd tertile of LST exposure, participants with lower SI exposure levels were 1.30 times more likely to experience incident cognitive decline than those with higher SI levels; and for those in the 3rd tertile of LST exposure, participants with lower SI exposure levels were 1.95 times more likely to experience incident cognitive impairment than those with higher SI levels. A similar analysis assessing the relationship between SI and other cognitive measures is underway for Word List Learning (assesses memory) and Animal Naming (assesses executive function).

4.3.2 *Acknowledgement*
The authors wish to acknowledge D. Quattrochi & D. Rickman (NASA); W. Crosson, S. Estes, M. Estes, S. Hemmings, M. Alhamdan & G. Wade (USRA); and S. Kent & L. McClure (UAB) for their contributions and support of the REGARDS project.

5 CONCLUSION

The purpose of this chapter is to broaden veteran remote sensing data users' understanding, and to inspire new remote sensing data users in the latest uses of satellite data for public health research. Experts in the field have come together to describe their work in their particular areas of research. In doing so, they have also provided insight to the data sets they use, the tools they build, the research methods they create, and the relationships they discover. It is hoped by the authors that this chapter will entice further public health research using remote sensing data and information services, and to tease out new ideas that further advance public health research.

REFERENCES

Adams, R.M. & Croker, T.D. 1989. The agricultural economics of environmental chance: Some lessons from air pollution. *J. Environ. Manage.* 28: 295–307.
Adimi, F.A., Soebiyanto, R.P., Safi, N. & Kiang, R. 2010. Towards malaria risk prediction in Afghanistan using remote sensing. *Malaria J.* 9: 125. doi:10.1186/1475-2875-9-125.
Al-Hamdan, M., Crosson, W., Limaye, A., Rickman, D., Quattrochi, D., Estes, M., Qualters, J., Sinclair, A., Tolsma, D., Adeniyi, K. & Niskar, A. 2009. Methods for characterizing fine particulate matter using ground observations and satellite remote-sensing data: Potential use for environmental public health surveillance. *J. Air & Waste Manag. Assoc.* 59: 865–881.

Anderson, M.C., Kustas, W.P. & Norman, J.M. 2007. Up-scaling flux observations from local to continental scales using thermal remote sensing. *Agron. J.* 99: 240–254.

Anenberg, S.C., Horowitz, L.W., Tong, D.Q. & West, J.J. 2010. An estimate of the gobal burden of anthropogenic ozone and fine particulate matter on premature human mortality using atmospheric modelling. *Environ. Health Perspect.* 118: 1189–1195.

Anyamba, A., Chretien, J-P, Small, J., Tucker, C.J., Formenty, P.B., Richardson, J.H., Britch, S.C., Schnabel, D.C., Erickson, R.L. & Linthicum, J. 2009. Prediction of a Rift Valley fever outbreak. *PNAS* 106(3): 955–959.

Aumann, H.H. & Pagano, R.J. 1994. Atmospheric infrared sounder on the Earth observing system. *Optical Engin.* 33: 776–784.

Avnery, S., Mauzerall, D.L., Liu, J. & Horowitz, L.W. 2011. Global crop yield reductions due to surface ozone exposure: 1. Year 2000 crop production losses & economic damage. *Atmos. Environ.* 45: 2284–2296.

Balcan, D., Colizza, V., Gocalves, B., Hu, H., Ramasco, J., Vespignani, A. 2009. Multiscale mobility networks and the spatial spreading of infectious diseases. *Proc. Nat. Acad. Sci.* 106(51): 21484–21489.

Balk, D., Brickman, M., Anderson, B., Pozzi, F. & Yetman, G. 2005. Mapping global urban and rural population distributions: Estimates of future global population distribution to 2015. UNFAO & CIESIN. Available from: http://sedac.ciesin.columbia.edu/gpw/docs/GISn.24_web_gpwAnnex.pdf [Accessed 19th January 2012].

Barker, C.M., Kramer V.L. & Reisen W.K. 2010. Decision support system for mosquito and arbovirus control in California. *Earthzine.* September 24th. Available from: http://www.earthzine.org/2010/09/24/decision-support-system-for-mosquito- and-arbovirus-control-in-california/ [Accessed 19th January 2012].

Beck, L.R., Rodriguez, M.H., Dister, S.W. Rodriguez, A.D., Rejmankva, E., Ulloa, A., Meza, R.A., Roberts, D.R., Paris, J.F., Spanner, M.A., Washino, R.K., Hacker, C. & Legters, J. 1994. Remote sensing as a landscape epidemiological tool to identify villages at high risk for malaria transmission. *Amer. J. Trop. Med. & Hyg.* 51(3): 271–280.

Beck, L.R., Lobitz, B.M. & Wood, B.L. 2000. Remote sensing and human health: New sensors and new opportunities. *Emerg. Infect. Dis.* 6(3): 217–226.

Beer, R., Glavich, T.A. & Rider, D.M. 2001. Tropospheric emission spectrometer for the Earth observing system's Aura satellite. *Appl. Optics* 40: 2356–2367.

Bell, M.L. 2006. The use of ambient air quality modelling to estimate individual and population exposure for human health research: A case study of ozone in the Northern Georgia region of the United States. *Environ. Int.* 32(5): 586–593.

Bell, M.L., McDermott, A., Seger, S.L., Samet, J.M. & Dominici, F. 2004. Ozone and short-term mortality in 95 US urban communities, 1987–2000. *JAMA* 292(19): 2372–2378.

Bell, M.L., Dominici, F. & Samet, J.M. 2005. A meta-analysis of time-series studies of ozone and mortality with comparison to the national morbidity, mortality, and air pollution study. *Epidemiol.* 16 (4): 436–445.

Benedict, K., Yang, P., Huang, Q. 2010. Project final report & feasibility report. NASA Applied Sciences Program cooperative agreement number NNX09AN53G.

Bernath, P.F., McElroy, C.T., Abrams, M.C., Boone, C.D., Butler, M., Camy-Peyret, C., Carleer, M., Clerbaux, C., Coheur, P.F., Colin, R., DeCola, P., DeMazière, M., Drummond, J.R., Dufour, D., Evans, W.F.J., Fast, H., Fussen, D., Gilbert, K., Jennings, D.E., Llewellyn, E.J., Lowe, R.P., Mahieu, E., McConnell, J.C., McHugh, M., McLeod, S.D., Michaud, R., Midwinter, C., Nassar, R., Nichitiu, F., Nowlan, C., Rinsland, C.P., Rochon, Y.J., Rowlands, N., Semeniuk, K., Simon, P., Skelton, R., Sloan, J.J., Soucy, M.-A., Strong, K., Tremblay, P., Turnbull, D., Walker, K.A., Walkty, I., Wardle, D.A., Wehrle, V., Zander, R. & Zou, J. 2005. Atmospheric Chemistry Experiment (ACE): Mission overview. *Geophys. Res. Lett.* 32, L15S01. doi:10.1029/ 2005GL022386.

Bishop, T.D., Miller III, H.L., Walker, R.L., Hurley, D.H., Menken, T. & Tilburg, C.E. 2010. Blue crab (*Callinectes sapidus* Rathbun, 1896) settlement at three Georgia (USA) estuarine sites. *Estuaries & Coasts* 33: 688–698.

Borden, K.A. & Cutter, S.L. 2008. Spatial patterns of natural hazards mortality in the United States. *Int. J. Health Geogra.* 7:64. doi:10.1186/1476-072X-7-64.

Bovensmann, H., Burrows, J.P., Buchwitz, M., Frerick, J., Noel, S., Rozanov, V.V., Chance, K.V., & Goede, A.P.H. 1999. SCIAMACHY: Mission objectives and measurement modes. *J. Atmos. Sci.* 56: 127–150.

Brandon, K. 2010. Maps: How mankind remade nature. Available from: http://www.wired.com/ wiredscience/2010/08/new-anthrome-maps/ [Accessed 13th April 2012].

Brooker, S., Leslie, T., Koaczinski, K., Mohsen, E., Mehboob, N., Saleheen, S., Khudonazarov, J., Freeman, T., Clements, A., Rowland, M. & Kolaczinski, J. 2006. Spatial epidemiology of *Plasmodium vivax*, Afghanistan. *Emerg. Infect. Dis.* 12(10): 1600–1602.

Burrows, J.P., Weber, M., Buchwitz, M., Rozanov, V., Ladstatter-Weissenmayer, A., Richter, A., DeBeek, R., Hoogen, R., Bramstedt, K., Eichmann, K.U. & Eisinger, M. 1999. The global ozone monitoring experiment (GOME): Mission concept and first scientific results. *J. Atmos. Sci.* 56: 151–175.

Callahan, C.M., Unverzagt, F.W., Hui, S.L., Perkins, A.J. & Hendrie, H.C. 2002. Six-item screener to identify cognitive impairment among potential subjects for clinical research. *Med. Care* 40(9): 771–781.

Caminade, C., Ndione, J.A., Kebe, C.M.F., Jones, A.E., Danuor, S., Tay, S., Tourre, Y.M., Lacaux, J.P., Vignolles, C., Duchemin, J.B., Jeanne, I. & Morse, A.P. 2011, Mapping Rift Valley fever and malaria risk over West Africa using climatic indicators. *Atmos. Sci. Lett.* 12: 96–103. doi:10.1002/asl.296.

Ceccato, P., Gobron, N., Flasse, S., Pinty, B. & Tarantola, S. 2002. Designing a spectral index to estimate vegetation water content from remote sensing data (Part 1: theoretical approach). *Rem. Sens. Environ.* 82 (2–3): 188–197.

Charland, K.M., Buckeridge, D.L., Sturtevant, J.L., Melton, F., Reis, B.Y., Mandl, K.D. & Brownstein, J.S., 2009. Effect of environmental factors on the spatio-temporal patterns of influenza spread. *Epidemiol. Infect.* 137(10): 1377–1387.

Chen, F., Mitchell, K., Schaake, J., Xue, Y., Pan, H.L., Koren, V., Duan, Q.Y., Ek, M. & Betts, A. 1996. Modeling of land surface evaporation by four schemes and comparison with FIFE observations. *J. Geophys. Res. Atmos.* 101(D3): 7251–7268.

Christophers, S.R. 1911. Malaria in the Punjab. *Sci. Mem, Med. & Sanit. Deps. India.* New Series 46: 197.

Clayton, D. & Kaldor J. 1987. Empirical Bayes estimates of age-standardized relative risks for use in disease mapping. *Biometrics* 43(3): 671–681.

ClearLead. 2011. Available from: http://www.clearleadinc.com/site/remote-sensing.html [Accessed 18th January 2012].

Clerbaux, C., George, M., Turquety, S., Walker, K.A., Barret, B., Bernath, P., Boone, C., Borsdorff, T., Cammas, J.P., Catoire, V., Coffey, M., Coheur, P.F., Deeter, M., De Maziére, M., Drummond, J., Duchatelet, P., Dupuy, E., de Zafra, R., Eddounia, F., Edwards, D.P., Emmons, L., Funke, B., Gille, J., Griffith, D.W.T., Hannigan, J., Hase, F., Höpfner, M., Jones, N., Kagawa, A., Kasai, Y., Kramer, I., Le Flochmoöen, E., Livesey, N.J., López-Puertas, M., Luo, M., Mahieu, E., Murtagh, D., Nédélec, P., Pazmino, A., Pumphrey, H., Ricaud, P., Rinsland, C.P., Robert, C., Schneider, M., Senten, C., Stiller, G., Strandberg, A., Strong, K., Sussmann, R., Thouret, V., Urban, J. & Wiacek. A. 2008. CO measurements from the ACE-FTS satellite instrument: Data analysis and validation using ground-based, airborne and spaceborne observations. *Atmos. Chem. & Physics* 8: 2569–2594.

CNES. 2011. Available from: http://www.cnes.fr/web/CNES-en/5073-monitoring-and-predicting-epidemics-with-satellites.php [Accessed 18th January 2012].

Cohen, A.J., Anderson, H.R., Ostro, B., Pandey, K.D., Krzyzanowski, M., Kuenzli, N., Gutschmidt K., Pope C.A., Romieu I., Samet J.M. & Smith, K.R. 2004. Mortality impacts of urban air pollution. In M. Ezzati, A.D. Lopez, A. Rodgers & C.J.L. Murray (eds.), *Comparative quantification of health risks: Global and regional burden of disease due to selected major risk factors, Vol.* 2: 1353–1433. Geneva: WHO.

Craig, M.H., Snow, R.W. & Sueur, D. 1999. A climate-based distribution model of malaria transmission in sub-saharan Africa. *Parasit. Today* 15(3): 105–111.

Curriero, F.C., Patz, J.A., Rose, J.B. & Lele, S. 2001. The association between extreme precipitation and waterborne disease outbreaks in the United States, 1948–1994. *Amer. J. Pub. Health* 91: 1194–1199.

Dai, Y.J., Zeng, X.B., Dickinson, R.E., Baker, I., Bonan, G.B., Bosilovich, M.G., Denning, A.S., Dirmeyer, P.A., Houser, P.R., Niu, G., Oleson, K.W., Schlosser, C.A. & Yang, Z.L. 2003. The common land model. *Bull. Amer. Meteorol. Soc.* 84(8): 1013–1023.

de la Beaujardiere, J. 2006 OpenGIS® web map server implementation specification. OGC Inc, Available from: http://www.opengeospatial.org/ [Accessed 19th January 2012].

Dinku, T., Ceccato, P., Grover-Kopec, E., Lemma, M., Connor, S.J. & Ropelewski, C.F. 2007 Validation and inter-comparison of satellite rainfall products over East African complex topography. *Int. J. Rem. Sens.* 28: 1503 1526. doi:10.1080/01431160600954688.

Dister, S.W., Beck, L.R., Wood, B.L., Falco, R. & Fish, D. 1993. The use of GIS and remote sensing technologies in a landscape approach to the study of Lyme disease transmission risk. In: *Proceedings of GIS '93: Geographic information systems in forestry, environmental and natural resource management.* Vancouver, B.C., Canada.

Dister, S.W., Fish, D., Bros, S., Frank, D.H. & Wood, B.L. 1997. Landscape characterization of peridomestic risk for Lyme disease using satellite imagery. *Amer. J Trop. Med. & Hyg.* 57: 687–92.

Dousset, B., Gourmelon, F., Laaidi, K., Zeghnoun, A., Giraudet, E., Bretin, P. & Vandentorren, S. 2009. Satellite monitoring of summer time heat waves in the Paris metropolitan area. 7th International Conference on Urban Climate, Yokohama, Japan.

Drummond, J.R. & Mand, G.S. 1996. The measurements of pollution in the troposphere (MOPITT) instrument: Overall performance and calibration requirements. *J. Atmos. & Oceanic Tech.* 13: 314–320.

EASTWeb. 2012. Available from: http://globalmonitoring.sdstate.edu/projects/eastweb/index.php [Accessed 19th January 2012].

Ebi, K.L. & Meehl, G.A. 2007. The heat is on: Climate change & heatwaves in the Midwest. Excerpted from *Regional impacts of climate change: Four case studies in the United States.* Arlington, VA: Pew Center on Global Climate Change.

Ebi, K.L. & Semenza, J.C. 2008. Community-Based Adaptation to the Health Impacts of Climate Change. *Amer. J. Prevent. Med.* 35(5): 501–507.

ECMWF. 2007. Available from: http://www.ecmwf.int/publications/newsletters/pdf/110_rev.pdf [Accessed 22nd March 2012].

ECOCAST. Available from: http://ecocast.arc.nasa.gov/topwp/ [Accessed 19th January 2012].

Ek, M.B., Mitchell, K.E., Lin, Y., Rogers, E., Grunmann, P., Koren, V., Gayno, G.& Tarpley, J.D. 2003. Implementation of Noah land surface model advances in the National Centers for Environmental Prediction operational mesoscale Eta model. *J. Geophys. Res. Atmos.* 108(D22). doi:10.1029 /2002JD003296.

Ellis, E.C. & Ramankutty, N. 2008. Putting people in the map: Anthropogenic biomes of the world. *Front. Ecol. & Environ.* 6(8): 439–447.

English, P.B., Sinclair, A.H., Ross, Z., Anderson, H., Boothe, V., Davis, C., Ebi, K., Kagey, B. Malecki, K., Shultz, R. & Simms, E. 2009. Environmental Health Indicators of Climate Change for the United States: Findings from the State Environmental Health Indicator Collaborative. *Environ. Health Perspect.* 117(11): 1673–1681.

ESA. 2011 Available from: http://www.esa.int/esaMI/Space_for_health/SEMNVMB474F_0.html [Accessed 18th January 2012].

Evans, J. & Geerken R. 2006. Classifying rangeland vegetation type and coverage using a Fourier component based similarity measure. *Rem. Sens of Env.* 105(1): 1–8.

Fairlie, T.D., Jacob, D.J. & Park, R.J. 2007. The impact of transpacific transport of mineral dust in the United States. *Atmos. Environ.* 41: 1251–1266.

Farr, T.G., Rosen, P.A., Caro, E., Crippen, R., Duren, R., Hensley, S., Kobrick, M., Paller, M., Rodriguez, E., Roth, L., Seal, D., Shaffer, S., Shimada, J., Umland, J., Werner, M., Oskin, M., Burbank, D. & Alsdorf, D. 2007. The shuttle radar topography mission. *Rev. Geophys.* 45: RG2004. doi:10.1029/ 2005RG000183.

Fayer, R., Dubey, J.P. & Lindsay, D.S. 2004. Zoonotic protozoa: From land to sea. *Trends in Parasitol.* 20: 531–536.

Felzer, B., Kicklighter, D., Melillo, J., Wang, C., Zhuang, Q. & Prinn, R. 2004. Effects of ozone on net primary production and carbon sequestration in the conterminous United States using a biogeochemistry model. *Tellus B* 56(3): 230–248.

Fielding, R. 2000. Architectural styles and the design of network-based software architectures. Ph.D. Dissertation. *Information and computer science* Univ. California, Irvine: Irvine, CA.

Fielding, R., Gettys, J., Mogul, J., Frystyk, H., Leach, P. & Berners-Lee, T. 1999. Hypertext Transfer Protocol HTTP/1.1 (RFC 2616). Available from: http://www.rfc-editor.org/rfc/rfc2068.txt [Accessed 10th April 2012].

Finkelman, B.S., Viboud, C., Koelle, K., Ferrari, M.J., Bharti, N. & Grenfell, B.T. 2007. Global patterns in seasonal activity of influenza A/H3N2, A/H1N1, and B from 1997 to 2005: Viral coexistence and latitudinal gradients. *PLoS One* 2(12): e1296.

Fischer, H., Birk, M., Blom, C., Carli, B., Carlotti, M., von Clarmann, T., Delbouille, L., Dudhia, A., Ehhalt, D., Endemann, M., Flaud, J.M., Gessner, R., Kleinert, A., Koopmann, R., Langen, J., Lopez-Puertas, M., Mosner, P., Nett, H., Oelhaf, H., Perron, G., Remedios, J., Ridolfi, M., Stiller, G. & Zander, R. 2008 MIPAS: An instrument for atmospheric and climate research *Atmos. Chem. Phys.* 8: 2151–2188.

Fishman, J., Creilson, J.K., Parker, P.A., Ainsworth, E.A., Vining, G.G., Szarka, J., Booker, F.L. & Xu, X. 2010. An investigation of widespread ozone damage to the soybean crop in the upper Midwest determined from ground-based and satellite measurements. *Atmos. Environ.* 44: 2248–2256.

Fox, R.M. 1957. *Anopheles gambiae* in relation to malaria and filariasis in coastal Liberia. *Amer. J. Trop. Med. & Hyg.* 6: 598–620.

Fraser, R.S. 1976. Satellite measurement of mass of Sahara dust in the atmosphere. *Appl. Opt.* 15: 2471–2479.

Fraser, R.S., Kaufman, Y.J. & Mahoney, R.L. 1984. Satellite measurements of aerosol mass and transport, *Atmos. Environ.* 18: 2577– 2584. doi:10.1016/0004-6981(84)90322-6.

Friedl, M.A., Sulla-Menashe, D., Tan, B., Schneider, A., Ramankutty, N., Sibley, A. & Huang, X. 2010. MODIS collection 5 global land cover: Algorithm refinements and characterization of new datasets. *Rem. Sens. Environ,* 114(1): 168–182.

Geerken, R., Zaitchik, B. & Evans, P. 2005. Classifying rangeland vegetation type and coverage from NDVI time series using Fourier Filtered Cycle Similarity. *Int. J. Rem. Sens.* 26(24): 5535–5554.

GEO. 2011a. Available from: http://www.earthobservations.org/about_geo.shtml [Accessed 18th January 2012].

GEO. 2011b. Available from: http://www.earthobservations.org/documents/work%20plan/GEO%202012-2015%20Work%20Plan_Rev1.pdf [Accessed 18th January 2012].

Gilles, H.M. & Warrell, D.A. (eds.). 1993. *Bruce-Chwatt's Essential Malariology*. London: Arnold Publishing.

Glantz, M. & Jamieson, D. 2000. Societal response to Hurricane Mitch and intra- versus intergenerational equity issues: Whose norms should apply? *Risk Anal.* 20: 869–882.

Golden, M.L., Downs, R.R. & Davis-Packard, K. 2005. Confidentiality issues and policies related to the utilization and dissemination of geospatial data for public health applications. Available from: http://www.ciesin.columbia.edu/pdf/SEDAC_ConfidentialityReport.pdf [Accessed 18th January 2012].

Golden, M.L., Yetman, G. & Chai-Onn, T. 2008. Assessment of populations in proximity to Superfund National Priorities List sites. Available from: http://sedac.ciesin.columbia.edu/eh/sfpop.html [Accessed 19th January 2012].

Gomez-Elipe, A., Otero, A., van Herp, M. & Aguirre-Jaime, A. 2007. Forecasting malaria incidence based on monthly case reports and environmental factors in Karuzi, Burundi, 1997–2003. *Malaria J.* 6: 129. doi:10.1186/1475-2875-6-129.

Goto, K., Nmor, J.C., Kurahashi, R., Minematsu, K., Yoda, T., Rakue, Y., Mizota, T. & Gotoh, K. 2010. Relationship between influx of yellow dust and bronchial asthma mortality using satellite data. *Scient. Res. & Essays* 5(24): 4044–4052.

Grover-Kopec, E., Kawano, M., Klaver, R.W., Blumenthal, B., Ceccato, P. & Connor, S.J. 2005. An online operational rainfall-monitoring resource for epidemic malaria early warning systems in Africa. *Malaria J.* 4: 1–6.

Hales, S., de Wet, N., Maindonald, J. & Woodward, A. 2002. Potential effect of population and climate changes on global distribution of dengue fever: An empirical model. *Lancet* 360: 830–834.

Haque, U. Hashizume, M., Glass, G.E., Dewan, A.M., Dewan, A.M., Overgaard, H.J. & Yamamoto, T. 2010. The role of climate variability in the spread of malaria in Bangladeshi highlands. *PLoS ONE* 5(12): e14341. doi:10.1371/journal.pone.0014341.

Harrington, L.C., Edman, J.D. & Scott, T.W. 2001. Why do female *Aedes aegypti* (Diptera: Culicidae) feed preferentially and frequently on human blood? *J. Med. Entomol.* 38(3): 411–422.

Hay, S.I., Packer, M.J. & Rogers, D.J. 1997. The impact of remote sensing on the study and control of invertebrate intermediate hosts and vectors for disease. *Int. J. Rem. Sens.* 18(14): 2899–2930.

Hay, S.I., Okiro, E.A., Gething, P.W., Patil, A.P., Tatem, A.J., Guerra, C.A. & Snow, R.W. 2010. Estimating the global clinical burden of *Plasmodium falciparum* malaria in 2007. *PLoS Med* 7(6): e1000290.

Herman, J., Bhartia, P., Torres, O., Hsu, C., Seftor, C. & Celarier, E. 1997. Global distribution of UV-absorbing aerosols from Nimbus-7/TOMS data. *J. Geophys. Res.* 102: 16,911–16,922.

Hertstein, U., Grunhage, L. & Jager, H.J. 1995. Assessment of past, present and future impacts of ozone and carbon dioxide on crop yields. *Atmos. Environ.* 29: 231–239.

Higurashi, A. & Nakajima, T. 2002. Detection of aerosol types over the East China Sea near Japan from four-channel satellite data. *Geophys. Res. Lett.*, 29(17): 1836. doi:10.1029/2002GL015357.

Holzer-Popp, T., Schroedter-Homscheidt, M., Breitkreuz, H., Martynenko, D. & Kluser, L. 2008. Improvements of synergetic aerosol retrieval for ENVISAT. *Atmos. Chem. Phys.* 8: 7651–7672.

Hoshen, M.B. & Morse, A.P. 2004. A weather-driven model of malaria transmission. *Malaria J.* 3: 32. doi:10.1186/1475-2875.

Hu, R.M., Sokhi, R.S. & Fisher, B.E.A. 2009. New algorithms and their application for satellite remote sensing of surface PM2.5 and aerosol absorption. *J. Aerosol Sci.* 40: 394–402.

Huete, A., Didan, K., Miura, T., Rodriguez, E.P., Gao, X. & Ferreira, L.G. 2002. Overview of the radiometric and biophysical performance of the MODIS vegetation indices. *Rem. Sens. Environ.* 83: 195–213.

Huffman, G.J., Adler, R.F., Morrissey, M., Bolvin, D.T., Curtis, S., Joyce, R., McGavock, B. & Susskind, J. 2001. Global precipitation at one-degree daily resolution from multi-satellite observations. *J. Hydrometeor.* 2(1): 36–50.

Ichii, K., Maruyama, M. & Yamaguchi, Y. 2003. Multi-temporal analysis of deforestation in Rondonia state in Brazil using Landsat MSS, TM, ETM+ and NOAA AVHRR imagery and its relationship to changes in the local hydrological environment. *Int. J. Rem. Sens.* 24(22): 4467–4479.

Igarashi, T. 2010. JAXA's concept on space initiatives for health. GEO Health and Environment Community of Practice Workshop. Centre National d'Études Spatiales (CNES) Paris, France 27–28 July. Available from: http://www.earthobservations.org/documents/cop/he_henv/20100727_France/16_ JAXA.pdf [Accessed 18th January 2012].

IPCC. 2007. Impacts, adaptation and vulnerability. Working Group II, Fourth Assessment Report. Cambridge: Cambridge UP.

Ito, K., De Leon, S.F. & Lippman, M. 2005. Associations between ozone and daily mortality: Analysis and meta-analysis. *Epidemiol.* 16: 446–457.

Jerrett, M., Burnett, R.T., Ma, R., Pope III, C.A., Krewski, D., Newbold, K.B., Thurston, G., Shi, Y., Finkel-stein, N., Calle, E.E. & Thun, M.J. 2005. Spatial analysis of air pollution and mortality in Los Angeles. *Epidemio.* 16(6) 727–736.

Jerrett, M, Burnett, RT, Pope, C.A.-III, Ito, K., Thurston, G., Krewski, D., Shi, Y., Calle, E., Thun, M. 2009. Long-term ozone exposure and mortality. *New Engl. J. Med.* 360(26): 1085–1095.

Jones, K.E., Patel, N.G., Levy, M.A., Storeygard, A., Balk, D., Gittleman, J.L. & Daszak, P. 2008. Global trends in emerging infectious diseases. *Nature* 451(7181): 990–993.

Joyce, R.J., Janowiak, J.E., Arkin, P.A. & Xie, P. 2004. CMORPH: A method that produces global precipitation estimates from passive microwave and infrared data at high spatial and temporal resolution. *J. Hydromet..* 5: 487–503.

Kahru, M. 1997. Using satellites to monitor large-scale environmental change in the Baltic Sea. In M. Kahru & C.W. Brown (eds.), *Monitoring algal blooms: New techniques for detecting large-scale environmental change*: 43–61. Berlin: Springer-Verlag.

Kalnay, E., Kanamitsu, M., Kistler, R., Collins, W., Deaven, D., Gandin, L., Iredell, M., Saha, S., White, G., Woollen, J., Zhu, Y., Leetmaa, A., Reynolds, R., Chelliah, M., Ebisuzaki, W., Higgins, W., Janowiak, J., Mo, K.C., Ropelewski, C., Wang, J., Jenne, R. & Joseph, D. 1996. The NCEP/NCAR 40-year reanalysis project. *Bull. Amer. Meteor. Soc.* 17(3): 437–471.

Kaufman, Y.J., Tanré, D. & Boucher, O. 2002. A satellite view of aerosols in the climate system. *Nature* 419: 215–223.

Kay, B.H., Fanning, I.D., Mottram, P. 1989. Rearing temperature influences flavivirus vector competence of mosquitoes. *Med. Vet. Entomol.* 3: 415–422.

Kelly-Hope, L.A., Hemingway J. & McKenzie, F.E. 2009. Environmental factors associated with the malaria vectors *Anopheles gambiae* and *Anopheles funestus* in Kenya. *Malaria J.* 8: 268.

Kiang, R.K., Adimi, F. & Soika, V. 2006. Meteorological, environmental remote sensing and neural network analysis of the epidemiology of malaria transmission in Thailand. *Geospat Health* 1: 71–84.

Kiang, R.K., Adimi, F. & Soebiyanto, R.P. 2011. Remote sensing-based modelling of infectious disease transmission. In T. Kass-Hout & X. Zhang (eds.), *Biosurveillance: Methods and case studies*. Boca Raton: CRC Press.

Knabb, R.D., Rhome, J.R & Brown, D.P. 2006. Tropical cyclone report: Hurricane Katrina: 23–30 August 2005. Available from: http://www.nhc.noaa.gov/pdf/TCR-AL122005_Katrina.pdf [Accessed 19th January 2012.]

Koster, R.D., Suarez, M.J., Ducharne, A., Stieglitz, M. & Kumar, P. 2000. A catchment-based approach to modeling land surface processes in a general circulation model 1. Model structure. *J. Geophys. Res. Atmos.* 105(D20): 24,809–24,822.

Kovats, R.S. & Hajat, S. 2008. Heat stress and public health: A critical review. *Annu. Rev. Pub. Health* 29: 41–55.

Kovats, R.S., Bouma, M.J., Hajat, S., Worrall, E. & Haines, A. 2003. El Niño and health. *Lancet* 362: 1481–1489.

Kowalczyk, E.Z., Wang, Y.P., Law, R.M., Davies, H.L., McGregor, J.L. & Abramowitz, G. 2006. The CSIRO Atmosphere Biosphere Land Exchange (CABLE) model for use in climate models and as an offline model. *CSIRO Marine & Atmos. Res.* Paper 013.

Krueger, A.J. & Jaross, G., 1999. TOMS ADEOS instrument characterization. *IEEE Trans.Geosci. & Rem. Sens.* 37: 1543–1549.

Kutser, T. 2004. Quantitative detection of chlorophyll in cyanobacterial blooms by satellite remote sensing. *Limnol. & Oceanogra.* 49: 2179–2189.

LBL. 2011. Available from: http://www.lbl.gov/Education/ELSI/Frames/pollution-health-effects-f.html [Accessed 18th January 2012].

Leptoukh, G., Csiszar, I., Romanov, P., Shen, S., Loboda, T. & Gerasimov, I. 2007. NASA NEESPI data center for satellite remote sensing data and services. *Environ. Res. Lett., 2.* doi:10.1088/1748-9326/2/4/045009.

Liang, X., Wood, E.F. & Lettenmaier, P. 1996. Surface soil moisture parameterization of the VIC-2L model: Evaluation and modification. *Global & Planetary Change* 13:(1–4): 195–206.

Linthicum, K.J., Anyamba, A., Tucker, C.J., Kelley, P.W., Myers, M.F. & Peters, C.J., 1999. Climate and satellite indicators to forecast Rift Valley Fever epidemics in Kenya. *Science* 285: 397–400.

Liu, Y., Sarnat, J. A., Coull, B. A., Koutrakis, P. & Jacob, D. J. 2004. Validation of Multiangle Imaging Spectro-radiometer (MISR) aerosol optical thickness measurements using Aerosol Robotic Network (AERONET) observations over the contiguous United States. *J. Geophys. Res.*, 109, D06205. doi:10.1029/2003JD003981

Liu, Y., Paciorek, C.J. & Koutrakis, P. 2009. Estimating regional spatial and temporal variability of PM2.5 concentrations using satellite data, meteorology, and land use information. *Environ. Health Perspec.* 117(6): 886–892.

Lynnes, C., Strub, R., Seiler, E., Joshi, T. & MacHarrie, P. 2009. Mirador: A simple, fast search interface for global remote sensing data sets. *IEEE Trans. Geosci. Rem. Sens.* 47(1): 92–96.

Lynnes, C., Olsen, E., Fox, P., Vollmer, B., Wolfe, R.E. & Samadi, S. 2010. A quality screening service for remote sensing data In: H. Salim & K. Keahey (general chairs). Proc. 19th ACM. *International symposium on high performance distributed computing (HPDC)*: 554–559. New York: ACM.

Lyons, W.A., Dooley, J.C. Jr. & Whitby, K.T. 1978. Satellite detection of long-range pollution transport and sulfate aerosol hazes. *Atmos. Environ.* 12: 621–631. doi:10.1016/0004-6981(78)90242-1.

Matricardi, E.A.T., Skole, D.L., Cochrane, M.A., Qi, J. & Chomentowski, W. 2005. Monitoring selective log-ging in tropical evergreen forests using Landsat: Multitemporal regional analyses in Mato Grosso, Brazil. *Earth Interactions* 9(24): 1–24. doi:10.1175/EI142.1.

McGranahan, G., Balk, D. & Anderson, B. 2007. The rising tide: Assessing the risks of climate change and human settlements in low elevation coastal zones. *Environ. & Urbaniza.* 19(1): 17–37.

Meehl, G.A. & Tebaldi, C. 2004 More intense, more frequent and longer lasting heat waves in the 21st Century. *Science.* 305(5686): 994–997.

Meng, C.J., Li, Z.L., Zhan, X., Xhi, J.C. & Liu, C.Y. 2009. Land surface temperature data assimilation and its impact on evapotranspiration estimates from the Common Land Model. *Water Res. Res.* 45(W02421). doi:10.1029/2008WR006971.

Mitchell, K.E., Lohmann, D., Houser, P.R., Wood, E.F., Schaake, J.C., Robock, A., Cosgrove, B.A., Sheffield, J., Duan, Q., Luo, L., Higgins, R.W., Pinker, R.T., Tarpley, J.D., Lettenmaier, D.P., Marshall, C.H., Entin, J.K., Pan, M., Shi, W., Koren, V., Meng, J., Ramsay, B.H. & Bailey, A.A. 2004. The multi-institution North American Land Data Assimilation System (NLDAS): Utilizing multiple GCIP products and partners in a continental distributed hydrological modelling system. *J. Geophys. Res.* 109: D07S90. doi:10.1029/2003JD003823.

Morain, S.A. & Budge, A.M. 2010. Suggested practices for forecasting dust storms and intervening their health effects. In O. Altan, R. Backhaus, P. Boccardo & S. Zlatanova (eds.), *Geoinformation for disaster and risk management*: 45–50. Copenhagen: Joint Board Geospatial Information Sciences & United Nation Office of Outer Space Affairs.

NASA. 2012. Available from: http://appliedsciences.nasa.gov/health-air.html [Accessed 24th January 2012].

Nemani, R., Hashimoto, H., Votava, P., Melton, F., Wang, W., Michaelis, A., Mutch, L., Milesi, C., Hiatt S. & White, M. 2009. Monitoring and forecasting ecosystem dynamics using the Terrestrial Observation and Prediction Systems (TOPS). *Rem. Sens. Environ.* 113: 1497–1509.

Nickovic, S., Kallos, G., Papadopoulos, A. & Kakaliagou, O. 2001. A model for prediction of desert dust cycle in the atmosphere. *J. Geophys. Res.* 106(D16): 18,113–18,130.

Njoku, E.G., Jackson, T.J., Lakshmi, V., Chan, T.K. & Nghiem, S.V. 2003. Soil moisture retrieval from AMSR-E. *IEEE Trans. Geosci. & Rem. Sens.* 41: 215–229.

Noor, A.M., Mutheu, J.J., Tatem, A.J., Hay, S.I. & Snow, R.W. 2009. Insecticide-treated net coverage in Africa: Mapping progress in 2000–07. *Lancet* 373(9657): 58–67.

NRC. 2001. Under the weather: Climate, ecosystems and infectious disease: Washington DC: National Academy Press.

NWS. 2012. Available from: http://www.nws.noaa.gov/os/heat/index.shtml [Accessed 22nd March 2012].

Oliveira, P.J., Asner, G.P., Knapp, D.E., Almeyda, A., Galvan-Gildemeister, R., Keene, S., Raybin, R.F. & Smith, R.C. 2007. Land-use allocation protects the Peruvian Amazon. *Science* 317(5842): 1233–1236.

O'Neill, M.S. Zanobetti. A. & Schwartz, J. 2003. Modifiers of the temperature and mortality association in seven US cities. *Amer. J. Epidemiol.* 157(12): 1074–1082.

Pampana, E. 1969. *A textbook of malaria eradication*. London: Oxford UP.

Patz, J. 2005. Satellite remote sensing can improve chances of achieving sustainable health. *Environ. Health Perspect.* 113(2): 84–85.

Patz, J.A., Daszak, P., Tabor, G.M., Aguirre, A.A., Pearl, M., Epstein, J., Wolfe, N.D., Kilpatrick, A.M., Foufopoulos, J., Molyneux, D. & Bradley, D.J. 2004. Unhealthy landscapes: Policy recommendations on land use change and infectious disease emergence. *Environ. Health Perspect.* 112: 1092–1098.

289

Pope, C.A. III, Burnett, R.T., Thun, M.J., Calle, E.E., Krewski, D., Ito, K. & Thurston, G.D. 2002. Lung cancer, cardiopulmonary mortality, and long-term exposure to fine particulate air pollution. *J. Amer. Med. Assoc.* 287: 1132–1141.

Prospero, J.M., Ginoux, P., Torres, O., Nicholson, S. & Gill, T. 2002. Environmental characterization of global sources of atmospheric soil dust identified with the NIMBUS 7 Total Ozone Mapping Spectrometer (TOMS) absorbing aerosol product. *Rev. of Geophys.* 41(1): 1–31.

Rahman, A., Kogan, F. & Roytman, L. 2006. Analysis of malaria cases in Bangladesh with remote sensing data. *Amer. J. Trop. Med. & Hyg.* 74(1): 17–19.

Raudenbush, S.W. & Bryk, A.S. 2002. *Hierarchical linear models: Applications and data analysis methods.* Thousand Oaks: Sage.

Rinner, C, Patychuk, D. Bassil, K., Nasr, S., Gower, S. & Campbell, M. 2010. The role of maps in neighbourhood-level heat vulnerability assessment for the City of Toronto. *Cart. & GIS* 37(1): 31–44.

Robine, J., Cheung, S., Le Roy, S., Van Oyen, H., Griffiths, C., Michel, J.P. & Herrmann, F. 2008. Death toll exceeded 70,000 in Europe during the summer of 2003. *C.R. Biol.* 331: 171–U175.

Rodell, M., Houser, P.R., Jambor, U., Gottschalck, J., Mitchell, K., Meng, C.J., Arsenault, K., Cosgrove, B., Radakovich, J., Bosilovich, M., Entin, J.K., Walker, J.P., Lohmann, D. & Toll, D. 2004. The global land data assimilation system. *Bull. Amer. Meteor. Soc.* 85(3): 381–394.

Roy, D.P., Ju, J., Kline, K., Scaramuzza, P.L., Kovalskyy, V. & Hansen, M. 2010. Webenabled Landsat data (WELD): Landsat ETM+ composited mosaics of the conterminous United States. *Rem. Sens. Environ.* 114(1): 35–49.

Sanchez, G.M. 2007. The application and assimilation of Shuttle Radar Topography Mission Version 1 data for high resolution dust modelling. MA Thesis, University of New Mexico, Department of Geography.

Schifano, P., Cappai, G., DeSario, M., Michelozzi, P., Marino, C., Bargagli, A.M. & Perucci, C.A. 2009. Susceptibility to heat wave-related mortality: A follow-up study of a cohort of elderly in Rome. *Environ. Health* 8:50. doi:10.1186/1476-069X-8-50.

Schlussel, P., Hultberg, T.H., Phillips, P.L., August, T. & Calbet, X. 2005. The operational IASI level 2 processor. In L. Burrows & J.P. Eichmann (eds.), *Atmospheric remote sensing: Earth's surface, troposphere, stratosphere and mesosphere I*: 982–988. Orlando: Elsevier.

Schwartz, J. 2005. Who is sensitive to extremes of temperature? A case-only analysis. *Epidemiol.* 16(1): 67–72.

SEDAC. 2007. National aggregates of geospatial data collection: Population, landscape and climate estimates (PLACE) Version II. Available from: http://sedac.ciesin.columbia.edu/place/methods jsp [Accessed 19th January 2012].

Semenza, J.C., Rubin, C.H., Falter, K.H., Selanikio, J.D., Flanders, W.D. Howe, H.L. & Wilhelm, J.L. 1996. Heat-related deaths during the July 1995 heat wave in Chicago. *N. Engl. J. Med.* 335: 84–90.

Shaman J., Pitzer, V.E., Viboud, C., Grenfell, B.T. & Lipsitch, M. 2010. Absolute humidity and the seasonal onset of influenza in the continental United States. *PLoS Biol* 8(2): e1000316.

Shryock, H.S., Siegel, J.S. & Stockwell, E.G. 1976. *The methods and materials of demography.* New York: Academic Press.

Soebiyanto, R.P., Adimi, F. & Kiang, R.K. 2010 Modelling and predicting seasonal influenza transmission in warm regions using climatological parameters. *PLoS ONE* 5(3): e9450.

Spinhirne, J.D., Palm, S.P., Hart, W.D., Hlavka, D.L. & Welton, E.J. 2005. Cloud and aerosol measurements from the GLAS space borne lidar: Initial results. *Geophys. Res. Lett.* 32: L22S03. doi:10.1029/2005GL023507.

Steininger, M.K., Tucker, C.J., Townshend, J.R.G., Killeen, T.J., Desch, A., Bell, V. & Ersts, P. 2001. Tropical deforestation in the Bolivian Amazon. *Environ. Conserv.* 28(2): 127–134.

Storeygard, A., Balk, D., Levy, M. & Deane, G. 2008. The global distribution of infant mortality: A subnational spatial view. *Popul. Space & Place* 14(3): 209–229.

Tanré, D., Bréon, F.M., Deuzé, J.L., Herman, M., Goloub, P., Nadal, F. & Marchand, A. 2001. Global observation of anthropogenic aerosols from satellite. *Geophys. Res. Lett.* 28(24): 4555–4558. doi:10.1029/2001GL013036.

Tatem, A.J., Smith, D.L., Gething, P.W., Kabaria, C.W., Snow, R.W. & Hay, S.I. 2010. Ranking of elimination feasibility between malaria-endemic countries. *Lancet* 376(9752): 1579–1591.

Teklehaimanot, A., McCord, G.C. & Sachs, J.D. 2007. Scaling up malaria control in Africa: An economic and epidemiological assessment. *Amer. J. Trop. Med. & Hyg.* 77(Supp6): 138–144.

Thompson, D.F., Malone, J.B., Harb, M., Faris, R., Huh, O.K., Buck, A.A. & Cline, B.L. 1996. Bancroftian filariasis distribution and diurnal temperature differences in the southern Nile delta. *Emerg. Infect. Dis.* 2: 234–235.

Thomson, M.C., Doblas-Reyes, F.J., Mason, S.J., Hagedorn, R., Connor, S.J., Phindela, T., Morse, A.P. & Palmer, T.N. 2006. Malaria early warnings based on seasonal climate forecasts from multi-model ensembles. *Nature* 439: 576–579.

Tong, D.Q. & Mauzerall, D.L. 2008. Summertime state-level source-receptor relationships between NOx emissions and downwind surface ozone concentrations over the continental United States. *Environ. Sci. & Tech.* 42(21): 7976–7984.

Tong, D., Mathur, R., Schere, K., Kang D. & Yu, S. 2007. The use of air quality forecasts to assess impacts of air pollution on crops: Methodology and case study. *Atmos. Environ.* 41(38): 8772–8794.

Tong, D.Q., Kan, H. & Yu, S. 2009. Ozone exposure and mortality. *New Engl. J. Med.* 360(26): 2787–2787.

Tucker, C.J. 1979. Red and photographic infrared linear combinations for monitoring vegetation. *Rem. Sens. Environ.* 8: 127–150.

Uppala, S, Dee, D., Kobayashi, S., Berrisford, P. & Simmons, A. 2008. Towards a climate data assimilation system: Status update of ERA-Interim. *ECMWF Newsl.* 115: 12–18.

Vandentorren, S., Bretin, P., Zeghnoun, A., Mandereau-Bruno, L., Croisier, A. & Cochet, C. 2006. August. 2003. Heat wave in France: Risk factors for death of elderly people living at home. *Eur. J. Pub. Health* 16: 583–591.

Van de Water, P., Main, C.E., Keever, T. & Levetin, E. 2003. An assessment of predictive forecasting *Juniperus ashei* pollen movement in the southern Great Plains. *Int. J. Biomet.* 48: 74–82.

van den Hurk, B.J.M.M., Viterbo, P., Beljaars, A.C.M. & Betts, A.K. 2000. Offline validation of the ERA40 surface scheme. *ECMWF Tech. Memo.* 295: 1–42.

van Donkelaar, A., Martin, R.V., Brauer, M., Kahn, R., Levy, R., Verduzco, C. & Villeneuve, P.J. 2010. Global estimates of ambient fine particulate matter concentrations from satellite-based aerosol optical depth: Development and application. *Environ. Health Perspect.* 118(6): 847–855.

Vittor, A.Y., Pan, W.K., Gilman, R.H., Tielsch, J., Glass, G., Shields, T., Sanchez-Lozano, W., Pinedo, V.V., Salas-Cobos, E., Flores, S. & Patz, J.A. 2009. Linking deforestation to malaria in the Amazon: Characterization of the breeding habitat of the principal malaria vector, *Anopheles darlingi*. *Amer.J. Trop. Med. & Hyg.* 81(1): 5–12.

Whiteside, A. & Evans, J.D. (eds.). 2006. Web Coverage Service (WCS) Implementation Specification, Version 1.1.0. *Open Geospat. Consort.* 06-083r8: 129.

Wilhelmi, O.V., Purvis, K.L. & Harriss, R.C. 2004. Designing a geospatial information infrastructure for mitigation of heat wave hazards in urban areas. *Nat. Haz. Rev.* 5: 147–158.

Witt, C.J., Richards, A.L., Masuoka, P.M., Foley, D.H., Buczak, A.L., Musila, L.A., Richardson, J.H., Colacicco-Mayhugh, M.G., Rueda, L.M., Klein, T.A., Anyamba, A., Small, J., Pavlin, J.A., Fukuda, M.M., Gaydos, J. & Russell, K.L. 2011. The AFHSC-Division of GEIS operations predictive surveillance program: A multidisciplinary approach for the early detection and response to disease outbreaks. *BMC Pub. Health* (Supp2): S10.

WHO 2009. Available from: http://www.ciesin.columbia.edu/docs/001-007/001-007.html [Accessed 18th January 2012].

WHO 2012. HealthMapper. Available from: http://www.who.int/health_mapping/tools/healthmapper/en/ [Accessed 23rd March 2012].

Wynne, T.T., Stumpf, R.P., Tomlinson, M.C., Warner, R.A., Tester, P.A., Dyble, J. & Fahnenstiel, G.L. 2008. Relating spectral shape to cyanobacterial blooms in the Laurentian Great Lakes. *Int. J. of Rem. Sens.* 29: 3665–3672.

Wynne, T.T., Stumpf, R.P., Tomlinson, M.C., Schwab, D.J., Watabayashi, G.Y. & Christensen, J.D. 2011. Estimating cyanobacterial bloom transport by coupling remotely sensed imagery and a hydrodynamic model. *Ecol. Appl.* 21: 2709–2721. doi:http://dx.doi.org/10.1890/10-1454.1.

Xiao X, Gilbert, M., Slingenbergh, J., Lei, F. & Boles, S. 2007. Remote sensing, ecological variables, and wild bird migration related to outbreaks of highly pathogenic H5N1 avian influenza. *J. Wldlfe. Dis.* 43: 540–546.

YCELP, CIESIN, World Economic Forum & the European Commission, Joint Research Centre. 2010. Available from: http://www.ciesin.columbia.edu/repository /epi/data/EPI _2010_ report.pdf [Accessed 18th January 2012].

Yu, H., Remer, L.A., Chin, M., Bian, H., Kleidman, R.G. & Diehl, T. 2008. A satellite-based assessment of transPacific transport of pollution aerosol. *J. Geophys. Res.* 113: D14S12. doi:10.1029/2007 JD009349.

Zaitchik, B.F., Rodell, M., Reichle, R. 2008. Assimilation of GRACE terrestrial water storage data into a land surface model: Results for the Mississippi River basin. *J. Hydrometeor.* 9(3): 535–548.

Zhang, H., Hoff, R.M. & Engel-Cox, J.A. 2009. The relation between moderate resolution imaging spectroradiometer (MODIS) aerosol optical depth and PM2.5 over the United States: A geographical comparison by US Environmental Protection Agency regions. *J. Air & Waste Manag. Assoc.* 59: 358–1369.

Chapter 7

Environmental modelling for health

S.A. Morain[1], S. Kumar[2] & T.J. Stohlgren[3] (Auth./eds.) with; O. Selinus[4], E. Steinnes[5],
M. Rosenberg[6] & M. Lo[7]

[1] *Earth Data Analysis Center, University of New Mexico, Albuquerque, NM, US*
[2] *Department of Ecosystem Science and Sustainability, Colorado State University, Fort Collins, CO, US*
[3] *US Geological Survey, Fort Collins Science Center, Fort Collins, CO, US*
[4] *Geological Survey of Sweden, Uppsala, Sweden*
[5] *Norwegian University of Science & Technology, Trondheim, Norway*
[6] *Queens University, Kingston, Ontario, Canada*
[7] *Jet Propulsion Laboratory, California Institute of Technology, Pasadena, CA, US*

ABSTRACT: This chapter highlights types of models and modelling strategies with a focus
on environment and human health applications that link environmental triggers and subsequent
disease exposures and risks. It reviews bio-geophysical modelling applications generally employed
by environmental and health scientists, and by policy and decision making authorities.

1 INTRODUCTION

Health and environmental monitoring is comprised of several substantially different modelling
approaches. In this book, health refers to human health and wellbeing, which in turn directs healthcare professionals to treat not only the factual, medical conditions of individuals and populations,
but also to consider the interplay of medical outcomes influenced by a patient's (or a population's)
social, economic, and demographic surroundings. As a consequence, healthcare providers search
for cause and effect relationships that are partly concrete and knowable, but that inevitably are also
soft and conditional (Patel *et al.*, Zhang 2005; Kahneman & Frederick 2005). Earth scientists, on
the other hand, are trained to explain how natural systems work based on measurable parameters
and to expand their scope of inquiry from simple to complex as the relationships among parameters
are better quantified. Thus it is that modelling strategies for healthcare are fundamentally different
from those employed by earth scientists. Spanning the strategic divide is one of the promises of the
21st Century. Specifically, the challenge is to find methods that link between extracted (measurable) environmental parameters and collected (statistical) health data. There are many definitions
and types of Models (Müller 2010).

Among the diversity of Earth sciences, the atmosphere and hydrosphere are more accurately
modelled than the lithosphere and biosphere. The physics of fluid motion in air and water lead to
deterministic (numerical) relationships among measurable parameters that are predictable. Properties in solid and semi-solid states (the lithosphere—geology and soils), must be sampled in the x, y,
and z domains to construct visualizations of what cannot otherwise be seen directly. The biosphere
is even more complex because no two biological specimens are exactly alike, and do not necessarily
react to environmental stimuli in the same ways. At different scales of reference, they are partly fluid
and partly solid; and, their properties change as their physical surroundings change (e.g. by fire,
drought, or extreme temperatures) Understanding these complexities has led the ICSU to inaugu-
rate a ten-year programme for researching health and wellbeing in changing urban environments.
The programme is based on a systems analysis approach that *aims to understand the complex*

relationships among components of a system including interrelationships between subsystems of a larger system. Such systems explicitly account for interactions and feedback between variables or processes (ICSU 2011).

1.1 Types of models and modelling systems

Models that link environment with health include those that are numerical (aka deterministic models), statistical, prognostic or diagnostic, empirical (i.e. based on repeated observations of parameters defining a system), probabilistic, heuristic, stochastic, conceptual, theoretical, kinematic (i.e. those describing the motion of objects and systems governed by Newton's second law, including systems that have geographic coordinates such as hydro-, bio- and geo-dynamics), forecast, and hindcast. The language of models and of modelling is complex and deeply rooted in philosophy (Frigg & Hartmann 2006; Müller 2010). Common forms are described here. Fortunately, much of the process involved in modelling can be expedited by a growing service industry based on user-supplied data sets selected from reliable sources to simulate complex systems and to deliver a variety of scenarios for decision makers (e.g. GoldSim 2012). Among these are services that link environmental models to possible human health outcomes; and many of these can be presented in 3-D real-world displays, or animated to show changing conditions through time. Ash flows from volcanic eruptions, atmospheric dust transport, toxic plume dispersal from industrial sources and slowly evolving vibrio events in coastal waters are among many mature applications.

Numerical models are based on equations whose parameters are measurable and solvable to mathematic precision. Systems of such equations, each addressing a different relationship among many parameters can be created to describe increasingly complex, interacting, phenomena. In mathematics, a deterministic model is one with no randomness that can predict future states of the system. Examples include air and water parameters used to simulate circulating atmospheric, hydrospheric systems; and, at least some non-fluid systems. As sophistication of these models mature, they may be coupled to associated and kindred models to form systems that address complex, multidisciplinary, and even societal applications with greater accuracy (Gobran & Clegg 1996; Frankenberger *et al.*, 1999; Xu *et al.*, 2010).

Statistical models are based also on known relationships among physical, cultural, economic, behavioural, and medical variables from which the probability of a given outcome can be derived. Such models consist of one or more equations that describe the behaviour of an object in terms of random variables and their associated probability distributions. One of the most basic models is the simple linear regression, which assumes a relationship between two random variables X and Y. For instance, one may want to explain child mortality in a given country by its gross domestic product (GDP). This is a statistical model because the relationships need not be perfect and the model may include disturbance terms that account for other effects on child mortality besides GDP.

Stochastic models are simulations that use a range of values for each variable. This indeterminate process evolves solutions over time or space such that future outcomes can be described by probability distributions. This means that even if the initial starting point is known, there are alternative future scenarios that might arise, some more probable than others. In a simple case, a stochastic process could use a time series of recurring, measurable phenomena, or a series of images to derive random functions whose arguments are drawn from a range of continuously changing values. One approach is to treat these as one or more deterministic arguments whose values are non-deterministic (single) quantities which have certain probability distributions (Papoulis & Unnikrishna 2001). Familiar examples of processes modelled as stochastic time series include signals derived from medical data, such as a patient's electrocardiogram; blood pressure or temperature; and random movement such as are found in image sequences of environmental or landscape changes.

Heuristic models are those that project an outcome based on acquired experience from a standard set of variables by creating a set, or system of rules that appear to govern an observed outcome. They are most useful in identifying the circumstances or attributes leading to an observed outcome even though observers do not know the exact interaction among the circumstances. Heuristic rules reduce a complex, continuously changing reality into a simpler, manageable problem. These rules are improved over time by refining or substituting attributes that more accurately simulate the observed outcomes. The heuristic method is used widely in healthcare, psychology, law, and philosophy, among others to simulate outcomes that are governed by time-dependent variations, rather than by precise mathematical relationships (Cioffi 1997; Kahneman & Frederick 2005; Ropella et al., 2005).

1.1.1 Earth science models

The Earth sciences contribute deterministic and statistical models that have applications for human health and wellbeing; the trick is to recognize how earth processes and health impacts might be linked given that results are dependent on the selection of model(s) and their applicability across various spatial and temporal resolutions. Table 1 is a list of selected natural and human-induced hazards and disasters (left column), typical earth science/health linkages (middle), and general spatio-temporal domains needed to generate model solutions (right column). Understanding the possible linkages influences both the kinds of input data needed and the interpretation of model outputs. One of the scientific challenges recognized by the EO2HEAVEN Programme in its 2011 European Commission research programme was to address *the methodological study of links between extracted environmental parameters (deterministic/numerical modellers) and collected health data (statistical, stochastic, heuristic modellers) in close collaboration with scenario leaders.* To achieve this goal requires experts from both arenas.

Avenues for geophysical modelling include observations and measurements of surface properties to assess conditions that are known empirically to have human health effects. Spectral measurements of surface properties are collected remotely by aerial and satellite sensors in digital raster format. These measured values are then assimilated into numerical models developed for specific air, water, soil, and related applications. When natural events or human-induced hazards alter surface conditions, new values for these can be modelled quickly and evaluated for their human health implications based on *a priori* knowledge that many diseases have associated environmental determinants.

Likewise there are numerous statistical and heuristic models to inform health communities about potential exposure levels occurring at the day-to-day level of health care. Statistical techniques commonly used by these communities also are being used increasingly by geoscientists in their effort to link physical phenomena to health outcomes. In part, this is made possible by the availability of long term, synoptic Earth observations that show patterns of changing surface properties on temporal scales ranging from days to decades. More than thirty-five years of continuous monitoring by an ever-growing number of satellite and sensor systems covering the globe make it possible to consider longitudinal data sets that link health to environment.

1.1.2 Data collection and modelling

Measurements of Earth's land, water, and atmospheric processes and conditions are collected by a host of sensors and instruments ranging from ground station monitors and ocean buoys to airborne and space-borne platforms. Data collected by these systems provide information on the changing conditions of environments for use in understanding possible future impacts on human health. Increasingly, these data are being assimilated into dynamic numerical models to improve model performance. Data for static elements of these models can be replaced (refreshed) every day, month, or season to ensure that model outputs represent the most recent conditions of environmental attributes.

1.1.2.1 Data collection

Data collected by atmospheric sensors provide information on air quality at the breathing level. There are sensors for monitoring atmospheric conditions like ozone and persistent organic pollutants. Earth observations also provide data for monitoring potentially hazardous conditions, both natural and anthropogenic. For example, dust, ash, and smoke plumes can be monitored and mapped to give early warning of risks to populations with respiratory ailments.

Table 1. Physical/anthropogenic forces, environmental linkages, and spatio-temporal model domains.

Natural and human hazards and disasters	Observed and/or modelled attributes	Spatio-temporal domain
Hydrometeorological hazards		
Floods and flash floods	Predicting human exposure; surface and groundwater flow; early warning systems	Hydrologic basins to watersheds; cities to regions; hourly to multi-year
Dust storms	Forecasting dust events for health alerts; potential human exposure	Cities to regions; hourly (event by event)
Drought	Forecasting extent and severity; impacts on water quantity and quality; spatial and temporal complexity	Hydrologic basins to watersheds; cities to regions
Geological hazards		
Earthquakes	Potential infrastructure vulnerability; actual infrastructure damage; transportation network mitigation	Pre-, real-time, and post assessment
Tsunamis	Travel speed; wave-height; potential damage areas; warning and mitigation	Ocean basins to cities; event by event
Landslides and mudflows	Areal extent and damage assessments; surface and groundwater assessments; infrastructure damage; relief optimization	Neighbourhood to cities; watershed to regional; as events occur
Human-caused hazards		
Air pollution	GHG, $PM_{2.5}$, smoke, ash concentrations; particle trajectories and dispersion; health advisories; inhalation exposures industrial emission monitoring	Point sources to cities; national to transnational; hourly
Soil pollution	Soil geochemistry; soil micro-nutrients; heavy metal pollutants; bio-infected soils (hygienic assessments); subsurface flow and transport	House-hold to cities; when needed
Human health issues		
Water supply	Areal assessment of degraded or abused land; clean-up monitoring; likelihood of diseases; likely contamination of public water supply	House-hold to neighbourhood and city; routinely or as needed
Food security and nutrition	Exposures to pesticides and chemicals; estimating daily dietary potential; risk levels	Individual to field; city to region; on demand to multi-year

Not surprisingly, most of the models and systems used address natural and human induced hazards and disasters. Many focus on air quality and pollution in densely populated areas. Others are used for regional and global modelling and monitoring at national, regional and global scales (e.g. dust storms, hurricanes, droughts, famine). The models fall into several categories: network models, object-oriented models, conceptual models, transport models, and forecast models. Several address water supply and sanitation; and systems are available also for identifying and tracking communicable and infectious diseases. Table 2 lists a few prominent examples and hundreds are available via the World Wide Web.

1.1.2.2 Data modelling

There are many avenues for geophysical modelling. Opportunities include observations and measurements of surface properties to assess conditions known to have health effects on individuals and

Table 2. Sample geophysical and biophysical models.

Application	Models/systems
Air pollution	CMAQ; HAPEM6; HEM-3; HYSPLIT; RAIMI; SHEDS/MENTOR-1A
Bio-terrorism*	SHEDS/MENTOR-2E
Communications and infectious disease	China Information System for Disease Control and Prevention; SYRIS; LMM
Contaminant risk	MENTOR-1A & 2E
Droughts	Drought Detection and Monitoring System
Dust forecasts	ADAM; DREAM; MASINGAR
Floods and flash floods	ITHACA Early Warning System for Floods
Food security and nutrition	DEPM; Lifeline Version 4.3
Groundwater	FEFLOW; FEHM; GMS; MODFLOW; Visual MODFLOW; ZOOMQ3D
Volcanic ash	HYSPLIT
Water supply and sanitation	GARR; CBPCALC; IONEX

*chemical and biological agents

populations. Spectral measurements of surface properties are collected in digital raster format and assimilated into numerical models developed for specific air, water, and soil applications. When natural events or human-induced hazards alter surface properties, observed new properties can be modelled quickly and evaluated for their human health implications based on *a priori* knowledge that many diseases have associated environmental determinants.

Geospatial and geospectral data are used widely by health and environmental monitoring COPs. As used here, *geospatial data* refers to values of environmental properties measured as points, lines, or areas in relational databases. These data are processed into products for visual interpretation, or fused with other data sets in a GIS to analyse, map, and manage multiple layers of data and information. Hardware and software systems are then programmed to integrate, manage and display data, or to query relationships, patterns and trends for representation as maps, charts, and reports. The technology is being used for many disease categories to map distributions of outbreak areas or point sources; or, to map relationships between diseases and possible environmental triggers. *Geospectral data* from satellite sensors are defined in terms of pixels. The physical dimensions of pixels are determined by a sensor's electro-optical-mechanical scanning design and the spectral values are determined by the sensitivity/detectivity of its detectors. These data can be assimilated into deterministic models and their outputs are often verified and validated by comparison to point data collected by *in-situ* ground-monitoring instruments. Harmonizing geospatial and geospectral data sets is essential when used jointly in modelling systems.

An important part of geoscience modelling involves interpolation techniques to represent data, often not on regular grids. The most widely used technique is kriging. It is a group of geostatistical techniques to interpolate the value of a point at an unobserved location from observations of its value at nearby locations. Kriging uses the spatial correlation among data points and constructs the interpolation via semi-variograms. For example, given an ordered set of measured grades of mineral concentrations, interpolation by kriging can predict concentrations at unobserved points. The technique is commonly used in mining, hydrogeology, natural resources, environmental sciences, geospectral sensing, and geospatial analyses.

There are numerous statistical and heuristic models to inform health communities about potential exposure levels occurring at the day to day level of health care. Statistical techniques commonly used by these communities also are being used increasingly by geoscientists in their effort to link Earth phenomena to health outcomes. In part, this is made possible by the availability of long term, synoptic Earth observations data on patterns of changing surface properties. More than thirty-five years of continuous monitoring by an ever growing number of satellite and sensor systems covering the globe make it possible now to consider longitudinal data sets linking health to environment.

Multivariate data analysis refers to any statistical technique that arises from more than one variable. This essentially models reality where each situation, product, or decision involves more than a single variable. The information age has resulted in masses of data in every field. Despite the quantum of data available, the ability to obtain a clear understanding of the situation and make intelligent decisions is a challenge. When data are stored in databases, multivariate analysis can be used to process the information in meaningful ways.

1.2   Survey of Earth science models relevant to human health and wellbeing

Synopses of biogeophysical phenomena having immediate consequences for human populations are itemized below (ICSU 2009; Davies 2010). General comments on methods and models used to assess specific incidences of these phenomena are given as an entrée to larger bodies of Earth science literature that guide multidisciplinary teams in forecasting events, developing mitigation plans in advance of an incident, or creating next generation model systems that link geospatial data with observed or emerging health outcomes.

1.2.1   Atmospheric processes

By virtue of its long history using satellite sensors, atmospheric modelling is arguably more mature than other Earth sciences modelling programmes. These processes lead to weather and climate oscillations, influence short- and long-term drought patterns, trigger extreme temperature events, and many other phenomena that link directly to human health, wellbeing, and economic conditions. Air quality at the breathing level is one area of immediate health concern.

1.2.1.1   Air quality

Air pollution models address applications ranging from assessing health risks for individuals and communities to dispersion of plumes at regional scales. The regional atmospheric climate model (RACMO) and the regional atmospheric modelling system (RAMS) represent higher resolution version of the global climate model (GCM). RACMO down-scales GCM results to regional and local levels to address physical phenomena in greater detail (CcSP 2012a). In addition to RACMO, RAMS is a multi-purpose system for medium resolution weather forecasting, photochemical ozone modelling, precursor transport, and nuclear emergency response. It can be used for predicting mesoscale pollution impacts in complex, time-dependent situations (CcSP 2012b). The regional air quality impact modelling initiative (RAIMI), and the modelling environment for total risk (MENTOR 1A & 2E) are risk-based modelling tools. MENTOR includes models for environmental and biological processes. RAIMI is fully integrated with Esri ArcView® and is designed to estimate potential health impacts associated with exposure to chemical pollutants. It can model risks associated with multiple levels of contaminants from multiple sources and multiple exposure pathways (US/EPA 2012a). MENTOR-2E uses a mechanistically consistent source-to-dose-to-response modelling framework to quantify inhalation exposure and doses resulting from emergencies. It models both chemical and biological agents and among other options uses a stochastic human exposure and dose simulation (SHEDS) approach (Georgopoulos & Lioy 2006; Georgopoulos 2008). Other models, such as CMAQ are deterministic, measuring trace gas concentrations and fine particulate matter. The HYSPLIT model is a complete system for computing air parcel trajectories and complex dispersion and deposition simulations. It can be used for tracking and forecasting radioactive material, volcanic ash, wildfire smoke, and pollutants from stationary and mobile emission sources.

   Modelling and forecasting dust and other aerosol episodes is becoming more widespread and recognized by health communities, especially in areas where extreme dust events occur (e.g. western China and Saharan Africa). These models typically rely upon weather forecast models as well as Earth observing data to produce forecasts and provide information about the intensity and timing of dust events. The DREAM and ADAM models forecast dust episodes at regional scales, while the model of aerosol species in the global atmosphere (MASINGAR) is a three-dimensional chemical transport model that provides information on the distribution of atmospheric aerosols and related trace species (Tanaka et al., 2003; Park et al., 2008).

   Air quality is one of the most monitored conditions in our environment. The US/EPA has a suite of models for analysing and monitoring a variety of conditions that affect the quality of air at the breathing level. In addition, there are models that use satellite data for monitoring types and concentrations of aerosols in the upper troposphere. There are models that estimate human inhalation exposures and health risks resulting from air pollution emissions. Systems have been developed that range from computing simple air parcel trajectories to complex dispersion and deposition simulations. Trace gas concentrations such as ozone, NO, and NO_2, and fine particulates such as $PM_{2.5}$ are significant contributors to health risks for persons with respiratory and cardiovascular diseases. These aerosols and particulates are modelled at local, regional and global scales, using both ground-based and satellite-based data.

For inhalation pathways US/EPA has developed a human exposure model (HEM) to assess sources of toxic emissions, usually producers or large users of air-borne toxic chemicals. HEM contains an atmospheric dispersion model, information on industrial sources, population data and other parameters that have to be customized for specific locations. Each source must be identified by latitude and longitude along with a description of release parameters including stack height, exit velocity, and emission rate. The model estimates the magnitude and distribution of ambient air concentrations in μgm^{-3} within a radius of fifty kilometres. These estimates are used as surrogates since exposure variables like duration of exposure, daytime *vs.* night time, human activity patterns, and residential occupancy are not addressed explicitly (US/EPA 2012b). There are two versions of HEM: The first, HEM-Screen, uses a long-term simplified industrial source complex model (ISCLT2) to derive dispersion. It is useful for screening-level assessments involving a large number of facilities. The second, HEM-3, uses either a short-term version of ISC (ISCST3) or the AERMOD to derive dispersion.

Another US/EPA model is called the hazardous air pollution exposure model (HAPEM) that is also available in two versions, HAPEM5 and HAPEM6 (US/EPA 2012c). These are designed to estimate inhalation exposures to various air-borne toxics for selected population groups. The models use ambient air concentration data, relational data for indoor/outdoor microenvironment concentrations, population data, and human activity patterns to estimate an expected range of inhalation exposures for groups of individuals.

Vulnerability of soils to aeolian erosion can be assessed by monitoring land cover changes on a seasonal or monthly basis, soil surface temperature, soil moisture patterns and precipitation amounts, topography, and surface roughness lengths. These need to be linked with information on land management practices. Experience exists in modelling dust storms together with pathogens and human health. Geoscientists are active in predicting some health risks, modelling airborne pathogens, and establishing early warning systems. Statistical models predict potential human exposure.

1.2.1.2 Extreme temperature events

Heat waves and prolonged cold snaps generally fall in the category of extreme temperature events. These are predicted reliably by deterministic weather forecast models from synoptic, satellite-based observations (Abaurrea *et al.*, 2007); but they seem to attract public attention when the broadcast media question the prospect of anthropogenic climate warning. They have been modelled also using Markov, statistical and stochastic procedures (Coles et al., 1994; Kysely 2002; Furrer et al., 2010). However, these models are limited in their ability to identify temperature variations within urban areas, and therefore the specific populations at risk of extreme heat, unless finer scale demographic, socio-economic, land cover, and urban attributes are integrated. When these finer-scaled data sets are assembled, weather forecasts become collaborative data for activating artificial neural networks that identify vulnerable neighbourhoods and local area residents at risk.

1.2.2 *Hydrologic processes*

Readers are directed to the alphabetical list of models, the computing systems they run on, model documentation, and related materials for water resources research and management (USGS 2012). Most of these can be integrated into other models to form systems that have relevance to human health applications.

1.2.2.1 Hurricanes and typhoons

Hurricanes and typhoons were detectable easily on early vintage geostationary satellite sensor imagery. Since the 1960s, technology has advanced to provide early warning of their formation in the eastern Atlantic, western Pacific, and eastern Pacific Ocean. Rates of movement, wind speeds, trajectories, rain rates, and storm surge magnitudes are the usual modelled parameters, although occasional storms still deviate significantly from projections of one or more of these elements. Two types of models are employed: statistical and dynamical; and, a combination of the two. Statistical models create forecasts from historical data sets that can be run on personal computers. Dynamical models solve equations for each point in a 3-D grid. However, the accuracy and precision of the calculations is dependent on three sources of error: initialization, the state of the atmosphere at any given point at the beginning of model process; resolution, the size of the 3-D cell around the point of calculation; and the basic understanding of the physics governing the atmosphere (Masters 2012). The statistical-dynamical approach was developed in the 1970s–1980s to create more reliable forecasts. The best known and mature models for hurricane forecasting are: ECMWF for medium range forecasts in the mid-latitudes; GFS; the NWS/Geophysical Fluid Dynamics Laboratory (GFDL) model; the UK Met Office model (UKMET); the NWS/Hurricane Weather Research Model (HWRF); and the US Navy's Navy Operational Global Prediction Center System (NOGAPS). There are also non-global models. The best forecasts are derived by combining outputs from several of the above models, called *consensus* forecasts.

Hurricane and typhoon modelling are mature technologies. For human health and wellbeing, the issues devolve more to individual human behaviour (i.e. those choosing not to evacuate impact areas), local decision making (i.e. authorities not optimising evacuation routes by reversing traffic flows), local alert mechanisms, and preparedness.

1.2.2.2 Droughts

Drought is defined as an extended period of time in which there is a significant deficiency of precipitation. Types of droughts include meteorological, agricultural, and hydrological, each of which has environmental and socio-economic impacts. Standard precipitation indices are used to model drought severity over a region. This information typically is used to monitor events using GIS. Agricultural drought models must consider precipitation shortages, evapotranspiration rates, soil moisture deficits, and ground water shortages. Hydrological droughts are driven by climatic changes that cause deficiencies in surface and sub-surface water supplies. These usually are modelled at watershed basin scales. Drought detection and monitoring systems are being developed that use satellite imagery and other spatial data as a basis for analysis. Results of these analyses are made available via web-based information systems (Mishra & Singh 2011).

1.2.2.3 Floods

Coastal and flood-plain flooding are vulnerable areas assessed primarily through observations of events and mapping the extent of alluvial deposits pre-dating the historical record. Event modelling is based on estimates of catchment precipitation and peak sea levels (for example, storm surges) used extensively as a basis for hazard and risk analyses. Flash flooding is less well studied but geomorphological mapping of constrained valleys and ravines and of alluvial fan deposits can be used to determine areas subject to potential hazards.

Groundwater flooding is often neglected even though this may cause surface floods or contribute to other events in areas with shallow groundwater tables and rising groundwater. Leakage from urban water supplies also may contribute to this hazard. The potential hazard can be assessed by monitoring groundwater levels. Modelling of groundwater behaviour is well developed and effective, provided enough primary survey data are available. Several models exist. Flooding causes serious health effects because of risks for polluted drinking water for people in cities.

The ITHACA early warning system for floods operates at regional and global scales and provides information and map products every three hours. Still under development, ITHACA's drought detection and monitoring system will produce information and map products of key drought indicators at regional and global scales via a web-enabled delivery scheme. Other types of systems are those primarily used by health professionals that address disease surveillance. These typically report, and in some cases map, incidences and outbreaks of communicable and infectious diseases. The China Information System for Diseases Control and Prevention and SYRIS are two examples.

1.2.3 *Seismic processes*

There are extensive data on locations, magnitudes/intensities, and frequency of seismic events. Records for developing countries often lack historical records, although seismic global monitoring networks have become increasingly effective over the past century. Retrospective modelling of historical events is relatively good, but indicators and prediction of future events remain elusive. The areas that most likely will suffer major earthquakes are well known. Most damage and fatalities arising from seismic events coincide with ground conditions in built environments, the quality of buildings and infrastructure, and secondary hazards such as landslides and dam collapses. Therefore, it is relatively straightforward to identify the areas that are most vulnerable to damage and to reduce risk by ensuring that structures are built to withstand the maximum events likely to occur in a given area. Historical maps of built environments would provide qualitative information on the probable building codes and regulations used at the time areas were developed; and since earthquakes cause serious damage to water, electrical, and heating distribution systems, maps and data on basic infrastructure should also become standards in earthquake hazard modelling.

1.2.3.1 Earthquakes

Earthquake modelling has surged since the 1960s and there is an extensive literature on their design and function, their software requirements, and their prediction capabilities in both 2-D and 3-D modes. Modern models can simulate complex numerical relationships among surface and subsurface parameters including the sources of uncertainties, which can be very high; and there is a growing amount of videography for engineering simulations (Bárdossy & Fodor 2004; Sijing & Marinos 1997; Ferguson *et al.*, 1998; Jaiswal *et al.*, 2009).

The locations of volcanic centres, and of the areas in which events are most frequent, are well known, but there is little appreciation that areas that have not been active within memory, or tradition, are still potentially active. The style of eruptions varies between types of volcanoes and even in the history of individual volcanoes. These differences give rise to hazards such as pyroclastic flows, lahars, lava flows, and in some cases instability and collapse of parts of the volcanic edifice. Volcanoes need to be studied in detail on a case by case basis to piece together the frequencies and styles of eruptions. The likely maximum extent of pyroclastic flows, lahars, and lava flows can be established by examining the local geomorphology to identify the most vulnerable areas where development should be avoided or where potential escape routes might be unusable. Warnings of possible increased activity can be detected by changes in heat flow monitored by *in-situ* and remote sensors. However there are many false alarms. The most active volcanoes near urbanized areas in developed countries are extensively monitored (e.g. seismic monitoring, tilt and inclinometry, heat and gas flow), but this is expensive and not always accurate in predicting eruptions. Few volcanoes in developing countries are monitored to the same extent in urban areas, although there are some notable exceptions like Lake Nyos, Cameroon, where a disaster resulted from emissions of carbon dioxide. Widespread modelling exists in volcanic areas using geological data and health data.

Mapping land movements via vintage aerial photography or on-the-ground establishes the extent and patterns of past landslides, which, when coupled with historical records give an idea of the frequency of events. Many prove to be linked to precipitation patterns. It is important to establish the potential for future movements in areas not yet affected based on the strength of materials compared with slope angles and other physical and engineering characteristics of the substrate. Data on ground water behaviour also can help to define areas where problems are most likely. Rock falls and avalanches in mountainous areas tend to be seasonal allowing warnings to be given in general terms. As the current interglacial age deepens, the frequencies of these events are increasing and may require re-evaluation of past experience. Heuristic models have a weakness in this regard because future predictions are tied to past events in a linear fashion and do not take account of non-linear tendencies. One health effect of landslides and mudflows is redistribution of polluted soils and effects on water distribution, for example disturbances in drinking water distribution.

Subsidence is associated with underground mines and shafts, natural caves, wells, cellars and similar cavities. Subsidence also occurs on collapsible soils like loess and, together with heave, is associated with the wetting and drying of certain clay types. Problems are widespread and, while often not dramatic, are costly; some events are large and sudden, and can lead to loss of life and property resulting from sinkholes and other collapses. Assessment depends on mapping the extent of previous uses of land and of limestone and gypsum karst layers liable to dissolution, collapse, or shrink/swell behaviour. While this is effective in outlining areas where subsidence is most likely to occur, ground investigations are needed to establish the actual presence or absence of a hazard and the need for designing remedial measures. Commonly, action is taken after an event occurs rather than identifying the need for action to prevent an event.

1.2.3.2 Tsunamis
Tsunamis are forecasted as a consequence of earthquakes and severe submarine geological shifts. These events are associated in the public mind with earthquakes but there is less appreciation that they also can be triggered by underwater and terrestrial landslides, in lakes as well as seas, or by volcanic cone collapses. There are warning systems in place for the Pacific, North Atlantic, and Indian Oceans; but, while these often allow time for escape to areas distant from the event's origin, they cannot prevent extensive damage in developed areas in vulnerable locations. The joint NASA/CNES JASON-1 satellite detects and monitors tsunamis progress and provides advance warning about probable wave height so populations can seek higher ground. Tsunami warnings associated with seismic and volcanic events are likely to be effective, but warnings related to massive mud and landslides are more difficult to assess. As with other sudden hydrologic disaster events, it is the aftermath of human suffering and loss of infrastructure that demand attention by health COPs.

1.2.4 *Decomposition processes*
Naturally toxic & deficient chemical elements and radiation are products of decomposition. Many naturally occurring elements affect human health if they are present in either toxic or deficient amounts. These concentrations are often transmitted via food to both human and domestic animal populations. To understand and verify the effects of nutrients on health, baseline geochemical mapping and statistical modelling are necessary. In countries having ancient residual soils like Australia, sub-Saharan Africa and India, medical problems have been related to micronutrient deficiencies in selenium (Se), iodine (I), and zinc (Zn) that play critical roles as catalysts in human biological processes. Deficiencies and excesses of macronutrients like phosphorus (P), sulphur (S), and iron (Fe) are equally bad (Davies 2010).

Chemical toxicities and deficiencies of lead (Pb), mercury (Hg), and others like polychlorinated biphenyls (PCBs) and asbestos, represent another dimension for health and wellbeing. Accumulations associated with chemical spills, automobile exhausts, lead in paint, mining operations near cities and unregulated recycling of electronic parts (aka *e-waste sites*) are a sinister threat for urban populations. They are especially dangerous in and around childrens' playgrounds. Sampling of both the geological and pedological substrates and of blood in children and adults is required to assess potential hazards. In most cases, statistical models are used and in some cases advanced multivariate geostatistical models have been used. Current research on urban environments has demonstrated linkages between lead contaminated urban soil, children's blood lead, and neurotoxic outcomes. It is suspected that, depending on the chemical nature and effect of a given hazardous substance, other toxins also will exhibit analogous exposure results and outcomes (Gobran & Clegg 1996; Chertov *et al.*, 2002; Shrestha *et al.*, 2007; Shamsad *et al.*, 2008; Davies 2010; Xu *et al.*, 2010). Cities are built environments that generate toxic hazards from birth to old age. The aim is to make them sustainable for present and future generations. Urban geochemistry and health is a subject requiring far more attention than the subject has received historically to make certain that future generations have the knowledge and physical ability to thrive in a changing world. In this

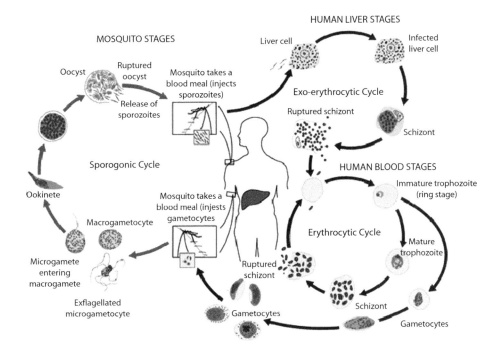

MOSQUITO STAGES

HUMAN LIVER STAGES

Oocyst

Ruptured oocyst

Mosquito takes a blood meal (injects sporozoites)

Release of sporozoites

Liver cell

Infected liver cell

Exo-erythrocytic Cycle

Ruptured schizont

Schizont

Sporogonic Cycle

HUMAN BLOOD STAGES

Ookinete

Mosquito takes a blood meal (injests gametocytes)

Macrogametocyte

Immature trophozoite (ring stage)

Erythrocytic Cycle

Mature trophozoite

Microgamete entering macrogamete

Ruptured schizont

Exflagellated microgametocyte

Gametocytes

Schizont

Gametocytes

Figure 1.   Malaria model linking parasite, mosquito and human infection and disease transmission dynamics (Modified from CDC). (see colour plate 59)

respect, baseline geochemical mapping and statistical modelling are necessary to keep pace with Earth observation technologies that can add spatial context to point sampling data.

Emissions of ionizing radiation are detected readily through radiometric monitoring, including aerial remote sensing. Radon monitoring is performed by placing detectors in properties and confined spaces where radon may present a hazard and where relatively straightforward remedial measures can be applied. In a number of countries, the geographical extents of geological formations that may emit radon have been mapped and the health risks are known. The range of geological/mineralogical circumstances where emissions may occur is well known to geoscientists, but is not widely known to others. As examples, there is a common awareness that granites may present a problem, but not that some granites are benign. Similarly, there is a lack of awareness that some shale and phosphate-rich deposits can have significant emissions. Distribution and health effects of radon can be modelled.

1.2.5   Biological processes

These include soil-borne, vector-borne, and water-borne diseases, the pathogens for which at some stage in their life cycle depend on soil or water for their existence and transmission. Geologists have traditionally studied those diseases that show a geographical distribution pattern related to an excess or deficiency of a specific element in the geochemical environment. What should be noted also is that there are other pathological states that are geo-medically relevant and that seem to be precipitated by environmental factors (e.g. climate, elevation, vegetation and land use patterns), which may themselves influence elemental distributions. Predicting the spread of infectious diseases will inevitably involve climate trends and variability issues (Stohlgren & Schnase 2006).

Part I of the Chapter has surveyed the conceptual and mathematical models typically used by health COPs. Schematic models like the one shown in Figure 1 for malaria convey state-of-the-art understanding about transmission pathways between parasites, mosquitoes and humans. Analytical models are mostly statistical, but there are increasing efforts to merge these with deterministic

Table 3. Life cycles, model approaches, research foci and selected references to 2007.

Model approaches	Research foci	Selected references (from MalaRis 2012)
A. Gonotrophic cycle*		
Deterministic; Mechanistic; probabilistic	Weather and climate change; database development	Hutchinson *et al.*, 1996; McMichael 1997; Hoshen & Morse 2004
Statistical; heuristic; simulations	Mosquito entomology; parisitology; genetics; ecological niches; phenology; popullation dynamics; distribution; mosquito behaviour; feeding cycles	Dietz *et al.*, 1974; Killeen *et al.*, 2000 Eichner *et al.*, 2001; Ishikawa *et al.*, 2003; Ahumada *et al.*, 2004; Depinay *et al.*, 2004; Chalvet-Monfrat *et al.*, 2007; Moffett *et al.*, 2007
B. Sporogonic cycle**		
Experimental model building	Epidemic processes; risk of human infection; effectiveness of indoor spraying; discrete event models; early warning systems; transmission intensity	Nedelman 1989; Moore 1992; Jetten *et al.*, 1996; McKenzie *et al.*, 1998; Patz *et al.*, 1998; Snow *et al.*, 1998; Craig *et al.*, 1999; McKenzie *et al.*, 1999; Hay *et al.*, 2001; Worrall *et al.*, 2001; Abeku *et al.*, 2004; Hoshen & Morse 2005; Omumbo *et al.*, 2005; Porphyre *et al.*, 2005; Le Menach *et al.*, 2007
Statistical; mathematic- simulations;	Human immunity; parameter estimates; eradication;	Nájera 1974; Pull & Grab 1974; Aron *et al.*, 1988; Smith *et al.*, 2006
Multi-model; probabilistic; deterministic	Quality assessments; transmission; burden of disease; early warning	Morse *et al.*, 2005; Thompson *et al.*, 2006
C. Erythrocytic cycle***		
Multiple and coupled models	Model inter-comparisons; assessments; evaluations of mosquito and human life cycle; inoculation and recovery rates	Jeffery *et al.*, 1959; Nedelman 1984; Martens *et al.*, 1995; Jacob *et al.*, 2001 K-1 Developers 2004; Palmer *et al.*, 2004; Randall *et al.*, 2007

*Egg development within the female mosquito depends on temperature, precipitation and humidity
**Parasite sporozoites infect mosquitoes. Development varies with temperature among species
***Humans liver cells infected with sporozoites which undergo several stages and re-infect mosquitoes

models for predicting disease outbreaks, or to merge them with GIS relational databases employing satellite data to assess environmental predictors of vector populations. An example is the malaria early warning systems (MEWS) described in Chapter 8. In addition, MalaRis, developed by the University of Köln provides a bibliographic history of malaria research and model-building (MalaRis 2012). In context of Figure 1, this history reveals the complexities of lifecycle stages in the mosquito and humans that permit the parasite to inhabit both forms, and to evade mosquito and human immune systems by passing through several stages, each having its own shape, structure and protein complement (NIH 2012; CDC 2012; Johns Hopkins 2012). Table 3 links a few of the entries in MalaRis to malaria life cycles in the sporogonic cycle of the mosquito and erythrocyte cycles of humans.

1.2.6 *Anthropogenic processes*
There are many facets to human health that are tied directly to how cultures occupy and use their land resources. Historically, humans have abused these resources and moved on, but such practices are no longer possible on a planet with diminishing resources and an ever-expanding population. The paragraphs below merely highlight some of the more obvious practices. Models and holistic model systems are needed to address these thoroughly.

Land degradation is largely an issue of how lands are managed. The geoscience dimension includes assessing soil vulnerability to erosion, compaction by livestock and vehicles (both can be readily mapped), and the availability of ground and surface water for which there are many sound modelling techniques available (Soil Erosion Site 2012). A late 1970s application of Landsat data used USDA's universal soil loss equation (USLE) to calculate the rate of soil loss in tons/acre/year as a means for estimating sheet erosion and fertilizer needs, and for designing strategies for land rejuvenation. All of the USLE parameters are amenable to modelling using EO-data, perhaps accounting for it having been modified many times from its original form by USDA to include mined lands, provide an agricultural production simulator, and related purposes. Degradation of agricultural land leading to smaller areas for food production is probably the most serious problem worldwide for human health and wellbeing.

Other categories of land degradation affecting health and wellbeing include: ecosystem contaminants, toxic waste, cultural conflicts, water supply and sanitation, and artisanal contaminants. For ecosystem contaminants and toxic waste, several modelling techniques are used for: 1) assessing rates of attenuation and dispersal of pollutants in both surface and ground water; 2) assessing the likelihood of contaminants being adsorbed or absorbed on sediment; and 3) assessing the effects of pollution on health. Complications often occur in geological materials that have incompletely understood or interconnected systems of macro and micro pores. While the assessment of the problem may be good, remediation is more problematical, especially with regard to those groundwater bodies that are subject to slow flow and replenishment. A relative newcomer in toxic disposal is imported commercial garbage like electronic equipment, computers, and retired ships. These are mostly unregulated activities that are not being adequately mapped or modelled for their health impacts. The number of such sites is growing in Africa and Asia, which accept these materials from global sources for their recycle value. Low cost, labor intensive *cottage industry* processes for extracting small amounts of precious metals and other useful products lead not only to health problems from inhalation but also to rising air, water and soil pollution, all of which have attendant health effects.

In general, soil pollution can be properly evaluated only on a site-by-site basis through sampling and testing. There are sound techniques for analysis, but despite the development of protocols for sampling based on statistical techniques to interpolate between sampling points, it is difficult to be sure that the whole range of contaminants has been properly evaluated throughout an area unless it is excavated. Since certain land uses are associated closely with particular types of pollution, evaluation of former land uses is helpful in identifying suspect areas. However, records often are deficient. It has been said of cities in the UK that all soils should be regarded as contaminated until proven otherwise because of their close proximity to long histories of industrial uses. This applies also to cities in developing countries affected by pollution and sanitary problems. The containment of toxic wastes in non-toxic landfills is well established in most developed countries but effective strategies depend on sound site location, high quality containment measures, and frequent (often remote) monitoring of facilities. The principles for site location also are well known, but mistakes are still made. A key remaining uncertainty is the very long term integrity of containment measures. Most modern facilities have been monitored only for decades. Landfills are not a preferred management option in most developed countries, but they are likely to be used elsewhere for many years to come.

Finally, many of the problems associated with artisanal mining are socio-economic hazards related to health and safety. Environmental problems include damage to the landscape and ecosystems through release of sediment into waterways and the use of toxic substances to extract economic materials (e.g. mercury in gold extraction). Resolution of these problems is essentially a matter of education and support.

These diseases and conditions are inherited or acquired. Inherited or acquired diseases may be cancers resulting from solar radiation, Type-1 diabetes, obesity, conditions resulting from nutritional (vitamin or chemical) deficiencies, malnutrition, and a host of others. Lifestyle diseases are acquired through eating habits (Type-2 diabetes, obesity), smoking (lung, lip and other cancers), occupational diseases related to industrial processes and mining environments (black lung disease), or other activities that place individuals in proximity to toxic materials. Though many have their roots in environmental phenomena, their impacts on individuals are mostly treated as societal issues related to wellbeing.

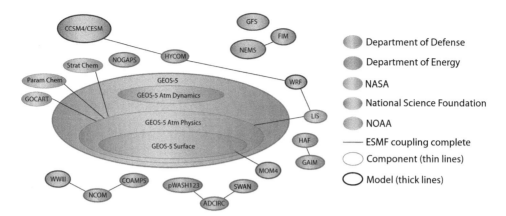

Figure 2. The ESMF is a collaborative, multiple US agency effort to couple over 50 geoscience models in such a way that they are interoperable and scalable (Adapted from O'Kuinghttons *et al.*, 2010). (see colour plate 60)

1.3 *Modelling systems*

The Earth system modelling framework (ESMF) program is assembling suites of mature models into a coherent, interoperable framework (Collins *et al.*, 2005). The idea behind ESMF is that complicated applications should be packaged into smaller pieces, called components. A component is a unit of software that has a coherent function and a standard calling interface and behaviour. Thus, ESMF is a component-based architecture for developing and assembling climate and related models. A virtual machine underlies the component-levels, providing both a foundation for performance portability and mechanisms for resource allocation and component sequencing. Components can be assembled to create multiple applications. Different implementations of a component may be available. In ESMF, a component may be a physical domain, or a function such as a coupler or I/O system (Hill *et al.*, 2004; Zhou 2006; O'Kuinghttons *et al.*, 2010).

In many research disciplines, models are the signature products of individual labs and investigators. Many components are documented poorly and require intensive experience before they can be used. However, the system offers promise for sharing models and model components more easily. This emergent system can enlarge the range of potential users, notwithstanding issues of trust, institutional structures, and other social dynamics that may be more problematic than technical concerns for effective implementation. ESMF includes many models that have been designed independently but which are known and respected for their high performance.

The ESMF assists model simulations across a wide range of applications from weather and climate to emergency response and to human health and well being. The framework allows scientists to build models quickly, reuse existing software rather than reinventing it, and to exchange components in a systematic way. This makes it easier for users to assemble complex models. The framework includes tools for re-gridding, data decomposition, and communication on parallel computers; and, for common modelling functions such as time management and message logging. Researchers using ESMF have a standard way to add new capabilities and swap-in different options. By helping scientists and engineers use common software to solve routine computational problems, ESMF will ultimately result in better research and accelerated progress in simulating the Earth's weather and climate systems. The goal is to increase scientific productivity and to promote new scientific opportunities.

Figure 2 shows how selected Earth science deterministic models are being coupled into the ESMF for multiple and interactive uses. Table 4 is a list of those commonly employed for atmospheric, hydrospheric and terrestrial parameters. The framework is structured around GEOS-5 Atmospheric General Circulation Model, built from the ground-up using ESMF protocols and philosophy. GEOS-5 is subdivided into several components forming subsets for atmospheric dynamics,

Table 4. Major components of the ESMF.

Acronym	Model or system name, or function	Lead US agency 2010
ADCIRC	Advanced circulation model	Department of Defense (DOD)
COAMPS	Coupled ocean-atmospheric prediction system	DOD
CCSM4	Community climate system model	National Science Foundation (NSF)
FIM	Flow-following finite volume icosahedral model	National Oceanic & Atmospheric Administration (NOAA)
GAIM	Global assimilation of ionospheric measurements model	DOD
GEOS	Goddard earth observing system	National Aeronautics & Space Administration (NASA)
GFS	Global forecast system	NOAA
GOCART	Goddard chemistry aerosol radiation & transport	NASA
HAF	Hakamada-Akasofu-Fry model	University
HYCOM	Hybrid coordinate ocean model	DOD
LIS	Land information system	NASA
MOM4	Modular ocean model	NOAA
NCOM	Naval Research Laboratory coastal ocean model	DOD
NEMS	National energy modelling system	NOAA
NoGAPS	Navy's operational global atmospheric prediction system	DOD
Param Chem	Parametrization of molecular force fields (Hamiltonians)	NSF
Strat Chem	Stratospheric chemistry and/or transport, especially O_3	NOAA
SWAN	Simulating waves near shore	DOD
pWASH123	Watershed modelling system	DOD
WRF	Weather research & forecasting	NSF

atmospheric physics, and surface models. Many of the ovals contain several model components, including couplers. The hierarchical tree of components can be snipped at different levels so that the whole physics package, or a single parameterization can be replaced. Each component is potentially interchangeable. The challenge is to couple biomedical models, environmental determinants of diseases, eco-physiological attributes of disease vectors, and a host of socioeconomic data and models into macro-scale systems.

2 MECHANISTIC MODELS

In this Section Doctor Lo focuses on the role of mechanistic models in human epidemiology. He further restricts the discussion to models based on differential equations. Readers will note the conspicuous absence of equations in this section since the goal is to keep the discussion at a higher level for a broader audience. References to technical papers and monographs that illustrate some of the key points and principles are provided.

Mechanistic models have their origins in the mechanistic concept of nature that became popular during the Age of Enlightenment in the 18th Century. The mechanistic theory asserts that the universe can be understood as a completely mechanical system, one that is composed entirely of matter in motion under a systematic set of natural laws. Newton's codification of the laws of motion in his *Principia* and his discovery of the law of gravitation helped to firmly cement this clockwork theory of the nature of the universe. The idea of mechanistic models assumes that complex systems are like machines which may be decomposed into individual mechanical parts. By understanding the workings of each individual part and the way they couple and work together, one can understand the complex systems completely. The modern physically based models simulate systems at the macro-scale, such as a bullet shattering glass, by modelling the behaviour of the bullet and glass down to the molecular and even quantum level. The main ingredients of mechanistic models are

307

algebraic and differential equations. The theoretical understanding and predictive power provided by these equations are the most attractive features of mechanistic models.

Although the fracture mechanics of glass is a complex problem, by comparison, biological and ecological systems are many orders of magnitude more complex. At the other extreme from mechanistic models are empirical models. The idea is that complex systems are too difficult to model and understand from first principles. Instead, by repeated observations and data collection, one is able to build a model based on observational data using statistical methods and models. Empirical models also have predictive powers, provided the conditions under which the system is operating are stable and do not change drastically. However, if these conditions change significantly, the models are unable to forecast the possible outcomes. A more serious short coming is their inability to explain the dynamics of the system, how it works and how it behaves.

In reality, no model is purely mechanistic or empirical in nature. It is a matter of degrees. Mechanistic models have parameters that are typically fitted from empirical data. The lack of such data and the associated empirical models frequently prevent mechanistic models from being useful at all in complex situations. Similarly, even the simplest empirical models require some understanding of the key components of the system and their interaction mechanism in order to collect the proper data and create an empirical model of the system under study.

The interaction of the environment with human health is an extremely diverse and complex set of problems. In general, very little is known about the mechanisms of these complex problems. In many cases, we do not even know what the relevant variables are for the problem; hence there is no way to set up a mathematical model describing the problem. Even when one is able to create a mathematical model of the problem, the parameters for it often require observations of environmental variables that are difficult and costly to obtain. Without such data, one cannot determine the parameters for the model. This is why empirical models are indispensable and will always be a critical tool for understanding the interaction between the environment and human health issues. Nevertheless, the power and potential of mechanistic models is so compelling and too important to ignore. Perhaps in the not too distant future, it may be possible to have multi-resolution mechanistic models of vector-borne diseases that take into account disparate model elements, from seasonal weather to the genetics of the vector and human host, to create epidemic forecasts for decision support at the local health level.

2.1 *Types of differential equations*

Differential equations measure the rate of change of one quantity against the rate of change of another. For example, the rate of change of the distance travelled against time, which gives us speed. In modeling natural phenomena, we are typically interested in the dynamics of the system. Dynamics is the study of how the different components of the system change with respect to one another. This is the reason why differential equations play such a major role in such problems.

2.1.1 *Ordinary and partial differential equations*

Differential equations are divided into two major categories: ordinary differential equations (ODE) and partial differential equations (PDE). Ordinary differential equations study the variation of the system with respect to a single variable such as time. The solutions of such a system are one-dimensional trajectories in the space of system variables. Partial differential equations, on the other hand, study the variation of the system with respect to multiple variables which may include time. For example, ocean waves are a natural phenomenon that change with respect to both space and time. The resulting PDE known as the *wave equation* measures the movement of the wave with respect to both spatial coordinates and the time coordinates. The resulting solutions are no longer one-dimensional trajectories, but instead, are multidimensional surfaces in the space of the system variables. One can easily see that the complexity of partial differential equations is several orders of magnitude above that of ordinary differential equations.

Modelling weather and climate relies heavily on partial differential equations. The movement of fluids is a key to these systems. Similarly, the movement of sand is also modelled as a fluid. Hence, the Navier-Stokes equations of fluid dynamics form the foundation of these systems. Thermodynamics also play a critical role in these problems. This is governed by another well-known partial differential equation known as the *heat equation* which describes the diffusion and transport of heat through an object. With a few changes in the parameters of the heat equation, it can be used to describe general diffusion processes in nature. For example, the spread of pollution in ground water and the transport of nutrients in plants are some processes which may be described by the heat equation.

As complex as the weather and climate may be to understand, these are processes which are governed by fundamental physical processes that are relatively well understood. By comparison, biological systems are infinitely more complex. There are no counterparts in biology to the fundamental physical laws such as gravity, electromagnetism, or quantum mechanics. Various aspects of biological processes may be modelled by differential equations, but the description of a biological system is so complex, it eludes our current attempts to capture it in any differential system to date. So when it comes to the ecosystem or the dynamics of vector-borne diseases, there is no general system of differential equations that can model and explain it all. Where it is possible to model such processes by differential equations, they must be hand crafted to suit each problem individually in its own environment with its own distinct peculiarities. For example, malaria may be transmitted by mosquitoes in the humid tropics as well as the very dry sub-Saharan regions. The behaviour of the different species of mosquitoes and the life style of humans in these two distinct ecosystems are entirely different. Consequently, one cannot model the transmission of malaria in these two very different regions by the same set of equations by merely changing parameters.

Partial differential equations may be used to describe theoretical aspects of ecological models. But when it comes to the actual simulations and forecasts, where does one get the values and ranges of the parameters in the equations? The initial data required by partial differential equations can be a set of data forming a multidimensional surface that must be obtained from field measurements. For example, the problem may require the temperature of the ocean to a depth of one kilometre over the Monterey Bay. The collection of such data, if possible at all, may be prohibitively expensive. Without such data, the model is essentially useless to the practitioners. The fact is that while there are many biological and ecological models built on partial differential equations, their predictive powers are limited by the amount of observational data available to properly parameterize these equations so as to be able to produce useful simulations and predictions.

2.1.2 *Ordinary differential equations*

Attention is turned now to ordinary differential equations and focus on epidemiological models with emphasis on vector-borne diseases. Although ordinary differential equations suffer the same issues as partial differential equations, their relative simplicity compared with partial differential equations make them much more tractable as modelling tools. Here the solutions are one-dimensional curves as opposed to the multidimensional solutions for partial differential equations. At the very least, one can visualize a solution curve in the space of variables relatively easily.

Differential equations for infectious disease are typically called compartmental models. This is because the population in the model is divided into compartments such as those individuals who are susceptible (S) to the disease, those who have been exposed (E) to the disease, those who are infectious (I) and can transmit the disease to others in the population, and those who are recovered or removed (R) from the disease. Here removal means either death or immune. The so-called SEIR model has the four compartments we just described. The differential equation relates these four groups in the population in a set of differential equations describing the rate of transfer from one compartment to another which captures the dynamics of the transmission and development of the disease from pre-infection to recovery/removal. The paper by (Mishra *et al.*, 2010) provides an excellent, succinct introduction to the terminology and the fundamentals of the compartmental models of infectious diseases. The actual equations that describe the relations between SEIR compartments are based on two well-known, fundamental equations in population dynamics: the logistic equation, and the predator prey equation.

2.1.3 The logistic equation

The logistic equation has its roots in Malthus' work on population growth. After reading it, Verhulst derived the logistic equation in 1838 by adding a correction term to Malthus' simple linear growth model for population (Verhulst 1838). Despite the many reservations and issues pointed out about mechanistic models, this simple equation is a grand example of how science strives to find mechanistic models for ecology and human health issues. Verhulst's simple logistic equation is able somehow to capture the dynamics of population growth from germs to humans despite the complexities and the enormous differences between microbes and humans. This model describes a closed population (with no migration) in an environment with limited resources, and is able to answer the question about the maximum population size sustainable within these resources. In 2010 this equation was used to fit the population of India since the 1960's into a model of malaria in India (Martcheva & Hoppensteadt 2010). The fitting is done using simple regression on the parameters of the differential equations, a typical approach used in mechanistic models.

An interesting side note regarding the logistic equation and its famous solution, the logistic function is the fact that this system is very unstable and exhibits deterministic chaos. The logistic function was one of the key models used to understand the behaviour of chaotic solutions starting in the 1960's. Amazingly, it remains as an active area of research. See Ausloos (2006) for the history of the logistic equation and its applications using the logistic map.

2.1.4 The predator-prey equation

The predator-prey equation (aka the Lotka-Volterra equation), is one of the main building blocks of biological systems and ecosystems. In 1910, Lotka wrote a paper on chemical reactions that are essentially Verhulst's logistic equation (Lotka 1910). In 1920, he extended it to biological systems and applied it to a model of plant and herbivore species (Lotka 1920). Finally in 1925, he derived his famous predator-prey model in his book on biomathematics (Lotka 1925). Volterra derived the same set of equations independently in his analysis of fish catches in the Adriatic Sea (Volterra 1926). The predator-prey equation is one of the earliest models of mathematical ecology, after the logistic equation.

Models such as SEIR or SIR, and hundreds of other variations of compartmental models use the logistic and predator-prey equations as their starting points to relate the behaviour of various populations and species involved. Terms are added and modified to tailor the equations to suit the circumstances. The growth rates are the key parameters in these differential equations which must be obtained from observations and field data collection. For example, for malaria the rate of infection depends on the number of mosquitoes present and the number of susceptible members in the population without immunity due to previous infections. These numbers must be collected before the model can be executed. When the parameters are obtained, remarkable things can be done even with the most rudimentary of mechanistic models. The paper by Martcheva & Hoppensteadt on modelling malaria in India for the last fifty years provides a wonderful example of what can be done. Their interesting observation was that, despite the availability of epidemiological data on malaria in India over the last six decades, no one ever bothered to construct the numerical model for malaria until 2010 (Martcheva & Hoppensteadt 2010).

2.2 Analysis methods for differential equations

In this section, some key concepts and tools used from dynamical systems theory used for analysing differential equations are described. These include equilibrium solutions, periodic solutions, stability, attractors, and invariant manifolds.

2.2.1 Equilibrium solution (fixed points)

An equilibrium solution is reached when all of the population compartments have reached a plateau and their rates of change are zero. This is also known as a fixed point of the differential equation because when one reaches such a solution, nothing changes and everything remains in a steady state. However, if this fixed point is not stable, the slightest perturbation will cause the solution to drift away from the vicinity of the fixed point. For ecosystems or populations, one usually likes steady states to remain constant and unchanging – provided it is a *good* steady state. For example, if humans can permanently eradicate malaria, that would be a good steady state; but if a state emerges where malaria remains constant, that would be an undesirable steady state.

Another situation is a periodic orbit. The equilibrium state just mentioned can be thought of as a periodic orbit with zero (0) period. A true periodic orbit would cycle through the same series of states repeatedly. According to Poincaré, who first developed many of these methodologies for studying these dynamical systems (another name for differential equations viewed from Poincaré's approach), the periodic orbits of a system of differential equations completely capture its geometry and dynamical behaviour (Poincaré 1993). Depending on the nature of the periodic orbit and the system being modelled, it may or may not be desirable to remain in a periodic orbit. In order to change the state from a periodic orbit, one must know the stability properties and trajectories forming *invariant manifolds* that may be associated with the periodic orbit. A few dynamical terms are defined to facilitate the discussion. A *manifold* is a high dimensional surface (e.g. a circle is a one-dimensional manifold, a sphere is a two-dimensional manifold, and a solid ball is a three-dimensional manifold with surface. The Euclidian 3-space is a 3-manifold without boundaries. The space of variables, where the differential equation operates, is called the *phase space*. Fixed points and periodic orbits are objects in the phase space. A manifold, M, in phase space is called *invariant* if starting from a point P on M, the trajectory of P remains always on M as the differential equations propagate. A fixed point and a periodic orbit are all examples of *invariant manifolds*.

2.2.2 *Delay differential equations*

In modelling vector-borne diseases like malaria, the development of the vectors such as insects must precede the actual spread of the diseases. Hence, there is a delay of the actual diseases transmission with respect to the development of the disease-carrying vectors. This is precisely what delay differential equations were invented to do. The introduction of the time delay greatly increases the complexity of these differential equations. Let us consider the simplest first order linear differential equation

$$x(t)' = kx(t), \tag{1}$$

where k is a constant, the delay version is $x(t)' = kx(t - \tau)$, here $\tau < 0$ is the time delay. The addition of this delay parameter causes two immediate problems: first, now the solution includes oscillatory motion; second, the initial value for the differential equation is no longer just the value of $x(t)$ at the initial time $t = 0$. Now, one must provide the value of $x(t)$ for the interval $[\tau, 0]$ (recall $\tau < 0$).

2.2.3 *Stochastic differential equations*

Stochastic differential equations are differential equations with stochastic parameters. The independent variables are now stochastic processes such as a Wiener process or a Levy process which follows certain probabilistic models. The following sub-Section describes a simple example. These equations require a different set of numerical algorithms for their solution than the standard integration algorithms such as Rung-Kutta or symplectic methods (Aubry & Chartier 1997). Stochasticity is an extremely important aspect to mechanistic models because natural processes are always filled with random effects such as noise or measurement and model uncertainties. In many models without the stochastic elements, the model will simply not work correctly.

2.3 *Examples of simple mechanistic models*

Perhaps nothing explains a concept better than actual examples. This section summarizes a series of papers on various mechanistic models of epidemiology to show how they are used in actual situations. An article by Earn (2008) on modelling measles in New York City is an excellent introduction to mechanistic models for biological systems. City health officials kept excellent records on measles epidemiology for many decades, which made this analysis possible. Getting access to the human health data records is one of the key problems for human health models. Earn began with the famous but simple SIR (Susceptible, Infected, Recovered) epidemiology model proposed in 1927 (Kermack & McKendrick 1927; Earn 2008). He compared the model step by step with the actual data. When the model was unable to explain a feature in the data, he augmented it to target that feature. By proceeding in steps, he was able to determine how to model the various features of the diseases and their relative importance and impacts on the model behaviour. The steps Earn followed were: 1) he showed that to obtain reasonable solutions to the SIR model, he had to begin with a good set of initial conditions for the variables S(t), I(t), R(t) at time t = 0 (the outbreak of a measles epidemic in NYC in 1962); 2) he set the susceptibles, S(0), to the total population of NYC in 1962; and 3), he set the infected, I(0), to the number of reported cases in September of 1962 (start of epidemic) times the infectious period of measles which is five days. With these conditions, the SIR model gave a reasonable prediction of the epidemic in 1962. However, measles is endemic and periodic. These initial conditions produced a single epidemic, with no recurrence for subsequent years. This is because individuals who recover from measles acquire immunity for life. For new cases of measles to occur in his model, it is necessary to introduce new susceptibles into it by adding births in the population. While this introduces the recurrence of measles in subsequent years, the epidemics show decrease in size year by year and eventually die out. Earn explains that this is actually a feature of the modified SIR model independent of the initial conditions and parameter values. To fix this problem, one must model the seasonality of the disease caused by the variation of the contact rate of children who are the key victims of this disease. The rate of contacts between children depends on the calendar of the school year. By including this seasonality in the contact rate in the model, it is now able to correctly predict an endemic periodic epidemiology for measles in NYC. A second article by Bauer (2008) on *Compartmental models in epidemiology* in the same book is an excellent summary of the subject.

Next a simple model for SARS by Ang (2007) based on the logistic equation discussed earlier is summarized. This is the two compartment SI model (Susceptible and Infected). Ang introduced a stochastic element into the model by using a standard Wiener process (also known as Brownian motion) to model the white noise in the model. He was able to fit the stochastic logistic equation to a portion of the SARS epidemiology data of Singapore in 2003. By adjusting the amount of stochasticity in the different times during the epidemic, a better fit could be obtained. However, this is somewhat artificial and does not provide insight into the actual epidemiology. The paper also introduces a standard numerical scheme based on the Euler method for solving stochastic differential equations known as the Euler-Maruyama method.

Lastly, a hybrid model on malaria in India by Martcheva & Hoppensteadt (2010) shows how a hybrid mechanistic and empirical model is constructed. The goal of this paper was to model the clinical cases of malaria in India from 1984 to 2009. They started with the famous Ross-MacDonald model for malaria transmission, which is an SIS model (Susceptible, Infected, Susceptible). Ross used this model to explain the transmission of malaria by mosquitoes and won a Nobel Prize for Medicine for this discovery (Ross 1911). His original equation only took humans into account. MacDonald extended the model by adding an SI model for the mosquito infection. However, for the period 1984 to 2009, the Indian government kept very good records of the malaria epidemiology in India. This enabled Martcheva & Hoppensteadt to add a new compartment (C) for those who had clinical symptoms. Thus, the SIS model became the SCIRS model.

The time series for malaria cases in India showed growth from 1984 to 1996 as there were no significant efforts to control malaria during this period. But in 1997, the Roll Back Malaria Program began in India (see also Chapter 8). Thus, from 1997 to 2009, the cases for malaria declined steadily. Another interesting development noted by the authors was that the infectious rate for malaria also declined steadily from twenty-one days in 1983 to just thirteen days in 2008. By adjusting the parameters and the terms in the SCIRS equations, the authors were able to fit the equations to the data using regression analysis.

For further reading on mechanistic models, readers should consider the lecture notes by Hoppensteadt (2011). This book presents various differential equation models including partial differential equations and stochastic differential equations. However, the author noted that obtaining observational data for the parameters of these equations is a serious challenge. Finally, Laneri *et al.*,

(2010) presents an analysis of the role of monsoon rains and malaria epidemics in northwest India. This paper presented an extremely sophisticated stochastic differential equation model for malaria. For an interesting application of invariant manifolds to modelling malaria, see Nakakawa (2011) on the relationship between malaria and the sickle cell gene that provides some protection against malaria, for reasons that are not well understood.

Mechanistic models like the Ross-MacDonald model and the Kermack-MacKendrik model have played significant roles in understanding and controlling malaria. However, for current epidemic risk assessments, sparseness of field data and knowledge prevent wider use of mechanistic models. For risk assessments, statistical empirical models are the most successful and broadly used. This is because differential equation models require many parameters and rates to be measured and modelled, which are often impossible to obtain with our current technology. Perhaps as remote sensing capabilities advance, the use of satellites and unoccupied aerial vehicles (UAVs) may provide some of the missing field data to help advance use of mechanistic models for epidemic risk assessments.

3   ECOLOGICAL NICHE MODELS FOR PUBLIC HEALTH

This Section by Doctors Kumar and Stohlgren reviews a variety of ecological niche models in context of their strengths and weaknesses for environmental change analyses at multiple spatial scales. Public health is referenced in context of environmental complexity and uncertainty given the enormous human effort to account for all physical and biological variables. The Section concludes with comments on ENMs and their potential to identify and map environments that may be evolving risks to human health.

3.1   *Introduction*

Ecological niche models (also called species distribution models, environmental matching models, and habitat suitability models) are increasingly being used to model and map invasive species distributions. Combining statistical algorithms with GIS, ecological niche models attempt to predict probability of occurrence of a species by using presence-only or presence-absence data in combination with environmental variables to predict the species' potential or actual distribution across a landscape (Elith & Leathwick 2009; Franklin 2009; Newbold 2010). These models are based on Hutchinson's classical niche concept: distributions of species are constrained by biotic interactions such as competition and predation, and by abiotic gradients such as elevation, temperature and precipitation (Hutchinson 1957; Elith & Leathwick 2009; Franklin 2009; Sinclair *et al.*, 2010). Although ecological niche models have become popular tools for invasive species ecology (Chen *et al.*, 2007; Evangelista *et al.*, 2008; Kumar *et al.*, 2009; Albright *et al.*, 2010; Stohlgren *et al.*, 2010a), they are increasingly being used for rare and endangered species (Guisan *et al.*, 2006; de Siqueira *et al.*, 2009), defining conservation priority areas (Pawar *et al.*, 2007; Fuller *et al.*, 2008; Kremen *et al.*, 2008), phytogeographical studies (Miller & Knouft 2006; Moritz *et al.*, 2009), public health (Costa *et al.*, 2002; Peterson *et al.*, 2008; Holt *et al.*, 2009), and the potential impacts of climate change (Thuiller *et al.*, 2007; Bradley *et al.*, 2009; Jarnevich & Stohlgren 2009). However, single species models and single model approaches have led to inconsistent uses and abuses of niche-based models. Thus, modelling potential to address multiple species and multiple stresses using climate change and land use change has yet to be realized.

There are still many unresolved issues regarding ENMs related to their potential predictive power and their ability to extrapolate or interpolate information from sampled areas to remaining un-sampled areas (Elith *et al.*, 2006; Marmion *et al.*, 2009; Sinclair *et al.*, 2010). The integration of different algorithms could allow users to integrate different types of data sets (e.g. presence-only vs. presence-absence) and predictive variables that are represented geospatially (i.e. shape files, grids, remote sensing). Some models have performed better for predicting potential distributions (e.g. BIOCLIM) (Busby 1991), while others are better at predicting realized distributions (e.g. Artificial

Neural Networks) (Jimenez-Valverde et al., 2008). Others are limited by sample size (Pearson et al., 2007), multi co-linearity (Neter et al., 1996; Graham 2003), species traits (e.g. classification and regression tree–CART) (Evangelista et al., 2008), and other unforeseen influences. For these reasons, careful consideration in model selection and their rigorous testing is critical.

Despite advances in ecological modelling, computing power, environmental and species occurrence databases, and statistical techniques, many challenges remain unresolved in the vital area of forecasting invasive species distributions using niche-based models. For example, niche models have difficulties predicting invasive species distributions that are far from equilibrium (Welk 2004). Also, species interactions, both negative and positive, are difficult to include because of limited data availability or poor distribution data, particularly in native ranges (Peterson 2003; Mitchell et al., 2006). ENMs are also affected by sampling bias and by spatial autocorrelation (Kadmon et al., 2004; Segurado et al., 2006; Veloz 2009). Because of these and other conceptual and practical problems, niche models are being haphazardly used in modelling native species distributions and forecasting biological invasions because of their simplicity and ease of use (Hampe 2004; Soberón & Peterson 2005; Graham et al., 2008; Menke et al., 2009).

Predicting the future is not that easy. Use of niche models for invasive species forecasting is challenging because they extrapolate species occurrence for novel environments or novel climates (in case of climate change studies). Their use for predicting invasive species is often criticized because they typically violate the assumption that an organism is in equilibrium with its environment (Vaclavik & Meentmeyer 2009). Different invasive species may have varying degrees of equilibrium with their environment; for example, well established invaders may be in relatively greater equilibrium with their environment than early invaders. Therefore, niche models that work well for established invaders may not work for early invaders; rigorous model testing is required.

A species' fundamental niche is defined as the region where the abiotic or environmental conditions are habitable, while its *realized* niche is more restricted to where both abiotic and biotic conditions are favourable and further confined by the dispersal and adaptation mechanisms (Pearson & Dawson 2003; Soberón & Peterson 2005). An important consideration for invasive species is that recent invaders may not have filled all suitable habitats, while species naturalized long ago may have filled a larger proportion of suitable habitat – but the differences between realized niche and fundamental niche may be uncertain for most species. Conceptually, niche models assume the fitted observational relationships to be an adequate representation of the realized niche of the species under a stable equilibrium or quasi-equilibrium constraint. As such, the ecological niche model result is only a first approximation of future distributions of individual species (Pearson & Dawson 2003), which are determined also by other processes such as dispersal, adaptation, competition, succession, fire and grazing pressure (Austin 2002). Still, an innovative integrated modelling approach may contribute considerably to a robust early warning system for decision makers to design more effective management and control strategies for harmful invasive species.

3.1.1 *Applications of niche models in public health*

Using ecological niche modelling for tracking disease vectors is not new. For example, GARP models were used to predict the potential distribution of stray dogs, and where the vector *Trypanosoma gerstaeckeri* might be expected to occur (Beard et al., 2003). The models indicated a potentially broader distribution of this species, and suggested additional areas of risk beyond those reported. Genetic algorithm for rule-set production (GARP) models have been used also to predict the geographic distribution of human monkeypox in Africa (Levine et al., 2007). Ecological niche models also have been used to model disease-carrying *Lutzomyia sandfly* species in the state of São Paulo, Brazil (Peterson & Shaw 2003). Maxent was used for predicting risk of malaria in Africa (Moffett et al., 2007) and for predicting potential distribution of *Culex tritaeniorhynchus*, a primary vector of Japanese encephalitis virus, in the republic of Korea (Masuoka et al., 2010). Despite these and other examples, more research is needed on model-comparison studies to refine the precision and accuracy of niche models for the protection of public health.

Table 5. Commonly used ecological niche models for predicting species distributions.

Model	Advantages*	Disadvantages
Maxent/Phillips et al., 2006	P; Nl; Np; NS-C; RIV; E; U	Pseudo-absence or background data
GARP/Stockwell & Nobel 1992; Stockwell & Peters 1999	P; Nl; Np; NS-C; U; NE	Pseudo-absence or background data; no RIV; Ne
Classification and Regression Tree (CART) Breiman et al., 1984	Np; Nl; P/A; E; RIV	Absence data needed
Random Forest/ Breiman 2001	Np; Nl; P/A; RIV	Absence data needed
Boosted Regression Tree/Friedman 2001, De'ath 2007	Np; Nl; P/A; RIV	Absence needed, limited spatial data; need more statistical details
Logistic Regression/ McCullagh & Nedler 1989	W; P/A	Absence data needed; sensitive to multicollinearity
BIOCLIM/ Busby 1991	P; simple	Needs absence data; less accurate than other niche models
DOMAIN/ Carpenter et al., 1993	P; simple	Needs absence data; less accurate than other niche models
Mahalanobis distance/ Farber & Kadmon 2003	P; Nl	No absence data
Environmental niche factor analyses or ENFA/ Hirzel et al., 2002	P	No absence data
Artificial Neural Network (ANN)/ Pearson et al., 2002	P/A	Absence data needed
Envelop/Jarnevich et al., 2007	P or A only can be run	All environmental factors are weighted equally

*Codes: P = species presence; P/A = Presence/Absence; Np = Nonparametric; Nl = Non-linear; NS-C = not sensitive to multi-colliniarity; RIV = relative importance of variables; CRV = continuous response variables; E = Efficiency; W = widely used. Presence-only algorithms. Note: A list of available software to implement these models can be found in Franklin (2009; pages 111–112)

3.2   Niche modelling algorithms

A variety of mathematical and statistical modelling algorithms is available to model spatial relations between ecological variables and environmental covariates. The choice of model algorithm may depend on the response variable. For example, if the response variable is presence/absence then LR, CART, BRT or RF may be appropriate. For presence-only data, algorithms like Maxent and GARP may be used.

With so many niche models to choose, users need to know how to select the appropriate model for their purposes (Table 5). Rare or newly invasive species may be more difficult to model than long-established species. Habitat (or host) specialists may require different models than habitat generalists. Other factors that influence the selection of models include: the number and distribution of occurrence points, the availability of true absence locations, the resolution of mapping layers (e.g. satellite data, climate data, soils information), and the specific model objectives. Because of the differences in their performance, it is suggested that users evaluate several niche-based species distribution models to find the one that most efficiently relates to their species of interest (Elith et al., 2006; Evangelista et al., 2008; Kumar et al., 2009). Available niche models can be broadly categorized into two categories: (1) presence-only models; and (2) presence-absence models. Most of these are available at no cost (Elith et al., 2006; Franklin 2009).

3.3   Presence-only niche models

For many species, presence-only data are more easily available than presence-absence data. Absence data, if available, may not be reliable. For example, the species may go undetected (i.e. low

315

probability of detection). Therefore, presence-only methods could be very useful when the user has access to only species presence data such as the data from published reports or from museums or herbaria. These methods predict the relative habitat suitability for a species. Commonly used presence-only niche models are: BIOCLIM (Busby 1986; 1991); DOMAIN (Carpenter *et al.*, 1993); Mahalanobis Distance (Farber & Kadmon 2003); environmental niche factor analysis (ENFA) (Hirzel *et al.*, 2002); maximum entropy model (Maxent) (Phillips *et al.*, 2004; 2006); and, the most widely used, GARP (Stockwell & Noble 1992).

BIOCLIM, a profile matching method, is one of the first species-distribution modelling softwares. It uses a "hyper-box" classifier (a hyper-rectangle or environmental envelope) to define the multi-dimensional environmental domain of a species (Busby 1986, 1991). It produces a grid-based habitat map in which cells are classified into different ranked classes of environmental suitability for the species. DOMAIN is a distance-based method and uses *a point-to-point similarity metric to assign a classification value to a candidate site based on the proximity in environmental space of the most similar record* (Carpenter *et al.*, 1993). Gower metric, a multivariate measure of similarity, is used in DOMAIN to make predictions of species habitat suitability. Mahalanobis distance is a multivariate method that defines perpendicular major and minor axes and is used to compare a single observation to a group of sites (Farber & Kadmon 2003). ENFA is also a multivariate method that is based on a multivariate description of species occurrence locations (Hirzel *et al.*, 2002). ENFA calculates the magnitude of the difference between a species' mean average conditions where a species is found and the entire range of environmental conditions within the background environment of the study area.

GARP is a superset of artificial intelligence modelling algorithms that uses a set of rules in a machine-learning environment to describe species' ecological niches (Stockwell & Noble 1992; Stockwell & Peters 1999). The rules are IF/THEN statements that describe non-random relationships between species occurrence data and environmental conditions. GARP utilizes four rule types: atomic, envelope or Bioclim, range, and logit. Atomic rules use single values of environmental variables (e.g. *if vegetation type is A then species is present*). Envelope rules use ranges of values of bioclimatic variables. Range rules are generalizations of Bioclim rules, and logit rules are an adaptation of logistic regression models (Stockwell & Peters 1999; Stockwell *et al.*, 2006). The rules are generated by sampling with replacement, and an equal number of presence and background points. The observations are divided into training and testing subsets and a large number of models are developed (usually 1000s). Due to the stochastic nature of GARP, it has been recommended that a set of best models be retained (e.g. 10 or 100). The most useful models are identified using the "best subsets" procedure (Anderson *et al.*, 2003). Next, the proportion of models predicting presence for a cell (or image pixel) can be interpreted as the probability of occurrence of a species.

Maxent was designed as a general-purpose, machine learning, predictive model that can be applied to incomplete data sets (Phillips *et al.*, 2004; Phillips *et al.*, 2006). Relatively new, the Maxent models are available at no cost from: www.cs.princeton.edu/~schapire/maxent/ [Accessed 28th March 2012]. Maxent operates on the principle of maximum entropy, making inferences from available data while avoiding unfounded constraints from the unknown (Phillips *et al.*, 2006). Entropy is the measure of uncertainty associated with a random variable. The greater the entropy, the greater will be the uncertainty. Maxent is a non-parametric, predictive model that uses presence-only data. This method estimates the probability distribution of a species by finding the probability distribution of maximum entropy, which is a probability that is closest to uniform (Phillips *et al.*, 2006). It automatically includes variable interactions and can handle continuous and categorical predictor variables. It uses a set of features (e.g. linear, quadratic, product, threshold and hinge) which are functions of environmental variables that constrain the geographical distribution of a species. It uses a regularization parameter, which is determined empirically, to control model overfitting. Maxent generates an estimate of probability of presence of the species that varies from zero (lowest probability) to one (highest probability). Maxent has fared well consistently in model comparison studies, and performs quite well even with small sample sizes (Elith *et al.*, 2006, Pearson *et al.*, 2007; Kumar *et al.*, 2009; Kumar & Stohlgren 2009). Finally, Maxent also generates response curves for each predictor variable and has a jack-knife option that allows estimation of relative influence of individual predictors (Phillips no date).

3.4 *Presence-absence models*

Presence-absence niche models include the type of modelling algorithms that can model binary response (1 and 0; one for the presence of species and zero for absence). Use of these niche models is recommended if presence-absence survey data are available. Most commonly used presence-absence niche models include generalized linear models (GLMs), generalized additive models (GAMs), CART, RF and BRT (Hastie & Tibshirani, 1986, 1990; De'ath & Fabricius 2000; Breiman 2001; De'ath 2007; Elith *et al.*, 2008). There are some modelling tools that include a combination of these algorithms. For example, BIOMOD includes the most widely used techniques for species distributions including GLMs, GAMS, CART and ANN (Thuiller 2003; Thuiller *et al.*, 2009).

3.4.1 *Generalized linear models*

Binomial logistic regression analyses are commonly used by ecologists for predictive modelling (Austin *et al.*, 1990; Pearce & Ferrier 2000). Based on discrete probability distribution, logistic regression analysis is a GLM that uses the logit (the natural log of the odds of the dependent occurring or not) as its link function (McCullagh & Nelder 1989). Requiring presence and absence data, logistic regression analysis is used to predict a dependent variable from a number of continuous or categorical independent variables ranking their relative importance and assessing the degrees of interaction. The dependent variable is usually dichotomous taking a value of one with a probability of success (p; present in a location) and value of zero with a probability of failure (1 – p; absent in a location). This method also applies a maximum likelihood estimation that quantifies the probability of a certain event occurring; in this case, the probability of occurrence by a species. The performance of a logistic regression model is generally assessed by a d-square value and goodness-of-fit tests (e.g. chi-square and Wald statistics). Prediction values are typically presented in tabular form as the percentage of data points correctly or incorrectly classified as present or absent. Only factors important in the model are given coefficients for mapping purposes. Modelled surfaces typically have values of probabilities of occurrence from 0 (absent) to 1 (present). Logistic regression analyses can be conducted with most statistical software packages and their results are easily displayed in GIS formats.

3.4.2 *Generalized additive models*

Generalized additive models or GAMs are a non-parametric extension of GLMs and are also commonly used in modelling species distributions (Hastie & Tibshirani 1986, 1990; Yee & Mitchell 1991; Austin 2007). They are excellent tools for identifying and describing non-linear relationships between species distributions and environmental predictors. Similar to GLMs, GAMs also use a link function to establish a relationship between the mean of the response variable and a 'smoothed' function of the explanatory variable(s) (Guisan *et al.*, 2002). GAMs have strong ability to model complex non-linear relationships between the response and a set of environmental predictors. The fit of GAM can be evaluated by testing the non-linearity of predictor by a model with linear fit for that predictor versus the non-parametric fit (the significance can be evaluated by a Chi-square test (Franklin 2009). GAMs are considered very powerful and flexible modelling tools for spatial prediction of species distributions and generally out perform GLMs (Austin 2002; Meynard & Quinn 2007).

3.4.3 *Decision tree-based methods*

Decision trees, also referred to as CART, are effective tools for exploring complex, non-linear relationships between response variables (e.g. presence-absence of a species) and multiple environmental variables (De'ath & Fabricius 2000). Classification tree analysis is a commonly used non-parametric model that predicts the response of a dependent variable through a series of simple

317

regression analyses (Breiman *et al.*, 1984). Unlike other regression approaches that conduct simultaneous analyses, classification trees statistically partition the dependent data into two homogenous groups at a node, repeating the procedure for each group in a continuing process that forms a hierarchal tree. Classification trees, which are used when the dependent data are categorical (e.g. presence or absence), function in a fashion similar to regression trees, which use continuous dependent data (e.g. per cent cover, species richness). There are several characteristics of this modelling approach that are appealing to researchers and resource managers. First, the analyses explicitly allows for nonlinear relationships between the dependent and independent variable. These methods make no *a priori* assumptions about the distribution of the data, the relationships among independent variables, or between the dependent and independent variables. Second, they are well suited to handle non-homogenous data sets (e.g. unbalanced sample sizes, high variability). Finally, the results are easily interpreted and the predictive strength of each independent variable is explicitly reported. Classification trees fit the most probable class as a constant while regression trees fit the mean response for observations in the homogenous group; both assume normally distributed errors. The results can be integrated easily with GIS and mapped (Hansen *et al.*, 2003). Classification tree analyses require presence and absence data while regression trees require continuous response data. CART models are relatively poor predictors and are subject to over-fitting and have other drawbacks (Evans *et al.*, 2011). These limitations have been addressed by ensemble learning approaches including bagging (Breiman 1996) and boosting (Friedman 2001; De'ath 2007).

3.4.4 *Random forests*
Random forests (Breiman 2001) build large numbers of de-correlated trees and combine them to produce more accurate classifications. It is a form of machine learning technique (bagging or bootstrap aggregation) that has no distributional assumptions about the predictor or response variables. The algorithm starts with the selection of many bootstrap samples from the data; approximately sixty-three per cent of the original observations occur at least once (Cutler *et al.*, 2007). A set of observations are held back (sometimes referred to as "out-of-bag" samples that are not part of the bootstrap sample) for estimating model error and variable selection or importance. A classification tree is fitted to each bootstrap sample and the trees are fully grown and used to predict the test (out-of-bag observations). Model performance and error rates are calculated for each observation using the out-of-bag predictions. The variable importance can be estimated in two ways: (1) based on the misclassification error rate calculated from the out-of-bag sample; and (2) based on in-the-bag training data by calculating the reduction in sum of squares (deviance) achieved by all splits of the tree that use that variable, averaged across all the trees (Prasad *et al.*, 2006; Franklin 2009). Random forests overcome the over-fitting by averaging predictions from a large number of models based on subsets of the data.

3.4.5 *Boosted regression trees*
Boosted regression trees (De'ath 2007; Elith *et al.*, 2008) are statistical learning methods that include two algorithms: boosting technique (similar to bagging), and a method for improving model accuracy; and regression trees, as described above. Boosting is a form of resampling in which weighted probability of a response is resampled based on previous classifications. In boosting, models are fitted iteratively to the training data and the weights are applied to observations, emphasizing poorly modelled ones. Boosted regression trees (BRT) is an additive model which is a linear combination of many trees (i.e. 100s to 1000s) that can be considered as a regression model in which each term is a tree (Elith *et al.*, 2008). BRT can handle: both continuous and categorical predictors allow missing data; are not sensitive to outliers; and automatically model variable interactions (Elith *et al.*, 2006). The relative importance of the predictor variables can be calculated based on the number of times a predictor variable is selected for splitting node and weighted based on the improvement to the model based on each split (Friedman & Meulman 2003). BRT models can be fitted to a variety of response types including Gaussian, Poisson and Binomial, and partial dependence plots can be used to visualize the effects of different predictors in the response variable. BRT are excellent predictors. They were ranked high in several model comparison studies, and

Table 6. Time series of NASA satellite products and other geospatial predictive layers that can be used in niche-based models for modelling and mapping disease vectors.

Data/variables	Description
EO data (MODIS AVHRR, 7 others)	Phenology from NDVI, Phenology from EVI, Phenology from LAI, Net Primary Production (NPP), Evapotranspiration, Soil Adjusted Vegetation Index (SAVI), Ratio Vegetation Index (RVI), TCI, per cent tree, herbaceous and bare ground cover, LST, surface albedo, and snow products.
Topographic, edaphic, geologic, and biogeochemical variables	Elevation, slope, aspect, solar radiation, head load, overland distance to water, snow potential index, compound topographic index, base flow index, flow direction, flow accumulation, surface geology, soil types, soil texture, nitrogen and carbon content.
Anthropogenic variables	LULUC from national sources like NLCD, fire, distance from human habitations, distance from roads and trails, human population density, human footprint/influence index (Sanderson *et al.*, 2002), global urban areas (Schneider *et al.*, 2010), human urbanization patterns in recent times (Stohlgren *et al.*, 2010b), and cattle density.
Climatic variables	Temperature (minimum, maximum, and mean), precipitation, 19 Bioclim variables (Nix *et al.*, 1986), humidity, growing degree days, and number of frost days.

were least affected by the changes in grain size (Elith *et al.*, 2006; Guisan *et al.*, 2007; Stohlgren *et al.*, 2010b).

3.5 Species and environmental data for niche models

Species occurrence data for niche models can be presence-only (location where a species was found present) or presence-absence collected during a field campaign or from existing sources. Several issues regarding the data collection on response variables should be considered. For example, sampling design and sampling bias can potentially affect the model results (Fortin *et al.*, 1989; Phillips *et al.*, 2009). Spatial locations of the field observations (X & Y coordinates) should be carefully recorded for spatial modelling because any spatial errors in the observations can result in high uncertainties in the model outputs (Graham *et al.*, 2004).

The choice of environmental variables considered in niche based models can have significant effects on their predictive power (Parolo *et al.*, 2008; Peterson & Nakazawa 2008). Environmental predictor variables can be broadly categorized into three categories: indirect, direct and resource variables (Austin 2002; 2007). Indirect variables are ones that have no direct physiological effect on species such as altitude, latitude and longitude. Direct variables include temperature and pH. The models developed using these variables are more biologically informative and more generalizable than models developed with indirect variables. Resource variables include the resources that are consumed by the species (e.g. light, water, and nutrients for plants) (Austin *et al.*, 2006; Austin 2002; 2007). The choice of environmental variables may also depend on the spatial extent of the study area (Mackey & Lindenmayer 2001) because different variables may affect species distributions at different spatial scales (grain and extent) (Pearson & Dawson 2003; Guisan *et al.*, 2007; Hortal *et al.*, 2010). Remotely sensed data have been widely used to predict species distributions and characterize habitat structure (Rogers *et al.*, 2002; Turner *et al.*, 2003; Asner & Vitousek 2005; Asner *et al.*, 2008) including forecasting distribution of vectors of malaria (*Anopheles gambiae* complex) in Africa.

Potential environmental variables that may be considered as predictors in niche models include climate variables (e.g. Worldclim, Daymet, PRISM), hydrological variables (e.g. flow direction, flow accumulation, and base flow index), remotely sensed data (e.g. Landsat and MODIS phenology products (Tan *et al.*, 2008; Morisette *et al.*, 2009), land use land cover types, and anthropogenic data layers (Table 6).

319

3.6 Multi co-linearity and variable reduction

With the greater availability of remotely sensed and other environmental data layers there is an increased tendency among researchers to consider a large number of environmental predictor variables in niche-based species distribution models. However, using too many variables in a niche model might result in over-fitting the model (Chatfield 1995). The predictor variables should be carefully chosen based on their biological significance and functional relevance to the species (Elith & Leathwick 2009). Most ecological environmental predictors are highly correlated (e.g. elevation is most often correlated with temperature). Multi co-linearity among predictor variables may result in high standard errors in parameters estimated (Graham et al., 2004). So, multi co-linearity among predictor variables should be tested (Neter et al., 1996) by examining cross-correlations among them. Only one variable from a set of highly correlated variables should be included in the model. The selection of this variable should be based on its biological relevance to the species and its ability to constrain species distribution.

Ecological niche models can also inform us about the environmental factors that may be associated with distribution of species. The variable importance in predicting species distributions can be determined using a number of criteria that have been suggested (e.g. Murray & Conner 2009). For example, AIC can be used to investigate the relative influence of predictors in GLMs (Akaike 1974; Burnham & Anderson 2002), and other methods such as Maxent 'jack-knife variables importance' can be used to investigate relative influence of different predictors in the model (Phillips et al., 2006).

3.7 Model evaluation and validation

Performance of the model should be evaluated using a number of threshold-dependent and threshold-independent metrics because comparing models using different statistics would allow for better overall evaluation (Lobo et al., 2010). Threshold-dependent metrics include Cohen's Kappa (Cohen 1960), sensitivity, specificity, per cent correct classification, odds ratio, and true skill statistic or TSS (Fielding & Bell 1997; Allouche et al., 2006; Franklin 2009; Mouton et al., 2010). The threshold-independent evaluation measure includes "area-under-the-receiver-operating-characteristic-curve", or ROC_AUC (Swets 1988), where sensitivity is plotted against specificity for all possible thresholds (Pearce & Ferrier 2000).

Since models are simplifications of reality, all models have prediction errors and uncertainty (Barry & Elith 2006; Franklin 2009); therefore, their performance should be evaluated. Ideally an independent data set from a new geographic location or from a different new field campaign from the same study area should be used for model validation (Fielding & Bell 1997; Barry & Elith 2006). If possible, independently collected field data should be used to validate the model performance. However, in cases where it is not feasible to collect new data because of logistical reasons "split sample approach" can be used which creates a quasi-independent data set for model validations (Fielding & Bell 1997; Guisan & Hofer 2003). A range of data partitioning methods has been suggested (e.g. re-substitution, cross-validation, 10-fold validation, jack-knife resampling, and bootstrap resampling (Verbyla & Litaitis 1989).

3.8 Model uncertainty and error

Uncertainty is inherent in species distribution models; almost all models have prediction errors and uncertainty (Barry & Elith 2006; Franklin 2009; Venette et al., 2010). Quantification of uncertainty and error in the model predictions can make models and maps of species distributions reliable and more useful and informative for managers in making critical management decisions. Uncertainty in species risk maps may be introduced by the input parameters. Parametric uncertainty caused by measurement and systematic errors, incomplete or sparse data, natural variability in the system and subjective decisions in the estimation of parameter values (Elith et al., 2002; Regan et al., 2002; Barry & Elith 2006; Venette et al., 2010). Other types of uncertainty arises from: (i) the

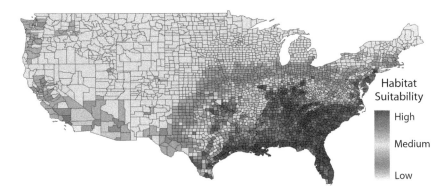

Figure 3. Potential habitat distribution of *Ae. albopictus* mosquito in the US. (see colour plate 61)

way in which the model is constructed and underlying models assumptions (Barry & Elith 2006); (ii) the selection of variables and the processes considered critical to species (Regan *et al.*, 2002); and (iii) the uncertainty in spatially explicit analyses (e.g. rescaling, aggregation, generalization, or extrapolation of inputs (Venette *et al.*, 2010). To address uncertainty, sensitivity analyses or Monte Carlo simulations can be conducted (Regan *et al.*, 2002; Johnson & Gillingham 2004). These methods measure the robustness of the model projections in the face of variability and uncertainty in model inputs and structure (Venette *et al.*, 2010). To reduce algorithmic or model-based uncertainty (Pearson *et al.*, 2006; Kumar *et al.*, 2009) the ensemble modelling approach can be used to combine predictions from several ENMs (Araujo & New 2007; Stohlgren *et al.*, 2010a).

3.9   A case study

The following example shows how species occurrence data can be integrated with remotely sensed, climatic, topographic, anthropogenic and other environmental data layers to produce disease vector risk maps using niche models (Figure 3). Potential habitat distribution for *Aedes albopictus* mosquitoes in the continental US was modelled using Maxent niche modelling methods. The Asian Tiger mosquito, *Ae. albopictus* (Skuse) is considered a great threat to public health in the US (Moore 1999). It is native to the tropical and sub-tropical parts of Southeast Asia but in the past few decades has invaded many countries throughout the world. It was first discovered in the US in 1985 and has now been reported from 1195 counties in thirty-six states. The case study considered thirty-four environmental variables, including nineteen bioclimatic variables calculated using the Daymet climate data set; humidity and other variables; MODIS vegetation continuous field variables; and topographic and anthropogenic variables like population density (Nix 1986; Hansen *et al.*, 2003; Daymet 2012). The number of variables in the final model was reduced to twenty-three after accounting for multi co-linearity. The average value of each predictor variable was extracted for all counties and used in the Maxent model. Twenty-five replicates were run using seventy per cent of the data for training the model and the remaining thirty per cent for testing model performance. The Maxent model performed far better than random with an average test AUC value of 0.81. Relative humidity, number of frost days, precipitation of driest month, and human population density (in order of their decreasing importance) were the top predictors of *Ae. albopictus* distribution in the continental US.

3.10   Caveats and challenges for ENM applications for public health

There are a number of factors that affect the accuracy of ENMs including specific assumptions of different modelling algorithms and the decisions made by the user during model calibration. The authors suggest: 1) taking an experimental approach when testing and using ENMs; 2) using a variety of appropriate models at various spatial scales; 3) reviewing all the model outputs carefully; 4) evaluating model uncertainty and interpreting results cautiously; 5) understanding and reporting model assumptions and limitations; and 6), taking an iterative approach to modelling as new data and techniques become available. The following includes a brief review of factors that can influence ENM performance.

3.10.1 *Background points*
Some of the models described above, especially presence-only models, are affected by the extent to which background or pseudo-absence points are drawn (Phillips 2008; Phillips *et al.*, 2009). Background or pseudo- absence data points, if used, should be drawn from within the environmental conditions in which the species were found to be present (i.e. area from which the species is making the habitat selection) because including the area far beyond the species' occurrence locations can result in misleading models. For inflated AUC values see VanDerWal (*et al.*, 2009). Minimum convex polygons (MCPs) around species' present locations can be used to define the extent for training the model and randomly drawing background or pseudo-absence points.

3.10.2 *Sample size and sampling bias*
The performance of most ENMs is affected by sample size (Stockwell & Peterson 2002; Hernandez *et al.*, 2006; Papes & Gaubert 2007; Pearson *et al.*, 2007; Wisz *et al.*, 2008). The models that are highly sensitive to sample size may not predict accurately any early invaders that have fewer samples. Therefore, use of a method like Maxent that performs well for even small sample sizes is advised.

Data on species distributions can often be biased (Graham *et al.*, 2004) and incomplete (Braunisch & Suchant 2010). This bias can be divided into four major categories: spatial, environmental, temporal and taxonomic (Soberón *et al.*, 2000). Biases can be introduced by several factors, including variable sampling efforts, spatial bias in sampling, focus on particular taxa (e.g. data on invasive plant species are more readily available compared to animals, pathogens and diseases), and sampling in a particular season (Peterson *et al.*, 1998). The location and intensity of field observations is heavily influenced by accessibility. For example, most species occurrence data are collected at convenient locations along roads, in flat spots, and near rest areas. Removing all biases from the species occurrence data is almost impossible because of the limited manpower, time and resources needed to collect field observations. However, there are ways to deal with such biases. For example using 'target group' data as background in Maxent can alleviate problems due to bias in data (Phillips *et al.*, 2009) or removing duplicates (to avoid pseudo-replication) and immediate neighbouring points or sample points within a certain distance based on assessment of spatial autocorrelation (Veloz 2009) can alleviate some of the problems due to biased data. Knowledge about the possible errors and biases associated with any species distribution data sets is critical because of its potential effects in predictive ability of niche-based models (Kadmon *et al.*, 2004). Despite these limitations, data with possible spatial errors or known biases still can be valuable when integrated with other reliable field data, and these data may assist in a better understanding of species distribution patterns. One must make best use of the available data and not wait for perfect unbiased data sets on invasive species or emerging disease vectors.

3.10.3 *Biotic interactions and dispersal*
Environmental niche models do not automatically address species interactions and dispersal limitations. Species interactions in niche-based models can be considered by including distribution layers of interacting species because interactions among species may play an important role in shaping species distributions (Grinnell 1917). Several studies on native species have found that including distributions of interacting species can improve the accuracy of distribution models (Leathwick & Austin 2001; Araujo & Luoto 2007; Ritchie *et al.*, 2009). Wherever possible, investigators should include dispersal vectors in the models because they may help predict species distributions and enhance predictability of niche models (Meentmeyer *et al.*, 2008; Vaclavik & Meentmeyer 2009).

3.11 *Discussion*

The invasion of non-native plants, animals and pathogens has escalated dramatically over the last few decades with the increase of trade, transportation and other elements of globalization. The economic costs and environmental consequences affect all levels of society negatively from indigenous cultures to world powers. Impacts include loss of native species and habitat, economic suppression, reduced food and water security, and direct threats to human health. Overall economic costs associated with invasive species in the US are estimated to exceed $120 billion per year (Pimentel *et al.*, 2000; 2005). The spatial and temporal distribution of vector mosquito distributions is controlled by environmental conditions such as temperature, precipitation, humidity, and land use land cover. Ecological niche modelling techniques can use these environmental factors and their interactions to predict probabilities of occurrence of disease vectors across multiple scales. State-of-the-art ecological niche models that can deal with presence-only and presence-absence data can be used to map current and future distributions of disease vectors and quantify those risks to inform climate change adaptation and public health intervention efforts to reduce current and future vulnerability of populations.

3.12 *Verification and validation*

In the bio-geosciences, numerical deterministic models always represent complex open systems that are only partially understood and for which there are only partial data inputs. Models are based on measurable properties of attributes and the standard approach to verifying and validating results is to compare model outputs with actual time-stamped measurements from often sparsely populated ground-based measurement networks. Differences between actual and modelled results must be explained in terms of: 1) uncertainties in the space-time domains of model parameters; 2) the realism of the model design; 3) the quality of input data; or 4), any number of other phenomena that cause discrepancies. Comparisons are only possible to the extent that observations of nature are complete; but they never are. If they were, there would be no need for modelling.

Verification and validation in the health sciences is frequently based on statistical relationships (an R^2) that measures the fraction of variability accounted for by the model. However, this statistic does not guarantee that the model fits the data well. If it does not, it cannot provide good answers to the questions being investigated. Consequently, an analysis of residuals is needed to measure the fit of data to the model. If the residuals appear to behave randomly, one interprets that the model does fit the data well; if non-random structure is evident in the residuals, there is probably a poor fit.

As important as modelling is for the growing array of health and wellbeing programmes, it is still true that verification, validation, and confirmation of numerical and statistical models for these purposes is difficult. Health and environmental systems are never closed, and model results are always non-unique (Oreskes *et al.*, 1994; Sterman 2006; Morain 2008). This fact takes added importance when public policy and population health are at stake. The pandemic H1N1 virus of 2009-10 is a recent example. Numerical models are used increasingly in the policy arena to justify difficult and controversial decisions.

REFERENCES

Abaurrea, J., Asin, J., Cebrian, A.C. & Centelles, A. 1992. Modeling and forecasting extreme hot events in the Ebro valley: A continental area. *Global & Planetary Change.* 57: 43–58.

Abeku, T.A., Hay, S.I., Ochola, S., Langi, P., Beard, B., deVlas, S.J. & Cox, J. 2004. Malaria epidemic early warning and detection in African highlands. *Trends in Parasit.* 20: 400–405.

Ahumada, J.A., Lapointe, D. & Samuel, M.D. 2004. Modeling the population dynamics of *Culex quinquefasciatus* (Diptera: Culicidae), along an elevational gradient in Hawaii. *J. Med. Entomol.* 41: 1157–70.

Akaike, H. 1974. A new look at statistical model identification. *IEEE Trans. Auto. Cont. AU-19*: 716–722.

Albright, T.P., Chen, H., Chen, L. & Guo, Q. 2010. The ecological niche and reciprocal prediction of the disjunct distribution of an invasive species: The example of *Ailanthus altissima*. *Biol. Invas.* 12: 2413–2427.

Allouche, O., Tsoar, A. & Kadmon, R. 2006. Assessing the accuracy of species distribution models: Prevalence, kappa and the true skill statistic (TSS). *J. Appl. Ecol.* 43(6): 1223–1232.

Anderson, R.P., Lew, D. & Peterson, A.T. 2003. Evaluating predictive models of species' distributions: Criteria for selecting optimal models. *Ecol. Modell.* 162: 211–232.

Ang, K. 2007 A simple stochastic model for an epidemic. *Electronic J. Math. & Techn.* 1(2): 116–127.

Araujo, M.B. & Luoto, M. 2007. The importance of biotic interactions for modelling species distributions under climate change. *Global Ecol. & Biogeog.* 16: 743–753.

Araujo, M.B. & New, M. 2007. Ensemble forecasting of species distributions. *Trends Ecol. & Evol.* 22: 42–47.

Aron, J.L. 1988. Mathematical modelling of immunity to malaria. *Mathemat. Biosci.* 90: 385–396.

Asner, G.P. & Vitousek, P.M. 2005. Remote analysis of biological invasion and biogeochemical change. *Proc. US Nat. Acad. Sci.* 102: 4383–4386.

Asner, G.P., Jones, M.O., Martin, R.E., Knapp, D.E. & Hughes, R.F. 2008. Remote sensing of native and invasive species in Hawaiian forests. *Rem. Sens. Environ.* 112: 1912–1926.

Aubry, A. & Chartier, P. 1997. Pseudo-simplectic Runge-Kutta methods. *BIT* 1–31.

Ausloos, M. & Dirickx, M. (eds.). 2006. *The logistic map and the route to chaos from the beginning to modern applications.* Berlin: Springer-Verlag.

Austin, M.P., Nicholls, A.O. & Margules, C.R. 1990. Measurement of the realized qualitative niche: Environmental niches of five *Eucalyptus* species. *Ecol. Monogr.* 60: 161–177.

Austin, M.P. 2002. Spatial prediction of species distribution: An interface between ecological theory and statistical modelling. *Ecol. Modell.* 157: 101–118.

Austin, M.P., Belbin, L., Meyers, J.A., Doherty, M.D. & Luoto, M. 2006. Evaluation of statistical models used for predicting plant species distributions: Role of artificial data and theory. *Ecol. Modell.* 199: 197–216.

Austin, M.P. 2007. Species distribution models and ecological theory: A critical assessment and some possible new approaches. *Ecol. Modell.* 200: 1–19.

Bárdossy, G. & Fodor, J. 2004. *Evaluation of uncertainties and risks in geology.* Heidelberg: Springer-Verlag.

Barry, S. & Elith, J. 2006. Error and uncertainty in habitat models. *J. Appl. Ecol* 43: 413–423.

Bauer, F. 2008. Compartmental models in epidemiology. In F. Brauer (ed.), *Mathematical epidemiology:* 19–79. New York: Springer-Verlag.

Beard, C.B., Pye, G., Steurer, F.J., Rodriguez, R., Campman, R., Peterson, A.T., Ramsey, J., Wirtz, R.A. & Robinson, L.E. 2003. Chagas disease in a domestic transmission cycle in southern Texas, USA. *Emerg. Infect. Dis.* 9: 103–105.

Bradley, B.A., Oppenheimer, M. & Wilcove, D.S. 2009. Climate change and plant invasions: Restoration opportunities ahead? *Global Change Biol.* 15: 1511–1521.

Braunisch, V. & Suchant, R. 2010. Predicting species distributions based on incomplete survey data: The trade-off between precision and scale. *Ecography* 33(5): 826–840.

Breiman, L. 2001. Random forests. *Mach. Learn.* 45: 5–32.

Breiman, L., Friedman, J.H., Olshen, R.A. & Stone, C.G. 1984. *Classification and regression trees.* Belmont: Wadsworth International Group.

Breiman, L. 1996. Bagging predictors. *Mach. Learn.* 24: 123–140.

Burnham, K.P. & Anderson, D.R. 2002. Model selection and multi-model inference: A practical information-theoretic approach. 2nd ed. New York: Springer.

Busby, J.R. 1986. A biogeoclimatic analysis of *Nothofagus cunninghamii* (Hook) Oerst in southeastern Australia. *Austra. J. Ecol.* 11: 1–7.

Busby, J.R. 1991. BIOCLIM - A Bioclimatic Analysis and Prediction System. In C.R. Margules & M.P. Austin (eds.), *Nature conservation: Cost effective biological surveys and data analysis:* 64–68. Canberra: CSIRO.

Carpenter, G., Gillison, A.N. & Winter, J. 1993. Domain: A flexible modelling procedure for mapping potential distributions of plants and animals. *Biodiv. & Conserv.* 2: 667–680.

CcSP. 2012a. Available from: http://www.knmi/~roode/racmo/racmo.htm [Accessed 31st March 2012].

CcSP. 2012b. Available from: http://atmet.com [Accessed 31st March 2012].

CDC. 2012. Available from: http://www.cdc.gov/malaria/about/biology/index.html [Accessed 19th April 2012].

Chalvet-Monfray, K., Sabatier, P. & Bicout, D.J. 2007. Downscaling modeling of the aggressiveness of mosquitoe vectors of diseases. *Ecol. Modell.* 204: 540–546.

Chatfield, C. 1995. Model uncertainty, data mining and statistical-inference. *J. Roy. Stat. Soc., Series a-Stats. in Soc.* 158: 419–466.

Chen, H., Chen, L.J. & Albright, T.P. 2007. Predicting the potential distribution of invasive exotic species using GIS and information-theoretic approaches: A case of ragweed (*Ambrosia artemisiifolia* L.) distribution in China. *Chinese Sci. Bull.* 52: 1223–1230.

Chertov, O.A., Nadporozhskaya, M., Bykhovets, S & Zudin, S. 2002. Simulation study in nitrogen supply in boreal forests using model of soil organic matter dynamics ROMUL. *Dev. Plant & Soil Sci.* 92(11): 900–901.

Cioffi, J. 1997. Heuristics, servants to intuition in clinical decision making. *J. Adv. Nursing* 26: 203–208.

Cohen, J. 1960. A coefficient of agreement of nominal scales. *Educat. & Psychol. Meas.* 20: 37–46.

Coles, S.G., Tawn, J.A. & Smith, R.L. 1994. A seasonal Markov model for extremely low temperatures. *Environmetrics* 5: 221–239.

Collins, N., Theurich, G., De Luca, C., Suarez, M., Trayanov, A., Balajl, V., Li, P., Yang, W., Hill, C. & da Silva, A. 2005. Design and implementation of components in the Earth system modeling framework. *Int. J. High Perf. Comp. Apps.* 19(3): 341–350.

Costa, J., Peterson, A.T. & Beard, C.B. 2002. Ecologic niche modelling and differentiation of populations of *Triatoma brasiliensis neiva*, 1911, the most important Chagas' disease vector in northeastern Brazil (hemiptcra, rcduviidae, triatominae). *Amer. J. Trop. Med. & Hyg.* 67: 516–520.

Craig, M.H., Snow, R.W. & le Sueur, D. 1999. A Climate-based distribution model of malaria transmission in sub-Saharan Africa. *Parasit. Today.* 15: 105–111.

Cutler, D.R., Edwards, T.C., Beard, K.H., Cutler, A. & Hess, K.T. 2007. Random forests for classification in ecology. *Ecology* 88(11): 2783–2792.

Daymet. 2012. Available from: http://www.daymet.org/ [Accessed 17th January 2012].

Davies, T.C. 2010. Geoscientific and environmental health issues in Africa. *Int. Geol. Rev.* 52(7-8): 873–897.

Depinay, J.-M.O., Mbogo, C.M., Killeen, G., Knols, B., Beier, J., Carlson, J., Dushoff, J., Billingsley, P., Mwambi, H., Githure, J., Toure, A.M. & McKenzie, F.E. 2004. A simulation model of African *Anopheles* ecology and population dynamics for the analysis of malaria transmission. *Malaria J.* 3: 29. doi:10.1186/1475-2875-3-29.

de Siqueira, M.F., Durigan, G., Junior, P.M. & Peterson, A.T. 2009. Something from nothing: Using landscape similarity and ecological niche modelling to find rare plant species. *J. Nat. Conserv.* 17: 25–32.

De'ath, G. & Fabricius, K.E. 2000. Classification and regression trees: A powerful yet simple technique for ecological data analysis. *Ecology* 81: 3178–3192.

De'ath, G. 2007. Boosted trees for ecological modelling and prediction. *Ecology* 88: 243–251.

Dietz, K., Molineaux, L. & Thomas, A.1974. A malaria model tested in the African savannah. *Bull. WHO* 50: 347–357.

Dole, R., Hoerling, M., Perlwitz, J., Eischeld, J., Pegion, P., Zhang, T., Quan, X.-W., Xu, T. & Murray, D. 2011. Was there a basis for anticipating the 2010 Russian heat wave. *Geophys. Res. Letts.* 38: doi:10.1029/2010GL046582.

Earn, D. 2008. A light introduction to modeling recurrent epidemics. In F. Brauer (ed.), *Mathematical epidemiology:* 3–17. New York: Springer Verlag.

Eichner, M., Diebner, H.H., Molineaux, L., Collins, W.E., Jeffery, G.M. & Dietz, K. 2001. Genesis, sequestration and survival of *Plasmodium falciparum* gametocytes: Parameter estimates from fitting a model to malaria therapy data. *Trans. Roy. Soc. Trop. Med. & Hyg.* 95: 497–501.

Elith, J., Burgman, M.A. & Regan, H.M. 2002. Mapping epistemic uncertainties and vague concepts in predictions of species distribution. *Ecol. Modell.* 157: 313–329.

Elith, J., Graham, C.H., Anderson, R.P, Dudik, M., Ferrier, S. Guisan, A., Hijmans, R.J., Huettmann, F., Leathwick, J.R., Lehmann, A., Li, J., Lohmann, L.G., Loiselle, B.A., Manion, G., Moritz, C., Nakamura, M., Yoshinori, N., Overton, J., McC.M., Peterson, T.A., Phillips, S.J., Richardson, K., Scachetti-Pereira, R., Schapire, R.E., Soberón, J., Williams, S., Wisz, M.S. & Zimmermann, N.E. 2006. Novel methods improve prediction of species' distributions from occurrence data. *Ecography* 29(2): 129–151.

Elith, J., Leathwick, J.R. & Hastie, T. 2008. A working guide to boosted regression trees. *J. Anim. Ecol.* 77: 802–813.

Elith, J. & Leathwick, J.R. 2009. Species distribution models: Ecological explanation and prediction across space and time. *Ann. Rev. Ecol. Evol. & System.* 40: 677–697.

Evangelista, P.H., Kumar, S., Stohlgren, T.J., Jarnevich, C.S., Crall, A.W., Norman III, J.B. & Barnett, D.T. 2008. Modelling invasion for a habitat generalist and a specialist plant species. *Divers. & Distrib.* 14: 808–817.

Evans, J.S., Murphy, M.A., Holden, Z.A. & Cushman, S.A. 2011. Modelling species distribution and change using random forest. In C.A. Drew, Y.F. Wiersma & F. Huettmann (eds.), *Predictive species and habitat modelling in landscape ecology: Concepts and applications:* 139–159. New York: Springer.

Farber, O. & Kadmon, R. 2003. Assessment of alternative approaches for bioclimatic modelling with special emphasis on the Mahalanobis distance. *Ecol. Model.* 160: 115–130.

Ferguson, C.D., Klein, W. & Rundle, J.B. 1998. Long-range earthquake fault models. *Computers in Physics* 12(1): 34–40.

Fielding, A.H. & Bell, J.F. 1997. A review of methods for the assessment of prediction errors in conservation presence/absence models. *Environ. Conserv.* 24: 38–49.

Fortin, M.J., Drapeau, P. & Legendre, P. 1989. Spatial auto-correlation and sampling design in plant ecology. *Vegetatio* 83(1–2): 209–22.

Frankenberger, J.R., Brooks, E.S., Walter, M.T., Walter, M.F. & Steenhuis, T.S. 1999. A GIS-based variable source area hydrology model. *Hydrolo. Process.* 13: 805–822.

Franklin, J. 2009. *Mapping species distributions: Spatial inference and prediction.* Cambridge: Cambridge UP.

Friedman, J.H. 2001. Greedy function approximation: A gradient boosting machine. *Ann. Stats.* 29(5): 1189–1232.

Friedman, J.H. & Meulman, J.J. 2003. Multiple additive regression trees with application in epidemiology. *Statis. Med.* 22: 1365–1381.

Frigg, R & Hartmann, S. 2006. Models in science. In Stanford encyclopaedia of philosophy. Stanford: Stanford UP: Available from: *http://plato.stanford.edu/entries/models-science/* [Accessed 26th December 2011].

Fuller, T., Morton, D.P. & Sarkar, S. 2008. Incorporating uncertainty about species' potential distributions under climate change into the selection of conservation areas with a case study from the arctic coastal plain of Alaska. *Biol. Conser.* 141: 1547–1559.

Furrer, E.M., Katz, R.W., Walter, M.D. & Furrer, R. 2010. Statistical modeling of hot spells and heat waves. *Climate Res.* 43: 191–205.

Georgopolos, P.G. & Lioy, P.J. 2006. From theoretical aspects of human exposure and dose assessment to computational model implementation. The modeling environment for total risk studies (MENTOR). *J. Toxic. & Environ. Health Part-B Critic. Revs.* 9(6): 457–483.

Georgopolos, P.G. 2008. A Multiscale approach for assessing the interactions of environmental and biological systems in a holistic health risk assessment framework. *Water, Air & Soil Pollut: Focus.* 8(1): 3–21.

Gobran, G.R. and Clegg, S. 1996. A conceptual model for nutrient availability in the mineral soil-root system. *Can. J. Soil Sci.* 76(2): 125–131.

GoldSim. 2012. Available from: http://www.goldsim.com/home/ [Accessed 26th March 2012].

Graham, M.H. 2003. Confronting multicollinearity in ecological multiple regression. *Ecol.* 84: 2809–2815.

Graham, C.H., Ferrier, S., Huettman, F., Moritz, C. & Peterson, A.T. 2004. New developments in museum-based informatics and applications in biodiversity analysis. *Trends Ecol. & Evol.* 19: 497–503.

Graham, C.H., Elith, J., Hijmans, R.J., Guisan, A., Peterson, A.T., Loiselle, B.A. & NCEAS. 2008. The influence of spatial errors in species occurrence data used in distribution models. *J. Appl. Ecol.* 45: 239–247.

Grinnell, J. 1917. The niche-relationships of the California Thrasher. *Auk* 34: 427–433.

Guisan, A. & Hofer, U. 2003. Predicting reptile distributions at the mesoscale: Relation to climate and topography. *J. Biogeog.* 30: 1233–1243.

Guisan, A., Edwards, T.C. & Hastie, T. 2002. Generalized linear and generalized additive models in studies of species distributions: Setting the scene. *Ecol. Modell.* 157: 89–100.

Guisan, A., Broennimann, O., Engler, R., Vust, M., Yoccoz, N.G., Lehmann, A. & Zimmermann, N.E. 2006. Using niche-based models to improve the sampling of rare species. *Conserv. Biol.* 20: 501–511.

Guisan, A., Graham, C.H., Elith, J., Huettmann, F. & NCEAS. 2007. Sensitivity of predictive species distribution models to change in grain size. *Divers. & Distrib.* 13: 332–340.

Hampe, A. 2004. Bioclimate envelope models: What they detect and what they hide. *Global Ecol. & Biogeo.* 13: 469–471.

Hansen, M.C., DeFries, R.S., Townshend, J.R.G., Carroll, M., Dimiceli, C. & Sohlberg, R.A. 2003. Global percent tree cover at a spatial resolution of 500 meters: First results of the MODIS vegetation continuous fields algorithm. *Earth Interac.* 7(paper 10): 1–15.

Hastie, T.J. & Tibshirani, R.J. 1986. Generalized additive models. *Stat. Sci.* 1: 297–318.

Hastie, T.J. & Tibshirani, R.J. 1990. *Generalized additive models.* London: Chapman & Hall.

Hay, S.I., Rogers, D.J., Shanks, G.D., Myers, M.F. & Snow, R.W. 2001. Malaria early warning in Kenya. *Trends in Parasit.* 17: 95–99.

Hernandez, P.A., Graham, C.H., Master, L.L. & Albert. D.L. 2006. The effect of sample size and species characteristics on performance of different species distribution modeling methods. *Ecography* 29(5): 773–785.

Hijmans, R.J. & Graham, C.H. 2006. The ability of climate envelope models to predict the effect of climate change on species distributions. *Global Change Biol.* 12(12): 2272–2281.

Hill, C., DeLuca, C., Balaji, V, Suarez, M. & DaSilva, A. 2004. The architecture of the Earth System Modeling Framework. *Comp. Sci. & Engin.* (Jan/Feb): 18–28.

Hirzel, A.H., Hausser, J. Chessel, D. & Perrin, N. 2002. Ecological-niche factor analysis: How to compute habitat-suitability maps without absence data? *Ecology* 83: 2027–2036.

Holt, A.C., Salkeld, D.J., Fritz, C.L., Tucker, J.R. & Gong, P. 2009. Spatial analysis of plague in California: Niche modelling predictions of the current distribution and potential response to climate change. *Int. J. Health Geogra.* 8: 1–14.

Hoppensteadt, F. 2011. *Mathematical methods for analysis of a complex disease.* Providence: AMS.

Hortal, J., Roura-Pascual, N., Sanders, N.J. & Rahbek, C. 2010. Understanding (insect) species distributions across spatial scales. *Ecography* 33: 51–53.

Hoshen, M.B. & Morse, A.P. 2004. A weather-driven model of malaria transmission. *Malaria J.* 3: 32. doi:10.1186/1475-2875- 3-32.

Hoshen, M.B. & Morse, A.P. 2005. A model structure for estimating malaria risk. In: W. Takken, P. Martens & R.J. Bogers (eds.), *Environmental change and malaria risk: Global and local implications.* 9: 41–50. Wageningen: Springer.

Hutchinson, G.E. 1957. Population studies – animal ecology and demography – concluding remarks. *Cold Spring Harbor symposia on quantitative biology* 22: 415–427.

Hutchinson, M.F., Nix, H.A., McMahon, J.P. & Ord, K.D. 1996. The development of a topographic and climate database for Africa. Proceedings of the 3rd international conference integrating GIS and environmental modeling, Santa Fe, New Mexico. Univ. Calif. Santa Barbara: NCGIA (Available on CDROM).

ICSU 2009. Health and wellbeing in a changing urban environment: A systems analysis approach. Contract report from the GeoUnions Joint Science Program Team for Health. Paris: ICSU.

Ishikawa, H., Ishii, A., Nagai, N., Ohmae, H., Harada, M., Suguri, S. & Leafasia, J. 2003. A mathematical model for the transmission of *Plasmodium vivax* malaria. *Parasit. Int.* 52: 81–93.

Jacob, D., van den Hurk, B.J.J.M., Andrae, U., Elgered, F.G.C., Graham, L.P., Jackson, S.D., Karstens, U., Koepken, C., Lindau, R., Podzun, R., Rockel, B., Rubel, F., Sass, B.H., Smith, R. & Yang, X. 2001. A comprehensive model intercomparison study investigating the water budget during the PIDCAP period. *Meteorol. & Atmosph. Phys.* **77**: 19–44.

Jaiswal, K.S., Wald, D.J., Earle, P.S., Porter, K.A. & Hearne, A. 2009. Earthquake casualty models within the USGS PROMPT Assessment of Global Earthquakes for Response (PAGER) system. Available from: http://earthquake.usgs.gov/learn/publications/pubs_tech.php [Accessed 28th March 2012].

Jarnevich, C.S. & Stohlgren. T.J. 2009. Near- term climate projections for invasive species distributions. *Biol. Invas.* 11: 1373–1379.

Jeffery, G.M., Young, M.D., Burgess, R.W. & Eyles, D.E. 1959. Early activity in sporozoite-induced *Plasmodium falciparum* infections. *Annals Trop. Med. & Parasit.* 53: 51–58.

Jetten, T.H., Martens, P. & Takken, W. 1996. Model simulations to estimate malaria risk under climate change. *J. Med. Ent.* 33: 361–371.

Jimenez-Valverde, A., Lobo, J.M. & Hortal, J. 2008. Not as good as they seem: The importance of concepts in species distribution modelling. *Divers. & Distrib.* 14: 885–890.

Johns Hopkins. 2012. Available from: http://www.malariasite.com/malaria/LifeCycle.htm [Accessed 19th April 2012].

Johnson, C.J. & Gillingham, M.P. 2004. Mapping uncertainty: Sensitivity of wildlife habitat ratings to expert opinion. *J. Appl. Ecol.* 41: 1032–1041.

Kadmon, R., Farber, O. & Danin, A. 2004. Effect of roadside bias on the accuracy of predictive maps produced by bioclimatic models. *Ecol. Appl.* 14: 401–413.

K-1 Model Developers. 2004. K-1 Coupled model (MIROC) Description. K-1: technical report 1. Center for Climate System Research. Tokyo: Tokyo Univ.

Kahneman, D. & Frederick, S. 2005. A model of heuristic judgement. In K.J. Holyoak & R.G. Morrison (eds.), *The Cambridge handbook of thinking and reasoning.* Chapter 12. New York: Cambridge UP.

Kermack, W. & McKendrick, A.G. 1927. A contribution to the mathematical theory of epidemics. *Proc. Roy. Soc. Lon.* 115: 700–721.

Killeen, G.F., McKenzie, F.E., Foy, B.D., Schieffelin, C., Billingsley, P.F. & Beier, J.C. 2000. A simplified model for predicting malaria entomologic inoculation rates based on entomologic and parasitologic parameters relevant to control. *Amer. J. Trop. Med. & Hyg.* 62: 535–544.

Kremen, C., Cameron, A., Moilanen, A., Phillips, S.J., Thomas, C.D., Beentje, H., Dransfield, J., Fisher, B.L., Glaw, F., Good, T.C., Harper, G.J., Hijmans, R.J., Lees, D.C., Louis, E., Nussbaum, R.A., Raxworthy, C.J., Razafimpahanana, A., Schatz, G.E., Vences, M., Vieites, D.R., Wright, P.C. & Zjhra, M.L. 2008.

Aligning conservation priorities across taxa in Madagascar with high-resolution planning tools. *Science* 320: 222–226.

Kumar, S. & Stohlgren, T.J. 2009. Maxent modeling for predicting suitable habitat for threatened and endangered tree Canacomyrica monticola in New Caledonia. *J. Ecol. & Nat. Environ.* 1(4): 94–98.

Kumar, S., Spaulding, S.A., Stohlgren, T.J., Hermann, K.A., Schmidt, T.S. & Bahls, L.L. 2009. Potential habitat distribution for the freshwater diatom *Didymosphenia geminata* in the continental US. *Front. Ecol. & Environ.* 7(8): 415–420.

Kysely, J. 2002. Probability estimates of extreme temperature events: Stochastic modelling approach vs. extreme value distributions. *Studia Geophys. et Geodaet.* 46: 93–112.

Laneri, K., Bhadra, A., Ionides, E., Bouma, M., Dhiman, R., Yadav, R. & Pascual, M. 2010 Forcing versus feedback: Epidemic malaria and monsoon rains in Northwest India. *PLos Comput. Biol.* 6(9).

Leathwick, J.R. & Austin. M.P. 2001. Competitive interactions between tree species in New Zealand's old-growth indigenous forests. *Ecology* 82: 2560–2573.

Le Menach, A.L., Takala, S., McKenzie, F.E., Perisse, A., Harris, A. Flahault A. & Smith, D.L. 2007. An elaborated feeding cycle model for reductions in vectorial capacity of nightbiting mosquitoes by insecticide-treated nets. *Malaria J.* 6: 10. doi:10.1186/1475-2875-6-10.

Levine, R.S, Peterson A., Yorita, K.L., Carroll, D., Damon, I.K. & Reynolds, M.G. 2007. Ecological niche and geographic distribution of human monkeypox in Africa. *PLoS ONE* 2(1): e176. doi:10.1371/journal.pone.0000176.

LMM. 2012. Available from: http://www.impetus.uni-koeln.de/malaris/images/lmm_scheme.png [Accessed 3rd April 2012].

Lobo, J.M., Jimenez-Valverde, A. & Hortal, J. 2010. The uncertain nature of absences and their importance in species distribution modelling, *Ecography* 33: 103–114.

Lotka, A.J. 1910 Contribution to the theory of periodic reaction. *J. Phys. Chem.* 14 (3): 271–274.

Lotka, A.J. 1920. Analytical note on certain rhythmic relations in organic systems. *Proc. Natl. Acad. Sci. US.* 6: 410–415.

Lotka, A.J. 1925. *Elements of physical biology.* Baltimore: Williams & Wilkins.

Mackey, B.G. & Lindenmayer, D.B. 2001. Towards a hierarchical framework for modelling the spatial distribution of animals. *J. Biogeo.* 28: 1147–1166.

MalaRis. 2012. Available from: http://www.impetus.uni-koeln.de/malaris/malariamodel_en.html [Accessed 3rd April 2012].

Marmion, M., Parviainen, M., Luoto, M.R., Heikkinen, K. & Thuiller, W. 2009. Evaluation of consensus methods in predictive species distribution modelling. *Divers. & Distrib.* 15: 59–69.

Martcheva, M. & Hoppensteadt, F. 2010. India's approach to eliminating *Plasmodium falciparum* malaria: A modeling perspective. *J. Biol. Syst.* 18(4): 867–891.

Martens,W.J.M., Jetten, T.H., Rottmans, J. & Niessen, L.W. 1995. Climate change and vector-borne diseases: A global modelling perspective. *Global Environ. Change.* 5: 195–209.

Masters, J. 2012. Available from: http://www.wunderground.com/hurricane/models.asp [Accessed 1st April 2012.

Masuoka, P., Klein, T.A., Kim, H.C., Claborn, D.M., Achee, N., Andre, R., Chamberlin, J., Small, J., Anyamba, A., Lee, D.K., Yi, S.H., Sardelis, M., Ju, Y.R. & Grieco, J. 2010. Modelling the distribution of *Culex tritaeniorhynchus* to predict Japanese encephalitis distribution in the Republic of Korea. *Geospat. Health* 5: 45–57.

McCullagh, P. & Nelder, J.A. 1989. *Generalized linear models.* London: Chapman & Hall.

McCune, B. 2006. Non-parametric habitat models with automatic interactions. *J. Veg. Sci.* 17: 819–830.

McKenzie, F.E., Wong, R.C. & Bossert, W.H. 1998. Discrete-event simulation models of *Plasmodium falciparum* malaria. *Simulation* 71: 250–261.

McKenzie, F.E., Wong, R.C. & Bossert, W.H. 1999. Discrete-event models of mixed-phenotype *Plasmodium falciparum* malaria. *Simulation.* 73: 213–217.

McMichael, A.J. 1997. Integrated assessment of potential health impact of global environmental change: Prospects and limitations. *Environ. Modell. Assess.* 2: 129–137.

Meentemeyer, R., Anacker, B.L., Mark, W. & Rizzo, D.M. 2008. Early detection of emerging forest disease using dispersal estimation and ecological niche modelling. *Ecol. Appl.* 18: 377–390.

Menke, S.B., Holway, D.A., Fisher, R.N. & Jetz, W. 2009. Characterizing and predicting species distributions across environments and scales: Argentine ant occurrences in the eye of the beholder. *Global Ecol. & Biogeo.* 18: 50–63.

Meynard, C.N. & Quinn, J.F. 2007. Predicting species distributions: A critical comparison of the most common statistical models using artificial species. *J. Biogeo.* 34: 1455–1469.

Miller, A.J. & Knouft, J.H. 2006. GIS-based characterization of the geographic distributions of wild and cultivated populations of the Mesoamerican fruit tree *Spondias purpurea* (Anacardiaceae). *Amer. J. Bot.* 93: 1757–1767.

Mishra, A.K. & Singh, V.P. 2011. Drought modeling: A review. *J. Hydrol.* 403(1–2): 157–175.

Mishra, S., Fisman, D. & Boily M. 2011. The ABC of terms used in mathematical models of infectious diseases. *J. Epidemiol Comm. Health.* 65: 87–94.

Mitchell, C.E., Agrawal, A.A., Bever, J.D., Gilbert, G.S., Hufbauer, R.A., Klironomos, J.N., Maron, J.L., Morris, W.F., Parker, I.M., Power, A.G., Seabloom, E.W., Torchin, M.E. & Vazquez, D.P. 2006. Biotic interactions and plant invasions. *Ecol. Lett.* 9: 726–740.

Moffett, A., Shackelford, N. & Sarkar, S. 2007. Malaria in Africa: Vector species' niche models and relative risk maps. *PLoS One* 2: e824.

Moore, P.S. 1992. Meningococcal meningitis in sub-Saharan africa: A model for the epidemic process. *Clinic. Infect. Dis.* 14: 515–525.

Morain, S.A. 2008. Improving public health services through space technology and spatial information systems. In R. Ragaini (ed.), *International seminar on nuclear war and planetary emergencies, 40th Session*: 199–218. Singapore: World Scientific.

Morisette, J.T., Richardson, A.D., Knapp, A.K., Fisher, J.I., Graham, E.A., Abatzoglou, J., Wilson, B.E., Breshears, D.D., Henebry, G.M., Hanes, J.M. & Liang, L. 2009. Tracking the rhythm of the seasons in the face of global change: Phenological research in the 21st century. *Front. Ecol. & Environ.* 7: 253–260.

Moritz, C., Hoskin, C.J., MacKenzie, J.B., Phillips, B.L., Tonione, M., Silva, N., VanDerWal, J., Williams, S.E. & Graham, C.H. 2009. Identification and dynamics of a cryptic suture zone in tropical rainforest. *Proc. Roy. Soc.: B-Biol. Sci.* 276: 1235–1244.

Moore, C.G. 1999. *Aedes albopictus* in the United States: Current status and prospects for further spread. *J. Amer. Mosq. Cont. Assoc.* 15(2): 221–227.

Morse, A.P., Doblas-Reyes, F.J., Hoshen, M.B., Hagedorn, R. & Palmer, T.N. 2005. A forecast quality assessment of an end-to-end probabilistic multi-model seasonal forecast system using a malaria model. *Tellus* 57A: 464–475.

Mouton, A.M., De Baets, B. & Goethals, P.L.M. 2010. Ecological relevance of performance criteria for species distribution models. *Ecol. Modell.* 221: 1995–2002.

Müller, R. 2010. *The concept of model: Definitions and types.* Switzerland: Müller Science. Available from: http://www.muellerscience.com/ENGLISH/Theconceptofmodel.definitions.htm [Accessed 26th December 2011].

Murray, K. & Conner, M.M. 2009. Methods to quantify variable importance: Implications for the analysis of noisy ecological data. *Ecology* 90(2): 348–55.

Nájera, J.A. 1974. A critical review of the field application of a mathematical model of malaria eradication. *Bull. WHO* 50: 449–457.

Nedelman, J. 1984. Inoculation and recovery rates in the malaria model of Dietz, Molineaux, and Thomas. *Mathemat. Biosci.* 69: 209–233.

Nedelman, J. 1989. Gametoctaemia and infectiousness in Falciparum malaria: Observations and models. *Advan. Dis. Vect. Res.* 6: 59–89.

Neter, J., Kutner, M.H., Nachtsheim, C.J. & Wasserman, W. 1996. *Applied linear statistical models: Regression, analysis of variance, and experimental designs.* 4th ed. Chicago: Irwin.

Newbold, T. 2010. Applications and limitations of museum data for conservation and ecology with particular attention to species distribution models. *Prog. Phys. Geog.* 34: 3–22.

NIH. 2012. Available from: http://www.niaid.nih.gov/topics/malaria/pages/lifecycle.aspx [Accessed 19th April 2012].

Nix, H.A. 1986. A biogeographic analysis of Australian elapid snakes. In: R. Longmore (ed.), *Australian flora and fauna.* Series 8. Canberra: Australian Government Publishing Service.

O'Kuinghttons, R., Oehmke, R. & DeLuca, C. 2010. Earth system modeling framework. *Workshop on coupling technologies for Earth system modelling: Today and tomorrow.* Available from: https://verc.enes.org/models/software-tools/oasis/general-information/events/workshop-on-coupling-technologies-for-earth-system-modelling-today-and-tomorrow-1/talks/ESMF.pdf [Accessed 5th April 2012].

Omumbo, J.A., Hay, S.I., Snow, R.W., Tatem, A.J. & Rogers, D.J. 2005. Modelling malaria risk in East Africa at high-spatial resolution. *Trop. Med. & Int. Health.* 10: 557–566.

Oreskes, N., Shrader-Frechette, K. & Belitz, K. 1994. Verification, validation, and confirmation of numerical models in the Earth sciences. *Science* 263(5147): 641–646.

Palmer, T.N., Alessandri, A., Andersen, U., Cantelaube, P., Davey, M., Délécluse, P., Déqué, M., Diez, E., Doblas-Reyes, F.J., Feddersen, H., Graham, R., Gualdi, S., Guérémy, J.-F., Hagedorn, R., Hoshen, M.,

Keenlyside, N., Latif, M., Lazar, A., Maisonnave, E., Marletto, V., Morse, A.P., Orfila, B., Rogel, P., Terres, J.-M. & Thomson, M.C. 2004. Development of a European multi-model ensemble system for seasonal to inter-annual prediction (DEMETER). *Bull. Amer. Meteorol. Soc.* 85: 853–872.

Papes, M. & Gaubert, P. 2007. Modelling ecological niches from low numbers of occurrences: Assessment of the conservation status of poorly known viverrids (Mammalia: Carnivora) across two continents. *Divers. & Distrib.* 13(6): 890–902.

Papoulis, A. & Unnikrishna, P.S. 2001. *Probability, random variables and stochastic processes.* New York: McGraw-Hill.

Park, S., Choe, A., Park, M. & Lee, E. 2008. Asian dust aerosol models (ADAM). *Asia Pacific Tech. Monitor* (Nov–Dec): 24–29.

Parolo, G., Rossi, G. & Ferrarini, A. 2008. Toward improved species niche modelling: *Arnica montana* in the Alps as a case study. *J. Appl. Ecol.* 45: 1410–1418.

Patel, V.L., Arocha, J.F. & Zhang, J.Z. 2005. Thinking and reasoning in medicine. In K.J. Holyoak & R.G. Morrison (eds.), *The Cambridge handbook of thinking and reasoning*: Chapter 30. New York: Cambridge UP.

Patz, J.A., Strzepec, K., Lele, S., Hedden, M., Green, S., Noden, B.S. Hay, I., Kalkstein, L. & Beier, J.C.1998. Predicting key malaria transmission factors, biting and entomological inoculation rates, using modelled soil moisture in Kenya. *Trop. Med. & Int. Health.* 3: 818–827.

Pawar, S., Koo, M.S., Kelley, C., Ahmed, M.F., Chaudhuri, S. & Sarkar, S. 2007. Conservation assessment and prioritization of areas in Northeast India: Priorities for amphibians and reptiles. *Biol. Conserv.* 136: 346–361.

Pearce, J. & Ferrier, S. 2000. Evaluating the predictive performance of habitat models developed using logistic regression. *Ecol. Modell.* 133(3): 225–245.

Pearson, R.G., Dawson, T.P., Berry, P.M. & Harrison. P.A. 2002. Species: A spatial evaluation of climate impact on the envelope of species. *Ecol. Modell.* 154: 289–300.

Pearson, R.G. & Dawson, T.P. 2003. Predicting the impacts of climate change on the distribution of species: Are climate envelope models useful? *Global Ecol. & Biogeo.* 12: 361–371.

Pearson, R.G., Thuiller, W., Araujo, M.B., Martinez-Meyer, E., Brotons, L., McClean, C., Miles, L., Segurado, P., Dawson, T.P. & Lees, D.C. 2006. Model-based uncertainty in species range prediction. *J. Biogeo.* 33: 1704–1711.

Pearson, R.G., Raxworthy, C.J., Nakamura M. & Peterson, A.T. 2007. Predicting species distribution from small numbers of occurrence records: A test case using cryptic geckos in Madagascar. *J. Biogeo.* 34: 102–117.

Peterson, A.T. 2003. Predicting the geography of species' invasions via ecological niche modelling. *Quart. Rev. Biol.* 78: 419–433.

Peterson, A.T., Navarro-Siguenza, A.G. & Benitez-Diaz, H. 1998. The need for continued scientific collecting: A geographic analysis of Mexican bird specimens. *Ibis* 140: 288–294.

Peterson, A.T. & Shaw, J. 2003. Lutzomyia vectors for cutaneous leishmaniasis in southern Brazil. Ecological niche models, predicted geographic distributions, and climate change effects. *Int. J. Parasit.* 33: 919–931.

Peterson, A.T. & Nakazawa, Y. 2008. Environmental data sets matter in ecological niche modelling: An example with *Solenopsis invicta* and *Solenopsis richteri*. *Global Ecol. & Biogeo.* 17: 135–144.

Peterson, A.T., Robbins, A., Restifo, R., Howell, J. & Nasci, R. 2008. Predictable ecology and geography of West Nile virus transmission in the central United States. *J. Vect. Ecol.* 33: 342–352.

Phillips, S.J. no date. Available from: ncep.amnh.org/linc/linc_download.php?component_id=39 [Accessed 28th March 2012.]

Phillips, S.J. 2008. Transferability, sample selection bias and background data in presence-only modelling: A response to Peterson *et al.*, 2007. *Ecography* 31: 272–278.

Phillips, S.J., Dudik, M. & Schapire, R.E. 2004. A maximum entropy approach to species distribution modelling. *Proc. 21st Intern. Conf. Mach. Learning.* Banff: 655–662.

Phillips, S.J., Anderson, R.P. & Schapire, R.E. 2006. Maximum entropy modelling of species geographic distributions. *Ecol. Modell.* 190: 231–259.

Phillips, S.J., Dudik, M., Elith, J., Graham, C.H., Lehmann, A., Leathwick, J. & Ferrier, S. 2009. Sample selection bias and presence-only distribution models: Implications for background and pseudo-absence data. *Ecol. Appli.* 19: 181–197.

Pimentel, D., Lach, L., Zuniga, R. & Morrison, D. 2000. Environmental and economic costs of nonindigenous species in the United States. *Biosci.* 50: 53–65.

Pimentel, D., Zuniga, R. & Morrison, D. 2005. Update on the environmental and economic costs associated with alien-invasive species in the United States. *Ecol. Econ.* 52: 273–288.

Poincaré , H. 1993. *New Methods of Celestial Mechanics*: 1(Ch. 3) 63–69. New York: Amer. Instit. Phys.

Porphyre, T., Bicout, D.J. & Sabatier, P. 2005. Modelling the abundance of mosquito vectors versus flooding dynamics. *Ecol. Modell.* 183: 173–181.

Prasad, A.M., Iverson, L.R. & Liaw, A. 2006. Newer classification and regression tree techniques: Bagging and random forests for ecological prediction. *Ecosys.* 9: 181–199.

Randall, D.A., Wood, R.A., Bony, S., Colman, R., Fichefet, T., Fyfe, J., Kattsov, V., Pitman, A., Shukla, J., Srinivasan, J., Stouffer, R.J., Sumiand, A. & Taylor, K.E. 2007. Climate models and their evaluation. In S. Solomon, D. Qin, M. Manning, Z. Chen, M. Marquis, K.B. Averyt, M. Tignor & H.L. Miller (eds.), *Climate change 2007: The physical science basis.* IPCC Fourth Assessment Report. Cambridge: Cambridge UP.

Regan, H.M., Colyvan, M. & Burgman, M.A. 2002. A taxonomy and treatment of uncertainty for ecology and conservation biology. *Ecol. Appl.* 12: 618–628.

Ritchie, E.G., Martin, J.K. Johnson, C.N. & Fox, B.J. 2009. Separating the influences of environment and species interactions on patterns of distribution and abundance: Competition between large herbivores. *J. Anim. Ecol.* 78: 724–731.

Rogers, D.J., Randolph, S.E., Snow, R.W. & Hay, S.I. 2002. Satellite imagery in the study and forecast of malaria. *Nature* 415: 710–715.

Ropella, G.E, Hunt, C.A. & Nag, D.A. 2005. Using heuristic models to bridge the gap between analytic and experimental models in biology. *38th Annual simulation symposium*: Spring simulation multiconference, San Diego.

Ross, R. 1911. *The Prevention of Malaria*. London: Murray

Sanderson, E.W., Jaiteh, M., Levy, M.A., Redford, K.H., Wannebo, A.V. & Woolmer, G. 2002. The human footprint and the last of the wild. *Biosci.* 52: 891–904.

Schneider, A., Friedl, M.A. & Potere, D. 2010. Mapping global urban areas using MODIS 500-m data: New methods and datasets based on 'urban ecoregions'. *Rem. Sens. Environ.* 114: 1733–1746.

Segurado, P., Araujo, M.B. & Kunin, W.E. 2006. Consequences of spatial autocorrelation for niche-based models. *J. Appl. Ecol.* 43: 433–444.

Shamsad, A., Leow, C.S., Ramlah, A., Wan Hussin, M.A. & Sanusi, A. 2008. Applications for a soil loss estimation and nutrient loading for Malaysian conditions. *Int. J. Appl. Earth Obs. Geoinf.* 10(3): 239–252.

Shrestha, S., Bastola, S., Babel, M.S., Dulal, K.N., Magome, J., Hapuarachchi, H.A.P., Kazama, F., Ishidaira, H. & Takeuchi, K. 2007. The assessment of spatial and temporal transferability of physically based distributed model parameters in different physiographic regions of Nepal. *J. Hydrol.* 347: 153–172.

Sijing, W & Marinos, P. 1997. Geologic risk models. In W. Sijing & P. Marinos (eds.), *Proc. 30th Int. Geol. Cong:* 30. The Netherlands: Ridderprint.

Sinclair, S.J., White, M.D. & Newell, G.R. 2010. How useful are species distribution models for managing biodiversity under future climates? *Ecol. & Soc.* 15: 1–13.

Smith, T., Killeen, G.F., Maire, N., Ross, A., Molineaux, L., Tediosi, F., Hutton, G., Utzinger, J., Dietz, K. & Tanner, M. 2006. Mathematical modeling of the impact of malaria vaccines on the clinical epidemiology and natural history of *Plasmodium falciparum* malaria: Overview. *Amer. J. Trop. Med. & Hyg.* 75(Supp2): 1–10.

Snow, R.W., Gouws, E., Omumbo, J., Rapuoda, B., Craig, M.H., Tanser, F.C., Sueur, D. le & Ouma, J. 1998. Models to predict the intensity of *Plasmodium falciparum* transmission: Applications to the burden of disease in Kenya. *Trans. Roy. Soc. Trop. Med. & Hyg.* 92: 601–606.

Soberón, J. & Peterson, A.T. 2005: Interpretation of models of fundamental ecological niche and species' distributional areas. *Biodivers. Inform.* 2: 1–10.

Soberón, J.M., Llorente, J.B. & Oñate, L. 2000. The use of specimen-label databases for conservation purposes: An example using Mexican Papilionid and Pierid butterflies. *Biodivers. & Conserv.* 9(10): 1441–1466.

Soil Erosion Site. 2012. Available from: http://soilerosion.net/doc/models_menu.html [Accessed 1st April 2012].

Sterman, J.D. 2006. Learning from evidence in a complex world. *Amer J. Pub. Health* 96(3): 505–514.

Stockwell, D.R.B. & Noble, I.R. 1992. Induction of sets of rules from animal distribution data: A robust and informative method of data analysis. *Math. & Comp. Simul.* 33: 385–390.

Stockwell, D.R.B. & Peters, D. 1999. The GARP modelling system: Problems and solutions to automated spatial prediction. *Int. J. Geogra. Inform. Sci.* 13(2): 143–158.

Stockwell, D.R.B. & Peterson, A.T. 2002. Effects of sample size on accuracy of species distribution models. *Ecol. Modell.* 148(1): 1–13.

Stockwell, D.R.B., Beach, J.H., Stewart, A., Vorontsov, G., Vieglais, D. & Pereira, R.S. 2006. The use of the GARP genetic algorithm and Internet grid computing in the Lifemapper world atlas of species biodiversity. *Ecol. Modell.* 195: 139–145.

Stohlgren, T.J. & Schnase, J. 2006. Biological Hazards: What we need to know about invasive species. *Risk Assess. J.* 26: 163–173.

Stohlgren, T.J., Ma, P., Kumar, S., Rocca, M., Morisette, J.T., Jarnevich, C.S. & Benson, N. 2010a. Ensemble habitat mapping of invasive plant species. *Risk Anal.* 30: 224–235.

Stohlgren, T.J., Jarnevich, C.S. & Giri, C.P. 2010. Modeling the human invader in the United States: *J. Appl. Rem. Sens.* 4(1): cit. # 043509. Available from: http://dx.doi.org/10.1117/1.3357386 [Accessed 28th March 2012].

Swets, J.A. 1988. Measuring the accuracy of diagnostic systems. *Science* 240(4857): 1285–1293.

Tan, B., Morisette, J.T., Wolfe, R.E., Gao, F., Ederer, G.A., Nightingale, J. & Pedelty, J.A. 2008. Vegetation phenology metrics derived from temporally smoothed and gap-filled MODIS data. *IGARSS* 3: 593–595.

Tanaka, T.Y., Orito, K., Sekiyama, T.T., Shibata, K. & Chiba, M. 2003. MASINGAR: A global tropospheric aerosol chemical transport model coupled with MRI/JMA98 GCM: model description. *Papers in Met. & Geophys.* 53(4): 119–138.

Thuiller, W., Richardson, D.M. & Midgley, G.F. 2007. Will climate change promote alien plant invasions? In W. Nentwig (ed.), *Biological Invasions*: 197–211. Berlin: Springer-Verlag.

Thuiller, W. 2003. BIOMOD – optimizing predictions of species distributions and projecting potential future shifts under global change. *Global Change Biol.* 9: 1353–1362.

Thuiller, W., Lafourcade, B., Engler, R., & Araujo, M.B. 2009. BIOMOD- a platform for ensemble forecasting of species distributions. *Ecography* 32: 369–373.

Thomson, M.C., Doblas-Reyes, F.J., Mason, S.J., Hagedorn, R., Connor, S.J., Phindela, T., Morse, A.P. & Palmer, T.N. 2006. Malaria early warnings based on seasonal climate forecasts from multi-model ensembles. *Nature* 439: 576–579.

Turner, W., Spector, S., Gardiner, N., Fladeland, M., Sterling, E. & Steininger, M. 2003. Remote sensing for biodiversity science and conservation. *Trends Ecol. & Evol.* 18: 306–314.

USEPA. 2012a. Available from: http://cfpub.epa.gov/crem/knowledge_base /crem_report.cfm?deid =74937 [Accessed 31st March 2012].

USEPA. 2012b. Available from: http:www.epa.gov/ttn/fera/human_hem.html [Accessed 1st April 2012].

USEPA. 2012c. Available from: http://www.epa.gov/ttn/fera/human_hapem.html [Accessed 1st April 2012].

USGS. 2011. Water resources alphabetical software. Available from: http://water.usgs.gov/software/lists/alphabetical [Accessed 26th March 2012].

USGS. 2012. Available from: http://earthquake.usgs.gov/research/modelling/ [Accessed 26th March 2012].

Vaclavik, T. & Meentemeyer, R.K. 2009. Invasive species distribution modelling (ISDM): Are absence data and dispersal constraints needed to predict actual distributions? *Ecol. Modell.* 220: 3248–3258.

VanDerWal, J., Shoo, L.P., Graham, C. & William, S.E. 2009. Selecting pseudo-absence data for presence-only distribution modelling: How far should you stray from what you know? *Ecol. Modell.* 220: 589–594.

Veloz, S.D. 2009. Spatially autocorrelated sampling falsely inflates measures of accuracy for presence-only niche models. *J. Biogeo.* 36: 2290–2299.

Venette, R.C., Kriticos, D.J., Magarey, R.D., Koch, F.H., Baker, R.H.A., Worner, S.P., Raboteaux, N.N.G., McKenney, D.W., Dobesberger, E.J., Yemshanov, D., De Barro, P.J., Hutchison, W.D., Fowler, G., Kalaris, T.M. & Pedlar, J. 2010. Pest risk maps for invasive alien species: A roadmap for improvement. *Biosci.* 60: 349–362.

Verbyla, D.L. & Litvaitis. J.A. 1989. Resampling methods for evaluating classification accuracy of wildlife habitat models. *Environ. Manage.* 13: 783–787.

Verhulst, P.H. 1838. Notice sur la loi que la population poursuit dans son accroissement. *Corresp. Mathémat. et physiq.* 10: 113–121.

Volterra, V. 1926. Variazioni e fluttuazioni del numero d'individui in specie animali conviventi. *Mem. Acad. Lincei Roma* 2: 31–113.

WHO EMRO RBM. 2012. Available from: http://www.emro.who.int/rbm/aboutmalaria-quickoverview .htm [Accessed 27th April 2012].

Worrall, E., Connor, S.J. & Thomson, M.C., 2007. A model to simulate the impact of timing, coverage and transmission intensity on the effectiveness of indoor residual spraying (IRS) for malaria control. *Trop. Med. & Int. Health.* 12: 75–88.

Wisz, M.S., Hijmans, R.J., Li, J., Peterson, A.T., Graham, C.H., Guisan, A. & NCEAS. 2008. Effects of sample size on the performance of species distribution models. *Diver. & Distrib.* 14: 763–773.

Xu, R., Huang, H., Luo, L., & Luo, H. 2010. Case study of distributed hydrologic model for soil and nutrient loss estimation in small watershed. doi: 10.1061/(ASCE)HZ.1944-8376.0000016.

Yee, T.W. & Mitchell, N.D. 1991. Generalized additive-models in plant ecology. *J. Veget. Sci.* 2(5): 587–602.

Zhou, S.J. 2006. Coupling climate models with the Earth System Modeling Framework and the Common Component Structure. *Concurrency Computat.: Pract. Exper.* 18: 203–213.

Environmental Tracking for Public Health Surveillance – Morain & Budge (eds)
© 2013 Taylor & Francis Group, London, ISBN 978-0-415-58471-5

Chapter 8

Early warning systems

P. Ceccato[1] & S.J. Connor[2] (Auth./eds.)
[1]*The International Research Institute for Climate and Society, The Earth Institute, Columbia University, Palisades, NY, US*
[2]*School of Environmental Sciences, University of Liverpool, Liverpool, UK*

ABSTRACT: This chapter presents the framework to develop advanced early warning systems (EWS) for human and animal health. Based on a malaria EWS case study in Africa, the chapter discusses the needs for developing and implementing effective EWS for several vector-borne diseases affected by climate and environmental factors. Combination of vulnerability assessment, climate forecast, climate monitoring based on remote sensing products and surveillance data compose the basis for creating an EWS. The remaining challenges to get the knowledge into practice and sustain it where it is needed are discussed to improve early warning and early action systems.

1 INTRODUCTION

Good health is one of the primary aspirations of human social development. Certain communicable and non-communicable diseases are associated with particular environmental and climate conditions. Droughts may lead to malnutrition, dust storms and smog can cause respiratory illnesses, and algal blooms contaminate seafood. Emerging and re-emerging infectious diseases, of both humans and animals, can spread wherever and whenever ecosystems change. Early warning systems for human and animal diseases are available in many countries. They consist of monitoring and forecasting environmental and climate conditions that often are harbingers of disease outbreaks. However, to be operational such systems must be integrated into a decision/action framework.

In sub-Saharan Africa, the greatest burden of disease morbidity and related mortality stems primarily from infectious diseases. Specifically, HIV-AIDS, Tuberculosis and Malaria, as a group, have come to be seen as a significant constraint to Africa's development prospects and massive investment in health services and control programs have been made in the most affected countries. WHO reports that ramped-up investments in malaria control over the last few years are having a significant impact (WHO 2010). However, the fact that these gains are considered to be fragile underscores the need to more clearly understand the roles that socioeconomic and environmental factors, including climate, play in the dynamics of Malaria.

Spatial information derived from EO sensor data or models is playing an increasingly important role in understanding the relationship between health and environmental factors, as well as locating and forecasting disease outbreaks. The ease of availability means they hold particular potential for efficient monitoring and forecasting of human and animal diseases, informing policies and interventions aimed at better controlling these diseases.

2 ELEMENTS OF MALARIA EARLY WARNING SYSTEMS

Malaria is a major public health problem in Africa. It is endemic throughout most of sub-Saharan Africa and its control is recognized as critical to achieving the Millennium Development Goals

(MDGs). The global strategic plan for roll back malaria states that *six out of eight Millennium Development Goals can only be reached with effective malaria control in place* (RBM 2005). Malaria is used here as the primary disease to illustrate how early warning systems can be used to inform health services in endemic and epidemic prone countries. Endemic and epidemic malaria are sometimes referred to as 'stable' and 'unstable' malaria respectively.

2.1 *Endemic malaria*

The greatest burden of malaria in Africa is borne by populations in regions where the disease pathogen is perennially present in the community. In these regions, the environment is conducive to interactions between the Anopheles mosquito, malaria parasites and human hosts because they contain surface water in which mosquitoes can lay their eggs, humid conditions which facilitate adequately long adult mosquito survival, and relative warmth which allows both the mosquito and the malaria parasite to develop rapidly. In addition, housing quality is generally poor and offers little protection from human-mosquito interaction. When malaria control measures are inadequate, as is the case in much of sub-Saharan Africa, then the disease distribution is closely linked with seasonal patterns of the climate and local environment. Those most vulnerable to endemic malaria are young children who have yet to acquire immunity to the disease, pregnant women whose immunity is reduced during pregnancy, and non-immune migrants or travellers.

2.1.1 *Risk maps and seasonal calendars for endemic malaria control planning*
In the absence of high-quality epidemiological data on malaria distribution in Africa, climate information has been used to develop malaria risk maps illustrating the boundaries of climatic suitability for endemic transmission. The best known of these are produced by the Pan-African based MARA Collaboration (MARA/ARMA 1998; Craig *et al.*, 1999). More recently malaria suitability maps have been produced in an online interactive format to enable graphic temporal information (seasonality) to be queried and displayed along with the spatial information (Grover-Kopec *et al.*, 2006). Climate suitability maps for malaria transmission are available from the International Research Institute (IRI 2012a). Seasonal information allows malaria control calendars to be developed. Such products enable health services to focus control activities such as drug procurement and anti-vector spraying more appropriately. Seasonal calendars may also help avoid misdiagnosis and inappropriate drug treatment, both of which are vital to reducing the development rate of parasite-drug resistance.

2.1.1.1 Inter-decadal variability and longer-term trends
We need to be aware however of some of the shortfalls in the climate data being used in such products. One climatology data set for Africa used in several mapping products, including those of MARA, is comprised of continental surfaces of temperature and rainfall compiled from interpolated observations collected during the period 1920–1980 (Hutchinson *et al.*, 1996). While a sixty-year data set is valuable for developing a climatology, it does not fully capture the more recent trends, such as the Sahelian drought, known to influence current disease distribution (Thomson *et al.*, 2004a; Thomson *et al.*, 2004b). Resulting disease distribution maps may therefore miss-represent the current spatial distribution. In the online products described above (Grover-Kopec *et al.*, 2006) the climatology is developed from observations over the period 1951–2000 (Mitchell *et al.*, 2004). This partly takes care of the problem of relating climate to more recent disease distributions.

However, the number of observations available from Africa to produce this climatology has been in decline over the past twenty years and this raises questions about the employed interpolation methods and the adequacy of the data (e.g. should the interpolation only be applied where there is an adequate number of observing stations?). Integrating temperature into disease risk models poses a particular challenge. This is very important in the highlands of east and southern Africa where temperature limits the extrinsic rate of malaria (and many other) parasite development. A fierce debate exists in the literature on the contribution of recently increasing temperatures to epidemic malaria in the highlands. The debate is confounded by the poor understanding that many health researchers have on the correct use and interpretation of climate information products and, conversely, the limited understanding of disease transmission dynamics within the climate community. Thus, caution needs to be exercised when national and sub-national maps are extracted from these types of continental-scale products because the results could be very misleading (Pascual *et al.*, 2006).

There are, however, considerably more observed data, and more recent data, in national meteorological service archives than is available to the global-level archives from which these climatology products are derived. Collating and incorporating these additional data into national climatology products would be highly beneficial to national development sectors, including health. Just recently, a project funded by Google.org and implemented by the IRI has developed an improved ten kilometre station and satellite merged rainfall product for Ethiopia demonstrating the possibility to enhance rainfall estimates based on 300 rain gauges stations. The Ethiopia enhanced national climate time series products are currently available from the National Meteorology Agency website (NMA 2012). Ideally, the development of national and sub-national distribution maps for malaria and other climate sensitive diseases should utilize these data. This is true also of non-health applications such as farming systems and food security zone maps which rely heavily on interpolated climate information. Otherwise, national health services are being supplied with sub-optimal information.

2.2 *Epidemic malaria*

Epidemic malaria tends to occur along the geographical margins of endemic regions, when the conditions supporting the equilibrium between the human, parasite and mosquito vector populations are disturbed. This often leads to a sharp but temporary increase in disease incidence. More than 124 million Africans live in such areas, and experience epidemics causing around twelve million malaria episodes and up to 310,000 deaths annually (Worrall *et al.*, 2004). All age groups are vulnerable to epidemic malaria because exposure to malaria is infrequent in these regions and, therefore, little acquired immunity to this life threatening disease is developed (Kizewski & Teklehaimanot 2004).

Epidemics can occur for a variety of reasons. Some epidemics are caused by human activities, such as an irrigation scheme in a warm, semi-arid environment, the displacement of non-immune populations between highland and lowland regions, or a breakdown in pre-existing levels of malaria control. Rarely, they may also follow the accidental introduction of an exotic mosquito species. These human-induced epidemics are less likely to return to the pre-existing equilibrium and may lead to endemic transmission (Figure 1).

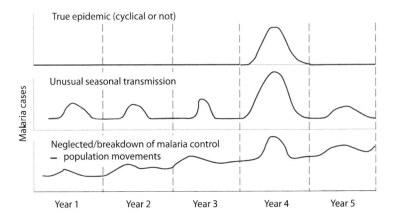

Figure 1. A conceptual diagram of three different epidemic scenarios derived from WHO 2002.

In the case of 'classic' or 'true' epidemics, the changes in suitability for transmission are brought about by natural causes such as climate anomalies in regions where the environment does not normally allow sufficient mosquito and parasite development to occur. These 'true' epidemics often occur in the desert-fringe or highland-fringe areas, which are typically too dry or cool, respectively, to support transmission. The climate anomalies are often periodic and temporary, and there is often a return to the original unstable state. Examples include the epidemics occurring in the semi-arid areas of southern Africa in 1996–1997, East Africa in 1997–1998, and the West African Sahel in 1999–2000, all of which were associated with large-scale unusually heavy rainfall (WHO 2004b). Being poorly prepared, health services often become rapidly overwhelmed, leading to perhaps ten times more malaria-related deaths than in non-epidemic years, across all age ranges (Kizewski & Teklehaimanot 2004; Brown et al., 1998). Epidemics also occur occasionally as enhanced seasonal peaks in regions where levels of seasonal transmission are normally low. Figure 1 illustrates three different types of epidemic situations confronting control services. The top represents a true epidemic, that is, an infrequent event occurring in areas where the disease does not normally occur. This type of epidemic is often associated with warm arid and semi-arid regions and may be cyclical in nature. The middle panel represents an unusually high peak in transmission in areas where malaria is normally present on a seasonal basis. This type of epidemic is often associated with hypoendemic or meso-endemic settings such as the highland fringes and may also be cyclical in nature. The bottom panel represents a resurgent outbreak where neglect or breakdown in control allows malaria to return to its higher pre-control level of endemicity. This third type of epidemic may be associated with more complex emergency situations involving political instability and displaced populations.

Analysis of previous epidemics provides some understanding of their underlying causes and offers opportunities to develop indicators for predicting and monitoring the conditions that are likely to give rise to new epidemics. This should enable authorities to development an epidemic early warning system, which, when combined with a flexible control plan, can lead to epidemic prevention (Connor et al., 1999).

2.2.1 Risk maps for epidemic prone regions

Climate anomalies are of primary interest when creating risk maps for epidemic prone regions. In warm semi-arid areas, temperature is rarely a limiting factor for malaria transmission. Insufficient rainfall, surface water and humidity, however, prevent malaria transmission from occurring on a regular seasonal basis. In highland fringe areas rainfall and humidity may be sufficient for malaria transmission, but relatively cool conditions normally slow down the parasite's reproductive cycle to the extent that it is longer than the life span of its mosquito host. Consequently, transmission does not often occur in these regions. Clearly, anomalies in these variables have important implications for allowing transmission among human populations with little or no immunity to the disease. Maps using deviations from mean climatology to demarcate epidemic prone regions (e.g. Figure 2) suffer from the same issues as endemic risk maps discussed above, including changes in which zones are epidemic prone as a result of inter-decadal variability and longer term climate trends.

2.2.2 Routine monitoring products for epidemic prone regions

To improve epidemic control in climate sensitive regions, the WHO has proposed a framework for developing integrated malaria early warning systems (MEWS) based on vulnerability monitoring, seasonal climate forecasting, environmental and meteorological monitoring, and epidemiological surveillance (Figure 3). This framework integrates climate forecast and climate/environment monitoring products that can provide an early warning to the occurrence of malaria epidemic from one to three months in advance (Ceccato et al., 2007).

Experience to date has shown that it is difficult in terms of availability, timing and cost to obtain meteorological observations from national meteorological services in Africa. National health services generally find the costs of purchasing these data prohibitive given their competing demands for resources across the spectrum of health service requirements. A number of those who have paid

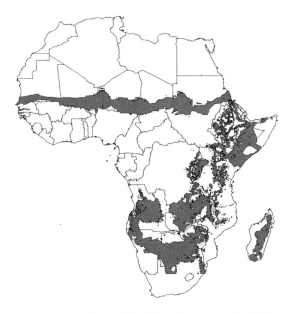

Figure 2.   Epidemic-prone regions in Africa modified from Ceccato (*et al.*, 2005).

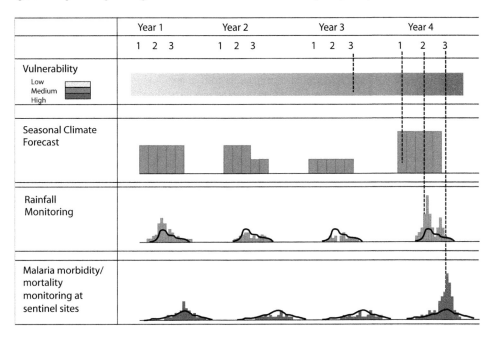

Figure 3.   MEWS integrated framework: gathering cumulative evidence for early and focused response (WHO 2004a). 1 represents assessment during pre-season, 2 represents assessment during the rainy season and 3 represents assessment during Malaria season every year. Year-3 Pre-season assessment – vulnerability increasing due to period of drought; Year-4 Pre-season assessment – vulnerability still increasing due to period of drought and seasonal forecast above normal – Flag-1; Year-4 Pre-season assessment – vulnerability remains high, weather monitoring indicates higher than normal rainfall – Flag-2; and, Year-4 Pre-season assessment – vulnerability remains high, rainfall higher than normal through much of season, malaria cases pass epidemic threshold – Flag-3. (see colour plate 62)

for meteorological data with the assistance of external project funding, have found their delivery to be too slow and irregular for use in an early warning application.

As with other sectors, health services have invested in their own hydro-meteorological network. The Onchercerciasis Control Programme, for example, established and maintained a comprehensive stream flow monitoring system across major river systems in West Africa as part of its vector control activities. Recent investments from the World Bank in malaria control have included the provision of meteorological stations at health facilities in Eritrea. In Niger, project funding has resulted in the creation of forty-four meteorological stations directly maintained and run by health staff. Rainfall, temperature and humidity data are recorded directly as part of electronic disease surveillance. These data are sent by satellite to the central levels of the Ministry of Health, and its partners, on a weekly basis. The use of this high technology however raises issues of sustainability and quality control.

Others have tried to overcome this problem of access by using proxies derived from meteorological satellites, which tend to be available free of charge, in near-real-time and therefore offer much promise for monitoring applications (Ceccato et al., 2005). The increasing capacity and affordability of personal computers means that these large and complicated sets of data can be stored, and analysed with relative ease. Furthermore, innovative methods of data capture, such as remote sensing and the visualization of data from multiple sources offered by spatial data handling software, such as geographic information systems, have greatly increased. These developments, combined with the possibilities for information sharing provided by the Internet, offer an unprecedented window of opportunity to increase their usefulness and applicability to developing country needs.

There are several satellite-derived rainfall estimates available. While some of these products are intended for research purposes, others are more appropriate for routine operational monitoring. However, temperature monitoring products currently remain mostly in the research realm and ground-based observations are essential for testing and validating sensors and estimation methods. This brings us once again to the problem of access to sufficient data and interpolation methods used, in this case, in the satellite products. If the data collected by national meteorological services were available to the developers of satellite estimation products, then these could be substantially improved. National meteorological services could better serve their national development partners in health and other climate sensitive sectors, by making these data available to them as well.

2.2.3 *Accessing climate and environmental information*

Despite the problems outlined above, studies have shown that monitoring rainfall anomalies has the potential to provide advanced warning of impending epidemics in several areas, including Kenya, Ethiopia, Eritrea, West Africa and Botswana (Thomson et al., 2006). In Eritrea, satellite imagery for monitoring rainfall is now part of the routine monitoring system. The National Malaria Control Program based in Asmara uses the MEWS interface developed by IRI staff. The MEWS enables users to gain a contextual perspective of the current rainfall season by comparing it to previous seasons and recent short-term averages. The rainfall estimates are updated every ten days through ADDS, which is maintained by the USGS and supported by US/AID.

The interface is available through the IRI data library in the form of an online *clickable map* (IRI 2012b). It displays the most recent decadal rainfall map over which national and district administrative boundaries and the epidemic risk zone can be overlaid. These visual features can be toggled on or off and the user can zoom in to any region for more clarity. In addition, the map can be downloaded in different formats compatible with common image analysis and GIS software such as ArcView® or *HealthMapper*, which is a no-cost GIS software developed by the WHO.

Decadal rainfall can be spatially averaged over a variety of user-selected areas, including administrative districts and 11 km^2, 33 km^2, 55 km^2 and 111 km^2 boxes. Upon selecting a sampling area and a specific location of interest through a click on the map, four time-series graphs are generated. These time-series provide an analysis of recent rainfall with respect to that of recent seasons and the recent short-term average. A description of the time-series figures, the data used and their source is also provided (Grover-Kopec et al., 2005).

The type and growth stage of vegetation also play an important role in determining mosquito abundance irrespective of their association with rainfall. The type of vegetation which surrounds the breeding sites, and thereby provides, food, potential resting and protection from climatic conditions are also important in determining the abundance of mosquitoes (Beck et al., 1994). Surface water provides the habitat for the juvenile stages (egg, larvae, pupae) of malaria vectors. Monitoring the state of vegetation and water bodies is therefore important to identify the source of malaria vectors. In order to monitor vegetation and water bodies, images from the satellite TERRA and sensor MODIS have been chosen because they provide frequent images at high spatial resolution (250 m) and are available free of charge (an important requirement considering the economic realities of the countries in the affected region). The products (vegetation indices 16-day L3 Global 250M SIN GRID V004) are automatically downloaded from the USGS Land-DAAC and provided to the user community via the IRI data library website (IRI 2012c). Users can download either the raw data, which are the single channels in the blue, red, NIR and SWIR wavelengths in different formats compatible with common image analysis and GIS software as well as NDVI and EVI (Huete et al., 2002). Using the IRI data library, users can: 1) combine the different spectral channels to create their own tailored vegetation indices for monitoring such as vegetation status in terms of moisture content by using a combination of the NIR-SWIR (Ceccato et al., 2002); 2) visualize a colour composite of the SWIR-NIR and Red channels (Red-Green-Blue) where the vegetation appears in green, the bare soils in brown and the water in blue; 3) integrate the colour composite into GIS software with ancillary data such as roads and villages; 4) extract weighted averages of the different indices per GIS layers such as district contours or any other shape file; and 5) create long-term time series of vegetation indices (Ceccato et al., 2006).

In Botswana, it was shown that further advanced warning of epidemic risk could be achieved by including seasonal climate forecasts into the malaria early warning system framework (Thomson et al., 2005; Thomson et al., 2006). The southern African region, which has a long history of malaria epidemics, now has the most advanced integrated approach to epidemic malaria control based on the evidence of the key determinants of epidemics in the region (Craig et al., 2004a; Craig et al., 2004b; Thomson et al., 2005). Experience and evidence for use of an integrated warning system approach within a national malaria control program has been demonstrated in Botswana over the past few years (Thomson et al., 2005). Other countries in the Southern African Development Community agree that this approach provides a useful framework for planning epidemic preparedness and response strategies. The WHO Southern Africa Inter-Country Programme for Malaria Control (SAMC) has supported them in exploring these tools further (DaSilva et al., 2004). As a result the region has for the second year running, prepared for the malaria season with a regional meeting (i.e. Malaria Outlook Forum) at which national and local vulnerability is assessed in the context of the pre-rainy season climate forecasts that have been tailored to the malaria community.

However, while such developments in computer and allied technologies are very welcome, their provision alone will do little to help solve the problems surrounding lack of capacity in sub-Saharan Africa. There is a massive need for institutional capacity strengthening, including the development and training of human resource capital. To get the relevant tools into practical use in routine decision making, a number of issues must be addressed. These are: 1) the problems identified and their potential solutions must stem from the needs and perspectives of the users and not the suppliers and developers, thus ensuring that the demand is real; 2) costs of developing and maintaining the tools and training curricula will ideally be met with monies over which national agencies have full command, even if they originate from donor sources, thus ensuring that the choices made and priority rankings are their own; 3) training needs to be trans-disciplinary and delivered through carefully selected national and regional institutions that can grow and be capable of maintaining the demand for specialized personnel (who may well move out of the original sector to take other jobs in the labour market as their skills base increase); 4) interfacing climate and health is a new disciplinary area that requires long-term investment in boundary institutions and the development of a hierarchy of curricula and training materials for use at different levels of the health sector; and 5) good career opportunities need to be apparent to attract those who are interested in working in this interface. These needs must be combined with a backdrop of problem-focused research activity that incorporates the latest health and climate knowledge.

Given the current status of institutions in sub-Saharan Africa, the above will undoubtedly need to be supported by development partners through the medium term. If national research and educational establishments are strengthened, then they will have a significant role to play. To ensure sustainability, national institutions that are accountable to their own constituencies and national development agendas must be leaders in this process.

3 FUTURE DIRECTIONS

The early warning framework presented has also been implemented for other human and animal diseases; namely, Rift Valley Fever and meningitis (Martin *et al.*, 2007; Thomson *et al.*, 2011). However, to be fully operational, the early warning systems must be integrated into a decision/action framework. There is currently a good deal of policy congruence through international, regional and local levels to support this effort. For example, the Global Framework for Climate Services (GFCS) aims are to develop more effective services to meet the increasing demand from climate sensitive sectors, including the health sector (WHO 2009). The remaining challenge is to get the knowledge into practice and sustaining it where it is needed. It is crucial that appropriate polices are developed and implemented to improve health system performance (Travis *et al.*, 2004). This may be helped by enhancing the workforces' ability to detect and treat diseases, monitor and predict spatio-temporal patterns and implement intervention and control strategies in a timely and cost-effective manner through the use of tools and analysis informed by climate data.

3.1 *The current policy context for the control of epidemic malaria*

In order to get research outcomes into policy and practice it is important to understand the context in which different policies are adopted and supported in a practical manner. Below is an example of how policies developed at the district and national levels connect to the larger political agenda of international policy makers. At the global scale, improved early warning, prevention and control of epidemics is one of the key technical elements of the current Global Strategy for Malaria Control (WHO 1993) and the RBM Partnership.

At the regional level, African Heads-of-State declared their support for the Roll Back Malaria initiative in Africa in April 2000 with the Abuja Targets (WHO 2000). In these targets, national malaria control services are expected to detect sixty per cent of malaria epidemics within two weeks of onset, and respond to sixty per cent of epidemics within two weeks of their detection. With the support of the WHO Regional Office for Africa, the WHO Inter-Country Programme Teams engage in the development of recommendations, guidelines and technical support to improve prevention and control of epidemics and transboundary/cross border within their various sub-regions (e.g. the Regional Economic Communities (RECS), the Economic Community of West African States (ECOWAS), the Intergovernmental Authority for Development (IGAD), and the Southern African Development Community (SADC) including collaborative activities with the African Development Bank.

As a consequence of these policy developments, nations that are prone to epidemics have: enhanced capabilities for delimiting epidemic/endemic-prone areas; established epidemic malaria surveillance systems in epidemic-prone areas; and strengthened their epidemic response capacities with the assistance of the global fund to fight AIDS, tuberculosis and malaria (GFATM 2012) and other donor support. In many national malaria control policy documents, countries now recognize that to achieve RBM and MDG targets, they need better information about where epidemics are most likely to occur, and some indication of when they are likely to happen. As a consequence, they have begun to explore the use of climate information in the development of integrated early warning systems. Thus, there is increasing congruence in policy initiatives from multilateral, bilateral, national and non-governmental agencies in relation to epidemic disease control and a growing demand for climate information and robust early warning systems to support these efforts. This is also reflected in the emerging GFCS.

3.2 *Linking health policies for climate-sensitive diseases to understand climate change*

This policy congruence extends to the current discussions on adaptation to climate change. Strengthened health systems are also seen as vital to improving the management of climate-sensitive disease in the context of climate change. The IPCC identified building public health infrastructure as *the most important, cost effective and urgently needed adaptation strategy*. Other measures endorsed by the IPCC include public health training programs, more effective surveillance and emergency response systems, and sustainable prevention and control programs. These measures are familiar to the public health community and are needed regardless of climate change and constitute what is the basis of a *no regrets* adaptation strategy (Grambsch & Menne 2003).

4 CONCLUSION

Early warning systems for human and animal diseases are in place and are used effectively in a number of countries. They are mostly composed of a number of information streams for monitoring and forecasting environmental and climatic factors that influence the risk of an outbreak for particular diseases based on seasonal climate forecasts/models, remote sensing technologies combined with epidemiological surveillance. The remote sensing and seasonal climate forecast inputs to these information streams have been subject to extensive research to validate their skill (Dinku, *et al.*, 2008; Vancutsem *et al.*, 2010; Barnston *et al.*, 2010; Ceccato *et al.*, 2012). However, to be fully operational, the systems must be integrated into an operational decision/action framework along with other essential socio economic information (DaSilva *et al.*, 2004).

The remaining challenge is to develop the essential partnerships to ensure effective and sustainable knowledge transfer and practice where they are most needed (Rogers 2011). One major opportunity for this is the emergence of the global framework for climate services (GFCS), an international programme to enhance applications of climate information and products in agriculture, food security, water, disasters and health sectors (WMO 2011). This will be accomplished through improved interactions between climate service providers and users at the global, regional and national levels. Early warning systems are expected to play an essential role in this initiative as the GFCS is developed and implemented.

REFERENCES

Barnston, A.G., Li, S., Mason, S.J., Dewitt, D.G., Goddard, L. & Gong, X. 2010. Verification of the first 11 years of IRI seasonal climate forecasts. *J. Appl. Meteor. Climat.* 49: 493–520.

Beck, L.R., Rodriguez, M.H., Dister. S.W., Rodriguez, A.D., Rejmankva, E., Ulloa, A., Meza, R.A., Roberts, D.R., Paris, J.F., Spanner, M.A., Washino, R.K., Hacker, C. & Legters, L.J. 1994. Remote sensing as a landscape epidemiological tool to identify villages at high risk for malaria transmission. *Amer. J. Trop. Med. & Hyg.* 51(3): 271–280.

Brown, V., Issak, M.A., Rossi, M., Barboza, P. & Paugam, A. 1998. Epidemic of malaria in north-eastern Kenya. *Lancet* 352: 1356–1357.

Ceccato, P., Gobron, N., Flasse, S., Pinty, B & Tarantola, S. 2002. Designing a spectral index to estimate vegetation water content from remote sensing data (Part 1: theoretical approach). *Rem. Sens. Environ.* 82(2–3): 188–197.

Ceccato, P., Connor, S.J., Jeanne, I. & Thomson, M.C. 2005. Application of geographical information system and remote sensing technologies for assessing and monitoring malaria risk. *Parasitol.* 47: 81–96.

Ceccato, P., Bell, N.A., Blumenthal, M.B., Connor, S.J., Dinku, T., Grover-Kopec, E.K., Ropelewski, C.F. & Thomson, M.C. 2006. Use of remote sensing for monitoring climate variability for integrated early warning systems: Applications for human diseases and desert locust management. *Proc. IGARSS-IEEE conference* Denver, Colorado, August 2006.

Ceccato, P., Ghebremeskel, T., Jaiteh, M., Graves, P.M., Levy, M., Ghebreselassie, S., Ogbamariam, A., Barnston, A.G., Bell, M., Del Corral, J., Connor, S.J., Fesseha, I., Brantly, E.P. & Thomson, M.C. 2007. Malaria stratification, climate and epidemic early warning in Eritrea. *Amer. J. Trop. Med. & Hyg.* 77: 61–68.

Ceccato, P., Vancutsem, C., Klaver, R., Rowland, J. & Connor, S.J. 2012. A vectorial capacity product to monitor changing malaria transmission potential in epidemic regions of Africa. *J. Trop. Med.* ID 595948. doi:10.1155/2012/595948.

Connor, S.J., Thomson, M.C. & Molyneux, D.H. 1999. Forecasting and prevention of epidemic malaria: New perspectives on an old problem. *Parasitol.* 41(1–3): 439–448.

Craig, M.H., Snow, R.W. & Le Sueur, D. 1999. A climate-based distribution model of malaria transmission in Sub-Saharan Africa. *Parasitol. Today* 15(3): 105–111.

Craig, M.H., Kleinschmidt, I., Nawn, J.B., Le Sueur, D. & Sharp, B. 2004a. Exploring 30 years of malaria case data in KwaZulu-Natal, South Africa: Part I. The impact of climatic factors. *Trop. Med. & Int. Health* 9(12): 1247–1257.

Craig, M.H., Kleinschmidt, I., Le Sueur, D. & Sharp, B. 2004b. Exploring 30 years of malaria case data in KwaZulu-Natal, South Africa: Part II. The impact of non-climatic factors. *Trop. Med. & Int. Health* 9(12): 1258–1266.

DaSilva, J., Garanganga, B., Teveredzi, V., Marx, S.M., Mason, S.J. & Connor, S.J. 2004. Improving epidemic malaria planning, preparedness and response in Southern Africa. *Malaria J.* 3: 37. doi:10.1186/1475-2875-3-37.

Dinku, T., Chidzambwa, S., Ceccato, P., Connor, S.J. & Ropelewski, C.F. 2008. Validation of high-resolution satellite rainfall products over complex terrain in Africa. *Int. J. Rem. Sens.* 29(14): 4097–4110.

GFATM. 2012. Available from: http://www.hipc-cbp.org/files/en/open/Guide_to_Donors/GFATM_11% 2009%202009.pdf [Accessed 3rd April 2012].

Grambsch, A. & Menne, B. 2003. Adaptation and adaptive capacity in the public health context. In: *Climate change and human health*: 220–236. Geneva: WHO.

Grover-Kopec, E.K., Kawano, M., Klaver, R.W., Blumenthal, M.B., Ceccato, P. & Connor, S.J. 2005. An online operational rainfall-monitoring resource for epidemic malaria early warning systems in Africa. *Malaria J.* 21: 4–6.

Grover-Kopec, E.K., Blumenthal, M.B., Ceccato, P., Dinku, T., Omumbo, J. & Connor, S.J. 2006. Web-based climate information resources for malaria control in Africa. *Malaria J.* 5: 38. doi: 10.1186/1475-2875-5-38.

Huete, A., Didan, K., Miura, T., Rodriguez, E.P., Gao, X. & Ferreira, L.G. 2002. Overview of the radiometric and biophysical performance of the MODIS vegetation indices. *Rem. Sens. Environ.* 83: 195–213.

Hutchinson, M.F., Nix, H.A. & McMahon, J.P. 1996. A topographic and climate data base for Africa Version 1.1 (CD-ROM). Canberra: The Australian National University.

IRI. 2012a. Available from: http://iridl.ldeo.columbia.edu/maproom/.Health/.Regional/.Africa/.Malaria/ .CSMT/ [Accessed 1st February 2012].

IRI. 2012b. Available from: http://iridl.ldeo.columbia.edu/maproom/.Health/.Regional/.Africa/.Malaria/ .MEWS/ [Accessed 1st February 2012].

IRI. 2012c. Available from: http://iridl.ldeo.columbia.edu/SOURCES/.USGS/.LandDAAC/.MODIS/ [Accessed 1st February 2012].

Kizewski, A. & Teklehaimanot, A. 2004. A review of the clinical and epidemiological burdens of epidemic malaria. *Amer. J. Trop. Med. & Hyg.* 71: 128–135.

MARA/ARMA. 1998. Towards an atlas of malaria risk in Africa: First technical report of the MARA/ARMA Collaboration. Durban, South Africa: MARA/ARMA Investigation Centre Medical Research Council. Available from: http://www.mara.org.za [Accessed 3rd April 2012].

Martin, V., De Simone, L., Lubroth, J., Ceccato, P. & Chevalier, V. 2007. Perspectives on using remotely-sensed imagery in predictive veterinary epidemiology and global early warning systems. *Geospat. Health* 2(1): 3–14.

Mitchell, T.D., Carter, T.R., Jones, P.D., Hulme, M. & New, M. 2004. A comprehensive set of high-resolution grids of monthly climate for Europe and the globe: The observed record (1901–2000) and 16 scenarios (2001–2100). In: Tyndall Centre for Climate Change Research Working Paper 55. Norwich. 30 pages Available from: http://www.tyndall.ac.uk/sites/default/files/wp55.pdf [Accessed 3rd April 2012].

NMA. 2012. Available from: http://213.55.84.78:8082/maproom/.NMA/ [Accessed 1st February 2012]

Pascual, M., Ahumada, J.A., Chaves, L.F., Rodó, X. & Bouma, M. 2006. Malaria resurgence in East African highlands: Tmperature trends revisited. *PNAS* 103(15): 5829–5834.

RBM. 2005 Available from: http://www.rollbackmalaria.org/forumV/docs/gsp_en.pdf [Accessed 1st February 2012].

Rogers, D.P. 2011. Partnering for health early warning systems. *Bull. WMO.* 61(1): 14–18.

342

Thomson, M.C., Erickson, P.J., Ben Mohamed, A. & Connor, S.J. 2004a. Land use change and infectious disease in West Africa. In: R. De Fries, G. Asner & R Houghton (eds.), *Ecosystems and land use change*: 169–187. Washington DC: American Geophysical Union.

Thomson, M.C., Connor, S.J., Ward, N. & Molyneux, D. 2004b. Impact of climate variability on infectious disease in West Africa. *Eco-Health* 1: 138–150.

Thomson, M.C., Mason, S.J., Phindela, T. & Connor, S.J. 2005. Use of rainfall and sea surface temperature monitoring for malaria early warning in Botswana. *Amer. J. Trop. Med. & Hyg.* 73(1): 214–221.

Thomson, M.C., Doblas-Reyes, F.J., Mason, S.J., Hagedorn, R., Connor, S.J., Phindela, T., Morse, A.P. & Palmer, T.N. 2006. Malaria early warnings based on seasonal climate forecasts from multi-model ensembles. *Nature* 439(7076): 576–579.

Thomson, M.C., Firth, E., Jancloes, M., Mihretie, A., Onoda, M., Nickovic, S., Broutin, H., Perea, W., Bertherat, E. & Hugonnet, S. 2011. Climate and public health: The MERIT initiative: A climate and health partnership to inform public health decision makers. Denver: World Climate Research Programme.

Travis, P., Bennett, S., Haines, A., Pang, T., Bhutta, Z., Hyder, A., Pielemeier, N., Mills, A. & Evans, T. 2004. Overcoming health-systems constraints to achieve the Millennium Development Goals. *Lancet* 364: 900–906.

Vancutsem, C., Ceccato, P., Dinku, T. & Connor, S.J. 2010. Evaluation of MODIS land-surface temperature data to estimate air temperature in different ecosystems over Africa. *Rem. Sens. Environ.* 114: 449–465.

WHO. 1993. A global strategy for malaria control. Geneva: WHO.

WHO. 2000. The Abuja Declaration on Roll Back Malaria Geneva: WHO.

WHO. 2002. Strategic framework to decrease the burden of TB/HIV. Geneva: WHO.

WHO. 2004a. World report on knowledge for better health. Geneva: WHO.

WHO. 2004b. Malaria epidemics: forecasting, prevention, early warning and control – From policy to practice. Geneva: WHO.

WHO. 2009. Available from http://www.climatesciencewatch.org/ [Accessed 6th March 2012].

WHO. 2010. World malaria report. Geneva: WHO Global Malaria Programme. ISBN 978 92 4 156410 6.

WMO. 2011. Report of the high-level taskforce on the global framework for climate services. Geneva: WMO.

Worrall, E., Rietveld, A. & Delacollette, C. 2004. The burden of malaria epidemics and cost-effectiveness of interventions in epidemic situations in Africa. *Amer. J. Trop. Med. & Hyg.* 71(Supp2): 136–140.

Chapter 9

Towards operational forecasts of algal blooms and pathogens

C.W. Brown[1] (Auth./ed.), with; D. Green[2], B.M. Hickey[3], J.M. Jacobs[4], L.W.J. Lanerolle[5], S. Moore[6], D.J. Schwab[7], V.L. Trainer[6], J. Trtanj[5], E. Turner[8], R.J. Wood[4] & T.T. Wynne[5]

[1] *National Oceanic and Atmospheric Administration (NOAA) Satellite Data and Information Service, College Park, MD, US*
[2] *NOAA, National Weather Service, Silver Spring, MD, US*
[3] *University of Washington, Seattle, WA, US*
[4] *NOAA, National Ocean Service, Oxford, MD, US*
[5] *NOAA, National Ocean Service, Silver Spring, MD, US*
[6] *NOAA, National Marine Fisheries Service, Seattle, WA, US*
[7] *NOAA Office of Oceanic and Atmospheric Research, Ann Arbor, MI, US*
[8] *NOAA, National Ocean Service, Durham, NH, US*

ABSTRACT: This chapter describes on-going regional projects sponsored by the US National Oceanic and Atmospheric Administration and its partners to forecast harmful algal blooms and water-borne pathogens in waters of the US coastal oceans and Great Lakes. Collectively, these provide an introduction to the general problems, the approaches employed, the products generated, and the on-going attempts to improve these ecological forecasts and transition them into operational products and services.

Disclaimer: the views, opinions, findings, conclusions, and recommendations contained in this document are those of the authors and are not official NSF, NOAA or US Government positions, policies, or decisions.

1 INTRODUCTION

The National Atmospheric and Oceanic Administration (NOAA) is well known for its operational weather forecasts, climate predictions, and biological forecasts associated with fish stock assessments for fisheries management. It is leveraging its observational capabilities and expertise in environmental prediction to provide forecasts of relevance to human and ecosystem health in various thematic and biogeographic domains. As our scientific understanding of ecosystem structure and function have matured, the ability to provide probabilistic forecasts is becoming viable for a number of environment, health, and safety issues in coastal waters. This understanding, along with access to growing streams of data from monitoring and observing networks, expanding computing power and increasing model sophistication, has permitted ecological forecasting to move from the research realm to proactive management application. Operational ecological forecasting requires a new era in interdisciplinary science that makes use of a wide range of information and links several different scientific disciplines and activities, such as meteorology, hydrology, climatology, data assimilation, ecological research, and environmental modelling, as well as a collaboration with policy and decision makers, and community based efforts to optimize a mix of technologies from national centres, regional networks, academia, and industry.

The American public, private corporations, coastal managers, and health and safety officials are demanding forecasts and early warnings of environmental changes and related impacts to save lives, protect our well being, reduce health risks, and to sustain ecosystem resources and maintain their economic benefits. This is especially true in coastal waters. These forecasts help coastal managers and others to answer questions that require immediate response or actions, such as *Is it safe to swim here?* or *Will this shellfish be safe to eat?*, as well as assist local, state and federal agencies in making sound management and policy decisions on large-scale, complex issues that have long-term consequences for the coastal environment. The National Environmental Policy Act (NEPA), in fact, requires the use of forecasts in selecting preferred alternatives for any contemplated federal action affecting the human environment.

NOAA's mission is to both understand and predict changes in Earth's environment and conserve and manage coastal and marine resources to meet US national economic, social, and environmental needs. As such, NOAA is the authoritative science-based, operational forecasting US Federal agency that collects, compiles, and disseminates data and information, modelling products, and forecasts that are vital for informing the public and private sectors about current and future environmental risks. It is important to note that it is not sufficient to strive for the best integrated science understanding, but in the context of commerce and health impact, it is necessary for NOAA to provide reliable and trusted services as well as promote environmental stewardship and ecosystem-based management. These activities reflect NOAA's goals of promoting a weather ready nation that is prepared for and responds to weather-related events including those of health concern, productive and healthy oceans that maintain and enhance fisheries, habitats and biodiversity, and resilient coastal communities that are environmentally and economically sustainable.

NOAA is well known for its weather forecasts, climate predictions, and biological forecasts associated with fish stock assessments for fisheries management. In response to these demands for predictions of human and ecosystem health in coastal waters (and elsewhere), the administration is leveraging its existing infrastructure and expertise in weather prediction, observation collection, advanced information processing and dissemination systems, and expanding its engagement with public health partners to develop and generate a suite of forecasts that focus on factors relevant to environmental and public health and span a range of time and space scales (Table 1). These ecological forecasts are intended to provide early warnings and scenario planning for ecosystem health, coastal management, and public health decision making. Just as weather and climate forecasts are used to adapt and mitigate hazard impacts, these health-relevant predictions are intended to inform management and policy decisions and enable moving from a reactive to a proactive mode. In much the same way that a weather forecast informs the public of the location and timing of severe weather, an ecological forecast provides pertinent information to local and state officials to fulfil their responsibilities in safe guarding the public. For instance, forecasts of the expected concentration of deleterious bacteria or the presence of toxic harmful algal can aid in generating advisories for beach and shellfish harvesting closure. Longer term forecasts help managers guide analysis of alternative management scenarios and develop plans to answer "what if" questions about the ocean and coastal environments. These ecological projections do not guarantee what is to come; instead, they offer scientifically based estimates and scenarios of what is likely to occur. Managers can use these forecasts to select the best options and take the appropriate actions to better manage the Nation's coastal resources and to protect human health.

Ecological forecasting, predicting the impacts of physical, chemical, biological, and human induced change on ecosystems and their components, is perhaps the most complex modelling challenge that NOAA faces. It requires a high-level of understanding and integration of several scientific fields of study, the infrastructure and observations to generate, validate and disseminate the forecasts, and collaboration between groups that specialize in each of these activities. As a consequence, ecological forecasting has become feasible only recently with our improved scientific understanding of ecosystem structure and function, and with improvements in integrating observations and modelling, computational capability, data management, telecommunications, and our ability to generate and issue operational ecological forecasts. Developing forecasts to meet the needs of the public health community is even more challenging, requiring the integration of complex public health surveillance or epidemiologic data to determine exposure pathways and related health outcomes. Furthermore, health related forecasts must be generated and disseminated through a sustained and durable operational framework upon which the users may rely. NOAA

346

Table 1. Examples of NOAA health related forecasting activities.

Event of concern	Common impact	NOAA forecast activities
Temperature extremes	Excessive heat or cold are among the top causes of US weather-related fatalities	Health messaging
Tropical cyclones tornadoes	Loss of life and property	Development, implementation and monitoring and prediction systems; early warning
Draught	Loss of livelihood	National Integrated Drought Information System (NIDIS) and early warning systems
Floods	Loss of life and property	Prediction and early warning
Allergens	Pollens, aerosols and air pollution, in concert with extreme temperature, changes in pressure or humidity can trigger asthma and other respiratory ailments	Health messaging
Harmful algal blooms and toxins	Fin- and shell-fish safety issues; human, animal and plant mortality	Development, implementation, and operations of monitoring and prediction systems; generation and dissemination of regional HAB bulletins and forecasts
Water-borne pathogens	Fin- and shell-fish safety issues; human, animal and plant mortality	Development and implementation of prototype forecast systems and products
Hypoxia/anoxia	Reduces habitat viability; kills sea life; inflicts economic losses and fish kills	Generation of forecasts in Gulf of Mexico; development of forecast system in Chesapeake Bay
Coral reef bleaching	Temporary stress to or death of coral	Global forecasts of potential bleaching

possesses the operational infrastructure and expertise in environmental prediction to enter into this expanding field of service.

In collaboration with its partners, NOAA is developing ecological forecasts that address several coastal issues relevant to its mission and its legislative mandates: fish recruitment, invasive species, shellfish closures, harmful algal blooms (HABs), beach and water quality, and transitioning them from the research realm to an operational setting in order to produce useful and reliable products that are routinely available in a timely manner. This chapter focuses on projects that generate local and regional forecasts of HABs and bacteria along US coasts and Great Lakes. It is not intended to be an exhaustive listing of projects on this subject, but to provide a cross-section of ecological forecasting activities currently conducted and funded by NOAA and other agencies, and represent studies in various geographic regions, approaches employed to generate the forecasts, and stages of development along the path to operations. The transitioning of a forecast (or any other product) from research to operations generally follows three, loosely-defined sequential steps: experimental, demonstration, and operational. Experimental products are in the research phase and are undergoing development and preliminary evaluation. Though scientists are the major participants in this stage, interaction with users often has been initiated and their needs documented. Demonstration products, which typically undergo enhancement and evaluation, are deemed worthwhile by the user and thereby applicable for operational transfer. Refining and testing forecasts in this phase may require several iterations with users. Pre-operational products are ready for transition to operations, but are awaiting resources for operational implementation and maintenance, and are likely distributed routinely by a non-operational office. This stage requires the involvement of the product developers as well as agency representatives to plan for resources. Operational products are generated reliably and issued routinely with a known measure of uncertainty from a well-controlled computational and networked environment with redundant infrastructure. Operational HAB forecasts, for example, are currently generated and issued by the Harmful Algal Bloom Operational Forecast System HAB-OFS, Center for Operational Oceanographic Products and Services (CO-OPS 2012). The skill of

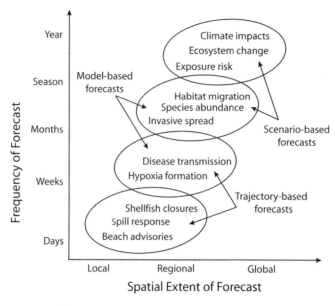

Figure 1. Management applications of ecological forecasts at representative spatial and temporal scales (Courtesy Scheurer, NOAA).

operational products is periodically assessed and their user outreach procedures are established. This final phase involves specific personnel trained as forecasters to interpret and transform results from observations and model guidance into forecasts. Therefore, strictly speaking, forecasts in NOAA parlance are only generated and issued by human forecasters. In this chapter, however, a more generic definition is employed, and *forecast* is considered to be synonymous with, and used interchangeably with *prediction*.

Coastal health-related forecasting activities in the aquatic environment can be grouped into three approaches: 1) trajectory-based forecasting, 2) model-based forecasting, and 3) scenario-based forecasting (Figure 1). In trajectory-based forecasting, the species, toxin, or an appropriate proxy is detected in observations and then is transported forward in time using velocity fields from a numerical hydrodynamic or transport model. Observations to identify the target are necessary, and can be acquired from various satellite and *in-situ* sampling sources. The second approach, model-based forecasting, can be either statistical or dynamical. In the statistical applications an empirical model relates the abundance or likelihood of a HAB, bacterium, toxin or event quantitatively, and relates them to environmental conditions. These types of models, such as a habitat suitability model, are constructed from available data of the target organism or event using a suite of statistical techniques that span from relatively simple data fitting models, such as logistic regression, to advanced computing techniques, such as artificial neural networks (Guisan & Zimmermann 2000; Austin 2002). The statistical model is then forced with the pertinent environmental conditions to predict the variable of interest. This statistical approach is useful when the target species or toxin cannot be detected readily with available observations and a process-oriented understanding of the factors regulating the organism or its toxin is incomplete. The dynamical modelling approach replicates physical and ecosystem processes based on numerical representations of physical, geochemical and biological principles. This approach, which is common in weather forecasting, is not currently available to generate any of the HAB and bacterial regional forecasts. However, ecological prediction using this mechanistic approach is the ultimate goal. Lastly, in scenario-based forecasting, the response of HABs, bacteria and associated toxic events (e.g. their intensity, frequency, distribution, and impacts) to environmental changes occurring over several decades (or longer) is approximated by applying their present environmental preferences or associations onto future climate projections like those provided by the IPCC. This approach is essentially an application of the statistical approach to obtain longer term projections. Scenario-based forecasting can test different management scenarios that may impact nutrient delivery or hydrologic forcing, and can evaluate the broader ecological implications of such actions.

2 FORECASTING HARMFUL ALGAL BLOOMS AND WATER-BORNE PATHOGENS

The case studies in this section, which are organized according to the three forecasting approaches discussed above, describe several projects dealing with the forecasting of water-borne pathogens and harmful algal blooms in waters of the US coastal ocean and Great Lakes, and provide an introduction to the general problem, the users, the forecast products, the challenges, and the on-going attempts to improve these ecological forecasts.

2.1 Trajectory-based models

The sub-sections of Section 2.1 describe trajectory-based HABs monitoring techniques for three coastal environments on the east coast, gulf coast, and Pacific Northwest coast of Oregon and Washington in the US. Doctors Wynne and Lanerolle open this discussion with a description of harmful species in the subtropical Gulf of Mexico featuring trajectory-based modelling outputs. This is followed by Doctors Wynne, Schwab and Lanerolle's description of blooms in the interior Great Lakes region using results from the Great Lakes coastal forecasting system (GLCFS) hydro-dynamic model, and introducing a new measure of vertical eddy-viscosity and its use to examine tracer and particle dispersion.

In the third example Doctors Trainer, Hickey and Moore describe bloom forecasts developed by the Pacific Northwest partnership (ECOHAB PNW). The blooms impact the fisheries and oceanography at the mouth of the Columbia River as well as ecology along the Washington/Oregon coasts to the south. At the end of Section 2.1, Doctor Schwab describes beach water quality monitoring in context of public safety. Three centres have been sponsored by NOAA to focus on developing predictive capabilities for beach water quality and harmful algal blooms.

2.1.1 Harmful algal blooms – Gulf of Mexico (US)

Algae occur in high concentration blooms in freshwater, estuarine, and marine waters where local, upstream, and offshore forcing promote rapid growth and accumulation of protists in the water column. Blooms are associated with the production of natural toxins, depletion of dissolved oxygen or other harmful effects, and are generally described as HABs and commonly referred to as *red tides*. The potential risk that HABs pose to human and marine mammal health, natural resources, and environmental quality has increased the urgency to monitor, detect, and forecast their occurrence and distribution.

Since October 2004, operational forecasts in the form of a HAB Bulletin are issued by the HAB-OFS to alert state, federal and local management agencies of the occurrence and impact of blooms of the dinoflagellate *Karenia brevis* in waters off the southwest coast of Florida. *K. brevis*, the major HAB species in the region, forms large monospecific blooms annually and produces neurological brevetoxin that can cause neurotoxic shellfish poisoning when shellfish that have ingested these algae are eaten by humans and marine mammals, and respiratory irritation when the brevetoxin is aerosolized and injected into the air after lysis, the disintegration or rupture of the cell membrane, resulting in the release of cell contents or the subsequent death of the cell. Forecasts of *K. brevis* blooms are generated twice weekly during a bloom event, using a combination of satellite derived image products, wind predictions, and a rule-based heuristic model based on previous observations and research (Stumpf *et al.*, 2009). Likely blooms of *K. brevis* are first detected in satellite imagery using a climatological chlorophyll anomaly product (Stumpf *et al.*, 2003; Tomlinson *et al.*, 2004). The anomaly is calculated by subtracting a moving sixty day average minus two-weeks to reduce the chance for persistent features in the imagery to bias the mean from the daily satellite-derived

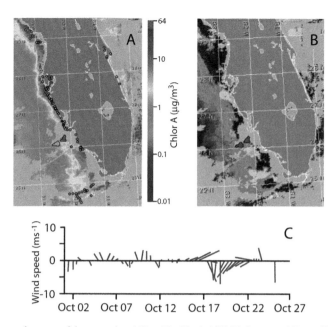

Figure 2.   The core elements of the operational West Florida shelf HAB forecast: (A) satellite-derived chloro-phyll image, with cell count information overlaid onto the image; (B) chlorophyll anomaly product, with confirmed HAB regions shown in red; and (C) local wind conditions. (see colour plate 63)

chlorophyll image. Any region with a difference equal to or more than one microgram per litre in chlorophyll concentration is flagged as a new bloom. The resulting anomaly image, combined with a set of expert rules, is used to determine whether the climatological anomaly is likely to be a *bona fide K. brevis* bloom.

Once the bloom is detected and confirmed by *in-situ* sampling, forecasts of intensification, transport, areal extent, and beach impact are generated by analysts using predictions of wind speed and direction from NOAA's Weather Services' marine weather forecasts, *in-situ* cell counts, and satellite-derived chlorophyll concentration (Stumpf *et al.*, 2009). The forecasts are published in operational HAB Bulletins that are sent twice a week via e-mail to registered users with natural resource management responsibilities (Figure 2). The bulletin displays the chlorophyll image, chlorophyll anomaly image, and wind speed. Additionally there is a two-part analysis performed by a forecaster. The first part is a *conditions report* that is available publically. This provides general statements regarding confirmed bloom activity and its likely impacts over the next three to four days. The second portion of the analysis is a more detailed section supplying more descriptive conditions and forecasting of bloom trajectory. Skill assessment of these four forecast parameters is conducted on a regular basis (Fisher *et al.*, 2006; Stumpf *et al.*, 2009). Archived HAB Bulletins, as well as documentation for understanding and interpreting them, are available (HAB-OFS 2012).

In 2010, the HAB-OFS was expanded to issue operational forecasts for HABs occurring in waters off the Gulf coast of Texas. Though the dinoflagellate *K. brevis* is also responsible for the HAB events along the Texas coast, the satellite-derived chlorophyll anomaly technique used in southwest Florida to detect their blooms had to be modified to account for regional differences. In Texas, HAB detection is complicated by resuspension of sediments and benthic, non-toxic diatoms in the water column that sometimes result in false positive HAB events. To improve the accuracy of detecting *K. brevis* blooms, a new technique was implemented that used satellite imagery to identify resuspension events and exclude them from the chlorophyll anomaly product (Wynne *et al.*, 2005). Once the *K. brevis* bloom is detected and confirmed, forecasts of the same four elements found in the southwest Florida HAB bulletin are generated, though the transport forecast is aided by a spill

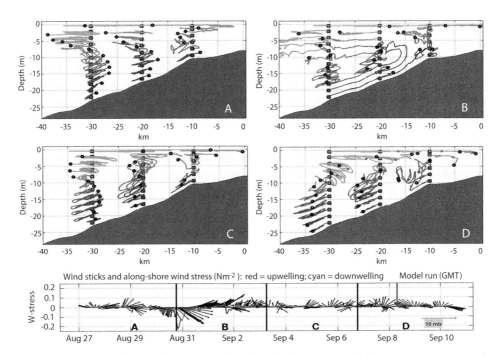

Figure 3. Output graphic from the two-dimensional upwelling transect model (Courtesy Lanerolle *et al.*, 2006). (see colour plate 64)

response trajectory model called the general NOAA operational modelling environment (GNOME) using coastal current forecasts generated by the Texas General Land Office, because the circulation is not wind driven as it is off southwest Florida.

To assess whether forecasting flow-fields would enable more accurate prediction of the transport and areal extent of HAB patches off southwest Florida, numerical modelling studies have been conducted (Lanerolle *et al.*, 2006; Stumpf *et al.*, 2008). For example, based on the hypothesis that upwelling contributes to HAB events on the west Florida shelf, a two-dimensional numerical ocean model was developed and implemented for a transect across this area (Lanerolle *et al.*, 2006). The model is based on Rutgers University's regional ocean modelling system (ROMS). It consists of a 400 point grid along the transect, nine points across and eighty vertical σ-levels, and it is capable of running as a nowcast/forecast system. ROMS is a free-surface, terrain-following, primitive equations model widely used by the scientific community for a diverse range of applications. In the model, HAB cells are simulated as mass-less and point-like Lagrangian particles that respond to the two-dimensional flow. The model and its output are useful for studying the effect of upwelling on HABs and their onset and could form a basis for enhancing the predictions from the HAB bulletins in the future (Figure 3). Each panel represents a 3.5 day segment of simulation and hence the full simulation is for fourteen days. Particles begin at blue squares and end at black circles and those corresponding to model initialization time are in magenta dots. Three sets of particle ensembles at 10km, 20km, and 30km distances from shore are employed and are in green, red and blue colours respectively. Deepest particles have darkest shades and shallowest the lightest. The upwelling and downwelling behaviours seen in the panels are closely correlated with the wind stress and wind stick plot shown in the lower panel where a negative wind stress in red colour signified upwelling. The wind sticks show the speed and direction of the wind. The magenta arrow indicates the model initialization time. As is evident in the Figure, especially panel B in the upper right on 31August 2006, particles respond to wind-driven upwelling/downwelling which cause them to move onshore or offshore. Current research is developing the downwelling index for incorporation into the operational forecast to provide managers with much earlier warning of HAB events.

2.1.2 Harmful algal blooms -Great Lakes (US)

Blooms of the toxic cyanobacterium *Microcystis aeruginosa* have become a common summer occurrence in certain areas of the Great Lakes, particularly western Lake Erie, and the ability to accurately delineate high concentration of *M. aeruginosa* is a public health priority. The blooms, which often form mats of surface scum, are aesthetically unappealing and are associated with various detrimental effects, including human respiratory irritation, unpleasant taste and odor of drinking water, human illness, and animal mortality (Wynne *et al.*, 2011).

In 2008, demonstration forecasts for blooms of these toxic cyanobacterial blooms were generated by the National Ocean Service's Center for Coastal Monitoring and Assessment. This system is more advanced than the one used off southwest Florida and Texas due to the nature of the image detection method and the implementation of a hydrodynamic model and a particle tracker to predict transport. For the detection of *M. aeruginosa* blooms, an algorithm was developed and applied to the satellite measurements of ESA's Medium-spectral Resolution Imaging Spectrometer (MERIS). This sensor has been available since 2003 at reduced spatial resolution (1200 metres) and since 2009 at full resolution (300 metres). The detection algorithm (Eq. 1) is based on a spectral shape based on the following wavelengths: 665, 681, and 709 nm (Wynne *et al.*, 2008).

$$SS(\lambda) = Rrs(\lambda) - Rrs(\lambda^-) - \{Rrs(\lambda^+) - Rrs(\lambda^-)\} * \frac{(\lambda - \lambda^-)}{(\lambda^+ - \lambda^-)}, \tag{1}$$

where SS is the spectral shape, Rrs is the remote sensing reflectance, and $\lambda = 681$ nm, $\lambda^+ = 709$ nm, and $\lambda^- = 665$ nm. The inverse of this relationship, coined the cyanobacterial index (CI), has been shown to identify blooms of cyanobacteria in Lake Erie (Wynne *et al.*, 2010). Based on the CI, field samples are collected for species identification and bloom extent.

Once the *M. aeruginosa* bloom has been detected and identified, its future location is forecast using the GNOME trajectory model with surface currents provided by the Great Lakes coastal forecasting system (GLCFS) hydrodynamic model (Wynne *et al.*, 2011). Seventy-two hour forecasts of surface current are generated and updated every six hours. Once the transport simulation is completed, the location of the bloom is produced as an image to create two separate scenarios; a *nowcast* and a *forecast*. The nowcast is defined as the simulation between satellite image acquisition time and the morning of the simulation run. The image simulated is often twenty-four hours old (time needed for data acquisition and processing), and sometimes is older depending on cloud cover and sensor orbit. The predicted currents are employed to transport the nowcast into the future to create a forecast for the following three days. Details of the techniques used in this forecasting system are available in Wynne (*et al.*, 2011).

This forecast information is disseminated to water quality managers, state officials and public water works in an experimental Lake Erie HAB bulletin (Figure 4). The satellite image used to detect the blooms, as well as the simulated nowcasts and forecasts, are illustrated in the bulletin, along with a *conditions* report and analysis, in much the same way as the Gulf of Mexico HAB bulletin. The bulletin also includes information on water temperature and wind stress, factors which can contribute to bloom maintenance, growth and dissipation (Wynne *et al.*, 2010; GLCFS 2012).

To better understand and predict the distribution of *M. aeruginosa* in the Great Lakes, several numerical modelling studies have been performed. One of these studies examined the effects of the vertical eddy-viscosity on the distribution of *Microcystis* cells. Although buoyant, the vertical distribution of cyanobacterial cells is dictated by the vertical eddy-viscosity of Lake-water, which is generated primarily by atmospheric conditions over the Lake where the bathymetry is shallow. To understand the eddy-viscosity generation, its distribution within the water column and its dispersive effects on *M. aeruginosa* cells, a one-dimensional numerical ocean modelling system based on ROMS was developed and a passive (inert) tracer and mass-less, point-like (and devoid of behaviour) Lagrangian particle simulation representing *M. aeruginosa* cells was conducted. The modelling configuration was designed to encompass a high degree of portability such that it could be applied easily to any open water body with a minimum number of inputs. The inputs to the model included a vertical profile of temperature and salinity, a representative bathymetric value,

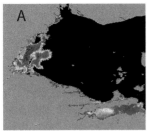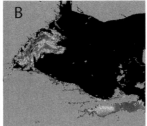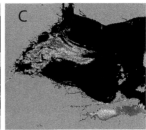

Figure 4. Example of the HAB forecast system for Lake Erie: (A) initial image of the CI; (B) nowcast; and (C) three-day forecast of likely bloom distribution. (see colour plate 65)

a representative Coriolis parameter, and a fixed-point time-series of meteorological variables to compute wind stresses and the net heat flux. The Coriolis parameter is defined as twice the vertical component of the Earth's angular velocity as a function of latitude. The tracer dispersion was modelled using the diffusion equation and its diffusion coefficient was taken to be the eddy-viscosity from the model. The particle dispersion was modelled assuming random walk behaviour.

The simple, quasi one-dimensional model proved useful in examining the influence of meteorologically induced eddy-viscosity on the mixing of *M. aeruginosa* cell mixing and distribution in the water column in Lake Erie. The simulated tracer and particles respond to the eddy-viscosity in the expected manner (i.e. high viscosity disperses the tracer/particles and low viscosity keeps them unaltered) (Figure 5). Furthermore, the tracer and particle pictures complement each other. This understanding and model may be incorporated into the forecast system to better predict the location and concentration of *M. aeruginosa* in Lake Erie in the future.

2.1.3 *Harmful algal blooms – Pacific Northwest (US)*

Two genera of harmful algal blooms concern shellfish safety in the Pacific Northwest – the diatom *Pseudo-nitzschia* and the dinoflagellate *Alexandrium*. Blooms of *Alexandrium* are responsible for the majority of HAB issues in the inland waters of Puget Sound, while blooms of *Pseudo-nitzschia* are responsible for the majority of HAB issues on the outer Washington-Oregon coast. The latter have been reported to cost as much as $22 million USD in lost revenue (Dyson & Huppert 2010). A series of multiyear studies, including the Olympic Region Harmful Algal Bloom (ORHAB) partnership, the Ecology and Oceanography in the Pacific Northwest (ECOHAB PNW) project, and the on-going Pacific Northwest Toxins (PNWTOX) project, have provided the scientific background needed to develop forecasts of the HABs in this region. Forecasting blooms of *Alexandrium* are described in Section 2.2.3 below.

Pseudo-nitzschia is a natural component of the marine ecosystem in the Pacific Northwest and is sometimes not toxic. Research has shown that the onset of toxicity and source of toxic cells are associated with two retentive oceanographic features, the seasonal Juan de Fuca eddy located off the Juan de Fuca Straight north of the Washington coast, and Heceta Bank located off the central Oregon coast (Hickey & Banas 2003; Trainer*et al.*, 2009). When toxic cells make their way to the coastal beaches, they can contaminate the Pacific razor clam, *Siliqua patula*, which has significant recreational, commercial and tribal value (Adams *et al.*, 2000; Trainer *et al.*, 2002; Hickey *et al.*, in press). Once razor clams are contaminated with domoic acid (DA), a nerve toxin occurring naturally in *Pseudo-nitzschia* spp., they can remain toxic for more than one year.

In 2007, the NOAA Northwest Fisheries Science Center developed a partnership with the CDC and the University of Washington to forecast toxic *Pseudo-nitzschia* events on the Washington coast based on the scientific insight into the ecology and oceanography of their blooms and toxicity gained during the studies mentioned above. The approach does not use satellite-derived chlorophyll to detect a toxic bloom as in the Gulf of Mexico and the Great Lakes because, even when *Pseudo-nitzschia* are abundant, the toxic cells make up only a small percentage of the total phytoplankton biomass (Trainer *et al.*, 2009). Forecasts are instead based on the likelihood that toxic cells from the Juan de Fuca eddy and Heceta Bank are transported to the coast.

353

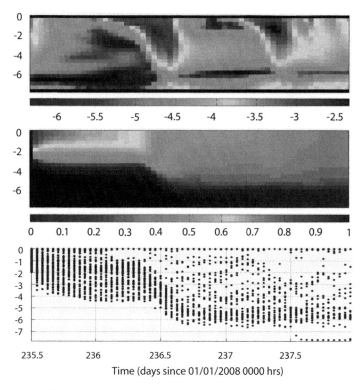

Figure 5. Modelled eddy-viscosity (top panel) and its use to examine tracer (middle panel) and particle dispersion (lower panel). (see colour plate 66)

Forecast generation involves deducing the likelihood of toxic cells in these retentive regions and assessing the likelihood that water parcels containing the toxic cells will arrive at coastal beaches. This includes locating the low-density plume of the Columbia River, which may act as a barrier to the onshore transport of toxic cells (Hickey & Banas 2003; Hickey *et al.*, 2005). Multiple types of environmental data are examined and synthesized by local experts with a detailed understanding of the physics of the retentive oceanographic features that promote toxicity of *Pseudo-nitzschia* cells and the conditions that advect shelf waters onto the coast in order to assess the risk of a toxic bloom occurring at beaches in Washington and Oregon. Periods when winds fluctuate between northward (retentive eddy) and southward for several days (release from the eddy and transport southward) as well as storm periods (northward winds), are considered particularly high risk especially if preceded by several days of good weather (southward winds).

The risk is communicated via the Pacific Northwest HAB bulletin and conveyed using a simple *traffic light* approach, where a green symbol indicates a low risk for a toxic HAB and a red symbol indicates a high risk. The PNW HAB bulletin also contains a summary of the rationale used to deduce these risks and the data examined. These data include: 1) maps of surface coastal velocities derived from high frequency radar and hydrodynamic models; 2) near-real time images of chlorophyll concentration from satellites; 3) SST from satellite measurements and numerical models; 4) a two-week time series of coastal winds from a National Data Buoy Center buoy in the region; 5) the location of the Columbia River plume and estimates of its runoff; 6) regional weather forecasts from the National Weather Service; 7) a cumulative upwelling index (CUI) calculated for the upwelling season using output from an atmospheric model; and 8) the number and type of *Pseudo-nitzschia* cells in samples collected at six beach monitoring sites by ORHAB (Trainer & Suddleson 2005). Chlorophyll and SST fields are used to determine whether the Juan de Fuca eddy is present, if the bloom is present offshore, and confirm if coastal upwelling has occurred. The

354

presence of colder temperatures near the coast indicates upwelling, movement away from the coast in surface layers and consequently a lower risk of toxic cells reaching the shore. Surface current velocities are used to predict whether the currents are likely to bring plankton from the Juan de Fuca eddy southward along the Washington shelf. The cells counts are identified and counted by light microscopy and grouped into large and small classes of *Pseudo-nitzschia*. When concentrations exceed a size-dependent threshold, specifically 50,000 cells/L for large cells and one million cells/L for small cells, toxin testing of seawater and shellfish by enzyme-linked immunosorbent assay is triggered (ELISA) (Litaker *et al.*, 2008).

The pilot bulletin provides several-day forecasts that are available only to managers for selected events coinciding with openings of the coastal razor clam beaches. They will likely become operational in two to five years, but the prototype is available for review (PNW HAB 2012). However, because of the complexity of species and toxicity variability, forecasts will necessarily be extremely conservative on the assumption that the retentive source regions always contain toxic cells. Until moored HAB sensors become readily available, this region would benefit greatly from weekly sample collections in the identified source regions of toxic cells over the summer season.

2.1.4 *Beach water quality*

The Center of Excellence for Great Lakes and Human Health (CEGLHH) is one of three centres sponsored by the oceans and human health initiative (OHHI) and has focused on developing a predictive capability for beach water quality and harmful algal blooms. CEGLHH has been successful in creating a mechanism for predicting the influence and movement of tributary plumes into the Great Lakes at several locations in the region. Through partnerships with several public health communities, CEGLHH continues to identify research needs that address public health needs for drinking and recreational water quality, and incorporate results into forecast models to better serve the public. Among these organizations are the Great Lakes beach association, the Great Lakes human health network, local drinking water utilities, the Michigan association for local public health, and Michigan environmental health association.

Current water quality monitoring involves a lag time between sample collection and water quality reporting. This may permit swimming at coastal beaches when bacterial levels could pose health threats or unregulated toxic algal blooms occur. Predictive models enable environmental and public health officials to notify the public of expected water quality one to two days in advance, thereby preventing beach closures when conditions are safe and avoiding negative local economic impacts. Current daily beach monitoring, which uses the 'persistence model' of using yesterday's *Escherichia coli* measurements to determine today's beach-bather water quality is not effective in predicting swimming conditions today. This model has been shown to be relatively ineffective and inaccurate, with error rates of up to fifty per cent. Nowcast models using statistical regression methods improve correct prediction of swimming conditions by twenty to thirty per cent or more over the current technology. This swimming information is provided by the afternoon of the same day. Forty-eight hour forecasts incorporate river and lake dynamics derived from process models, as well as forecasted atmospheric variables (rainfall, wind velocity and direction, and wave height) to enable beach managers to better inform the public of beach water quality, assist in the development of remediation projects to reduce bacterial sources to beaches, and provides families with data for recreational planning. Beach water quality nowcast and forecast models can be useful screening tools in determining where human health threats may be impacting recreational beaches.

One approach to improving beach forecasting capabilities, particularly for beaches that might be impacted by a nearby tributary, was to develop near-shore hydrodynamic models that can simulate and predict near-shore circulation and plume trajectories (Figure 6). The coloured contours in the figure indicate the concentrations (blue for low, red for high) of a hypothetical pathogen carried from the respective tributary (Burns Ditch, Indiana on the left, Clinton River, Michigan on the right) into the lake and potentially onto nearby swimming beaches (Indiana Dunes and Metro Beach). CEGLHH has developed several near-shore models and is currently evaluating their utility for beach quality forecasting. The purpose of the mechanistic river plume models is to provide information about the fate of river-borne pathogenic material once it enters the lake. Models have

355

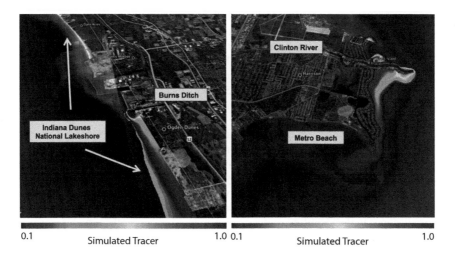

0.1   Simulated Tracer   1.0 0.1   Simulated Tracer   1.0

Figure 6.   River plume transport simulations at Burns Ditch, IN and Clinton River, MI. (see colour plate 67)

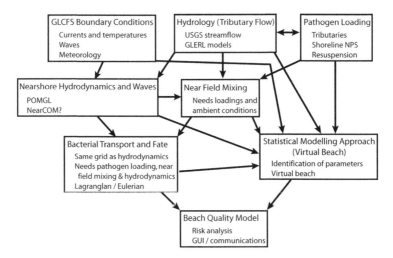

Figure 7.   Combined statistical and deterministic beach water quality modelling approach.

been developed for the Milwaukee River, the Grand River, and Burns Ditch in Lake Michigan; for the Saginaw River in Lake Huron; and for the Clinton River in Lake St. Clair. The current two-dimensional approach to plume concentration prediction will soon be replaced by a fully three-dimensional particle-based approach, which has been shown to be more accurate and versatile in simulating and predicting the fate of river plumes. The particle-based approach has the capability to simulate dispersion and decay of bacteria in a more realistic manner than the two-dimensional concentration-based approach.

A critical component of the mechanistic modelling framework is an estimate of bacterial concentration at the river mouth. To address this need, a new model for predicting bacteria loading rates from the contributing watershed will be added to the modelling framework. The proposed watershed-loading model will combine GIS-based land use assessments and hydrological model algorithms to forecast bacteria loadings and concentrations that will serve as inputs to the mechanistic river plume models (Figure 7).

2.2 Empirical model-based forecasts

Using spectral measurements of environmental parameters, both from space platforms and from instrumented ground networks is a common approach for monitoring or forecasting out-breaks of disease-causing organisms. Knowing how, when, and how often environments change, and observing the outcomes of those changes provides powerful, common sense interventions for curbing the impact of potentially deadly outcomes. Section 2.2 explores these capabilities for monitoring HABs. Doctors Jacobs, Wood, Brown, Trainer, Hickey and Moore discuss empirical approaches for forecasting V*ibrio spp.* in Chesapeake Bay and the Pacific Northwest, US.

2.2.1 Forecasting the likelihood of Vibrio vulnificus in Chesapeake Bay

The Cooperative Oxford Laboratory has been working with federal, state, and academic partners since 2005 to: 1) monitor potentially pathogenic species of bacteria in the Chesapeake region; 2) determine environmental factors which govern their distribution; 3) develop models and forecasts for use by public health officials; and 4), understand the implications for human and living resource health (Jacobs *et al.*, 2009a; Jacobs *et al.*, 2009b; Jacobs *et al.*, 2009c; Jacobs *et al.*, 2009d; Stine *et al.*, 2009; Jacobs *et al.*, 2010; Matsche *et al.*, 2010). A primary focus of this effort is the estuarine bacteria of the genus *Vibrio.*

One of these species, *Vibrio vulnificus*, is of particular concern because it is responsible for ninety-five per cent of all seafood related mortalities in the US (Oliver & Kaper 2001). Primary septicemia associated with seafood consumption and wound infections are the most common type of *V. vulnificus* infection in humans, with gastroenteritis occurring relatively infrequently (Strom & Paranjpye 2000). Cases which become septic have as high as a fifty per cent mortality rate (Rippey 1994; Oliver & Kaper 2001). Previous reports by the Centers for Disease Control and Prevention estimated *V. vulnificus* infections annually at ninety-seven total cases in the US, with forty-eight associated deaths (Mead *et al.*, 2000). Thus, it is extremely important to predict the presence and abundance of *V. vulnificus* in the interest of human health.

Several efforts have attempted to examine correlations of abundance or presence of *V. vulnificus* with environmental factors as a means to predict their occurrence (O'Neill *et al.*, 1992; Wright *et al.*, 1996; Lipp *et al.*, 2001; Heidelberg *et al.*, 2002; Pfeffer *et al.*, 2003; Randa *et al.*, 2004). While various parameters have been reported as being correlated to *V. vulnificus* abundance, water temperature, and to a lesser extent salinity, are consistently identified variables. In general, growth and abundance are positively correlated with water temperature when greater than 15°C. Salinity may also govern abundance, but the relationship is not as clear. Several researchers have reported estuarine waters of ten to fifteen parts per thousand as the preferred salinity for *Vibrios* (Lipp *et al.*, 2001; Randa *et al.*, 2004). The inconsistencies seen in the response of *V. vulnificus* to salinity gradients may result from regional and strain differences between these various studies.

Efforts are increasingly focusing on the development and application of empirical habitat suitability models of organisms in support of environmental forecasting (Brown *et al.*, 2002; Decker *et al.*, 2007; Constantin de Magny *et al.*, 2010; Jacobs *et al.*, 2010). In the Chesapeake Bay, the Chesapeake Bay ecological prediction system (CBEPS) is being developed and implemented by scientists at NOAA, the University of Maryland system, the Chesapeake Research Consortium, and the Maryland Department of Natural Resources. The CBEPS generates Bay-wide nowcasts and three-day forecasts of several environmental variables, including temperature and salinity (Brown *et al.*, in press). In simulation, these environmental variables are used to drive empirical habitat models of target organisms to make first order predictions of their likelihood of occurrence. Here we describe a predictive model developed to forecast the likelihood of *V. vulnificus* presence in surface waters of Chesapeake Bay.

The *V. vulnificus* habitat suitability model currently used is a logistic regression model that predicts the likelihood of its presence/absence based on ambient sea-surface temperature and salinity. In Chesapeake Bay, *V. vulnifius* was observed to occur most often in a narrow salinity range of ten to fifteen parts per thousand and is most prevalent at water temperatures above 15°C (Figure 8). The empirical model, developed by the Cooperative Oxford Laboratory, was constructed

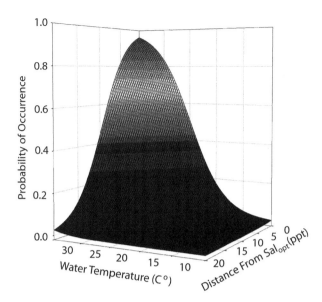

Figure 8.   Graphic representation of logit model using temperature and salinity to predict the probability of occurrence of *V. vulnificus* in Chesapeake Bay. (see colour plate 68)

from *V. vulnificus* counts enumerated using quantitative polymerase chain reaction (Panicker *et al.*, 2004; Panicker & Bej 2005) and contemporaneously sampled water quality data collected by the MD DNR and Virginia Department of Environmental Quality according to accepted protocols. The combination of water temperature and salinity in the habitat model correctly classifies the presence of *V. vulnificus* ninety-three per cent of the time in Chesapeake Bay.

The *V. vulificus* habitat model can be used to hindcast, nowcast and forecast the likelihood of *V. vulnificus* presence throughout Chesapeake Bay by forcing it with surface temperature and salinity simulated using the CBEPS. CBEPS uses a version of the ROMS adapted for Chesapeake Bay. This version uses near-real time observation data to provide model forcing, such as atmospheric momentum and heat fluxes, river outflow and ocean sea level, to simulate salinity, temperature, and other physical and biogeochemical variables in the Bay. For example, daily freshwater discharge data for nine major tributaries are acquired from the USGS stream water monitoring website and atmospheric forcing quantities (including three hourly wind, air temperature, relative humidity, and pressure) are obtained from NCEP/nmm. The Chesapeake version of ROMS has considerable skill in simulating temperature and salinity in the Bay. For the period from 2007–2009, both sea-surface temperature and salinity fields match relatively well with observations with an RMSE of one to two degrees centigrade for temperature and a RMSE of two to three for salinity, with model performance generally better in the main stem of the Bay than in its tributaries. A full assessment of the Chesapeake version of the ROMS temperature and salinity hindcast skill, and details about it, are reported in Xu (*et al.*, 2012).

Current forecasting products (Figure 9) are available from a restricted access web platform to Maryland and Virginia public and environmental health agencies, which are currently assessing how they can make use of the *Vibrio spp.* nowcasts and forecasts to streamline their monitoring efforts and aid in providing public warnings and determining beach closures. Dissemination of these types of sensitive products is an issue that needs to be addressed by the public health community; it is not clear that making these products available to the public is the best way to protect public health. Rather, it may be necessary to keep these products out of public view and provide them only to public health officials.

This effort represents one of the most spatially intensive sampling programs for potential human pathogens ever attempted in Chesapeake Bay. The efforts are culminating in a suite of models and

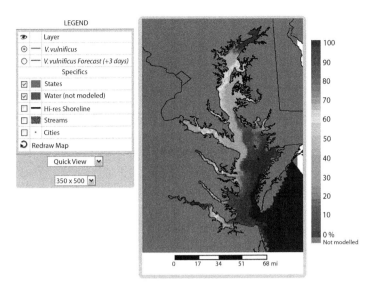

Figure 9. *Vibrio vulnificus* model output for 11 August 2010 and web interface provided to state and county health officials. Scale represents probability of occurrence from 0 (blue) to 100% (red). No data are available for areas shaded grey. (see colour plate 69)

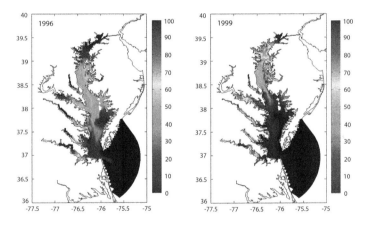

Figure 10. Hindcast depicting probability of occurrence of *V. vulnificus* in wet (1996) and dry (1999) years. Both figures represent conditions present on 1 August. (see colour plate 70)

forecasts. The *V. vulnificus* model represents the first of these forecasts and others are being constructed (e.g. *V. parahaemolyticus*). Because these variables are modified by climate, large changes in the distribution of *V. vulnificus* can occur annually and perhaps over longer time frames with potential global climate change. For example, hindcasts of *V. vulnicifus* suggest climate variability (wet-cool years vs. warm-dry) is expected to play a major role in their spatial and temporal distribution (Figure 10).

2.2.2 *Harmful algal blooms in Chesapeake Bay*
Similar to the monitoring and prediction of *Vibrio* species, the forecasting of HABs is also of interest in the Chesapeake Bay. Numerous HAB species, many of which are seasonal, are present in Chesapeake Bay and can achieve high concentrations. The most prevalent and potentially toxic HAB

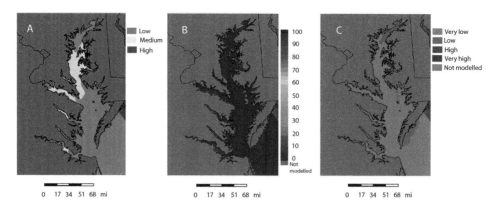

Figure 11. Examples of harmful algal bloom forecasts for the Chesapeake Bay using a hybrid mechanistic-empirical approach: (a) relative abundance of *Karlodinium veneficum*, (b) likelihood of a *Prorocentrum minimum* bloom, and (c) likelihood of a *Microcystis aeruginosa* bloom. (see colour plate 71)

species include *Alexandrium monilatum, Cochlodinium polykrikoides, Microcystis aeruginosa, Prorocentrum minimum,* and *Karlodinium veneficum*. Blooms of these species may impact regional finfish and shellfish populations, degrade water quality conditions, and affect water based industries and human health. The Maryland Department of Natural Resources and the Virginia Department of Environmental Quality require short (one to three day) and longer-term (seasonal) predictions of HAB events to protect fisheries and human health, and to direct sampling of toxin and precursor indicators in potential bloom areas, as well as meet a statutory requirement to assess the presence of HABs.

With the possible exception of the cyanobacterium *M. aeruginosa*, the species responsible for HAB events in the Chesapeake Bay cannot be remotely identified in satellite imagery and another method to detect and forecast them is required. It is known that HABs are characterized by the proliferation and dominance of a particular species of potentially toxic or otherwise potentially harmful algae. This proliferation and toxin production are the result of a suite of complex interactions between habitat, physical processes, life cycles, community structures, and growth and grazing behaviours. These bio-physical interactions all combine to regulate the dynamics of the HAB event. The exact mechanisms and environmental conditions that give rise to most HABs are poorly understood and it is therefore not currently possible to construct mechanistic models to predict blooms of HAB species. Although the specific mechanisms and environmental interactions that give rise to HABs are not fully understood, general habitat preferences are known for a number of species and statistically significant empirical relationships have been established between HAB occurrence and environmental factors, such as temperature, salinity, and nutrient concentrations.

As a consequence, the approach used to forecast the HABs in Chesapeake Bay is the same as described above for *V. vulnificus* and involves forcing multivariate empirical habitat suitability models for each HAB species with numerical model simulations of habitat-relevant, environmental variables.

Starting in 2011, forecasts of the relative abundance or the likelihood of a bloom of three HAB species in the Chesapeake Bay were produced experimentally using this hybrid mechanistic – empirical approach: the dinoflagellates *Karlodinium veneficum* and *Prorocentrum minimum* and the cyanobacterium *M. aeruginosa* (Brown *et al.*, in press; Brown *et al.*, unpubl.). These most current short-term forecasts are staged and available on the Mapping Harmful Algal Blooms in the Chesapeake Bay website (NOAA 2012) (Figure 11).

The habitat models of the three HAB species were constructed using different statistical and advanced computing techniques (Table 2). Standard regression techniques are not appropriate to define habitat models for HABs due to a large proportion of zero density observations. Threshold concentrations for defining bloom levels for *P. minimum* and *M. aeruginosa* are based on living

Table 2. Summary of harmful algal bloom habitat suitability models used in the Chesapeake Bay HAB prediction system.

Species	Model type	Input variables*	Output	Model accuracy**
Karlodininum veneficum	Neural network	SST, SSS, month	Relative abundance	84%
Prorocentrum minimum	Logistic regression	Chl*a*, NH4, TON TSS, month	Probability of bloom occurrence	88%
Microcystis aeruginosa	Hierarchical decision tree	Chl*a*, DIN, DO Kd, NH4, SSS TN, TSS	Probability of bloom occurrence	90%

*Acronyms: Chl*a* = chlorophyll-a; DIN = dissolved inorganic nitrogen; DO = dissolved oxygen; Kd = diffuse attenuation coefficient; NH4 = ammonium; SSS − sea surface salinity; SST − sea surface temperature; TN = total nitrogen; TON = total organic nitrogen; TSS = total suspended solids
**Correct forecasts ÷ N

resource and human health effects and therefore the forecasts are relevant to coastal managers. A habitat suitability model is used to predict the likelihood of *Pseudo-nitzschia* spp., the species complex that produces the potent DA neurotoxin responsible for amnesic shellfish poisoning (ASP), yet some of the environmental variables used by the habitat model at present cannot be simulated, and the likelihood of their blooms cannot be forecast in Chesapeake Bay (Anderson *et al.*, 2010).

Both abiotic and biotic environmental variables are used by the habitat suitability model (Table 2) and are simulated with CBEPS, the same prediction system used in forecasting *V. vulnificus* as described above. The HAB habitat models perform well when tested against a subset of *in situ* data that were not used to train them, with estimates of accuracy ranging from eighty-four to ninety per cent (Table 2). However, this model validation does not provide an assessment of the integrated forecasts because it does not take into account errors associated with estimating the forcing variables (e.g. temperature and salinity) generated by the Chesapeake Bay version of ROMS.

Evaluation of the integrated HAB forecasts has been initiated but remains rudimentary at this stage. The comparison of nowcasts and *in-situ* observations has only been performed adequately for the dinoflagellate *K. veneficum*. The results indicate the nowcasts perform reasonably well, though not as well as the accuracy statistics estimated during model validation. Comparison of nowcasts of *K. veneficum* relative abundances to those observed at twenty-four Chesapeake Bay Program stations for a three-year period from January 2007 to January 2009 indicates that the predictions performed better than chance (thirty-three per cent), with an overall accuracy of correct classification of fifty-nine per cent in the Bay. A considerable amount of work still needs to be done to fully assess the skill of these forecasts. In future, the Chesapeake Bay version of ROMS will assimilate satellite-derived estimates and *in-situ* measurements to improve the forecast skill of the physical variables (e.g. salinity) and forecasts will use satellite imagery to improve HAB detection.

2.2.3 *Harmful algal blooms in Puget Sound (US)*

As mentioned previously, blooms of *Alexandrium* are responsible for the majority of HAB issues in the inland waters of Puget Sound, with shellfish harvesting closures due to unsafe levels of toxins in shellfish tissues increasing in recent years and now occurring almost annually throughout the Sound (Trainer *et al.*, 2003; Moore *et al.*, 2009). An empirical approach is under development to forecast their presence. The NOAA West Coast Center for Oceans and Human Health identified and validated the components of a forecast using a fifteen year record of paralytic shellfish toxin levels in the blue mussel, *Mytilus edulis*, from 1993–2007 (Moore *et al.*, 2009). The forecast uses a *window of opportunity* approach, whereby periods of time with an increased risk of a toxic event are identified based on the presence of a specific combination of environmental conditions. A wider window of opportunity translates into an increased risk for *Alexandrium* blooms. In general, warm

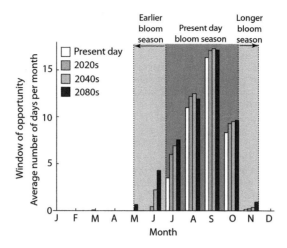

Figure 12.   Average monthly values of the *window of opportunity* for *Alexandrium* in Puget Sound for the 1980s and in the future for the 2020s, 2040s, and 2080s. The dark grey shaded area represents the present day bloom season (2010–2019) and the lighter grey shaded area represents a wider future bloom season that begins earlier in the year and persists longer (Modified from Moore *et al.*, 2011). (see colour plate 72)

air and water temperatures, weak winds, low stream flow, and small tidal variability were found to increase the risk of blooms. This approach is similar in concept to the one employed in forecasting *Vibrio* and HABs in the Chesapeake Bay, though the details and implementation differ.

The NOAA Northwest Fisheries Science Center and the National Weather Service are currently investigating the possibility of providing several-day forecasts for *Alexandrium* blooms in Puget Sound using the window of opportunity approach. This timescale is most useful for shellfish growers and managers to implement mitigation strategies and better target monitoring practices. For example, strategies that shellfish growers can implement include selectively harvesting different growing areas or increasing pre-bloom harvests to partially offset losses during bloom periods, and managers can better allocate limited resources to monitoring by targeting hot spot locations during time periods with increased risk for a bloom, or close growing areas during bloom periods more selectively than they otherwise could without the forecast. Pilot forecasts of the window of opportunity for *Alexandrium* in Puget Sound on a three to five day timescale will likely be available in two years.

2.3   Scenario-based forecasts

2.3.1   Harmful algal blooms in Puget Sound

In order to forecast the possible effects of climate change on the risk of *Alexandrium* blooms in Puget Sound, the *window of opportunity* approach used to generate short-term forecasts was applied to three future time periods (the 2020s, 2040s, and 2080s) using climate change projections for the Pacific Northwest region. The future scenarios for the window of opportunity were calculated using the IPCC greenhouse gas and sulphate aerosol emissions scenario A1B (Nakicenovic 2000). This scenario describes a world with rapid economic growth and a global population that peaks in the mid-21st Century with a rapid increase in carbon dioxide emissions early in the century and stabilization after mid-century due to the rapid introduction of new and efficient technologies. Of the ensemble of global climate models (GCMs) with archived data, it has been found that twenty simulations of 21st century climate with the A1B scenario were able to capture past climate variations in the Pacific Northwest (Mote & Salathé 2010). Average changes to the environmental conditions that comprise the HAB forecast were calculated from these GCMs for each future time period (Moore *et al.*, 2011). The resulting forecast indicates that by the end of the 21st Century, *Alexandrium* blooms may begin up to two months earlier in the year and persist for one month later compared to second decade of the century (Figure 12). Changes to the duration of the bloom season appear to be imminent and are detectable within thirty years using this forecast. The extended lead time offered by these projections will allow managers to put mitigation measures in place faster and more effectively to protect human health against these toxic outbreaks.

3 DISCUSSION AND NEXT STEPS

The HAB and water-borne pathogen forecasts briefly described in the previous section represent a subset of those currently under development and operational by NOAA and its partners for US coastal waters. More forecasts at various stages of development exist, such as nowcasts of the concentration of the bacterium *V. parahaemolyticus* in oysters of the Gulf of Mexico (Grimes, pers. comm.) and predictions of blooms of the dinoflagellate *A. fundyense* in the Gulf of Maine (Anderson *et al.*, 2005; McGillicuddy *et al.*, 2005; McGillicuddy *et al.*, 2011), with the number of HAB forecasts currently exceeding those of water-borne pathogens.

Nevertheless, examination of even this limited number of forecasts reveals several important characteristics. Foremost of these is that forecasts are regionally centric. They are designed to meet regional differences in user needs, data availability, and the HAB or water-borne pathogen of interest. From providing short-term forecasts of the likely transport, expansion and impact of *K. brevis* blooms in southwest Florida by first detecting them in satellite imagery and then having an analyst predict their movement based on wind velocity and other factors, to supplying long-term projections of the risk of toxic *Alexandrium* blooms in Puget Sound several decades into the future by applying environmental conditions associated with their present occurrence to climate change scenarios, the projects tailored their approaches based on the needs of local users, the availability of observations and model output, and the ecology and attributes of the target organism for the region.

As a consequence, a diversity of techniques that employ a range of data has evolved to generate the predictions in the different regions. Some techniques and approaches work well in one region but not in another, and alternative methods and data must be exploited. Satellite measurements, for example, are essential for initially detecting *K. brevis* in waters of Florida's West Coast but are not of significant help in identifying blooms of *P. minimum* in Chesapeake Bay. As a result, an empirical approach is used to detect these blooms.

A difference in methodologies between regions does not imply that one technique cannot be applied elsewhere. The techniques developed in one region can often be applied to other locations, either directly or with slight modification, thereby augmenting the existing method(s). For example, maps of cyanobacterial index estimated from satellite measurements, as developed and applied in the Great Lakes, can be applied to detect *M. aeruginosa* in Chesapeake Bay. Combining this satellite-derived information with predictions of the likelihood of *M. aeruginosa* blooms generated using the mechanistic – empirical habitat model approach should improve the accuracy of detecting and identifying these blooms in Chesapeake Bay by providing additional information into the analysis process. Though the extent of exchanging methods between regions will be limited by the characteristics of the target organism or event, it is expected that the adoption of techniques and products developed and implemented in other regions is likely to become widespread when the same modelling and observational capabilities are available in all regions.

Though not detailed in the examples of the previous section, the involvement and inclusion of local users and stakeholders throughout the process, even after it has become operational, is critical for the success of creating and maintaining a useful forecast. Developers need to work with the user(s) to define and refine a forecast to make it locally relevant. Many of the forecasts under development are likely candidates to be transitioned to operational products and services based on their adoption by users, whereas it is too early to tell for others, which first have to be vetted by users and possibly modified to meet their needs. For example, the presence of *K. veneficum* blooms does not necessarily signify high levels of toxins associated with this HAB species and the presence of *V. vulnificus* does not equate to the risk of infection and illness. Both the toxin and risk assessments may be of more use to the users, such as state health officials, and the present forecasts would therefore have to be modified to provide the relevant information.

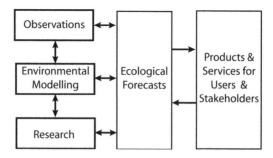

Figure 13.   Notional framework for an ecological forecasting system.

Recognition of regional differences and the concept of tailoring products to local user needs are vital in developing operational HAB and pathogen forecasts, yet it is also important to note their commonalities. All forecast projects comprise a common set of component activities in their prediction system that, when linked together, define the general framework of an operational eco-logical forecasting system (Figure 13). The key activities of this system include: 1) conducting research in multiple fields of study, from physical to social sciences, to understand and quantify the cause(s) of the problem and their impacts; 2) performing environmental modelling on local to global scales in order to generate appropriate meteorological, physical, geochemical and ecological output; 3) collecting, compiling, archiving and disseminating observations of physical, geochemi-cal, and biological variables, both real-time and historical, for numerical model initialization, data assimilation, and product validation, enhancement and skill assessment; 4) generating ecological model guidance and forecasts at appropriate space and time scales; and 5) using the forecasts to make educated short- to long-term decisions. The environmental modelling and observations, which benefit from advances in research and technological development, are used to generate the ecological predictions. These predictions, either used directly or interpreted by a trained ana-lyst, provide the informational products and services employed in the decision making process. Refinement and improvement of the ecological forecasts through successive iterations with the user community will also drive additional research and enhancements in environmental modelling and observational infrastructure and technology.

With the exception of actually making the decisions based on the forecasts provided, which lies entirely in the purview of the user, the elements of this integrated ecological forecast framework (Figure 13) exist within current capabilities of NOAA. Supplementing these components would thereby strengthen existing health related and other ecological forecasts and development of new ones. In regards to research, many avenues can be pursued. For example, there is a need to better understand the factors involved in the regulation of HAB and pathogen presence, abundance, and toxicity/virulence (Dyble *et al.*, 2008). This knowledge, in turn, can be used to develop desperately needed mechanistic models (e.g. Gregg *et al.*, 2003; Moore *et al.*, 2004; Le Quéré *et al.*, 2005) to generate the forecasts and avoid the limitations imposed by empirical approaches. Empirical models, for instance, are restricted to the observational domain, both temporal and spatial, from which the data were collected. Applying an empirical habitat model beyond its observational domain should be done with caution, such as extrapolating over long time scales to predict long-term changes. Furthermore, empirical models do not necessarily provide significant insights into the various processes determining the distribution pattern of species, such as top down control, as would a mechanistic approach. Funding for basic and applied research into human and ecosystem related health in coastal regions, such as through NOAA's Oceans and Human Health Initiative and Harmful Algal Bloom Program, provides these new and innovative technologies and approaches.

Timely access to reliable observations and numerical model output is also a critical component of an operational ecological forecasting system (Jochens *et al.*, 2010). Towards that end, attempts by many federal agencies are underway to ensure that observational platforms and networks pro-vide a full suite of interoperable capabilities, with many of the relevant issues addressed through coordination and alignment with the integrated ocean observing system (IOOS) and various local, regional and global observing networks. This capability will improve access to existing observations as well as enable the rapid broadcasting and use of new measurements once they become available. In addition, the development of new sensors, such as *in-situ* genomic sensors, and new platforms, such as unmanned aerial and underwater vehicles with adaptive sampling capabilities, offers the opportunity to acquire new types of information at spatial and temporal scales not currently sampled.

As the same environmental variables are used to generate many of the regional ecological forecasts, such as sea-surface temperature and current velocity, NOAA is considering the construction of an environmental modelling backbone to provide a consistent set of forcing and boundary conditions, including meteorological, hydrology, nutrient and sediment supply. This could integrate NOAA's operational modelling capacity that exists in the National Ocean Service and the National Weather Service, and provide uniform modelling approaches that could be applied to several ecological forecasts. This approach would support regional modelling and recognize that the sensors and physical modelling backbone required may vary with the ecological problem (site) of concern. Furthermore, by coupling validated ecological and biogeochemical models with real time weather forecasts and assimilation of real-time marine and coastal-ocean data acquired through an integrated ocean observing system as described above, it will be possible to provide more accurate and relevant operational forecasts of the location, time, and duration of ecosystems hazards and long term threats.

Ideally, the modelling prediction system in a region would ultimately assume the shape of a regional Earth System model, a fully integrated, coupled model of the regional ecosphere, which includes the atmosphere, land, and ocean and the biosphere, and the anthroposphere, which includes human activity and its impact on the ecosphere (Murtugudde, 2009). Such a model could be used for both real time applications, such as short-term of marine organisms and ocean health, and climate research, such as projecting the effect of climate change on the health of coastal marine ecosystems. Though the generation of an ecological prediction by a regional Earth System model is unlikely to be achieved any time soon, its explicit modelling of human behaviour and their effect on the ecosystem is an important aspect that has not received sufficient attention to date and should be pursued.

4 CONCLUSIONS

In summary, ecological forecasting has made great strides as scientific understanding of ecosystem structure and function has matured. These advances, along with improvements in integrated observations and modelling, computational power, data management, IT and telecommunications, have made it possible for a number of critical coastal environment, health, and safety issues to be addressed and forecast. NOAA, through its collaborative relationships with other agencies and institutions, as well as local users and stakeholders, is leveraging and building upon its existing technical capabilities and scientific expertise to develop, generate and issue operational forecasts of HAB and water-borne pathogens for US coastal waters and make these ecological forecasts an integral part of its environmental prediction portfolio.

ACKNOWLEDGMENTS

The author/editor thanks Michelle Hawkins for supplementing Table 1, Michelle Tomlinson and Kathleen Fisher for elucidating the finer points of the NOAA HAB-OFS, and Quay Dortch and Marc Suddleson for improving an earlier draft of the Chapter. The models developed for *Vibrio* spp. could not have been completed without the extensive efforts of the Maryland Department of Natural Resources and Virginia Department of Environmental Quality's respective water quality monitoring programs, and the technical skill and expertise of Matt Rhodes (JHT Inc./NOAA Oxford

Laboratory) in sample processing. Support for this work was provided by the NOAA Center for Satellite Applications and Research, the NOAA Center for Sponsored Coastal Ocean Research (NA09NOS4780180) and the National Science Foundation (OCE0942675) as part of the PNWTOX project (BH), and the West Coast Center for Oceans and Human Health as part of the NOAA OHHI. This is GLERL Contribution No. 1618.

REFERENCES

Adams, N.G., Lesoing, M. & Trainer, V.L. 2000. Environmental conditions associated with domoic acid in razor clams on the Washington coast. *J. Shellfish Res.* 19(2): 1007–1015.

Anderson, C.R., Sapiano, M.R.P., Prasad, M.B.K., Long, W., Tango, P.J., Brown, C.W. & Murtugudde, R. 2010. Predicting potentially toxigenic *Pseudo-nitzschia* blooms in the Chesapeake Bay. *J. Marine Syst.* 83(3–4): 127–140.

Anderson, D.M., Keafer, B.A., McGillicuddy, Jr., D.J., Mickelson, M.J., Keay, K.E., Scott, L.P., Manning, J.P., Mayo, C.A., Whittaker, D.K., Michael Hickey, J., He, R., Lynch, D.R. & Smith, K.W. 2005. Initial observations of the 2005 *Alexandrium fundyense* bloom in southern New England: General patterns and mechanisms. *Deep Sea Res. Part 2 Top. Stud. Oceanogr.* 52(19–21): 2856–2876.

Austin, M.P. 2002. Spatial prediction of species distribution: An interface between ecological theory and statistical modelling. *Ecol. Model.* 157(2–3): 101–118.

Brown, C.W., Hood, R.R., Li, Z., Decker, M.B., Gross, T., Purcell, J.E. & Wang, H. 2002. Forecasting system predicts presence of sea nettles in Chesapeake Bay. *Eos, Trans. Amer. Geophys. Union* 83(30): 321, 325–326.

Brown, C.W., Hood, R.R., Long, W., Jacobs, J.M., Ramers, D.L., Wazniak, C., Wiggert, J.D., Wilson, D., Wood, R. & Xu, J. (in press). Ecological forecasting in Chesapeake Bay: Using a mechanistic-empirical modelling approach. *J. Marine Syst.*

Brown, C.W., Ramers, D.L., Hood, R.R., Wazniak, C., Long, W. & Tango, P. (unpublished). Predicting the relative abundance of the dinoflagellate *Karlodinium veneficum* in the Chesapeake Bay.

Constantin de Magny, G., Long, W., Brown, C.W., Hood, R.R., Huq, A., Murtugudde, R. & Colewell R.R. 2010. Predicting the distribution of *Vibrio spp.* in the Chesapeake Bay: *Vibrio cholerae* case study. *Ecohealth* 6(6): 378–389.

CO-OPS 2012. Available from: http://tidesandcurrents.noaa.gov/hab/ [Accessed 20th March 2012].

Decker, M.B., Brown, C.W., Hood, R.R., Purcell, J.E., Gross, T.F., Matanoski, J.C., Bannon, R.O. & Setzler-Hamilton, E.E. 2007. Predicting the distribution of the scyphomedua *Chrysaora quinquecirrha* in Chesapeake Bay. *Marine Ecol. Prog. Series* 329: 99–113.

Dyble, J., Bienfang, P., Dusek, E., Hitchcock, G., Holland, F., Laws, E., Lerczak, J., McGillicuddy, D., Minnett, P., Moore, S., O'Kelly, C., Solo-Gabriele, H. & Wang, J. 2008. Environmental controls, oceanography and population dynamics of pathogens and harmful algal blooms: Connecting sources to human exposure. *Environ. Health* 7(Supp2): S5

Dyson, K. & Huppert, D.D. 2010. Regional economic impacts of razor clam beach closures due to harmful algal blooms (HABs) on the Pacific coast of Washington. *Harmful Algae.* 9(3): 264–271.

Fisher, K.M., Allen, A.L., Keller, H.M., Bronder, Z.E., Fenstermacher, L.E. & Vincent, M.S. 2006. Annual report of the Gulf of Mexico Harmful Algal Bloom Operational Forecast System (GOM HAB-OFS). *NOAA Technical Report.* Silver Spring: NOAA NOS CO-OPS.

GLCFS. 2012. Available from: http://www.glerl.noaa.gov/res/Centers/HABS/lake_erie_hab/lake_erie_hab.html [Accessed 20th March 2012].

Gregg, W.W., Ginoux, P., Schopf, P.S. & Casey, N.W. 2003. Phytoplankton and iron: Validation of a global three-dimensional ocean biogeochemical model. *Deep-Sea Research II* 50(22–26): 3143–3169.

Guisan, A. & Zimmermann, N.E. 2000. Predictive habitat distribution models in ecology. *Ecol. Model.* 135(2–3): 147–186.

HAB-OFS. 2012. Available from: http://tidesandcurrents.noaa.gov/hab/bulletins.html [Accessed 30th March 2012].

Heidelberg J.F., Heidelberg, K.B. & Colwell, R.R. 2002. Seasonality of Chesapeake Bay bacterioplankton species. *Appl. & Environ. Microbio.* 68(11): 5488–5497.

Hickey, B.M. & Banas, N.S. 2003. Oceanography of the US Pacific Northwest coastal ocean and estuaries with application to coastal ecology. *Estuaries* 26(4B): 1010–1031.

Hickey, B.M., Geier, S., Kachel, N. & Macfadyen, A. 2005. A bi-directional river plume: The Columbia in summer. *Continental Shelf Res.* 25(14): 1631–1636.

Hickey, B.M., Trainer, V.L., Kosro, P.M., Adams, N.G., Connolly, T.P., Kachel, N.B. & Geier, S.L. (in press). Seasonal differences in sources of toxic *Pseudo-nitzschia* cells on Washington's razor clam beaches. *J. Geophys. Res.*

Jacobs, J., Rhodes, M., Sturgis, B. & Wood, B. 2009a. Influence of environmental gradients on the abundance and distribution of *Mycobacterium* spp. in a coastal lagoon estuary. *Appl. & Environ. Microbio.* 75(23): 7378.

Jacobs, J.M., Howard, D.W., Rhodes, M.R., Newman, M.W., May, E.B. & Harrell, R.M. 2009b. Historical presence (1975–1985) of mycobacteriosis in Chesapeake Bay striped bass *Morone saxatilis*. *Dis. Aquatic Org.* 85(3): 181–186.

Jacobs, J.M., Rhodes, M.R., Baya, A., Reimschuessel, R., Townsend, H. & Harrell, R.M. 2009c. Influence of nutritional state on the progression and severity of mycobacteriosis in striped bass *Morone saxatilis*. *Dis. Aquatic Orgs.* 87(3): 183–197.

Jacobs, J.M., Stine, C.B., Baya, A.M. & Kent, M.L. 2009d. A review of mycobacteriosis in marine fish. *J. Fish Dis.* 32(2): 119–130.

Jacobs, J.M., Rhodes, J., Brown, C.W., Hood, R.R., Leigh, A., Long, W. & Wood, R. 2010 Predicting the distribution of *Vibrio vulnificus* in Chesapeake Bay. *NOAA Technical Memorandum*. NOAA National Center for Coastal Ocean Science, Center for Coastal Environmental Health and Biomolecular Research: Oxford Laboratory.

Jochens, A.E., Malone, T.C., Stumpf, R.P., Hickey, B.M., Carter, M., Morrison, R., Dyble, J., Jones, B. & Trainer, V.L. 2010. Integrated ocean observing system in support of forecasting harmful algal blooms. *Mar. Technol. Soc. J.* 44(6): 99–121.

Lanerolle, L.W.J., Tomlinson, M.C., Gross, T.F., Aikman, F., III, Stumpf, R.P., Kirkpatrick, G.J. & Pederson, B.A. 2006. Numerical investigation of the effects of upwelling on harmful algal blooms off the west Florida coast. *Estuarine Coastal & Shelf Sci.* 70(4): 599–612.

Le Quéré, C., Harrison, S.P., Prentice, C., Buitenhuis, E.T., Aumonts, O., Bopp, L., Claustre, H., Da Cunha, L.C., Geider, R.J., Giraud, X., Klaas, C., Kohfeld, K.E., Legendre, L., Manizza, M., Platt, T., Rivkin, R.B., Sathyendranath, S., Uitz, J., Watson, A.J. & Wolf-Gladrow, D. 2005. Ecosystem dynamics based on plankton functional types for global ocean biogeochemistry models. *Global Change Biol.* 11(11): 2016–2040.

Lipp, E.K., Rodriguez-Palacios, C. & Rose, J.B. 2001. Occurrence and distribution of the human pathogen *Vibrio vulnificus* in a subtropical Gulf of Mexico estuary. *Hydrobiologia* 460(1): 165–173.

Litaker R.W., Stewart T. N., Eberhart B.T., Wekell J.C., Trainer V.L., Kudela R.M., Miller P.E., Roberts A., Hertz C., Johnson T.A. Frankfurter G., Smith G.J., Schnetzer A., Schumacker J., Bastian J.L, Odell A., Gentien P., Dominique L.G., Hardison D.R. & Tester, P.A. 2008. Rapid enzyme-linked immunosorbent assay for the detection of the algal toxin domoic acid. Harmful Algae 27(5):1301–1310.

Matsche, M.A., Overton, A., Jacobs, J., Rhodes, M.R. & Rosemary, K.M. 2010. Low prevalence of splenic mycobacteriosis in migratory striped bass *Morone saxatilis* from North Carolina and Chesapeake Bay, USA. *Dis. Aquatic Orgs.* 90(3): 181–189.

McGillicuddy Jr., D.J., Anderson, D.M., Solow, A.R. & Townsend, D.W. 2005. Interannual variability of Alexandrium fundyense abundance and shellfish toxicity in the Gulf of Maine. *Deep Sea Res. Part 2 Top. Stud. Oceanogr.* 52(19–21): 2843–2855.

McGillicuddy Jr., D.J., Townsend, D.W., He, R., Keafer, B.A., Kleindinst, J.L., Li, Y., Manning, J.P., Mountain, D.G., Thomas, M.A. & Anderson, D.M. 2011. Suppression of the 2010 Alexandrium fundyense bloom by changes in physical, biological, and chemical properties of the Gulf of Maine. *Limno.l Oceanogr.* 56(6): 2411–2426.

Mead, P.S., Slutsker, L., Dietz, V., McCaig, L.F., Bresee, J.S., Shapiro, C., Griffin, P.M. & Tauxe, R.V. 2000. Food-related illness and death in the United States. *J. Environ. Health* 62(7): 607–625.

Moore, J.K., Doney, S.C. & Lindsay, K. 2004. Upper ocean ecosystem dynamics and iron cycling in a global three-dimensional model. *Global Biogeochem. Cycles*, 18(4): GB4028. doi:10.1029/ 2004GB 0022220.

Moore, S.K., Mantua, N.J., Hickcy, B.M. & Trainer, V.L. 2009. Recent trends in paralytic shellfish toxins in Puget Sound, relationships to climate, and capacity for prediction of toxic events. *Harmful Algae* 8(3): 463–477.

Moore, S.K., Mantua, N.J. & Salathé, E.P.J. 2011. Past trends and future scenarios for environmental conditions favoring the accumulation of paralytic shellfish toxins in Puget Sound shellfish. *Harmful Algae*, 10(5): 521–529.

Mote, P.W. & Salathé, E.P.J. 2010. Future climate in the Pacific Northwest. *Climate Change* 102(1–2): 29–50.

Murtugudde, R. 2009. Regional Earth system prediction: A decision-making tool for sustainability? *Curr. Opin. Environ. Sustainability*. 1(1): 37–45.

Nakicenovic, N. & Swart, R. (eds.). 2000. IPCC Special Report on Emissions Scenarios. Cambridge: Cambridge UP.

NOAA. 2012. Available from: http://155.206.18.162/cbay_hab/ [Accessed 20th March 2012].

O'Neill, K.R., Jones, S.H. & Grimes, D.J. 1992. Seasonal incidence of *Vibrio vulnificus* in the Great Bay estuary of New Hampshire and Maine. *Appl. & Environ. Microbiol.* 58(10): 3257–3262.

Oliver, J.D. & Kaper, J.B. 2001. *Vibrio* species. In: M.P. Doyle (ed.), *Food microbiology: Fundamentals and frontiers:* 263–300. Washington DC: ASM Press.

Panicker, G. & Bej, A.K. 2005. Real-time PCR detection of *Vibrio vulnificus* in oysters: Comparison of oligonucleotide primers and probes targeting *vvhA*. *Appl. & Environ. Microbiol.* 71(10): 5702–5709.

Panicker, G., Call, D.R., Krug, M.J. & Bej, A.K. 2004. Detection of pathogenic *Vibrio* spp. in shellfish by using multiplex PCR and DNA microarrays. *Appl. & Environ. Microbiol.* 70(12): 7436–7444.

Pfeffer, C.S., Hite, M.F. & Oliver, J.D. 2003. Ecology of *Vibrio vulnificus* in estuarine waters of eastern North Carolina. *Appl. & Environ. Microbiol.* 69(6): 3526–3531.

PNW HAB. 2012. Available from: http://www.pnwhabs.org/pnwhabbulletin/issue20070831.php [Accessed 20th March 2012].

Randa, M.A., Polz, M.F. & Lim, E. 2004. Effects of temperature and salinity on *Vibrio vulnificus* population dynamics as assessed by quantitative PCR. *Appl. & Environ. Microbiol.* 70(9): 5469–5476.

Rippey, S.R. 1994. Infectious diseases associated with molluscan shellfish consumption. *Clinic. Microbiol. Revs.* 7(4): 419–425.

Stine, C.B., Jacobs, J.M., Rhodes, M.R., Overton, A., Fast, M. & Baya, A.M. 2009. Expanded range and new host species of *Mycobacterium shottsii* and *M. pseudoshottsii*. *J. Aqua. Anim. Health* 21(3): 179–183.

Strom, M.S. & Paranjpye, R.N. 2000. Epidemiology and pathogenesis of *Vibrio vulnificus*. *Microbes & Infect.* 2(2): 177–188.

Stumpf, R.P., Culver, M.E., Tester, P.A., Tomlinson, M., Kirkpatrick, G.J., Pederson, B.A., Truby, E., Ransibrahmanakul, V. & Soracco, M. 2003. Monitoring *Karenia brevis* blooms in the Gulf of Mexico using satellite ocean color imagery and other data. *Harmful Algae* 2(2): 147–160.

Stumpf, R.P., Litaker, R.W., Lanerolle, L. & Tester, P.A. 2008. Hydrodynamic accumulation of *Karenia* off the west coast of Florida. *Continental Shelf Res.* 28(1): 189–213.

Stumpf, R.P., Tomlinson, M.C., Calkins, J.A., Kirkpatrick, B., Fisher, K., Nierenberg, K., Currier, R. & Wynne, T.T. 2009. Skill assessment for an operational algal bloom forecast system. *J. Marine Syst.* 76(1–2): 151–161.

Tomlinson, M.C., Stumpf, R.P., Ransibrahmanakul, V., Truby, E.W., Kirkpatrick, G.J., Pederson, B.A., Vargo, G.A. & Heil, C.A. 2004. Evaluation of the use of SeaWiFS imagery for detecting *Karenia brevis* harmful algal blooms in the eastern Gulf of Mexico. *Rem. Sens. Environ.* 91(3–4): 293–303.

Trainer, V.L. & Suddleson, M. 2005. Monitoring approaches for early warning of domoic acid events in Washington State. *Oceanogr.* 18(2): 228–237.

Trainer, V.L., Hickey, B.M. & Homer, R.A. 2002. Biological and physical dynamics of domoic acid production off the Washington coast. *Limn. & Oceanogr.* 47(5): 1438–1446.

Trainer, V.L., Eberhart, B.T.L., Wekell, J.C., Adams, N.G., Hanson, L., Cox, F. & Dowell J. 2003. Paralytic shellfish toxins in Puget Sound, Washington State. *J. Shellfish Res.* 22(1): 213–223.

Trainer, V.L., Hickey, B.M., Lessard, E.J., Cochlan, W.P., Trick, C.G., Wells, M.L., MacFadyen, A. & Moore, S.K. 2009. Variability of *Pseudo-nitzschia* and domoic acid in the Juan de Fuca eddy region and its adjacent shelves. *Limn. & Oceanogr.* 54(1): 289–308.

Wright, A.C., Hill, R.T., Johnson, J.A., Roghman, M.C., Colwell, R.R. & Morris, J.G. Jr. 1996. Distribution of *Vibrio vulnificus* in the Chesapeake Bay. *Appl. & Environ. Microbiol.* 62(2): 717–724.

Wynne, T.T., Stumpf, R.P., Tomlinson, M.C., Ransibrahmanakul, V. & Villareal, T.A. 2005. Detecting *Karenia brevis* blooms and algal resuspension in the western Gulf of Mexico with satellite ocean color imagery. *Harmful Algae* 4(6): 992–1003.

Wynne, T.T., Stumpf, R.P., Tomlinson, M.C., Warner, R.A., Tester, P.A., Dyble, J. & Fahnenstiel, G.L. 2008. Relating spectral shape to cyanobacterial blooms in the Laurentian Great Lakes. *Int. J. Rem. Sens.* 29(2): 3665–3672.

Wynne, T.T., Stumpf, R.P., Tomlinson, M.C. & Dyble, J. 2010. Characterizing a cyanobacterial bloom in western Lake Erie using satellite imagery and meteorological data. *Limn. & Oceanogr.* 55(5): 2025–2036.

Wynne, T.T., Stumpf, R.P., Tomlinson, M.C., Schwab, D.J., Watabayashi, G.Y. & Christensen, J.D. 2011. Estimating cyanobacterial bloom transport by coupling remotely sensed imagery and a hydrodynamic model. *Ecolog. Appl.* 21(7): 2709–2721.

Xu, J., Long, W., Wiggert, J.D., Lanerolle, L.W.J., Brown, C.W., Murtugudde, R. & Hood, R.R. 2012. Climate forcing and salinity variability in the Chesapeake Bay, USA. *Estuar. & Coasts* 35(1): 237–261.

Chapter 10

Information and decision support systems

W. Hudspeth (Auth./ed.)[1] with; W.K. Reisen[2], C.M. Barker[2], V. Kramer[3], M. Caian[4],
V. Crăciunescu[4], H.E. Brown[5], A.C. Comrie[5], A. Zelicoff[6], T.G. Ward[7], R.M. Ragain[8];
G. Simpson[9]; W. Stanhope[10]; T.A. Kass-Hout[11], A. Scharl[12], A.L. Sonricker[13,14] &
J.S. Brownstein[13,14]

[1] *Earth Data Analysis Center, University of New Mexico, Albuquerque, New Mexico, US*
[2] *School of Veterinary Medicine, University of California, Davis, Davis, CA, US*
[3] *California Department of Public Health, Sacramento, CA, US*
[4] *National Meteorological Administration, Bucharest, Romania*
[5] *School of Geography & Development, University of Arizona, Tucson, AZ, US*
[6] *Saint Louis University School of Public Health, St. Louis, MO, US*
[7] *City of Lubbock Health Department (retired), Lubbock, TX, US*
[8] *Saint Louis University School of Public Health, St Louis, MO, US*
[9] *New Mexico Department of Health (retired), Albuquerque, NM, US*
[10] *Saint Louis University School of Public Health, St Louis, MO, US*
[11] *Centers for Disease Control & Prevention, Atlanta, GA, US*
[12] *MODUL University, Vienna, Austria*
[13] *Massachusetts Institute of Technology, Boston, MA, US*
[14] *Children's Hospital, Boston, MA, US*

ABSTRACT: This chapter describes various approaches for extending environmental observations into actionable public health decisions.

1 INTRODUCTION

Decision support systems (DSS) have been defined as a class of computer-based information systems, and knowledge-based systems that support business and organizational decision making activities. Properly designed, a DSS is an interactive, software-based system that allows decision makers to compile information from raw data, documents, personal knowledge and/or business models to identify and solve problems (Wikipedia 2011). They belong to an environment with multidisciplinary foundations, including database research, artificial intelligence, human-computer interaction, simulation methods, software engineering and telecommunications.

While numerous taxonomies have been devised to classify decision support systems, Power's described criteria for distinguishing them. They include communication-, data-, document-, knowledge-, and model-driven DSSs (Power 2000). Public health DSSs can usually be described as both data-driven and model-driven, in which time series data, a detailed understanding of the environmental links to human health, and access to a simulation model guides decision makers.

All decision support systems share a number of core features. Three fundamental components of a DSS architecture are: 1) the database (or knowledge base); 2) the model (e.g. the decision context and user criteria); and 3) the user interface. There are also considerations for the development framework employed, which can be iterative in nature, such that the DSS can be changed and redesigned at various intervals. Once the system is designed it will need to be tested and revised for the desired outcome. For a public health DSS, the types of information normally required include: 1) an assessment of all current information assets; 2) understanding of the models and associated techniques used to integrate EO data with observed human health outcomes; 3) examples of historical cases of these linkages; 4) data on those current and forecasted environmental conditions known to affect human health outcomes; 5) the capability to predict health outcomes from current and forecasted environmental conditions; 6) a means for assessing relative degree of risk to health outcomes given environmental conditions; 7) a means to distil simplified risk warnings from complex results and conveying them to decision makers; 8) a means for assessing the utility and predictive power that actions taken will confer to outcomes; 9) production and dissemination of data and analytical outputs that support decisions and 10) implementing appropriate, evidence-based remediation efforts among decision makers.

Public health is an important outcome of decision support tools involving air quality, water management, energy management, and agricultural efficiency among others. Equally important have been the effects of weather and climate cycles on infectious disease systems. As discussed in earlier chapters of this book, these cycles impact several factors that trigger disease prevalence and transmission. Glass notes that public health communities have been slow to adopt EO data and observations. One of the most obvious barriers to more rapid adoption is that satellite platforms are not designed to monitor disease risk directly. Ideally, any system would have to be based on a proven history of verified and validated data that record changes in population distribution and environmental conditions associated with observed disease outcomes. Moreover, observations need to have demonstrable resolution and temporal frequency to actually identify when environmental and population changes occur. There are also privacy issues and restrictions to consider. Finally, Glass argues that aggregating health data to finer and finer geographic scales produces an ecological fallacy in linking cause and effect relationships. Moving responsibility for integrating decision support systems to higher levels of authority might ameliorate some of the challenges.

Decision support solutions for infectious diseases fall into two fundamentally different categories (Adler 2005). The first includes epidemiological models that simulate the progression of disease outbreaks and projects the rates of transmission among subsets of the population. The infectious disease outbreak decision support system (IDODSS), developed by DecisionPath®, is designed for organizations around the world to raise alarms about potential pandemics caused by cross-species migration of avian influenza (H1N1). IDODSS was stimulated by threats of bio-terrorism and US preparedness to respond. The system is designed to address preparedness at multiple levels from national to local, and to have operational strategies that allow responses to outbreaks in real-time. The syndromic reporting information system (SYRIS), described in Section 5, is one such; especially because it is designed around individual patient/doctor interviews and observations networked across multiple health care jurisdictions. In contrast, many IDODSS' employ analytical techniques such as differential equations, Monte Carlo simulations and dynamic network models (Lewis *et al.*, 2002; Lee *et al.*, 2006; Jain & McLean 2008; Brandeau *et al.*, 2009; Lee *et al.*, 2009).

The second category consists of crisis management for tracking and reporting those infected by a disease. Such systems include capacity planning and resource allocation such as hospital staffing and stocking medical supplies, continuity and recovery planning, and disaster preparedness and simulations. Crisis management tools are often situational analysis tools. They include database archives, analytical tools and capabilities, and related visualization and summary tools. They typically provide snapshots of current status, past performance of tool sets, and overviews of short term trends. Although true decision support systems enable active decision making, crisis management tools are a necessary prerequisite. Decision support systems generate alarms that require attention by public health authorities to act (Mnatsakyan & Lombardo 2008). Among this category of DSSs are those for applications requiring tracking and reporting. Often called business intelligence (BI), they include databases, data warehouses, analytics, dashboards, and other summary/visualization tools. BI tools provide insight into current status, historical performance, and short-term trends. BI is a necessary prerequisite to decision support, but used alone are not sufficient.

The support generated by a DSS can also be targeted at three distinct, interrelated entities: personal, group, and organizational. System components are often classified in four categories as: *Inputs*: factors, numbers, and characteristics; *User Knowledge and Expertise*: inputs requiring manual analysis by the user; *Outputs*: transformed data from which DSS decisions are generated; and *Decisions*: results generated by the system based on user criteria.

Public health is an important outcome component of decision support tools involving air quality, water management, energy management and agricultural efficiency issues. There was a time when public health COPs were slow to adopt Earth observations technology; but this has changed dramatically since 2006 when advances in data discovery, access and delivery mechanisms began to appear (ICSU 2012). When monitoring human disease, there needs to be a long history of data recording to provide information on changes in population distribution and environmental conditions associated with outbreaks of disease; and these need to be of sufficient resolution and frequency to enable identification of changing conditions (see Chapter 6). As such, many support tools being developed have substantial integration of Earth observations, but lack end-to-end public health outcomes. Privacy requirements, restrictions on the distribution of public health data, and lack of spatial resolution in data collection are common problems that complicate establishing relationships between monitored environmental conditions and human health outcomes.

1.1 *Participatory design and collaborative prototyping*

Participatory design (PD) and collaborative prototyping are techniques that can assist development of computer software (or applications) than meet the ends of end users (Dredger *et al.*, 2007). While PD emphasizes mutuality and reciprocity, traditional PD methods tend to be uni-directional, wherein developers collect data and analyse the requirements from users, and then deliver a system to users. An alternative means of DSS development emphasizes a third space, or a hybrid realm that overlaps the work domains of software professionals and the end users. Such an exercise facilitates capacity building, where an exchange of knowledge, skills, resources and infrastructure better allows developers to understand and integrate the types of tasks users need to perform. It also allows users to gain an understanding of the development process and computational limits of their requests.

A significant challenge in designing a decision support system that includes EO data and/or imagery is integrating them into a GIS structure (Dredger *et al.*, 2007). Maps as visual data are recognized by public health communities as providing an ability to assess large amounts of data in a single view. The ensuing challenge is the need to train public health users of GIS to manipulate and interpret spatial data products properly for health applications; and realizing this suggests a collaborative prototyping design to ensure robust public health applications (see Section 5, SYRIS). An alternative approach overlaps the work domains of software and system developers with intended end users. This facilitates capacity building, exchange of knowledge skills and resources leading to a better understanding by development teams of the types of data, analyses, and models that users need. This approach permits end users to better understand the development process and computational capabilities. Typically, it has been found that developers need more exact information on specific functions needed for decision making, and users become more knowledgeable about the specific questions that can be answered by a DSS.

Dredger (*et al.*, 2007) found that, when developing a web-based, GIS-enabled DSS, the design process would have been more streamlined had there been a stronger focus on spatial literacy and more research into the decision making process. They argued that developers required more explicit information regarding the functionality that users needed, and that users required a better understanding of the types of questions a GIS can help answer.

1.2 *Assessing DSS performance*

Verification and validation can be an on-going process, or can be thought of iteratively, wherein feedback from historical performance can continuously guide the use of the decision support system as a type of early warning system. To get public health communities to accept biosurveillance systems, it is necessary to make complex information more manageable and to achieve performances that are more robust (Mnatsakanyan & Lomardo 2008). Many algorithms used in syndromic surveillance systems can employ univariate statistical anomaly detection techniques, and can identify potentially large disease outbreak events quickly, though with a high number of false positives. Their utility improves when users can make informed decisions based on reliable situational-awareness information; hence, the role of developers should be to create advanced tools to manage and summarize the diversity of information related to situational awareness (Mnatsakanyan & Lomardo 2008). Multivariate analyses of data from multiple sources can be fused with algorithms that provide early-warnings. While temporal anomaly detection algorithms can show high sensitivity on the single data streams, information fusion algorithms can rule out epidemiologically irrelevant patterns, thus refining the specificity of the system (Mnatsakanyan & Lomardo 2008).

1.3 *What follows?*

Section 2 by Doctors Reisen, Barker and Kramer is both an early warning and a real-time information system for forecasting environmental and biological conditions that spawn WNV outbreaks in California. Section 3 by Doctors Caian and Crăciunescu describes *AirAware Romania*, a system for monitoring *in-situ* air quality, traffic patterns, and hourly weather parameters in a multifaceted GIS framework. The objective is to provide local authorities with advisories on air pollution levels on a street-by-street basis across Bucharest. Section 4 by Doctors Brown and Comrie is a translational social science and physical science approach to concerns about rising health care costs and incidence of Coccidioidomycosis (aka Cocci) in Arizona. A scheme is presented to integrate iterative interactions among many concerned fields of science and the feedback mechanisms that inform decisions makers. Section 5 by Doctor Zelicoff describes a commercial syndromic surveillance system (SYRIS) that informs doctors, veterinarians, ER-responders, school nurses, health officials, and others ranging from schools to counties/districts, states/provinces and national levels. Section 6 by Doctor Kass-Hout concludes the Chapter with a semantic approach to health information focusing on crowdsourcing, social networks, and near real-time access to media and related alerts from regional, national, and international sources.

2 INTERVENTING MOSQUITO-BORNE ENCEPHALITIS VIRUS IN CALIFORNIA

The North American mosquito-borne encephalitides are viral zoonoses maintained and amplified in nature by transmission among culicine mosquitoes and passerine birds. Humans and equines are infected tangentially but develop mild to severe disease. Although equine vaccines are available, integrated mosquito management programmes remain the only option to protect public health. Deciding when and where to apply what level of mosquito control for effective intervention requires careful surveillance coordinated through an effective decision support system. The current chapter describes how the California State Mosquito-borne Virus Surveillance and Response Plan collects, collates and distributes surveillance information to provide a real time decision support system for participating mosquito control agencies and the California Department of Public Health.

The North American mosquito-borne encephalitides in the Flaviviridae and Togaviridae are zoonoses maintained in nature by horizontal transmission among culicine mosquitoes and passerine birds. Equids and humans are dead-end hosts for these viruses and usually become infected after enzootic amplification exceeds thresholds that enable tangential transmission. Equids and some endangered or valuable avian hosts are protected by annual vaccination; however, there currently are no vaccines or effective therapeutics licensed for humans. Therefore, public health intervention relies solely on integrated mosquito management and health education programmes. Because of the extensive spatial and temporal variation in enzootic amplification, effective mosquito management relies upon detailed surveillance of vector population abundance and virus activity to direct control operations. In California, special mosquito and vector control districts supported by local taxes attempt to maintain mosquito populations below levels where there is high risk for tangential transmission of West Nile (WNV), Saint Louis encephalitis (SLEV) and western equine encephalomyelitis (WEEV) viruses to humans. Surveillance and control efforts are coordinated through the Mosquito and Vector Control Association of California (MVCAC) and the Vector-borne Disease Section of the California Department of Public Health (CDPH). Surveillance diagnostics and data management systems are provided, in part, by the Center for Vector-borne Diseases (CVEC) within the School of Veterinary Medicine at the University of California, Davis, which also conducts collaborative basic and applied research on arbovirus epidemiology and control.

Background surveillance programmes and intervention responses follow The California State Mosquito-borne Virus Surveillance and Response Plan (CMVSRP) that was developed jointly by CDPH, CVEC and MVCAC, and is reviewed and updated annually (Kramer 2009). Because most precipitation in California occurs during the winter months, either as rain at lower elevations or snow in the Sierra Nevada, the prior year's mosquito abundance (Barker 2008) and winter weather conditions are associated with succeeding vernal mosquito abundance (Reisen et al., 2008b), which, in turn, affects summer arbovirus activity (Barker et al., 2009). However, vernal viral amplification may be constrained by factors, such as peridomestic passerine immunity (Wilson et al., 2006), avian host depopulation (Wheeler et al., 2009), and effective early-season mosquito control (Lothrop et al., 2008; Reisen et al., 2008e). Therefore, nowcasts utilizing enzootic surveillance data in real-time are critical for intervention decision support (Reisen & Barker 2008).

The Surveillance Gateway is an internet data collection, management, analysis, storage, visualization and reporting system conceptualized through the CMVSRP (Reisen & Barker 2008), developed as a project within NASA's ASD (Park et al., 2008). When coupled with high-throughput diagnostics (Dannen et al., 2007; Feiszli et al., 2009), the Gateway allows almost real-time surveillance and auto-generates weekly risk assessments based on enzootic virus monitoring. Previously, SLEV and WEEV risk was estimated by averaging quintile scores assigned to temperature, mosquito abundance, mosquito infection incidence and sentinel chicken seroconversions. This system successfully tracked risk during representative years showing no, low and high enzootic activity (Barker et al., 2003). Following the invasion of California by WNV (Reisen et al., 2004), test results on dead birds reported by the public were added as an additional risk factor. At present, WNV seemingly has displaced SLEV in California since 2003 (Reisen et al., 2008d), and effective mosquito control seems to have limited WEEV amplification, despite the fact the virus has remained virulent for both mosquitoes and avian hosts (Reisen et al., 2008c). Therefore, the remainder of our presentation will focus on WNV decision support.

Input data measures of environmental conditions, mosquito abundance, and virus activity are combined to estimate the risk of tangential transmission of virus to humans. Equine cases are not considered, because most animals now are protected by annual vaccination against both WEEV and WNV. Human case reports are not included, because the time between the date of infection or the onset of disease to confirmatory diagnosis and reporting typically is too long to provide adequate early warning to initiate intervention. In retrospective analyses, the inclusion of human data as a risk factor dampened risk estimates and reduced lead time (Barker et al., 2003) and were more appropriately used as the outcome metric to assess an enzootic risk model.

2.1 Environmental factors

Although precipitation and snow pack have been correlated with vernal mosquito abundance (Reisen et al., 2008b), the distribution of surface water during spring and summer in California is under the control of a variety of water agencies. As a result, the data are complex and difficult to assess spatially (Barker 2008), and therefore environmental risk in our current model has been limited to temperature. Temperature surfaces for California with one km^2 resolution are derived from EO-data and ground-based data provided by TOPS (Melton et al., 2008; Nemani et al., 2003) and downloaded daily into the CVEC computing system. Risk is scored into quintiles (Figure 1, Table 1) based on the duration of the extrinsic incubation period of WNV in *Culex tarsalis* mosquitoes (Reisen et al., 2006). This time period was found to be similar among *Culex* species (Goddard et al., 2003) and driven by temperature. Warm years enable virus activity by elongating the transmission season, reducing the duration of the extrinsic incubation period, and increasing the frequency of vector-host contact (Reisen et al., 2006).

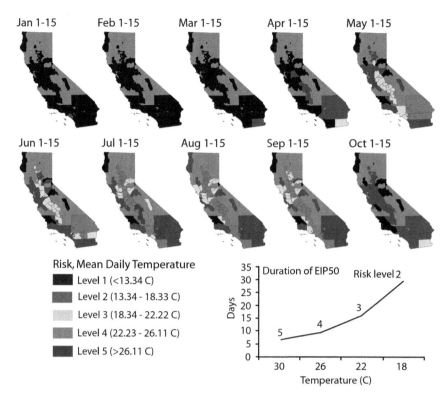

Figure 1. Temperature risk for each mosquito control agency in California in 2009. Level of risk based on the median duration of the extrinsic incubation period of WNV in *Cx. tarsalis*. (see colour plate 73)

2.2 *Mosquitoes*

Culex tarsalis and members of the *Cx pipiens* complex are the primary vectors of WNV throughout California (Goddard *et al.*, 2002; Reisen *et al.*, 2008a) and therefore surveillance focuses on these species. *Cx. thriambus* and *Cx. stigmatosoma* are highly competent laboratory vectors, but are limited in abundance and distribution, feed mostly on birds, and therefore were not included in our model. *Culex* abundance is monitored at fixed sites on a weekly or biweekly schedule by dry ice-baited traps (Newhouse *et al.*, 1966) and by gravid female traps in urban habitats (Cummings 1992). Abundance is ranked into quintiles based on the amplitude of the current anomaly relative to the five year historical average for that collection period (Table 1). Prevalence of infection is estimated by testing pools of approximately fifty female mosquitoes of each species from each site for WNV RNA using real time RT-PCR (Shi & Kramer 2003), calculated using the maximum likelihood estimate (Biggerstaff 2003), and ranked into quintiles.

2.3 *Birds*

Transmission of WNV to birds is monitored by sequentially bleeding sentinel chickens to detect seroconversions and by testing dead birds reported by the public for WNV RNA. Over 200 flocks of five to ten white leghorn hens are deployed throughout the state each spring, bled weekly or biweekly, and the blood samples screened using an enzyme immunoassay (EIA). Chickens were selected as sentinel birds, because they are relatively refractory to infection but produce elevated antibody levels (Langevin *et al.*, 2001; Reisen *et al.*, 2005). Risk is related to the number of chickens (i.e. intensity) and the number of flocks (i.e. spatial extent) showing evidence of infection since the previous sampling (Table 1). Dead birds reported by the public to the state-wide hot line

Table 1.   Form showing surveillance factors and risk assessment values from the CMVSRP.

WNV surveillance factor	Assessment value	Benchmark
1. Environmental conditions. Rural transmissions may favour El Niño conditions, whereas urban transmission may favour La Niña conditions.*	1	Average daily temperature, preceding month <56°F
	2	Average daily temperature, preceding month 57–65°F
	3	Average daily temperature, preceding month 66–74°F
	4	Average daily temperature, preceding month 75–83°F
	5	Average daily temperature, preceding month >83°F
2. Adult *Culex tarsalis* and *Cx. pipiens* complex abundance. Determined by trapping adults, identifying them to species, and comparing numbers to those previously documented for an area for the current time period.	1	Vector abundance well below 50%
	2	Vector abundance below average 50–90%
	3	Vector abundance average 50–150%
	4	Vector abundance above average 150–300%
	5	Vector abundance well above average 300%
3. Virus infection rate in *Culex tarsalis* and *Cx. pipiens* complex mosquitoes tested in pools of 50. Test results expressed as minimum infection rate (MIR) per 1000 female mosquitoes tested, or per 20 pools.	1	MIR/1000 = 0
	2	MIR/1000 = 0–1.0
	3	MIR/1000 = 1.1–2.0
	4	MIR/1000 = 2.1–5.0
	5	MIR/1000 = >5.0
4. Sentinal chicken seroconversion. Number of chickens in a flock that develop antibodies to WNV. If more than one flock is present in a region, number of flocks with seropositive chickens is an additional consideration. Typically 10 chickens per flock	1	No seroconversions
	2	One seroconversion in single flock over broad region
	3	1–2 seroconversions in a single flock in specific region
	4	More than 2 seroconversions in single flock, or 1–2 seroconversions in specific region
	5	More than 2 seroconversions per flock in multiple flocks in specific region
5. Dead bird infection. Includes zoo collections.	1	No WNV positive dead birds
	2	One WNV positive dead bird in broad region
	3	One WNV positive dead bird in specific region
	4	2-5 WNV positive dead birds in specific region
	5	More than 5 positive dead birds and multiple reports of dead birds in specific region
Response level/average rating: Normal season 1.0 to 2.5 Emergency planning 2.6 to 4.0 Epidemic 4.1 to 5.0		TOTAL = AVERAGE =

*Temperature data link available from: http://www.ipm.ucdavis.edu:/WEATHER/wxretrieve.html [Accessed 12th April 2012]

(McCaughey *et al.*, 2003) are recovered by local agencies, and if suitable for testing, shipped to the California Animal Health and Food Safety laboratory at UC Davis for necropsy. Oral swabs from American crows and kidney snips from other species are tested for WNV RNA by RT-PCR. Similar to chickens, risk is related to the number of birds positive from different geographical areas (Table 1).

Risk of human infection Quintile scores from each of the five risk factors described above are arithmetically averaged to calculate an overall risk score, which provides a metric of tangential risk for virus transmission to humans. This score is interpreted based on the CMVSRP as normal season (1.0-2.5), emergency planning (2.6-4.0) or epidemic conditions (4.1-5.0; Table 1). Calculated levels of risk are related directly to recommended intervention responses in the CMVSRP, which range from routine larval control during a normal season to emergency adult mosquito control during

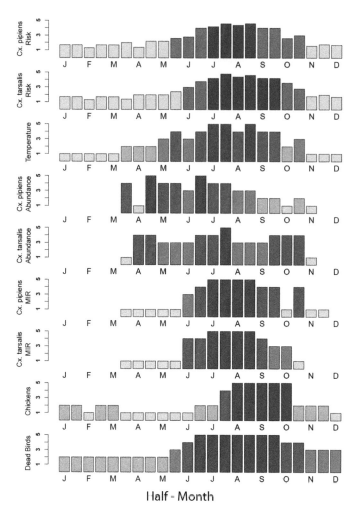

I Ialf - Month

Figure 2. Risk of WNV tangential transmission summarized by half month for Kern County, California, 2009. yellow = normal season (risk = 1 – 2.5); orange = epidemic planning (risk = 2.6 – 4.0); red = epidemic (risk = 4.1 – 5.0). (see colour plate 74)

an epidemic (Kramer 2009). Surveillance methods and response activities in the CMVSRP are in general agreement with published CDC Guidelines (Moore *et al.*, 2002), but response is graded in accordance to risk score levels. These values can be utilized by health planners to visualize the spatial and temporal distribution of risk throughout the state of California.

Risk estimates can be calculated manually using the tools available in the Gateway to estimate mosquito species-specific abundance anomalies or infection prevalence, but also are auto-generated weekly from the Gateway based on the previous half-month of data and sent via email as PDF files to all participating agencies. Representative data are shown for Kern County during 2009; note that not all risk factors are available during every half-month due to the seasonality of sampling (Figure 2). In California the sampling period or virus season becomes progressively shorter as a function of increasing latitude. Aggressive mosquito control in response to risk values during 2009 limited the number of human cases reported from Kern County to nineteen, the second lowest annual total since the arrival of WNV in 2004.

Risk based on environmental and enzootic factors has been correlated temporally with the occurrence of human cases in rural and urban areas (Figure 3). Importantly, emergency planning

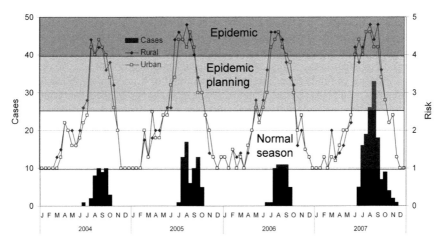

Figure 3. Risk levels from 1–5 estimated from environmental and enzootic measures for rural and urban locations and the numbers of reported human cases reported by date of onset plotted as a function of half month from 2004–2007 in Kern County, California (Reisen *et al.*, 2009). (see colour plate 75)

levels of risk have been recognized several weeks prior to the onset of human cases, allowing control agencies to recognize the eminent threat of an outbreak and initiate intervention. Control of urban mosquito larval sources such as abandoned swimming pools (Reisen *et al.*, 2008e) and aggressive focal adult and larval control in response to increased risk during 2009 seemed to limit tangential transmission of WNV factors that enable successful intervention.

Successful intervention to reduce vector-borne pathogen transmission in both developed and developing countries requires similar essential elements to enable successful surveillance programmes. An important precursor to successful intervention is a thorough understanding of the ecology and epidemiology of a pathogen's persistence and amplification in time and space. Process-based mechanistic models in combination with statistical correlative models are useful in understanding what factors drive amplification. Forecasts provide early-season warning to health planners, whereas nowcasts provide verification of these forecasts and are useful for operational response. Satellite and aerial data fused or assimilated into a GIS provide important interpolative and extrapolative approaches to estimating risk over heterogeneous landscapes.

2.4 *Surveillance*

After sound epidemiological research outlines key factors and delineates amplification patterns in time and space, a statistically sound population-based sampling scheme is needed to provide information on critical surveillance factors such as vector abundance and infection prevalence and transmission to vertebrate hosts. Usually a spatially delimited, fixed sampling grid and historical data are necessary to understand where and when amplification occurs, and when enzootic transmission is approaching outbreak levels. These point estimates can be extrapolated spatially using remotely sensed landscape measures to indicate when and where outbreaks are likely. Rapid, effective response to early warnings of amplification can be successful in arresting transmission and limiting human infection (Carney *et al.*, 2008; Elnaiem *et al.*, 2008; Nielsen *et al.*, 2008).

2.5 *Sustainability*

WNV is now endemic throughout the New World from Canada to Argentina and focal outbreaks are likely when and where vectors, hosts and pathogens intersect within a suitable environment. This is also true for other anthroponoses (e.g. malaria and dengue) and zoonoses (e.g. Rift Valley fever and Japanese encephalitis) throughout the world. Zoonoses with wildlife hosts are especially difficult to eliminate, but can be managed using population-based surveillance and vector control. Detection and effective response requires continued fiscal input to sustain surveillance and control programmes. Long-term sustainability relies upon local funds and cannot be accomplished by the inconsistent infusion of funds from international, national or even state agencies. In California, local communities have voted to provide annual revenues in support of special mosquito and vector control programmes that provide protection against vector-borne pathogens and enhance the quality of life by reducing biting nuisance.

3 AIRAWARE-ROMANIA

3.1 *A regional air quality DSS*

Bucharest is the capital and the biggest city in Romania. The urban population is roughly two million in an area of approximately $228\,Km^2$. The development strategy for the municipal and metropolitan area of Bucharest has been elaborated for a horizon of time up to 2025. The vision of this strategy is that Bucharest will become a European metropolis, having a well-defined regional, continental and inter-continental role. Unfortunately, presently the Bucharest Municipality has serious problems in ensuring a healthy life environment for its citizens, the air quality being an environmental aspect negatively influenced by multiple pollution sources. The air pollution in Bucharest has a complex character because of the multiple sources, types (traffic, thermal power plants, industry, and extensive building construction) and conditions and spatial distribution.

In this context, the AIRAWARE project, funded in the EU LIFE framework, is aiming to build a pilot air quality monitoring and forecasting system to ensure a sustainable development of the rapidly expanding urban areas in Bucharest, minimizing and preventing the air pollution impact on human health. The AIRAWARE system has a distributed architecture with dedicated sub-systems for: (1) air quality monitoring; (2) numerical modeling and forecast; (3) geospatial portal for data integration, visualization, query and analysis; and (4) slow-flow and rapid-flow feedback.

The AIRAWARE system users list includes the main local and national authorities in the domain of air quality monitoring, forecasting, air pollution abatement and mitigation of impacts, such as: the National Meteorological Administration, the Regional Environmental Protection Agency of Bucharest, the Direction for Public Health of Bucharest, the Centre for Urban Planning of Bucharest, the Institute of Biology of the Romanian Academy and Meteo-France as strategic European partner.

3.2 *The AIRAWARE system*

The AIRAWARE architecture was designed to be a distributed system. All the partners and users are able to access the system using a thin web browser (Internet Explorer or Mozilla Firefox) or a thick desktop client to store, display, query, analyse, and retrieve data and information about Bucharest's air quality (Figure 4). Figure 4 shows the role of each project partner and the information flow between data providers, end users, and project partners. The partners' short names are shown in red. REPA-B and NAM are using mobile laboratories and sensors to monitor concentrations for the most important pollutants. NAM and Meteo France use the data for model-based air quality forecasts. The integrated results are used by REPA-B, IB-RA and UMPC-B to assess the impact on humans, plants and animals.

3.3 *AIRAWARE monitoring system*

The system includes fixed ground monitoring stations measuring the most important pollutants (NO_2, SO_2, O_3, PM_{10}, $PM_{2.5}$, Pb, benzene, CO), mobile laboratories and a complex LiDAR/DIAL spatial measurement of emission/ambient air contamination data, allowing both 3-D modelling of diffusive and non-point source emissions, as well as complex 3-D spatial ambient air contamination information.

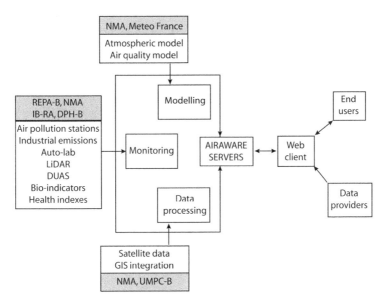

Figure 4.   AIRAWARE system information flow.

3.3.1 *Industrial emission monitoring network*

Point pollutant sources with major health impacts have been selected and their main sources identified and interconnected in the AIRAWARE emission data collection network. This network receives daily data (CO, SO_2, NO_2, PM) from ten automatic monitors and daily consumption data from another twenty-one sources. Data are also used from another set of sixty-two climatological emission sources computed from annual consumption data. These are further processed through the European Environment Agency's COReINventoryAIR (CORINAIR) Programme into hourly data. Altogether in 2004 there were ninety-three sources covering ninety per cent of total point source emissions in Bucharest. The daily flow of information is established through agreements signed by EPA-Bucharest with the thirty-one main air polluters in Bucharest.

3.3.2 *Traffic emission network*

This network was defined on a regular grid, initially at two kilometre resolution and further improved, with two information-fields: 1) the roads area fraction; and 2) traffic/vehicle type and flux (with hourly, and week-day variations derived from traffic measurements and parking facility databases. Based on these data, emission coefficients are derived for each grid point, hour and week-day for four vehicle types through a statistical downscaling method. The model was validated on a seasonal basis during the project (first campaign was made on 26Apr2006). These computations are currently updated through online continuous traffic flux measurements at 120 monitoring points across Bucharest, and through a continuously updated GIS road map from CUMPC-B, as shown in Figure 4. The analysis of these data is made in conformity with European air quality norms and methods.

3.3.3 *Two dimensional ground emission data*

These data are obtained on an hourly basis from eight automated monitoring stations in Bucharest, measuring concentrations of CO, NO2, SO2, O3, C6H6, and transmitted hourly to the EPA server. For particles (PM_{10} or $PM_{2.5}$) and lead, these stations have automatic samplers (24 hours sampling on each filter) and the filters are analysed in the laboratory. The emission data are transmitted daily to NAM at 8UTC. They are used also to calibrate the models, to diagnose the current condition of air quality, to correlate with health outcomes, and to serve as bio-indicators. The historical emission

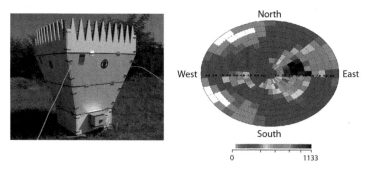

Figure 5.   (Left) SODAR equipment; (Right) sound emission matrix. (see colour plate 76)

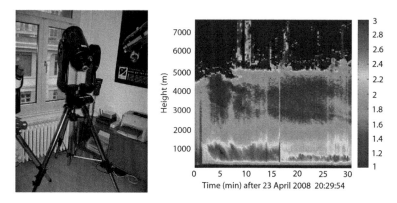

Figure 6.   LiDAR equipment (left) is used for studying atmospheric composition, structure, clouds, and aerosols; (right) is a sample of range-corrected LiDAR signals through the Earth's boundary layer. (see colour plate 77)

database is available since Model validation implies also emission monitoring campaigns at the forecasted maximum concentrations locations. For this purpose, an auto-laboratory was equipped with automatic analysers for NO_2, SO_2, CO, O_3, PM_{10} and benzene.

3.3.4   *Three dimensional emission data*
These data are obtained from a system comprised of gas analysers and vertically distributed air samplers on a 33 m tower at 3 m, 15 m, and 30 m for O_3, SO_2, CH_4, NMHC, and NOx (Figure 5 left). The system is located at the Afumati meteorological station. The 3D wind profile is measured by a sonic detection and ranging (SODAR) with a twenty-four sound-emission points-matrix (Figure 5 right). All data are made available online from the central AIRAWARE system, and all operational meteorological data are available from the NAM network and GIS processed for analysis during the AIRAWARE feedback chains. The meteorological database is available and updated at six hour and twenty-four hour intervals.

   Data from a ground-based LiDAR system are also used operationally to measure the level and diurnal evolution of the aerosol layer, boundary layer height, cloud base and other properties. Software has been developed to process these LiDAR data (Figure 6). They are used to correlate with the 2-dimensional air pollution monitoring network and for model validation. MODIS satellite data are used operationally for integrated aerosol optical depth and aerosols parameters.

3.3.5   *Human health indicators*
All data for Bucharest have been analysed and three main types of providers have been identified. Vital statistics are provided by the department of public health-Bucharest. Acute care hospital data

are provided by two university clinics for infants and children; two infectious disease clinics; and two general emergency clinics. Ambulance services are provided by the municipality of Bucharest. Health data can be obtained according to provisions in a written protocol of collaboration. This protocol has five steps: in Step 1, weekly data received from the distributed sources are aggregated using a software developed specifically for AIRAWARE; in Step 2, daily frequencies for each single/group of ischaemic heart disease (IHD) codes are calculated; in Step 3, time series data sets indexed by calendar day are compiled; in Step 4, analyses are performed; and in Step 5, the AIRAWARE database is updated weekly.

In step 4 two types of assessments are performed: 1) a general assessment of the existence and nature of associations between outdoor air pollutant levels and health event frequencies through correlation analysis on time series, where the independent variable is the pollutant level and the dependent variable is the daily frequency of health event, respectively; and 2) an operational assessment estimate of attributable number of events prevented through pollution control scenarios, based on relative risks calculated by epidemiological studies for each 10 units of change in pollutant levels. Note: as a derivative of time series analysis based on correlation coefficients, a bidirectional symmetric case-crossover analysis is performed randomly to avoid bias induced by autocorrelation, personal characteristics, weather indexes and other factors. Health indicators based on historical morbidity caused by respiratory diseases, heart diseases, tumours, some infectious diseases and mortality data from cardio-respiratory diseases are monitored. Regular reports are produced on these air pollution related health-indicators to monitor trends and propose mitigation measures. Integrated recommendations from the status-warning health codes (SWHC) characterizing the current and forecasted states are issued as a result of the daily analysis of the georeferenced health related information layers and forecasted pollution risk levels. This index is based on forecasts and on historical-statistical data mapped at town level, covering six city-hall districts.

3.3.6 *Bio-indicators*
The workflow for bio-indicators has four steps: 1) identifying ground survey sites based on patterns documented in measured air pollution data; 2) inventorying plants and animals from the selected sites; 3) analysing correlated data-bases and defining bio-indicators related to air pollution; and 4) integrating and analysing bio-indicator data integration, analysis.

Step 2 involved GIS techniques to select ground sites and to code them according to their probable polluted status based on soil chemistry, the presence of fungi, lichens, mosses, indicator plants, other organisms, and invertebrate forms (Acari, Collembola, Chilopoda) known to be useful bio-indicators of degraded sites. Data are obtained through periodic surveys started in 2006 from five fixed sources: three polluted parks in Bucharest (Izvor, Unirii, and Cismigiu) and two control (clean) sites in similar ecosystems (Balotesti and Baneasa). Climatology and air pollution have synergetic/antagonistic effects on bio-indicators, so integrated monitoring and data analysis are necessary for these studies.

3.3.7 *Urban planning indicators*
Urban planning indicators are complex measures aimed at creating liveable environments for city dwellers. They include the quality of road surfaces, variety of efficient transportation modes, the quality and efficiency of water and power infrastructure, location of city parks, historical markers, cultural centres, and many more. Infrastructure and aesthetics are two of the most important sets of critcria for urban planners because they impact the quality of health and wellbeing for all inhabitants.

3.3.7.1 Methodology
There are three steps to address planning indicators. In step 1, under the authority of the Urban Technical Commission of Bucharest Municipality, data and information are being collected for an urban database to support proposals for urban development as elaborated by the new zone urban plans. These are then approved by other institutions charged by law to verify the proposals and by the Bucharest City Council. The objective is to retrieve proposed urban indicators that include

aesthetic qualities like land-use, building density (land use per cent) and skyline (height regime and functional land-use). In step 2, surveys are undertaken to verify the height of new buildings located in areas where air pollution is, or could become, a health issue in context of human exposure levels. Step 2 also reviews the urban database and assesses new development proposals; and loads this information monthly into the database at Bucharest City Hall. Step 3 creates thematic maps about exposed areas.

3.3.7.2 Numerical modelling and forecast component

Air quality numerical modelling consists in several multiple-nested, coupled (MNC) systems of atmospheric models and air quality models. The multiple nesting approach includes a meso-scale, country-wide domain at ten kilometre horizontal resolution; a regional scale urban and urban fringe domain at 3.5 km resolution; an urban scale domain at 1.5 km resolution for dynamical adaptation; and finally a street-level domain for detailed air quality monitoring. There are two MNC runs daily: a preliminary one that uses emission data collected from the previous 48 hours; and an updated one using current-day estimated emissions from economic agents integrated over twenty-four hours.

Pollution modeling involves chemical (photolytic) processes, 3-D transport, convection, and dispersion of pollutants based on the forecasted state of the atmospheric boundary layer by the atmospheric model. It provides the 3-D air pollutant evolution for various time ranges over the coming two days. Several models are used for this system component. The atmospheric model Aire Limitee Adaptation Dynamique INitialisation (ALADIN) is a complex 3-D primitive equations, non-hydrostatic, state-of-the-art numerical model used as an operational weather forecast model in more than fifteen countries. It has been developed through international cooperation between France and Romania and refreshed by the very latest modelling improvements. A dynamical down-scaling procedure has been implemented to provide short term high resolution forecasts. The entire modelling chain is updated continuously. There is also an option to migrate ALADIN to a more complex, two-way interactive Applications of Research to Operations at Mesoscale (AROME) model, if desired.

Multiple nesting models for air pollution use the two-way interaction chemistry and transport Multi-scale Chemistry and Transport Model (MOCAGE) at meso-scale, and at the regional scale it uses the operationelle meteorologiske luftkvalitsmodeller (OML) model for pollutant dispersion and chemical transformation at urban scale, and the operational street pollution model (OSPM) for pollutant dispersion and photolytic transformations at major transportation intersections. Direct and inverse forecasted trajectories are computed for predefined initial and final points, and frequency statistics are computed based on trajectories and stored over the area of interest as indicators of contamination probability.

Standard scenarios are computed daily following protocols with decision making authorities. These scenarios are used to re-direct traffic patterns, reduce industrial emissions, and mitigate individual source contributions. Output data from the full modelling chain are processed using GIS technology to serve each sub-system's needs and user requirements. Current and forecasted pollution hats are used for two levels of information flow (rapid and slow flow). Required scenarios are performed during the analysis step by decision making authorities. The models are validated and improved continuously through measures from meteorological networks, acquired tools (SODAR and LiDAR), emission data (2-D and 3-D) and other sources.

3.4 *AIRAWARE geoportal*

A geospatial portal is a human interface to a collection of online geospatial information resources, including data sets and services. The AIRAWARE geoportal represents the public face of the AIRAWARE's on-line system, created to facilitate data and knowledge sharing of geospatial air quality outputs. The free and open source (FOSS) GIS space includes products to fill every level of the OpenGIS spatial data infrastructure stack. The main advantages of FOSS software are: 1) the availability of source code and the right to modify and use the software in any way; 2) not tied to a single vendor; 3) big community to support; 4) good security, reliability & stability; 5) very good standard compliancy; and 6), lower implementation cost. The entire AIRAWARE geoportal and back-end modules are based on the existing FOSS and FOSS4G applications.

3.4.1 *Data fusion and routing*

The system needs to handle different types of geospatial data (scanned maps, satellite images, vector files, digital elevation models, model outputs) in different file formats and coordinate systems (Stereographic 1970 – official Romanian coordinate system, Gauss-Kruger, UTM and others), processed by the project partners on different computing platforms and software environments. To make all the work easily available to participants and end-users, a detailed specification package based mostly on the geospatial data abstraction library (GDAL) and simple feature library (OGR) have been developed and ingested into a central application. These ensure that every piece of information is parsed and represented correctly in the portal.

The application is also able to automatically route the data from the project participants to the system (mainly the mathematical models) and to web-mapping applications. For this purpose, a task and stay resident (TSR) application was constructed. When a project participant sends new data to the system, a configuration file is created. The TSR application checks the configuration files. When changes are detected, the application forwards the new file through FTP connections to the appropriate processing application or mathematical model. The TSR application waits for the output file to be created and notifies users to check the results. If necessary, more processing steps are initiated. The TSR has been constructed such that it is able to monitor every ten seconds for the arrival of a triggering file. Multiple requests can be handled simultaneously. All the performed operations are recorded in a file log. Based on this log, users and project partners are able to track system activity to detect any malfunctions.

3.4.2 *Portal structure*

The content is managed by Text pattern, a powerful and flexible, open source content management system (CMS) application. For supplementary, specific functionality, custom modules were built. Other free applications are providing server-side functionality (e.g. MySQL relational database management system; PHP; Python; Java server-side scripting languages; Apache webserver; Tomcat servlet container; phpMyAdmin; and phpPgAdmin web clients for database management.

For geospatial data management, top open source applications were also integrated in the website. These include: PostGIS (geospatial data storage); GeoNetwork Opensource (geospatial data catalogue and metadata editor); Geoserver (standard geospatial server for serving data via WMS and WFS); OpenLayers (client web-mapping application); and TileCache (Python-based WMS/TMS server with pluggable caching mechanisms and rendering back-ends).

The information flow between the various server side applications and the front end graphical interface is determined by interaction among portal users and their requests, as shown in Figure 7. Table 2 lists the set of user requests and server responses.

3.4.3 *Geoportal interface*

The website interface was carefully designed, respecting the existing World Wide Web Consortium (W3C) standards and separating the structure from the presentation by using strict XHTML mark-up and cascade style sheets. New web technologies, like asynchronous JavaScript and XML (AJAX), were also used to increase the interactivity. The goal was to obtain a simple, friendly and accessible environment for air quality data management. From the user's perspective, when a button is clicked, an operation is performed, and a result appears on the screen. This summarizes a complex process of communication between the viewer and the server.

The portal is divided into several functional sections. Each contains a predefined type of information. The most important one is the map section. Here the users can view and explore the cartographic representation of the spatial data stored in the GIS database. Both raster and vector data can be displayed. Maps can also include labels for the generated spatial entities based on the represented database attributes. The cartographic symbols and labels are automatically adjusted according to the presentation scale. Maps are refreshed after each action performed by the user.

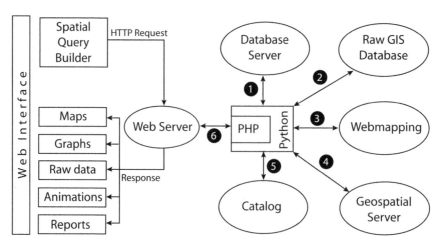

Figure 7. AIRAWARE geoportal structure. Numbers in black circles correspond to types of user requests.

Table 2. Typical user requests and web server responses.

Types of request	Request	Web server response
1	SQL query	Set
2	File	via FTP, HTTP
3	Map query	Map
4	Get map	via WMS, WFS, GeoRSS
5	Search	Set
6	Php	HTML

The geospatial information, represented as map layers, can be divided into two categories: 1) the background database; and 2) an air quality database. A typical background database might include the transportation grid, buildings, drainage network, land use, points of interest, an elevation model and satellite data) In addition to the background data, users can chose to display OpenStreetMap, Google Maps, Yahoo Maps or Microsoft Maps as background data. Air quality data might include mathematical model outputs such as weather forecast, temperature, pressure, wind speed & direction, relative humidity; pollutant forecast concentrations for PM_{10}, CO, NO, NO_2, SO_2, O_3, benzene, pollution measured status; and bio and health indicators. Examples of the map section graphical interface are shown in Figures 8 and 9.

Users with who require more advanced GIS functionality, or want to use their own data processing algorithms, can access the AIRAWARE system content using a desktop (thick) client. This is possible due to the standard data access protocols and methods implemented on the AIRAWARE server-side. Also, users with minimal or zero GIS knowledge are able to access some of the AIRAWARE products within popular, easy to use, applications like Google Earth.

Other sections on the AIRAWARE portal allow the user to display the air quality information as graphs (Figure 7), tables or download the source data in popular file formats like Esri® shapefile, keyhole mark-up language (KML), MS Excel, or as comma separated values (CSV).

3.5 Conclusions

The initial project proposal took into consideration only a few pollution sources like large power plants, and traffic volume. During the model implementation process it eventuated that these have very quick dynamics and require frequent updates. It is necessary also to cover a sufficient range of emissions to develop a correct forecast. Thus, there were cooperation protocols established

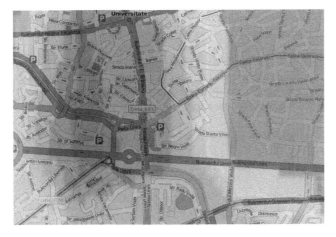

Figure 8. NO$_X$ concentration forecast displayed using the AIRAWARE system. (see colour plate 78)

Figure 9. Pollutant trajectories forecast displayed using the AIRAWARE system. (see colour plate 79)

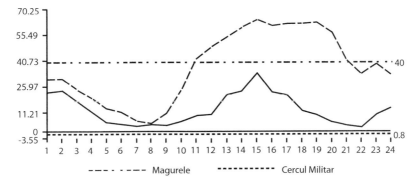

Figure 10. April 2008: O$_3$ recorded values at two monitoring stations.

385

among the main air polluters. Special software had to be developed and implemented specifically at each location for automatic online data emission or consumption collection and pre-processing, data validation, and use of CORINAIR for consumption data conversions. Deterministic tools have been developed and implemented to forecast and analyse results in such a manner to function operationally online while adapting to a large number of user requirements. This led to large database creation to allow systematic analysis. The innovative pilot system provides useful information daily for a variety of users, especially those in environmental and health communities, and those in urban planning and decision making. As the system matures, it will be poised to extend the daily observations and analysis to long-term assessments of health outcomes from climate interactions with air quality consequences.

4   CLIMATE, COCCIDIOIDOMYCOSIS AND CO-PRODUCTION OF SCIENCE

4.1   *Introduction*

This section describes translational research in response to concerns of public health decision makers to increasing rates of coccidioidomycosis (Cocci) incidence, a fungal disease endemic to the south western US. First, the aetiology of cocci is explained, followed by a review of climate-based models of inter-annual incidence in Arizona as a basis for improved understanding of disease risk. The discussion then turns to data quality and how data are addressed to decrease model complexity and increase model utility for the community stakeholders. The section ends by describing the authors' experience with co-production of science in support of public health decision making.

4.2   *Disease aetiology*

Cocci (aka Valley Fever), is a fungal disease common to the south western US. The disease has been endemic in the region since historic times, but recent increases in incidence have renewed investigations. In just five years from 1990 to 1995, the incidence of cocci more than doubled in Arizona to 14.9 cases per 100,000 population (Ampel *et al.*, 1998). Just over a decade later, in 2008, 4768 cases (73/100K population) were reported to the Arizona department of health services (Tsang *et al.*, 2010). As with many other diseases, the very young and the very old typically represent the more severe forms of disease, and immuno-compromised individuals are more susceptible (Ampel *et al.*, 1998; Galgiani *et al.*, 2005). Though most cases are mild, a recent study in Arizona showed annual hospital costs for coccidioidomycosis were $86 million USD or an average of $30,000 USD per hospital visit (Tsang *et al.*, 2010).

Cocci is not a contagious disease and most survivors develop life-long immunity (Stevens 1995; Galgiani *et al.*, 2005). After an incubation period of one to three weeks, sixty per cent of infected individuals present no, or mild, symptoms (Smith *et al.*, 1946b; Stevens 1995). Fewer than one per cent experience severe disease in which the pathogen enters the bloodstream and causes disseminated infection. This more life threatening form of the disease involves other body parts and occasionally coccidioidal meningitis with almost certain death (Stevens 1995).

Two fungal species are known to cause cocci; *Coccidioides immitis*, which occurs primarily in California, and *Coccidioides posadasii*, which occurs more broadly across the American Southwest, southern California, Mexico, and parts of South America (Fisher *et al.*, 2002). The *Coccidioides spp.* life cycle involves both a saprophytic and a parasitic phase. The saprophytic phase consists of soil-living mycelia. The mature hyphae are structured as alternating functional and sterile *arthroconidia* (Kolivras *et al.*, 2001). The sterile cells are fragile and provide convenient breaking points that breach easily and allow the *arthroconidia* to aerosolize (Stevens 1995; Kolivras *et al.*, 2001).

The parasitic phase manifests subsequent to inhalation of mature *arthroconidia* by an animal host. In this phase, *arthroconidia* enlarge to form spherical, double-walled cells containing up to several hundred endospores (Stevens 1995). When the mature spherule ruptures the endospores are released and each may form a new spherule resulting in exponential growth of the fungus.

4.3 Model Development

As reviewed by Kolivras, climate was long believed to play a role in the seasonal incidence of coccidioidomycosis, but empirical evidence was limited (Kolivras 2001). Clinical reports of coccidioidomycosis incidence in Army recruits to southern California showed distinct seasonality with cases occurring in the dry summer and fall (Smith *et al.*, 1946a). Moreover, more cases were reported following one particularly wet winter. Other reports from California also showed seasonality where cases were more common in the windy, dusty periods (Maddy 1965). Finally, a bimodal peak in hospital admissions for coccidioidomycosis was reported in Maricopa County, AZ and again it was linked to temperature, rainfall, and dust (Hugenholtz 1957).

According to the so-called grow and blow hypothesis, moisture is required for fungal growth in the soil followed by a drying period when the *arthroconidia* are aerosolized and can then be inhaled (Comrie & Glueck 2007). In the *grow period* of the hypothesis, moisture is critical for fungal growth. Precipitation has a two-fold influence where it is required for the fungus to grow in the soil but it has the immediate effect of removing dust and spores from the air. In the *blow period*, dry and windy weather aerosolizes the mature *arthroconidia* (Kolivras *et al.*, 2001). The nuances of when and how severe of a weather event must occur to meet the requirements of the *grow and blow* hypothesis remain to be determined.

The Applied Climate for Environment and Society (ACES) laboratory at the University of Arizona set about to determine the role of climate in the observed seasonality and inter-annual incidence of cocci. The models from ACES focused on Arizona in part due to the recent increased incidence, but also because this area has a characteristic seasonal pattern and has been known since the 1950s to be the most infective area (Maddy 1958; Kolivras *et al.*, 2001).

Focusing on data from Pima County, which includes Tucson, the major southern urban centre within Arizona's endemic region, Kolivras & Comrie found long-term (one year or more) and winter meteorological conditions were more important in predicting monthly incidence (Kolivras & Comrie 2003). Their findings provided support for the hypothesis of precipitation requirements for the grow period and a drying period for *arthroconidia* dispersal. Data from Maricopa County, which includes Phoenix as another major urban centre in the endemic zone, showed the importance of rainfall, temperature and dust in the incidence of cocci (Park *et al.*, 2005). Further evidence for the grow and blow hypothesis was established recently by looking at both Pima and Maricopa counties and finding a significant autocorrelation among the monthly incidence data that ends abruptly with the onset of the summer monsoon precipitation (Tamerius & Comrie 2012).

4.4 Data Quality Concerns

As with many investigations utilizing human incidence data, the quality of the data posed noteworthy challenges. Reporting requirements and case definitions changed in the 1990s, affecting the consistency and comparability of annual incidence (Ampel *et al.*, 1998; Park *et al.*, 2005). It has been found that de-trending the data helped to account for the increased reporting in a data set spanning 1992-2005 for Pima County (Comrie & Glueck 2007). They also addressed issues relating to incidence data by conducting a sensitivity analysis. They showed that the time spent cleaning messy surveillance data was not sufficient to improve data and was thus unnecessary.

The data for cocci in Arizona were also affected by the transient winter season population; that is, people moving to Arizona for the winter half-year. Mild winter conditions mean that the population of susceptible individuals increases in the winter but they are not counted in the national census counts. Park and associates dealt with this issue by adjusting the census population estimates to reflect winter visitors (Park *et al.*, 2005).

Which data were recorded also presented a potential bias and had to be addressed. Case reports may include a date for symptom onset, diagnosis date, or report date which had to be converted to an estimated exposure date. Report dates were available and probably accurate for most cases, but tend to be delayed with respect to the exposure date. Conversely, onset dates were less frequently reported and less reliable, but are more useful for estimating date of exposure. Comrie investigated onset-to-diagnosis and onset-to-report lags and found the best approximation of onset dates was achieved by first averaging the available data by month and applying a three-month moving average algorithm to smooth the data (Comrie 2005). Recent results from enhanced surveillance conducted by the Arizona Department of Health Services (ADHS) further simplifies this issue with their finding that the median time from symptom onset to diagnosis is fifty-five days (Tsang *et al.*, 2010).

4.5 *Decision support*

The research described above into climate influences on cocci was initiated in response to a need identified by the ADHS to address increased disease incidence. Its success is measured not only in the scientific merit of identifying climate drivers of cocci incidence and the improved ability to predict case dynamics, but also in its usefulness to public health decision makers; a type of translational science. This dual scientific-societal benefit was achieved through the iterative interaction among an interdisciplinary group of researchers and public health decision makers.

The interdisciplinary nature of the research team ensured that the science produced was useable to stakeholders (Lemos & Morehouse 2005). With respect to the models described here, the team consisted of researchers in the ACES laboratory, epidemiologists at the ADHS, and members at the Valley Fever Center for Excellence (VFCE) at the University of Arizona (Kolivras & Comrie 2003). The collaborative process meant the ACES researchers were able to build models that ADHS could use through a better understanding of how they would be applied in a public health setting. For example, using lagged climatic predictors enables forecasts of disease risk. This serves not only as an early warning mechanism for primary care personnel to consider cocci in their differential diagnosis, but also provides ADHS with information which can be utilized to determine resource allocation.

Involving stakeholders (physicians and public health decision makers) throughout the process facilitated flow of knowledge between researchers and users. The two-way flow of information between researchers and users fosters user-oriented experiments with operational outcomes (Ray *et al.*, 2007). This also helps to ensure that users are aware of limitations and opportunities for the application of the research and thus more thoughtful useable science is generated which can be applied to achieve societal benefits (Ray *et al.*, 2007).

Ray (*et al.*, 2007) presented a schematic for the benefits of interactive iterative science which summarizes the ACES experience with the co-production of science (Figure 11). Straight arrows in the Figure indicate the iterative interaction between scientific communities, while the dashed lines indicate processes of feedback from stakeholders informing research questions and assessment activities. Rather than a single tangible tool, this interdisciplinary approach engaging both scientific and user communities facilitated an application focus to the climate-driven cocci models. Assessing model sensitivity to the data quality helped to reduce unnecessary complexity and to clarify the message of the models, thus promoting their utility to decision makers. Seasonal associations identified by this work can be used by ADHS to inform physicians and patients about cocci aetiology. The more knowledgeable the patient is about the disease, the more likely they to be tested for and the earlier the diagnosis (Tsang *et al.*, 2010) thus reducing the societal burden of cocci. Much of the progress made toward elucidating climatic predictors of cocci can be attributed to interactions that result from the co-production of science.

4.6 *Acknowledgement*

Funds for developing the surveillance data set and the Surveillance Gateway programme were provided by NOAA, NASA, the CDC, the California Department of Public Health, and the Mosquito and Vector Control Association of California.

5 SYNDROME-BASED DISEASE SURVEILLANCE SYSTEMS

The key to mitigating epidemics is situational awareness, both before an outbreak occurs and after an outbreak is reported. Syndromic surveillance is among many public health reporting and alerting

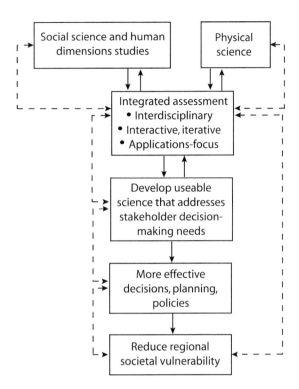

Figure 11.   Schematic describing the co-production of science (Amended from Ray *et al.*, 2007).

systems designed around situational awareness, and these are being implemented at local, state, national, and international levels (Chen *et al.*, 2010). Chen and colleagues describe thirteen national syndromic systems; twenty state-level systems; seven industrial systems; ten international systems; and six systems for special events like national and international sporting events, disasters, and financial economic summits. The WHO also has initiated the Epidemic and Pandemic Alert and Response programme as an effort toward global syndromic surveillance, though this is more policy and administrative than technical in scope.

Public Health officials (PHOs) have long recognized the critical importance of comprehensive reporting of infectious disease in both animals and humans. It is estimated that more than sixty per cent of 1400 or so human infectious agents are, in fact, zoonotic in origin, and recent events such as the arrival of WNV into North America, and the subsequent spread of the virus to bird and mosquito populations across the continental US, underscore the importance of early recognition of unusual signs and symptoms in wildlife, domestic animals and fowl, and humans to identify novel diseases and their scope.

Unfortunately, as the WNV and other surprising outbreaks (e.g. monkeypox in prairie dogs in Midwestern states in 2004) make clear, there is virtually no near real-time communication between veterinary and human medical practitioners, and precious little training in medical schools in recognizing zoonotic diseases in humans. Further, public health officials who would ostensibly evaluate clinical reports and provide advice and initiate contact investigations in general have no method for reliably and rapidly communicating with clinicians in their jurisdictions. For example, very few physicians subscribe to, or read, state public health offices' Health Alert Networks (HAN). The number of veterinarians, emergency management and ambulance systems (EMS) who use local HANs is unknown, but is likely to be a very small percentage of these professionals. Further, environmental factors such as dust and pollen can mimic infectious disease syndromes, and there is no integration of such data into medical alert or disease reporting systems.

The current reportable disease paradigm that has long dominated public health reporting is inherently limited by: 1) the predictable and indivisible duration in laboratory analysis of clinical samples; 2) failure to recognize signs and symptoms of classical infectious diseases that have now become rather rare (e.g. measles, diphtheria, mumps, and tularaemia); 3) failure to obtain and handle laboratory samples correctly; and 4) poor compliance on the part of physicians to report *per se* even when reportable conditions are diagnosed, either because of lack of awareness of the requirement to do so, or the time-consuming process of notifying public health officials via telephone or FAX (Doyle *et al.*, 2002). While there is no question that laboratories are valuable sources of clinical test results that are diagnostic of key reportable conditions, they are for all intents and purposes the main source of infectious disease data for most PHOs. Substantial time may be required for processing samples and generating actionable results. By definition, truly novel diseases or disease syndromes (such as the causative Coronavirus of SARS) elude straightforward laboratory detection because of the absence of reagents for identifying the presence of the organism.

Finally, the threat of bioterrorism along with the potential for rapid spread of naturally occurring communicable diseases of animals and humans, places an extraordinary premium on rapid diagnosis and situational awareness of population health. Many exercises conducted by the Departments of Defense, Homeland Security and Health and Human Services lead inevitably to the conclusion that "hours matter" in a bioterrorism event or in some naturally occurring diseases, particularly in the domesticated large animal population. Multiple disease propagation models utilizing agent-based and deterministic approaches also confirm these lessons.

Syndrome-based disease surveillance systems (SBDSS) are defined as data-gathering approaches focusing on clinical descriptors like signs and symptoms instead of specific diseases. In theory, SBDSSs by virtue of their timeliness and volume of information flows could assist in meeting these central public health responsibilities. However, in practice the specific scientific approach, design and underlying technical features like ease of use are dramatically different across the dozens of SBDSSs currently available. Some of these designs have been implemented only in narrowly defined demographic settings or have other limiting features. The promise is often not met in real-world use.

All SBDSS fall into two basic categories: *passive* and *active* (Brevata *et al.*, 2004). Some authors choose to reverse the definitions of these terms. *Active* is sometimes associated with exploitation of existing data streams such as hospital emergency room chief complaints. *Passive* systems are, by contrast, based on direct reporting from healthcare workers. However, the authors here reverse the above definitions because doing so makes them more intuitive and because reversing them correctly characterizes the role of the clinician who is *actively* involved only when making the decision to report a case that may warrant further investigation. The first category is automated or *passive* surveillance systems that exploit existing data streams and employ various statistical algorithms to detect infectious disease. Some of the data sources exploited by these passive systems include: pharmacy sales (including over-the-counter medications); total volume of nurse *hot-line* calls; brief *chief complaint* summaries from emergency room logs and ambulance logs; and absenteeism (i.e. schools and workplaces). A 2004 report called into question the utility and timeliness of *data-mining* approaches (Brevata *et al.*, 2004). Data-mining methods, though superficially attractive because they obviate the need for additional effort on the part of the clinician to enter data, appear to have unacceptable signal-to-noise ratios. False positives are common because some of the data gathered, such as billing codes, are non-specific.

The second category of SBDSSs as defined above consists of clinical or *active* surveillance systems that depend on selected reporting from physicians, veterinarians, EMS services and other health care providers based on the clinical judgment when assessing severity of illness among patients (whether animal or human). It is of course difficult to quantify the value of clinical judgment, but it is nonetheless a key component of daily parlance and processes of medical and veterinary practice (Brevata *et al.*, 2004). It is also important to note that the overwhelming majority of SBDSS data gathering features focus solely on human patients, despite the fact that in all significant outbreaks of novel diseases over the past decade or more in North America, animals were the primary source of the diseases. These outbreaks include: monkeypox in the Midwestern US in 2003; avian influenza H5N1 in 1996, mostly in SE Asia; haemorrhagic *E. Coli* throughout the US in 2007; food-borne botulism in at least four states in the US in 2007; and H1N1 worldwide in 2009.

The authors have used several data-mining systems, and two clinician-driven reporting systems in the epidemiologically complicated, mixed urban, rural and agricultural environments of west Texas and eastern New Mexico. Based on these experiences they were instrumental in developing a syndrome reporting information that is clinician-based and cost effective.

5.1 *The syndrome reporting information system (SYRIS™)*

One of the systems summarized by Chen (*et al.*, 2010), among the many proliferating since 2000 is the Rapid Syndrome Validation Project later commercialized as the Syndrome Reporting Information System (SYRIS) (ARES 2012). It supports two-way disease information reporting and data sharing for medical professionals using fast, reliable, portable methods to report suspicious or novel symptoms that may be part of a known disease or disease-complex. Reporting is based on symptom complexes known as syndromes. These can be defined with a high degree of specificity like haemorrhagic fever syndromes, or can be made more general to address more common medical care. The system functions mainly as a data integration tool using data supplied by users and summarized for PHOs to view as temporal graphs and map layers. This architecture facilitates identification of epidemic disease factors and locations, and provides a means for distributing medical alerts. The design protects patient confidentiality for all types of data reporting because it does not use patient-specific identifiers. Public health officials may contact a physician or other health professional privately to obtain patient-specific information if they believe such information is needed to protect public health. Among other contagious and communicable diseases SYRIS detects many high-risk respiratory diseases including, but not limited to: anthrax, influenza, SARS, West Nile Virus, and avian influenza. It addresses six syndromes specifically: flu-like illness such as hantavirus pulmonary syndrome (HPS), fever with skin rash, fever with altered mental status, acute bloody diarrhoea, acute hepatitis, and acute respiratory distress.

Based on this experience the US Department of Energy funded a project called the rapid syndrome validation program (RSVP). Beginning in 2003, two of its developers (Zelicoff & Simpson) evolved RSVP's scope into what is now SYRIS. This system is a fully-integrated animal and human SBDSS owned by ARES Corporation. Prototype versions have been used by the City of Lubbock's Health Department and the State of Texas' Health Service Region-I, a separate public health jurisdiction comprising the forty counties around the City of Lubbock. In addition, SYRIS is in use by the State of California and CDC's border infectious disease surveillance system (BIDSS). In total, more than one million people, and countless animals, are in the SYRIS coverage areas, with the participation of approximately 120 physicians and nurse-practitioners, twenty five veterinarians, twenty five school nurses, multiple emergency medical service companies, and animal control officials. Wildlife-rehabilitators who are often the first to recognize disease in bird and free-ranging mammalian species also report to SYRIS.

5.2 *Rationale*

Since the turn of the 21st Century, human and animal populations around the globe have suffered from a large and unpredictable array of new diseases of obscure origin. Among these affecting humans are SARS-coronavirus, at least two new bird influenza strains, monkeypox, and Mad Cow disease. There is no part of the human-inhabited world that is safe from infectious diseases, many of which were either unknown or thought to be irrelevant to human health in the 20th Century. Many of these cannot be treated with medications, so prevention seems to be the best approach for limiting their impact.

Second, as a result of jet-age transit of people and animal life forms across the globe, diseases can spread rapidly to virtually any major city in a matter of hours. Concentrations of animals and birds in market places create perfect environments for organisms to jump from infected life forms to humans, and vice versa. As pointed out in Part II of this book, it seems inevitable that new diseases, many of them potentially deadly, will spread to environments that are now sparsely populated.

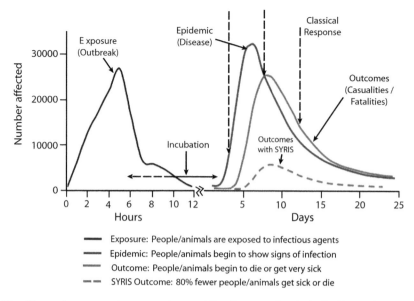

Exposure: People/animals are exposed to infectious agents
Epidemic: People/animals begin to show signs of infection
Outcome: People/animals begin to die or get very sick
SYRIS Outcome: 80% fewer people/animals get sick or die

Figure 12. The syndrome reporting approach to avoiding disease outbreaks in the first twelve hours vs. the classical identification and response approach that result in epidemics. SYRIS utilizes GIS and EO-data in the system design (Courtesy ARES Corporation). (see colour plate 80)

Arguably, the best reason for syndromic information systems is that the existing public health infrastructure is inadequate to deal with large numbers of human and animal disease carriers at the geographic point of infection. Much of this situation has arisen because of inadequate funding at local levels to provide effective health services quickly, when and where they are needed. The situation is further exacerbated by complacency following the supposed eradication of diseases like smallpox in the 1960s and '70s, and by extensive vaccination programmes for polio. SYRIS is designed to provide timely and actionable information to health officials and policy makers to recognize and respond to evolving disease outbreaks at their inception when intervention is most needed to avoid epidemics.

The developers of SYRIS recognize that few physicians are trained in statistics or epidemiology, and that very few have regular contact with public health offices. Furthermore, very few of the nearly eighty reportable diseases across Canada and the US are ever reported by physicians or primary care specialists. Recent history shows that food-borne outbreaks of E. coli, for example, can take weeks to trace to their sources even given the resources available to national health monitoring agencies. SYRIS is designed for qualified health providers to share syndromes at the time and place of patient observation, many of whom may be first cases in an ensuing outbreak; and, to receive information from colleagues who may be observing the same or similar symptoms in their service areas.

Like all syndromic systems, SYRIS aims to detect exposures to suites of illnesses represented by each syndrome within one to six hours in their incubation stages; and to implement identification and response measures before an outbreak occurs (Figure 12). Throughout history, and without early warning obtainable through social and information networks, classical epidemics have blossomed in anywhere from five to twenty-five days of a few people becoming infected; and classical responses have not been able to prevent them.

5.3 Software design and operational requirements

The basic design requirements for the SYRIS software are: 1) complete platform independence. It is written in JAVA and works on all Windows, Macintosh, LINUX and other Java-enabled operating systems (including tablet PCs); 2) local public health officials serve as data analysts for all information reported from the clinical communities listed above. PHOs also register users via an administrative page in SYRIS; 3) data entry in less than thirty seconds based on syndromic classifications for animals or humans suspected of having symptoms and signs indicating serious infectious disease based on clinical judgment; 4) immediate updating of geographic maps and temporal graphs so that all users can visualize local, regional and national reporting levels quickly; 5) veterinarian, physician, school nurse and EMS reports are colour coded for easy interpretation. 6) the specific details of individual case reports (for example, reporting clinician, specific clinical findings, and approximate patient age) is under the control of local PHOs but can be shared easily as a spread-sheet file readable by almost all statistics software packages; 6) GIS-based analysis tools for public health officials and clinicians complemented by statistical tools for rapidly determining unusual events; 6) free-text advice and analysis from public health officials; 7) automated alarm criteria, set on an ad hoc basis by public health officials such that any report meeting those criteria generates immediate notification of on-call PHOs via e-mail and/or digital pager; 8) a highly secure database (SQL-server) that meets all HL-7 and NEDSS requirements, and which can import other data streams easily (e.g. laboratory data); 9) easy installation and automatic updating of software; 10) on-line training via videos and manuals; and 11) low cost, that is, approximately $0.15 USD per capita for annual licensing fee.

5.4 Functions

A *user guide* is available online (ARES 2012). Upon login, the map displays the geographic area associated with the user's jurisdiction. When DETAILS are selected for a syndrome, the map view changes to display the geographic area where cases have occurred that are associated with the syndrome. Navigation buttons allow users to zoom in or out, or to move up, down, left and right. The map can be viewed as a series of layers. Some are more or less transparent, while others are collared in various shades, and still others represent borders of counties, states, and zip codes. Various features of the map can be turned on or off in "groups" or individually. The community of users includes physicians, veterinarians, school nurses, animal control officers, lab technicians, emergency medical services, medical investigators, and public health officials. Once a user logs onto the system, they can view instantaneous syndromic information, maps, public health alerts, and contact information. The intention is to provide a mechanism whereby disparate networks of health users inform themselves about current and rapidly evolving situations at the grass-roots level.

Data entry in SYRIS is via "intelligent" pull-down menus which automatically change depending on the specific situation and radio-buttons. There are nine veterinary syndromes in six human syndromes. An example of a veterinary submission is seen in Figure 13.

Note that data fields include a "site descriptor" (zoo vs. feedlot vs. domestic/residential). Veterinary infectious disease experts who advised us in the design of SYRIS noted that this stratification was particularly important given the unique diseases that are found in each of these subpopulations of animals. A wide variety of animal species are listed (and more can easily be added), and the veterinarian can enter their clinical impression in addition to categorizing the case into a specific syndrome.

After entering the case by clicking on "Submit Report", the data are transferred securely to a central database. Within a few seconds, the veterinarian receives the screen shown in Figure 14. The map shows a case of neurological lameness just reported by a veterinarian in zip code 87042 near Albuquerque NM, US. Both the map and the graph report data as a thirty-day rolling window in time (this is an arbitrary parameter that can be set by the public health official). The colour coding on the map reflects the syndromes reported: horizontal stripes represent human syndromes, vertical stripes represent veterinary syndrome reports such as the one just entered, and diagonal stripes indicate school nurse, EMS or animal control syndrome reports. In any given public health jurisdiction (e.g. county or city health departments) all reports from users registered in that jurisdiction appear on the map.

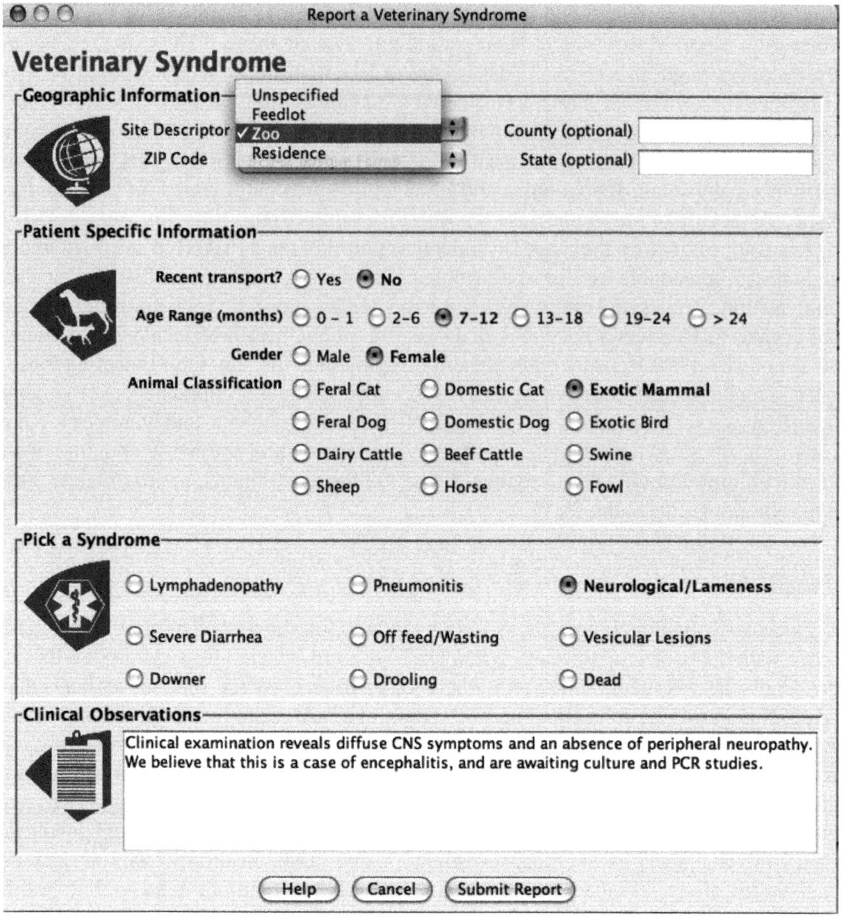

Figure 13. Reporting screen for veterinarians.

Data query is similarly straightforward. The map can be panned and zoomed in real-time; maximum resolution is at the zip code level (which has been assessed as sufficiently precise for analysis purposes while at the same time protecting patient confidentiality), and entire regions can be visualized and queried. For example, by zooming out and right-clicking on a state, SYRIS summarizes all case reports by syndrome from the data-base in a few seconds (Figure 15, left panel).

Finally, clinicians and public health officials can use the GIS tools beneath the map to quickly ascertain the overlap(s) of syndromes that may help narrow differential diagnoses. The icons beneath the map represent geographic features (such as major highways, political boundaries, and postal codes), each of the six human syndromes (Fever with CNS findings, acute hepatitis, Adult Respiratory Distress Syndrome, Influenza-like illness, Fever with skin rash and severe diarrhoea) and the nine veterinary syndromes. If users are concerned about encephalitis cases, they can turn off all syndromes except animal neurologic syndrome and the human syndrome of "Fever with CNS findings" (Figure 16). In this case there is a geographical and temporal overlap of veterinary and human reports, and as local PHOs would be aware, the specific areas involved are both urban and rural and are also known to be heavily infested with mosquitoes as the Zip Code boundaries span the Rio Grande in the centre of Albuquerque.

Public health officials can easily determine which clinicians have reported cases of interest that may warrant further discussion with the health-care provider. When a public health official logs

394

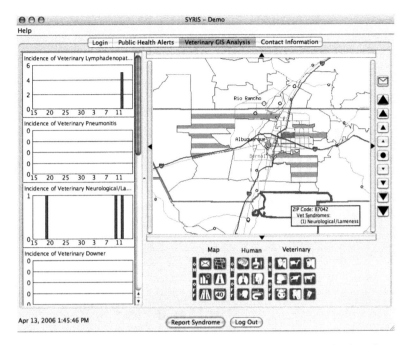

Figure 14. Screen after a case report showing human and veterinary cases in the Albuquerque data (notional data).

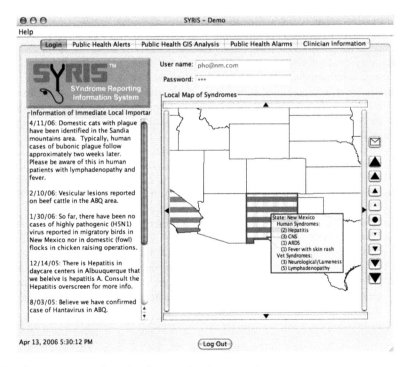

Figure 15. Summary screen for regional or state-level case counts.

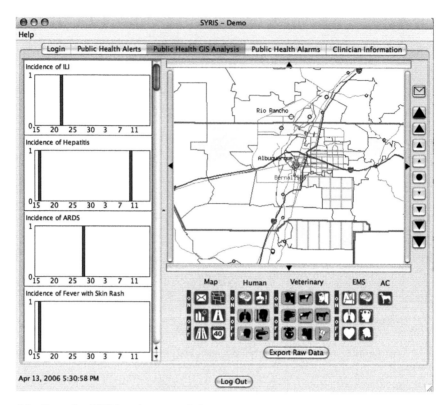

Figure 16.   Example of GIS-based case correlations.

into SYRIS, they can perform the same GIS analysis. SYRIS organizes the data by clinician type (i.e. physicians, veterinarians, school nurses) and then by date within each category. Included in the database is the name and phone number of the reporting clinician. Thus, initiating a contact investigation is streamlined and allows the "index of suspicion" of PHOs to be rapidly assessed. PHOs can easily view the database in almost any spread sheet or statistics programme, facilitating further analysis.

Further, PHOs can set specific criteria, syndromes, and signs and symptoms within those syndromes, for which they wish to be notified immediately, even if they are away from their computers. For example, if a PHO was interested in knowing about all cases of international travellers over the age of seventeen with severe diarrhoea, they simply go to the "Alarm" tab, initiate a "New Alarm" and select criteria from intelligent pull-down menus (Figure 17).

This process makes operational, the public health principle of "case definitions" which are, in practice, almost impossible to communicate to clinicians. Thus, when a clinician enters a case that meets the criteria of the case definition, SYRIS notifies the PHO with an e-mail and/or digital page specifying the name of the clinician, their telephone number, the syndrome reported, and any other clinical details reported. The PHO can then make the decision to contact the clinician or not. Should the criteria be too general such that too many alarms are generated, the PHO can at any time refine the criteria of interest.

5.5   Early successes with SYRIS

SYRIS has become the primary public health tool for routine daily communication among clinicians, animal control officials, and PHOs in the City of Lubbock and the surrounding forty counties

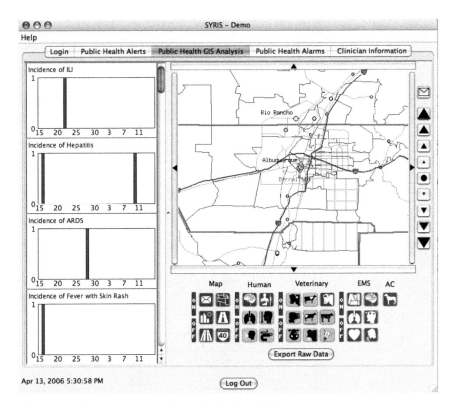

Figure 17.   Alarm criteria screen as used by public health officials.

of northwest Texas. In addition to providing regular, timely summaries of advice to clinicians SYRIS was also employed during the Katrina disaster by physicians pressed into service to staff two medical clinics at the evacuee centre established for approximately 850 people in a mostly abandoned air force base approximately twenty miles (32+km) west of Lubbock. Because volunteer physicians were working in shifts and because the evacuees were at high risk for unusual infectious disease (e.g. *Vibrio vulnificus cellulitis*, meningococcal disease, and enteric infections) with which most physicians had little experience, PHOs and doctors installed SYRIS in two laptop computers within a few minutes of the arrival of evacuees. Local physicians volunteered to staff the intake sessions for evacuees and then staffed an on-going clinic over a two week period. Many of these physicians ordinarily worked in private practice or as faculty members at Texas Tech School of Medicine. None of these doctors had any formal training with the SYRIS programme or syndromic surveillance. The system was so intuitive in its design that little training was required for these physicians for them to use it. A single page hand-out to guide physicians was employed. Data collection was very simple and users were able to collect syndromic surveillance data successfully.

The data from SYRIS allowed for planning and preventing the outbreak of communicable disease with other evacuees and the Lubbock community. The display of data from the SYRIS system is easy to read and allowed physicians working with the evacuees to assess and react to important outbreak information in a timely manner. In the system, a single screen summarizes syndromic data for easy interpretation.

The system also aided the care of individual patient with symptoms by helping to guide diagnosis. SYRIS was added to the Health Director's station in Lubbock's Emergency Operations Center (EOC) when the EOC was activated for the Katrina response. He was able to monitor SYRIS from the EOC 20 miles away, and determine if resources would be allocated to meet the health needs of the evacuees. Based on information observed in SYRIS, the director was able to identify with

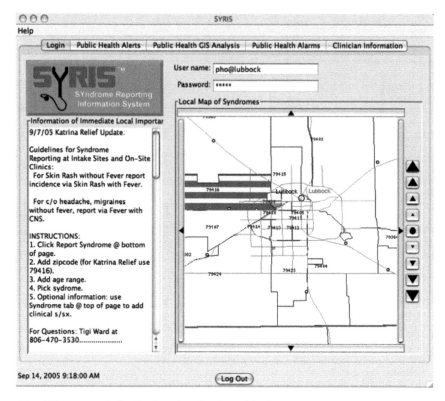

Figure 18.   SYRIS in use during Katrina relocation (actual data).

reasonable certainty the most prevalent health complaints of the evacuees. Ordering unnecessary supplies and unneeded resources were thereby reduced significantly.

Because of the concern about unusual disease and associated misdiagnosis and incorrect treatment, the local community wanted to be prepared for outbreaks of communicable diseases should they occur either at the evacuee housing site or into the surrounding area. SYRIS addressed both problems easily. For example, on the main SYRIS login page, continuously updated advice was provided by public health officials and the SYRIS map displayed the incidence and location of a variety of infectious disease syndromes (Figure 18).

At intake, numerous cases of diarrhoea, skin rash and headache, with or without fever (fever with CNS findings) were reported, but rapidly waned, reassuring physicians, evacuees and residents that there was little likelihood of communicable disease outbreak (Figure 9). Few, if any, unnecessary diagnostic tests were ordered, and there were no complications from missed diagnoses or inappropriate treatment. Public health officials at multiple levels were apprised of clinical assessments, dramatically reducing costs and waste of limited resources. Because of its intuitive interface and on-line training, clinicians were able to train themselves in the use of the system with minimal assistance from busy PHOs.

In 2008, there was an outbreak of Shigella among school age children in Lubbock, TX. Although to the best of our knowledge this outbreak was not reported via the CDC's Morbidity and Mortality Weekly report, nor through the Health Alert Network, school nurses, public health officials and physicians using SYRIS were able to keep track of the outbreak. Further, *SYRIS facilitated the dissemination of information on the time course of the outbreak via the* initial screen (Figure 19).

Finally, during the H1N1 2009 pandemic, SYRIS apprised local healthcare workers and public health officials on the incidence of disease in near real-time. Lubbock experienced a low morbidity, and apparently relatively low infection rate, as well (Figure 20). The "Information of Immediate

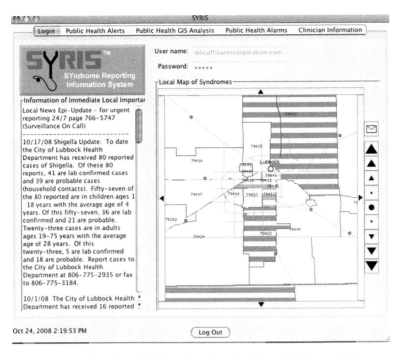

Figure 19.   Screen during a Sept-Oct 2008 Shigella outbreak in Lubbock, TX (actual data).

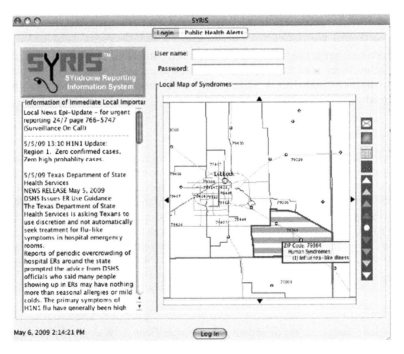

Figure 20.   Screen during H1N1 outbreak in early May 2009 (actual data).

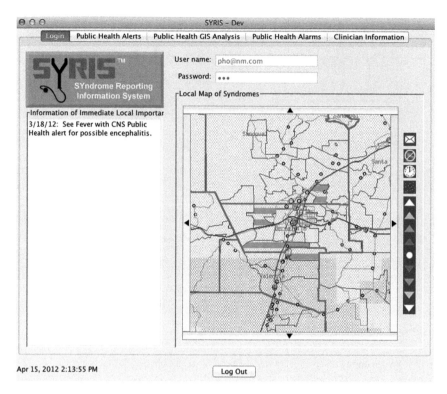

Figure 21.   Screen showing two dust concentrations from DREAM/eta model for PM_{10} on 15 April 2012. Different concentrations are indicated by intensity of grey shading with the heavier concentration toward the bottom of the scene.

Local Importance" on the front page of SYRIS included information on the local pandemic situation (zero confirmed cases as of early May 2009), as well as copies of essential information from the Texas Department of State Health Services that might otherwise not have reached providers in clinics and schools.

5.6   Integration of atmospheric data into SYRIS

Because respiratory syndromes similar to influenza can occur with environmental dust exposures in susceptible individuals (e.g. those with pre-existing chronic lung disease), SYRIS includes a display of DREAM output, updated daily to include current conditions and predictions for dust levels stratified by size into roughly 2.5 μm and 10 μm bins up to 24 hours in the future (depending on the time in the daily modelling cycle when users access the dust data). The dust map is superimposed on the SYRIS GIS display of syndromic information with a click on the dust 'icon' on the right hand side of the SYRIS screen (Figure 21). Thus, clinicians can be aware of current as well as likely future dust concentrations, possibly narrowing the differential diagnosis of respiratory symptoms in patients with asthma, COPD or other chronic lung diseases.

5.7   Summary

PHOs have long recognized the desirability of high-specificity, high sensitivity data characterizing the population health of animals and humans in near real-time to provide actionable information to identify and interrupt infectious disease outbreaks. SYRIS has met this essential function and provided an additional, unexpected benefit of strengthening the relationships between public health and clinical practice professionals. In west Texas and parts of New Mexico, health COPs have evaluated multiple data-mining and clinician-driven syndromic surveillance systems, but to date, no system has met the ease of use, low cost, low false positive rate and broadly-based clinician acceptance as has SYRIS.

Unlike some other electronic syndrome-based systems, there have been no violations of patient confidentiality after many hundreds of case reports, no false positive results (defined here as cases that led to formal public health investigations into insignificant events) nor false negatives (defined as cases of classical reportable disease that were identified by laboratory examination but had no awareness from clinical syndromic data entry from clinicians). The authors believe that physicians, in particular, utilize SYRIS because it provides them with a real-time assessment of infectious disease epidemiology, thus fostering efficient use of laboratory tests and appropriate antibiotic prescribing.

In short, SYRIS makes it possible to practice "evidence-based" medicine across all age groups who present with signs and symptoms of acute, multi-system infectious disease. In addition, and, to our knowledge unique among any functioning disease reporting systems, SYRIS integrates environmental dust data of key importance in pulmonary and cardiovascular disease diagnosis. Our experience to date is limited by the absence of controlled trial comparing SYRIS to other reporting systems, including notifying public health officials of standard "reportable diseases" (usually defined by state PHOs). Such a trial is now indicated, given the limited utility of electronic reporting systems that have received significant support to date (Witkop *et al.*, 2010; Sokolow *et al.*, 2005).

5.8   *Conflict of Interest Disclosure*

Authors declare that they have no financial or ownership interest in ARES Corporation or the SYRIS software system, either singly or jointly.

6   A SEMANTIC APPROACH TO HEALTH INFORMATION

The challenge for both data providers and users is to make data exchange and data discovery more effective while tackling the problems of information overload, data interoperability, and data integration for distributed, disparate sources that are tailored to spatial data. A holistic approach to geospatial and temporal searching for the environmental tracking enterprise is needed that enables communities of users to work together. This approach needs to synthesize a variety of emerging technologies at each stage of the information collection and search process. This section provides an overview to a holistic search capability and outlines the major capabilities needed at each stage of the process. Secondly, the key technologies needed at each of these stages are summarized. The Chapter closes with two case studies that are shaping the future of this capability. Some of the key issues for an enterprise approach for geospatial and temporal data search are outlined below. Four terms and their definitions according to Wikipedia are given for this Section.

Crowdsourcing is a neologism for the act of taking tasks by a group (crowd) of people or community in the form of an open call. The term has become popular for leveraging the mass collaboration. *WSDL-based* is the web services description language (WSDL, pronounced wuz-dul). It is an XML-based language that provides a model for describing Web services software designed to support interoperable machine-to-machine interaction over a network. *RESTful* is a style of software architecture for distributed hypermedia systems such as the World Wide Web. The term *representational state transfer* was introduced and defined in 2000 by the principal authors of the hypertext transfer protocol (HTTP) specification. *Folksonomy* is a system of classification derived from the practice and method of collaboratively creating and managing tags to annotate and categorize content; this practice also is known as collaborative tagging, social classification, social indexing, and social tagging. Folksonomy is a portmanteau of folks and taxonomy. Folksonomies became popular on the Web around 2004 as part of social software applications such as social bookmarking and photograph annotation. Tagging, which is one of the defining characteristics of Web 2.0 services, allows users to classify and find information collectively. Some websites include tag clouds as a way to visualize tags in a folksonomy. A good example of a social website that utilizes folksonomy is 43 Things. An empirical analysis of tagging systems dynamics has shown that consensus around stable distributions and shared vocabularies emerges, even in the absence of a central controlled vocabulary (Halpin *et al.*, 2007). For content to be searchable, it should be categorized and grouped like keywords in a journal article.

6.1 *Elements of the semantic approach*

6.1.1 *Data from multiple sources and owners*
The environmental community will often need to act as a data broker for data owned by many contributors, agencies and organizations. The Federation of Earth Science Information Partners (ESIP) addressed this challenge and developed DataFed (ESIP 2012). ESIP has accomplished the initial step in creating an environment for information discovery and sharing. In crisis situations, it will be necessary to Crowd-source data from a variety of volunteer sources. Although this reduces the cost and effort needed to make data available for environmental tracking, it does introduce additional data integration and data quality challenges. The user base includes environmental specialists and researchers, policy making and regulatory organizations, as well as the public and media. Each group requires a different context to ensure that proper information is communicated and is actionable.

6.1.2 *Varied support of metadata and tagging*
Data differ widely to the degree to which it is geo-referenced, tagged, and identified with metadata that describe who provided the data, when they are valid, and what the semantic meaning of the data is. Tagging is a resource-intensive process. Data sources may also have different levels of trust associated with them, and the relevancy of data to a given search is not always easy to determine. Furthermore, since a fully centralized repository of relevant data would be cost-prohibitive to develop and maintain, a federated approach to cataloguing and retrieving data will be essential. Source schemas, metadata and capabilities (functions supported by the source) can be exported. This provides the capability of registering data sources to a particular platform so that they can be made available via a single web service registry. This service can semantically enhance the discovery of both traditional (WSDL-based) geospatial services and non-traditional (RESTful) OGC geospatial services. While there have been efforts to standardize geospatial data and related ontologies, derived analytical data have not benefited from similar standardization.

6.1.3 *Optimal user experience, collaboration with limited resources*
End users require data provided in a meaningful context, amenable to their native language. They will likely request that data be presented in a viewer client such as Google Earth which is inexpensive, highly intuitive, and allows one to overlay additional data as needed. They will also need a notion of search that is much broader than a keyword search or query. They will need to be able to explore relationships between data based on location, time window, or themes and concepts. They will also need to subscribe to notification systems so that they can be notified of key events matching their criteria of interest.

Users will need to work together to share data, their interpretations of the data, their hypotheses, and to coordinate their actions. Dedicated communities of interest will need to be supported so that concerned individuals can unite around the same tactical mission. These distributed communities can jointly develop and evaluate hypotheses, evaluate the relevance of data sources and artefacts, and coordinate actions.

Many environmental agencies do not have the time or resources needed to integrate and maintain all possible data feeds personally, but they also need to ensure quality and usability for those provided by other sources. It is also important to identify freeware and open-ware solutions that provide for the maximum amount of capability without a crippling amount of license fees.

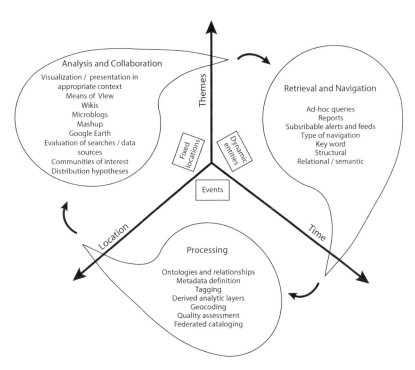

Figure 22.   A unified approach to environmental data search.

6.1.4   *Geospatial or temporal enterprise search*

The following terms represent the rapidly expanding lexicon of user-led innovation: .Open source software; citizen journalism; crowdsourcing; user-generated content; social networks; the sharing economy; peer production; multi-user virtual environments; participatory media; and, collaborative creativity. Central to each term is a common concept, that of the co-operative active user. Digital technologies are making it possible for audiences to cease being merely passive consumers of information and, as well, become active producers. These developments represent the beginnings of what is shaping up to be an immense and enduring shift in the relational patterns encompassing the production and consumption of information. It is a shift as audiences collaborate to produce content for themselves, and for each other; to devise their own ways to connect; and to distribute what is relevant to them. These innovations will lead to a unified approach to an enterprise search of environmental data, where a myriad of sources, retrieval and collaboration techniques will coexist seamlessly, as shown in Figure 22. The search space for environment data consists of three major axes. The geospatial axis addresses the relevant locations and elevations for specific elements. The temporal axis addresses the relevant time window for a given datum, such as time and date when an observation was made, or the time period when a given facility was producing waste. The thematic axis addresses the types of information that are being displayed, whether they are atmospheric observations, water contamination measures, organizational structures or regulations. Individual elements in the search space can be fixed locations, such as a township, a protected area, or a geographical feature which are likely to remain stable over time; or, transient entities that include resources such as hazmat vans, individuals and task groups, meteorological conditions, and specific environmental phenomena such as plumes.

Events may have hypotheses or interpretations associated with them such as, Why is the vegetation in this area showing signs of stunted growth? Elements in the search space are also described by their relationships with each other. These could be described in formal ontologies (a tightly controlled mapping of concepts and their inherent relationships), or be developed ad-hoc as users define relationships using Tags (folksonomies). Relationships are crucial for navigation and discovery within specific users' contexts, since they allow users to move between elements along many possible paths, and not be limited to a traditional keyword search or query. It is necessary for Earth science data and models to be accessible in semantic frameworks because Earth system science data originate from many disciplines that have different standards and terminologies, and that provide data in a variety of mark-up languages and formats (Raskin 2004). The ESMF described in Chapter 7 is an excellent example. One can anticipate that success in the semantic ESMF efforts will accrue huge benefits to initiatives like WMO's Thorpex and the ICSU's SHWB programmes as they continue to evolve (ICSU 2011).

6.2 HealthMap: Tool for environmental health tracking

HealthMap was created in 2006. It is an established global leader in utilizing online informal sources for disease outbreak monitoring, with over a million users a year. The website, freely available at http://healthmap.org, delivers real-time intelligence on a broad range of emerging infectious diseases for a diverse audience including local health departments, governments, and international travellers. HealthMap brings together disparate data sources to achieve a unified and comprehensive view of the current global state of infectious diseases and their effect on human and animal health (Brownstein & Freifeld 2007). HealthMap relies on a variety of electronic media sources, including online news aggregators, eyewitness reports, expert-curated discussions and validated official reports. Through an automated process, updating 24/7/365, the system monitors, organizes, integrates, filters, visualizes and disseminates online information about emerging diseases in seven languages, facilitating early detection of global public health threats (Brownstein *et al.*, 2008).

In February 2009 a new category (environmental) was added to the HealthMap system in an effort to distinguish alerts of potential public health importance due to natural or man-made ecological occurrences. Whether natural or man-made, instances of human disease have been linked to environmental factors and events such as earthquakes, floods, hurricanes, fires, chemical spills, and acute air pollution. Following a natural or man-made environmental event, a noticeable increase in risk factors for the emergence of epidemics may occur (immediately or gradually) including population displacement, lack of adequate shelter, overcrowding, disruption to the power supply, and lack of access to safe drinking water (Watson *et al.*, 2007). HealthMap posts environmental occurrences of concern to its website in an effort to increase awareness on the potential for outbreaks of human illness and other effects on public health to users from the general public as well as public health professionals around the globe.

6.3 Case studies

6.3.1 Haiti

Immediately following the January 12, 2010 earthquake that devastated Haiti, HealthMap created a dedicated page focusing on the country (www.*HealthMap* org/haiti), and began a more active surveillance approach to identify infectious disease events. Utilizing similar methods for enhanced surveillance as those developed previously for mass-gathering events, additional feeds were monitored to increase the sensitivity of the automated system. All incoming reports were manually reviewed to ensure specificity (Khan *et al.*, 2010). As early as January 21st, media reports of increased cases of tetanus, measles, respiratory illnesses, diarrhoea, and dengue emerged. On October 20, 2010 reports were received of a diarrheal disease outbreak in Saint-Marc, Artibonite, Haiti that was later confirmed to be cholera (Washington Post 2010). In partnership with Humanity Road and Crisis Mappers, HealthMap added additional layers to the focused Haiti map to assist aid workers in identifying such things as health facilities, cholera treatment centres, and new safe water installations. This partnership also enabled HealthMap to begin posting reports from text messages and Twitter posts, which collectively provided an additional data source for tracking the spread of cholera in Haiti.

6.3.2 *Pakistan*

Flooding that began in July 2010 quickly worsened in Pakistan, and by mid-August the massive flooding had affected approximately one quarter of the country. Much of the population was displaced, and clean water became scarce, raising the risk for the spread of diseases such as dysentery, cholera, and diarrheal illnesses in a country with an already weakened public health infrastructure (Solberg 2010). In response to the events in Pakistan, HealthMap created a QuickView pre-set, a shortcut tab that users are able to customize to view specific data of interest, titled Pakistan Flood-Related Alerts. The pre-set was also the HealthMap website default view during the height of the flooding in an effort to increase awareness of the on-going public health crisis in Pakistan, and to allow users to view easily any new infectious disease events that were occurring in the region.

These two examples demonstrate the capability of the HealthMap system to detect near real-time instances of human illness following environmental disasters. While some environmental occurrences lead to outbreaks of human illness, other events recorded by the system have caused devastation by way of economic or ecological losses such as: 1) the eruption of Iceland's Eyjafjal-lajökull volcano in April 2010 that led to massive disruptions in European air travel, and economic losses (BBC News 2010); 2) the September 2010 grasshopper outbreak that devastated alfalfa and other grass crops in Wyoming, US (Billings Gazette 2010); and 3) coral bleaching due to increased ocean temperatures, affecting an estimated ninety per cent of coral in the Gulf of Thailand and the Andaman Sea leading to the temporary closure of eighteen popular dive sites in February 2011 (Wilson 2011).

HealthMap continues to track environmental occurrences with the utilization of the environ-mental classification for alerts on the system's main page visualization to increase awareness on the potential for outbreaks of human illness, a weakened public health infrastructure, or economic or ecological losses that may follow natural and man-made environmental events.

6.4 *The climate change collaboratory*

The climate change collaboratory (CCC) is a project funded by the Austrian climate and energy fund within the Austrian climate research programme. It aims to strengthen the relations between Austrian scientists, policy makers, educators, environmental NGOs, news media and corporations – stakeholders who recognize the need for adaptation and mitigation, but differ in world views, goals and agendas. The collaboratory manages expert knowledge and provides a technology platform for effective communication and collaboration. It assists networking with leading international orga-nizations, bridges the science-policy gap and promotes rich, self-sustaining community interaction to translate knowledge into coordinated action. Innovative survey instruments in the tradition of games with a purpose (Rafelsberger & Scharl 2009) create shared meaning and leverage networking platforms to capture indicators of environmental attitudes, lifestyles and behaviours.

Despite credible forecasts and warnings from the scientific community about anthropogenic climate change, greenhouse gas emissions have continued to grow. Scientists studying the issue predict more adverse consequences unless stronger actions are taken. From the policy-making level to personal voting and purchasing decisions, however, observable actions have not been commensurate with the threat of climate change. Governments and societies remain far short of undertaking emission reductions that some scientists say are required to forestall dangerous interference in the climate system (Abbasi 2006). Although public concern about climate change has risen in the past few years, a much smaller percentage is actually taking action. Reasons for this discrepancy include: 1) on the micro level, the widespread perception of climate change as a risk that will predominantly impact geographically and temporally distant people and places; and the lack of personal efficacy (belief that their own actions will make a difference and one's voice will be heard), a critical motivating factor in behavioural change that can be supported by Web-based applications to share knowledge and coordinate action; and 2) on the meso- and macro-levels, a gap between policies and research needed to promote and support adaptation, and also mitigation (and their interrelation), and what is currently available.

The over-arching goal of the collaboratory is to build capacity among policy makers, scientists, educators, environmental NGOs, news media and corporations to close this gap and translate increased awareness into behavioural change on the local, regional, national and international levels.

Triple-C is an interdisciplinary initiative to encourage and study discourse and critical debate that lead to a shared understanding of climate change issues on all political levels, ranging from inter-individual communication and local communities to global campaigns and treaties. By investigating communicative strategies and processes that function between disciplines and stakeholders, the Triple-C project aims to unearth hidden assumptions and misconceptions about climate change, contribute to a mutual understanding of existing problems, and suggest priorities for research and policy development. Participants of the collaboratory will benefit from a synergy of skills and resources, the constitution and dynamic maintenance of shared knowledge, flexible and non-hierarchical modes of cooperation, and mechanisms for distributed decision making.

Environmental Web resources such as documents and best-practice examples are often being created through processes of cooperation and social exchange. They depend on and benefit from a synergy of skills, the dynamic maintenance of shared knowledge, flexible and non-hierarchical portfolios of services, and distributed decision making. Triple-C recognizes and supports the social construction of meaning via distributed information services that aim to improve the quality of decisions, build trust and help resolve conflicts among competing interests. It will provide matchmaking services for ad-hoc team composition and a range of Web-enabled communication and collaboration tools. Facilitating the collaboration between stakeholders will require a tight integration of heterogeneous services. System connectivity, contextualization and semantic interoperability to achieve this integration are at the core of Triple-C. Collaborative ontology building, for example, ensures that recent findings are understood by all members of a virtual community.

The envisioned collaboration platform will draw upon the lessons learned from building the Media Watch on Climate Change (MWCC) (Hubmann-Haidvogel et al., 2009) (ECOresearch.net 2012). This award-winning news aggregator provides geographic and semantic visualizations based on multiple coordinated view technology. It will be extended by communication and collaboration tools such as messaging services, Wikis, Web-based discussion forums, multi-language support, and a layered security model to distinguish between public and private information. Geographic mapping will play a central role, using the virtual globe technology of NASA World Wind to integrate different types of data objects (documents, best-practice examples, expert profiles, etc.), and put them into a regional context.

REFERENCES

Abbasi, D.R. 2006. *Americans and climate change: Closing the gap between science and action*. New Haven: Yale School of Forestry & Environmental Studies.

Adler, R. 2005. Executive summary: Infectious disease outbreak decision support system (IDODSS). Available from: http://www.decpath.com/ForeTell%20IDODSS%20Exec%20Summary.pdf [Accessed 21st February 2012].

Ampel, N.M., Mosley, D.G., England, B., Vertz, P.D., Komatsu, K. & Hajjeh. R.A. 1998. Coccidioidomycosis in Arizona: Increase in incidence from 1990 to 1995. *Clin. Infect. Dis.* 27: 1528–1530.

ARES. 2012. Available from: http://syris.arescorporation.com/demo/ [Accessed 25th January 2012]

Barker, C.M. 2008. Spatial and temporal patterns in mosquito abundance and virus transmission in California. PhD Diss. Davis: University of California.

Barker, C.M., Reisen, W.K. & Kramer, V.L. 2003. California State mosquito-borne virus surveillance and response plan: Retrospective evaluation using conditional simulations. *Am. J. Trop. Med. Hyg.* 68: 508–518.

Barker, C.M., Reisen, W.K., Eldridge, B.F., Park, B. & Johnson, W.O. 2009. *Culex tarsalis* abundance as a predictor of western equine encephalomyelitis virus transmission. *Proc. Mosq. Vect. Cont. Assoc. Calif.* 77: 65–68.

BBC News. 2010. Iceland volcano cloud: The economic impact. Available from: *http://news.bbc. co.uk/2/hi/8629623.stm* [Accessed 16th February 2012].

Biggerstaff, B.J. 2003. Pooled infection rate. Available from: http://www.cdc.gov/ncidod/dvbid/westnile/software.htm [Accessed 16th February 2012].

Billings Gazette. 2010. Grasshopper outbreak brings disaster declaration in western Wyoming. Available from: http://billingsgazette.com/news/state-and-regional/wyoming/article_39ca8834-085f-11e0-a77f-001cc4c03286.html. [Accessed 16th February 2012].

Brandeau, M.L., McCoy, J.H., Hupert, N., Holty, J-E. & Bravata, D.M. 2009. Recommendations for modeling disaster responses in public health and medicine: A position paper of the Society for Medical Decision Making. *Med. Dec. Making* 29: 438–460.

Bravata, D.M., McDonald, K.M., Smith, W.M., Rydzak, C., Szeto, H., Buckeridge, D.L, Haberland, C. & Owens, D.K. 2004. Systematic review: Surveillance systems for early detection of bioterrorism-related diseases *Ann. Intern. Med.*140: 910–922.

Brownstein, J.S. & Freifeld, C.C. 2007. HealthMap: The development of automated real-time internet surveillance for epidemic intelligence. *Euro Surveil.* 12(48): pii=3322.

Brownstein, J.S., Freifeld, C.C., Reis, B.Y. & Mandl, K.D. 2008. Surveillance sans frontieres: Internet-based emerging infectious disease intelligence and the HealthMap project. *PLoS Med.* 5(7): 6. doi:10.1371/journal.pmed.0050151.

Carney, R.M, Husted, S., Jean, C., Glaser, C. & Kramer, V. 2008. Efficacy of aerial spraying of mosquito adulticide in reducing incidence of West Nile Virus, California, 2005. *Emerg. Infect. Dis.* 14: 747–754.

Chen, H., Zeng, D. & Yan, P. 2010. Public health syndromic surveillance systems. In: H. Chen, D. Zeng & P. Yan (eds.), *Infectious Disease Informatics for Public Health & Biodefense*: 9–21. New York: Springer Science.

Comrie, A.C. 2005. Climate factors influencing coccidioidomycosis seasonality and outbreaks. *Environ. Health Perspec.* 113: 688–692.

Comrie, A.C. & Glueck, M.E. 2007. Assessment of climate-coccidioidomycosis model: Model sensitivity for assessing climatologic effects on the risk of acquiring coccidioidomycosis. *Coccidioidomycosis: 6th Intern. Symp.* 1111: 83–95.

Cummings, R.F. 1992. Design and use of a modified Reiter gravid mosquito trap for mosquito-borne encephalitis surveillance in Los Angeles County, California. *Proc. Mosq. Vect. Cont. Assoc. Calif.* 60: 170–176.

Dannen, M., Simmons, K., Chow, A., Park, B. & Fang, Y. 2007. Surveillance 2006: Overview, changes and improvements in turnaround time. *Proc. Mosq. Vect. Cont. Assoc. Calif.* 75: 37.

Doyle, T.J., Glynn, M.K. & Groseclose, S.L. 2002. Completeness of notifiable infectious disease reporting in the United States: An analytical literature review. *Amer. J. Epidemiol.* 155: 866–74.

Dredger, S., Michelle, A.K., Morrison, J., Sawada, M., Crighton, E.J. & Graham, I.D. 2007. Using participatory design to develop (public) health decision support systems through GIS. *Int. J. Health Geog.* 6: 53. doi:10.1186/1476-072X- 6-53.

ECOresearch.net. 2012. Available from: http://www.ecoresearch.net/climate [Accessed 6th March 2012].

Elnaiem, D.E., Kelley, K., Wright, S., Laffey, R., Yoshimura, G., Reed, M., Goodman, G., Thiemann, T., Reimer, L., Reisen, W.K. & Brown, D. 2008. Impact of aerial spraying of pyrethrin insecticide on *Culex pipiens* and *Culex tarsalis* (Diptera: Culicidae) abundance and West Nile Virus infection rates in an urban/suburban area of Sacramento County, California. *J Med. Entomol.* 45: 751–757.

ESIP. 2012. Available from: http://datafedwiki.wustl.edu/index.php/DataFed [Accessed 27th January 2012].

Feiszli, T., Husted, S., Park, B., Eldridge, B.F., Fang, Y., Reisen, W.K., Jean, C., Parker, E. & Kramer, V.L. 2009. Surveillance for mosquito-borne encephalitis virus activity in California, 2008. *Proc. Mosq. Vect. Cont. Assoc. Calif.* 77: 84–97.

Fisher, M.C., Koenig, G.L. White, T.J. & Taylor, J.W. 2002. Molecular and phenotypic description of *Coccidioides posadasii* sp nov., previously recognized as the non-California population of *Coccidioides immitis*. *Mycologia* 94:73–84.

Galgiani, J.N., Ampel, N.M., Blair, J.E., Catanzaro, A., Johnson, R.H., Stevens, D.A. & Williams, P.L. 2005. Coccidioidomycosis. *Clin. Infect. Dis.* 41: 1217–1223.

Goddard, L.B., Roth, A., Reisen, W.K. & Scott, T.W. 2003. Extrinsic incubation period of West Nile Virus in four California *Culex* (Diptera: Culicidae) species. *Proc. Mosq. Vect. Cont. Assoc. Calif.* 71: 70–75.

Goddard, L.B., Roth, A.E., Reisen, W.K. & Scott, T.W. 2002. Vector competence of California mosquitoes for West Nile Virus. *Emerg. Infect. Dis.* 8: 1385–1391.

Halpin, H., Robu, V. & Sheperd, H. 2007. The complex dynamics of collaborative tagging, Proc.16th Int. World Wide Web Conf. (WWW'07). Banff, Canada Available from: http://socialmedia.scribblewiki.com/Halpin_et_al%2C_WWW_2007 [Accessed 4th April 2012].

Hubmann-Haidvogel, A., Scharl, A. & Weichselbraun, A. 2009. Multiple coordinated views for searching and navigating web content repositories. *Inform. Sci.* 179(12): 1813–1821.

Hugenholtz, P. 1957. Climate and coccidioidomycosis. In *Symposium on Coccidioidomycosis*: 136–143. Phoenix, AZ: Public Health Services.

ICSU. 2011. Report of the ICSU planning group on health and wellbeing in the changing urban environment: A systems analysis approach. Paris: ICSU.

ICSU. 2012. Available from: http://www.icsu-wds.org/ [Accessed 4th April 2012].

Jain, S. and McLean, C.R. 2008. Components of an incident management simulation and gaming framework and related developments. Simulation 84: 3. Doi:10.1177/0037549708088956. Available from: http//sim.sagepub.com/content/84/1/3 [Accessed 4th April 2012].

Khan, K., Freifeld, C., Wang, J., Mekaru, S., Kossowsky, D., Sonricker, A. 2010. Preparing for infectious disease threats at mass gatherings: The case of the 2010 Vancouver winter olympic games. *Can. Med. Assoc.* 182(6): 579–83.

Kolivras, K.N. & Comrie, A.C. 2003. Modeling valley fever (coccidioidomycosis) incidence on the basis of climate conditions. *Int. J. Biomet.* 47: 87–101.

Kolivras, K.N., Johnson, P.S., Comrie, A.C. & Yool, S.R. 2001. Environmental variability and coccidioidomycosis (valley fever). *Aerobiol.* 17: 31–42.

Kramer, V.L. 2009. California State mosquito-borne virus surveillance and response plan. Available from: http://westnile.ca. gov/resources.php [Accessed 8th July 2012].

Langevin, S.A., Bunning, M., Davis, B. & Komar, N. 2001. Experimental infection of chickens as candidate sentinels for West Nile Virus. *Emerg. Infect. Dis.* 7: 726–729.

Lee, E.K., Maheshwary, S., Mason, J. & Glisson, W. 2006. Large-scale dispensing for emergency response to bioterrorism and infectious-disease outbreak. *Interfaces* 36(6): 591–607.

Lee, E.K., Chen, C.-H., Pietz, F. & Beneke, B. 2009. Modeling and optimizing the public-health infrastructure for emergency response. *Interfaces* 39(5): 476–490.

Lemos, M.C. & Morehouse, B.J. 2005. The co-production of science and policy in integrated climate assessments. *Global Environ. Change-Human & Policy Dimens.* 15: 57–68.

Lewis, M.D., Pavlin, J.A., Mansfield, J.L., O'Brien, S., Boomsma, L.G., Elbert, Y. & Kelly, P.W. 2002. Disease outbreak detection system using syndromic data in the greater Washington, DC area. *Amer. J. Prev. Med.* 23(3): 180–186.

Lothrop, H.D., Lothrop, B.B., Gomsi, D.E. & Reisen, W.K. 2008. Intensive early season adulticide applications decrease arbovirus transmission throughout the Coachella Valley, Riverside County, California. *Vect. Borne Zoon. Dis.* 8: 475–489.

Maddy, K.T. 1958. The geographic distribution of *Coccidiodes immitis* and possible ecological implications. *Ariz. Med.* 15: 178–188.

Maddy, K.T. 1965. Observations on *Coccidioides immitis* found growing naturally in soil. *Ariz. Med.* 22: 281–288.

McCaughey, K., Miles, S.Q., Woods, L., Chiles, R.E., Hom, A., Kramer, V.L., Jay-Russel, M., Sun, B., Reisen, W.K., Scott, T.W., Hui, L.T., Steinlein, D.B., Castro, M., Houchin, A. & Husted, S. 2003. The California West Nile Virus dead bird surveillance program. *Proc. Mosq. Vect. Cont. Assoc. Calif.* 71: 38–43.

Melton, F., Nemani, R.R., Michaelis, A., Barker, C.M., Park, B. & Reisen, W.K. 2008. Monitoring and modeling enviornmental conditions related to mosquito abundance and virus transmission risk with the NASA Terrestrial Observation and Prediction System. *Proc. Mosq. Vect. Cont. Assoc. Calif.* 76: 89–93.

Mnatsakanyan, Z.R. & Lombardo, J.S. 2008. Decision support models for public health informatics. *Johns Hopkins APL Tech. Digest* 27: 332–339.

Moore, C.G., McLean, R.G., Mitchell, C.J., Nasci, R.S., Tsai, T.F., Calisher, C.H., Marfin, A.A., Moore, P.S. & Gubler, D.J. 2002. *Guidelines for arbovirus surveillance programs in the United States.* Ft. Collins, CO: DVBID, CDC, PHS: US Department Health & Human Services.

Nemani, R.R., Keeling, C.D., Hashimoto, H., Jolly, W.M., Piper, S.C., Tucker, C.J., Myneni, R.B. & Running, S.W. 2003. Climate-driven increases in global terrestrial net primary production from 1982 to 1999. *Science* 300: 1560–1563.

Newhouse, V.F., Chamberlain, R.W., Johnston, J.G. Jr. & Sudia, W.D. 1966. Use of dry ice to increase mosquito catches of the CDC miniature light trap. *Mosq. News* 26: 30–35.

Nielsen, C.F., Armijos, M.V., Wheeler, S., Carpenter, T.E., Boyce, W.M., Kelley, K., Brown, D., Scott, T.W. & Reisen, W.K. 2008. Risk factors associated with human infection during the 2006 West Nile Virus outbreak in Davis, a residential community in Northern California. *Amer. J. Trop. Med. & Hyg.* 78: 53–62.

Park, B.J., Sigel, K., Vaz, V., Komatsu, K., McRill, C., Phelan, M., Colman, T., Comrie, A.C., Warnock, D.W., Galgiani, J.N. & Hajjeh, R.A. 2005. An epidemic of coccidioidomycosis in Arizona associated with climatic changes, 1998–2001. *J. Infect. Dis.* 191: 1981–1987.

Park, B., Eldridge, B.F., Barker, C.M. & Reisen, W.K. 2008. Building upon California's Surveillance Gateway. *Proc. Mosq. Vector Control Assoc. Calif.* 76: 27–28.

Power, D.J. 2000. Web-based and model-driven decision support systems: Concepts and issues. *Proc. Americas Conference on Information Systems.* Long Beach, California.

Rafelsberger, W. & Scharl, A. 2009. Games with a purpose for social networking platforms. In: 21st ACM Conference on Hypertext and Hypermedia: 193–197. Torino, Italy: ACM.

Raskin, R. 2004. Enabling semantic interoperability for Earth science data. Available from: http://esto.nasa.gov/conferences/estc2004/papers/a5p1.pdf [Accessed 5th April 2012].

Ray, A.J., Garfin, G.M., Wilder, M., Vasquez-Leon, M., Lenart, M. & Comrie, A.C. 2007. Applications of monsoon research: Opportunities to inform decision making and reduce regional vulnerability. *J. Clim.* 20:1608–1627.

Reisen, W.K., Lothrop, H.D., Chiles, R.E., Madon, M.B., Cossen, C., Woods, L., Husted, S., Kramer, V.L. & Edman, J.D. 2004. West Nile Virus in California. *Emerg. Infect. Dis.* 10: 1369–1378.

Reisen, W.K., Fang, Y. & Martinez, V.M. 2005. Avian host and mosquito (*Diptera: Culicidae*) vector competence determine the efficiency of West Nile and St. Louis Encephalitis Virus transmission. *J. Med. Entomol.* 42: 367–375.

Reisen, W.K., Fang, Y. & Martinez, V.M. 2006. Effects of temperature on the transmission of West Nile Virus by *Culex tarsalis* (*Diptera: Culicidae*). *J. Med. Entomol.* 43: 309–317.

Reisen, W.K. & Barker, C.M. 2008. Use of climate variation in vectorborne disease decision support systems. In *Global climate change and extreme weather events: Understanding the potential contributions to the emergence, reemergence, and spread of infectious disease:* 198–218. Washington DC: Institute of Medicine.

Reisen, W.K., Barker, C.M., Fang, Y. & Martinez, V.M. 2008a. Does variation in *Culex vector* competence enable outbreaks of West Nile Virus in California? *J. Med. Entomol.* 45: 1126–1138.

Reisen, W.K., Cayan, D., Tyree, M., Barker, C.M., Eldridge, B.F. & Dettinger, M. 2008b. Impact of climate variation on mosquito abundance in California. *J. Soc. Vect. Ecol.* 33: 89–98.

Reisen, W.K., Fang, Y. & Brault, A.C. 2008c. Limited interdecadal variation in mosquito (*Diptera: Culicidae*) and avian host competence for Western Equine Encephalomyelitis Virus (*Togaviridae: Alphavirus*). *Amer. J. Trop. Med. & Hyg.* 78: 681–686.

Reisen, W.K., Lothrop, H.D., Wheeler, S.S., Kensington, M., Gutierrez, A., Fang, Y., Garcia, S. & Lothrop, B. 2008d. Persistent West Nile Virus transmission and the displacement St Louis Encephalitis Virus in southeastern California, 2003–2006. *J. Med. Entomol.* 45: 494–508.

Reisen, W.K., Takahashi, R.M., Carroll, B.D. & Quiring, R. 2008e. Delinquent mortgages, neglected swimming pools and West Nile Virus in California, USA. *Emerg. Infect. Dis.* 14: 1747–1749.

Reisen, W.K., Carroll, B.D., Takahashi, R., Fang, Y., Garcia, S., Martinez, V.M. & Quiring, R. 2009. Repeated West Nile Virus epidemic transmission in Kern County, California, 2004–2007. *J. Med. Entomol.* 46: 139–157.

Shi, P.Y. & Kramer, L.D. 2003. Molecular detection of West Nile Virus RNA. *Expert. Rev. Mol. Diagn.* 3: 357–366.

Smith, C.E., Beard, R.R., Rosenberg H.G. & Whiting E.G. 1946a. Effect of season and dust control on coccidioidomycosis. *J. Amer. Med. Assoc.* 132: 833–838.

Smith, C.E., Beard, R.R., Whiting E.G. & Rosenberg H.G. 1946b. Varieties of coccidioidal infection in relation to the epidemiology and control of the diseases. *Amer. J. Pub. Health* 36: 1394–1402.

Solberg, K. 2010. Worst floods in living memory leave Pakistan in paralysis. *Lancet* 376(9746):2.

Sokolow, L.Z., Grady, N., Rolka, H., Walker, D., McMurray, P., English-Bullard, R. & Loonsk, J. 2005 Deciphering data anomalies in BioSense. *MMWR* 26(54 Supp) :133–139. Stevens, D.A. 1995. Coccidioidomycosis. *N. Engl. J. Med.* 332: 1077–1082.

SYRIS 2012. Available from: http://www.arescorporation.com & http://www.arescorporation.com/ products. aspx?style=2&%20pict_id=189&menu_id=103&id=87 [Accessed 16th April 2012].

Tamerius, J.D. & Comrie, A.C. 2012. Coccidioidomycosis incidence in Arizona predicted by seasonal precipitation. *PLOS ONE* 6(6): e21009. doi:10.1371/journal pone.0021009.

Tsang, C.A., Anderson, S.M., Imholte, S.B., Erhart, L.M., Chaen, S., Park, B.J., Christ, C., Komatsu, K., Chiller, T. & Sunenshine, R. 2010. Enhanced surveillance of coccidioidomycosis, Arizona, USA, 2007–2008. *Emerg. Infect. Dis.* 16: 1738–1744.

Washington Post. 2010. Officials probe possible outbreak in rural Haiti. Available from: *http://www.washington post.com/wp-dyn/content/article/2010/10/20/AR2010102006051.html.* [Accessed 16th February 2012].

Watson, J.T., Gayer, M. & Connolly, M.A. 2007. Epidemics after natural disasters. *Emerg. Infect. Dis.* 13(1): 1–5.

Wheeler, S.S., Barker, C.M., Armijos, M.V., Carroll, B.D., Husted, S.R. & Reisen, W.K. 2009. Differential impact of West Nile Virus on California birds. *The Condor* 111: 1–20.

Wikipedia. 2011. Decission support system Available from: http://en.wikipedia.org/wiki/Decision_support_system [Accessed 2nd February 2012].

Wilson, D. 2011. Coral bleaching outbreak in Thailand shutting dive sites and slowing tourism. Available from: http://www.reuters.com/article/2011/02/08/idUS52935383320110208. [Accessed 16th February 2012].

Wilson, J.L., Reisen, W.K. & Madon, M.B. 2006. Three years of West Nile Virus in Greater Los Angeles County. *Proc. Mosq. Vect. Cont. Assoc. Calif.* 74: 9–11.

Witkop, C.T., Duffy, M.R., Macias, E.A., Gibbons, T.F., Escobar, J.D., Burwell, K.N. & Knight, K,K. 2010. Novel influenza A (H1N1) outbreak at the U.S. Air Force Academy: Epidemiology and viral shedding duration. *Amer. J. Prev. Med.* 38(2):121–126.

Author index

Subject index

Giovanni 268; for health applications 257–68; IRI data library for 248; for poverty mapping 259–60; Mirador tool for 268; screening for 268; search services for 266–8; from SEDAC 257-64; for superfund sites 260; IT systems for 252-5

data analysis systems: and I-HEAT 250–52; and IT systems 252–55; land data analysis for 278; and MERRA 164; and REGARDS 280-283; *see also* AIRAWARE

decision support systems: assessing performance of 369–70; business intelligence tools for 370–1; communities of interest for 402; and crisis management 370; collaborative prototyping for 371; definition of 369–70; design criteria for 369–70; and infectious diseases 370; models for 370; participatory design of 371; performance of 372; and public health 370-71; for situational awareness 370, 372, 388–9; *see also* AIRAWARE; *see also* early warning systems; *see also* mosquito-borne encephalitis; *see also* coccidioidomycosis; *see also* syndromic surveillance; *see also* semantic approach to health information

dengue virus: and *Aedes aegypti* 200; climate effects on 201–3; DALYS for 36; emergence of 200, 204; mortality from 201; range expansion of 201, 204; remote sensing of 39–40; role of urbanization on 200–1, 203–4; serotypes of 200; transmission of 38–39, 200–1; vector biology of 36

desert dust: anthropogenic sources for 145–6; in Asia 146–7; and climate cycles 146; data sources for 255–6; dust source classification of 166–8; entrainment of 148–9, 156; geography of 144–5; and human health 147–50; long range transport of 146, 243–4; microorganisms in 150–52; modelling of 252–7; mortality from 148; in North Africa 145–6; nutrients in 146; prediction of 239–40; resuspension of 156; saltation of 145; viral diseases from 151; and warning system for 168; *see also* emerging sensors; *see also* climate cycles; *see also* modelling

differential equations: analysis methods for 310–11; dynamical behaviour of 311; and epidemiology 309, 311–13; fixed points for 310–11; infectious disease 309; and logistic equation 309–10; and mechanistic models 308–13; and predator-prey equation 309–10; and SEIR model 309–10, 312; types of 308, 310–12; uses for 308–10

disability-adjusted life years (DALYS): and burden of disease 112–18; calculation of 114–15; strengths and weaknesses of 115–18; and water-borne diseases 116

disease: categories of 10–14; and climate sensitivity 10; and forecasting outbreaks 21; habitat suitability for 21; weather and climate links to 245

early warning systems: and Australian Tsunami Early Warning System 97; future directions for 340–1; and malaria 232, 234, 278–9; 275–80, 333–40; NDVI for 102; and tsunamis 96; for west Nile virus 234; *see also* malaria

ebola virus: *see* filoviruses

ecological niche modelling (ENM): and algorithms 315; applications of 314–15; and bias in 322; case study of 321; caveats and challenges for 321–323; evaluation and validation of 320, 323; history and development of 313–14; and equilibrium 314; and multi co-linearity 320; and presence-only 315–317; and species presence/absence 317–19; public health applications of 314–15; satellite sensors for 319; species and environmental data for 319; uncertainty and error in 320–1; *see also* modelling

ecological forecasting: examples of 347; and harmful algal blooms 347; and national framework for 364; predicting impacts of 346; spatial and temporal scales for 348; types of 348

El Niño Southern Oscillation (ENSO): and plague 55–58; and rift valley fever 48–51; *see also* climate cycles

emerging diseases: abiotic drivers of 188; background for 187–90; biotic drivers of 187; concept of 187–9; definition of 187; drivers of 187–8; facilitators for 187–8; GEIS for 231–2; information system for 231–41; transition pathways for 188–9

emerging sensors: and AOD gaps 165; and A-Train sensors 161–2; and dust source classification 166; and machine learning 159, 164; and public health implications 165–7; and self-organizing maps 166

414

empirical mode decomposition: and filovirus outbreaks 196–7; and plague 55–6

empirical modelling: concept for 308; and harmful algal blooms 357; in Chesapeake Bay 357–62; *see also* modelling

encephalitis: bird transmission by 374; and *Bunyaviridae* 44; in California 373–7; environmental factors for 373; and *Flaviviridae 41*; intervention of 372–8; mosquito-borne types of 40–4, 372–3; and Nipah virus 205; and North American encephalitides 372–3; and primary amoebic meningoencephalitis 93–4; remote sensing of 45; risk of 376; surveillance of 377; sustainability of 377; tangential transmission of 372; and *Togaviridae* 43; vectors for 374; and West Nile virus 372–8; *see also* mosquito-borne diseases; *see also* *Naegleria fowleri*; *see also* Nipah virus

enhanced vegetation index (EVI): calculation of 104; and rotavirus infections 104; *see also* rotavirus infections

enteric infections: burden of 97–8; causes of 98; and *Cryptosporidium* 97; and *Giardia* 97; seasonal fluctuation in 98; surveillance systems for 98; *see also* water-borne diseases

environment: and climate cycles 9; and extreme heat events 299; and health consequences *ix*; and weather 9

environmental data: and cohort data 280; and air quality modelling 296–9; and hydrological modelling 299–301

environmental modelling: for air quality 298–9; and anthropological processes 304–5; and atmospheric processes 298–9; and biological processes 303–4; and decomposition processes 302–3; and earth science models 298–305; and earth science modelling framework (ESMF) 306–7; equations for 278; examples of 278; for extreme heat events 299; and hydrological processes 299–301; land use/land use change for 275–9; and machine learning 316, for seismic processes 301–2; and types of models 294–8; *see also* mechanistic models; *see also* modelling

filoviruses (*Filoviridae*): and climate cycles 196; and cross species transmission 191–2; early warning system for 194; and

Ebola 192–4; empirical mode decomposition of 197; Marburg 192–4; prediction framework for 197; and NAO signal for 197; NDVI behaviour of 194; outbreak hypothesis for 194–8; and outbreak processes 196; research questions for 194–6; tracking of 192–3; *see also* modelling

floods: and emergency alert 96; modelling of 300–1; mortality from 96

forecasting: and ecological 345–8; empirical models for 357–62; scenario models for 362–3

Haemorrhagic fever: in sub-Saharan Africa 231–2

harmful algal blooms: and *Alexandrium monilatum* 360–3; bacterial species forming 95, 360; in Chesapeake Bay 359–61; and *Cochlodinium polykrikoides* 360; and cyanobacterial blooms 237, 360; and habitat suitability 361; and management response 96; and *Mytilis edulis* 361; predictive modelling for 96; and *Prorocentrum minimum* 360–1; and *Pseudo-nitzschia* 95, 361; in Puget Sound 361; *see also* forecasting

health programmes: and disaster monitoring 7, 132; and HealthMap 112; for health & wellbeing 8; and I-HEAT 250–2; for PROMED 112; for system of systems 8; for weather forecasting 9

heat waves: and health impacts 249–50; I-HEAT system for 251–2

hydrological disasters: and floods 96, 300–1; and hurricane/typhoon modelling 300; and tsunamis 96

image classification: and malaria 276–80; *see also* modelling

infectious diseases: diarrhoeal diseases 91; early history of 21; and human ecosystems 245; mitigating risks of 21; spread of 21

influenza: and avian 236; climate & seasonality of 68–70, 236; modelling of 71–3; risk of 234–6; types & characteristics of 68, 236

information systems: and environmental tracking 401–2; data for 269; holistic approach to 401; and precipitation events 269–75; RESTful-based 401–2; and reduced water quality events (RWQEs) 269–72; and predicting RWQEs 273–4;

abstraction for 104–6; African reference sites for 105

satellite observations: as data for human health 3; and epidemiology for 246–7; for expanding health monitoring 6–9; metadata records for 247; obstacles facing 246–7; physical basis for 14; recent history of 3–9

scenario-based forecasting: and harmful algal blooms 362–3; *see also* mechanistic models

semantic approach to health information: and case studies 404–5; and climate change collaboratory 405–6; elements of 402–4; and HealthMap 404; rationale for 401; unified schematic for 403

sensor technology: role of 4–6; trends in 3

stepwise optimal hierarchical clustering: and plague 55

surveillance systems: developments in 112; environmental monitoring for 98–9; and global burden of disease 115; and HealthMap 112; integrated systems for 99; international systems for 112; and malaria 247–8; monitoring rainfall for 248; monitoring breeding sites for 248–9; and PROMED 112; and REGARDS 280–3; role of climate in 247–8; systems for 388–401

swimming pool safety: and DPSEEA framework 88; and early warning risk for 89

syndromic surveillance: about 388–91; functions of 393–400; future designs for 118–9; rationale for 392; sample system for (SYRIS) 391–401; *see also* decision support

Togaviridae: and encephalitis 43

trajectory based models: and beach water quality 355–6; in Great Lakes (US) 352–3; in Gulf of Mexico 349–51;and harmful algal blooms 349; in Pacific Northwest (US) 353–5

transmission pathways: and diarrhoeal diseases 91

tsunamis: warning system for 97; communities at risk of 97; drownings from 96–97; and seismic processes 302

valley fever: *see* Coccidioidomycosis

vector-borne diseases: spectral data sets for 231; health studies using 231–7; and

Haemorrhagic fever 231; spectral data sets for 231;

Vibrio fulnificus: forecasting presence of 357–61; hindcasting of 359

volcanic ash: and air traffic safety 137; and chemical composition 133; ash detection 137–9, and dispersal modelling 137, 140–1; human exposure to 132–3, 135–7; and Eyjafjallajökull 144, 168, 405; and Guagua Pichincha 135–7, 143; health impacts from 134–5, 143–4; health limitations of 136–7; and integration with GIS 140–1; and International Decade for Natural Disaster Reduction (IDNDR) 132; linking health data to 142–144; remote data acquisition for 138; resuspension of 144; satellite monitoring of 133, 135, 137; and serum immunoglobulins 134–5; and SO_2 detection of 141–2; societal impacts of 133–4; toxicities from 134–5

water-borne diseases: definition of 108–10; diarrhoeal burden of 108–9, 115–18; and exposure pathways 88; and haemolytic uraemic syndrome 91, 110; health studies about 110–12, 237–9; and Legionnaires' disease 92–3; outbreaks of 91, 96; and oyster *Norovirus* 238; pathogens for 91, 349; and primary amoebic meningoencephalitis (PAM) 93–4; public health actions for 90; quantification of 109–10; spectral data sets for 237; as surrogate for ecosystem health 97; surveillance of 110; transmission of 92–3; types of 109–10; and water cycle 89; *see also* harmful algal blooms; *see also* cryptosporidium; *see also* information access

water quality: and arsenic 94; chemical contaminants in 94–5; and *cryptosporidium* 97; and disinfectant by-products 94; *E.coli* in 91, 110–11; environmental monitoring of 92; fluoride in 95; and harmful algal blooms 95, 146; and hydrological disasters 96; and Legionnaires' disease 92; poor quality consumption of 91; and primary amoebic meningoencephalitis 93–4; and recreational waters 92; and rotavirus infections 104; and tsunamis 96; and water treatment facilities 91; and Trihalomethanes 94

ISPRS Book Series

1. Advances in Spatial Analysis and Decision Making (2004)
   Edited by Z. Li, Q. Zhou & W. Kainz
   ISBN: 978-90-5809-652-4 (HB)

2. Post-Launch Calibration of Satellite Sensors (2004)
   Stanley A. Morain & Amelia M. Budge
   ISBN: 978-90-5809-693-7 (HB)

3. Next Generation Geospatial Information: From Digital Image Analysis to Spatiotemporal
   Databases (2005)
   Peggy Agouris & Arie Croituru
   ISBN: 978-0-415-38049-2 (HB)

4. Advances in Mobile Mapping Technology (2007)
   Edited by C. Vincent Tao & Jonathan Li
   ISBN: 978-0-415-42723-4 (HB)
   ISBN: 978-0-203-96187-2 (E-book)

5. Advances in Spatio-Temporal Analysis (2007)
   Edited by Xinming Tang, Yaolin Liu, Jixian Zhang & Wolfgang Kainz
   ISBN: 978-0-415-40630-7 (HB)
   ISBN: 978-0-203-93755-6 (E-book)

6. Geospatial Information Technology for Emergency Response (2008)
   Edited by Sisi Zlatanova & Jonathan Li
   ISBN: 978-0-415-42247-5 (HB)
   ISBN: 978-0-203-92881-3 (E-book)

7. Advances in Photogrammetry, Remote Sensing and Spatial Information Science. Congress
   Book of the XXI Congress of the International Society for Photogrammetry and Remote
   Sensing, Beijing, China, 3–11 July 2008 (2008)
   Edited by Zhilin Li, Jun Chen & Manos Baltsavias
   ISBN: 978-0-415-47805-2 (HB)
   ISBN: 978-0-203-88844-5 (E-book)

8. Recent Advances in Remote Sensing and Geoinformation Processing for Land Degradation
   Assessment (2009)
   Edited by Achim Röder & Joachim Hill
   ISBN: 978-0-415-39769-8 (HB)
   ISBN: 978-0-203-87544-5 (E-book)

9. Advances in Web-based GIS, Mapping Services and Applications (2011)
   Edited by Songnian Li, Suzana Dragicevic & Bert Veenendaal
   ISBN: 978-0-415-80483-7 (HB)
   ISBN: 978-0-203-80566-4 (E-book)

10. Advances in Geo-Spatial Information Science (2012)
    Edited by Wenzhong Shi, Michael F. Goodchild, Brian Lees & Yee Leung
    ISBN: 978-0-415-62093-2 (HB)
    ISBN: 978-0-203-12578-6 (E-book)

11. Environmental Tracking for Public Health Surveillance (2012)
    Edited by Stanley A. Morain & Amelia M. Budge
    ISBN: 978-0-415-58471-5 (HB)
    ISBN: 978-0-203-09327-6 (E-book)

Colour plates

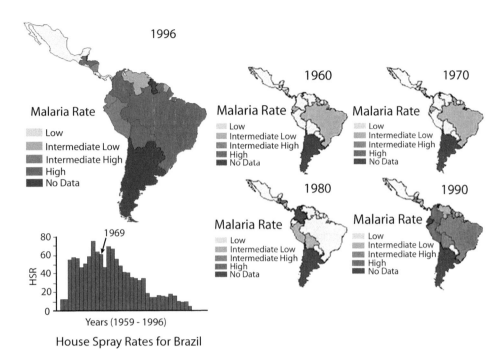

Plate 1.   Changes in malaria prevalence distribution within South America from the 1960's–1990 (Re-created from Roberts *et al.*, 2002b) (See page 24).

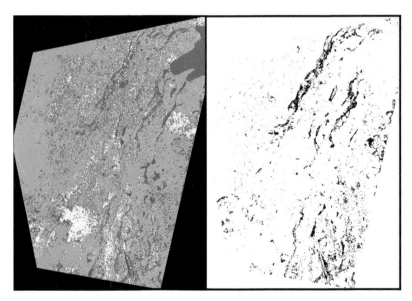

Plate 2.   Classified SPOT and Radarsat imagery of marshes in northern Belize (Permission *ESA.*, 2005) (See page 28).

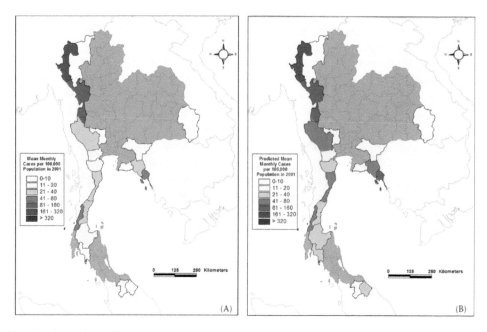

Plate 3.   Comparison of actual and modelled malaria case rates (per 100,000 population) in 2001. Training of neural networks was performed using 1994–2000 data: (a) actual; (b) modelled case rates (See page 33).

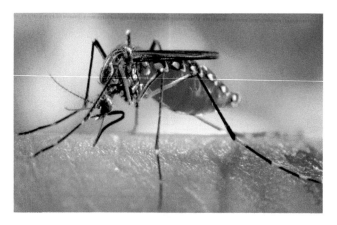

Plate 4.   *Aedes aegypti*, the primary vector of dengue viruses (Source, CDC Public Health Image Library; photo credit, James Gathany) (See page 37).

Plate 5.   Larval breeding sites of *Aedes aegypti* in Thailand: A. indoor bathroom basins; B. ant trap under table leg; C. outdoor animal water vessel; D. Indoor and outdoor water storage jars (See page 37).

Plate 6.   Aerial photograph of dambo habitats in Eastern Free State province, South Africa (Photo courtesy R. Swanepoel) (See page 46).

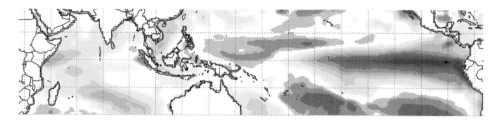

Plate 7a.   Sea surface temperature anomalies during December 1997–February 1998.

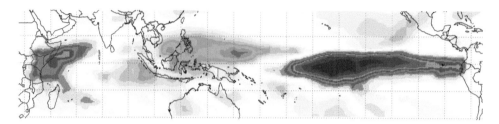

Plate 7b.   Outgoing long wavelength radiation anomalies during December 1997–February 1998 (See page 50).

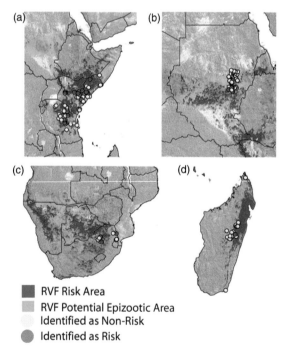

■ RVF Risk Area
▨ RVF Potential Epizootic Area
□ Identified as Non-Risk
● Identified as Risk

Plate 8.   Summary RVF risk areas for: (A) eastern Africa, Sep. 2006 to May 2007; (B) Sudan, May 2007 to Dec. 2007; (C) southern Africa, Sep. 2007 to May 2008; and (D) Madagascar, Sep. 2007 to May 2008 (Courtesy Anyamba) (See page 52).

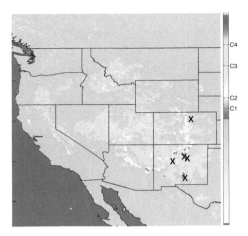

Plate 9.   A stepwise optimal hierarchical clustering approach was applied to identify training samples that contribute the most in the cluster classification and optimize criterion functions. The climatic, landscape and ecological properties of these samples were used to identify and generalize temporal characteristics of plague outbreaks in the region (Pinzon *et al.*, unpublished) (See page 56).

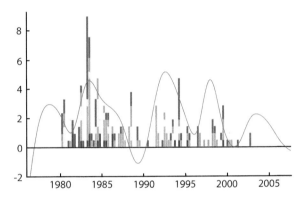

Plate 10.   Three intrinsic mode functions (IMF_{4-6}) of the EMD of the multivariate ENSO index (MEI) are combined into an almost ten year-cycle wave (blue line), and lagged twelve months to overlap the number of human cases of plague in each quarter of the year. The number of cases is clustered according to the five SOHC classes. Notice that incidence of plague occurred when IMF_{4-6} is positive (Pinzon *et al.*, unpublished) (See page 56).

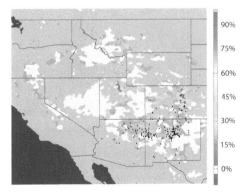

Plate 11.   Plague risk maps masked with a density population map of the Four Corners region. The plague endemic area is thus concentrated on peridomestic regions that constitute about eighty-five per cent of the cases (Pinzon *et al.*, unpublished) (See page 57).

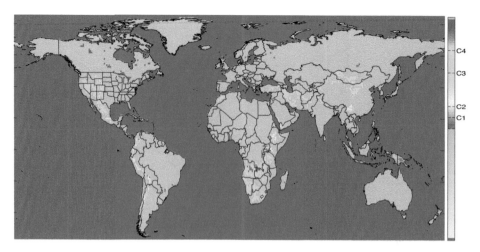

Plate 12. Global extension of the SOHC model based on five clusters (See page 57).

Plate 13. Global distribution of CL (left) and VL (right) (Courtesy WHO) (See page 58).

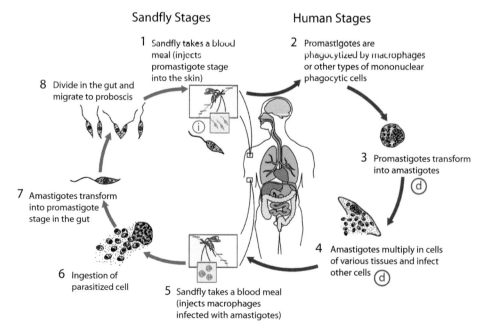

Plate 14. Life cycle of *Leishmania* parasites (Modified from CDC). The infectious stage (i) in red circle occurs in stage 1; diagnostic stages (d) in blue circles occur in stages 3 and 4 (Courtesy CDC) (See page 59).

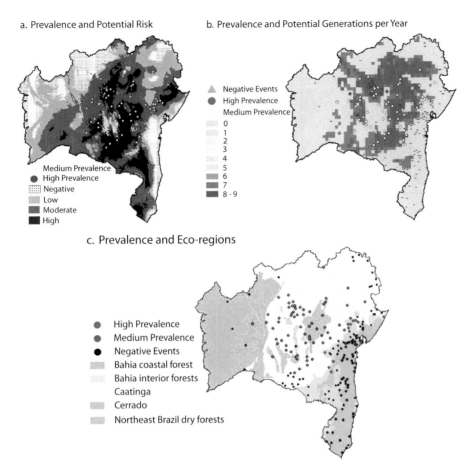

a. Prevalence and Potential Risk

b. Prevalence and Potential Generations per Year

▲ Negative Events
● High Prevalence
● Medium Prevalence
  0
  1
  2
  3
  4
  5
  6
  7
  8 - 9

Medium Prevalence
● High Prevalence
▦ Negative
  Low
  Moderate
  High

c. Prevalence and Eco-regions

● High Prevalence
● Medium Prevalence
● Negative Events
  Bahia coastal forest
  Bahia interior forests
  Caatinga
  Cerrado
  Northeast Brazil dry forests

Plate 15. (a) Ecological niche and (b) GDD-water budget models predicted a similar distribution and abundance pattern for vector-parasite transmission. Highest transmission risk was predicted in the Caatinga region, and no risk was predicted in the coastal forest for (c) ecological regions of Bahia State (Modified from Nieto *et al.*, 2006) (See page 66).

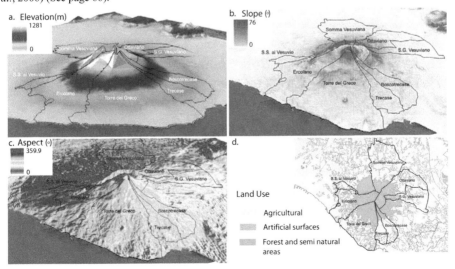

a. Elevation(m)
1281
0

b. Slope (°)
76
0

c. Aspect (°)
359.9
0

d.

Land Use

  Agricultural
  Artificial surfaces
  Forest and semi natural areas

Plate 16.   Elevation, slope, aspect and land use classes formed the data layers used to construct a GIS database (Modified from Rossi *et al.*, 2007) (See page 67).

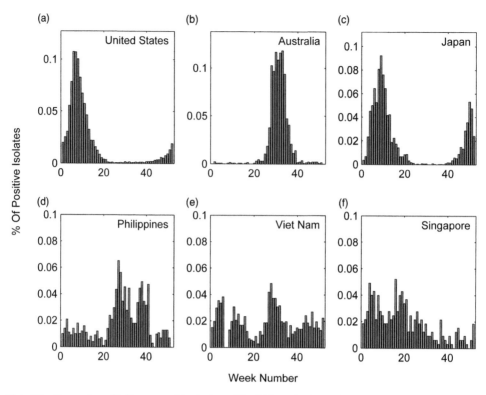

Plate 17.   Proportion of influenza positive isolates (Y-axis) by calendar week in 2007 (X-axis) for countries representing the Northern Hemisphere (A), Southern Hemisphere (B), and East Asia from temperate (C) to tropical (F) countries (data from FluNet, WHO 2010b) (See page 69).

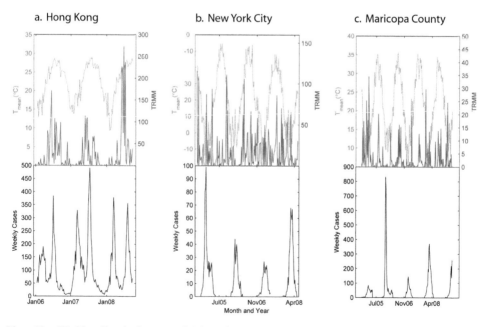

Plate 18.   Weekly climatic factors and lab-confirmed influenza positive isolates for (A) Hong Kong, (B) New York City and (C) Maricopa County. In the top panel for each location, the green line mean temperature (°C) compared to the blue line showing TRMM precipitation rate (mm/hr.); bottom panel for each location is the weekly number of influenza cases for each location (See page 70).

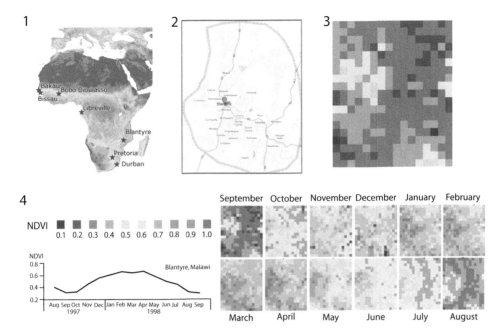

Plate 19.   Schematic representation of steps in the data abstraction process for monthly vegetation index data from remote sensing imagery. Large numbers represent sequential steps in the process (See page 100).

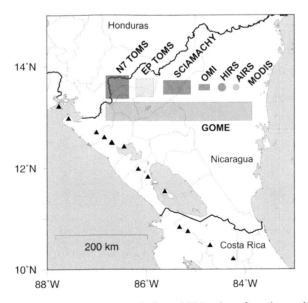

Plate 20.   The nadir footprint of SO$_2$ sensors, excluding ASTER, whose footprint would not register at this scale. GOES has a slightly larger footprint than MODIS (Courtesy Carn, University of Maryland, Baltimore County) (See page 139).

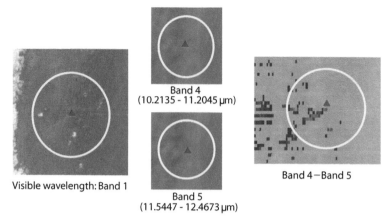

Band 4
(10.2135 - 11.2045 μm)

Band 5
(11.5447 - 12.4673 μm)

Visible wavelength: Band 1

Band 4−Band 5

Plate 21.   GOES-8 imagery, Tungurahua Volcano, Ecuador 14:45 UTC 16 September 2001 (See page 139).

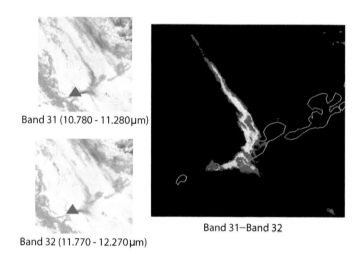

Band 31 (10.780 - 11.280 μm)

Band 32 (11.770 - 12.270 μm)

Band 31−Band 32

Plate 22.   MODIS Imagery, Mount Cleveland Volcano, Alaska 23:10 UTC 19 February 2001 (See page 140).

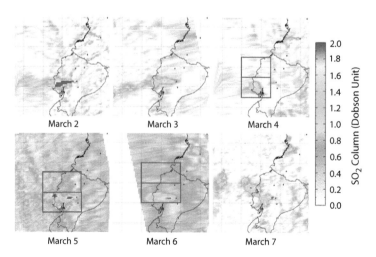

March 2

March 3

March 4

March 5

March 6

March 7

SO_2 Column (Dobson Unit)

2.0
1.8
1.6
1.4
1.2
1.0
0.8
0.6
0.4
0.2
0.0

Plate 23.   OMI SO_2 imagery for Tungurahua volcano, Ecuador, March 2007 (See page 141).

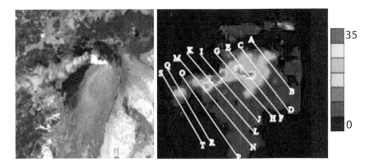

Plate 24. Emissions from Pacaya volcano, Guatemala are similar to emissions from Tungurahua (Courtesy Watson, University of Bristol) (See page 141).

Plate 25. Aerial photograph of Quito, Ecuador. Pichincha volcano lies west of Quito in the central background. A narrow zone of haze stretches from left to right across the city (Geophysical Institute of Ecuador) (See page 143).

Plate 26. An atmospheric bridge of dust observed on 8 August 2001 extends along the arrow from its source in North Africa to the Caribbean (Courtesy NASA & ORBIMAGE) (See page 146).

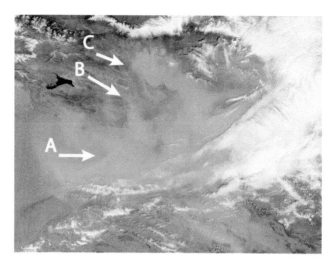

Plate 27. A large dust storm originating in western China on 24 April 2010. This storm resulted from dust converging from three locations. Arrows indicate direction of dust movement from each source region (Courtesy MODIS image, NASA) (See page 147).

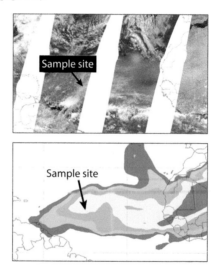

Plate 28. African dust-borne microorganisms transported across the Atlantic Ocean between 22 May and 30 June 2003: Ocean Drilling Program, Leg 209. The black bars are CFUs and the Y-axis on the left shows their values in μg/litre^3 of air; the dotted line shows modelled NAPPS dust concentrations on the right side Y-axis in μg/litre^3 of air (See page 149).

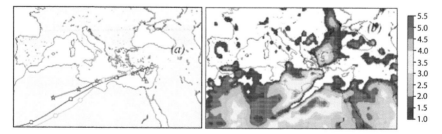

Plate 29. (Left): Air-mass back-trajectory from Erdemli, Turkey; (Right): Earth Probe TOMS satellite data (trajectory data courtesy ECMWFTOMS; TOMS image courtesy NASA) (See page 151).

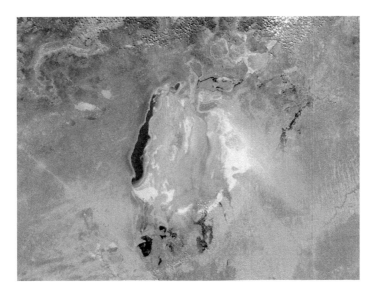

Plate 30.   MODIS image for 26 March 2010 showing a dust event originating in the Aral Sea region and moving to the southeast, along the Kazakhstan/Uzbekistan border (Courtesy NASA Earth-observatory) (See page 153).

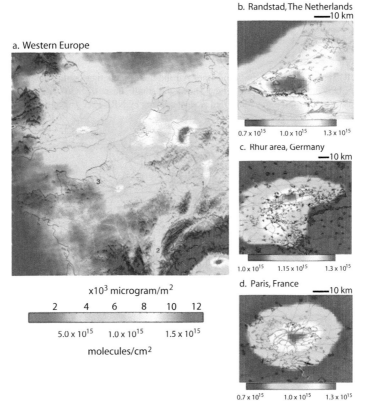

Plate 31.   Mean tropospheric NO_2 from OMI over Western Europe December 2004-November 2005. Data from Boersma *et al.*, 2007 (Courtesy J.P. Veefkind) (See page 159).

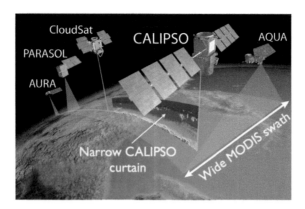

Plate 32.   Visualization of the two complimentary types of aerosol information obtained by NASA's A-TRAIN constellation, detailed vertical information from CALIPSO curtains and global coverage of the total optical depth from MODIS (See page 161).

Plate 33.   (Left) shows a region centred on Iraq. All the land pixels that could contain dust sources were classified by a SOM into 1000 classes. (Right) shows the small subset of SOM classes (shown in red) that have the largest overlap with regions identified as dust sources by NRL (See page 167).

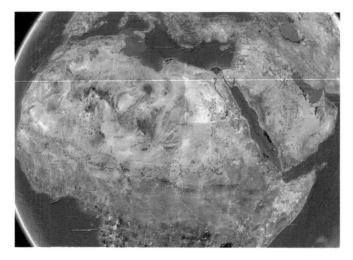

Plate 34.   The shaded pixels (blue and cyan) show all the land surface pixels in the SOM classes identified as dust sources. These classes were the same SOM classes displayed in Figure 19 (right) (See page 168).

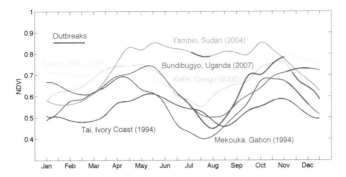

Plate 35.   Time series behaviour of the NDVI data from the documented outbreak sites of EBOV HF. Outbreak periods are denoted by the thick brown sections of the timelines. Note that the outbreaks tend to occur toward the middle of the second dry season (See page 194).

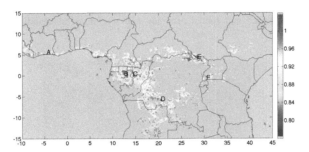

Plate 36.   Regions where EBOV HF is endemic. The colour bar at right shows levels of ecological similarity between known Ebola outbreak sites and other areas in central Africa. Red (1) = high ecological similarity. Blue (0.1) = low ecological similarity. Sites of known Ebola outbreaks are denoted as: A = Tai, Côte d'Ivoire; B = Mekuoka, Gabon; C = Kelle, Dem. Rep. Congo; D = Luebo, Dem. Rep. Congo; E = Nzara, Sudan; and F = Bundibugyo, Uganda. Grey areas: areas with low ecologic similarity (<0.75). Red and orange areas outside the known outbreak sites are possible future *Ebola hot zones* (Updated from Pinzon *et al.*, 2004) (See page 195).

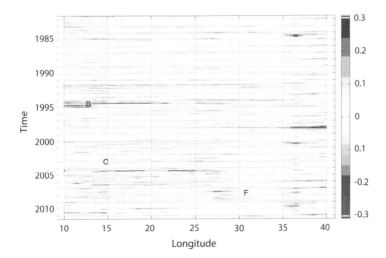

Plate 37.   Longitudinal Hovmoller anomaly NDVI image of the *Ebola hot zone* linked to outbreak sites B, C and F from Figure 3. Colour bar at right denotes NDVI readings. Green denotes above average NDVI levels, or rain. Brown denotes below average NDVI levels, or no rain. The linear coloured markings over a geographic area (width) and time (height) show markedly drier environmental conditions in 1991, 1994, 2000–2001 and 2004 (See page 195).

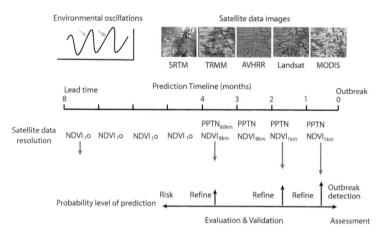

Plate 38. Model prediction framework: Developing a multi-level risk map with dynamic decreasing uncertainty and increasing temporal and spatial accuracy that reinforce expandability and accessibility (See page 197).

Plate 39. MODIS true colour image showing a bloom in the western basin of Lake Erie, August 2009 (See page 237).

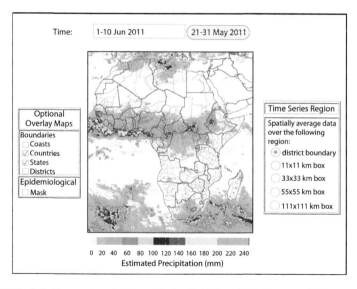

Plate 40. MEWS *clickable map* for rainfall monitoring 1–10 June 2011 (See page 249).

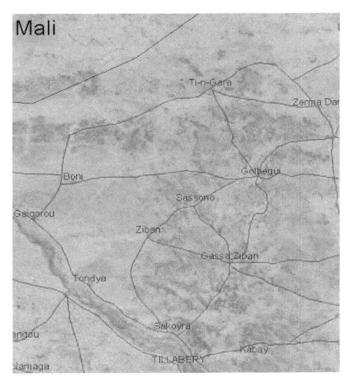

Plate 41. TERRA-MODIS colour composite RGB where the SWIR channel is displayed in red, the NIR channel in green and the visible red channel in blue. This image was acquired during the rainy season in north Niamey, Niger. It is roughly 200 km^2 in area. Small water bodies in blue are breeding sites for mosquitoes and locations where nomads water their cattle (See page 250).

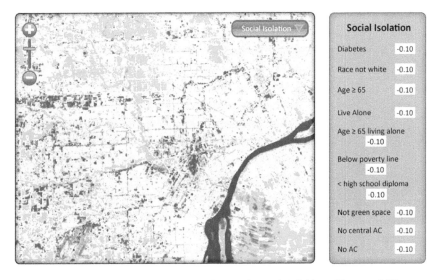

Plate 42. The I-HEAT interface showing a heat-risk map of Detroit, Michigan (See page 253).

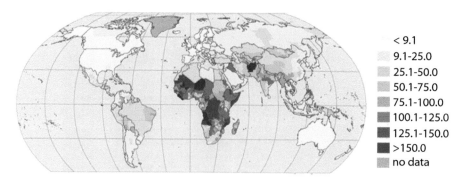

Plate 43. Subnational infant mortality rate data serves as one proxy in mapping global poverty (See page 260).

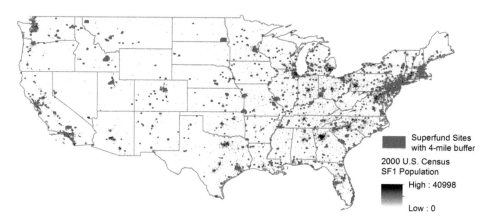

Plate 44. US population and NPL superfund sites with four mile buffers (See page 261).

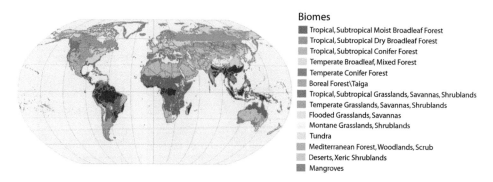

Plate 45. PLACE II provides population and land area distribution for classes of several different demographic, physical, biological, and climatic variables, including Biomes (See page 261).

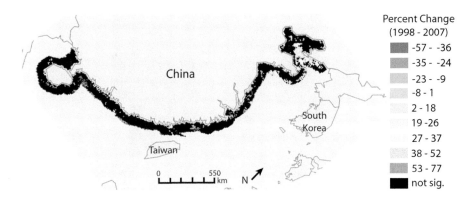

Percent Change
(1998 - 2007)

- -57 - -36
- -35 - -24
- -23 - -9
- -8 - 1
- 2 - 18
- 19 -26
- 27 - 37
- 38 - 52
- 53 - 77
- not sig.

China

South Korea

Taiwan

0   550
km   N

Plate 46. Percentages of change in chlorophyll concentrations as an indicator of algal biomass along the China coast, 1998–2007 (See page 262).

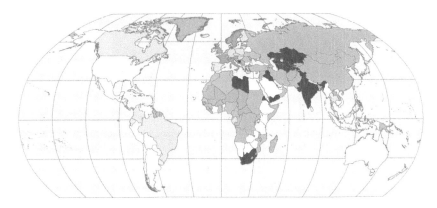

Plate 47. 2012 Environmental performance index scores. Dark green: strongest index; light green, strong; ivory, modest; orange, weaker; red, weakest; and grey, no index (Courtesy CIESIN & YCELP) (See page 263).

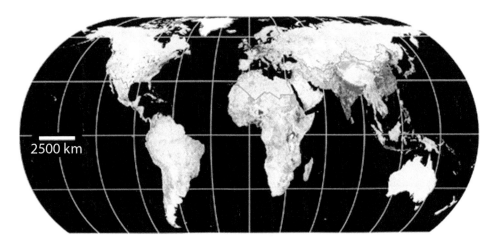

2500 km

Plate 48. Anthromes based on population density, land use and vegetation cover. Blue, purple and red colours define heavily cultivated regions in Asia and Europe; greens and yellows are semi-natural and croplands; pale to darker orange colours are defined as rangelands; and pale green to grey areas are "wild" lands (Courtesy Brandon, 2010) (See page 264).

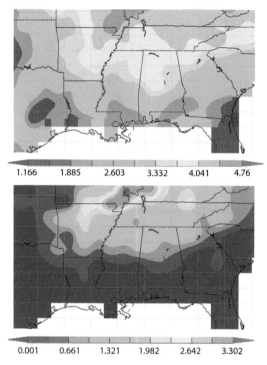

Plate 49. (Top) GLDAS monthly soil moisture and (Bottom) monthly surface runoff for the south eastern US between April and June 2011. Soil moisture ranges from <1.2 in purple and blue areas to more than 4.5 in the orange and red areas. Surface runoff values range from <0.6 to >2.6 (See page 269).

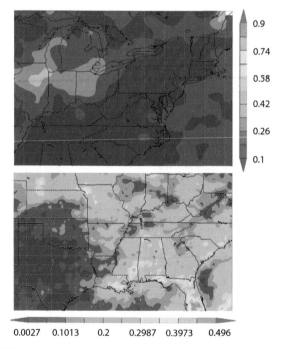

Plate 50. (Top) MODIS AOD values range from lowest in purple (<0.1) to highest in pale green (>0.3). (Bottom) Average TRMM precipitation rate at 00:00 UTC 22 June to 22 August 2011. Purple, blue, turquois represent <.0.002 to 0.2; green, yellow represent >0.2 <0.39; and orange, red >0.39 <0.49 mm/hr (See page 270).

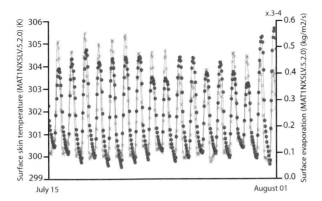

Plate 51.   Hourly area-averaged temperature time series along the Texas/Louisiana coast 15July to 1August 2011. The blue line is surface skin temperature ranging from 299°K to 306°K, and the red line is surface evaporation ranging from 0.0 to 1.2 kg/m² (See page 271).

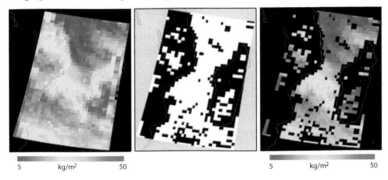

Plate 52.   Total column precipitable water (kg/m²) from AIRS near Madagascar on 4 June 2009. (Left) raw data; (Center) quality mask; (right) post screen out-put. Red is ±5 km²; blue is ±50 kg/m² (See page 271).

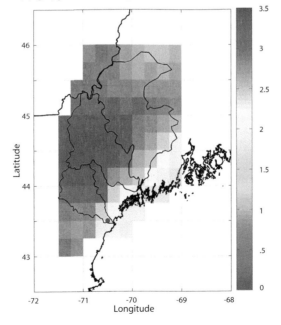

Plate 53.   Mean daily precipitation (mm/day) observed for TRMM grid cells between 1998 & 2009. Black lines represent boundaries of three different watersheds. Blue circle is a water sampling site in Saco River Watershed (See page 272).

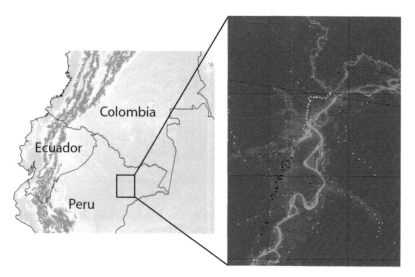

Plate 54. Portion of the Peruvian Amazon, shown on a 432-RGB Landsat TM composite from 22 August 2005. Red and yellow points show mosquito collection sites along the Nauta-Iquitos and Iquitos-Mazan roads, respectively. Green points are settlements (See page 276).

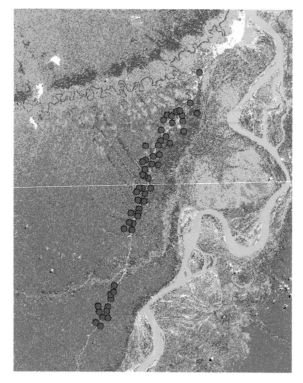

Plate 55. Collection sites reported along the Nauta-Iquitos road, overlain on a supervised classification of a Landsat image. Iquitos is located in the northeast corner of the image (See page 277).

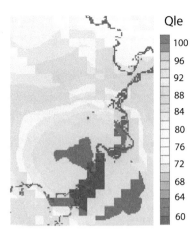

Plate 56.   Monthly averaged latent heat flux (W/m²) for February 2009 over a portion of the PAFLDAS domain. Data cells have different sizes due to differing resolution of input parameters and forcing data, but can be improved as LDAS incorporates additional satellite data sets and downscaling techniques (See page 279).

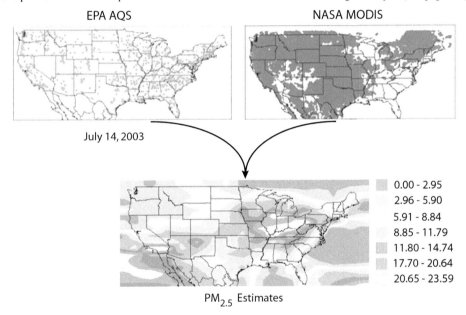

Plate 57.   A schematic showing how EPA/AQS data are merged with NASA MODIS data to form a smoothed PM$_{2.5}$ surface (Courtesy McClure & Alhamdan, USRA) (See page 282).

Plate 58.   The REGARDS participants; those in red are whites, those in blue are African Americans (See page 282).

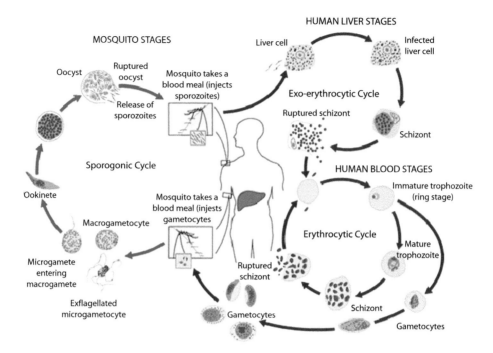

Plate 59.   Malaria model linking parasite, mosquito and human infection and disease transmission dynamics (Modified from CDC) (See page 303).

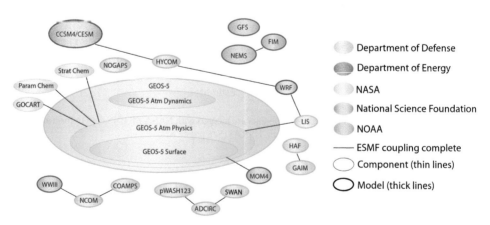

Plate 60.   The ESMF is a collaborative, multiple US agency effort to couple over 50 geoscience models in such a way that they are interoperable and scalable (Adapted from O'Kuinghttons *et al.*, 2010) (See page 306).

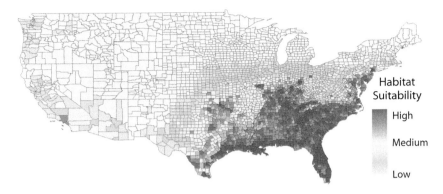

Plate 61. Potential habitat distribution of *Ae. albopictus* mosquito in the US (See page 321).

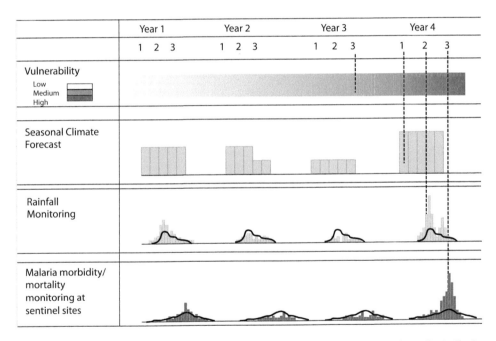

Flag 1 - Flag 2 - Flag 3

Plate 62. MEWS integrated framework: gathering cumulative evidence for early and focused response (WHO 2004a). 1 represents assessment during pre-season, 2 represents assessment during the rainy season and 3 represents assessment during Malaria season every year. Year-3 Pre-season assessment – vulnerability increasing due to period of drought; Year-4 Pre-season assessment – vulnerability still increasing due to period of drought and seasonal forecast above normal – Flag-1; Year-4 Pre-season assessment – vulnerability remains high, weather monitoring indicates higher than normal rainfall – Flag2; and, Year-4 Pre-season assessment – vulnerability remains high, rainfall higher than normal through much of season, malaria cases pass epidemic threshold – Flag-3 (See page 337).

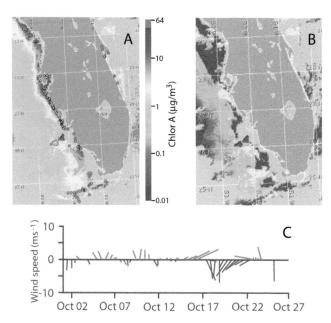

Plate 63.   The core elements of the operational West Florida shelf HAB forecast: (A) satellite-derived chloro-phyll image, with cell count information overlaid onto the image; (B) chlorophyll anomaly product, with confirmed HAB regions shown in red; and (C) local wind conditions (See page 350).

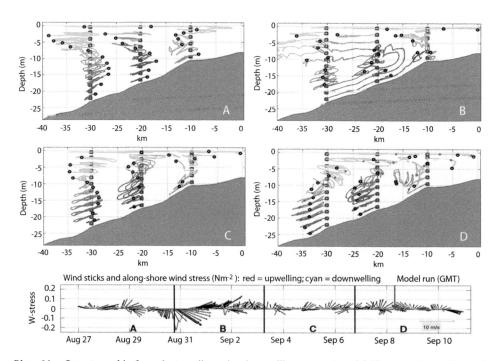

Plate 64.   Output graphic from the two-dimensional upwelling transect model (Courtesy Lanerolle *et al.*, 2006) (See page 351).

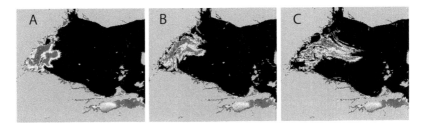

Plate 65.   Example of the HAB forecast system for Lake Erie: (A) initial image of the CI; (B) nowcast; and (C) three-day forecast of likely bloom distribution (See page 353).

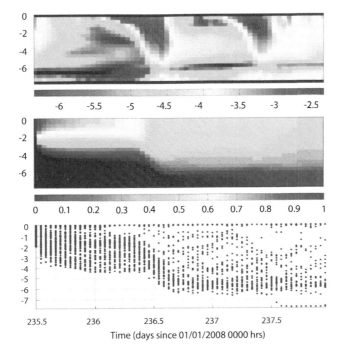

Plate 66.   Modelled eddy-viscosity (top panel) and its use to examine tracer (middle panel) and particle dispersion (lower panel) (See page 354).

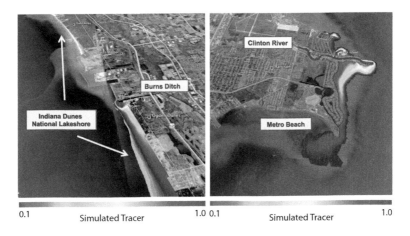

Plate 67.   River plume transport simulations at Burns Ditch, IN and Clinton River, MI (See page 356).

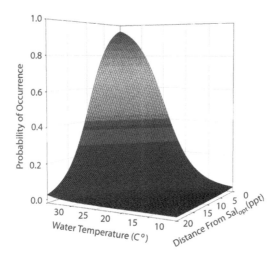

Plate 68.   Graphic representation of logit model using temperature and salinity to predict the probability of occurrence of *V. vulnificus* in Chesapeake Bay (See page 358).

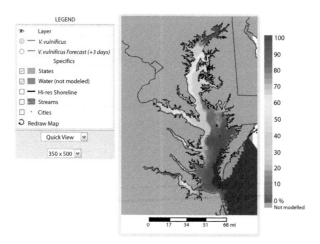

Plate 69.   *Vibrio vulnificus* model output for 11 August 2010 and web interface provided to state and county health officials. Scale represents probability of occurrence from 0 (blue) to 100% (red). No data are available for areas shaded grey (See page 359).

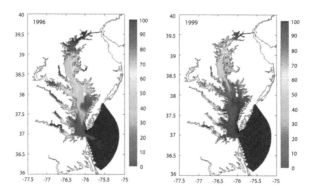

Plate 70.   Hindcast depicting probability of occurrence of *V. vulnificus* in wet (1996) and dry (1999) years. Both figures represent conditions present on 1 August (See page 359).

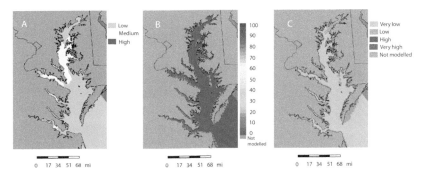

Plate 71. Examples of harmful algal bloom forecasts for the Chesapeake Bay using a hybrid mechanistic-empirical approach: (a) relative abundance of *Karlodinium veneficum*, (b) likelihood of a *Prorocentrum minimum* bloom, and (c) likelihood of a *Microcystis aeruginosa* bloom (See page 360).

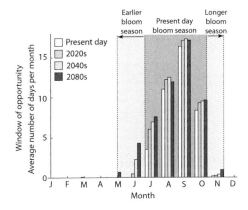

Plate 72. Average monthly values of the *window of opportunity* for *Alexandrium* in Puget Sound for the 1980s and in the future for the 2020s, 2040s, and 2080s. The dark grey shaded area represents the present day bloom season (2010–2019) and the lighter grey shaded area represents a wider future bloom season that begins earlier in the year and persists longer (Modified from Moore *et al.*, 2011) (See page 362).

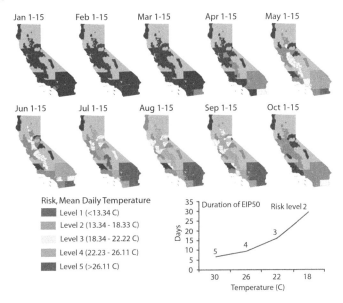

Plate 73. Temperature risk for each mosquito control agency in California in 2009. Level of risk based on the median duration of the extrinsic incubation period of WNV in *Cx. tarsalis* (See page 374).

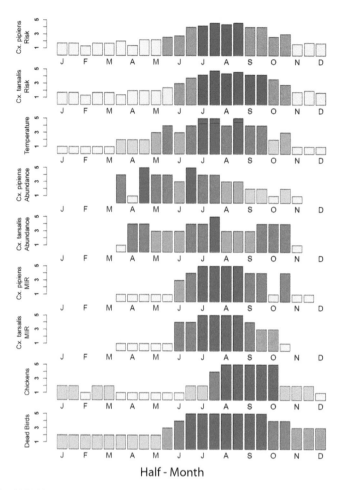

Plate 74. Risk of WNV tangential transmission summarized by half month for Kern County, California, 2009. yellow = normal season (risk = 1 – 2.5); orange = epidemic planning (risk = 2.6 – 4.0); red = epidemic (risk = 4.1 – 5.0) (See page 376).

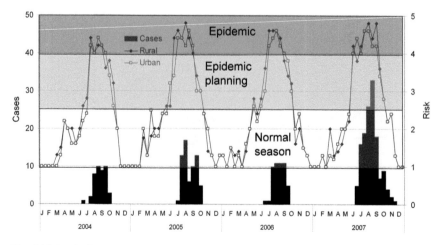

Plate 75. Risk levels from 1-5 estimated from environmental and enzootic measures for rural and urban locations and the numbers of reported human cases reported by date of onset plotted as a function of half month from 2004-2007 in Kern County, California (Reisen *et al.*, 2009) (See page 377).

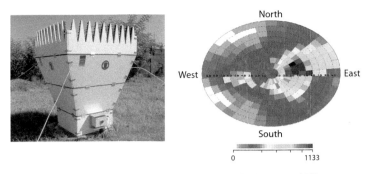

Plate 76. (Left) SODAR equipment; (Right) sound emission matrix (See page 380).

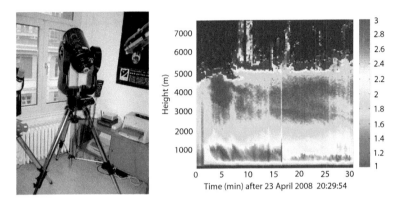

Plate 77. LiDAR equipment (left) is used for studying atmospheric composition, structure, clouds, and aerosols; (right) is a sample of range-corrected LiDAR signals through the Earth's boundary layer (See page 380).

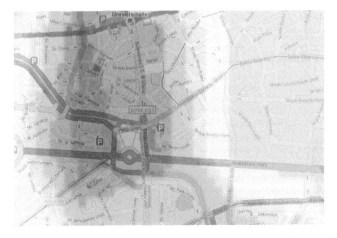

Plate 78. NO$_X$ concentration forecast displayed using the AIRAWARE system (See page 385).

Plate 79. Pollutant trajectories forecast displayed using the AIRAWARE system (See page 385).

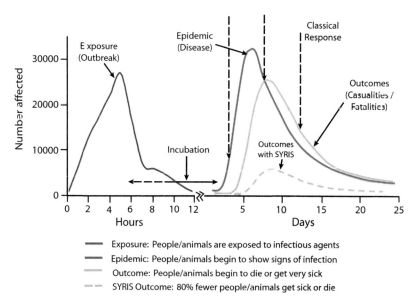

Plate 80.   The syndrome reporting approach to avoiding disease outbreaks in the first twelve hours vs. the classical identification and response approach that result in epidemics. SYRIS utilizes GIS and EO-data in the system design (Courtesy ARES Corporation) (See page 392).